ISBN 978-0-282-31801-7
PIBN 10065068

1 MONTH OF
FREE
READING

at

www.ForgottenBooks.com

By purchasing this book you are eligible for one month membership to ForgottenBooks.com, giving you unlimited access to our entire collection of over 700,000 titles via our web site and mobile apps.

To claim your free month visit:

www.forgottenbooks.com/free65068

English
Français
Deutsche
Italiano
Español
Português

www.forgottenbooks.com

Mythology Photography **Fiction**
Fishing Christianity **Art** Cooking
Essays Buddhism Freemasonry
Medicine **Biology** Music **Ancient**
Egypt Evolution Carpentry Physics
Dance Geology **Mathematics** Fitness
Shakespeare **Folklore** Yoga Marketing
Confidence Immortality Biographies
Poetry **Psychology** Witchcraft
Electronics Chemistry History **Law**
Accounting **Philosophy** Anthropology
Alchemy Drama Quantum Mechanics
Atheism Sexual Health **Ancient History**
Entrepreneurship Languages Sport
Paleontology Needlework Islam
Metaphysics Investment Archaeology
Parenting Statistics Criminology
Motivational

AMERICAN MACHINISTS' HANDBOOK

Published by the

McGraw-Hill Book Company
New York

Successors to the Book Departments of the
McGraw Publishing Company   Hill Publishing Company

Publishers of Books for

Electrical World   The Engineering and Mining Journal
Engineering Record   American Machinist
Electric Railway Journal   Coal Age
Metallurgical and Chemical Engineering   Power

American Machinists' Handbook

AND

DICTIONARY OF SHOP TERMS

A REFERENCE BOOK OF MACHINE SHOP AND
DRAWING ROOM DATA, METHODS AND
DEFINITIONS

BY

FRED H. COLVIN, A.S.M.E.

Associate editor of the *American Machinist*, Author of "*Machine
Shop Arithmetic*," "*Machine Shop Calculations*,"
"*The Hill Kink Books*," etc., etc.

AND

FRANK A. STANLEY

Associate editor of the *American Machinist*, Author of "*Accurate
Tool Work*," "*Automatic Screw Machines*,"
"*The Hill Kink Books*," etc.

EIGHTH IMPRESSION — TWENTY-FOURTH THOUSAND
CORRECTED

McGRAW-HILL BOOK COMPANY
239 WEST 39TH STREET, NEW YORK
6 BOUVERIE STREET, LONDON, E.C.
1909

DEPARTMENT OF ENGINEERING.

Copyright, 1908, BY THE HILL PUBLISHING COMPANY

First Printing, October, 1908
Second Printing, February, 1909
Third Printing, May, 1909
Fourth Printing, November, 1909
Fifth Printing, April, 1910
Sixth Printing, October, 1910
Seventh Printing, March, 1911
Eighth Printing, November, 1911

The Plimpton Press Norwood Mass. U.S.A.

PREFACE

EVERY man engaged in mechanical work of any kind, regardless of his position in the shop or drawing room, frequently requires information that is seldom remembered and is not usually available when wanted.

With this in mind it has been our endeavor to present in convenient form such data as will be of value to practical men in the various branches of machine work. While some of the matter included may seem elementary, it was considered necessary in order to make the work complete. Much of the information has never before been available to the mechanic without tiresome search and consultation.

We believe that the Dictionary section will be found of service to the younger mechanics and in helping to establish standard names for various parts which are now more or less confused in different sections of the country.

Our indebtedness to various manufacturers and individuals is hereby acknowledged, and in the back of the book will be found a list of such authorities with page references to the information furnished by them.

We dare not hope that no errors will be found and we shall be glad to have them pointed out and to receive any suggestions as to additions or other changes which may add to the value of the book.

THE AUTHORS.

CONTENTS

SCREW THREADS

Cutting Screw Threads

PAGE

Stud and Lead Screw Gears on Lathe 1
Examples in Finding Gears in Screw Cutting 1
Cutting Fractional Threads 1
Diagram of Simple and Compound Gearing for Screw Cutting. 2
Condensed Rules for Screw Cutting 3
Gears for Screw Cutting 3
Following the Motion of the Train of Gears 4
Effect of Compound Gearing in the Train 4
A Screw Thread Angle Table 4
Side Clearance of the Tool for Thread Helix 4
Formulas for Finding the Angle of Helix 5
Use of the Protractor in Getting the Side Clearance on the
  Tool 6
Multiple Thread Cutting 6
Table of Distances to Move Carriage in Multiple Thread
  Cutting 7
Opening and Closing the Lead Screw Nut 7
Face Plate for Multiple Thread Cutting 8
Cutting Diametral Pitch Worms in the Lathe 8
Table of Change Gears for Diametral Pitch Worms . . . 9
Examples in Finding Gears for Worm Cutting 10
The Brown & Sharpe 29-degree Worm Thread and the Acme
  29° Screw Thread 10
Full Size Sections of B. & S. Worm and Acme Screw Threads,
  1-inch Pitch 11
Measurements of V Tools 11
The Gear Tooth Caliper as Used for Thread Measurements . 12
Table for V Tool Angle Measurements 13
Grinding the Flat on Thread Tools 14, 13
Table for Grinding Flat on Tools for U. S. Form of Thread 14, 15

Standard Proportions of Screw Threads

Table of United States Standard Screw Threads 16
Table of Sharp "V" Standard Screw Threads 17
Table of Whitworth Standard Screw Threads 18
Table of British Association Standard Screw Threads . . 19
Table of French (Metric) Standard Screw Threads . . . 20
Table of International Standard Screw Threads 21
Table of Acme 29-degree Standard Screw Threads . . . 22
Table of Acme 29 degree Tap Threads . . . . ~~

Measuring Screw Threads

PAGE

B. & S. Screw Thread Micrometer Readings for U. S. Threads . . 24
B. &. S. Screw Thread Micrometer Readings for Sharp "V"
Threads 25
B. &. S Screw Thread Micrometer Readings for Whitworth
Threads 26
Explanation of Screw Thread Micrometer Caliper . . . 26, 27
Measuring Thread Diameters with Micrometer and Wires . . 27
Measuring Threads of Special Diameter 27
Formulas for U. S. Thread Measurement with Micrometer and
Wires 28
Table for U. S. Thread Measurement with Micrometer and
Wires 29
Formulas for Sharp "V" Thread Measurement with Microm-
eter and Wires 30
Table for Sharp "V" Thread Measurement with Micrometer
and Wires 31
Formulas for Whitworth Thread Measurement with Microm-
eter and Wires 32
Table for Whitworth Thread Measurement with Micrometer
and Wires 33
Measuring Fine Pitch Screw Threads with Micrometer and
Wires 34
Constants for 3-wire and Micrometer System of Measuring
Threads 34
Measuring Acme 29-degree Screw and Tap Threads with Mi-
crometer and Wires 35
Table of Wire Sizes for Measuring Acme 29-degree Screw and
Tap Threads 35
Brown & Sharpe 29-degree Worm Thread Formulas . . . 36
Table of B. & S. 29-degrees Worm Thread Parts 36
Measuring B. &. S. 29-degree Worm Thread with Microm-
eter and Wires 37
Table of Wire Sizes for Measuring B. & S. 29-degree Worm
Threads 37
Worm Wheel Hobs 38

PIPE AND PIPE THREADS

Briggs Standard Pipe Threads 39
Taper of Pipe End and Form of Thread in the Briggs System . 39
Longitudinal Section of Pipe Thread 39
Perfect and Imperfect Threads 40
Dimensions of Wrought Iron Welded Tubes, Briggs Standard . 40
Whitworth Pipe Sizes and Threads 41
Table of Tap Drill Sizes for Briggs and Whitworth Pipe Taps . 41
The Pipe Joint in the Briggs System 42
Relation of Reamer, Tap, Die and Gages in the Briggs System 42
Forming the Joint in the Briggs System 43
Table of Diameters, Thread Lengths, Gaging Dimensions, etc.,
in the Briggs Joint 43, 44

Gage Sets for Briggs Pipe and Fittings 45
Briggs Standard-, Working Allowance-, and Inspection
    Allowance Gages 45
Relation and Application of the Gages 45, 46
National Standard Hose Coupling Dimensions 46

TWIST DRILLS AND TAPS

Angle of Spiral 47
Clearance or Relief 47
Grooves for Best Results 47
Grinding or Sharpening 48
Angle of Clearance 48
Feeds and Speeds for Different Material 49-50
Reasons for Chipping and Splitting 50
Drill Pointers and Troubles 50
Special Drills and Their Uses 50
Size of Drills, Decimal 53-54
Size of Drills, Letter 55
Sizes of Drills, Decimal Equivalent 56
Tap Drills for Regular Threads 57
Tap Drills for Machine Screw Taps 58
Drills for Dowel Pins 59
Double Depth of Threads 60

TAPS

Dimensions of Machine Screw Taps, old Standard 62
Dimensions of Machine Screw Taps, A. S. M. E. 63
Dimensions of Hand Taps 64
Dimensions of Taper Taps 65
Dimensions of Pipe Taps 66
Dimensions of Stove Bolt Taps 66
Dimensions of Taper Die Taps 67
Dimensions of Sellers Hobs 68
Dimensions of Square Thread Taps 69

FILES

Measurement of Files 70
Methods of Designating 70
Terms Used by File Makers 70
Hight of Work 70
Pickling Bath for Work to be Filed 70
Actual Spacing of Teeth 71
Teeth per Inch . . . . . : 72
Shapes and Grades of Files 72-74
When a File Cuts Best 74
File Tests 7
Small Files for Fine Work . . . .
Needle Files and Riffles . . . .

WORK BENCHES

PAGE

Filing and Assembling Benches 76
Benches for Average Shop Work 76
Locating Bench Away from Wall 76
Modern Designs for Benches 76
High and Low Cost Benches 77
Material for Benches 77
Building Benches from Small Blocks 78
Hight of Work Benches 78
Width and Thickness 78

SOLDERING

Cleaning the Joint 78
Strength of Soldered Joint 78
The Proper Heat for Soldering 78
Fluxes for Different Metals 78
Soldering Salts 78
Fluxes for Sheet Tin 78
Fluxes for Lead 79
Fluxes for Lead Burning 79
Fluxes for Brass 79
Fluxes for Copper 79
Fluxes for Zinc 79
Fluxes for Galvanized Iron 79
Fluxes for Wrought Iron or Steel 79
Making the Fluxes 80
Cleaning and Holding the Work 81
Soldering with Tin Foil 81
Soldering Cast Iron 81
Cold Soldering Metals, etc. 82

GEARING

Gear Teeth, Shapes of 83
Gear Teeth, Parts of Teeth 83
Circular Pitch of Gears, Table 84
Diametral Pitch of Gears, Table 85
Diametral and Circular Pitch, Table 86
Chordal Pitch and Spur Gear Radius 87
Table of Constants for Chordal Pitch 87
Tables of Tooth Parts, Diametral Pitch 88–89
Tables of Tooth Parts, Circular Pitch 90–91
Diagram for Cast Gear Teeth 92
Laying out Spur Gear Blanks 93
Actual Sizes of Diametral Pitches 94
Laying out Single Curve Teeth 96
Pressure Angles 96
Stub-tooth Gears 97
Fellows Stub-tooth Dimensions 98
Nuttall Stub-tooth Dimensions 98

                                                        PAGE
Tables for Turning Gear Blanks 99–100
Gear Tooth Cutters 101
Table of Depth and Thickness of Tooth 101
Block Indexing in Cutting Gear Teeth 101
Block Indexing, Tables for 102
Metric Pitch, Formula and Tables 103
Sprocket-wheels for Block Center Chains 104
Sprocket-wheels for Roller Chain 105
Bevel Gears 106
Bevel Gears, Names of Angles and Parts 106
Laying out Bevel Gear Blanks 106
Cutters for Bevel Gears 107
Bevel Gear Tables 108, 109
Examples of Use of Table 108–109
Miter Gear Tables 110
Spiral Gears 111
Spiral Gears, Formulas and Rules 111, 112
Spiral Gears, Table and Its Uses 113–114
Thread of Worms 115
Width of Face of Worm Wheels 115
Table of Worm Threads and Wheels 117

MILLING AND MILLING CUTTERS

Milling Machine Feeds and Speeds

Cutter Speeds for Steel, Cast Iron and Brass 118
Thickness of Chip 118
Effect of Overspeeding Cutters 118
The Question of Power 119
Cutters with Three and Five Teeth in Contact 119
Comparison of Fine and Coarse Pitch Cutters 119
Test in Cast Iron with Three- and Five-Tooth Contact . . 119
Advantages of Coarse Pitch Cutters for Heavy Milling . . 120
Finer Tooth Cutters for Finish Cuts 120
Clearance Angles for Roughing and Finishing Cutters . . 120
Lubrication for Cutters Milling Steel 120
Examples of Rapid Milling 120

Cam Milling

Milling Heart-shaped Cams 121
Method of Laying Out Cam 121
Selecting the Cutter 121
Locating Cam and Cutter at the Start 121
Selecting the Correct Index Plate 121
Operating the Table for Successive Cuts on the Cam . . 122
Milling Cams by Gearing up the Dividing Head . . . 122
Diagram for Determining the Angle of Index Head . . 122
Gearing the Machine for Cam Lobes of Different Leads . . 12.
Method of Feeding the Work to the Cutter . . . . . 12.

Tables for Use with the Dividing Head

PAGE

Table for Cutting Spirals on the Universal Milling Machine . 124–127
Method of Finding the Angle for Work Larger than Given in
Table 124
Table for Plain and Differential Indexing on Brown & Sharpe
Milling Machines 128–137
Dividing Head Arranged for Differential Indexing . . . 128
Method of Computing Gears for Differential Indexing . . . 128

Milling Cutter, Reamer and Tap Flutes

No. of Teeth in End Mills, Straight and Spiral Flute . . . 138
No. of Teeth in inserted Tooth Cutters 138
Pitch of Metal Slitting Cutters 138
Pitch of Screw Slotting Cutters 138
No. of Teeth in Plain Milling Cutters 139
Form of Cutter for Milling Teeth in Plain Milling Cutters . 139
No. of Teeth in Side or Straddle Mills 139
Angular Cutter for Milling Teeth in Side or Straddle Mills . 139
No. of Teeth in Corner Rounding, Concave and Convex Cutters, 140
Angular Cutter for Milling Teeth in " " " " 140
No. of Teeth in Single and Double Angle and Spiral Mill
Cutters 140
Cutter for Milling Teeth in Double Angle and Spiral Mill . 140
. No. of Flutes in Taps — Hand, Machine Screw, Tapper, Nut,
and Screw Machine 141
Tap Fluting Cutters 141
No. of Flutes in Taper and Straight Pipe Taps 142
Fluting Cutters for Taper and Straight Pipe Taps . . . 142
No. of Flutes in Pipe Hobs, Sellers Hobs and Hob Taps . . 142
Fluting Cutter for Hobs 142
No. of Flutes in Shell Reamers 143
Cutters for Fluting Reamers 143
No. of Flutes in Chucking and Taper Reamers 144
Diameters of Straddle Mills for Fluting Center Reamers . . 145
Cutter Keyways, Square and Half-round 145
Table of Standard T-Slot Cutter Dimensions 146
Table of Largest Squares that can be Milled on Round Stock . 146
Table of Divisions Corresponding to Given Circumferential Dis-
tances 147

GRINDING AND LAPPING

Grinding Wheels and Grinding

The Commercial Abrasives; Emery, Corundum, Carborundum
and Alundum 148
Grading of Wheels 148
Selection of Suitable Wheels 149
The Combination Grit Wheel 149
Hard Wheels 149
Wheel Grades for Given Classes of Work 150

PAGE

Speed and Efficient Cutting 150
Action of Wheels that are Too Hard or Too Soft 150
When a Wheel is Sharp 151
Contact of Wheel on Different Diameters, Flat Surfaces, and
    Internal Work 151
Selecting Wheels According to Contact 151
The Contact Area of a Wheel 152
Wheel Pressure and Wear 152
Wearing Effect of High Work Speeds 152
Grinding Allowances 153
Grinding Hardened Work 153
Undercut Corners for Shoulders to be Ground 153
Use of Water 154
Methods of Setting Diamonds 154
The Use of Diamonds 154
Preservation of the Diamond 154
Setting Diamonds 155
Speed Table for Grinding Wheels of Various Diameters . . 155
Grading Abrasive Wheels 156
The Norton Co.'s Grade Marks 156
The Carborundum Co.'s Grade Marks 156
Grade List of the Safety Emery Wheel Co. 156
Table of Grades of Wheels Usually Furnished for Different
    Classes of Work 157

Lapping

The Common Classes of Laps 158
A Lapping Plate for Flat Work 158
Speed of Diamond Laps 158
Lapping Flat Surfaces 159
The Lubricant Used 159
Laps for Holes 159
Adjustable Laps 159
Advantages of Lead Laps 160
Various Types of Internal Laps 160
How to do Good Lapping 160
Using Cast Iron, Copper and Lead Laps 161
Ring Gage and Other Work 161
A Lap for Plugs 161
Abrasives for Different Kinds of Laps 161
Diamond Powder in the Machine Shop 162
The Kind of Diamond Used 162
Reducing the Diamond to Powder 162
Settling Diamond Powder in Oil 162
Table for Settling Diamond 162
Rolling the Diamond Powder into Laps 162
Diamond Laps 163
Tools used in Charging Laps 163
Diamond Lap for Grinding Small Drills 163
Grinding Holes in Hard Spindles 164
Diamond used on Boxwood Laps . . . .

Reamer and Cutter Grinding

PAGE

Reamer Clearances 164
Chucking Reamer Blade for Cast Iron and Bronze . . . 165
Shape of Reamer Blade for Steel 165
Clearance of Reamer Blades 165
Grinding the Clearances on Various Kinds of Reamers . . . 165
Table for Grinding Clearances on Different Sizes of Reamers, 166, 167
Cup Wheel Clearance Table, Giving Tooth Rest Settings for
    Desired Clearance 168
Disk Wheel Clearance Table 168

SCREW MACHINE TOOLS, SPEEDS AND FEEDS

Types of Tools and Their Construction

Box Tools and Cutters 169
Roughing Box Tool with Tangent Cutter 169
Clearance for Box Tool Cutters 169
Sizes of Steel Recommended for Box Tool Cutters . . . 170
Finishing Box Tool with Radial Cutter 170
Hollow Mills 170
Location of Cutting Edge and Rake for Hollow Mills . . . 170
Hollow Mill Dimensions 171
Clamp Collar Proportions for Hollow Mills 171
Dies and Taps 171
Tapping Out Spring Dies 171
Spring Die Dimensions 172
Sizing Work for Threading 172
Table of Over- and Under-size Allowances for Tapping and
    Threading 173
Tap Lengths, Number of Flutes and Width of Lands . . . 173
Circular and Dovetail Forming Tools 174
Cutting Clearances on Forming Tools 175
Diameters of Circular Tools and Amount Usually Cut Below
    Center 175
Getting the Tool Diameters at Different Points to Produce a
    Given Form 176
Finishing a Circular Tool to Correct Outline 176
Formulas for Obtaining Depths to Finish Circular Tool on
    Center Line 176
Dovetail Tool Depths for Producing Correct Outline on Work, 177
Finishing a Dovetail Forming Tool 177
Location of Master Tool in Finishing Circular and Dovetail
    Tools 177
Circular Tools for Conical Points 178
Finishing a Circular Tool to Produce a 60-degree Cone . . . 178

Speeds and Feeds for Screw Machine Work

Table of Speeds and Feeds for Turning Screw Stock and Brass 179
Table of Speeds and Feeds for Turning with Finish Box Tool, 180
Table of Speeds and Feeds for Forming 181

PAGE
Table of Speeds and Feeds for Drilling 182
Table of Speeds and Feeds for Reaming 183
Table of Speeds and Feeds for Threading 183
Rate of Feed for Counterboring 183

PUNCH PRESS TOOLS

Method of Finding Diameters of Shell Blanks 184
Diagrams and Formulas for Blank Diameters for Plain,
  Flanged, Hemispherical and Taper Shells 185
Punch and Die Allowances for Accurate Work 184
Governing Size of Work by Punch and Die 186
Table of Clearance between Punch and Die for Different
  Metals 187
Clearance for Punches and Dies for Boiler Work 187
Lubricants for Press Tools 188
Oiling Copper and German Silver Sheets for Punching . . 188
Mixture for Drawing Steel Shells 188
Preparations for Drawing Brass, Copper, etc. 188

BOLTS, NUTS AND SCREWS

U. S. Standard Bolts and Nuts 189-191
Shearing and Tensile Strength of Bolts from $\frac{1}{4}$ to 3 inches
  Diameter 189
Dimensions U. S. Standard Rough Bolts and Nuts 190
Dimensions U. S. Standard Finished Bolts and Nuts . . . 191
Sizes of Machine Bolts with Manufacturers' Standard Heads . 192
Set Screw Dimensions 193
Hartford Machine Screw Co.'s Standard Set Screw Dimensions, 193

Tables of Cap and Machine Screw Dimensions

Hexagon and Square Head Cap Screws 194
Collar Head or Collar Screws 195
Fillister Head Cap Screws (P. &. W Std.) 195
Flat, Round and Oval Fillister Head Cap Screws 196
Button Head Cap Screws 197
Flat and Oval Countersunk Head Cap Screws 197
Flat and Round Head Machine Screws (American Screw Co.'s
  Standard) 198
Fillister Head Machine Screws (American Screw Co.'s Std.) . 199
Threads per Inch on Machine Screws (American Screw
  Co.'s Standard) 200
A. S. M. E. Standard Form of Thread, Pitch Formula, Etc. . 200
Diagrams of Basic Maximum and Minimum Screw and Tap
  Threads 201

Tables of A. S. M. E. Standard Machine Screw Dimensions

Outside, Root and Pitch Diameters of Standard Screws . . 202
Diameters of Taps for Standard Screws 203
Outside, Root and Pitch Diameters of Special Screws . . . 2

PAGE

Diameters of Taps for Special Screws 205
Dimensions of Oval Fillister Heads 206
Dimensions of Flat Fillister Heads 207
Dimensions of Flat Countersunk Heads 208
Dimensions of Round or Button Heads 209

Nut and Bolt Tables

U. S. Standard Hot Pressed and Cold Punched Nuts 210
Cold Punched Check and Jam Nuts 210
Manufacturers' Standard Hot Pressed and Forged Nuts . . 211
Manufacturers' Standard Cold Punched Nuts. 212
Manufacturers' Standard Narrow Gage Hot Pressed Nuts . . 213
Button Head Machine, Carriage and Loom Bolts 214
Length of Bolts 214
Lengths of Threads Cut on Bolts 214
Round and Square Countersunk Head Bolts 215
Tap Bolts 215
Stove Bolt Diameters and Threads 215
Automobile Bolt and Nut Standards Adopted by the A. L. A. M. 216
Planer Nuts 217
Coupling Bolts 217
Planer Head Bolts, Nuts and Washers 217

Miscellaneous Tables

Depths to Drill and Tap for Studs 218
Bolt Heads for Standard T-Slots 218
Eye Bolts 219
Spring Cotters 219
U. S. Standard Washers 220
Narrow Gage and Square Washers 220
Cast Iron Washers 221
Riveting Washers 221
Machine and Wood Screw Gage Sizes 221
Coach and Lag Screws 222
Lengths of Threads on Coach and Lag Screws 222
Lag Screw Test 222
Wood Screws 223
Boiler and Tank Rivet Heads 224

CALIPERING AND FITTING

The Vernier and How to Read it 225
The Vernier Graduations 225
Principle of the Vernier Scales 226
Reading the Micrometer 226
The Micrometer Parts 226
The Ten Thousandth Micrometer 227
Micrometer Graduations 227
Measuring Three-Fluted Tools with Micrometer and V-Block,
227, 228

Press and Running Fits

PAGE

Parallel Press, Drive and Close Fits 228
Parallel Running Fits 228
Table of Limits for Press, Drive and Hand Fits 229
Table of Limits for Close, Free and Loose Running Fits . 230
Shrink Fit Allowances 231
Limits in Shop Gages for Various Kinds of Fits 231, 232
Limits in Plug Gages for Standard Holes 231
Allowances Over Standard for Force Fits. 232
Allowances Over Standard for Driving Fits 232
Allowances Below Standard for Push or Keying Fits . . . 232
Clearances of Running Fits 232
Making Allowances with the Calipers for Various Kinds of
  Fits 233–236
Side Play of the Calipers when Measuring for Running Fits . 233
Table of Reduced Diameters Indicated by Side Play of Calipers, 233
Axial Inclination of Calipers in Measuring for Shrink or Press
  Fits 234
Table of Caliper Inclination for Allowances for Shrink or
  Force Fits 234
Side Play of Calipers in Boring Holes Larger than a Piece of
  Known Diameter 235
Rule for Finding Variation in Size of Hole Corresponding to
  Given Amount of Side Play 235
Allowing for Running and Driving Fits 236

Dimensions of Keys and Key-Seats

Rules for Key and Keyway Proportions 236
Key and Keyway Dimensions 237
Dimensions of Straight Keys 238
Square Feather Keys and Straight Key Sizes 238, 239
Barth Keys 239
Whitney Keys and Cutters 240, 241
Proportions of Key Heads 241
Table for Finding Total Keyway Depths 242, 243
Table of Amount of Taper for Keys of Various Lengths. . . 244

TAPERS AND DOVETAILS

Measuring Tapers

An Accurate Taper Gage 245
Applications of the Taper Gage 246
Setting the Adjustable Gage Jaws by Means of Disks . . . 246
Formulas for Use with Taper Gage 247
Finding Center Distances between the Gage Disks . . . 247
Finding the Disk Diameters 248
Finding the Amount of Taper per Foot 248
Finding the Width of Opening at the Ends of the Gage Jaws 24'

Tables of Standard Tapers

PAGE

Brown & Sharpe Standard Tapers 250, 251
Morse Standard Tapers 252, 253
Reed Standard Tapers 254
Jarno Standard Tapers 254, 255
Sellers Standard Tapers 256, 257
Taper Pins and Reamers 257
Table Giving Total Amount of Taper for Work Tapering from
o to 1¼ in. per Foot and Ranging up to 24 inches long . . 258
Table of Tapers per Foot in Inches and Corresponding
Angles 259
Table for Computing Tapers Corresponding to any Given
Angle 260, 261
Explanation of Table for Computing Tapers 262
Table for Dimensioning Dovetail Slides and Gibs . . . 262, 263.
Measuring External and Internal Dovetails 264
Diagrams of Various Classes of Dovetails 264
Table of Constants for Measuring Dovetails with Plugs . . 265
Examples of Uses of the Table of Constants 265

SHOP AND DRAWING ROOM STANDARDS

Standard Jig Parts

Drill Bushings 266, 267
Dimensions of Loose and Fixed Bushings 266
Dimensions of Fixed Bushings for Tools Having Stop Collars . 267
Dimensions of Collar Head Jig Screws 267
Dimensions of Winged Jig Screws 267
Binding Screws 268, 269
Supporting and Locking Screws 268, 269
Dimensions of Square Head Jig Screws 268
Dimensions of Headless Jig Screws 268
Dimensions of Nurled Head Jig Screws 269
Dimensions of Locking Jig Screws 269
Sizes of Straps for Jigs 269

Tables of Dimensions of Standard Machine Parts

Hand Wheels 270
Handles for Hand Wheels 271
Knobs 271
Ball Handles 272
Binder Handles 272
Single End Ball Handles 273
Ball Lever Handles 273
Machine Handles 274
Wing Nuts 274
Thumb Nuts 275
Hook Bolts 275

Miscellaneous Tables

PAGE

Standard Plug and Ring Gages 276
Counterbores with Inserted Pilots 277
Two-point Ball Bearing Dimensions 278, 279
Four-point Ball Bearing Dimensions 280
Integral Right-angle Triangles for Erecting Perpendiculars . 281
Chords of Arcs from 10 Minutes to 90 degrees . . . 281–283
Construction of Angles from Table of Chords 281
Table for Spacing Holes in Circles; Diameters 1 to 12, Holes 3
   to 32 284–285
Explanation of Table for Spacing Holes 286
Table of Sides, Angles and Sines for Spacing 3 to 500 Holes or
   Sides in a Circle 286–291
Actual Cutting Speeds of Planers with Various Return Ratios, 292
Stock Allowed for Standard Upsets 292
Stock Required to Make Bolt Heads, Mfgrs. Standard Sizes . 293
Stock Required to Make Bolt Heads, U. S. Standard Sizes . 293
Quick Way of Estimating Lumber for a Pattern 294
Table of Proportionate Weight of Castings to Weight of
   Pattern 294
Degrees Obtained by Opening a Two-foot Rule 294
Weight of Fillets 295
Table of Areas or Volumes of Fillets 295

WIRE GAGES AND STOCK WEIGHTS

Twist Drill and Steel Wire Gage Sizes 296
Stubs' Gages 296
Different Standards for Wire Gages in Use in the United States, 297
Wire and Drill Sizes Arranged Consecutively 298, 299
Stubs' Steel Wire Sizes and Weights 300
Music Wire Sizes 301
Weights of Sheet Steel and Iron, U. S. Standard Gage . . 301
Weights of Steel, Iron, Brass and Copper Plates, B. &. S. Gage, 302
Weights of Steel, Iron, Brass and Copper Plates, B'ham. Gage, 303
Weights of Steel, Iron, Brass and Copper Wire, B. & S. Gage . 304
Weights of Steel, Iron, Brass and Copper Wire, B'ham. Gage . 305
Weights of Steel and Iron Bars per Linear Foot 306
Weights of Brass, Copper and Aluminum Bars per Linear Foot, 307
Weights of Flat Sizes of Steel 308
Weights of Seamless Brass and Copper Tubing 309

BELTS AND SHAFTING

Belt Fastenings 310
Belt Hooks 310
Belt Lacings 310
Belt Splices 310
Belt Studs 310
Lacing Belts with Leather 3ᵀ

PAGE

Lacing Belts with Wire 310
Strength of Lacings 311
Tension on Belts 311
Alining Shafting by Steel Wire 312, 313
Table of Wire Sag for Lining Shafting 312, 313
Speeds of Pulleys and Gears 314
Rules for Speeds of Pulleys and Gears 314

STEEL AND OTHER METALS

Heat Treatment of Steel 315
Molecular Changes in Cooling 315
Safe Temperatures for Steel 315
Methods of Heating 315
Furnaces for Different Fuels 315
Heating in Liquids 315
Baths for Heating 316
Gas as a Fuel 316
Cooling the Steel 316
Baths for Cooling and Hardening 316
Annealing 317
Hardening Bath 317
High Speed Steels 318
Case Hardening 318
Harveyizing 318
Carbonization or Case Hardening 318
Penetration of Carbon 319
Carbonizing Materials 319
Action of Wood Charcoal 320
Tests of Carbon Penetration 319–321
Carbonizing with Gas 321
Effect of Composition on Strength 322
Effect of Hardening on Strength 322
Mechanical Properties, Annealed 322
Mechanical Properties, Hardened 322
Fahrenheit and Centigrade Thermometer 323
Converting one Thermometer to the Other 323
Properties of Metals 324
Alloys for Coinage 324
Composition of Bronzes 325
Bearing Metals 325
Bismuth Alloys, Fusible Metals 325
Alloys 326
Shrinkage of Castings 326
Aluminum, Properties of 327
Aluminum, Melting, Polishing and Turning 327

GENERAL REFERENCE TABLES

Common Weights and Measures 328, 329
Water Conversion Factors 329

CONTENTS

PAGE

Convenient Multipliers 330
The Metric System 330
Metric Weights and Measures 330, 331
Metric and English Conversion Tables 331
Miscellaneous Conversion Factors 332
Decimal Equivalents of Fractions of Millimeters, Advancing
   by $\frac{1}{100}$ mm 332
Decimal Equivalents of Fractions of Millimeters, Advancing
   by $\frac{1}{50}$ mm. 333
Equivalents of Inches in Millimeters . . . . ·334
Decimal Equivalents of Fractions of an Inch, Advancing by
   64ths 335
Decimal Equivalents of Fractions of an Inch, Advancing by
   8ths, 16ths, 32ds, and 64ths 335
Decimal Equivalents of 6ths, 7ths, 8ths, 12ths, 16ths, 24ths, etc.
                                         336, 337
Equivalents of Inches in Decimals of a Foot 338, 339
Squares, Cubes, Square and Cube Roots of Fractions from
   $\frac{1}{64}$ to 1 inch 340, 341
Circumferences and Areas of Circles from $\frac{1}{64}$ to 1 inch . . 340, 341
Circumferences and Areas of Circles from 1 to 100 . . . 342-347
Squares, Cubes, Square and Cube Roots of Numbers from
   1 to 520 . . . . ·348-360
Circumferences and Circular Areas of Numbers from 1 to 520,
                                       348-360
Circumferences and Diameters of Circles from 1 to 200 . . 361

SHOP TRIGONOMETRY

Explanations of terms 362, 363
Finding Depth of V-Thread 364
Finding Diagonal of Bar 364
Finding Square for Taps 364
Spacing Bolt Circles 365
Laying out Jigs 365
Trigonometry Formulas. 366
Use of Formulas 366
Table of Regular Polygons 367
Practical Examples 368
Finding Radius without Center 368
Properties of Regular Figures: Circle, Triangle, Square, Hexa-
   gon and Octagon 369, 370
Table of Tangents and Co-tangents 371-382
Table of Sines and Co-sines 382-393
Table of Secants and Co-secants 394-405

DICTIONARY OF SHOP TERMS

Definitions and illustrations of shop terms 406-49

THE AMERICAN MACHINISTS' HANDBOOK

SCREW THREADS

CUTTING SCREW THREADS

NEARLY all lathes are geared so that if gears having the same number of teeth are placed on both stud and lead screw, it will cut a thread the same pitch as the lead screw. This is called being geared "even." If the lathe will not do this, then find what thread will be cut with even gears on both stud and lead screw and consider that as the pitch of lead screw. In speaking of the pitch of lead screw it will mean the thread that will be cut with even gears.

In cutting the same thread with even gears, both the work and the lead screw are turning at the same rate. To cut a faster thread, the lead screw must turn faster than the work, so the larger gear goes on the stud and the smaller on the lead screw. To cut a slower thread (finer-pitch or less lead), the larger gear goes on the screw and the smaller on the stud.

Calling the lead screw 6 to the inch, what gears shall we use to cut an 8 thread?

Multiply both the lead screw and the thread to be cut by some number (the same number for both) that will give two gears you have in the set. If the gears vary by 4 teeth, try 4 and get 24 and 32 as the gears. If by 5, you get 30 and 40 as the gears. Then as 8 is slower than ·6· the large gear goes on the lead screw and the small one on the stud.

Cut an 18 thread with a 5-pitch lead screw and gears varying by 5 teeth. $5 \times 5 = 25$ and $5 \times 18 = 90$. There may not be a 90 gear, but you can use a 2 to 1 compound gear and use a 45 gear instead. That is, put the 25 gear on the stud, use any 2 to 1 combination between this and the 45 gear on the screw.

The 25 gear must drive the large gear of the 2 to 1 combination and the small gear drive the 45-tooth gear, either directly or through an intermediate.

In cutting fractional threads the same rule holds good. To cut $11\frac{1}{2}$ threads with gears that change by 4 teeth, use $4 \times 6 = 24$ and $4 \times 11\frac{1}{2} = 46$, with the 24 gear on the stud and the 46 on the screw. With gears changing by 5 this is not so easy, as $5 \times 11\frac{1}{2} = 57\frac{1}{2}$, an impossible gear. Multiplying by 10 would give 60 and 115, not much better. Multiply by 6 and get $6 \times 6 = 36$ and $6 \times 11\frac{1}{2} = 69$, neither of which is in the set. It seems as though 35 and 70 would come pretty near it, but they will cut a 12 thread instead.

To find what thread any two gears will cut, multiply the pitch of lead screw and the gear which goes on it and divide this by the gear on the stud. Suppose we try 40 on the stud and 75 on the

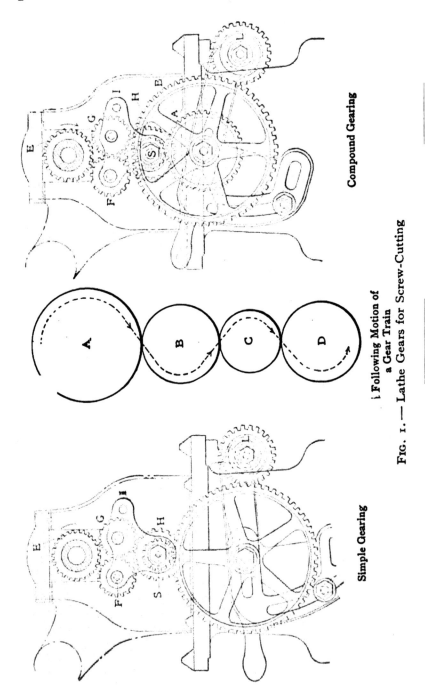

Compound Gearing

Simple Gearing

Following Motion of
a Gear Train

FIG. 1. — Lathe Gears for Screw-Cutting

screw. Multiply 75 by 6 = 450 and divide by 40 which gives 11¼ as the thread that will be cut. Try 45 and 80. 6 × 80 = 480; divided by 45 = 10⅔, showing that the 40 and 75 are nearest and that to cut it exactly a special gear will have to be added to the set. In reality the gears would not change by 5 teeth with a 6-pitch lead screw.

Rules for screw cutting may be summed up as follows, always remembering that the lead screw is the thread that will be cut when gears having the same number of teeth are placed on both screw and stud.

Having	To Find	Rule
A = True lead of screw and B = Thread to be cut	C = Gear for stud and D = Gear for screw	Multiply both A and B by any one number that will give gears in the set. Put gear A on stud and gear B on lead screw.
A = True lead of screw B = Thread to be cut C = Gear for stud	D = Gear for screw	Multiply B by C and divide by A.
A = True lead of screw B = Thread to be cut D = Gear for screw	C = Gear for stud	Multiply A by D and divide by B
A = True lead of screw C = Gear for stud D = Gear for screw	B = Thread that will be cut	Multiply A by D and divide by C

GEARS FOR SCREW-CUTTING

GEAR trains for screw-cutting are usually arranged similarly to the illustration, Fig. 1. If the gear E on the lathe spindle has the same number of teeth as the gear H on the stud S, the lathe is geared even, i.e., gears having the same teeth placed on both stud and lead screw will cut a thread like the lead screw. As shown, the gears are out of mesh because the tumbler gears F and G do not mesh with E; but moving the handle I down throws F into mesh with E so the drive is through E, F, G, H, S and intermediate to L, driving it so as to cut a right-hand screw if it is a right-hand thread, as is usually the case. Raising handle I cuts out F entirely and reverses the direction of the lead screw.

To follow the motion of a train of gears, take a stick (or your finger if they are *not* running) and trace the motion from the driver to the end as shown by the dotted lines in *A*, *B*, *C* and *D*.

When a lathe is compound geared the stud gear drives an auxiliary gear as *A*, which multiplies or reduces the motion as the case may be. It will readily be seen, if the stud drives *A* and *B* drives *L*, the motion will be reduced one-half because *A* has one-half the number of teeth in *B*.

A SCREW-THREAD ANGLE TABLE

THE accompanying table gives the angle of helix of various pitches and diameters with respect to a line perpendicular to the axis. These angles were worked out with the idea of using them for grinding thread tools for threads of various pitches upon different diameters of work. This table will enable one to set the protractor at the proper angle of side clearance for the work in hand and grind the thread tool correctly without guesswork.

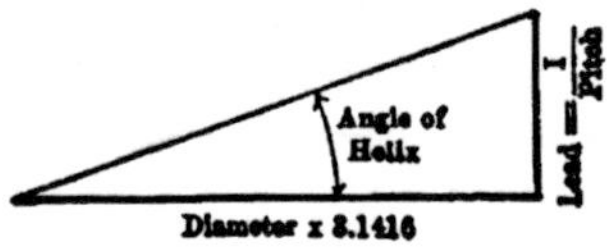

THREAD ANGLE TABLE

THREADS PER INCH = *P*

Diam. of Work	1	2	3	4	5	6	7	8	9	10	11	12
1/4"	50°-54	32°-31	22°-59	17°-39	14°-18	11°-59	10°-19	9°- 2	8°- 3	7°-58	7°-11	6°- 3
3/8"	40°-23	23°- 1	15°-48	12°-16	9°-39	8°- 8	7°-13	6°-37	5°-23	5°-20	4°-49	4°-24
1/2"	32°-30	17°-41	11°-58	9°- 3	7°-16	6°-37	5°-40	4°-33	4°- 3	4°- 1	3°-37	3°- 3
5/8"	27°- 2	14°-18	9°-38	7°-15	5°-49	5°- 8	4°-10	3°-52	3°-15	3°-13	2°-54	2°-26
3/4"	23°- 5	12°- 1	8°- 8	6°- 4	4°-52	4°- 3	3°-52	3°- 3	2°-43	2°-41	2°-19	2°- 2
7/8"	20°- 4	10°-20	7°- 1	5°-12	4°-10	3°-28	2°-59	2°-37	2°-19	2°-18	2°- 4	1°-50
1 "	17°-39	9°- 2	6°- 2	4°-33	3°-39	3°- 2	2°-36	2°-17	2°- 2	2°-	1°-48	1°-31
1 1/8"	15°-49	8°- 4	5°-23	4°- 4	3°-15	2°-42	2°-19	2°- 2	1°-48	1°-47	1°-37	1°-21
1 1/4"	14°-10	7°-12	4°-48	3°-39	2°-55	2°-26	2°- 2	1°-50	1°-44	1°-37	1°-27	1°-13
1 3/8"	13°- 4	6°-37	4°-25	3°-19	2°-40	2°-13	1°-54	1°-36	1°-29	1°-28	1°-19	1°- 6
1 1/2"	11°-59	6°- 4	4°- 3	3°- 3	2°-26	2°- 2	1°-44	1°-31	1°-21	1°-20	1°-13	1°- 1
1 5/8"	11°- 6	5°-36	3°-44	2°-49	2°-15	1°-52	1°-36	1°-21	1°-15	1°-14	1°- 7	56
1 3/4"	10°-26	5°-16	3°-29	2°-37	2°- 5	1°-44	1°-29	1°-18	1°-10	1°- 8	1°- 2	53
1 7/8"	9°-39	4°-52	3°-15	2°-26	1°-57	1°-37	1°-23	1°-13	1°- 5	1°- 4	58′	40
2 "	9°- 4	4°-34	3°- 3	2°-18	1°-50	1°-31	1°-18	1°- 8	1°- 1	1°- 0	54′	47
2 1/4"	8°- 8	4°- 9	2°-42	2°- 2	1°-37	1°-21	1°- 8	1°- 1	54′	53′	49′	41
2 1/2"	7°-15	3°-39	2°-26	1°-49	1°-28	1°-16	1°- 3	55′	49′	48′	43′	37
2 3/4"	6°-37	3°-19	2°-13	1°-40	1°-22	1°- 7	57′	50′	45′	44′	40′	33
3 "	6°- 4	3°- 3	2°- 2	1°-31	1°-13	1°- 1	52′	46′	41′	40′	36′	30

While the table is worked out for single threads, it can be used for double or triple threads by considering the lead equal to the advance of the work in one revolution instead of $\frac{1}{p}$, as given in the table.

It is customary in many shops to have several thread tools in stock to cut these various thread angles, each cutting within a certain range of angles. This table will be useful in determining the best range for each thread tool.

$$P = \text{Pitch} = \text{Threads per inch.} \quad \frac{1}{P} = \text{Lead} = L$$

$$D = \text{Diameter of work in inches.} \quad \pi = 3.1416 = \frac{C}{D}$$

$$C = \text{Circumference of Work in inches} = \pi D$$

$$\frac{L}{C} = \frac{\text{Lead}}{\text{Circumference of Work}} = \frac{\frac{1}{P}}{\pi D} = \text{Tangent of Angle}$$

Find Angle in Table of Tangents

THREAD ANGLE TABLE

THREADS PER INCH = P

Diam. of Work	13	14	15	16	18	20	22	24	26	28	30	32
	5°-36	5°-12	4°-51	4°-33	4°- 3	3°-39	3°-19	3°- 1	2°-45	2°-36	2°-25	2°-17
	3°-45	3°-28	3°-14	3°- 3	2°-43	2°-26	2°-13	2°- 1	1°-53	1°-44	1°-37	1°-31
	2°-49	2°-36	2°-26	2°-17	2°- 2	1°-49	1°-40	1°-31	1°-24	1°-18	1°-13	1°- 8
	2°-12	2°- 5	1°-57	1°-48	1°-37	1°-28	1°-20	1°-13	1°- 8	1°- 3	59'	54'
	1°-53	1°-44	1°-37	1°-31	1°-21	1°-13	1°- 7	1°- 1	57	53'	49'	46'
	1°-37	1°-30	1°-24	1°-18	1°-10	1°- 3	57'	53'	49	45'	42'	39'
1″	1°-24	1°-18	1°-13	1°- 8	1°- 1	55'	50'	45'	45	39'	36'	34'
1⅛″	1°-15	1°-11	1°- 5	1°- 1	54'	49'	45'	40'	42	35'	32'	30'
1¼″	1°- 7	1°- 3	59'	54'	49'	44'	40'	37'	34	31'	29'	27'
1⅜″	1°- 1	57'	53'	50'	44'	40'	36'	33'	31	28'	27'	25'
1½″	56'	52'	49'	46'	40'	36'	33'	30'	28	26'	24'	23'
	52'	48'	45'	42'	37'	34'	31'	28'	26	24'	23'	21'
	48'	45'	42'	40'	35'	31'	28'	26'	24	22'	21'	19'
	45'	42'	39'	37	33'	29'	27'	24'	23	21'	19'	18'
	42'	40'	36'	34'	31'	27'	25'	22'	22	19'	18'	17'
	37'	35'	32'	30'	27'	24'	22'	20'	19	18'	16'	15'
	34'	31'	29'	27'	24'	22'	20'	18'	17	15'	14'	14'
2″	31'	28'	26'	25'	22'	20'	18'	17'	16	14'	13'	13'
	28'	26'	24'	23'	20'	18'	17'	15'	14	13'	12'	11'

Figs. 2 and 3 show side and front elevations of the thread tool
and of the protractor as applied to obtain the proper angle of side
clearance to cut a right-hand screw thread. The front edge of the
thread tool is used to determine the angle of side clearance. Fig. 4
shows a section taken along the line $a\ b$, Fig. 2. It will be noticed
that line $e\ f$ is shorter than $G\ H$ to give clearance to the cutting
edges of the thread tool, and also that $G\ R$ is equal to $H\ R$ and $e\ S$
is equal to $f\ S$. The angle of the helix at half the depth of the
thread, Fig. 5, can be used, if desired, and can be approximated to
from the table, or figured exactly by the method given at the top of
the table.

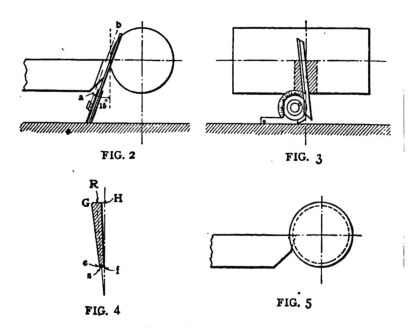

FIG. 2   FIG. 3

FIG. 4   FIG. 5

The Use of the Protractor

MULTIPLE THREAD CUTTING

THE accompanying table will be found useful when cutting multiple threads. When one thread is cut, the feed nut may be opened
(the spindle of course being stopped) and the carriage moved along
by hand the distance given in the table; the nut is then closed on
the screw and the next thread cut. This is a quick and sure method
of starting the second, third or fourth thread where the lead screw
of the lathe is of the pitch given in the table.

TABLE FOR MULTIPLE THREAD CUTTING

Cut	Thread on Lead Screw	Move Carriage
DOUBLE		
1	Even	½ inch
1¼	Any	2 inch
1½	Any	1 inch
2	4	¼ inch
2¼	Any	2 inch
2½	Any	1 inch
3	Even	½ inch
3¼	Any	2 inch
3½	Any	1 inch
4	8	⅛ inch
4¼	Any	2 inch
4½	Any	1 inch
5	Even	½ inch
5½	Any	1 inch
TRIPLE		
1	6	⅓" or 2 threads on lead screw
1¼	6	1⅓" or 8 threads on lead screw
1½	6	⅔" or 4 threads on lead screw
2	6	⅙" or 1 thread on lead screw
2¼	6	1⅓" or 8 threads on lead screw
2½	6	⅔" or 4 threads on lead screw
QUADRUPLE		
1	4	¼ inch
1¼	Any	1 inch
1½	Even	½ inch
2	8	⅛ inch
2¼	Any	1 inch
2½	Even	½ inch

Say we wish to cut a 3½-pitch double-thread screw: the lathe must be geared the same as for a single, triple or quadruple thread. The tool will of course have to be the same width and the depth of cut the same as for a 7 per inch screw. After the first thread is cut it will appear very shallow and wide. With the lathe spindle idle, the nut is opened and the carriage moved (in either direction) 1 inch; the nut is then closed on the lead screw and the tool is in the proper position to make the second cut.

If the carriage were moved 2 inches, the tool could follow exactly the first groove cut. In the case of a triple-thread screw, if the carriage were moved 3 inches, the tool would follow its original path and it would do the same in the case of a quadruple thread if mov 4 inches.

The carriage can, of course, be moved 1 inch and the nut closed no matter what the pitch of the lead screw may be (unless it is fractional), but in order to close the nut after moving ½ inch, the screw must have some even number of threads per inch.

As will be seen by referring to the table, a lead screw with any even number of threads per inch is used in a number of cases, while in several other instances the screw may be of any pitch — either odd or even. In certain cases 4 and 8 per inch lead screws are specified; and in cutting triple threads a 6 per inch screw is required.

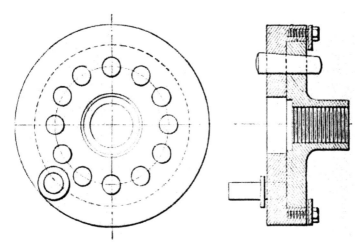

FIG. 6. — Face-Plate for Multiple Thread Cutting

FACE-PLATE FOR MULTIPLE THREAD CUTTING

FIG. 6 shows a face-plate fixture used on various numbers of threads. On an ordinary driving plate is fitted a plate having, as shown, twelve holes enabling one to get two, three, four or six leads if required. This ring carries the driving stud, and is clamped at the back of the plate by two bolts as an extra safeguard. All that is necessary in operation is to slack off the bolts, withdraw the index pin, move the plate the number of holes required, and re-tighten the bolts. It is used on different lathes, as occasion requires, by making the driving plates alike and drilling a hole for the index pin. It is found that the index pin works best when made taper, and a light tap is sufficient to loosen or fix it.

CUTTING DIAMETRAL PITCH WORMS IN THE LATHE

THE accompanying table is to be used in cases where fractional worm-thread cutting is necessary for diametrical pitch worm threads to mesh into diametral-pitch worm gears.

Table of Change Gears for Diametral Pitch Worms

Diametral Pitch	Single Depth	Width of Tool Point A	Width of Top of Thread B	Pitch of Lead Screw							
				2	3	4	5	6	7	8	10
2	1.078″	.487″	.526″								
2½	.862″	.390″	.421″								
3	.719″	.325″	.350″								
3½	.616″	.278″	.300″								
4	.540″	.243″	.263″								
5	.431″	.195″	.210″								
6	.360″	.162″	.175″								
7	.308″	.139″	.150″								
8	.270″	.122″	.131″								
9	.240″	.108″	.117″								
10	.216″	.097″	.105″								
11	.196″	.088″	.096″								
12	.180″	.081″	.088″								
14	.154″	.069″	.075″								
16	.135″	.061″	.066″								
18	.120″	.054″	.058″								
20	.108″	.048″	.053″								
24	.090″	.040″	.044″								
28	.077″	.034″	.038″								
32	.067″	.030″	.033″								
40	.054″	.024″	.026″								
48	.045″	.020″	.022″								

Formula: $\dfrac{22 \times \text{Lead Screw}}{7 \times \text{Diametral Pitch}} \Big\} = $ Ratio of Wheels.

In the first column is found the diametral pitch to be cut. In the second column is found the corresponding single depth of the worm thread. Under the third column is found the width of the tool at the point, the tool being the regular 29-degree included angle. In the fourth column is found the width at the top of the worm thread.

The next heading in the chart is "Pitch of lead screw," and here are found different pitches of lead screws from 2 to 10.

Example: Suppose it is desired to cut a worm thread of 4 diametral pitch on a single-geared lathe having a 6-pitch lead screw. Now, opposite 4 in the first column find the single depth of worm thread, or 0.540 inch; and continuing in the same direction from left to right, under the next column find the width of the worm-thread tool at the point or end, which is 0.243 inch, and so on to the next column where is found the width of the worm thread at the top, which is 0.263 inch. Say there is a 6-pitch lead screw on the lathe. Then follow right on in the same direction until coming to the square under 6, and the gear, will be in the ratio of $\frac{87}{7}$. Of course there is no 7 gear on the lathe, so simply bring the fraction $\frac{87}{7}$ to higher denominations, say, $\frac{87}{7} \times \frac{3}{3} = \frac{99}{21}$: that is, put the 99 gear on the spindle or stud, and the 21 gear on the screw. Then use a gear of any convenient size to act as an intermediate gear, and thus connect the gear on the spindle with the gear on the screw. Taking the fraction $\frac{87}{7}$ and multiplying the numerator and denominator by 4 would give $\frac{132}{28}$ as the two gears to be used. It will be seen that this last fraction simply changes the number of teeth in the gears, but does not change the value of the fraction; thus there is the same ratio of gears.

Take another case: Suppose it is desired to cut a 20-diametral pitch worm thread in a lathe having a 4-pitch lead screw. What would be the necessary gears to cut the desired thread? Next to 20 in the first column is found the single depth of the worm thread, which is 0.108 inch. Continuing on, reading from left to right as in the first case, and 0.048 inch is found as the width of the tool at the point. In the next column is found the width at the top of the worm thread, which in this case is 0.053 inch. Under column 4, and opposite 20, are found the gears necessary for cutting a 20-diametral pitch worm thread in a lathe with a 4-pitch lead screw. The gear thus found, namely, $\dfrac{22 \text{ stud}}{35 \text{ screw}}$ may not be in the regular set of gears furnished with the lathe. In that case double up on both and make it $\dfrac{44 \text{ stud}}{70 \text{ screw}}$, which is the same in value. The two examples thus worked out could have been cut on lathes with lead screws having any number of threads per inch, with the same result. One point in cutting these threads is that the tool must be of exact dimension all over, for if it is not exactly 29 degrees included angle, or the point is not as it should be for width, then there will be an error in the worm thread all around.

THE BROWN & SHARPE 29-DEGREE WORM THREAD
AND THE ACME 29-DEGREE STANDARD
SCREW THREAD

THERE seems to be some confusion among mechanics regarding the 29-degree Acme standard screw thread and the Brown & Sharpe 29-degree worm thread.

The sketches, Figs. 7 and 8, show plainly the difference between threads of the same pitch in the two systems. The sections views are of threads of one-inch linear pitch drawn to scale to the proportions given by the thread formulas in connection with the complete tables of the two systems of threads as given on pages following.

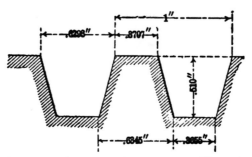

FIG. 7. — Acme 29-Degree Screw Thread

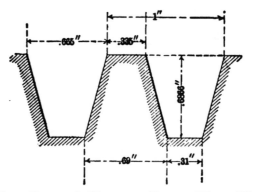

FIG. 8. — Brown & Sharpe 29-Degree Worm Thread

MEASUREMENT OF V-TOOLS

THE accompanying table of angle measurements should prove of convenience to all who make tools for cutting angles or make the gages for these tools.

The principle here adopted is that, on account of the difficulty and in some cases the impossibility of measuring the tool at its point, the measurement is taken on the angle of the tool at a given distance from the point. In this case the true measurement will be less than the actual measurement by an amount equal to twice the tangent of half of the angle multiplied by the distance of the line of measurement from the point.

For making the measurement the Brown & Sharpe gear-tooth caliper may be used. Fig. 9 shows this tool in position for measuring. The depth vernier A is set to a given depth h, and the measurement is taken by means of the vernier B. The width of the tool point x is equal to the measurement on the line $a\,b$ less $2h\left(tan.\dfrac{C}{2}\right)$. To use the table, h is always taken to be $\frac{1}{16}$ inch, which is found to be a convenient depth for most work. If a greater depth is required, all that is necessary is to multiply the figures given by the ratio of the required depth to $\frac{1}{16}$ inch. For instance, if the depth is required

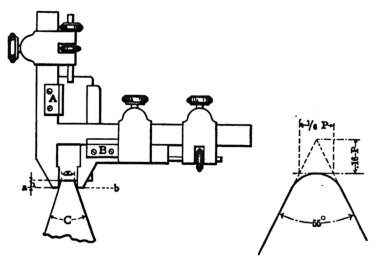

FIG. 9. — Measuring Thread Tools

to be $\frac{1}{8}$ inch, the figures given are multiplied by 2. In the great majority of cases, $\frac{1}{16}$ will be found a suitable value for h, when to find the width of the point x it is merely necessary to deduct the value of $\dfrac{tan.\dfrac{C}{2}}{8}$ for the angle required, which can be obtained at a glance from the table.

In the case of the Sellers or United States standard thread, the point of the tool should be one-eighth of the pitch of the screw, while in the Whitworth standard, as shown, the point of the tool would be one-sixth of the pitch if it were not rounded. By using these figures in combination with the table, it can be determined when sufficient has been ground from the point of the tool.

The table is called "Table for Angle Measurements," because if a sharp angle, that is, one without the point ground away, is measured as above, this measurement, by reference to the table, will give the angle direct.

TABLE FOR V-TOOL ANGLE MEASUREMENTS

Degrees	$\dfrac{tan.\frac{C}{2}}{8}$	Degrees	$\dfrac{tan.\frac{C}{2}}{8}$	Degrees	$\dfrac{tan.\frac{C}{2}}{8}$
1	0.0011	31	0.0346	61	0.0736
2	0.0022	32	0.0358	62	0.0751
3	0.0033	33	0.0370	63	0.0766
4	0.0044	34	0.0382	64	0.0781
5	0.0055	35	0.0394	65	0.0796
6	0.0066	36	0.0406	66	0.0811
7	0.0077	37	0.0418	67	0.0827
8	0.0088	38	0.0430	68	0.0843
9	0.0099	39	0.0442	69	0.0859
10	0.0110	40	0.0454	70	0.0875
11	0.0121	41	0.0466	71	0.0891
12	0.0132	42	0.0489	72	0.0908
13	0.0143	43	0.0492	73	0.0925
14	0.0154	44	0.0505	74	0.0942
15	0.0165	45	0.0518	75	0.0959
16	0.0176	46	0.0531	76	0.0976
17	0.0187	47	0.0544	77	0.0994
18	0.0198	48	0.0557	78	0.1012
19	0.0209	49	0.0570	79	0.1030
20	0.0220	50	0.0583	80	0.1048
21	0.0231	51	0.0596	81	0.1067
22	0.0242	52	0.0609	82	0.1086
23	0.0253	53	0.0623	83	0.1105
24	0.0264	54	0.0637	84	0.1125
25	0.0275	55	0.0651	85	0.1145
26	0.0286	56	0.0665	86	0.1165
27	0.0298	57	0.0679	87	0.1186
28	0.0310	58	0.0693	88	0.1207
29	0.0322	59	0.0707	89	0.1228
30	0.0334	60	0.0721	90	0.1250

GRINDING THE FLAT ON THREAD TOOLS

To facilitate grinding the correct width of flat for the single-point inserted tool to cut United States standard form of threads the accompanying table on pages 14 and 15 has been prepared. The distance from the point of the tool to the back is first measured with the micrometer, then the point of the tool may be ground off until the micrometer measurement from the back is equal to the whole depth minus dimension A, when we may be sure, without undertaking the difficult job of measuring it directly, that the flat B has the proper width. The dimensions A and B for pitches from 1 to 64 threads per inch are included in the table.

TABLE FOR GRINDING FLAT END OF TOOL FOR CUTTING U. S. FORM OF THREAD

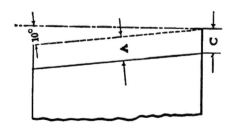
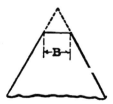

Threads per Inch	Pitch	A	B	C	Double Depth	Depth
1	1.000	.1064	.125	.1082	1.299	.6495
2	.5000	.0532	.0625	.0541	.6495	.3247
3	.3333	.0355	.0416	.0360	.433	.2165
4	.2500	.0266	.0312	.0270	.3247	.1623
5	.2000	.0213	.0250	.0216	.2598	.1299
6	.1666	.0177	.0208	.0180	.2165	.1082
7	.1428	.0152	.0178	.0154	.1855	.0927
8	.1250	.0133	.0156	.0135	.1623	.0812
9	.1111	.0118	.0138	.0120	.1443	.0721
10	.1000	.0106	.0125	.0108	.1299	.0649
11	.0909	.00963	.0113	.0098	.1180	.0592
12	.0833	.00886	.0104	.0090	.1082	.0541
13	.0769	.00818	.0096	.0083	.0999	.0499
14	.0714	.00758	.0089	.0077	.0920	.0460
15	.0666	.00707	.0083	.0071	.0866	.0433
16	.0625	.00673	.0079	.0068	.0812	.0406
17	.0588	.00620	.0073	.0063	.0764	.0382
18	.0555	00588	.0069	.0059	.0721	.0360
19	.0526	.00554	.0065	.0056	.0683	.0341
20	.0500	.00530	.0062	.0054	.0649	.0324
21	.0476	.00503	.0059	.0051	.0618	.0309
22	.0454	.0048	.0056	.0049	.0590	.0295
23	.0431	.00451	.0053	.0046	.0564	.0282
24	.0416	.00433	.0052	.0045	.0541	.0270
25	.0400	.00426	.0050	.0043	.0519	.0259
26	.0384	.00409	.0048	.0041	.0491	.0245
27	.0370	.00393	.0046	.0040	.0481	.0240
28	.0357	.00375	.0044	.0038	.0463	.0231
29	.0344	.00366	.0043	.0037	.0447	.0223
30	.0333	.00354	.0041	.0036	.0433	.0216
31	.0322	.00341	.0040	.0035	.0419	.0209
32	.0312	.00332	.0039	.0034	.0405	.0202

TABLE FOR GRINDING FLAT END OF TOOL FOR CUTTING U. S FORM OF THREAD

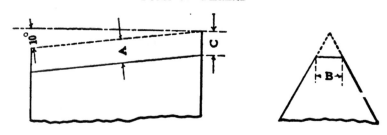

Threads per Inch	Pitch	A	B	C	Double Depth	Depth
33	.0303	.00315	.0037	.0032	.0393	.0196
34	.0294	.00307	.0036	.0031	.0382	.0191
35	.0285	.00295	.0035	.0030	.0370	.0185
36	.0277	.00289	.0034	.00295	.0360	.0180
37	.0270	.00281	.0033	.00286	.0350	.0175
38	.0263	.00272	.00325	.00282	.0341	.0170
39	.0256	.00268	.00320	.00277	.0333	.0166
40	.0250	.00264	.00312	.00270	.0324	.0162
41	.0243	.00255	.00303	.00262	.0319	.0159
42	.0238	.00251	.00295	.00257	.0309	.01545
43	.0232	.00247	.00290	.00251	.0302	.01520
44	.0227	.00238	.00283	.00245	.0295	.0147
45	.0222	.00233	.00277	.00240	.0290	.0145
46	.0217	.00230	.00271	.00235	.0282	.0141
47	.0212	.00225	.00265	.00230	.0274	.0137
48	.0208	.00221	.00260	.00225	.0270	.0135
49	.0204	.00217	.00255	.00220	.0263	.0131
50	.0200	.00213	.00250	.00216	.0258	.0129
51	.0196	.00208	.00245	.00212	.0254	.0127
52	.0192	.00204	.00240	.00208	.0249	.01245
53	.0188	.00200	.00235	.00203	.0245	.01225
54	.0185	.00196	.00231	.00200	.02405	.01202
55	.0181	.00192	.00226	.00196	.0236	.0118
56	.0178	.00189	.00222	.00192	.0232	.0116
57	.0175	.00185	.00218	.00189	.0228	.0114
58	.0172	.00184	.00215	.00186	.0223	.01115
59	.0169	.00180	.00211	.00183	.02201	.0110
60	.0166	.00177	.00208	.00180	.02165	.01082
61	.0163	.00173	.00203	.00177	.02119	.01059
62	.0161	.00172	.00202	.00175	.02095	.01047
63	.0158	.00169	.00198	.00171	.02061	.01030
64	.0156	.00167	.00196	.00169	.02029	.0101/

TABLE OF U. S. STANDARD SCREW THREADS

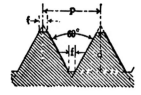

$$\text{Formula} \begin{cases} p - \text{Pitch} = \dfrac{1}{\text{No. Threads per Inch}} \\ d - \text{Depth} = p \times .6495 \\ f - \text{Flat} = \dfrac{p}{8} \end{cases}$$

Diam. of Screw	Threads to Inch	Pitch	Depth of Thread	Diam. at Root of Thread	Width of Flat
1/4	20	.0500	.0325	.185	.0063
5/16	18	.0556	.0361	.2403	.0069
3/8	16	.0625	.0405	.2936	.0078
7/16	14	.0714	.0461	.3447	.0089
1/2	13	.0769	.0499	.4001	.0096
9/16	12	.0833	.0541	.4542	.0104
5/8	11	.0909	.0591	.5069	.0114
3/4	10	.1000	.0649	.6201	.0125
7/8	9	.1111	.0721	.7307	.0139
1	8	.1250	.0812	.8376	.0156
1 1/8	7	.1429	.0928	.9394	.0179
1 1/4	7	.1429	.0928	1.0644	.0179
1 3/8	6	.1667	.1082	1.1585	.0208
1 1/2	6	.1667	.1082	1.2835	.0208
1 5/8	5 1/2	.1818	.1181	1.3888	.0227
1 3/4	5	.2000	.1299	1.4902	.0250
1 7/8	5	.2000	.1299	1.6152	.0250
2	4 1/2	.2222	.1444	1.7113	.0278
2 1/4	4 1/2	.2222	.1444	1.9613	.0278
2 1/2	4	.2500	.1624	2.1752	.0313
2 3/4	4	.2500	.1624	2.4252	.0313
3	3 1/2	.2857	.1856	2.6288	.0357
3 1/4	3 1/2	.2857	.1856	2.8788	.0357
3 1/2	3 1/4	.3077	.1998	3.1003	.0385
3 3/4	3	.3333	.2165	3.3170	.0417
4	3	.3333	.2165	3.5670	.0417
4 1/4	2 7/8	.3478	.2259	3.7982	.0435
4 1/2	2 3/4	.3636	.2362	4.0276	.0455
4 3/4	2 5/8	.3810	.2474	4.2551	.0476
5	2 1/2	.4000	.2598	4.4804	.0500
5 1/4	2 1/2	.4000	.2598	4.7304	.0500
5 1/2	2 3/8	.4210	.2735	4.9530	.0526
5 3/4	2 3/8	.4210	.2735	5.2030	.0526
6	2 1/4	.4444	.2882	5.4226	.0556

TABLE OF SHARP " V " SCREW THREADS

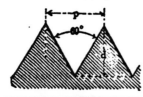

Formula $\begin{cases} p = \text{Pitch} = \dfrac{1}{\text{No. Threads per Inch}} \\ d = \text{Depth} = p \times .8663 \end{cases}$

Diam. of Screw	No. Threads per Inch	Pitch	Depth of Thread	Diam. at Root of Thread
¼	20	.0500	.0433	.1634
5/16	18	.0556	.0481	.2163
⅜	16	.0625	.0541	.2667
7/16	14	.0714	.0618	.3140
½	12	.0833	.0722	.3557
9/16	12	.0833	.0722	.4182
⅝	11	.0909	.0787	.4676
11/16	11	.0909	.0787	.5301
¾	10	.1000	.0866	.5768
13/16	10	.1000	.0866	.6393
⅞	9	.1111	.0962	.6826
15/16	9	.1111	.0962	.7451
1	8	.1250	.1083	.7835
1⅛	7	.1429	.1237	.8776
1¼	7	.1429	.1237	1.0026
1⅜	6	.1667	.1443	1.0864
1½	6	.1667	.1443	1.2114
1⅝	5	.2000	.1733	1.2784
1¾	5	.2000	.1733	1.4034
1⅞	4½	.2222	.1924	1.4902
2	4½	.2222	.1924	1.6152
2⅛	4½	.2222	.1924	1.7402
2¼	4½	.2222	.1924	1.8652
2⅜	4½	.2222	.1924	1.9902
2½	4	.2500	.2165	2.0670
2⅝	4	.2500	.2165	2.1920
2¾	4	.2500	.2165	2.3170
2⅞	4	.2500	.2165	2.4420
3	3½	.2857	.2474	2.5052
3⅛	3½	.2857	.2474	2.6301
3¼	3½	.2857	.2474	2.7551
3⅜	3¼	.3077	.2666	2.8418
3½	3¼	.3077	.2666	2.9668
3⅝	3¼	.3077	.2666	3.0918
3¾	3	.3333	.2886	3.1727
3⅞	3	.3333	.2886	3.2977
4	3	.3333	.2886	3.4227

Table of Whitworth Standard Screw Threads

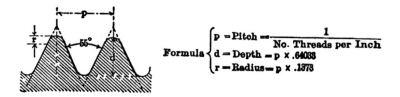

$$\text{Formula}\begin{cases} p = \text{Pitch} = \dfrac{1}{\text{No. Threads per Inch}} \\ d = \text{Depth} = p \times .64033 \\ r = \text{Radius} = p \times .1373 \end{cases}$$

Diam. of Screw	No. of Threads per inch	Pitch	Depth of Thread	Diam. at Root of Thread
¼	20	.0500	.0320	.1860
5/16	18	.0556	.0356	.2414
⅜	16	.0625	.0400	.2950
7/16	14	.0714	.0457	.3460
½	12	.0833	.0534	.3933
9/16	12	.0833	.0534	.4558
⅝	11	.0909	.0582	.5086
11/16	11	.0909	.0582	.5711
¾	10	.1000	.0640	.6219
13/16	10	.1000	.0640	.6844
⅞	9	.1111	.0711	.7327
1	8	.1250	.0800	.8399
1⅛	7	.1429	.0915	.9420
1¼	7	.1429	.0915	1.0670
1⅜	6	.1667	.1067	1.1616
1½	6	.1667	.1067	1.2866
1⅝	5	.2000	.1281	1.3689
1¾	5	.2000	.1281	1.4939
2	4½	.2222	.1423	1.7154
2¼	4	.2500	.1601	1.9298
2½	4	.2500	.1601	2.1798
2¾	3½	.2857	.1830	2.3841
3	3½	.2857	.1830	2.6341
3¼	3¼	.3077	.1970	2.8560
3½	3¼	.3077	.1970	3.1060
3¾	3	.3333	.2134	3.3231
4	3	.3333	.2134	3.5731
4½	2⅞	.3478	.2227	4.0546
5	2¾	.3636	.2328	4.5343
5½	2⅝	.3810	.2439	5.0121
6	2½	.4000	.2561	5.4877

TABLE OF BRITISH ASSOCIATION SCREW THREADS

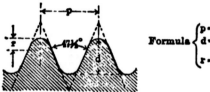

Formula $\begin{cases} p = \text{Pitch} \\ d = \text{Depth} = p \times .6 \\ r = \text{Radius} = \dfrac{2 \times p}{11} \end{cases}$

Number	Diam. of Screw mm.	Approximate Diam. Inches	Pitch mm.	Approximate Pitch Inches	Depth of Thread mm.	Diam. at Root of Thread mm.
0	6.0	.236	1.0	.0394	.6	4.8
1	5.3	.209	.9	.0354	.54	4.22
2	4.7	.185	.81	.0319	.485	3.73
3	4.1	.161	.73	.0287	.44	3.22
4	3.6	.142	.66	.0260	.395	2.81
5	3.2	.126	.59	.0232	.355	2.49
6	2.8	.110	.53	.0209	.32	2.16
7	2.5	.098	.48	.0189	.29	1.92
8	2.2	.087	.43	.0169	.26	1.68
9	1.9	.075	.39	.0154	.235	1.43
10	1.7	.067	.35	.0138	.21	1.28
11	1.5	.059	.31	.0122	.185	1.13
12	1.3	.051	.28	.0110	.17	.96
13	1.2	.047.	.25	.0098	.15	.9
14	1.0	.039	.23	.0091	.14	.72
15	.9	.035	.21	.0083	.125	.65
16	.79	.031	.19	.0075	.115	.56
17	.70	.028	.17	.0067	.10	.50
18	.62	.024	.15	.0059.	.09	.44
19	.54	.021	.14	.0055	.085	.37
20	.48	.019	.12	.0047	.07	.34
21	.42	.017	.11	.0043	.065	.29
22	.37	.015	.10	.0039	.06	.25
23	.33	.013	.09	.0035	.055	.22
24	.29	.011	.08	.0031	.05	.19
25	.25	.010	.07	.0028	.04	.17

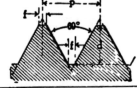

Formula $\begin{cases} p - \text{Pitch} \\ d - \text{Depth} = p \times .6495 \\ f - \text{Flat} = \dfrac{p}{8} \end{cases}$

Diameter of Screw mm.	Pitch mm.	Diameter at Root of Thread mm.	Width of Flat mm.
3	0.5	2.35	.06
	0.75	3.03	.09
5	0.75	4.03	.09
6	1.0	4.70	.13
7	1.0	5.70	.13
8	1.0	6.70	.13
8	1.25	6.38	.16
9	1.0	7.70	.13
9	1.25	7.38	.16
10	1.5	8.05	.19
11	1.5	9.05	.19
12	1.5	10.05	.19
12	1.75	9.73	.22
14	2.0	11.40	.25
16	2.0	13.40	.25
18	2.5	14.75	.31
20	2.5	16.75	.31
22	2.5	18.75	.31
22	3.0	18.10	.38
24	3.0	20.10	.38
26	3.0	22.10	.38
27	3.0	23.10	.38
28	3.0	24.10	.38
30	3.5	25.45	.44
32	3.5	27.45	.44
33	3.5	28.45	.44
34	3.5	29.45	.44
36	4.0	30.80	.5
38	4.0	32.80	.5
39	4.0	33.80	.5
40	4.0	34.80	.5
42	4.5	36.15	.56
44	4.5	38.15	.56
45	4.5	39.15	.56
46	4.5	40.15	.56
48	5.0	41.51	.63
50	5.0	43.51	.63
52	5.0	45.51	.63
56	5.5	48.86	.69
60	5.5	52.86	.69
64	6.0	56.21	.75
68	6.0	60.21	.75
72	6.5	63.56	.81
76	6.5	67.56	.81
80	7.0	70.91	.88

TABLE OF INTERNATIONAL STANDARD SCREW THREADS
DIMENSIONS IN MILLIMETERS

$$\text{Formula}\begin{cases} p = \text{Pitch} \\ d = \text{Depth} = p \times .6495 \\ f = \text{Flat} = \dfrac{p}{8} \end{cases}$$

Diam. of Screw	Pitch	Diam. of Screw	Pitch	Diam. of Screw	Pitch	Diam. of Screw	Pitch
6	1.00	18	2.50	39	4.00	68	6.00
7	1.00	20	2.50	42	4.50	72	6.50
8	1.25	22	2.50	45	4.50	76	6.50
9	1.25	24	3.00	48	5.00	80	7.00
10	1.50	27	3.00	52	5.00	88	7.50
11	1.50	30	3.50	56	5.50	96	8.00
12	1.75	33	3.50	60	5.50	116	9.00
14	2.00	36	4.00	64	6.00	136	10.00
16	2.00						

The "International Standard" is the same, with modifications noted, as that now in general use in France.

INTERNATIONAL STANDARD THREADS

At the "Congress International pour L'Unification des Filetages," held in Zurich, October 24, 1898, the following resolutions were adopted:

The Congress has undertaken the task of unifying the threads of machine screws. It recommends to all those who wish to adopt the metric system of threads to make use of the proposed system. This system is the one which has been established by the "Society for the Encouragement of National Industries," with the following modification adopted by this Congress.

1. The clearance at the bottom of thread shall not exceed $\frac{1}{16}$ part of the hight of the original triangle. The shape of the bottom of the thread resulting from said clearance is left to the judgment of the manufacturers. However, the Congress recommends rounded profile for said bottom.

3. The table for Standard Diameters accepted is the one which has been proposed by the Swiss Committee of Action. (This table is given above.) It is to be noticed especially that 1.25 mm. pitch is adopted for 8 mm. diameter, and 1.75 mm. pitch for 12 mm. diameter. The pitches of sizes between standard diameters indicated in the table are to be the same as for the next smaller standard diameter.

ACME 29° SCREW THREADS

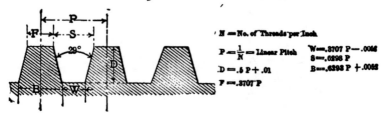

N = No. of Threads per Inch

$P = \frac{1}{N}$ = Linear Pitch

$D = .5 P + .01$

$F = .3707 P$

$W = .3707 P - .0052$

$S = .6293 P$

$B = .6293 P + .0052$

The Acme standard thread is an adaptation of the most commonly used style of Worm Thread and is intended to take the place of the square thread.

It is a little shallower than the worm thread, but the same depth as the square thread and much stronger than the latter.

The various parts of the Acme standard thread are obtained as follows:

Width of Point of Tool for Screw Thread =

$$\frac{.3707}{\text{No. of Threads per inch}} - .0052.$$

$$\text{Width of Screw or Nut Thread} = \frac{.3707}{\text{No. of Threads per inch}}.$$

Diameter of Screw at Root =

$$\text{Diameter of Screw} - \left(\frac{1}{\text{No. of Threads per inch}} + .020\right).$$

$$\text{Depth of Thread} = \frac{1}{2 \times \text{No. of Threads per inch}} + .010.$$

TABLE OF ACME 29° SCREW THREAD PARTS

N	P	D	F	W	S	B
Number of Threads per Inch	Pitch of Single Thread	Depth of Thread	Width of Top of Thread	Width of Space at Bottom of Thread	Width of Space at Top of Thread	Thickness at Root of Thread
1	1.0	.5100	.3707	.3655	.6293	.6345
1½	.750	.3850	.2780	.2728	.4720	.4772
2	.500	.2600	.1853	.1801	.3147	.3199
3	.3333	.1767	.1235	.1183	.2098	.2150
4	.250	.1350	.0927	.0875	.1573	.1625
5	.200	.1100	.0741	.0689	.1259	.1311
6	.1666	.0933	.0618	.0566	.1049	.1101
7	.1428	.0814	.0529	.0478	.0899	.0951
8	.125	.0725	.0463	.0411	.0787	.0839
9	.1111	.0655	.0413	.0361	.0699	.0751
10	.10	.0600	.0371	.0319	.0629	.0681

ACME 29° TAP THREADS

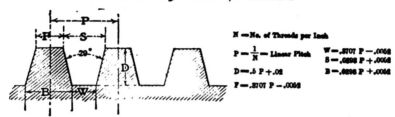

THE Acme standard tap-thread is cut with the same width of tool as the screw-thread and the diameter at the root is the same for tap and screw. Clearance at bottom of thread between screw and nut is obtained by boring the nut blank .020 oversize.

The outside diameter of the tap is made .020 larger than the screw to give clearance between top of screw-thread and bottom of nut.

Width of Point of Tool for Tap-Thread =
$$\frac{.3707}{\text{No. of Threads per Inch}} - .0052.$$

$$\text{Width of Thread} = \frac{.3707}{\text{No. of Threads per Inch}} - .0052$$

Diameter of Tap = Diameter of Screw + .020.

Diameter of Tap at Root =
$$\text{Diameter of Tap} - \left(\frac{1}{\text{No. of Threads per Inch}} + .040.\right)$$

$$\text{Depth of Thread} = \frac{1}{2 \times \text{No. of Threads per Inch}} + .020.$$

TABLE OF ACME STANDARD 29° TAP-THREAD PARTS

N.	P	D	F	W	S	B
Number of Threads per Inch	Pitch of Single Thread	Depth of Thread	Width of Top of Thread	Width of Space at Bottom of Thread	Width of Space at Top of Thread	Thickness at Root of Thread
1	1.0	.5200	.3655	.3655	.6345	.6345
1½	.750	.3950	.2728	.2728	.4772	.4772
2	.500	.2700	.1801	.1801	.3199	.3199
3	.3333	.1867	.1183	.1183	.2150	.2150
4	.250	.1450	.0875	.0875	.1625	.1625
5	.200	.1200	.0689	.0689	.1311	.1311
6	.1666	.1033	.0566	.0566	.1101	.1101
7	.1428	.0914	.0478	.0478	.0951	.0951
8	.125	.0825	.0411	.0411	.0839	.0839
9	.1111	.0755	.0361	.0361	.0751	.0751
10	.10	.0700	.0319	0319	.0681	.0681

BROWN & SHARPE SCREW THREAD MICROMETER CALIPER READINGS

READING OF CALIPER

$$\text{For U. S. Threads} = D - \frac{.6495}{P}$$

U. S. STANDARD THREADS

Diam.	Pitch	Caliper Reading		Diam.	Pitch	Caliper Reading	
D	P	$D - \dfrac{.6495}{P}$	$\dfrac{.6495}{P}$	D	P	$D - \dfrac{.6495}{P}$	$\dfrac{.6495}{P}$
	64		.0101	1/4	20	.2176	.0324
	62		.0105	5/16	18	.2765	.0360
	60		.0108	3/8	16	.3344	.0406
	58		.0112	7/16	14	.3911	0464
	56		.0116	1/2	13	.4501	.0499
	54		.0120	9/16	12	.5084	.0541
	52		.0125	5/8	11	.566	.0590
	50		.0130	3/4	10	.6851	.0649
	48		.0135	7/8	9	.8029	.0721
	46		.0141	1	8	.9188	.0812
	44		.0148	1 1/8	7	1.0322	.0928
	42		.0155	1 1/4	7	1.1572	.0928
	40		.0162	1 3/8	6	1.2668	.1082
	38		.0171	1 1/2	6	1.3918	.1082
	36		.0180	1 5/8	5 1/2	1.507	.1180
	34		.0191	1 3/4	5	1.6201	.1299
	32		.0203	1 7/8	5	1.7451	.1299
	30		.0217	2	4 1/2	1.8557	.1443
	28		.0232	2 1/4	4	2.3376	.1624
	26		.0250	3	3 1/2	2.8145	.1855
	24		.0271	3 1/2	3 1/4	3.3002	.1998
	22		.0295	4	3	3.7835	.2165

As there is no standard of diameter for the finer pitches, the columns for diameter and caliper reading are left blank. The column on the right gives the number to be subtracted from the diameter to obtain the caliper reading.

For explanation of screw thread micrometer caliper, refer to page 26.

Brown & Sharpe Screw Thread Micrometer Caliper Readings

READING OF CALIPER

$$\text{For ``V'' Threads} = D - \frac{.866}{P}$$

"V" Threads

Diam.	Pitch	Caliper Reading		Diam.	Pitch	Caliper Reading	
D	P	$D-\dfrac{.866}{P}$	$\dfrac{.866}{P}$	D	P	$D-\dfrac{.866}{P}$	$\dfrac{.866}{P}$
	64		.0135	$\frac{1}{4}$	24	.2139	.0361
	62		.0140	$\frac{1}{4}$	20	.2067	.0433
	60		.0144	$\frac{5}{16}$	20	.2692	.0433
	58		.0149	$\frac{5}{16}$	18	.2644	.0481
	56		.0155	$\frac{3}{8}$	18	.3269	.0481
	54		.0160	$\frac{3}{8}$	16	.3209	.0541
	52		.0167	$\frac{7}{16}$	16	.3834	.0541
	50		.0173	$\frac{7}{16}$	14	.3756	.0619
	48		.0180	$\frac{1}{2}$	14	.4381	.0619
	46		.0188	$\frac{1}{2}$	13	.4334	.0666
	44		.0197	$\frac{1}{2}$	12	.4278	.0722
	42		.0206	$\frac{9}{16}$	14	.5006	.0619
	40		.0217	$\frac{9}{16}$	12	.4903	.0722
	38		.0228	$\frac{5}{8}$	11	.5463	.0787
	36		.0241	$\frac{5}{8}$	10	.5384	.0866
	34		.0255	$\frac{11}{16}$	10	.6009	.0866
	32		.0271	$\frac{3}{4}$	10	.6634	.0866
	30		.0289	$\frac{7}{8}$	9	.7788	.0962
	28		.0309	1	8	.8918	.1082
	26		.0333	$1\frac{1}{8}$	8	1.0168	.1082
				$1\frac{1}{4}$	7	1.1263	.1237
				$1\frac{1}{2}$	6	1.3557	.1443

As there is no standard of diameter for the finer pitches, the columns for diameter and caliper reading are left blank. The column on the right gives the number to be subtracted from the diameter to obtain the caliper reading.

For explanation of screw thread micrometer caliper, refer to page 26.

BROWN & SHARPE SCREW THREAD MICROMETER CALIPER READINGS

READING OF CALIPER

For Whitworth Threads $= D - \dfrac{.640}{P}$

WHITWORTH STANDARD THREADS

Diam.	Pitch	Caliper Reading	
D	P	$D - \dfrac{.640}{P}$	$\dfrac{.640}{P}$
$\frac{1}{4}$	20	.2180	.0320
$\frac{5}{16}$	18	.2769	.0355
$\frac{3}{8}$	16	.3350	.0400
$\frac{7}{16}$	14	.3918	.0457
$\frac{1}{2}$	12	.4467	.0533
$\frac{9}{16}$	12	.5092	.0533
$\frac{5}{8}$	11	.5668	.0582
$\frac{11}{16}$	11	.6293	.0582
$\frac{3}{4}$	10	.6860	.0640
$\frac{13}{16}$	10	.7485	.0640
$\frac{7}{8}$	9	.8039	.0711
$\frac{15}{16}$	9	.8664	.0711
1	8	.9200	.0800
$1\frac{1}{8}$	7	1.0336	.0914
$1\frac{1}{4}$	7	1.1586	.0914
$1\frac{3}{8}$	6	1.2684	.1066
$1\frac{1}{2}$	6	1.3934	.1066
$1\frac{5}{8}$	5	1.4970	.1280
$1\frac{3}{4}$	5	1.6220	.1280
$1\frac{7}{8}$	$4\frac{1}{2}$	1.7328	.1422
2	$4\frac{1}{2}$	1.8578	.1442
$2\frac{1}{8}$	$4\frac{1}{2}$	1.9828	.1422

SCREW-THREAD MICROMETER CALIPER

THE Brown & Sharpe thread micrometer is fitted with pointed spindle and "V" anvil as in Fig. 10, to measure the actual thread on the cut surface. Enough of the point is removed and the bottom of the "V" is carried low enough so that the anvil and spindle clear the top and bottom of the thread and rest directly on the sides of e thread.

As it measures one-half of the depth of the thread from the top, on each side, the diameter of the thread as indicated by the caliper, or the pitch diameter, is the full size of the thread less the depth of one thread.

This depth may be found as follows;

Depth of   V   threads = .866 ÷ number of threads to 1″
   "   " U. S. Std.   "   = .6495 ÷   "   "   "   "   "
   "   " Whitworth   "   = .64   ÷   "   "   "   "   "

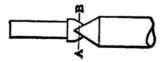

FIG. 10. — Spindle and Anvil of Thread Micrometer

As the U. S. thread is flatted ⅛ of its own depth on top, it follows that the pitch diameter of the thread is increased ⅛ on each side, equaling ¼ of the whole depth and instead of the constant .866 we use the constant .6495, which is ¾ of .866.

When the point and anvil are in contact the o represents a line drawn through the plane *A B*, Fig. 10, and if the caliper is opened, say to .500, it represents the distance of the two planes .500″ apart. The preceding tables are used in connection with the micrometer.

MEASURING EXTERNAL SCREW-THREAD DIAMETERS WITH MICROMETERS AND WIRES

It is frequently necessary, especially in making a tap or thread-plug gage, to measure the thread diameter on the thread angle in addition to measuring on top of the thread and at the bottom of the thread groove, and unless calipers made expressly for such work are at hand, the measurement on the thread angle is not made with any degree of accuracy or is omitted entirely. The accompanying sketches, Figs. 11, 12 and 13, formulas, and tables, are worked out for convenience in screw-thread inspection, so that by using ordinary micrometer calipers and wire of the diameter called for in the table the standard threads can be compared with the figures given.

Threads of Special Diameter

For threads of special diameter the values of x, x_1 or x_2 can be readily computed from the formula corresponding to the method of measuring to be used. The method shown in Fig. 11 at x is liable to lead to an error unless care be taken that the diameter on top of the thread is correct, and also that the flatted surface on the top of the threads is concentric with the rest of the thread. The concentricity of the flatted surface can be tested by measuring, as at x, Fig. 11, at several points on a plane through the axis and at right ʳ⁻ to it. The wire used must be round and of uniform diamet

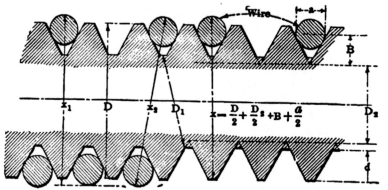

FIG. 11. — Measuring U. S. Standard Threads

D = outside diameter of thread.
D_1 = root diameter measured in thread groove.
n = number of threads per inch of length.
d = depth of thread.
p = distance from center to center of adjacent threads.
f = width of flat on U. S. Standard thread.
a = diameter of wire used.
B = distance from apex of thread angle at root, to center of wire.
D_2 = diameter of cylinder touched by apexes of thread angles.
x = diameter from top of threads on one side of tap or bolt, to top of wire laid in thread groove on opposite side.

U. S. STANDARD THREAD

p = lead = $\dfrac{1}{n}$, for single threads.

$d = p \times .6495 = \dfrac{.6495}{n}.$

$D_1 = \sqrt{(D - 2d)^2 + \left(\dfrac{\text{lead}}{2}\right)^2}.$

$f = \dfrac{p}{8}.$

a = from p, max; to $p \times .505$, min.

$B = \dfrac{a}{2} \div \sin 30° = a.$

$D_2 = D - \dfrac{1.5155}{n}.$

$x = \dfrac{D}{2} + \dfrac{D_2}{2} + B + \dfrac{a}{2}.$

$x_1 = D_2 + 2B + a.$

$x_2 = \sqrt{(D_2 + 2B)^2 + \left(\dfrac{\text{lead}}{2}\right)^2} + a.$

TABLE FOR MEASURING U. S. STANDARD THREADS WITH MICROMETERS AND WIRES

D	n	D_1	D_2	a	B	$\left(\frac{\text{lead}}{2}\right)^2$	x	x_1	x_2
1/4"	20	.1867	.1742	.04	.04	.000625	.2721	.2942	.2955
5/16"	18	.2419	.2283	.04	.04	.000771	.3304	.3483	.3495
3/8"	16	.2954	.2803	.04	.04	.000976	.3876	.4003	.4016
7/16"	14	.3465	.3292	.04	.04	.001274	.4433	.4492	.4507
1/2"	13	.4019	.3834	.06	.06	.001479	.5317	.5634	.5647
9/16"	12	.4561	.4362	.06	.06	.001735	.5893	.6162	.6177
5/8"	11	.5089	.4872	.06	.06	.002065	.6461	.6672	.6681
11/16"	11	.5712	.5497	.06	.06	.002065	.7086	.7297	.7312
3/4"	10	.6221	.5984	.06	.06	.0025	.7643	.7784	.7801
13/16"	10	.6844	.6609	.06	.06	.0025	.8267	.8409	.8425
7/8"	9	.7327	.7066	0.10	0.10	.003086	.9408	1.0066	1.0083
15/16"	9	.7950	.7691	0.10	0.10	.003086	1.0033	1.0691	1.0706
1"	8	.8399	.8105	0.10	0.10	.003906	1.0553	1.1105	1.1124
1 1/8"	7	.9421	.9085	0.10	0.10	.005102	1.1667	1.2085	1.2107
1 1/4"	7	1.0668	1.0335	0.10	0.10	.005102	1.2917	1.3335	1.3355
1 3/8"	6	1.1614	1.1224	0.10	0.10	.006944	1.3987	1.4224	1.4250
1 1/2"	6	1.2862	1.2474	0.10	0.10	.006944	1.5237	1.5474	1.5497
1 5/8"	5½	1.3917	1.3494	0.15	0.15	.008263	1.7122	1.7994	1.8019
1 3/4"	5	1.4935	1.4469	0.15	0.15	.010	1.8234	1.8969	1.8997
1 7/8"	5	1.6182	1.5719	0.15	0.15	.010	1.9484	2.0219	2.0245
2"	4½	1.7149	1.6632	0.15	0.15	.012343	2.0566	2.1132	2.1163
2 1/8"	4½	1.8393	1.7882	0.15	0.15	.012343	2.1816	2.2382	2.2411
2 1/4"	4½	1.9641	1.9132	0.15	0.15	.012343	2.3066	2.3632	2.3667
2 3/8"	4	2.0540	1.9961	0.15	0.15	.015625	2.4105	2.4461	2.4495
2 1/2"	4	2.1787	2.1211	0.15	0.15	.015625	2.5355	2.5711	2.5742
2 3/4"	4	2.4284	2.3711	0.15	0.15	.015625	2.7855	2.8211	2.8240
3"	3½	2.6326	2.5670	0.20	0.20	.020392	3.0835	3.1670	3.1704
3 1/4"	3½	2.8823	2.8170	0.20	0.20	.020392	3.3335	3.4170	3.4200
3 1/2"	3¼	3.1041	3.0337	0.20	0.20	.023654	3.5668	3.6337	3.6368
3 3/4"	3	3.3211	3.2448	0.20	0.20	.02775	3.7974	3.8448	3.8486
4"	3	3.5708	3.4948	0.20	0.20	.02775	4.0474	4.0948	4.0983
4 1/4"	2½"	3.8019	3.7228	0.20	0.20	.03024	4.2864	4.3228	4.3264
4 1/2"	2½"	4.0318	3.9489	0.20	0.20	.03305	4.5244	4.5500	4.5530
4 3/4"	2⅝"	4.2592	4.1728	0.20	0.20	.03625	4.7614	4.7728	4.7767
5"	2½"	4.4848	4.3938	0.20	0.20	.040	4.9970	4.9938	4.9980
5 1/4"	2½"	4.7346	4.6438	0.20	0.20	.040	5.2470	5.2438	5.2477
5 1/2"	2½"	4.9574	4.8619	0.20	0.20	.04431	5.4810	5.4619	5.4661
5 3/4"	2½"	5.2072	5.1119	0.20	0.20	.04431	5.7310	5.7119	5.7160
6"	2¼"	5.4271	5.3264	0.20	0.20	.049373	5.9032	5.9264	5.9307

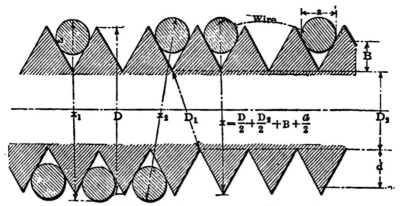

FIG. 12. — Measuring 60-Degree V-Threads

D = outside diameter of thread.
D_1 = root diameter measured in thread groove. ·
n = number of threads per inch of length.
d = depth of thread.
p = distance from center to center of adjacent threads.
a = diameter of wire used.
B = distance from apex of thread angle at root, to center of wire.
D_2 = diameter of cylinder touched by apexes of thread angles.
x = diameter from top of threads on one side of tap or bolt, to top of wire laid in thread groove on opposite side.

60° V Thread

$$p = \text{lead} = \frac{1}{n}, \text{ for single threads.}$$

$$d = p \times .866, = \frac{.866}{n}.$$

$$D_1 = \sqrt{(D - 2d)^2 + \left(\frac{\text{lead}}{2}\right)^2}.$$

$$a = \frac{p}{.866}, \text{ max; to } p \times .577, \text{ min.}$$

$$B = \frac{a}{2} \div \sin 30° = a.$$

$$D_2 = D - \frac{1.732}{n}.$$

$$x = \frac{D}{2} + \frac{D_2}{2} + B + \frac{a}{2}.$$

$$x_1 = D_2 + 2B + a.$$

$$x_2 = \sqrt{(D_2 + 2B)^2 + \left(\frac{\text{lead}}{2}\right)^2} + a.$$

TABLE FOR MEASURING 60-DEGREE V-THREADS WITH MICROMETERS AND WIRES

D	n	D_1	D_2	a	B	$\left(\dfrac{\text{lead}}{2}\right)^2$	x	x_1	x_2
1/4″	20	.1653	.1634	0.04	0.04	.000625	.2667	.2834	.2846
5/16″	18	.2180	.2163	0.04	0.04	.000771	.3244	.3363	.3375
3/8″	16	.2685	.2667	0.04	0.04	.0009765	.3808	.3867	.3881
7/16″	14	.3158	.3138	0.06	0.06	.0001274	.4656	.4938	.4957
1/2″	12	.3580	.3557	0.06	0.06	.001735	.5178	.5357	.5375
9/16″	12	.4202	.4182	0.06	0.06	.001735	.5803	.5982	.5998
5/8″	11	.4697	.4676	0.06	0.06	.0020657	.6363	.6476	.6492
11/16″	11	.5319	.530	0.06	0.06	.0020657	.6987	.7100	.7115
3/4″	10	.5789	.5768	0.10	0.10	.0025	.8134	.8768	.8784
13/16″	10	.6412	.6393	0.10	0.10	.0025	.8759	.9313	.9413
7/8″	9	.6847	.6826	0.10	0.10	.003086	.9288	.9826	.9843
15/16″	9	.7470	.7450	0.10	0.10	.003086	.9912	1.045	1.0466
1″	8	.7859	.7835	0.10	0.10	.003906	1.0417	1.0835	1.0854
1 1/8″	7	.8803	.8776	0.10	0.10	.005102	1.1513	1.1776	1.1800
1 1/4″	7	1.0050	1.0026	0.10	0.10	.005102	1.2763	1.3026	1.3047
1 3/8″	6	1.0895	1.0863	0.15	0.15	.006944	1.4556	1.5363	1.5388
1 1/2″	6	1.2141	1.2113	0.15	0.15	.006944	1.5806	1.6613	1.6635
1 5/8″	5	1.2825	1.2786	0.15	0.15	.010	1.6768	1.7286	1.7317
1 3/4″	5	1.4071	1.4036	0.15	0.15	.010	1.8018	1.8536	1.8565
1 7/8″	4½	1.4941	1.490	0.15	0.15	.012343	1.9075	1.9400	1.9434
2″	4½	1.6188	1.615	0.15	0.15	.012343	2.0325	2.0650	2.0682
2 1/8″	4½	1.7435	1.740	0.15	0.15	.012343	2.1575	2.1900	2.1930
2 1/4″	4½	1.8683	1.8651	0.15	0.15	.012343	2.2825	2.3150	2.3178
2 3/8″	4½	1.9930	1.990	0.15	0.15	.012343	2.4075	2.440	2.4426
2 1/2″	4	2.0707	2.067	0.20	0.20	.015625	2.5835	2.670	2.6670
2 3/4″	4	2.3203	2.317	0.20	0.20	.015625	2.8335	2.917	2.9196
3″	3½	2.5089	2.505	0.20	0.20	.020392	3.0525	3.105	3.1085
3 1/4″	3½	2.7587	2.755	0.20	0.20	.020392	3.3025	3.355	3.3582
3 1/2″	3¼	2.9711	2.967	0.20	0.20	.023654	3.5335	3.567	3.5705
3 3/4″	3	3.1770	3.1727	0.20	0.20	.02775	3.7613	3.7727	3.7765
4″	3	3.4266	3.4227	0.20	0.20	.02775	4.0113	4.0227	4.0263

WATCH SCREW THREADS

WATCH screw threads are of sharp V-form and generally 45-degree angle for screws used in nickel and brass; though 60 degrees for use in steel. The Waltham Watch Company and others use the centimeter as the unit for all measurements with the exception of the pitch, which is based on the inch; the Waltham threads being 110, 120, 140, 160, 170, 180, 200, 220, 240, 254, per inch and the diameters ranging from 0.120 to 0.035 cm.

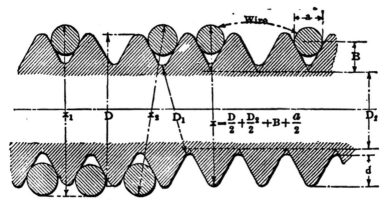

FIG. 13. — Measuring Whitworth Threads

D = outside diameter of thread.
D_1 = root diameter measured in thread groove.
n = number of threads per inch of length.
d = depth of thread.
p = distance from center to center of adjacent threads.
r = radius on Whitworth thread.
a = diameter of wire used.
B = distance from apex of thread angle at root, to center of wire.
D_2 = diameter of cylinder touched by apexes of thread angles.
x = diameter from top of threads on one side of tap or bolt, to top of wire laid in thread groove on opposite side.

WHITWORTH THREAD

p = lead = $\dfrac{1}{n}$, for single threads.

$$d = p \times .64033 = \frac{.64033}{n}.$$

$$D_1 = \sqrt{(D - 2d)^2 + \left(\frac{\text{lead}}{2}\right)^2}.$$

$r = p \times .1373.$

$a = p \times .84,\ \text{max; to } p \times .454,\ \text{min.}$

$$B = \frac{a}{2} \div \sin 27^\circ\ 30' = \frac{a}{.9235}.$$

$$D_2 = D - \frac{1.600825}{n}.$$

$$x = \frac{D}{2} + \frac{D_2}{2} + B + \frac{a}{2}.$$

$x_1 = D_2 + 2B + a.$

$$x_2 = \sqrt{(D_2 + 2B)^2 + \left(\frac{\text{lead}}{2}\right)^2} + a.$$

TABLE FOR MEASURING WHITWORTH THREADS WITH MICROMETERS AND WIRES

D	n	D_1	D_2	a	B	$\left(\dfrac{\text{lead}}{2}\right)^2$	x	x_1	x_2
1/4″	20	.1875	.1699	0.04	.04331	.000625	.2733	.2965	.2977
5/16″	18	.2428	.2235	0.04	.04331	.000771	.3313	.3501	.3514
3/8″	16	.2965	.2749	0.04	.04331	.000976	.3883	.4015	.4029
7/16″	14	.344	.3231	0.04	.04331	.001274	.4436	.4497	.4512
1/2″	12	.3953	.3666	0.06	.06496	.001735	.5232	.5563	.5582
9/16″	12	.4576	.4291	0.06	.06496	.001735	.5907	.6190	.6204
5/8″	11	.5105	.4794	0.06	.06496	.002065	.6372	.6693	.6710
11/16″	11	.5728	.5420	0.06	.06496	.002065	.7097	.7319	.7334
3/4″	10	.6239	.5899	0.06	.06496	.0025	.7649	.7798	.7815
13/16″	10	.6862	.6524	0.06	.06496	.0025	.8274	.8423	.8438
7/8″	9	.7348	.6971	0.06	.06496	.003086	.8810	.8870	.8882
15/16″	9	.797	.7596	0.06	.06496	.003086	.9435	.9495	.9512
1″	8	.8422	.7999	0.10	.10839	.003906	1.0583	1.1167	1.1185
1 1/8″	7	.9447	.8963	0.10	.10839	.005102	1.169	1.2131	1.2153
1 1/4″	7	1.0693	1.0213	0.10	.10839	.005102	1.294	1.3381	1.340
1 3/8″	6	1.1644	1.1082	0.10	.10839	.006944	1.400	1.4250	1.4276
1 1/2″	6	1.2892	1.2332	0.10	.10839	.006944	1.525	1.5500	1.5523
1 5/8″	5	1.3726	1.3048	0.15	.16242	.010	1.7023	1.7796	1.7826
1 3/4″	5	1.497	1.4298	0.15	.16242	.010	1.8273	1.9046	1.9074
1 7/8″	4 1/2	1.5942	1.5193	0.15	.16242	.012343	1.9345	1.9941	1.9973
2″	4 1/2	1.7185	1.6443	0.15	.16242	.012343	2.0595	2.1191	2.1221
2 1/8″	4 1/2	1.8437	1.7693	0.15	.16242	.012343	2.1845	2.2441	2.2470
2 1/4″	4	1.9338	1.8498	0.15	.16242	.015625	2.2873	2.3246	2.328
2 3/8″	4	2.0585	1.9750	0.15	.16242	.015625	2.4123	2.4498	2.453
2 1/2″	4	2.1833	2.100	0.15	.16242	.015625	2.5373	2.5748	2.5778
2 3/4″	3 1/2	2.3882	2.2926	0.20	.21567	.020392	2.837	2.9240	2.9276
3″	3 1/2	2.6397	2.5426	0.20	.21567	.020392	3.087	3.1740	3.1773
3 1/4″	3 1/4	2.860	2.7574	0.20	.21567	.023654	3.3194	3.3887	3.3924
3 1/2″	3 1/4	3.1098	3.0074	0.20	.21567	.023654	3.5694	3.6387	3.642
3 3/4″	3	3.327	3.2164	0.20	.21567	.027755	3.799	3.8477	3.8515
4″	3	3.5768	3.4664	0.20	.21567	.027755	4.049	4.0977	4.1012
4 1/4″	2 7/8	3.808	3.693	0.20	.21567	.030241	4.287	4.3243	4.328
4 1/2″	2 7/8	4.0582	3.943	0.20	.21567	.030241	4.537	4.5743	4.578
4 3/4″	2 5/8	4.2878	4.168	0.20	.21567	.033051	4.7746	4.7993	4.8025
5″	2 5/8	4.5376	4.418	0.20	.21567	.033051	5.0245	5.0493	5.0524
5 1/4″	2 1/2	4.7658	4.640	0.20	.21567	.036252	5.2607	5.2713	5.275
5 1/2″	2 1/2	5.0156	4.890	0.20	.21567	.036252	5.5107	5.5213	5.5248
5 3/4″	2 1/2	5.2415	5.110	0.20	.21567	.040	5.7455	5.7413	5.7446
6″	2 1/2	5.4913	5.360	0.20	.21567	.040	5.9955	5.9013	5.9944

MEASURING FINE PITCH SCREW-THREAD DIAMETERS

THE accompanying table should be of service to those using the three-wire system of measurement as the constants cover the finer pitches and may be easily applied to screw threads of any diameter. The diagrams, Fig. 14, make the method of application so plain that no description appears necessary.

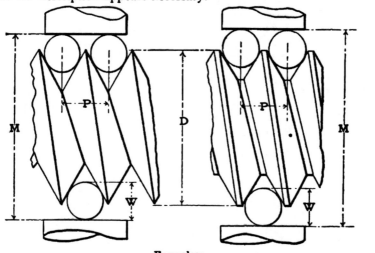

Formulas:

For V Thread
$$D = M - 3W + 1.732P.$$
$$M = D - 1.732P. + 3W.$$

For Sellers Thread
$$D = M - 3W + 1.5155P.$$
$$M = D - 1.5155P. + 3W.$$

FIG. 14. — Measuring Fine Pitch Threads

CONSTANTS FOR USE WITH THE 3-WIRE SYSTEM OF MEASURING SCREW THREADS

Threads per Inch	For V Thread $1.732\ P =$	For Sellers Thread $1.5155\ P =$	Threads per Inch	For V Thread $1.732\ P =$	For Sellers Thread $1.5155\ P =$
8	.21650	.18943	25	.06928	.06062
9	.19244	.16839	28	.06185	.05412
10	.17320	.15155	32	.05412	.04736
11	.15745	.13777	36	.04811	.04210
12	.14433	.12629	40	.04330	.03789
13	.13323	.11658	48	.03608	.03157
14	.12371	.10825	50	.03464	.03031
16	.10825	.09472	56	.03093	.02706
18	.09622	.08419	64	.02706	.02368
20	.08660	.07578	80	.02165	.01894
22	.07872	.06889	100	.01732	.01516
24	.07216	.06314			

MEASURING ACME 29-DEGREE THREADS

THE diameter of a wire which will be flush with tops of thread on tap when laid in the Acme thread groove, Fig. 15, will be found as follows:

Rad. of wire section = side opp. = side adj. × tan. 37° 45′ =

$$\frac{p \times .6293 + .0052}{2} \times .77428.$$

Diam. of wire = $(p \times .6293 + .0052) .77428.$

Wires of the diameter given in the table come flush with the tops of tap threads and project .010 above the top of screw threads.

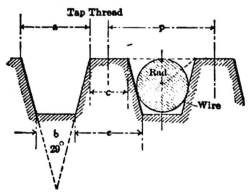

FIG. 15. — Measuring Acme Threads

TABLE OF WIRE SIZES FOR MEASURING ACME STANDARD 29° SCREW
AND TAP THREADS

Threads per Inch	Pitch	Diam. of Wire
1	1.	0.4913
1⅓	.750	0.3694
1½	.6666	0.3288
1¾	.5774	0.2824
2	.500	0.2476
2½	.400	0.1989
3	.3333	0.1664
4	.250	0.1278
5	.200	0.1014
6	.1666	0.0852
7	.1428	0.0736
8	.125	0.0649
9	.1111	0.0581
10	.100	c.0527

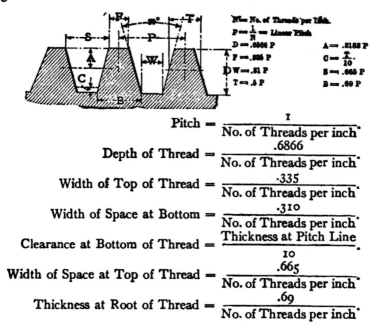

$$\text{Pitch} = \frac{1}{\text{No. of Threads per inch}}.$$

$$\text{Depth of Thread} = \frac{.6866}{\text{No. of Threads per inch}}.$$

$$\text{Width of Top of Thread} = \frac{.335}{\text{No. of Threads per inch}}.$$

$$\text{Width of Space at Bottom} = \frac{.310}{\text{No. of Threads per inch}}.$$

$$\text{Clearance at Bottom of Thread} = \frac{\text{Thickness at Pitch Line}}{10}.$$

$$\text{Width of Space at Top of Thread} = \frac{.665}{\text{No. of Threads per inch}}.$$

$$\text{Thickness at Root of Thread} = \frac{.69}{\text{No. of Threads per inch}}.$$

TABLE OF BROWN & SHARPE 29° WORM THREAD PARTS

	P	D	F	W	T	A	C	S	B
Number of Threads Per Inch	Pitch of Single Thread	Depth of Thread	Width of Top of Thread	Width of Space at Bottom	Thickness of Thread at Pitch Line	Thread Above Pitch Line	Clearance at Bottom of Thread	Width of Space at Top	Thickness at Root of Thread
1	1.0	.6866	.3350	.3100	.5000	.3183	.05	.665	.69
1¼	.8	.5492	.2680	.2480	.4000	.2546	.04	.532	.552
1½	.6666	.4577	.2233	.2066	.3333	.2122	.0333	.4433	.4599
2	.5	.3433	.1675	.1550	.2500	.1592	.0250	.3325	.345
2½	.4	.2746	.1340	.1240	.2000	.1273	.0200	.2660	.276
3	.3333	.2289	.1117	.1033	.1666	.1061	.0166	.2216	.2299
3½	.2857	.1962	.0957	.0886	.1429	.0909	.0143	.1901	.2011
4	.250	.1716	.0838	.0775	.1250	.0796	.0125	.1637	.1725
4½	.2222	.1526	.0744	.0689	.1111	.0707	.0111	.1478	.1533
5	.2	.1373	.0670	.0620	.1000	.0637	.0100	.1330	.138
6	.1666	.1144	.0558	.0517	.0833	.0531	.0083	.1108	.115
7	.1428	.0981	.0479	.0443	.0714	.0455	.0071	.095	.0985
8	.125	.0858	.0419	.0388	.0625	.0398	.0062	.0818	.0862
9	.1111	.0763	.0372	.0344	.0555	.0354	.0055	.0739	.0766
10	.10	.0687	.0335	.0310	.0500	.0318	.005	.0665	.069
12	.0833	.0572	.0279	.0258	.0416	.0265	.0042	.0551	.0575
16	.0625	.0429	.0209	.0194	.0312	.0199	.0031	.0409	.0431
20	.050	.0343	.0167	.0155	.0250	.0159	.0025	.0332	.0345

MEASURING BROWN & SHARPE 29-DEGREE WORM THREADS

THE diameter of wire for Brown & Sharpe worm thread, Fig. 16, for each pitch, that will rest in the thread groove on the thread angle and be flush with the tops of the finished threads, is found as follows; Rad. of wire section (see table) = side opp. = side adj. × tan.

$$37° 46' = \frac{0.665\ P}{2} \times 0.77428 = 0.257448\ P \text{ and diam. of wire} = 0.5149\ P.$$

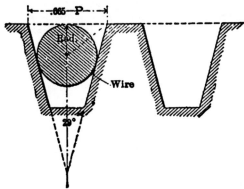

FIG. 16. — Measuring Brown & Sharpe Worm Threads.

TABLE OF WIRE SIZES FOR MEASURING B. & S. 29° WORM THREADS

Threads per Inch	Pitch	Diam. of Wire
½	2.	1.0298
⅝	1.750	0.9010
¾	1.500	0.7723
⅘	1.250	0.6436
1	1.0	0.5149
1½	.6666	0.3432
2	.5	0.2574
2½	.4	0.2060
3	.3333	0.1716
3½	.2857	0.1471
4	.250	0.1287
4½	.2222	0.1144
5	.2	0.1030
6	.1666	0.0858
7	.1428	0.0735
8	.125	0.0643
9	.1111	0.0572
10	.10	0.0515
12	.0833	0.0429
16	.0625	0.0322
20	.050	0.0257

WORM WHEEL HOBS

Hobs are made larger in diameter than the worm they are used with by the amount of two clearances. The Brown & Sharpe method is to make the clearance one-tenth of the thickness of the tooth on the pitch line or .05 inch for a worm of one pitch. If the worm was 3 inches outside diameter, which would be a fair proportion for this pitch, the outside diameter of the hob would be $3 + (2 \times .05) = 3.1$ inches. The thread tool would be .31 inch wide at the point and would cut $.6360 + .1 = .7366$ deep, leaving the top of the thread the same thickness as the bottom, which is different from the worm.

The land L should be made as near the proportions given as possible.

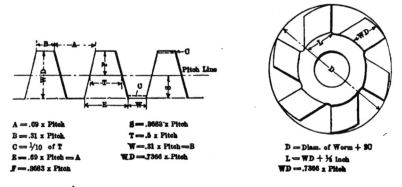

A = .09 x Pitch	S = .3683 x Pitch
B = .31 x Pitch	T = .5 x Pitch
C = 1/10 of T	W = .31 x Pitch = B
E = .69 x Pitch = A	WD = .7366 x Pitch
F = .3683 x Pitch	

D = Diam. of Worm + 2C
L = WD + ½ inch
WD = .7366 x Pitch

FIG. 17. — Section of Hob Thread   FIG. 18. — End View of Hob

The diagram Fig. 17 shows the shape and proportions of the thread of a worm hob, and Fig. 18 shows the proportions for the depth of tooth, the lead and the outside diameter. In these diagrams:

A = Width of space at top of tooth.
B = Width of thread at top.
C = Clearance or difference between the hob and worm.
D = Diameter of hob.
E = Width of tooth at bottom.
F = Hight above pitch line.
L = Width of land or tooth at bottom.
S = Depth below pitch line.
T = Width at pitch line.
W = Width of space at bottom.
WD = Whole depth of tooth.

PIPE AND PIPE THREADS

BRIGGS STANDARD PIPE THREADS

THE particulars in the following paragraph regarding this system of pipe standards are from a paper by the late Robert Briggs, C.E., read in 1882, before the Institution of Civil Engineers of Great Britain.

The taper employed has an inclination to 1 in 32 to the axis. The thread employed has an angle of 60 degrees; it is slightly rounded off, both at the top and at the bottom, so that the hight or depth of the thread, instead of being exactly equal to the pitch $\times$ .866 is only four-fifths of the pitch, or equal to $0.8\dfrac{1}{n}$, if n be the number of

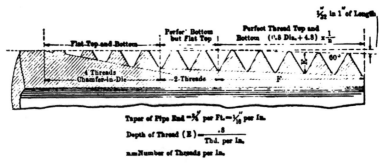

FIG. 1. — Longitudinal Section of Briggs Pipe Thread

threads per inch. For the length of tube-end throughout which the screw-thread continues perfect, the formula used is $(0.8\,D + 4.8)$ $\times \dfrac{1}{n}$, where D is the actual external diameter of the tube throughout its parallel length, and is expressed in inches. Further back, beyond the perfect threads, come two having the same taper at the bottom, but imperfect at the top. The remaining imperfect portion of the screw-thread, furthest back from the extremity of the tube, is not essential in any way to this system of joint; and its imperfection is simply incidental to the process of cutting the thread at a single operation.

Thread Section

The threads as produced at the pipe end in the Briggs system are represented clearly in the longitudinal section, Fig. 1.

Here the threads that are perfect at top and bottom are shown at F, the depth being indicated at E. Back of the perfect threads are represented the two threads with perfect bottom and flat tops and behind these are the imperfect threads produced by the chamfer or bell mouth of the threading die. A table giving the general dimensions of wrought iron tubes in the Briggs system will be found on page 40, while complete data pertaining to the thread depths, lengths of perfect and imperfect portions, allowances for making the joint in screwing the pipe into the fitting, gaging allowances, etc., are contained in the tables on pages 43 and 44.

In cutting pipe threads with a lathe tool as in threading taper work in general, the tool should be set at right angles to the axis of the piece and not square with the conical surface.

STANDARD DIMENSIONS OF WROUGHT-IRON WELDED TUBES
BRIGGS STANDARD

Diameter of Tubes			Thickness of Metal Inches	Screwed Ends	
Nominal Inside Inches	Actual Inside Inches	Actual Outside Inches		Number of Threads per Inch	Length of Perfect Thread Inches
$\frac{1}{8}$	0.270	0.405	0.068	27	0.19
$\frac{1}{4}$	0.364	0.540	0.088	18	0.29
$\frac{3}{8}$	0.494	0.675	0.091	18	0.30
$\frac{1}{2}$	0.623	0.840	0.109	14	0.39
$\frac{3}{4}$	0.824	1.050	0.113	14	0.40
1	1.048	1.315	0.134	$11\frac{1}{2}$	0.51
$1\frac{1}{4}$	1.380	1.660	0.140	$11\frac{1}{2}$	0.54
$1\frac{1}{2}$	1.610	1.900	0.145	$11\frac{1}{2}$	0.55
2	2.067	2.375	0.154	$11\frac{1}{2}$	0.58
$2\frac{1}{2}$	2.468	2.875	0.204	8	0.89
3	3.067	3.500	0.217	8	0.95
$3\frac{1}{2}$	3.548	4.000	0.226	8	1.00
4	4.026	4.500	0.237	8	1.05
$4\frac{1}{2}$	4.508	5.000	0.246	8	1.10
5	5.045	5.563	0.259	8	1.16
6	6.065	6.625	0.280	8	1.26
7	7.023	7.625	0.301	8	1.36
8	7.982	8.625	0.322	8	1.46
*9	9.000	9.688	0.344	8	1.57
10	10.019	10.750	0.366	8	1.68

* By the action of the manufacturers of wrought-iron pipe and boiler tubes, at a meeting held in New York, May 9, 1839, a change in size of actual outside diameter of 9-inch pipe was adopted, making the latter 9.625 instead of 9.688 inches, as given in the table of Briggs standard pipe diameters.

WHITWORTH PIPE THREADS

THE table below shows the practice in Great Britain in regard to pipe and pipe threads. The Whitworth pipe thread, which is the standard, is cut both straight and taper, the Engineering Standards Committee recommending, for taper threads, 1/4 inch per foot as in the Briggs standard.

WHITWORTH PIPE THREADS

Size In.	Diameter	Dia. at Bottom of Thread	No. of Threads per In.	Size In.	Diameter	Dia. at Bottom of Thread	No. of Threads per In.
1/8	0.3825	0.3367	28	1 5/8	2.021	1.905	11
1/4	0.518	0.4506	19	1 3/4	2.047	1.9305	11
3/8	0.6563	0.5889	19	1 7/8	2.245	2.1285	11
1/2	0.8257	0.7342	14	2	2.347	2.2305	11
5/8	0.9022	0.8107	14	2 1/4	2.5875	2.471	11
3/4	1.041	0.9495	14	2 1/2	3.0013	2.8848	11
7/8	1.189	1.0975	14	2 3/4	3.247	3.1305	11
1	1.309	1.1925	11	3	3.485	3.3685	11
1 1/8	1.492	1.3755	11	3 1/4	3.6985	3.582	11
1 1/4	1.65	1.5335	11	3 1/2	3.912	3.7955	11
1 3/8	1.745	1.6285	11	3 3/4	4.1255	4.009	11
1 1/2	1.8825	1.765	11	4	4.339	4.223	11

TAP DRILLS FOR PIPE TAPS

The sizes of Twist Drills to be used in boring holes, to be reamed with Pipe Reamers, and Threaded with Pipe Taps, are as follows:

Size Pipe Tap	BRIGGS		WHITWORTH		Size Pipe Tap	BRIGGS		WHITWORTH	
	Thread	Drill	Thread	Drill		Thread	Drill	Thread	Drill
1/8	27	21/64	28	5/16	1 3/4			11	1 15/16
1/4	18	27/64	19	27/64	2	11 1/2	2 3/16	11	1 5/32
3/8	18	9/16	19	9/16	2 1/4			11	2 3/8
1/2	14	11/16	14	11/16	2 1/2	8	2 9/16	11	2 25/32
5/8			14	23/32	2 3/4			11	3 1/32
3/4	14	29/32	14	27/32	3	8	3 3/16	11	3 9/32
7/8			14	1 1/16	3 1/4			11	3 1/2
1	11 1/2	1 1/8	11	1 1/8	3 1/2	8	3 11/16	11	3 3/4
1 1/4	11 1/2	1 15/32	11	1 15/32	3 3/4			11	4
1 1/2	11 1/2	1 23/32	11	1 23/32	4	8	4 3/16	11	4 1/4

THE PIPE JOINT IN THE BRIGGS SYSTEM

THE illustrations below and the tables on pages 43 and 44, represent the relation of the reamer, tap, die and testing gages in the preparation of the Briggs pipe end and fitting preliminary to making up the joint.

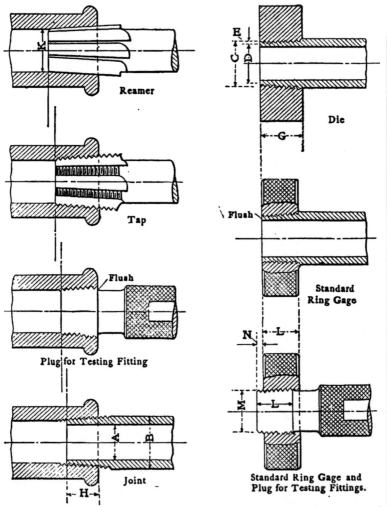

FIG. 2. — Reamer, Tap, Die and Gages for Briggs Pipe Standard

The illustrations to the left in Fig. 2 show the relative distances that the pipe reamer, tap, testing plug and pipe end are run into the fitting in making the joint; while at the right are shown the die and ring gage on the pipe end, and the relative diameters of the standard ring gage and the testing plug for the fittings.

In pipe fitting the end of the pipe should always be cut to fit the Briggs standard pipe gage. The fitting should be tapped small in order to insure a tight joint. Theoretically the joint should be tight when the pipe end has been screwed into the fitting a distance represented at H in the diagrams, Fig. 2 and following tables. However, to allow for errors the thread on the pipe is actually cut two threads beyond H. Similarly the fitting should be tapped two threads deeper than distance H.

The following table used in conjunction with the illustrations in Fig. 2, contains information as to length and number of perfect and imperfect threads; distance and number of turns the pipe screws into fitting by hand and with wrench, or the total length and number of threads of joint; ring and plug gage data for testing tools; besides general pipe dimensions, drill and reamer sizes, etc.

BRIGGS PIPE THREAD TABLE (See page 42)

Dia. of Pipe		Actual Outside	No. of Threads per Inch	Dia. at End of Pipe	Dia. at Bottom of Th'd	Depth of Thread	Length of Perfect Threads	No. of Perfect Threads
Nominal Inside	Actual Inside							
A	B			C	D	E	F	
⅛″	.270″	.405″	27	.393″	.334″	.029″	.19″	5.13
¼″	.364″	.540″	18	.522″	.433″	.044″	.29″	5.22
⅜″	.494″	.675″	18	.656″	.568″	.044″	.30″	5.4
½″	.623″	.840″	14	.815″	.701″	.057″	.39″	5.46
¾″	.824″	1.050″	14	1.025″	.911″	.057″	.40″	5.6
1 ″	1.048″	1.315″	11½	1.283″	1.144″	.069″	.51″	5.87
1¼″	1.380″	1.660″	11½	1.626″	1.488″	.069″	.54″	6.21
1½″	1.610″	1.900″	11½	1.866″	1.728″	.069″	.55″	6.33
2 ″	2.067″	2.375″	11½	2.339″	2.201″	.069″	.58″	6.67
2½″	2.468″	2.875″	8	2.819″	2.619″	.100″	.89″	7.12
3 ″	3.067″	3.500″	8	3.441″	3.241″	.100″	.95″	7.6
3½″	3.548″	4.000″	8	3.938″	3.738″	.100″	1.00″	8.0
4 ″	4.026″	4.500″	8	4.434″	4.234″	.100″	1.05″	8.4
4½″	4.508″	5. ″	8	4.931″	4.731″	.100″	1.10″	8.8
5 ″	5.045″	5.563″	8	5.490″	5.290″	.100″	1.16″	9.28
6 ″	6.065″	6.625″	8	6.546″	6.346″	.100″	1.26″	10.08
7 ″	7.023″	7.625″	8	7.540″	7.340″	.100″	1.36″	10.88
8 ″	7.982″	8.625″	8	8.534″	8.334″	.100″	1.46″	11.68
9 ″	9.000″	9.625″	8	9.527″	9.327″	.100″	1.57″	12.56
10 ″	10.019″	10.750″	8	10.645″	10.445″	.100″	1.68″	13.44

(Table Continued on Page 44)

BRIGGS PIPE THREAD TABLE—Continued. (See page 42)

Dia. Pipe	G — Total Length of Th'd and Thickness of Die	No. of Turns Pipe Screws into Fitting by Hand	Distance Pipe Screws into Fitting by Hand	No. of Turns Made with Wrench	Distance Pipe is Screwed in with Wrench	Total No. of Turns Pipe Screws into Fitting	H — Total Distance Pipe Screws into Fitting	Dia. Drill to be used with Pipe Reamer	K — Dia. End Pipe Before Chamfering Reamer	L — Length of Th'd on Plug and Ring Gage	No. of Th'ds on Plug and Ring Gage	M — Dia. of End of Plug for Testing Fittings	No. of Th'ds Inspection Plug Projects Through Ring	N — Distance Inspection Plug Projects Through Ring
1/8"	.412"	4	.148"	1.13	.042"	5.13	.19"	21/64"	.320"	.264"	7.13	.386"	3.13	.116"
1/4"	.624"	4	.222"	1.22	.068"	5.22	.29"	7/16"	.414"	.401"	7.22	.511"	3.22	.179"
3/8"	.634"	4	.222"	1.4	.078"	5.40	.30"	9/16"	.539"	.411"	7.4	.644"	3.4	.189"
1/2"	.818"	4	.285"	1.46	.105"	5.46	.39"	11/16"	.648"	.533"	7.46	.800"	3.46	.247"
3/4"	.828"	4	.285"	1.6	.115"	5.60	.40"	29/32"	.867"	.543"	7.6	1.009"	3.6	.257"
1"	1.03"	4½	.391"	1.37	.119"	5.87	.51"	1-9/64"	1.101"	.684"	7.87	1.265"	3.37	.293"
1 1/4"	1.06"	5	.435"	1.21	.105"	6.21	.54"	1-31/64"	1.449"	.714"	8.21	1.608"	3.21	.279"
1 1/2"	1.07"	5	.435"	1.33	.115"	6.33	.55"	1-23/32"	1.695"	.724"	8.33	1.848"	3.33	.289"
2"	1.10"	5	.435"	1.67	.145"	6.67	.58"	2-3/16"	2.164"	.754"	8.67	2.319"	3.67	.319"
2 1/2"	1.64"	5	.625"	2.12	.265"	7.12	.89"	2-9/16"	2.554"	1.14"	9.12	2.787"	4.12	.515"
3"	1.70"	5	.625"	2.6	.325"	7.60	.95"	3-3/16"	3.180"	1.2"	9.6	3.405"	4.6	.575"
3 1/2"	1.75"	5½	.688"	3.0	.375"	8.00	1.00"	3-11/16"	3.680"	1.25"	10.0	3.899"	5.0	.625"
4"	1.80"	5½	.688"	2.9	.362"	8.40	1.05"	4-3/16"	4.180"	1.3"	10.4	4.396"	4.9	.612"
4 1/2"	1.85"	5½	.688"	3.3	.412"	8.80	1.10"			1.35"	10.8	4.890"	5.3	.662"
5"	1.91"	6	.750"	3.78	.472"	9.28	1.16"			1.41"	11.28	5.445"	5.78	.722"
6"	2.01"	7	.875"	4.08	.510"	10.08	1.26"			1.51"	12.08	6.499"	6.08	.760"
7"	2.11"	8	1.000"	3.88	.485"	10.88	1.36"			1.61"	12.88	7.494"	5.88	.735"
8"	2.21"	9	1.125"	3.68	.460"	11.68	1.46"			1.71"	13.68	8.490"	5.68	.710"
9"	2.32"			3.56	.445"	12.56	1.57"			1.82"	14.56	9.484"	5.56	.695"
10"	2.43"	10	1.250"	3.44	.430"	13.44	1.68"			1.93"	15.44	10.603"	5.44	.680"

GAGE SETS FOR BRIGGS PIPE AND FITTINGS

THE gages manufactured by the Pratt & Whitney Company for makers and users of pipe and fittings include three distinct sets for each size of pipe, and these are illustrated in Fig. 3. Set No. 1 consists of a ring and plug conforming in all dimensions to the Briggs standard, and is known as the standard reference set. The plug screws into the ring with faces flush — as indicated by the position of the two gages. The flat milled on the plug shows the depth to which the latter should enter the fitting to allow for screwing up with tongs to make a steam-tight joint; the ring, of course, screws on to the pipe flush with the end.

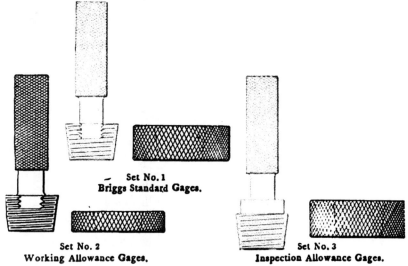

Set No. 1
Briggs Standard Gages.

Set No. 2
Working Allowance Gages.

Set No. 3
Inspection Allowance Gages.

FIG. 3. — Briggs Pipe Thread Gages

Set No. 2 — the working allowance set — consists of the plug already described and a ring whose thickness is equal to the standard ring less the allowance for screwing up the joint. As the plug and ring threads are of the same diameter at the small end, the bottom surfaces come flush when the two members are screwed together. It will be noted that, as the plug enters the fitting only to the bottom of the flat at the side, and the ring screws on to the pipe only far enough to bring the outer face flush with the pipe end, there are a few threads on, or in, the work beyond the reach of the gages; hence with this type of gage a reasonable amount of wear may be permitted at the end of the tap or the mouth of the die without causing the rejection of the work.

The plug and ring in set No. 3 are inspection allowance gages, the ring being the same in all particulars as the standard gage in set No. 1, while the plug is longer than Nos. 1 and 2 by an amount equal to the allowance for screwing up for a tight joint, this extra length being represented by the cylindrical portion at the rear o' the thread cone. When the gages are screwed together the back of the cylindrical section comes flush with the ring face and the threaded end of the plug projects through the ring, as indicated, a distance equal to the length of the cylinder, or the screwing-up allowance. This plug will enter a perfect fitting until the back of the threaded section is flush with the end of the fitting, thus testing the full depth of the tapped thread in the same way that the standard ring gage covers the thread on the pipe end, and at the same time showing that the fitting is tapped to right diameter to allow the joint to screw up properly.

NATIONAL STANDARD HOSE COUPLING

This standard for fire hose couplings was adopted by the National Fire Protection Association May 26, 1905 and has since been approved and adopted by various other organizations.

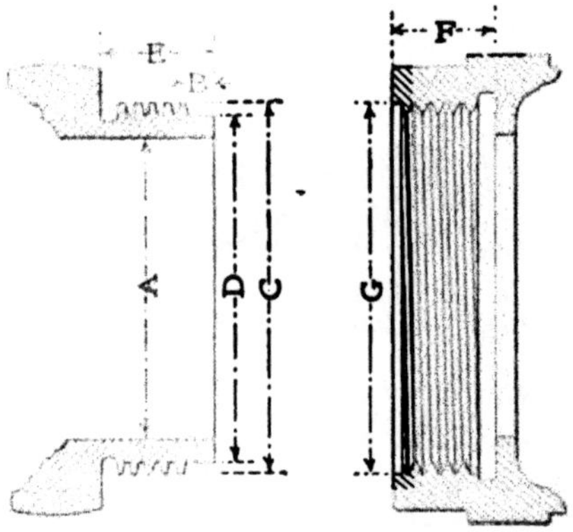

FIG. 4.—National Standard Hose Coupling

DIMENSIONS OF NATIONAL STANDARD HOSE COUPLINGS

A	Inside Diameter of Hose Couplings...	2½	3	3½	4½
B	Length of Blank End on Male Part...	¼	¼	¼	¼
C	Outside Diameter of Thread, Finished	3 1/16	3⅝	4¼	5¾
D	Diameter at Root of Thread.........	2.8715	3.3763	4.0013	5.3970
E	Total Length of Male End..........	1	1⅛	1⅛	1⅛
	Number of Threads per inch........	7½	6	6	4
F	Length of Female Thread...........	⅞	1	1	1¼
G	Diameter of Top of Female Thread..	3.0925	3.6550	4.28	5.80

NOTE:—The above to be of the 60-deg. V-thread pattern with one-hundredth inch cut off the top of thread and one-hundredth inch left in the bottom of the 2½-inch, 3-inch, and 3½-inch couplings, and two hundredths inch in like manner for the 4½-inch couplings, and with one-quarter inch blank end on male part of coupling in each case; female ends to be cut ⅛-inch shorter for endwise clearance. They should also be bored out .03 inch larger in the 2½-inch, 3 and 3½-inch sizes, and .05 inch larger on the 4½-inch size in order to make up easily and without jamming o sticking.

TWIST DRILLS

THE twist-drill is perhaps one of the most efficient tools in use as, although one half is cut away in the flutes, it has a very large cutting surface in proportion to its cross-sectional area. This is made possible by the fact that the work helps to support the drill and the feed pressure on the drill tends to force the point into a cone-shaped hole which centers it.

In addition to the radial relief or backing-off behind the cutting edge, twist-drills have longitudinal clearance by decreasing the diameter from the point toward the shank, varying from .00025 to .0015 per inch of length. This prevents binding and is essential in accurate drilling.

To increase the strength the web is increased gradually in thickness from the point toward the shank by drawing the cutters apart. This decreases chip room and to avoid this defect the spiral is increased in pitch and the flute widened to make up the chip room.

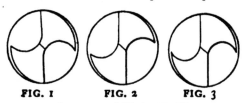

FIG. 1   FIG. 2   FIG. 3

Grooves of Twist Drills

The shape of the groove affects the power and the shape of the chip and experiments by the Cleveland Twist-Drill Company are interesting. The groove in Fig. 1 does not give a good cutting edge, especially near the center, as it does not allow a full curl to the chip. Fig. 2 is a very free cutting-groove, the chips curl up to the full size of the groove and this reduces the power required to bend the chips. Fig. 3 is an even better form as it rolls a chip with each turn conical so that one lays inside the other and makes a much shorter chip from the same depth of hole.

The angle of spirals varies from 18 to 35 degrees according to the ideas of the maker. In theory the finer the pitch or the greater the angle, the easier it should be to cut and curl the chip. But this gives a weak cutting edge and reduces the ability to carry off the heat, and it does not clear itself of chips so well. After a long series of tests the same firm adopted $27\frac{1}{2}$ degrees for the spiral. This angle makes the spiral groove of all drills start at the point with a pitch equal to six diameters of the blank, the increase in twist being a constant function of the angular movement of rotation of the drill blank. This angle is based on holes from one to three diameters deep. For deeper holes a smaller angle might be advisable and greater angle for holes of less depth. There is practically no difference in torsional stress with the angle between 25 and 30 degre

SHARPENING DRILLS

Drills should be sharpened so as to cut the right size and with as little power as possible. To cut the right size both lips must be the same length and the same angle. A gage as shown in Fig. 4 will help both to get the angle and to grind them central. This gives the usual lip edge of 59 degrees. Fig. 5 shows how you can see if both lips are ground alike, but does not give the angle. Fig. 6 is a suggestion by Professor Sweet of relieving the drill back of the cutting edge, making it similar to a flat drill in this respect.

For drilling brass or for any thin stock where the drill goes clear through, it is best to grind the cutting edge parallel with the axis of

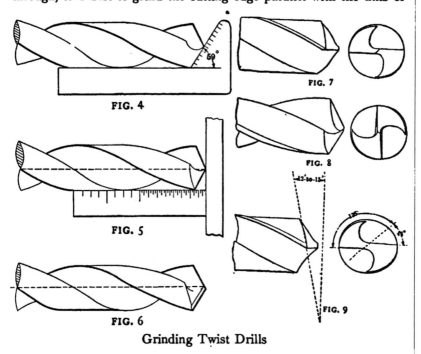

FIG. 4

FIG. 7

FIG. 8

FIG. 5

FIG. 6

FIG. 9

Grinding Twist Drills

the drill. This does away with the tendency to draw into the work. Fig. 7 shows how this is done.

It is sometimes necessary to thin the point of the drill to get best results. This requires care in grinding but can be done as shown in Fig. 8.

The best all-around clearance angle is 12 degrees, though for softer metals 15 degrees can be used. The 12 degrees is the angle at the cutting edge, but this should increase back of the cutting edge so that the line across the web should be 45 degrees, with the cutting edges. This is important, as it not only saves power but prevents splitting in hard service. The point of the drill should look like Fig. 7 or Fig. 8. Fig. 9 shows the clearance angle and the right angles for the drill point.

SPEED OF DRILLS

Learn to run drills at their proper speed to secure the most work with fewest grindings and breakages. The best practise is to use a speed that will give 30 feet a minute cutting speed for steel, 35 feet for cast iron and 60 feet for brass. This means that the cutting edge must run fast enough to make these speeds. For drilling steel with a $\frac{1}{16}$-inch drill this means 1834 revolutions a minute, while for brass it would be 3668 revolutions. The table gives the speeds without any figuring for all drills up to 3 inches. These speeds require plenty of lubricant. This is for carbon steel drills.

These speeds may have to be reduced with very hard material, but should not be exceeded in ordinary cases. High-speed steel drills will stand about double these speeds. It is easy to memorize the speeds of the drills most in use.

TABLE OF DRILL FEEDS

Diam. of Drill	Inches of Feed per Minute at Cutting Speed of								
	30 Feet—Steel			35 Feet—Iron			60 Feet—Brass		
	Rev. per Minute	Feed .004-.007		Rev. per Minute	.004-.007		Rev. per Minute	.004-.007 per Revolution	
$\frac{1}{16}$	1834	7.33	12.83	2140	8.56	14.97	3668	14.66	25.76
$\frac{1}{8}$	917	3.66	6.41	1070	4.28	7.49	1834	7.33	12.83
$\frac{3}{16}$	611	2.44	4.27	713	2.85	4.99	1222	4.88	8.58
$\frac{1}{4}$	458	1.83	3.20	535	2.14	3.74	917	3.66	6.44
		Feed .007	.015		.007	.015		.007	.015
$\frac{5}{16}$	367	2.57	5.5	428	3	6.42	733	5.14	11
$\frac{3}{8}$	306	2.14	4.6	357	2.5	5.35	611	4.28	9.2
$\frac{7}{16}$	262	1.83	3.9	306	2.14	4.58	524	3.66	7.8
$\frac{1}{2}$	229	1.60	3.43	268	1.87	4.	459	3.20	6.86
$\frac{5}{8}$	184	1.28	2.75	214	1.50	3.21	367	2.57	5.5
$\frac{3}{4}$	153	1.07	2.3	178	1.25	2.67	306	2.14	4.6
$\frac{7}{8}$	131	.91	1.95	153	1.07	2.29	262	1.88	3.93
1	115	.80	1.71	134	.93	2	229	1.60	3.43
$1\frac{1}{8}$	102	.71	1.53	119	.83	1.79	204	1.43	3.06
$1\frac{1}{4}$	91.8	.64	1.37	107	.75	1.61	183	1.28	2.75
$1\frac{3}{8}$	83.3	.58	1.25	97.2	.68	1.45	167	1.17	2.51
$1\frac{1}{2}$	76.3	.53	1.15	89.2	.62	1.38	153	1.07	2.3
$1\frac{5}{8}$	70.5	.49	1.05	82.2	.57	1.23	141	.99	2.11
$1\frac{3}{4}$	65.5	.45	.97	76.4	.53	1.15	131	.94	1.96
$1\frac{7}{8}$	61.1	.42	.92	71.3	.50	1.07	122	.85	1.81
2	57.3	.40	.85	66.9	.46	1.	115	.80	1.73
$2\frac{1}{4}$	51	.36	.71	59.4	.41	.89	102	.71	1.53
$2\frac{1}{2}$	45.8	.32	.68	53.5	.37	.80	91.7	.64	1.37
$2\frac{3}{4}$	41.7	.29	.62	48.6	.34	.73	83.4	.58	1.21
3	38.2	.27	.57	44.6	.31	.67	76.4	.53	1.15

FEED OF DRILLS

The feed of drills is usually given in parts of an inch per revolution, 0.004 to 0.007 inch for drills of ¼ inch and smaller and 0.007 to 0.015 inch for larger drills being recommended. This has been worked out into the table for the standard speeds to show inches of feed per minute for the three speeds given, which is more convenient. This is not an iron-clad rule but should be used with judgment. For high-speed steel these figures can be just about doubled.

DRILL TROUBLES

Twist-drills will stand more strain in proportion to their size and weight than almost any other tool, and when a good drill gives trouble it is pretty safe to say some of the conditions are wrong.

If it chips on the edge, the lip clearance is too great and fails to support the cutting edge or the feed is too heavy. Ease off on the feed first and then watch the grinding.

If it splits in the web it is either ground wrong, *i.e.*, does not have the center lip at the angle of 45 degrees or the feed is altogether too heavy.

If the outer corner wears, it shows that the speed is too great. This is particularly noticeable on cast iron.

DRILL POINTERS

In most cases it is better to use high speeds almost to the point where the drill corners commence to wear with a light feed than to use slower speed and heavy feed.

This is specially true of drilling in automatic machines where the holes are not more than twice as deep as the diameter where drills are flooded with lard oil. With deeper holes the chips are harder to get rid of and it is better to use slower speeds and heavier feeds as the drilled hole gets deeper.

Watch the drill chip and try to grind so that it will come out in a small compact roll. It is better to have this continuous clear to bottom of hole if possible.

In drilling brass use a heavier feed especially on automatic machines, as it helps to work out the chips. If you lubricate at all, flood the work. Twist-drills ground as for steel often catch and "hog in" on brass, especially at the bottom of the hole, where it breaks through. To avoid this, grind the lead or rake from the cutting edge.

In drilling hard material use turpentine as a lubricant.

Drills feed easier by thinning the extreme point if this is carefully done. This is important in hand feeding.

SPECIAL DRILLS AND THEIR USES

Ratchet-drills have a square taper shank, are used in hand-ratchet braces and in air-driven drills. Used in bridge building, structural and repair work.

The shell drill, Fig. 10, is used after a two-groove drill in chucking out cored holes or for enlarging holes that have been made with a two-groove drill. It has a taper hole and a number of sizes can be used on the same arbor.

Wire drills and jobbers or machinists drills both have round shanks and only differ in size. Wire drills are made to a twist-drill gage and the others to a jobbers or fractional gage.

Blacksmith drills all have a ½-inch shank 2½ inches long, so as to all fit the same holder. There is a flat on the shank for set-screw.

The straightway or Farmers drill has the same clearance as a twist-drill but the flutes are straight. It is used mostly in drilling brass and soft metals or in drilling cross holes or castings where blow holes may be found, as it is less likely to run than the twist-drill.

Oil drills have the advantage of the cutting edge being kept cool and of the chips being forced back through the grooves which reduces friction to a minimum. They are used for all kinds of drilling, mostly deep hole work. In cast-iron drilling air is sometimes used to blow out the chips and keep the drill cool. They are generally used in a screw or chucking machine or a lathe fitted for this work. Where the drill is held stationary and the work revolves, the oil is pumped to the connection and flows through the holes in the drill as in Fig. 11.

Where the drill revolves as in a drill press, the oil is pumped into a collar which remains stationary while the drill socket revolves, as in Fig. 12. An oil groove around the socket and holes through to the drill connects with the holes in the drill itself. Other types are shown in Figs. 13 and 14. The latter is used mostly

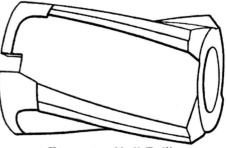

FIG. 10. — Shell Drill

in screw or chucking machine turrets where the oil is pumped into the center of the turret and into the large hole in the shank of the drill.

The hollow drill shown in Fig. 15 is used for deep drilling or long holes and is used in a lathe or some similar machine fitted for the purpose. It has a hole lengthwise through the shank connecting with the grooves of the drill. The shank can be threaded and fitted to a metal tube of such length as desired. The outside of the drill has a groove the whole length of the body. The lubricant is conveyed to the point of the drill on the outside through these grooves, while the hollow tube admits of the passage of oil and chips from the point. In using this drill the hole is first started with a short drill the size of the hole desired and drilled to a depth equal to the length of the body of the hollow drill to be used. The body of the hollow drill acts as a packing, compelling the oil to follow the grooves and the chips to flow out through the hollow shank.

Three and four groove drills are used for chucking out cored holes or enlarged holes that are first drilled with a two-groove drill. They are much better than a two-groove drill for use in cored holes or to follow another drill. The ends of the drills, Fig. 16 and 17, indicate that they are not made to drill from solid stock but for enlarging a hole already made.

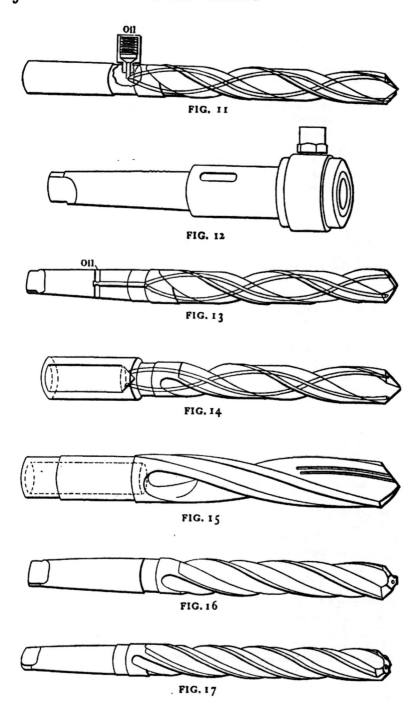

FIG. 11

FIG. 12

FIG. 13

FIG. 14

FIG. 15

FIG. 16

FIG. 17

The following tables give standard drill sizes in various ways, each being very convenient for certain classes of work:

DECIMAL EQUIVALENTS OF NOMINAL SIZES OF DRILLS

Inch	M.M.	Wire Gage	Decimals of an Inch	Inch	M.M.	Wire Gage	Decimals of an Inch	Inch	M.M.	Wire Gage	Decimals of an Inch
		80	.0135		1.2		.047244			37	.104
		79	.0145		1.3		.051181		2.7		.1063
$\frac{1}{64}$			.015625			55	.052			36	.1065
	.4		.01574			54	.055	$\frac{7}{64}$			.109375
		78	.016		1.4		.055118			35	.11
		77	.018		1.5		.05905		2.8		.11024
	.5		.01968			53	.0595			34	.111
		76	.020	$\frac{1}{16}$			.0625			33	.113
		75	.021		1.6		.06299		2.9		.11417
		74	.0225			52	.0635			32	.116
	.6		.02362		1.7		.066929		3		.11811
		73	.024			51	.067			31	.12
		72	.025			50	.07		3.1		.12205
		71	.026		1.8		.070866	$\frac{1}{8}$			.125
	.7		.02756			49	.073		3.2		.12598
		70	.028		1.9		.0748			30	.1285
		69	.02925			48	.076		3.3		.12992
		68	.031	$\frac{5}{64}$			.078125		3.4		.13386
$\frac{1}{32}$			.03125			47	.0785			29	.136
	.8		.031496		2		.07874		3.5		.1378
		67	.032			46	.081			28	.1405
		66	.033			45	.082	$\frac{9}{64}$			.140625
		65	.035		2.1		.082677		3.6		.14173
	.9		.03543			44	.086			27	.144
		64	.036		2.2		.086614		3.7		.14567
		63	.037			43	.089			26	.147
		62	.038		2.3		.09055			25	.1495
		61	.039			42	.0935		3.8		.14961
	1		.03937	$\frac{3}{32}$			.09375			24	.152
		60	.04		2.4		.09448		3.9		.15354
		59	.041			41	.096			23	.154
		58	.042			40	.098	$\frac{5}{32}$			.15625
		57	.043		2.5		.098425			22	.157
	1.1		.043307			39	.0995		4		.15748
		56	.0465			38	.1015			21	.159
$\frac{3}{64}$			.046875		2.6		.102362			20	.161

DECIMAL EQUIVALENTS OF NOMINAL SIZES OF DRILLS, *Contin*

Inch	M.M.	Wire Gage	Decimals of an Inch	Inch	M.M.	Letter Sizes	Decimals of an Inch	Inch	M.M.	Letter Sizes	Decimals of an Inch
	4.1		.16142			A	.234			P	.323
	4.2		.16536	15/64			.234375	21/64			.328125
		19	.166		6		.23622			Q	.332
	4.3		.16929			B	.238		8.5		.33465
		18	.1695		6.1		.24015		8.6		.33859
11/64			.171875			C	.242			R	.339
		17	.173		6.2		.2441	11/32			.34375
	4.4		.17323			D	.246		8.8		.34646
		16	.177		6.3		.24803			S	.348
	4.5		.17717	1/4		E	.25		9		.35433
		15	.18		6.4		.25197			T	.358
	4.6		.1811		6.5		.25591	23/64			.359375
		14	.182			F	.257		9.2		.36221
		13	.185		6.6		.25984			U	.368
	4.7		.18504			G	.261		9.5		.37402
3/16			.1875		6.7		.26377	3/8			.375
	4.8		.18898	17/64			.265625			V	.377
		12	.189			H	.266		9.6		.37796
		11	.191		6.8		.26772		9.8		.38583
	4.9		.19291		6.9		.27165			W	.386
		10	.1935			I	.272	25/64			.390625
		9	.196		7		.27559		10		.3937
	5		.19685			J	.277			X	.397
		8	.199		7.1		.27952			Y	.404
	5.1		.20079			K	.281	13/32			.40625
		7	.201	9/32			.28125			Z	.413
13/64			.203125		7.2		.28347		10.5		.4134
		6	.204		7.3		.2874	27/64			.421875
	5.2		.20473			L	.29		11		.43307
		5	.2055		7.4		.29133	7/16			.4375
	5.3		.20866			M	.295		11.5		.45276
		4	.209		7.5		.29528	29/64			.453125
	5.4		.2126	19/64			.296875	15/32			.46875
		3	.213		7.6		.29922		12		.47244
	5.5		.21654			N	.302	31/64			.484375
7/32			.21875		7.7		.30314		12.5		.4921
	5.6		.22047		7.8		.30709	1/2			.5
		2	.221		7.9		.31102				
	5.7		.22441	5/16			.3125				
		1	.228		8.		.31496				
	5.8		.22835			O	.316				
	5.9		.23228		8.2		.32284				

Decimal Equivalents of Nominal Sizes of Drills, *Continued*

Inch	M.M.	Decimals of an Inch	Inch	M.M.	Decimals of an Inch	Inch	M.M.	Decimals of an Inch
	13	.51181	43/64		.671875	27/32		.84375
33/64		.515625	11/16		.6875		21.5	.84646
17/32		.53125		17.5	.689	55/64		.859375
	13.5	.5315	44/64		.703125		22	.86614
35/64		.546875		18	.70866	7/8		.875
	14	.55118	23/32		.71875		22.5	.88583
9/16		.5625		18.5	.72835	57/64		.890625
	14.5	.57087	47/64		.734375		23	.90551
37/64		.578125		19	.74803	29/32		.90625
	15	.59055	3/4		.75			.921875
19/32		.59375	49/64		.765625		23.5	.9252
		.609375		19.5	.76772	15/16		.9375
	15.5	.61024	25/32		.78125		24	.94488
5/8		.625		20	.7874	61/64		.953125
	16	.62992	51/64		.796875		24.5	.9646
41/64		.640625		20.5	.8071	31/32		.96875
	16.5	.6496	13/16		.8125		25	.98425
21/32		.65625		21	.82677	63/64		.984375
	17	.66929	53/64		.828125	1		1.

Letter Sizes of Drills

Diameter Inches	Decimals of 1 Inch	Diameter Inches	Decimals of 1 Inch
A 15/64	.234	N	.302
B	.238	O 5/16	.316
C	.242	P 21/64	.323
D	.246	Q	.332
E 1/4	.250	R 11/32	.339
F	.257	S 11/32	.348
G	.261	T 23/64	.358
H 17/64	.266	U	.368
I	.272	V 3/8	.377
J	.277	W 25/64	.386
K 9/32	.281	X	.397
L	.290	Y 13/32	.404
M 19/64	.295	Z	.413

DECIMAL EQUIVALENTS OF DRILL SIZES FROM $\frac{1}{2}''$ TO No. 80

Size	Decimal Equivalent	Size	Decimal Equivalent	Size	Decimal Equivalent
$\frac{1}{2}$	0.500	3	0.213	$\frac{3}{32}$	0.0937
$\frac{31}{64}$	0.4843	4	0.209	42	0.0935
$\frac{15}{32}$	0.4687	5	0.2055	43	0.089
$\frac{29}{64}$	0.4531	6	0.204	44	0.086
$\frac{7}{16}$	0.4375	$\frac{13}{64}$	0.2031	45	0.082
$\frac{27}{64}$	0.4218	7	0.201	46	0.081
Z	0.413	8	0.199	47	0.0785
$\frac{13}{32}$	0.4062	9	0.196	$\frac{5}{64}$	0.0781
Y	0.404	10	0.1935	48	0.076
X	0.397	11	0.191	49	0.073
$\frac{25}{64}$	0.3906	12	0.189	50	0.070
W	0.386	$\frac{3}{16}$	0.1875	51	0.067
V	0.377	13	0.185	52	0.0635
$\frac{3}{8}$	0.375	14	0.182	$\frac{1}{16}$	0.0625
U	0.368	15	0.180	53	0.0595
$\frac{23}{64}$	0.3593	16	0.177	54	0.055
T	0.358	17	0.173	55	0.052
S	0.348	$\frac{11}{64}$	0.1718	$\frac{3}{64}$	0.0468
$\frac{11}{32}$	0.3437	18	0.1695	56	0.0465
R	0.339	19	0.166	57	0.043
Q	0.332	20	0.161	58	0.042
$\frac{21}{64}$	0.3281	21	0.159	59	0.041
P	0.323	22	0.157	60	0.040
O	0.316	$\frac{5}{32}$	0.1562	61	0.039
$\frac{5}{16}$	0.3125	23	0.154	62	0.038
N	0.302	24	0.152	63	0.037
$\frac{19}{64}$	0.2968	25	0.1495	64	0.036
M	0.295	26	0.147	65	0.035
L	0.290	27	0.144	66	0.033
$\frac{9}{32}$	0.2812	$\frac{9}{64}$	0.1406	$\frac{1}{32}$	0.0312
K	0.281	28	0.1405	67	0.032
J	0.277	29	0.136	68	0.031
I	0.272	30	0.1285	69	0.029
H	0.266	$\frac{1}{8}$	0.125	70	0.028
$\frac{17}{64}$	0.2656	31	0.120	71	0.026
G	0.261	32	0.116	72	0.025
F	0.257	33	0.113	73	0.024
E-$\frac{1}{4}$	0.250	34	0.111	74	0.0225
D	0.246	35	0.110	75	0.021
C	0.242	$\frac{7}{64}$	0.1093	76	0.020
B	0.238	36	0.1065	77	0.018
$\frac{15}{64}$	0.2343	37	0.104	$\frac{1}{64}$	0.0156
A	0.234	38	0.1015	78	0.016
1	0.228	39	0.0995	79	0.0145
2	0.221	40	0.098	80	0.0135
$\frac{7}{32}$	0.2187	41	0.096		

TAP DRILL SIZES FOR REGULAR THREADS

THESE sizes give an allowance above the bottom of thread on sizes 3/16 to 2; varying respectively as follows: for "V" threads, .010 to 055 inch; for U. S. S. and Whitworth threads, 005 to 027 inch. These are found by adding to the size at bottom of thread, 1/4 of the pitch for "V" threads, and 1/8 of the pitch for U. S. S. and Whitworth, the pitch being equal to 1 inch divided by the number of threads per inch. In practice it is better to use a larger drill if the exact size called for cannot be had.

Size Tap	No. of Threads	Size of Drill			Size Tap	No. of Threads	Size of Drill		
		U.S.S.	V	W			U.S.S	V	W
3/16	24	.128	.111	.129	15/16	9	.808	.790	.810
1/4	20	.191	.184	.192	1	8	.854	.832	.856
5/16	18	.248	.239	.249	1 1/16	8	.917	.894	.919
3/8	16	.302	.293	.303	1 1/8	7	.957	.932	.960
7/16	14	.354	.345	.355	1 1/4	7	1.082	1.057	1.085
1/2	13	.409	.399	.410	1 3/8	6	1.179	1.144	1.182
1/2	12	.402	.391	.403	1 1/2	6	1.304	1.269	1.307
9/16	12	.465	.453	.466	1 5/8	5 1/2	1.412	1.372	1.416
5/8	11	.518	.506	.520	1 5/8	5	1.390	1.347	1.394
11/16	11	.581	.568	.583	1 3/4	5	1.515	1.472	1.519
3/4	10	.632	.618	.634	1 7/8	5	1.640	1.597	1.644
13/16	10	.695	.680	.697	1 7/8	4 1/2	1.614	1.566	1.619
7/8	9	.745	.728	.747	2	4 1/2	1.739	1.691	1.744

A very simple rule, which is good enough in many cases, is: Subtract the pitch of one thread from the diameter of the tap.

A 3/8-inch tap 16-thread would be 3/8 minus 1/16 = 5/16 drill; a 3/4-inch tap, ten-thread, would be 3/4 minus 1/10 = 75/100 − 10/100 or 0.75 − 0.10 = 65/100 or 0.65, or a little over 5/8 of an inch, so a 5/8-inch drill will do nicely. With a 1-inch tap we have 1 − 1/8 = 7/8-inch drill, which is a little large but leaves enough thread for most cases.

TAP DRILLS

For Machine Screw Taps

THESE drills will give a thread full enough for all practical purposes. but not a *full* thread as this is very seldom required in practical work, Further data along this line will be found in the tables which follow.

TAP DRILLS

Sizes of Taps	No. of Threads	Sizes of Drills	Sizes of Taps	No. of Threads	Sizes of Drills
2	48	48	12	24	19
2	56	46	13	20	17
2	64	45	13	24	15
3	40	48	14	20	14
3	48	47	14	22	13
3	56	45	14	24	11
4	32	45	15	18	12
4	36	43	15	20	10
4	40	42	15	24	7
5	30	41	16	16	10
5	32	40	16	18	7
5	36	38	16	20	5
5	40	36	16	24	1
6	30	39	17	16	7
6	32	37	17	18	4
6	36	35	17	20	2
6	40	33	18	16	2
7	28	32	18	18	1
7	30	31	18	20	B
7	32	30	19	16	C
8	24	31	19	18	D
8	30	30	19	20	E
8	32	29	20	16	E
9	24	29	20	18	E
9	28	27	20	20	F
9	30	26	22	16	H
9	32	24	22	18	I
10	24	26	24	14	K
10	28	24	24	16	L
10	30	23	24	18	M
10	32	21	26	14	O
11	24	20	26	16	P
11	28	19	28	14	R
11	30	18	28	16	S
12	20	21	30	14	T
12	22	19	30	16	U

DIMENSIONS FOR TWIST DRILLS

FOR BORING HOLES TO BE THREADED WITH U. S. FORM OF THREAD TAPS $\frac{1}{16}$ to $\frac{1}{4}$ INCH DIAMETER

Diameter Inches	No. of Threads to the Inch	Exact Diameter Bottom of Thread Inches	Gage No. of Drill	Diameter Inches	No. of Threads to the Inch	Exact Diameter Bottom of Thread Inches	Gage No. of Drill
$\frac{1}{16}$	60	.041	57	$\frac{1}{4}$	26	.200	6
$\frac{1}{16}$	64	.042	56	$\frac{5}{64}$	56	.055	53
$\frac{3}{32}$	48	.067	50	$\frac{5}{64}$	60	.056	53
$\frac{3}{32}$	50	.068	50	$\frac{7}{64}$	40	.077	46
$\frac{3}{32}$	56	.071	49	$\frac{7}{64}$	44	.080	45
$\frac{3}{32}$	60	.072	48	$\frac{7}{64}$	48	.082	44
$\frac{1}{8}$	40	.093	41	$\frac{9}{64}$	32	.100	38
$\frac{1}{8}$	44	.096	40	$\frac{9}{64}$	36	.105	36
$\frac{1}{8}$	48	.098	39	$\frac{9}{64}$	40	.108	34
$\frac{5}{32}$	32	.116	31	$\frac{11}{64}$	32	.131	29
$\frac{5}{32}$	36	.120	31	$\frac{11}{64}$	36	.136	28
$\frac{5}{32}$	40	.124	30	$\frac{11}{64}$	40	.139	28
$\frac{3}{16}$	24	.133	29	$\frac{11}{64}$	24	.149	24
$\frac{3}{16}$	28	.141	27	$\frac{11}{64}$	28	.157	21
$\frac{3}{16}$	30	.144	26	$\frac{11}{64}$	32	.162	19
$\frac{3}{16}$	32	.147	25	$\frac{11}{64}$	36	.167	18
$\frac{3}{16}$	36	.152	23	$\frac{13}{64}$	24	.180	13
$\frac{7}{32}$	24	.164	19	$\frac{13}{64}$	28	.188	10
$\frac{7}{32}$	28	.172	16	$\frac{15}{64}$	32	.194	8
$\frac{7}{32}$	32	.178	14	$\frac{15}{64}$	36	.198	7
$\frac{7}{32}$	36	.183	12	$\frac{17}{64}$	18	.193	9
$\frac{1}{4}$	18	.178	14	$\frac{17}{64}$	20	.201	5
$\frac{1}{4}$	20	.185	12	$\frac{17}{64}$	24	.211	3
$\frac{1}{4}$	22	.190	10	$\frac{17}{64}$	26	.216	2
$\frac{1}{4}$	24	.196	8	$\frac{17}{64}$	32	.225	1

DRILLS AND REAMERS FOR DOWELL PINS

Sizes of Rod		Drills and Reamers for Drive Fits			Drills for Clearance	
No. of Gage (Stubbs Steel Wire)	Dia.	Size of Drill	Dia. of Drill	Dia. of Reamer	Size of Drill	Dia. of Drill
54	.055	No. 55	.052		No. 54	.055
45	.081	" 47	.0785		" 46	.081
33	.112	" 36	.1065	.110	" 33	.113
30	.127	" 31	.120	.125	" 30	.1285
21	.157	" 24	.152	.155	" 22	.157
10	.191	" 13	.185	.189	" 11	.191
	.252	C	.242	.250 −.2505	F	.257
	.315	$\frac{5}{16}$ Reamer Drill	.307	.3125−.313	O	.316
V	.377	$\frac{3}{8}$ "	.366	.375 −.3755	V	.377
	.439	$\frac{7}{16}$ "	.427	.4375−.438		
	.503	$\frac{1}{2}$ "	.489	.500 −.5005		
	.628	$\frac{5}{8}$ "	.616	.625 −.6255		
	.753	$\frac{3}{4}$ "	.734 ($\frac{3}{4}$)	.750 −.7505		

DOUBLE DEPTH OF THREADS

Threads per in.	V Threads D D	U. S. St'd D D	Whit. St'd D D	Threads per in.	V Threads D D	U.S. St'd D D	Whit. St'd D D
2	.86650	.64950	.64000	28	.06185	.04639	.04571
2¼	.77022	.57733	.56888	30	.05773	.04330	.04266
2⅜	.72960	.54694	.53894	32	.05412	.04059	.04000
2½	.69320	.51960	.51200	34	.05097	.03820	.03764
2⅝	.66015	.49485	.48761	36	.04811	.03608	.03555
2¾	.63019	.47236	.46545	38	.04560	.03418	.03368
2⅞	.60278	.45182	.44521	40	.04330	.03247	.03200
3	.57733	.43300	.42666	42	.04126	.03093	.03047
3¼	.53323	.39966	.39384	44	.03936	.02952	.03136
3½	.49485	.37114	.36571	46	.03767	.02823	.02782
4	.43300	.32475	.32000	48	.03608	.02706	.02666
4½	.38488	.28869	.28444	50	.03464	.02598	.02560
5	.34660	.25980	.25600	52	.03332	.02498	.02461
5½	.31490	.23618	.23272	54	.03209	.02405	.02370
6	.28866	.21650	.21333	56	.03093	.02319	.02285
7	.24742	.18557	.18285	58	.02987	.02239	.02206
8	.21650	.16237	.16000	60	.02887	.02165	.02133
9	.19244	.14433	.14222	62	.02795	.02095	.02064
10	.17320	.12990	.12800	64	.02706	.02029	.02000
11	.15745	.11809	.11636	66	.02625	.01968	.01939
11½	.15069	.11295	.11121	68	.02548	.01910	.01882
12	.14433	.10825	.10666	70	.02475	.01855	.01828
13	.13323	.09992	.09846	72	.02407	.01804	.01782
14	.12357	.09278	.09142	74	.02341	.01752	.01729
15	.11555	.08660	.08533	76	.02280	.01714	.01673
16	.10825	.08118	.08000	78	.02221	.01665	.01641
18	.09622	.07216	.07111	80	.02166	.01623	.01600
20	.08660	.06495	.06400	82	.02113	.01584	.01560
22	.07872	.05904	.05818	84	.02063	.01546	.01523
24	.07216	.05412	.05333	86	.02015	.01510	.01476
26	.0661	.04996	.04923	88	.01957	.01476	.01454
27	.06418	.04811	.04740	90	.01925	.01443	.01422

This gives the depth to allow for a full thread in a nut or similar piece of work for threads for 2 to 90 per inch, regardless of the diameter. A special nut for a 2-inch bolt, 20 threads per inch, U. S. Standard would have a hole 2. − .06495 = 1.93505 inches in diameter bored in it.

SIZES OF TAP DRILLS FOR TAPS WITH "V" THREAD

Diam. Tap, in Inches	Threads per Inch	Size of Drill, Number	Diam. Tap, in Inches	Threads per Inch	Size of Drill	Diam. Tap, in Inches	Threads per Inch	Size of Drill, Inches	Diam. Tap, in Inches	Threads per Inch	Size of Drill, Inches
$\frac{3}{32}$	48	50	$\frac{7}{32}$	24	No. 20	$\frac{19}{32}$	12	$\frac{31}{64}$	$1\frac{7}{32}$	7	$1\frac{1}{64}$
$\frac{3}{32}$	52	50	$\frac{7}{32}$	28	No. 17	$\frac{19}{32}$	14	$\frac{1}{2}$	$1\frac{7}{32}$	8	$1\frac{8}{64}$
$\frac{3}{32}$	54	49	$\frac{7}{32}$	30	No. 16	$\frac{5}{8}$	10	$\frac{31}{64}$	$1\frac{1}{4}$	7	$1\frac{8}{64}$
$\frac{3}{32}$	56	49	$\frac{7}{32}$	32	No. 15	$\frac{5}{8}$	11	$\frac{1}{2}$	$1\frac{9}{32}$	7	$1\frac{6}{64}$
$\frac{3}{32}$	60	48	$\frac{1}{4}$	24	No. 16	$\frac{5}{8}$	12	$\frac{33}{64}$	$1\frac{5}{16}$	7	$1\frac{10}{64}$
$\frac{7}{64}$	32	50	$\frac{1}{4}$	28	No. 12	$\frac{21}{32}$	10	$\frac{43}{64}$	$1\frac{5}{16}$	7	$1\frac{6}{64}$
$\frac{7}{64}$	36	49	$\frac{1}{4}$	32	No. 10	$\frac{21}{32}$	11	$\frac{9}{16}$	$1\frac{3}{8}$	6	$1\frac{3}{16}$
$\frac{7}{64}$	40	47	$\frac{1}{4}$	18	No. 17	$\frac{21}{32}$	12	$\frac{37}{64}$	$1\frac{13}{32}$	6	$1\frac{5}{32}$
$\frac{7}{64}$	48	44	$\frac{1}{4}$	20	No. 14	$\frac{11}{16}$	11	$\frac{19}{32}$	$1\frac{7}{16}$	6	$1\frac{3}{16}$
$\frac{7}{64}$	56	43	$\frac{1}{4}$	24	No. 9	$\frac{11}{16}$	12	$\frac{39}{64}$	$1\frac{15}{32}$	6	$1\frac{7}{32}$
$\frac{1}{8}$	32	44	$\frac{5}{16}$	16	No. 10	$\frac{23}{32}$	11	$\frac{5}{8}$	$1\frac{1}{2}$	6	$1\frac{1}{4}$
$\frac{1}{8}$	36	43	$\frac{9}{32}$	18	$\frac{11}{64}$ in.	$\frac{23}{32}$	12	$\frac{41}{64}$	$1\frac{17}{32}$	6	$1\frac{9}{32}$
$\frac{1}{8}$	40	42	$\frac{9}{32}$	20	No. 3	$\frac{3}{4}$	10	$\frac{5}{8}$	$1\frac{9}{16}$	6	$1\frac{5}{16}$
$\frac{1}{8}$	42	41	$\frac{5}{16}$	16	No. 1	$\frac{3}{4}$	11	$\frac{41}{64}$	$1\frac{19}{32}$	6	$1\frac{11}{32}$
$\frac{1}{8}$	48	39	$\frac{5}{16}$	18	$\frac{13}{64}$ in.	$\frac{3}{4}$	12	$\frac{21}{32}$	$1\frac{5}{8}$	5	$1\frac{13}{32}$
$\frac{9}{64}$	30	41	$\frac{11}{32}$	16	F	$\frac{25}{32}$	10	$\frac{43}{64}$	$1\frac{5}{8}$	$5\frac{1}{2}$	$1\frac{7}{16}$
$\frac{9}{64}$	32	40	$\frac{11}{32}$	18	$\frac{17}{64}$ in.	$\frac{25}{32}$	11	$\frac{11}{16}$	$1\frac{21}{32}$	5	$1\frac{7}{16}$
$\frac{9}{64}$	36	37	$\frac{3}{8}$	14	J	$\frac{25}{32}$	12	$\frac{45}{64}$	$1\frac{11}{16}$	$5\frac{1}{2}$	$1\frac{1}{2}$
$\frac{9}{64}$	40	34	$\frac{3}{8}$	16	L	$\frac{13}{16}$	10	$\frac{23}{32}$	$1\frac{23}{32}$	5	$1\frac{1}{2}$
$\frac{5}{32}$	30	33	$\frac{3}{8}$	18	$\frac{19}{64}$ in.	$\frac{13}{16}$	10	$\frac{23}{32}$	$1\frac{3}{4}$	$5\frac{1}{2}$	$1\frac{17}{32}$
$\frac{5}{32}$	32	32	$\frac{13}{32}$	14	N	$\frac{7}{8}$	9	$\frac{49}{64}$	$1\frac{25}{32}$	5	$1\frac{9}{16}$
$\frac{5}{32}$	36	31	$\frac{13}{32}$	16	P	$\frac{7}{8}$	10	$\frac{25}{32}$	$1\frac{13}{16}$	5	$1\frac{19}{32}$
$\frac{11}{64}$	40	30	$\frac{13}{32}$	18	$\frac{21}{64}$ in.	$\frac{15}{16}$	9	$\frac{13}{16}$	$1\frac{7}{8}$	5	$1\frac{21}{32}$
$\frac{11}{64}$	32	30	$\frac{7}{16}$	14	R	$\frac{15}{16}$	9	$\frac{53}{64}$	$1\frac{29}{32}$	5	$1\frac{11}{16}$
$\frac{11}{64}$	36	29	$\frac{7}{16}$	16	S	$\frac{15}{16}$	9	$\frac{13}{16}$	$1\frac{15}{16}$	5	$1\frac{23}{32}$
$\frac{11}{64}$	40	28	$\frac{15}{32}$	14	$\frac{3}{8}$ in.	1	8	$\frac{27}{32}$	$1\frac{31}{32}$	5	$1\frac{3}{4}$
$\frac{3}{16}$	24	29	$\frac{15}{32}$	16	W	$1\frac{1}{32}$	8	$\frac{57}{64}$	2	$4\frac{1}{2}$	$1\frac{47}{64}$
$\frac{3}{16}$	28	28	$\frac{1}{2}$	12	$\frac{25}{64}$ in.	$1\frac{1}{16}$	8	$\frac{29}{32}$	2	5	$1\frac{49}{64}$
$\frac{3}{16}$	30	27	$\frac{1}{2}$	13	X	$1\frac{1}{16}$	8	$\frac{29}{32}$	2	$4\frac{1}{2}$	$1\frac{47}{64}$
$\frac{3}{16}$	32	26	$\frac{1}{2}$	14	$\frac{13}{32}$ in.	$1\frac{1}{8}$	7	$\frac{61}{64}$	2	5	$1\frac{13}{16}$
$\frac{3}{16}$	36	24	$\frac{17}{32}$	12	$\frac{27}{64}$ in.	$1\frac{1}{8}$	8	$\frac{63}{64}$	2	$4\frac{1}{2}$	$1\frac{51}{64}$
$\frac{13}{64}$	24	26	$\frac{17}{32}$	13	$\frac{7}{16}$ in.	$1\frac{5}{32}$	7	$\frac{63}{64}$	2	5	$1\frac{27}{32}$
$\frac{13}{64}$	28	22	$\frac{17}{32}$	14	$\frac{7}{16}$ in.	$1\frac{3}{16}$	8	$\frac{63}{64}$	2	$4\frac{1}{2}$	$1\frac{51}{64}$
$\frac{13}{64}$	32	20	$\frac{9}{16}$	12	$\frac{29}{64}$ in.	$1\frac{3}{16}$	7	$1\frac{3}{16}$	2	5	$1\frac{27}{32}$
$\frac{13}{64}$	36	18	$\frac{9}{16}$	14	$\frac{31}{32}$ in.	$1\frac{1}{16}$	8	$1\frac{6}{64}$	2	$4\frac{1}{2}$	$1\frac{53}{64}$

This table gives similar information but in a way that would be more convenient in some cases.

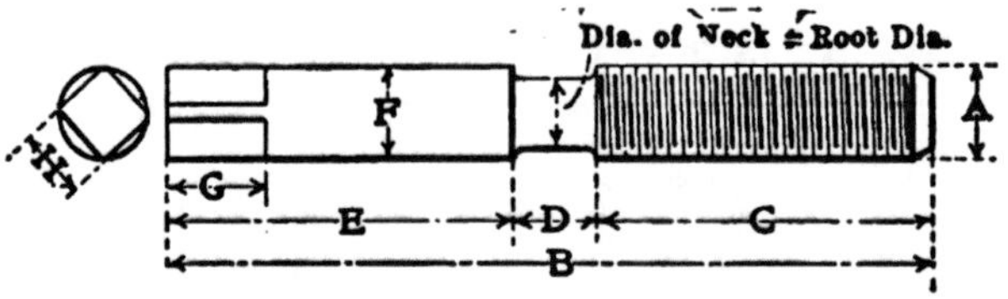

Dimensions of Machine Screw Taps

Number of Tap	Diameter of Tap	Number of Threads per Inch	Total Length	Length of Thread	Length of Neck	Length of Shank	Diameter of Shank	Length of Square	Size of Square	No. of Flutes
	A		B	C	D	E	F	G	H	
1	.071	64	1 3/4	9/16	No neck used on these.	1 3/16	.125	3/16	3/32	3
1 1/2	.081	56	1 3/4	9/16		1 3/16	.125	3/16	3/32	3
2	.089	56	1 3/4	9/16		1 3/16	.125	3/16	3/32	3
3	.101	48	1 7/8	5/8		1 1/4	.125	3/16	3/32	3
4	.113	36	2	11/16		1 5/16	.125	3/16	3/32	3
5	.125	36	2 1/8	3/4		1 3/8	.125	7/32	3/32	3
6	.141	32	2 1/8	3/4		1 3/8	.141	7/32	7/64	3
7	.154	32	2 1/8	3/4		1 3/8	.154	7/32	7/64	3
8	.166	32	2 1/4	13/16	1/8	1 5/16	.166	7/32	1/8	4
9	.180	30	2 1/4	7/8	1/8	1 1/4	.180	1/4	1/8	4
10	.194	24	2 1/4	7/8	1/8	1 1/4	.194	1/4	5/32	4
11	.206	24	2 3/8	7/8	1/8	1 1/4	.206	1/4	5/32	4
12	.221	24	2 3/8	15/16	1/8	1 5/16	.221	9/32	5/32	4
13	.234	22	2 1/2	1	3/16	1 5/16	.234	9/32	3/16	4
14	.246	20	2 5/8	1 1/16	3/16	1 3/8	.246	9/32	3/16	4
15	.261	20	2 3/4	1 1/8	3/16	1 7/16	.261	5/16	3/16	4
16	.272	18	2 3/4	1 1/8	3/16	1 7/16	.272	5/16	7/32	4
18	.298	18	2 3/4	1 1/8	3/16	1 7/16	.298	5/16	7/32	4
20	.325	16	3	1 1/4	1/4	1 1/2	.325	11/32	1/4	4
22	.350	16	3	1 1/4	1/4	1 1/2	.350	11/32	1/4	4
24	.378	16	3 1/4	1 1/4	5/16	1 11/16	.378	3/8	9/32	4
26	.404	16	3 1/4	1 1/4	5/16	1 11/16	.404	3/8	5/16	4
28	.430	14	3 1/2	1 3/8	5/16	1 13/16	.430	13/32	5/16	4
30	.456	14	3 1/2	1 3/8	5/16	1 13/16	.456	7/16	11/32	4

These are for the American Screw Company's Standard screws that have been in use for many years.

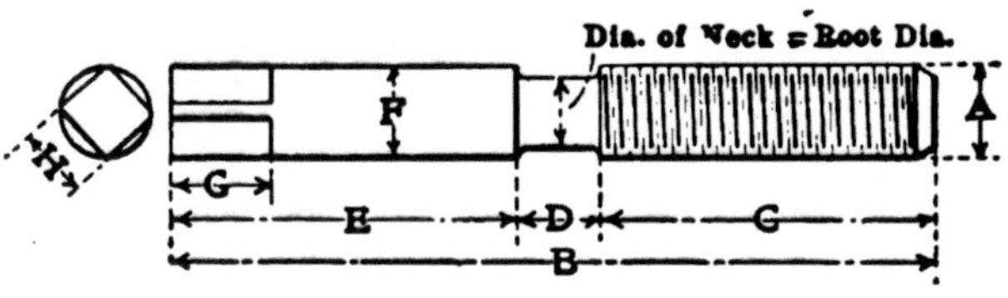

Dimensions of Machine Screw Taps

No. of Tap	Dia. of Tap	Threads per Inch	Total Length	Length of Thread		Length of Shank	Diameter of Shank	Length of Square	Size of Square	No. of Flutes
	A					E	F	G	H	
0	.060	80	$1\frac{3}{4}$	$\frac{9}{16}$	No Neck used on these	$1\frac{3}{16}$	.125	$\frac{3}{16}$	$\frac{3}{32}$	3
1	.073	72	$1\frac{3}{4}$	$\frac{9}{16}$		$1\frac{3}{16}$	.125	$\frac{3}{16}$	$\frac{3}{32}$	3
2	.086	64	$1\frac{3}{4}$	$\frac{9}{16}$		$1\frac{3}{16}$	.125	$\frac{3}{16}$	$\frac{3}{32}$	3
3	.099	56	$1\frac{7}{8}$	$\frac{5}{8}$		$1\frac{1}{4}$	.125	$\frac{3}{16}$	$\frac{3}{32}$	3
4	.112	48	2	$\frac{11}{16}$		$1\frac{5}{16}$	.125	$\frac{3}{16}$	$\frac{3}{32}$	3
5	.125	44	$2\frac{1}{8}$	$\frac{3}{4}$		$1\frac{3}{8}$	.125	$\frac{7}{32}$	$\frac{3}{32}$	3
6	.138	40	$2\frac{1}{8}$	$\frac{3}{4}$		$1\frac{3}{8}$	.138	$\frac{7}{32}$	$\frac{7}{64}$	3
7	.151	36	$2\frac{1}{8}$	$\frac{3}{4}$		$1\frac{3}{8}$	.151	$\frac{7}{32}$	$\frac{7}{64}$	3
8	.164	36	$2\frac{1}{4}$	$\frac{13}{16}$	$\frac{1}{8}$	$1\frac{5}{16}$	.164	$\frac{7}{32}$	$\frac{1}{8}$	4
9	.177	32	$2\frac{1}{4}$	$\frac{7}{8}$	$\frac{1}{8}$	$1\frac{1}{4}$	.177	$\frac{1}{4}$	$\frac{1}{8}$	4
10	.190	30	$2\frac{1}{4}$	$\frac{7}{8}$	$\frac{1}{8}$	$1\frac{1}{4}$	.190	$\frac{1}{4}$	$\frac{5}{32}$	4
12	.216	28	$2\frac{3}{8}$	$\frac{7}{8}$	$\frac{1}{8}$	$1\frac{1}{4}$	.216	$\frac{1}{4}$	$\frac{5}{32}$	4
14	.242	24	$2\frac{3}{8}$	$\frac{15}{16}$	$\frac{1}{8}$	$1\frac{5}{16}$	.242	$\frac{9}{32}$	$\frac{5}{32}$	4
16	.268	22	$2\frac{1}{2}$	1	$\frac{3}{16}$	$1\frac{5}{16}$	.268	$\frac{9}{32}$	$\frac{3}{16}$	4
18	.294	20	$2\frac{5}{8}$	$1\frac{1}{16}$	$\frac{3}{16}$	$1\frac{3}{8}$	.294	$\frac{9}{32}$	$\frac{3}{16}$	4
20	.320	20	$2\frac{3}{4}$	$1\frac{1}{8}$	$\frac{3}{16}$	$1\frac{7}{16}$	.320	$\frac{5}{16}$	$\frac{3}{16}$	4
22	.346	18	$2\frac{3}{4}$	$1\frac{1}{8}$	$\frac{3}{16}$	$1\frac{7}{16}$	.346	$\frac{5}{16}$	$\frac{7}{32}$	4
24	.372	16	3	$1\frac{1}{4}$	$\frac{1}{4}$	$1\frac{1}{2}$	.372	$\frac{11}{32}$	$\frac{1}{4}$	4
26	.398	16	$3\frac{1}{4}$	$1\frac{1}{4}$	$\frac{5}{16}$	$1\frac{11}{16}$	.398	$\frac{3}{8}$	$\frac{9}{32}$	4
28	.424	14	$3\frac{1}{2}$	$1\frac{3}{8}$	$\frac{5}{16}$	$1\frac{13}{16}$	.424	$\frac{13}{32}$	$\frac{5}{16}$	4
30	.450	14	$3\frac{1}{2}$	$1\frac{3}{8}$	$\frac{5}{16}$	$1\frac{13}{16}$	.450	$\frac{7}{16}$	$\frac{11}{32}$	4

This table covers the sizes adopted by the American Society of Mechanical Engineers in June, 1907, and now known as the A.S.M.E. Standard for machine screw sizes.

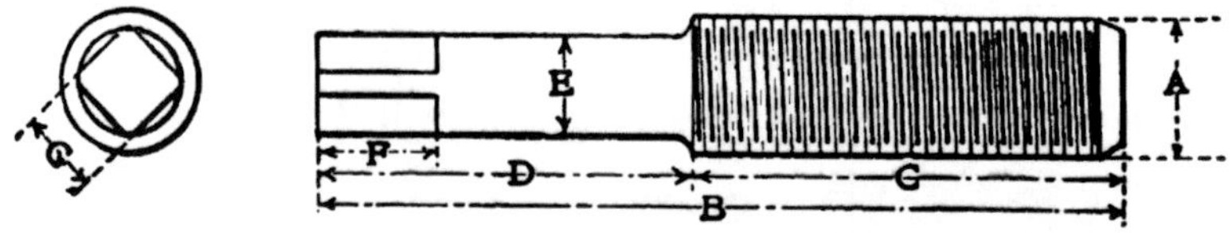

Dimensions of Hand Taps

Diameter of Tap	Number of Threads per Inch		Total Length	Length of Thread	Length of Shank	Diameter of Shank E		Length of Square	Size of Square	No. of Flutes
A	U.S.St'd	V. St'd	B	C	D	U.S.St'd	V. St'd	F	G	
3/16	32	24	2 1/4	13/16	1 7/16	0.187	0.187	3/16	9/64	4
1/4	20	20	2 3/4	1 1/8	1 5/8	0.250	0.250	1/4	3/16	4
5/16	18	18	3	1 1/4	1 3/4	0.225	0.200	5/16	5/32	4
3/8	16	16	3 1/4	1 3/8	1 7/8	0.280	0.250	3/8	3/16	4
7/16	14	14	3 1/2	1 1/2	2	0.330	0.300	3/8	1/4	4
1/2	13	12	3 3/4	1 9/16	2 3/16	0.385	0.340	7/16	9/32	4
9/16	12	12	4	1 11/16	2 5/16	0.440	0.400	1/2	5/16	4
5/8	11	11	4 1/4	1 13/16	2 7/16	0.490	0.455	9/16	11/32	4
11/16	11	11	4 1/2	1 15/16	2 9/16	0.555	0.515	9/16	13/32	4
3/4	10	10	4 3/4	2	2 3/4	0.605	0.560	5/8	7/16	4
13/16	10	10	4 7/8	2 1/8	2 3/4	0.670	0.625	11/16	1/2	4
7/8	9	9	5 1/8	2 3/16	2 15/16	0.715	0.675	3/4	1/2	4
15/16	9	9	5 1/4	2 5/16	2 15/16	0.780	0.730	3/4	9/16	4
1	8	8	5 1/2	2 3/8	3 1/8	0.825	0.770	13/16	5/8	4
1 1/16	7	8	5 3/4	2 1/2	3 1/4	0.860	0.830	7/8	5/8	4
1 1/8	7	7	5 7/8	2 9/16	3 5/16	0.925	0.860	7/8	11/16	4
1 3/16	7	7	5 7/8	2 9/16	3 5/16	0.980	0.920	7/8	11/16	4
1 1/4	7	7	6 1/4	2 3/4	3 1/2	1.050	0.985	15/16	3/4	4
1 5/16	6	7	6 1/4	2 3/4	3 1/2	1.080	1.050	15/16	13/16	4
1 3/8	6	6	6 5/8	2 7/8	3 3/4	1.145	1.070	15/16	13/16	4
1 7/16	6	6	6 5/8	2 7/8	3 3/4	1.20	1.130	1	7/8	4
1 1/2	6	6	7	3	4	1 270	1.200	1	15/16	4
1 5/8	5 1/2	5	7 1/4	3 3/16	4 1/16	1.375	1.265	1 1/16	1	4
1 3/4	5	5	7 5/8	3 5/16	4 5/16	1.475	1.390	1 1/16	1 1/16	4
1 7/8	5	4 1/2	7 7/8	3 7/16	4 7/16	1.600	1.475	1 1/8	1 1/8	6
2	4 1/2	4 1/2	8 1/4	3 5/8	4 5/8	1.700	1.600	1 3/16	1 1/4	6
2 1/8	4 1/2	4 1/2	8 1/2	3 3/4	4 3/4	1.810	1.710	1 3/16	1 5/16	6
2 1/4	4 1/2	4 1/2	8 3/4	3 7/8	4 7/8	1.945	1.845	1 1/4	1 7/16	6
2 3/8	4	4 1/2	9	4	5	2.030	1.975	1 5/16	1 1/2	6
2 1/2	4	4	9 1/4	4 1/8	5 1/8	2.160	2.040	1 5/16	1 9/16	6
2 5/8	4	4	9 1/2	4 1/4	5 1/4	2.285	2.175	1 3/8	1 11/16	6
2 3/4	4	4	9 3/4	4 5/16	5 7/16	2.400	2.290	1 7/16	1 3/4	6
2 7/8	3 1/2	4	10	4 7/16	5 9/16	2.485	2.425	1 7/16	1 13/16	6
3	3 1/2	3 1/2	10 1/8	4 9/16	5 9/16	2.600	2.480	1 1/2	1 7/8	8

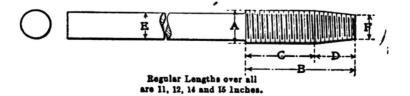

Regular Lengths over all
are 11, 12, 14 and 15 Inches.

DIMENSIONS OF TAPPER TAPS

Diameter of Tap	Number of Threads per Inch		Length of Thread	Length of Straight Part	Length of Chamfered Part	Diameter of Shank E		Diameter of Point F		No. of Flutes
A	U.S. St'd	V. St'd	B	C	D	U.S. St'd	V. St'd	U.S. St'd	V. St'd	
1/4	20	20	1 3/4			0.170	0.150	0.170	0.158	4
5/16	18	18	2			0.225	0.200	0.234	0.210	4
3/8	16	16	2			0.280	0.250	0.287	0.261	4
7/16	14	14	2 1/4			0.330	0.300	0.338	0.307	4
1/2	13	12	2 1/4			0.385	0.340	0.393	0.348	4
9/16	12	12	2 1/2			0.440	0.400	0.446	0.411	4
5/8	11	11	2 1/2			0.490	0.455	0.499	0.462	4
11/16	11	11	2 1/2			0.555	0.515	0.561	0.523	4
3/4	10	10	2 1/4			0.605	0.560	0.611	0.570	4
13/16	10	10	2 1/4			0.670	0.625	0.673	0.631	4
7/8	9	9	3			0.720	0.675	0.722	0.675	4
15/16	9	9	3			0.780	0.730	0.783	0.736	4
1	8	8	3 1/2			0.820	0.770	0.828	0.775	4
1 1/8	7	7	3 1/2			0.925	0.860	0.928	0.869	4
1 1/4	7	7	3 1/2			1.050	0.985	1.053	0.993	4
1 3/8	6	6	4			1.145	1.070	1.147	1.075	4
1 1/2	6	6	4			1.270	1.195	1.272	1.200	4

NOTE. — Tapper taps differ from machine taps in not having a square on the end of the shank. They are used in nut tapping machines, the nuts being run over the tap on to the shank and when full the tap is removed and the nuts slid off. The tap is then replaced for another lot of nuts.

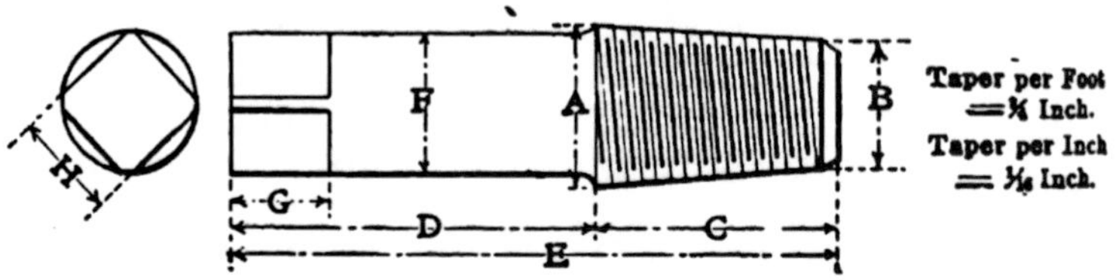
H
G
F
A
B
D
C
E
Taper per Foot = 3/4 Inch.
Taper per Inch = 1/16 Inch.

Briggs Standard Pipe Taps

Nominal Dia. of Tap	Threads per Inch	Dia. Large End	Dia. of Small End Before Chamfering	Length of Thread	Length of Shank	Total Length	Diameter of Shank	Length of Square	Size of Square	Number of Flutes
		A	B	C	D	E	F	G	H	
1/8	27	.443	.381	1	1 1/2	2 1/2	7/16	9/16	5/16	4
1/4	18	.575	.505	1 1/8	1 5/8	2 3/4	1/2	9/16	3/8	4
3/8	18	.718	.640	1 1/4	1 3/4	3	9/16	5/8	7/16	4
1/2	14	.887	.793	1 1/2	1 7/8	3 3/8	3/4	11/16	9/16	4
3/4	14	1.104	.993	1 5/8	2 1/8	3 3/4	15/16	3/4	11/16	4
1	11 1/2	1.366	1.257	1 3/4	2 3/8	4 1/8	1 1/8	13/16	13/16	5
1 1/4	11 1/2	1.717	1.599	1 7/8	2 5/8	4 1/2	1 5/16	15/16	1	5
1 1/2	11 1/2	1.963	1.838	2	2 7/8	4 7/8	1 1/2	1	1 1/8	5
2	11 1/2	2.453	2.312	2 1/4	3 1/2	5 3/4	1 7/8	1 3/16	1 3/8	7
2 1/2	8	2.961	2.781	2 7/8	4	6 7/8	2 1/4	1 3/8	1 11/16	8
3	8	3.605	3.402	3 1/4	4 1/2	7 3/4	2 5/8	1 1/2	1 15/16	9
3 1/2	8	4.125	3.899	3 5/8	4 9/16	8 3/16	2 13/16	1 9/16	2 1/8	11
4	8	4.629	4.395	3 3/4	4 5/8	8 3/8	3	1 5/8	2 1/4	11

Stove Bolt Taps

	Diameters					
Threads per Inch	5/32	3/16	7/32	1/4	5/16	3/8
Present Standard......	28	24	22	18	18	16
Old Standard.........	30	24	24	18	18	18

NOTE. — These have no fixed standard form of thread, being usually something like an Acme thread in general appearance.

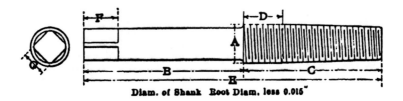

Diam. of Shank   Root Diam. less 0.015″

DIMENSIONS OF TAPER DIE TAPS

Diameter of Tap	Length of Shank	Length of Thread	Length of Straight Thread	Total Length	Length of Square	Size of Square	Number of Flutes
A	B	C	D	E	F	G	
$\frac{1}{4}$	$1\frac{1}{2}$	2	$\frac{1}{4}$	$3\frac{1}{2}$	$\frac{9}{16}$	$\frac{1}{8}$	5
$\frac{5}{16}$	$1\frac{1}{2}$	$2\frac{1}{2}$	$\frac{5}{16}$	4	$\frac{5}{8}$	$\frac{5}{32}$	5
$\frac{3}{8}$	$1\frac{1}{2}$	3	$\frac{3}{8}$	$4\frac{1}{2}$	$\frac{11}{16}$	$\frac{3}{16}$	5
$\frac{7}{16}$	$1\frac{3}{4}$	$3\frac{1}{4}$	$\frac{7}{16}$	5	$\frac{11}{16}$	$\frac{7}{32}$	5
$\frac{1}{2}$	2	$3\frac{1}{2}$	$\frac{1}{2}$	$5\frac{1}{2}$	$\frac{3}{4}$	$\frac{15}{64}$	5
$\frac{9}{16}$	$2\frac{1}{4}$	$3\frac{3}{4}$	$\frac{9}{16}$	6	$\frac{13}{16}$	$\frac{17}{64}$	5
$\frac{5}{8}$	$2\frac{1}{2}$	4	$\frac{5}{8}$	$6\frac{1}{2}$	$\frac{13}{16}$	$\frac{19}{64}$	5
$\frac{11}{16}$	$2\frac{3}{4}$	$4\frac{1}{4}$	$\frac{11}{16}$	7	$\frac{7}{8}$	$\frac{21}{64}$	6
$\frac{3}{4}$	3	$4\frac{1}{2}$	$\frac{3}{4}$	$7\frac{1}{2}$	$\frac{7}{8}$	$\frac{7}{16}$	6
$\frac{13}{16}$	$3\frac{1}{4}$	$4\frac{3}{4}$	$\frac{13}{16}$	8	$1\frac{1}{16}$	$\frac{1}{2}$	6
$\frac{7}{8}$	$3\frac{1}{2}$	5	$\frac{7}{8}$	$8\frac{1}{2}$	1	$\frac{9}{16}$	6
$\frac{15}{16}$	$3\frac{1}{2}$	$5\frac{1}{4}$	$\frac{15}{16}$	$8\frac{3}{4}$	1	$\frac{5}{8}$	6
1	$3\frac{1}{2}$	$5\frac{1}{2}$	1	9	$1\frac{1}{16}$	$\frac{5}{8}$	6
$1\frac{1}{8}$	$3\frac{1}{2}$	$5\frac{3}{4}$	$1\frac{1}{8}$	$9\frac{1}{4}$	$1\frac{1}{8}$	$1\frac{1}{16}$	6
$1\frac{1}{4}$	$3\frac{1}{2}$	6	$1\frac{1}{4}$	$9\frac{1}{2}$	$1\frac{1}{16}$	$\frac{3}{4}$	7
$1\frac{3}{8}$	$3\frac{3}{8}$	$6\frac{1}{4}$	$1\frac{3}{8}$	$9\frac{3}{4}$	$1\frac{3}{16}$	$1\frac{3}{16}$	7
$1\frac{1}{2}$	$3\frac{3}{8}$	$6\frac{3}{8}$	$1\frac{1}{2}$	10	$1\frac{1}{4}$	$1\frac{3}{16}$	7
$1\frac{5}{8}$	$3\frac{3}{8}$	$6\frac{5}{8}$	$1\frac{5}{8}$	$10\frac{1}{4}$	$1\frac{7}{16}$	1	7
$1\frac{3}{4}$	$3\frac{3}{8}$	$6\frac{3}{4}$	$1\frac{3}{4}$	$10\frac{1}{2}$	$1\frac{1}{2}$	$1\frac{1}{16}$	8
$1\frac{7}{8}$	$3\frac{3}{8}$	$7\frac{1}{4}$	$1\frac{7}{8}$	$10\frac{3}{4}$	$1\frac{5}{8}$	$1\frac{1}{8}$	8
2	$3\frac{3}{8}$	$7\frac{3}{8}$	2	11	$1\frac{11}{16}$	$1\frac{1}{4}$	8

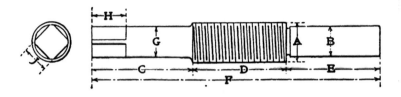

DIMENSIONS OF SELLERS HOBS

Diameter of Hob	Number of Threads per Inch		Diameter of Pilot B	Length of Shank	Length of Thread	Length of Pilot	Total Length	Diameter of Shank G		Length of Square	Size of Square	No. of Flutes
A	U.S. St'd	V. St'd	U.S. St'd	C	D	E	F	U.S.St'd	V. St'd	H	J	
1/4	20	20	3/16	2	1 1/8	1 1/8	4 1/4	0.170	0.150	3/8	1/8	6
5/16	18	18	3/16	2	1 1/4	1 1/4	4 1/2	0.225	0.200	3/8	5/32	6
3/8	16	16	1/4	2 1/8	1 7/16	1 7/16	5	0.280	0.250	3/8	3/16	6
7/16	14	14	5/16	2 1/8	1 9/16	1 9/16	5 1/2	0.330	0.300	13/16	1/4	6
1/2	13	12	3/8	2 1/8	1 11/16	1 11/16	5 3/4	0.385	0.340	13/16	5/32	8
9/16	12	12	3/8	2 1/4	1 7/8	1 7/8	6	0.440	0.400	13/16	5/16	8
5/8	11	11	1/2	2 1/4	2 1/8	2 1/8	6 1/2	0.490	0.455	1	13/32	8
11/16	11	11	1/2	2 1/4	2 3/8	2 3/8	7	0.555	0.515	7/8	13/32	8
3/4	10	10	1/2	2 1/2	2 1/2	2 1/2	7 1/2	0.605	0.560	7/8	3/16	8
13/16	10	10	1/2	2 1/2	2 3/4	2 3/4	8	0.670	0.625	15/16	1/2	8
7/8	9	9	11/16	2 1/2	3	3	8 1/2	0.715	0.675	15/16	1/2	8
15/16	9	9	11/16	2 1/2	3 1/4	3 1/4	9	0.780	0.730	1	9/16	10
1	8	8	11/16	2 5/8	3 7/16	3 7/16	9 1/2	0.825	0.770	1	5/8	10
1 1/8	7	7	7/8	2 5/8	3 9/16	3 9/16	9 3/4	0.925	0.860	1 1/16	11/16	10
1 1/4	7	7	7/8	2 5/8	3 11/16	3 11/16	10	1.050	0.985	1 1/16	3/4	10
1 3/8	6	6	1 1/16	2 5/8	3 15/16	3 15/16	10 1/2	1.145	1.070	1 1/8	13/16	10
1 1/2	6	6	1 1/16 1	2 5/8	4 3/16	4 3/16	11	1.270	1.200	1 3/16	15/16	10
1 5/8	5 1/2	5	1 1/16 1	2 3/4	4 3/8	4 3/8	11 1/2	1.375	1.265	1 1/4	1	12
1 3/4	5	5	1 1/16 1	2 3/4	4 5/8	4 5/8	12	1.475	1.390	1 1/4	1 1/16	12
1 7/8	5	4 1/2	1 1/2 1	2 3/4	4 7/8	4 7/8	12 1/2	1.600	1.475	1 5/16	1 1/8	12
2	4 1/2	4 1/2	1 1/2 1	2 3/4	5 1/8	5 1/8	13	1.700	1.600	1 3/8	1 1/4	12

NOTE. — The Sellers hob is designed to be run on centers, the work, such as hand or die chasers, being held against it and fed along by the lathe carriage.

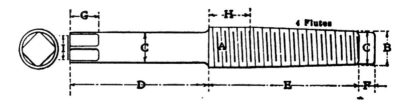

STANDARD SQUARE-THREAD TAPS

Size	A	B	C	D	E	F	G	H	I
Diameter $\frac{5}{8}''$..(1)	$\frac{35}{64}$	$\frac{1}{2}$	$\frac{1}{2}$	$3\frac{5}{8}$	$3\frac{1}{4}$	$\frac{1}{2}$	$\frac{9}{16}$	$\frac{5}{8}$	$\frac{7}{16}$
(2)	$\frac{19}{32}$	$\frac{17}{32}$	$\frac{1}{2}$	$3\frac{5}{8}$	$3\frac{1}{4}$	$\frac{1}{2}$	$\frac{9}{16}$	$\frac{5}{8}$	$\frac{7}{16}$
Pitch 8......(3)	$\frac{5}{8}$	$\frac{37}{64}$	$\frac{1}{2}$	$3\frac{5}{8}$	$3\frac{1}{4}$	$\frac{1}{2}$	$\frac{9}{16}$	$\frac{5}{8}$	$\frac{7}{16}$
Diameter $\frac{3}{4}''$..(1)	$\frac{41}{64}$	$\frac{37}{64}$	$\frac{37}{64}$	$3\frac{3}{4}$	$3\frac{3}{8}$	$\frac{1}{2}$	$\frac{3}{4}$	$\frac{3}{4}$	$\frac{7}{16}$
(2)	$\frac{45}{64}$	$\frac{5}{8}$	$\frac{37}{64}$	$3\frac{3}{4}$	$3\frac{3}{8}$	$\frac{1}{2}$	$\frac{3}{4}$	$\frac{3}{4}$	$\frac{7}{16}$
Pitch 6......(3)	$\frac{3}{4}$	$1\frac{1}{16}$	$\frac{37}{64}$	$3\frac{3}{4}$	$3\frac{3}{8}$	$\frac{1}{2}$	$\frac{3}{4}$	$\frac{3}{4}$	$\frac{7}{16}$
Diameter $\frac{7}{8}''$..(1)	$\frac{3}{4}$	$\frac{41}{64}$	$\frac{41}{64}$	4	$3\frac{1}{2}$	$\frac{1}{2}$	$\frac{13}{16}$	$\frac{7}{8}$	$\frac{1}{2}$
(2)	$1\frac{5}{16}$	$\frac{47}{64}$	$\frac{41}{64}$	4	$3\frac{1}{2}$	$\frac{1}{2}$	$\frac{13}{16}$	$\frac{7}{8}$	$\frac{1}{2}$
Pitch $4\frac{1}{2}$.....(3)	$\frac{7}{8}$	$\frac{41}{64}$	$\frac{41}{64}$	4	$3\frac{1}{2}$	$\frac{1}{2}$	$\frac{13}{16}$	$\frac{7}{8}$	$\frac{1}{2}$
Diameter $1''$..(1)	$\frac{57}{64}$	$\frac{51}{64}$	$\frac{51}{64}$	$4\frac{1}{4}$	$4\frac{1}{4}$	$\frac{1}{2}$	$\frac{15}{16}$	1	$\frac{5}{8}$
(2)	$\frac{61}{64}$	$\frac{7}{8}$	$\frac{51}{64}$	$4\frac{1}{4}$	$4\frac{1}{4}$	$\frac{1}{2}$	$\frac{15}{16}$	1	$\frac{5}{8}$
Lead D'BL $\frac{3}{8}''$(3)	1	$1\frac{15}{16}$	$\frac{51}{64}$	$4\frac{1}{4}$	$4\frac{1}{4}$	$\frac{1}{2}$	$\frac{15}{16}$	1	$\frac{5}{8}$
Diameter $1\frac{1}{4}$..(1)	$\frac{61}{64}$	$\frac{55}{64}$	$\frac{55}{64}$	$4\frac{3}{4}$	$4\frac{3}{8}$	$\frac{1}{2}$	1	$1\frac{1}{4}$	$1\frac{1}{16}$
(2)	$1\frac{1}{32}$	$1\frac{5}{8}$	$\frac{55}{64}$	$4\frac{3}{4}$	$4\frac{3}{8}$	$\frac{1}{2}$	1	$1\frac{1}{4}$	$1\frac{1}{16}$
(3)	$1\frac{5}{64}$	$1\frac{1}{64}$	$\frac{55}{64}$	$4\frac{3}{4}$	$4\frac{3}{8}$	$\frac{1}{2}$	1	$1\frac{1}{4}$	$1\frac{1}{16}$
Pitch $3\frac{1}{2}$......(4)	$1\frac{1}{8}$	$1\frac{1}{16}$	$\frac{55}{64}$	$4\frac{3}{4}$	$4\frac{3}{8}$	$\frac{1}{2}$	1	$1\frac{1}{8}$	$1\frac{1}{16}$
Diameter $1\frac{1}{2}''$.(1)	$1\frac{15}{64}$	$1\frac{1}{8}$	$1\frac{1}{8}$	$5\frac{1}{4}$	$4\frac{5}{8}$	$\frac{5}{8}$	$1\frac{1}{8}$	$1\frac{1}{2}$	$\frac{7}{8}$
(2)	$1\frac{9}{32}$	$1\frac{1}{16}$	$1\frac{1}{8}$	$5\frac{1}{4}$	$4\frac{5}{8}$	$\frac{5}{8}$	$1\frac{1}{8}$	$1\frac{1}{2}$	$\frac{7}{8}$
(3)	$1\frac{21}{64}$	$1\frac{17}{64}$	$1\frac{1}{8}$	$5\frac{1}{4}$	$4\frac{5}{8}$	$\frac{5}{8}$	$1\frac{1}{8}$	$1\frac{3}{8}$	$\frac{7}{8}$
Lead D'BL $\frac{1}{2}''$(4)	$1\frac{3}{8}$	$1\frac{5}{16}$	$1\frac{1}{8}$	$5\frac{1}{4}$	$4\frac{5}{8}$	$\frac{5}{8}$	$1\frac{1}{8}$	$1\frac{3}{8}$	$\frac{7}{8}$

NOTE. — While in theory the thread and the space are both one half the pitch in practice it is necessary to make the thread a little more than half in order to allow clearance for the screw that goes into the threaded hole. The amount of this clearance depends on the character of the work and varies from .001 inch up. Some also make the tap so that the screw will only bear on the top or bottom and the sides.

FILES

FILES are designated both by the spacing of their teeth and the shape or cross-section of steel on which the teeth are cut; the size always referring to their length which is measured from the point cutting to the end of the file proper but the measurement never includes the tang which fits into the handle.

TERMS USED

The back of a file is the convex or rounding side of half-round, cabinet and other files having a similar shape.

A file is Bellied when it is full or large in the center.

A Blunt file is the same size its whole length instead of being tapered.

An Equalling file is one which looks blunt but which has a slight belly or curve from joint to tang.

A Float file is a coarse single cut made for use on soft metals or wood and frequently used by plumbers.

A Safe-edge is an edge left smooth or blank so that the file will not cut if it strikes against the side of a slot or similar work.

The Tang is the small pointed end forged down for fitting into the handle.

Three square files are double cut and have teeth only on the sides, while taper saw files are usually single cut and have teeth on the edge as well as the sides. This makes the taper saw files broad on the edge or without sharp corners, while the three square files have very sharp corners.

A special angle tooth file is made for brass work. The first cut is square across the file, while the second is at quite an acute angle, about 60 degrees from the first cut.

Doctor files are very similar to these except that the first cut is about 15 degrees instead of being square across the file.

A lock file has safe edge and the teeth only go about one third the way across from each side leaving the center blank. The teeth are single cut.

HIGHT OF WORK

The work should be at a convenient hight which will usually vary from 40 to 44 inches for most men with an average of. 42 inches. This means the hight of the work, not the bench.

PICKLING BATH

A good pickle to soften and loosen the scale on cast iron before filing is made of two or three parts of water to one part of sulphuric acid. Immerse castings for a short time.

For brass castings use a pickle of five parts water to one part nitric acid.

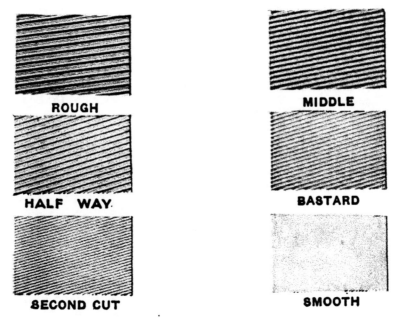

Actual Tooth Spacing of Single Cut Files

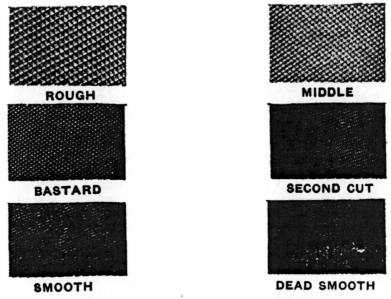

Actual Tooth Spacing of Double Cut Files

The Teeth of Files

The cut of a file or the number of teeth per inch vary with the length of the file itself and the kind of a file, and is a little confusing, as a rough cut in a small file may be as fine as a second cut of a larger size. The cuts used on regular 12-inch files are shown in the illustration and represents the practice of Henry Disston & Sons. The same makers also supply the table of cuts per inch used on their machines, which are as follows:

Regular Taper Files

Length, inches. — 2½, 3, 3½, 4, 4½, 5, 5½, 6, 6½, 7, 8, 9, 10,
Teeth per inch — 64, 56, 52, 50, 48, 46, 44, 42, 42, 40, 38, 36, 34
Slim Tapers —   64, 64, 60, 58, 56, 52, 50, 50, 46, 46, 44, 40, 38.

Mill File, Bastard Cut

Length — 4, 5, 6, 7, 8, 9, 10, 11, 12, 13, 14, 15, 16, 17, 18, 20 in.
Teeth — 56, 50, 48, 46, 44, 42, 40, 38, 36, 34, 32, 30, 28, 26, 24, 22 per inch.

Flat File, Bastard Cut

48, 42, 38, 36, 32, 30, 26, 24, 22, 20, 20, 18, 18, 16, 16, 14.

Single cut files usually have teeth at about 25 degrees and in double cut files the other cut is usually from 45 to 50 degrees. Fine machinists files are made in ten numbers from oo to 8.

The Shapes of Files

In the following pages the shapes of standard files are shown. The names are as follows:

1. Metal saw — blunt.
2. Three-square or tri-angular.
3. Barrette.
4. Slitting.
5. Square.
6. Round or rat-tail.
7. Pippin.
8 Knife.
9. Crossing.
10. Half-round.
11. Crochet.
12. Warding.
13. Extra narrow pillar.
14. Narrow pillar.
15 Pillar.
16 Hand.

Files are cut in two ways, single and double The first has but a single line of cuts across the surface, at an angle with the file body but parallel to each other. The double-cut file has two lines of cuts, at an angle with each other, and the second cut being usually finer than the first. Some prefer the single cut, for filing in the lathe. Rasps have single teeth forced up with a punch.

The old method of designating the cuts were rough, coarse, bastard, second cut, smooth and dead smooth. Some makers are now using a series of numbers — usually eight to ten — instead of the six designations by name formerly employed. The uses of the various cuts depend on the shop in question and must be learned from observation and experience in each case.

The grades of cut used by them run from No oo to No. 8, and while it is hard to exactly compare them with the old-style designations, it will be found that No. oo is about the same as a bastard, No. 1 as a second cut, No. 2 or 3 with a smooth, and Nos. 6 to 8 with a dead smooth file.

The Standard Shapes of Files

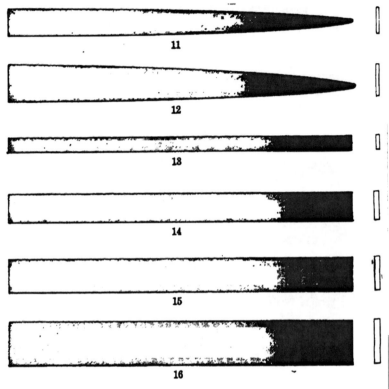

The Standard Shapes of Files

WHEN A FILE CUTS BEST

One who has given the matter careful attention, and has built file-testing machines, Edward G. Herbert, of Manchester, England, has come to the conlcusion that a file does not cut best when it is new but after it has been used for some little time, say 2500 strokes or the filing away of one cubic inch of metal. Another curious feature is that its usefulness seems to come to a sudden instead of a gradual end.

A bastard file having 25 teeth to the inch, operating on a surface one inch square with a pressure of 30 pounds, which is about equal to heavy hand filing, gives 25 cutting edges about one inch long, which likens it somewhat to a broad cutting tool in a planer.

In cutting a file the metal is forced up in a sort of a bur, and occasionally the top of the tooth slopes over backward which is the reason that a file often cuts better after these are broken or worn off. Then, too, when a file is new all the teeth are not of the same hight and only a few points cut. As they wear down more teeth come into ccntact and do more work.

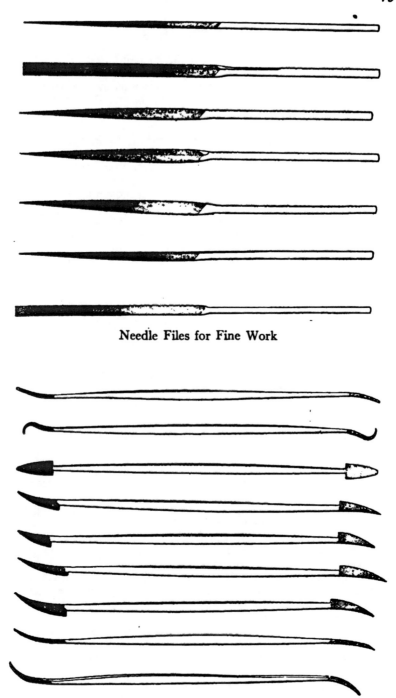

Needle Files for Fine Work

Die Sinkers Files or Riffles

WORK BENCHES

THE duties of a bench vary with the shop in which it is located according to the work that is to be done on it or at it. If it is simply a filing bench, the main requirement is that it support a vise firmly and at the proper hight. If an assembling bench, these are not the important features, and just what it does need depends on the kind of work being handled.

For the average shop work we want a bench that is rigid; that will stand chipping and filing; that can be used in testing work on a surface plate or in handling jigs and fixtures; that will not splinter badly nor yet injure a tool should it happen to drop on it. For the toolmaker the cast-iron bench top has many advantages, but both the bench and the tool are very liable to be marred by dropping the tool on it, so that for general use we rely on wood as in the days of old, except that a bench with solid 2- or 3-inch planking the whole width is now too expensive to consider. We no longer want the bench braced up against the side of the shop but set it out from

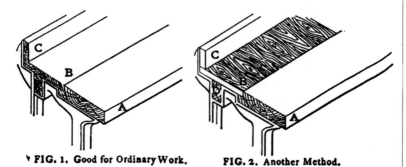

▸ FIG. 1. Good for Ordinary Work.   FIG. 2. Another Method.

the wall to allow the heat to rise and the air to circulate, as well as giving the sprinklers a chance to get at a fire on the floor near the walls.

The use of a lighter board at the back has become so common that the New Britain Machine Company's design for a bench leg is made for this construction as shown in Fig. 1. This also shows the back-board B rabbeted to the plank A, which supports it all along the front edge, and it is also supported by the stringer D, which runs the whole length of the bench. These supports, in addition to the cross bearing of the legs every 6 or 8 feet, give the backboard a stiffness that was unknown where they are simply laid flush and not rabbeted and the stringer is absent.

Benches made without these supports are open to the serious objection that the backboard springs down when a heavy weight, such as a jig or surface plate, is put on the bench and throws them out of level.

All cracks are more or less of a nuisance in bench work, but in this case any shrinkage can be taken up by wedging against the ⸱⸱⸱on support of the board C and the edge of the backboard B.

Another style bench with this same leg is shown in Fig. 2. Here the front plank *A* and the backboard *C* are the same as before, but instead of having one backboard, this part of the bench is made up of narrow strips as *B*, fitting into rabbet in plank *A* and supported by the stringer *D* as before. These narrow boards can be either tongued and grooved hardwood flooring, or can be square edges, as preferred; in either case any shrinkage can be taken up by forcing the boards together.

A cheaper form of bench is shown in Fig. 3, where the heavy planking is entirely dispensed with and the boards *B* run the full width of the bench as shown. Running along the front, underneath the main boards is a soft plank *A* which supports the edge of the bench where the most work comes, and under the back is the 2 × 6-inch stringer as before. Here, too, the boards can be either notched or square edge, each having its advocates; the objection raised against the tongue and groove being that the edges are apt to split off from heavy articles dropping on them. An advantage claimed for the boards running this way is that work going on or off the

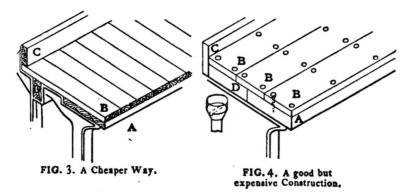

FIG. 3. A Cheaper Way.

FIG. 4. A good but expensive Construction.

bench is always in the direction of the grain of the wood and that fewer splinters are formed on that account. In either Figs. 2 or 3 any local wear can be remedied by replacing the worn board with a new one. Some object to the end of the grain at the front of a bench.

The material used in any of these can be varied to suit the individual requirements. Maple is generally considered the best wood for a bench, while others prefer ash. For the backboards hard pine is often used and even cheaper woods will answer if necessary, although it probably pays to use maple all through if you can afford it.

Still another style of bench is shown in Fig. 4 and one which was designed to be serviceable and have a long life without so much regard to first cost as the others. The bench leg was flat on top, the first layer of maple planks *A* and *D* and on top, narrower boards of the same material. These were fastened with long wood screws, holes being countered and plugged as shown.

The theory of this construction is that the boards are sure to be more thoroughly seasoned than the planks, consequently the plank will shrink the most and tend to draw the top boards closer together. It certainly makes a solid bench, but the first cost is rather high.

Benches are also occasionally built up from small blocks so as to present an end grain on the top, the same as butchers' blocks. One shop we know of surfaces these when worn by putting them on a Daniels' planer and substituting a circular saw for the swinging knives. This saws the top very smooth and leaves a good surface. Others glue up strips on edge and plane down to a smooth surface so as to do away with all cracks. Zinc or even heavy paper covers are often used where fine work is being assembled to prevent its finding its way into cracks and crevices.

The usual work bench is from 33 to 35 inches from the floor to the top, about 29 or 30 inches wide, and has the front plank 3 inches with backboards 1 inch thick. A cast-iron leg of this type weighs about 50 pounds.

SOLDERING

ALMOST every one thinks he can solder, yet if we examine the work carefully we will find that only about 10 per cent of the work is really done as it should be. Thorough soldering is frequently referred to as sweating, and it is remarkable the difference in strength between a well-fitted and "sweated" junction of the metals and one as ordinarily soldered.

A point frequently overlooked is the important one of properly cleaning the surfaces to be joined. This is too often left for the flux to correct. Another neglected point is the selection of the flux to be used, although nearly all of the metals can be joined by the use of the same flux. The after effects resulting from improper cleaning after soldering are frequently worse than the good effects of the soldering. This is particularly noticeable in electrical work.

For strength, fit the parts accurately. The more accurate the fitting the stronger the result. Use a solder with as high a melting point as possible. Apply the proper heat as it should be. The nearer the temperature of work to be joined is brought to the fusing point of the solder the better will be the union, since the solder will flow more readily.

Fluxes for Different Metals

There are on the market a number of fluxes or soldering salts that are giving good satisfaction. A form that is non-corrosive and very popular with electrical workers is the soldering stick in which the ingredients are molded into stick form about 1 inch diameter and 6 inches long.

The action and use of a flux in soldering are to remove and prevent the formation of an oxide during the operation of soldering, and to allow the solder to flow readily and to unite more firmly with the surfaces to be joined.

For sheet tin, on the best work rosin or colophony is used; but owing to the ease of applying and rapidity of working, zinc chloride or acid is more generally used. Beeswax can also be used, as also almost any of the pastes, fats or liquids prepared for the purpose.

For lead, a flux of oil and rosin in equal parts works very well. Tallow is also a good flux. Rosin or colophony is much used, and zinc chloride will keep the surfaces in good condition.

Lead burning is a different operation from soldering, and at the present time almost a lost art. The surfaces must be bright and free from oxide; solder is not used as a flux, but a piece of lead and rosin or oil.

For brass, zinc chloride or almost any of the soldering preparations is used. Care must be taken to remove any scale or oxide if a good joint is wanted. On new metal this is not much trouble, but on old or repair work it is sometimes exceedingly difficult. This is particularly noticeable on metal patterns that have been in use for some time. The scraper must be brought into use to remove it. Many use with considerable success an acid dip such as is commonly used by electro platers, for removing the oxide. Oily or greasy work can be cleaned by the use of potash or lye, but care must be exercised that the brass is not left too long in the solution, especially if it contains any joints previously soldered, since the action set up will dissolve the solder entirely or roughen up the joint to such an extent as to require refinishing.

For copper, the same fluxes as for brass are used. On old work it is almost always necessary to scrape the parts to be joined to get the solder to hold. A particularly difficult piece of work to solder is an old bath tub. The grease and soap form a layer that is impervious to any of the fluxes, and it must be carefully removed entirely if good work is wanted.

For zinc, use muriatic acid almost full strength or chloride of zinc solution. Zinc is the metal that has a "critical temperature" more than any other metal except the softer alloys. If the iron is overheated, the zinc is melted and a hole burned in the metal; even if this does not occur, the surface of the metal is roughened and there is formed on the soldering copper an alloy that will not flow but simply makes a pasty mass. At the correct heat the solder will flow readily and unite firmly with the metal. Especially if the work is to be painted, care should be taken to neutralize and wash off any excess of acid or soldering solution, as it is impossible to cause paint to adhere properly unless this is done.

For galvanized iron, use muriatic acid, chloride of zinc solution or rosin, and be sure to see that the acid is neutralized if the work is to be painted. Many cornices and fronts are made of this metal and are very unsightly in a short time after being painted, particularly at the joints, owing to lack of care in removing the excess flux.

An action that is not usually taken into consideration in the joining of galvanized iron or zinc with copper, as is sometimes done, is the electrical action set up by the metals if any moisture is present. This is very noticeable in cities where the acid from the atmosphere assists in the action. It will nearly always be found that the zinc or galvanized iron has been greatly injured at the places joined.

For wrought iron or steel, zinc chloride is best. The iron or steel, to make good work, should be previously freed from scale or oxide and tinned before joining. Where the oxide is not very heavy, the iron can be cleaned by brushing with muriatic acid and rubbing with a piece of zinc.

The Fluxes Themselves

The above paragraphs give the fluxes adapted to the various metals; the fluxes themselves are as follows:

Hydrochloric or muriatic acid. The ordinary commercial acid is much used in full strength or slightly diluted to solder zinc, particularly where the zinc is old or covered with an oxide.

Rosin or colophony, powdered, is commonly used for copper, tin and lead, and very generally by canneries and packing houses on account of its non-poisonous qualities. It is also used mixed with common olive oil. Turpentine can also be used as a flux. Beeswax is a good but expensive flux. Tallow is also used for lead pipe, but is more frequently mixed with rosin.

Palm or cocoa oil will work well, but is more generally used in the manufacture of tin plate. The common green olive oil works very well with the more fusible solders.

As expedients I have used a piece of common stearine candle or a piece of common brown rosin soap or cheap furniture varnish, which is largely composed of rosin. Paraffin, vaseline and stearine are recommended for use with some of the alloys for soldering aluminum.

Chloride of zinc, acid, or soldering liquid, is the most commonly used of all the fluxes; as usually prepared, simply dissolve as much scrap zinc in the ordinary commercial muriatic acid as it will take up. But if it is diluted with an equal quantity of water and a small quantity of sal ammoniac is added it works much better and is less likely to rust the articles soldered. If they are of iron or steel, about 2 ounces to the pint of solution is about the proper quantity of powdered sal ammoniac to add.

In preparing this solution, a glass or porcelain vessel should be used; owing to the corrosive fumes, it should be done in a well-ventilated place. Use a vessel of ample capacity, since there is considerable foaming or boiling of the mixture.

Soldering liquid, non-corrosive, is also prepared by dissolving the zinc in the acid as above and adding one fourth of the quantity of aqua ammonia to neutralize the acid, then diluting with an equal quantity of water.

Soldering liquid, neither corrosive nor poisonous. Dissolve $1\frac{1}{2}$ parts glycerin, 12 parts water, and add $1\frac{1}{2}$ parts lactic acid.

Soldering paste. When a solution of chloride is mixed with starch paste, a syrupy liquid is formed which makes a flux for soldering.

Soldering fat or paste. Melt 1 pound of tallow and add 1 pound of common olive oil. Stir in 8 ounces of powdered rosin; let this boil up and when partially cool, add with constant stirring, $\frac{1}{2}$ pint of water that has been saturated with powdered sal ammoniac. Stir constantly until cool. By adding more rosin to make it harder, this can be formed into sticks. A very good acid mixture for cleaning work to be soldered is equal parts of nitric and sulphuric acid and water. *Never pour the water into the acid.*

Cleaning and Holding Work

For copper work a dilute sulphuric acid is best. Articles of lead and zinc can be cleaned with a potash solution, but care must be exercised as the alkalies attack these metals. For zinc, a dilute solution of sulphuric or muriatic acid will clean the surface.

For cleaning or removing the oxide or other foreign material, scrapers and files are frequently used. An old file bent at the ends and with the corners flaring makes a handy tool. Grind the edge sharp and make as hard as possible.

To enable difficult points to be "filled," sometimes a small piece of moist clay pressed into shape to form the desired shape can be used to advantage as a guide for the solder. Another use of the clay is to embed the parts in, to hold them in position for soldering.

Plaster of paris is also used for this purpose, but is sometimes difficult to remove, especially in hollow pieces. A dilute solution of muriatic acid will help to get this out, however.

Castings containing aluminum are always harder to solder than other alloys. In some instances where the percentage of aluminum is high, it is necessary to copperplate the parts to be joined before a satisfactory joint can be made. In nearly every instance the work can be "stuck" together, but not actually soldered.

In metal-pattern making too little attention is given to both the fitting of the parts and the selection of the solder to joint the work. A good grade should always be used, and it must be borne in mind that the higher the melting point of the solder the stronger the joint.

A very good job of soldering can be done on work that will permit of it by carefully fitting the parts, laying a piece of tin foil, covered on both sides with a flux, between the parts to be joined and pressing them tightly together. Heat until the foil is melted. This is very good in joining broken parts of brass and bronze work. If they fit well together, they can frequently be joined in this manner so that the joint is very strong and almost imperceptible.

Soldering Cast Iron

For cast iron, the flux is usually regarded as a secret. A number of methods are in use; one of the oldest and least satisfactory is to brush the surfaces thoroughly with a brass scratch brush. Brush until the surface is coated with brass, then tin this surface and solder as usual. If plating facilities are to be had, copperplate the parts and solder together. This method has been used very successfully for a number of years.

A fair substitute for the above is to clean the surfaces thoroughly and copperplate them with a solution of sulphate of copper: about 1 ounce sulphate of copper, ½ pint water, ½ ounce sulphuric acid. Brush this solution on or dip into the solution, rinse off and dry it well before soldering.

Another method is to tin the cast iron. To do this, first remove all scale until the surface is clean and bright. The easiest way to do this is with the emery wheel. Dip in a lye to remove any grease, and rinse the lye off; then dip into muriatic acid of the usual strength

Then go over the surface with rosin and a half and half solder. It may be necessary to dip into the acid several times to get the piece thoroughly tinned. Rubbing the surface of the iron with a piece of zinc while the acid is still on it will facilitate the tinning.

Another method of soldering cast iron is to clean the surface as in the previous operation and then brush over with chloride of zinc solution and sprinkle powdered sal ammoniac on it; then heat until the sal ammoniac smokes. Dip into melted tin and remove the surplus; repeat if not thoroughly tinned. Half tin and half lead works well as a solder for this.

Commutator wires and electrical connections should never be soldered by using an acid solution, owing to the corrosive action afterward. A good flux is an alcoholic solution of rosin.

Cold Soldering for Metals, Glass, Porcelain, etc.

Precipitate the copper from a solution of the sulphate by putting in strips of zinc. Place the copper powder in a porcelain or wedgewood mortar and mix it with from 20 to 30 parts of sulphuric acid of 1.85 degrees Baumé. Then add 70 parts of mercury when well mixed, wash well with water to remove the excess of acid and allow it to cool. To use it, heat it and pound it well in an iron mortar until it becomes plastic. It can then be used and adheres very firmly when cold. No flux is needed, but the surfaces must be clean. This is used where heat cannot be used, or to join metal parts to glass or porcelain.

A solution of copper for copperplating steel or cast iron before soldering will work by simply immersing the work in it. This is also useful to copper the surface of dies and tools to enable the mechanic to "lay out" or scribe the work so that the lines can be readily seen. Take copper sulphate $3\frac{1}{2}$ ounces, sulphuric acid $3\frac{1}{2}$ ounces, water 1 to 2 gallons. Dissolve the copper in the water and add the acid.

GEARING

GEAR. TEETH — SHAPES OF

Cycloidal or Epicycloidal. — A curved tooth generated by the point of a circle rolling away from the gear wheel or rack.

Involute. — A curved tooth generated by unwinding a tape or string from a cylinder. The rack tooth has straight sides.

Involute Standard. — The standard gear tooth has a 14½ degree pressure angle which means that the teeth of a standard rack have straight sides 14½ degrees from the vertical.

Involute — Stubbed. — A tooth shorter than the standard and usually with a 20-degree pressure angle.

GEARS — TEETH AND PARTS

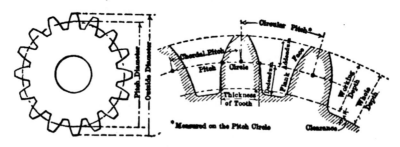

FIG. 1. — Part of Gear Teeth

Addendum. — Length from pitch line to outside.

Chordal Pitch. — Distance from center to center of teeth in a straight line.

Circular Pitch. — Distance from center to center of teeth measured on the pitch circle.

Clearance. — Extra depth of space between teeth.

Dedendum. — Length from pitch line to base of tooth.

Diametral Pitch. — Number of teeth divided by the pitch diameter or the teeth to each inch of diameter.

Face. — Working surface of tooth outside of pitch line.

Flank. — Working surface of tooth below pitch line.

Outside Diameter. — Total diameter over teeth.

Pitch Diameter. — Diameter at the pitch line.

Pitch Line. — Line of contact of two cylinders which would have the same speed ratios as the gears.

Linear Pitch. — Sometimes used in rack measurement. Same as circular pitch of a gear.

Having	To Get	Rule	Formula
The Diametral Pitch	The Circular Pitch	Divide 3.1416 by the Diametral Pitch	$P' = \dfrac{3.1416}{P}$
The Pitch Diameter and the Number of Teeth ...	The Circular Pitch	Divide Pitch Diameter by the product of .3183 and Number of Teeth	$P' = \dfrac{D'}{.3183\,N}$
The Outside Diameter and the Number of Teeth ...	The Circular Pitch	Divide Outside Diameter by the product of .3183 and Number of Teeth plus 2	$P' = \dfrac{D}{.3183\,N+2}$
The Number of Teeth and the Circular Pitch	Pitch Diameter	The continued product of the Number of Teeth, the Circular Pitch and .3183	$D' = N P' \,.3183$
The Number of Teeth and the Outside Diameter ...	Pitch Diameter	Divide the product of Number of Teeth and Outside Diameter by Number of Teeth plus 2	$D' = \dfrac{N D}{N+2}$
The Outside Diameter and the Circular Pitch	Pitch Diameter	Subtract from the Outside Diameter the product of the Circular Pitch and .6366	$D' = D - (P'.6366)$
Addendum and the Number of Teeth ...	Pitch Diameter	Multiply the Number of Teeth by the Addendum	$D' = N s$
The Number of Teeth and the Circular Pitch	Outside Diameter	The continued product of the Number of Teeth plus 2, the Circular Pitch and .3183	$D = (N+2)\,P'\,.3183$
The Pitch Diameter and the Circular Pitch	Outside Diameter	Add to the Pitch Diameter the product of the Circular Pitch and .6366....	$D = D' + (P'.6366)$
The Number of Teeth and the Addendum........	Outside Diameter	Multiply Addendum by Number of Teeth plus 2	$D = s\,(N+2)$
The Pitch Diameter and the Circular Pitch	Number of Teeth	Divide the product of Pitch Diameter and 3.1416 by the Circular Pitch	$N = \dfrac{D'\,3.1416}{P'}$
The Circular Pitch	Thickness of Tooth	One half the Circular Pitch	$t = \dfrac{P'}{2}$
The Circular Pitch	Addendum	Multiply the Circular Pitch by .3183 or $s = \dfrac{D'}{N}$	$s = P'\,.3183$
The Circular Pitch	Root	Multiply the Circular Pitch by .3683	$s + f = P'\,.3683$
The Circular Pitch	Working Depth	Multiply the Circular Pitch by .6366	$D'' = P'\,.6366$
The Circular Pitch	Whole Depth	Multiply the Circular Pitch by .6866	$D'' = P'\,.6866$
The Circular Pitch	Clearance	Mutliply the Circular Pitch by .05	$f = P'\,.05$
Thickness of Tooth	Clearance	One tenth the Thickness of Tooth at Pitch Line	$f = \dfrac{t}{10}$

Having	To Get	Rule	Formula
The Circular Pitch	The Diametral Pitch	Divide 3.1416 by the Circular Pitch	$P = \dfrac{3.1416}{P'}$
The Pitch Diameter and the Number of Teeth ...	The Diametral Pitch	Divide Number of Teeth by Pitch Diameter	$P = \dfrac{N}{D}$
The Outside Diameter and the Number of Teeth	The Diametral Pitch	Divide Number of Teeth plus 2 by Outside Diameter	$P = \dfrac{N+2}{D}$
The Number of Teeth and the Diametral Pitch	Pitch Diameter	Divide Number of Teeth by the Diametral Pitch	$D' = \dfrac{N}{P}$
The Number of Teeth and the Outside Diameter	Pitch Diameter	Divide the Product of Outside Diameter and Number of Teeth by Number of Teeth plus 2	$D' = \dfrac{DN}{N+2}$
The Outside Diameter and the Diametral Pitch ...	Pitch Diameter	Subtract from the Outside Diameter the quotient of 2 divided by the Diametral Pitch..................	$D' = D - \dfrac{2}{P}$
Addendum and the Number of Teeth ...	Pitch Diameter	Multiply Addendum by the Number of Teeth	$D' = s\,N$
The Number of Teeth and the Diametral Pitch	Outside Diameter	Divide Number of Teeth plus 2 by the Diametral Pitch..................	$D = \dfrac{N+2}{P}$
The Pitch Diameter and the Diametral Pitch	Outside Diameter	Add to the Pitch Diameter the quotient of 2 divided by the Diametral Pitch	$D = D' + \dfrac{2}{P}$
The Pitch Diameter and the Number of Teeth ...	Outside Diameter	Divide the Number of Teeth plus 2 by the quotient of Number of Teeth and by the Pitch Diameter	$D = \dfrac{N+2}{\dfrac{N}{D'}}$
The Number of Teeth and Addendum ..	Outside Diameter	Multiply the Number of Teeth plus 2 by Addendum	$D = (N+2)\,s$
The Pitch Diameter and the Diametral Pitch	Number of Teeth	Multiply Pitch Diameter by the Diametral Pitch	$N = D'\,P$
The Outside Diameter and the Diametral Pitch	Number of Teeth	Multiply Outside Diameter by the Diametral Pitch and subtract 2	$N = DP - 2$
The Diametral Pitch	Thickness of Tooth	Divide 1.5708 by the Diametral Pitch	$t = \dfrac{1.5708}{P}$
The Diametral Pitch	Addendum	Divide 1 by the Diametral Pitch or $s = \dfrac{D'}{N}$	$s = \dfrac{1}{P}$
The Diametral Pitch	Root	Divide 1.157 by the Diametral Pitch	$s + f = \dfrac{1.157}{P}$
The Diametral Pitch	Working Depth	Divide 2 by the Diametral Pitch..................	$D'' = \dfrac{2}{P}$
The Diametral Pitch	Whole Depth	Divide 2.157 by the Diametral Pitch	$D'' + f = \dfrac{2.157}{P}$
The Diametral Pitch	Clearance	Divide .157 by the Diametral Pitch..................	$f = \dfrac{.157}{P}$
Thickness of Tooth	Clearance	Divide Thickness of Tooth at pitch line by 10	$f = \dfrac{t}{10}$

TABLE OF CORRESPONDING DIAMETRAL AND CIRCULAR PITCHES

TABLE No. 1		TABLE No. 2	
Diametral Pitch	Circular Pitch	Circular Pitch	Diametral Pitch
1¼	2.5133	2	1.571
1½	2.0944	1⅞	1.676
1¾	1.7952	1¾	1.795
2	1.571	1⅝	1.933
2¼	1.396	1½	2.094
2½	1.257	1 7/16	2.185
2¾	1.142	1⅜	2.285
3	1.047	1 5/16	2.394
3½	.898	1¼	2.513
4	.785	1 3/16	2.646
5	.628	1⅛	2.793
6	.524	1 1/16	2.957
7	.449	1	3.142
8	.393	15/16	3.351
9	.349	7/8	3.590
10	.314	13/16	3.867
11	.286	3/4	4.189
12	.262	11/16	4.570
14	.224	5/8	5.027
16	.196	9/16	5.585
18	.175	1/2	6.283
20	.157	7/16	7.181
22	.143	3/8	8.378
24	.131	5/16	10.053
26	.121	1/4	12.566
28	.112	3/16	16.755
30	.105	1/8	25.133
32	.098	1/16	50.266
36	.087		
40	.079		
48	.065		

No. 1 table shows the diametral pitches with the corresponding circular pitches.

No. 2 table shows the circular pitches with the corresponding diametral pitches.

It is most natural to think of gears in circular or linear pitch and we soon get to know the size of any pitch, as 12, as being a little over ¼ inch from center to center. But the diametral system has many advantages in figuring gear blanks, center distances, etc.

CONSTANTS FOR DETERMINING CHORDAL PITCH AND RADIUS OF SPUR GEARS

P = Chordal Pitch of Teeth.
R = Radius of Pitch Circle.
N = Number of Teeth.
C = Constant. (See table below.)

$$\text{Chordal pitch} = \frac{\text{Radius of pitch circle}}{\text{Constant for number of teeth}}.$$

Radius of pitch circle = Constant × chordal pitch.

$$\text{Constant for any number of teeth} = \frac{\text{Radius of pitch circle}}{\text{Chordal pitch of teeth}}.$$

EXAMPLES: 1. What is radius of pitch circle of a gear having 45 teeth, 1¼ inch pitch? Follow 40 in table to column 5 (making 45 teeth), and find 7.163. Multiply by pitch, 1¾ inch, and get 12.53 inches radius or 25.06 pitch diameter.

2. What is the chordal pitch of a gear 32 inches pitch diameter, 67 teeth? Follow 60 in table to column 7 and find 10.665. Divide radius (½ of 32 = 16 inches) by constant. 16 ÷ 10.665 = 1.5 inch pitch.

3. What number of teeth has a gear of 1.5 inch chordal pitch and pitch diameter 32 inches? Divide by 2 to get radius. Divide this by chordal pitch which will give constant. 16 ÷ 1.5 = 10.665. Look in table for this constant which will be found to represent 67 teeth.

TABLE OF CONSTANTS

N	0	1	2	3	4	5	6	7	8	9
0	0.000	0.159	0.318	0.477	0.636	0.795	0.955	1.114	1.273	1.432
10	1.591	1.750	1.910	2.069	2.229	2.387	2.546	2.706	2.865	3.024
20	3.183	3.342	3.501	3.661	3.820	3.979	4.138	4 297	4.457	4.616
30	4.775	4.934	5.093	5.252	5.412	5.571	5.730	5.889	6.048	6.208
40	6.367	6.526	6.685	6.844	7.003	7.163	7.322	7.481	7.640	7.799
50	7.959	8.118	8.277	8.436	8.595	8.754	8.914	9.073	9.232	9.391
60	9.550	9.709	9.869	10.028	10.187	10.346	10.505	10.665	10.824	10.983
70	11.142	11.301	11.460	11.620	11.779	12.938	12.097	12.256	12.416	12.575
80	12.734	12.893	13.052	13.211	13.371	13.530	13.689	13.848	14.007	14.167
90	14.326	14.485	14.644	14.803	14.962	15.122	15.281	15.440	15.599	15.758
100	15.918	16.077	16.236	16.395	16.554	16.713	16.873	17.032	17.191	17.350
110	17.509	17.668	17.828	17.987	18.146	18.305	18.464	18.624	18.783	18.942
120	19.101	19.260	19.419	19.579	19.738	19.897	20.056	20.215	20.375	20.534
130	20.693	20.852	21.011	21.170	21.330	21.489	21.648	21.807	21.966	22.126
140	22.285	22.444	22.603	22.762	22.921	23.081	23.240	23.399	23.558	23.717
150	23.877	24.036	24.195	24.354	24.513	24.672	24.832	24.991	25.150	25.309
160	25.468	25.627	25.787	25.946	26.105	26.264	26.423	26.583	26.742	26.901
170	27.060	27.219	27.378	27.538	27.697	27.856	28.015	28.174	28.334	28.493
180	28.652	28.811	28.970	29.129	29.289	29.448	29.607	29.766	29.925	30.085
190	30.242	30.403	30.562	30.721	30.880	31.040	31.199	31.358	31.517	31.676
200	31.830	31.989	32.148	32.307	32.446	32.625	32.785	32.944	33.103	33.262
210	33.427	33.586	33.746	33.905	34.064	34.223	34.382	34.542	34.701	34.860
220	35.019	35.178	35.337	35.497	35.656	35.815	35.974	36.133	36.293	36.452
230	36.611	36.770	36.929	37.088	37.248	37.407	37.566	37.725	37.884	38.044
240	38.203	38.362	38.521	38.680	38.839	38.999	39.158	39.317	39.476	39.635
250	39.795									

Gear Wheels

TABLE OF TOOTH PARTS — DIAMETRAL PITCH IN FIRST COLUMN

Diametral Pitch	Circular Pitch	Thickness of Tooth on Pitch Line	Addendum and $\frac{1}{P}$	Working Depth of Tooth	Depth of Space below Pitch Line	Whole Depth of Tooth
P	P'	t	s	D''	$s+f$	$D''+f$
$\frac{1}{2}$	6.2832	3.1416	2.0000	4.0000	2.3142	4.3142
$\frac{3}{4}$	4.1888	2.0944	1.3333	2.6666	1.5428	2.8761
1	3.1416	1.5708	1.0000	2.0000	1.1571	2.1571
$1\frac{1}{4}$	2.5133	1.2566	.8000	1.6000	.9257	1.7257
$1\frac{1}{2}$	2.0944	1.0472	.6666	1.3333	.7714	1.4381
$1\frac{3}{4}$	1.7952	.8976	.5714	1.1429	.6612	1.2326
2	1.5708	.7854	.5000	1.0000	.5785	1.0785
$2\frac{1}{4}$	1.3963	.6981	.4444	.8888	.5143	.9587
$2\frac{1}{2}$	1.2566	.6283	.4000	.8000	.4628	.8628
$2\frac{3}{4}$	1.1424	.5712	.3636	.7273	.4208	.7844
3	1.0472	.5236	.3333	.6666	.3857	.7190
$3\frac{1}{2}$	.8976	.4488	.2857	.5714	.3306	.6163
4	.7854	.3927	.2500	.5000	.2893	.5393
5	.6283	.3142	.2000	.4000	.2314	.4314
6	.5236	.2618	.1666	.3333	.1928	.3595
7	.4488	.2244	.1429	.2857	.1653	.3081
8	.3927	.1963	.1250	.2500	.1446	.2696
9	.3491	.1745	.1111	.2222	.1286	.2397
10	.3142	.1571	.1000	.2000	.1157	.2157
11	.2856	.1428	.0909	.1818	.1052	.1961
12	.2618	.1309	.0833	.1666	.0964	.1798
13	.2417	.1208	.0769	.1538	.0890	.1659
14	.2244	.1122	.0714	.1429	.0826	.1541

To obtain the size of any part of a diametral pitch not given in the table, divide the corresponding part of 1 diametral pitch by the pitch required.

As it is natural to think of gear pitches as the distance between teeth the same as threads, it is well to fix in the mind the approximate center distances of the pitches most in use. Or it is easy to remember that if the diametral pitch be divided by $3\frac{1}{7}$ we have the teeth per inch on the pitch line. By this method we easily see that in a 10 diametral pitch gear there are approximately 3 teeth per inch while in a 22 diametral pitch there will be just 7 teeth to the inch.

TABLE OF TOOTH PARTS — *Continued*

DIAMETRAL PITCH IN FIRST COLUMN

Diametral Pitch	Circular Pitch	Thickness of Tooth on Pitch Line	Addendum and $\frac{1}{P}$	Working Depth of Tooth	Depth of Space below Pitch Line	Whole Depth of Tooth
P	P'	t	s	D''	$s+f$	$D''+f$
15	.2094	.1047	.0666	.1333	.0771	.1438
16	.1963	.0982	.0625	.1250	.0723	.1348
17	.1848	.0924	.0588	.1176	.0681	.1269
18	.1745	.0873	.0555	.1111	.0643	.1198
19	.1653	.0827	.0526	.1053	.0609	.1135
20	.1571	.0785	.0500	.1000	.0579	.1079
22	.1428	.0714	.0455	.0909	.0526	.0980
24	.1309	.0654	.0417	.0833	.0482	.0898
26	.1208	.0604	.0385	.0769	.0445	.0829
28	.1122	.0561	.0357	.0714	.0413	.0770
30	.1047	.0524	.0333	.0666	.0386	.0719
32	.0982	.0491	.0312	.0625	.0362	.0674
34	.0924	.0462	.0294	.0588	.0340	.0634
36	.0873	.0436	.0278	.0555	.0321	.0599
38	.0827	.0413	.0263	.0526	.0304	.0568
40	.0785	.0393	.0250	.0500	.0289	.0539
42	.0748	.0374	.0238	.0476	.0275	.0514
44	.0714	.0357	.0227	.0455	.0263	.0490
46	.0683	.0341	.0217	.0435	.0252	.0469
48	.0654	.0327	.0208	.0417	.0241	.0449
50	.0628	.0314	.0200	.0400	.0231	.0431
56	.0561	.0280	.0178	.0357	.0207	.0385
60	.0524	.0262	.0166	.0333	.0193	.0360

To obtain the size of any part of a diametral pitch not given in the table, divide the corresponding part of 1 diametral pitch by the pitch required.

As it is natural to think of gear pitches as the distance between teeth the same as threads, it is well to fix in the mind the approximate center distances of the pitches most in use. Or it is easy to remember that if the diametral pitch be divided by 3½ we have the teeth per inch on the pitch line. By this method we easily see that in a 10 diametral pitch gear there are approximately 3 teeth per inch while in a 22 diametral pitch there will be just 7 teeth to the inch.

Gear Wheels

TABLE OF TOOTH PARTS — CIRCULAR PITCH IN FIRST COLUMN

Circular Pitch	Threads or Teeth per Inch Linear	Diametral Pitch	Thickness of Tooth on Pitch Line	Addendum and Module	Working Depth of Tooth	Depth of Space below Pitch Line	Whole Depth of Tooth	Width of Thread-Tool at End	Width of Thread at Top
P'	$\frac{1''}{P'}$	P	t	s	D''	$s+f$	$D''+f$	$P' \times .31$	$P' \times .335$
2	$\frac{1}{2}$	1.5708	1.0000	.6366	1.2732	.7366	1.3732	.6200	.6700
$1\frac{7}{8}$	$\frac{8}{15}$	1.6755	.9375	.5968	1.1937	.6906	1.2874	.5813	.6281
$1\frac{3}{4}$	$\frac{4}{7}$	1.7952	.8750	.5570	1.1141	.6445	1.2016	.5425	.5863
$1\frac{5}{8}$	$\frac{8}{13}$	1.9333	.8125	.5173	1.0345	.5985	1.1158	.5038	.5444
$1\frac{1}{2}$	$\frac{2}{3}$	2.0944	.7500	.4775	.9549	.5525	1.0299	.4650	.5025
$1\frac{7}{16}$	$\frac{16}{23}$	2.1855	.7187	.4576	.9151	.5294	.9870	.4456	.4816
$1\frac{3}{8}$	$\frac{8}{11}$	2.2848	.6875	.4377	.8754	.5064	.9441	.4262	.4606
$1\frac{1}{3}$	$\frac{3}{4}$	2.3562	.6666	.4244	.8488	.4910	.9154	.4133	.4466
$1\frac{5}{16}$	$\frac{16}{21}$	2.3936	.6562	.4178	.8356	.4834	.9012	.4069	.4397
$1\frac{1}{4}$	$\frac{4}{5}$	2.5133	.6250	.3979	.7958	.4604	.8583	.3875	.4188
$1\frac{3}{16}$	$\frac{16}{19}$	2.6456	.5937	.3780	.7560	.4374	.8156	.3681	.3978
$1\frac{1}{8}$	$\frac{8}{9}$	2.7925	.5625	.3581	.7162	.4143	.7724	.3488	.3769
$1\frac{1}{16}$	$\frac{16}{17}$	2.9568	.5312	.3382	.6764	.3913	.7295	.3294	.3559
1	1	3.1416	.5000	.3183	.6366	.3683	.6866	.3100	.3350
$\frac{15}{16}$	$1\frac{1}{15}$	3.3510	.4687	.2984	.5968	.3453	.6437	.2906	.3141
$\frac{7}{8}$	$1\frac{1}{7}$	3.5904	.4375	.2785	.5570	.3223	.6007	.2713	.2931
$\frac{13}{16}$	$1\frac{3}{13}$	3.8666	.4062	.2586	.5173	.2993	.5579	.2519	.2722
$\frac{4}{5}$	$1\frac{1}{4}$	3.9270	.4000	.2546	.5092	.2946	.5492	.2480	.2680
$\frac{3}{4}$	$1\frac{1}{3}$	4.1888	.3750	.2387	.4775	.2762	.5150	.2325	.2513
$\frac{11}{16}$	$1\frac{5}{11}$	4.5696	.3437	.2189	.4377	.2532	.4720	.2131	.2303
$\frac{2}{3}$	$1\frac{1}{2}$	4.7124	.3333	.2122	.4244	.2455	.4577	.2066	.2233
$\frac{5}{8}$	$1\frac{3}{5}$	5.0265	.3125	.1989	.3979	.2301	.4291	.1938	.2094
$\frac{3}{5}$	$1\frac{2}{3}$	5.2360	.3000	.1910	.3820	.2210	.4120	.1860	.2010
$\frac{4}{7}$	$1\frac{3}{4}$	5.4978	.2857	.1819	.3638	.2105	.3923	.1771	.1914
$\frac{9}{16}$	$1\frac{7}{9}$	5.5851	.2812	.1790	.3581	.2071	.3862	.1744	.1884

To obtain the size of any part of a circular pitch not given in the table, multiply the corresponding part of 1″ pitch by the pitch required.

As an example take a gear having 21 diametral pitch to find the various tooth parts. Take 1 diametral pitch and divide 3.1416 by 21 to find the corresponding circular pitch, which is .14951. The tooth thickness is 1.5708 ÷ 21 = .748; the addendum is 1. ÷ 21 = .04761; the working depth is 2. ÷ 21. = .09522; the depth below

Table of Tooth Parts

CIRCULAR PITCH IN FIRST COLUMN

Circular Pitch	Threads or Teeth per Inch Linear	Diametral Pitch	Thickness of Tooth on Pitch Line	Addendum and Module	Working Depth of Tooth	Depth of Space below Pitch Line	Whole Depth of Tooth	Width of Thread-Tool at End	Width of Thread at Top
P'	$\frac{1''}{P'}$	P	t	s	D''	$s+f$	$D''+f$	$P' \times .31$	$P' \times .335$
$\frac{1}{2}$	2	6.2832	.2500	.1592	.3183	.1842	.3433	.1550	.1675
$\frac{4}{9}$	$2\frac{1}{4}$	7.0685	.2222	.1415	.2830	.1637	.3052	.1378	.1489
$\frac{7}{16}$	$2\frac{2}{7}$	7.1808	.2187	.1393	.2785	.1611	.3003	.1356	.1466
$\frac{3}{7}$	$2\frac{1}{3}$	7.3304	.2143	.1364	.2728	.1578	.2942	.1328	.1436
$\frac{2}{5}$	$2\frac{1}{2}$	7.8540	.2000	.1273	.2546	.1473	.2746	.1240	.1340
$\frac{3}{8}$	$2\frac{2}{3}$	8.3776	.1875	.1194	.2387	.1381	.2575	.1163	.1256
$\frac{4}{11}$	$2\frac{3}{4}$	8.6394	.1818	.1158	.2316	.1340	.2498	.1127	.1218
$\frac{1}{3}$	3	9.4248	.1666	.1061	.2122	.1228	.2289	.1033	.1117
$\frac{5}{16}$	$3\frac{1}{5}$	10.0531	.1562	.0995	.1989	.1151	.2146	.0969	.1047
$\frac{3}{10}$	$3\frac{1}{3}$	10.4719	.1500	.0955	.1910	.1105	.2060	.0930	.1005
$\frac{2}{7}$	$3\frac{1}{2}$	10.9956	.1429	.0909	.1819	.1052	.1962	.0886	.0957
$\frac{1}{4}$	4	12.5664	.1250	.0796	.1591	.0921	.1716	.0775	.0838
$\frac{2}{9}$	$4\frac{1}{2}$	14.1372	.1111	.0707	.1415	.0818	.1526	.0689	.0744
$\frac{1}{5}$	5	15.7080	.1000	.0637	.1273	.0737	.1373	.0620	.0670
$\frac{3}{16}$	$5\frac{1}{3}$	16.7552	.0937	.0597	.1194	.0690	.1287	.0581	.0628
$\frac{2}{11}$	$5\frac{1}{2}$	17.2788	.0909	.0579	.1158	.0670	.1249	.0564	.0609
$\frac{1}{6}$	6	18.8496	.0833	.0531	.1061	.0614	.1144	.0517	.0558
$\frac{2}{13}$	$6\frac{1}{2}$	20.4203	.0769	.0489	.0978	.0566	.1055	.0477	.0515
$\frac{1}{7}$	7	21.9911	.0714	.0455	.0910	.0526	.0981	.0443	.0479
$\frac{2}{15}$	$7\frac{1}{2}$	23.5619	.0666	.0425	.0850	.0492	.0917	.0414	.0446
$\frac{1}{8}$	8	25.1327	.0625	.0398	.0796	.0460	.0858	.0388	.0419
$\frac{1}{9}$	9	28.2743	.0555	.0354	.0707	.0409	.0763	.0344	.0372
$\frac{1}{10}$	10	31.4159	.0500	.0318	.0637	.0368	.0687	.0310	.0335
$\frac{1}{16}$	16	50.2655	.0312	.0199	.0398	.0230	.0429	.0194	.0209
$\frac{1}{20}$	20	62.8318	.0250	.0159	.0318	.0184	.0343	.0155	.0167

To obtain the size of any part of a circular pitch not given in the table, multiply the corresponding part of 1″ pitch by the pitch required.

pitch line in 1.1571 ÷ 21 = .0551 and the whole depth is 2.1571 ÷ 21 = .1027 inches. These could also have been obtained by splitting the difference between the figures for 20 and 22 pitch. Th[illegible] same can be done for circular pitch except that we multiply inst[illegible] of divide.

DIAGRAM FOR CAST–GEAR TEETH

THE accompanying diagram (Fig. 2) for laying out teeth for cast gears will be found useful by the machinist, patternmaker and drafts-man.   The diagram for circular pitch gears is similar to the one given by Professor Willis, while the one for diametral pitch was obtained by using the relation of diametral to circular pitch.

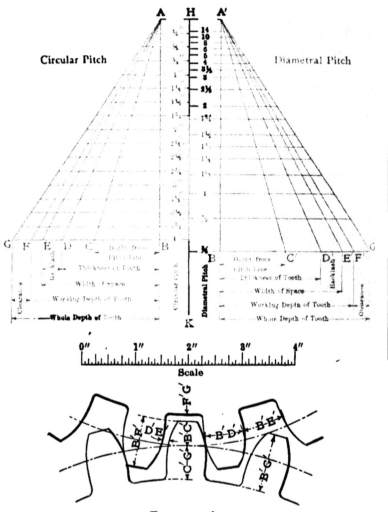

FIGS. 2 and 3.

By the diagram the relative size of a tooth may be easily deter-mined.   For example, if we contemplate using a gear of 2 diametral pitch, by referring to line $H K$, which shows the comparative distance between centers of teeth, on the pitch line, it will be observed that

2 diametral pitch is but little greater than 1½ inches circular pitch, or exactly 1.57 inches circular pitch. This result is obtained by dividing 3.1416 by the diametral pitch (3.1416 divided by 2 equals 1.57). In similar manner, if the circular pitch is known, the diametral pitch which corresponds to it is found by dividing 3.1416 by the circular pitch; for example, the diametral pitch which corresponds to 3 inches circular pitch is by the line *H K* a little greater than 1 diametral pitch, or exactly 1.047 (3.1416 divided by 3 equals 1.047).

The proportions of a tooth may be determined for either diametral or circular pitch by using the corresponding diagram.

Continue, for illustration, the 2 diametral pitch. We have found, above, the distance between centers of teeth on the pitch line to be a little more than 1½ inches (1.57 inches). The hight of tooth above pitch line *B′ C′* will be found on the horizontal line corresponding to 2 pitch. The distance between the lines *A′ B′* and *A′ C′* on this line may be taken in the dividers and transferred to the scale below. Thus we find the hight of the tooth to be 1½ inch. In the same manner the thickness of tooth *B′ D′*, width of space *B′ E′*, working depth *B′ F′* and whole depth of tooth *B′ G′* may be determined.

The backlash or space between the idle surfaces of the teeth of two gear wheels when in mesh is given by the distance *D′ E′*. The clearance or distance between the point of one tooth and the bottom of space into which it meshes is given by the distance *F′ G′*. The backlash and clearance will vary according to the class of work for which the gears are to be used and the accuracy of the molded product. For machine molded gears which are to run in enclosed cases, or where they may be kept well oiled and free from dirt, the backlash and clearance may be reduced to a very small amount, while for gears running where dirt is likely to get into the teeth, or where irregularities due to molding, uneven shrinkage, and like causes, enter into the construction, there must be a greater allowance. The diagram is laid out for the latter case. Those who have more favorable conditions for which to design gears should vary the diagram to suit their conditions. This can be done by increasing *B D* and decreasing *B E*, and by increasing *B C* or decreasing *B G*, or both, to get the clearance that will best meet the required conditions. The same kind of diagram could be laid out for cut gears, but as tables are usually at hand which give the dimensions of the parts of such gears, figured to thousandths of an inch, it would be as well to consult one of these.

LAYING OUT SPUR GEAR BLANKS

DECIDE upon the size wanted, remembering that 12-pitch teeth are $\frac{2}{12}$ deep and 8-pitch — as in the drawing — ⅜ deep, etc. Should it be 8 pitch, as shown in the cut, draw a circle measuring as many eighths of an inch in diameter as there are to be teeth in the gear. This circle is called the Pitch Line. Then with a radius ⅛ of an inch larger, draw another circle from the same center, which will give the outside diameter of the gear, or ⅛ larger than the pitch circle. Thus we have for the diameter of an 8-pitch gear of 24 teeth, $\frac{26}{8}$. Should there be 16 teeth, as in the small spur gear in the cut, th outside diameter would be $\frac{18}{8}$, the number of teeth being always two less than there are eighths — *when it is 8 pitch* — in the outside diameter.

The distance from the pitch line to the bottom of the teeth is the same as to the top, excepting the clearance, which varies from $\frac{1}{8}$ of the pitch to $\frac{1}{10}$ of the thickness of the tooth at the pitch line. This latter is used by Brown & Sharpe and many others, but the clearance being provided for in the cutters the two gears would be laid out to mesh together just $\frac{7}{8}$.

These rules apply to all pitches, so that the outside diameter of a 5-pitch gear with 24 teeth would be $\frac{26}{5}$; if a 3-pitch gear with 40 teeth it would be $\frac{42}{3}$. Again, if a blank be $4\frac{1}{6}$ ($\frac{25}{6}$) in diameter, and cut 6 pitch, it should contain 23 teeth.

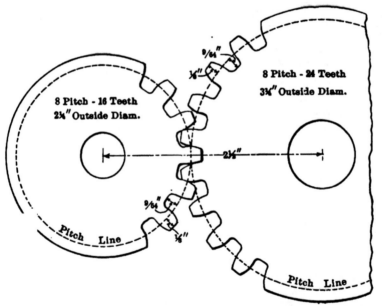

FIG. 4. — Laying out a Pair of Gears

ACTUAL SIZE OF DIAMETRAL PITCHES

It is not always easy to judge or imagine just how large a given pitch is when measured by the diametral system. To make it easy to see just what any pitch looks like the actual sizes of twelve diametral pitches are given on the following page, ranging from 20 to 4 teeth per inch of diameter on the pitch line, so that a good idea of the size of any of these teeth can be had at a glance.

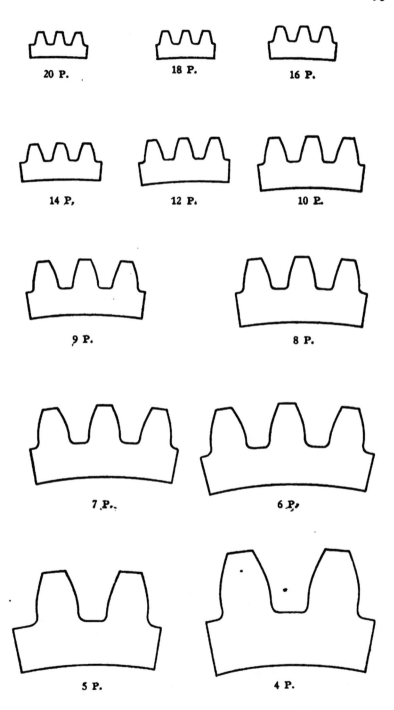

LAYING OUT SINGLE CURVE TOOTH

A VERY simple method of laying out a standard tooth is shown in Fig. 5, and is known as the single curve method. Having calcu lated the various proportions of the tooth by rules already given, draw the pitch, outside, working depth and clearance or whole depth circles as shown. With a radius one half the pitch radius draw the semicircle from the center to the pitch circle. Take one quarter the pitch radius and with one leg at top of pitch circle strike arc cutting the semicircle. This is the center for the first tooth curve and locates the base circle for all the tooth arcs. Lay off the tooth thickness and space distances around the pitch circle and draw the tooth curves through these points with the tooth curve radius already found. The fillets in the tooth corners may be taken as one seventh of the space between the tops of the teeth.

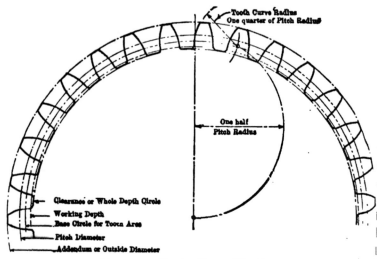

FIG. 5. — Single Curve Tooth

PRESSURE ANGLES

WE next come to pressure angles of gear teeth, which means the angle at which one tooth presses against the other and can best be shown by the pinion and rack, Figs. 6 and 7.

The standard tooth has a 14½ degree pressure angle, probably because it was so easy for the millwright to lay it out as he could obtain the angle without a protractor by using the method shown for laying out a thread tool (see Fig. 14). As the sides of an involute rack tooth are straight, and at the pressure angle from the perpendicular, draw the line of pressure at 14½ degrees from the pitch line. The base circle of the tooth arcs can be found by drawing a line from the center of the gear to the line of pressure and at right angle to it as shown, or by the first method, and working from this the to curve can be drawn by either the single-curve method or, as is more usual, by stopping the curve from two or more points on this same circle.

The difference between the 14½- and 20-degree pressure angles can be seen by comparing Figs. 6 and 7. Not only is the tooth shorter, but the base is broader. The base circle for the tooth arcs is found in the same way as before.

This form of tooth is largely used in automobile transmission and similar work. William Sellers & Co. use a 20 degree pressure angle with a tooth of standard length.

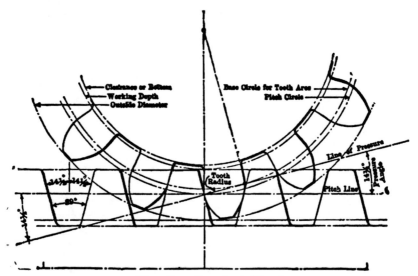

FIG. 6. — Standard Tooth

STUB-TOOTH GEARS

ANY tooth shorter than the regular standard length is called a "stub" tooth, but like the bastard thread there have been many kinds. In 1899 the Fellows Gear Shaper Company introduced a short tooth with a 20-degree pressure angle instead of the usual 14½-degree. This gives a broader flank to the tooth and makes a stronger gear, especially for small pinions where strength is needed. While the Fellows tooth is shorter than the standard tooth there is no fixed relation between them, as, on account of the tooth depth graduations of the gear shaper, it was thought best to give the new tooth depth in the same scale which is shown in the following table. This means that if the pitch is 4 it has the depth of a 5-pitch standard tooth divided as shown. The clearance is one-quarter the addendum or ledendum.

TABLE OF TOOTH DIMENSIONS OF THE FELLOWS STUB-TOOTH GEAR

Cutters Marked Pitch	Stub Tooth Pitch	Has Depth of Standard Tooth	Thickness on Pitch Line	Addendum	Dedendum and Clearance
	4	5	.3925	.200	.250
	5	7	.314	.1429	.1785
	6	8	.2617	.125	.1562
	7	9	.2243	.111	.1389
	8	10	.1962	.100	.125
	9	11	.1744	.0909	.1137
	10	12	.157	.0833	.1042
	12	14	.1308	.0714	.0893

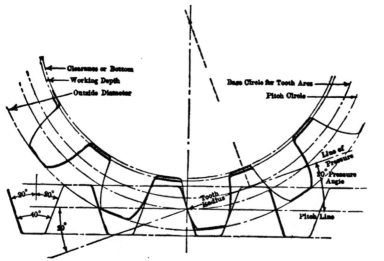

FIG. 7. — Stubbed Tooth

The Nuttall Company also use a 20-degree stub tooth, but have a fixed length or depth in the following proportions.

Addendum = .25 × circular pitch instead of .3683.
Dedendum = .30 × circular pitch instead of .3683.
Working depth = .50 × circular pitch instead of .6366.
Clearance = .05 × circular pitch same as standard.
Whole depth = .55 × circular pitch instead of .6866.

TABLE FOR TURNING AND CUTTING GEAR BLANKS

FOR STANDARD LENGTH TOOTH

Pitch	16	12	10	8	Pitch	16	12	10	8
Depth of Tooth	.135	.180	.216	.270	Depth of Tooth	.135	.180	.216	.270
No. of Teeth	Outside Diameter				No. of Teeth	Outside Diameter			
10	$\frac{3}{4}$	1	$1\frac{2}{10}$	$1\frac{1}{2}$	51	$3\frac{5}{16}$	$4\frac{5}{12}$	$5\frac{3}{10}$	$6\frac{5}{8}$
11	$\frac{13}{16}$	$1\frac{1}{12}$	$1\frac{3}{10}$	$1\frac{5}{8}$	52	$3\frac{3}{8}$	$4\frac{6}{12}$	$5\frac{4}{10}$	$6\frac{3}{4}$
12	$\frac{7}{8}$	$1\frac{2}{12}$	$1\frac{4}{10}$	$1\frac{3}{4}$	53	$3\frac{7}{16}$	$4\frac{7}{12}$	$5\frac{5}{10}$	$6\frac{7}{8}$
13	$\frac{15}{16}$	$1\frac{3}{12}$	$1\frac{5}{10}$	$1\frac{7}{8}$	54	$3\frac{1}{2}$	$4\frac{8}{12}$	$5\frac{6}{10}$	7
14	1	$1\frac{4}{12}$	$1\frac{6}{10}$	2	55	$3\frac{9}{16}$	$4\frac{9}{12}$	$5\frac{7}{10}$	$7\frac{1}{8}$
15	$1\frac{1}{16}$	$1\frac{5}{12}$	$1\frac{7}{10}$	$2\frac{1}{8}$	56	$3\frac{5}{8}$	$4\frac{10}{12}$	$5\frac{8}{10}$	$7\frac{1}{4}$
16	$1\frac{1}{8}$	$1\frac{6}{12}$	$1\frac{8}{10}$	$2\frac{1}{4}$	57	$3\frac{11}{16}$	$4\frac{11}{12}$	$5\frac{9}{10}$	$7\frac{3}{8}$
17	$1\frac{3}{16}$	$1\frac{7}{12}$	$1\frac{9}{10}$	$2\frac{3}{8}$	58	$3\frac{3}{4}$	5	6	$7\frac{1}{2}$
18	$1\frac{1}{4}$	$1\frac{8}{12}$	2	$2\frac{1}{2}$	59	$3\frac{13}{16}$	$5\frac{1}{12}$	$6\frac{1}{10}$	$7\frac{5}{8}$
19	$1\frac{5}{16}$	$1\frac{9}{12}$	$2\frac{1}{10}$	$2\frac{5}{8}$	60	$3\frac{7}{8}$	$5\frac{2}{12}$	$6\frac{2}{10}$	$7\frac{3}{4}$
20	$1\frac{3}{8}$	$1\frac{10}{12}$	$2\frac{2}{10}$	$2\frac{3}{4}$	61	$3\frac{15}{16}$	$5\frac{3}{12}$	$6\frac{3}{10}$	$7\frac{7}{8}$
21	$1\frac{7}{16}$	$1\frac{11}{12}$	$2\frac{3}{10}$	$2\frac{7}{8}$	62	4	$5\frac{4}{12}$	$6\frac{4}{10}$	8
22	$1\frac{1}{2}$	2	$2\frac{4}{10}$	3	63	$4\frac{1}{16}$	$5\frac{5}{12}$	$6\frac{5}{10}$	$8\frac{1}{8}$
23	$1\frac{9}{16}$	$2\frac{1}{12}$	$2\frac{5}{10}$	$3\frac{1}{8}$	64	$4\frac{1}{8}$	$5\frac{6}{12}$	$6\frac{6}{10}$	$8\frac{1}{4}$
24	$1\frac{5}{8}$	$2\frac{2}{12}$	$2\frac{6}{10}$	$3\frac{1}{4}$	65	$4\frac{3}{16}$	$5\frac{7}{12}$	$6\frac{7}{10}$	$8\frac{3}{8}$
25	$1\frac{11}{16}$	$2\frac{3}{12}$	$2\frac{7}{10}$	$3\frac{3}{8}$	66	$4\frac{1}{4}$	$5\frac{8}{12}$	$6\frac{8}{10}$	$8\frac{1}{2}$
26	$1\frac{3}{4}$	$2\frac{4}{12}$	$2\frac{8}{10}$	$3\frac{1}{2}$	67	$4\frac{5}{16}$	$5\frac{9}{12}$	$6\frac{9}{10}$	$8\frac{5}{8}$
27	$1\frac{13}{16}$	$2\frac{5}{12}$	$2\frac{9}{10}$	$3\frac{5}{8}$	68	$4\frac{3}{8}$	$5\frac{10}{12}$	7	$8\frac{3}{4}$
28	$1\frac{7}{8}$	$2\frac{6}{12}$	3	$3\frac{3}{4}$	69	$4\frac{7}{16}$	$5\frac{11}{12}$	$7\frac{1}{10}$	$8\frac{7}{8}$
29	$1\frac{15}{16}$	$2\frac{7}{12}$	$3\frac{1}{10}$	$3\frac{7}{8}$	70	$4\frac{1}{2}$	6	$7\frac{2}{10}$	9
30	2	$2\frac{8}{12}$	$3\frac{2}{10}$	4	71	$4\frac{9}{16}$	$6\frac{1}{12}$	$7\frac{3}{10}$	$9\frac{1}{8}$
31	$2\frac{1}{16}$	$2\frac{9}{12}$	$3\frac{3}{10}$	$4\frac{1}{8}$	72	$4\frac{5}{8}$	$6\frac{2}{12}$	$7\frac{4}{10}$	$9\frac{1}{4}$
32	$2\frac{1}{8}$	$2\frac{10}{12}$	$3\frac{4}{10}$	$4\frac{1}{4}$	73	$4\frac{11}{16}$	$6\frac{3}{12}$	$7\frac{5}{10}$	$9\frac{3}{8}$
33	$2\frac{3}{16}$	$2\frac{11}{12}$	$3\frac{5}{10}$	$4\frac{3}{8}$	74	$4\frac{3}{4}$	$6\frac{4}{12}$	$7\frac{6}{10}$	$9\frac{1}{2}$
34	$2\frac{1}{4}$	3	$3\frac{6}{10}$	$4\frac{1}{2}$	75	$4\frac{13}{16}$	$6\frac{5}{12}$	$7\frac{7}{10}$	$9\frac{5}{8}$
35	$2\frac{5}{16}$	$3\frac{1}{12}$	$3\frac{7}{10}$	$4\frac{5}{8}$	76	$4\frac{7}{8}$	$6\frac{6}{12}$	$7\frac{8}{10}$	$9\frac{3}{4}$
36	$2\frac{3}{8}$	$3\frac{2}{12}$	$3\frac{8}{10}$	$4\frac{3}{4}$	77	$4\frac{15}{16}$	$6\frac{7}{12}$	$7\frac{9}{10}$	$9\frac{7}{8}$
37	$2\frac{7}{16}$	$3\frac{3}{12}$	$3\frac{9}{10}$	$4\frac{7}{8}$	78	5	$6\frac{8}{12}$	8	10
38	$2\frac{1}{2}$	$3\frac{4}{12}$	4	5	79	$5\frac{1}{16}$	$6\frac{9}{12}$	$8\frac{1}{10}$	$10\frac{1}{8}$
39	$2\frac{9}{16}$	$3\frac{5}{12}$	$4\frac{1}{10}$	$5\frac{1}{8}$	80	$5\frac{1}{8}$	$6\frac{10}{12}$	$8\frac{2}{10}$	$10\frac{1}{4}$
40	$2\frac{5}{8}$	$3\frac{6}{12}$	$4\frac{2}{10}$	$5\frac{1}{4}$	81	$5\frac{3}{16}$	$6\frac{11}{12}$	$8\frac{3}{10}$	$10\frac{3}{8}$
41	$2\frac{11}{16}$	$3\frac{7}{12}$	$4\frac{3}{10}$	$5\frac{3}{8}$	82	$5\frac{1}{4}$	7	$8\frac{4}{10}$	$10\frac{1}{2}$
42	$2\frac{3}{4}$	$3\frac{8}{12}$	$4\frac{4}{10}$	$5\frac{1}{2}$	83	$5\frac{5}{16}$	$7\frac{1}{12}$	$8\frac{5}{10}$	$10\frac{5}{8}$
43	$2\frac{13}{16}$	$3\frac{9}{12}$	$4\frac{5}{10}$	$5\frac{5}{8}$	84	$5\frac{3}{8}$	$7\frac{2}{12}$	$8\frac{6}{10}$	$10\frac{3}{4}$
44	$2\frac{7}{8}$	$3\frac{10}{12}$	$4\frac{6}{10}$	$5\frac{3}{4}$	85	$5\frac{7}{16}$	$7\frac{3}{12}$	$8\frac{7}{10}$	$10\frac{7}{8}$
45	$2\frac{15}{16}$	$3\frac{11}{12}$	$4\frac{7}{10}$	$5\frac{7}{8}$	86	$5\frac{1}{2}$	$7\frac{4}{12}$	$8\frac{8}{10}$	11
46	3	4	$4\frac{8}{10}$	6	87	$5\frac{9}{16}$	$7\frac{5}{12}$	$8\frac{9}{10}$	$11\frac{1}{8}$
47	$3\frac{1}{16}$	$4\frac{1}{12}$	$4\frac{9}{10}$	$6\frac{1}{8}$	88	$5\frac{5}{8}$	$7\frac{6}{12}$	9	$11\frac{1}{4}$
48	$3\frac{1}{8}$	$4\frac{2}{12}$	5	$6\frac{1}{4}$	89	$5\frac{11}{16}$	$7\frac{7}{12}$	$9\frac{1}{10}$	$11\frac{3}{8}$
49	$3\frac{3}{16}$	$4\frac{3}{12}$	$5\frac{1}{10}$	$6\frac{3}{8}$	90	$5\frac{3}{4}$	$7\frac{8}{12}$	$9\frac{2}{10}$	$11\frac{1}{2}$
50	$3\frac{1}{4}$	$4\frac{4}{12}$	$5\frac{2}{10}$	$6\frac{1}{2}$	91	$5\frac{13}{16}$	$7\frac{9}{12}$	$9\frac{3}{10}$	$11\frac{5}{8}$

TABLE FOR TURNING AND CUTTING GEAR BLANKS

FOR STANDARD LENGTH TOOTH

Pitch	16	12	10	8	Pitch	16	12	10	8
Depth of Tooth	.135	.180	.216	.270	Depth of Tooth	.135	.180	.216	.270
No. of Teeth	Outside Diameter				No. of Teeth	Outside Diameter			
92	$5\frac{7}{8}$	$7\frac{10}{12}$	$9\frac{4}{10}$	$11\frac{2}{4}$	133	$8\frac{7}{16}$	$11\frac{3}{12}$	$13\frac{5}{10}$	$16\frac{7}{8}$
93	$5\frac{15}{16}$	$7\frac{11}{12}$	$9\frac{5}{10}$	$11\frac{7}{8}$	134	$8\frac{1}{2}$	$11\frac{4}{12}$	$13\frac{6}{10}$	17
94	6	8	$9\frac{6}{10}$	12	135	$8\frac{9}{16}$	$11\frac{5}{12}$	$13\frac{7}{10}$	$17\frac{1}{8}$
95	$6\frac{1}{16}$	$8\frac{1}{12}$	$9\frac{7}{10}$	$12\frac{1}{8}$	136	$8\frac{5}{8}$	$11\frac{6}{12}$	$13\frac{8}{10}$	$17\frac{1}{4}$
96	$6\frac{1}{8}$	$8\frac{2}{12}$	$9\frac{8}{10}$	$12\frac{1}{4}$	137	$8\frac{11}{16}$	$11\frac{7}{12}$	$13\frac{9}{10}$	$17\frac{3}{8}$
97	$6\frac{3}{16}$	$8\frac{3}{12}$	$9\frac{9}{10}$	$12\frac{3}{8}$	138	$8\frac{3}{4}$	$11\frac{8}{12}$	14	$17\frac{1}{2}$
98	$6\frac{1}{4}$	$8\frac{4}{12}$	10	$12\frac{1}{2}$	139	$8\frac{13}{16}$	$11\frac{9}{12}$	$14\frac{1}{10}$	$17\frac{5}{8}$
99	$6\frac{5}{16}$	$8\frac{5}{12}$	$10\frac{1}{10}$	$12\frac{5}{8}$	140	$8\frac{7}{8}$	$11\frac{10}{12}$	$14\frac{2}{10}$	$17\frac{3}{4}$
100	$6\frac{3}{8}$	$8\frac{6}{12}$	$10\frac{2}{10}$	$12\frac{3}{4}$	141	$8\frac{15}{16}$	$11\frac{11}{12}$	$14\frac{3}{10}$	$17\frac{7}{8}$
101	$6\frac{7}{16}$	$8\frac{7}{12}$	$10\frac{3}{10}$	$12\frac{7}{8}$	142	9	12	$14\frac{4}{10}$	18
102	$6\frac{1}{2}$	$8\frac{8}{12}$	$10\frac{4}{10}$	13	143	$9\frac{1}{16}$	$12\frac{1}{12}$	$14\frac{5}{10}$	$18\frac{1}{8}$
103	$6\frac{9}{16}$	$8\frac{9}{12}$	$10\frac{5}{10}$	$13\frac{1}{8}$	144	$9\frac{1}{8}$	$12\frac{2}{12}$	$14\frac{6}{10}$	$18\frac{1}{4}$
104	$6\frac{5}{8}$	$8\frac{10}{12}$	$10\frac{6}{10}$	$13\frac{1}{4}$	145	$9\frac{3}{16}$	$12\frac{3}{12}$	$14\frac{7}{10}$	$18\frac{3}{8}$
105	$6\frac{11}{16}$	$8\frac{11}{12}$	$10\frac{7}{10}$	$13\frac{3}{8}$	146	$9\frac{1}{4}$	$12\frac{4}{12}$	$14\frac{8}{10}$	$18\frac{1}{2}$
106	$6\frac{3}{4}$	9	$10\frac{8}{10}$	$13\frac{1}{2}$	147	$9\frac{5}{16}$	$12\frac{5}{12}$	$14\frac{9}{10}$	$18\frac{5}{8}$
107	$6\frac{13}{16}$	$9\frac{1}{12}$	$10\frac{9}{10}$	$13\frac{5}{8}$	148	$9\frac{3}{8}$	$12\frac{6}{12}$	15	$18\frac{3}{4}$
108	$6\frac{7}{8}$	$9\frac{2}{12}$	11	$13\frac{3}{4}$	149	$9\frac{7}{16}$	$12\frac{7}{12}$	$15\frac{1}{10}$	$18\frac{7}{8}$
109	$6\frac{15}{16}$	$9\frac{3}{12}$	$11\frac{1}{10}$	$13\frac{7}{8}$	150	$9\frac{1}{2}$	$12\frac{8}{12}$	$15\frac{2}{10}$	19
110	7	$9\frac{4}{12}$	$11\frac{2}{10}$	14	151	$9\frac{9}{16}$	$12\frac{9}{12}$	$15\frac{3}{10}$	$19\frac{1}{8}$
111	$7\frac{1}{16}$	$9\frac{5}{12}$	$11\frac{3}{10}$	$14\frac{1}{8}$	152	$9\frac{5}{8}$	$12\frac{10}{12}$	$15\frac{4}{10}$	$19\frac{1}{4}$
112	$7\frac{1}{8}$	$9\frac{6}{12}$	$11\frac{4}{10}$	$14\frac{1}{4}$	153	$9\frac{11}{16}$	$12\frac{11}{12}$	$15\frac{5}{10}$	$19\frac{3}{8}$
113	$7\frac{3}{16}$	$9\frac{7}{12}$	$11\frac{5}{10}$	$14\frac{3}{8}$	154	$9\frac{3}{4}$	13	$15\frac{6}{10}$	$19\frac{1}{2}$
114	$7\frac{1}{4}$	$9\frac{8}{12}$	$11\frac{6}{10}$	$14\frac{1}{2}$	155	$9\frac{13}{16}$	$13\frac{1}{12}$	$15\frac{7}{10}$	$19\frac{5}{8}$
115	$7\frac{5}{16}$	$9\frac{9}{12}$	$11\frac{7}{10}$	$14\frac{5}{8}$	156	$9\frac{7}{8}$	$13\frac{2}{12}$	$15\frac{8}{10}$	$19\frac{3}{4}$
116	$7\frac{3}{8}$	$9\frac{10}{12}$	$11\frac{8}{10}$	$14\frac{3}{4}$	157	$9\frac{15}{16}$	$13\frac{3}{12}$	$15\frac{9}{10}$	$19\frac{7}{8}$
117	$7\frac{7}{16}$	$9\frac{11}{12}$	$11\frac{9}{10}$	$14\frac{7}{8}$	158	10	$13\frac{4}{12}$	16	20
118	$7\frac{1}{2}$	10	12	15	159	$10\frac{1}{16}$	$13\frac{5}{12}$	$16\frac{1}{10}$	$20\frac{1}{8}$
119	$7\frac{9}{16}$	$10\frac{1}{12}$	$12\frac{1}{10}$	$15\frac{1}{8}$	160	$10\frac{1}{8}$	$13\frac{6}{12}$	$16\frac{2}{10}$	$20\frac{1}{4}$
120	$7\frac{5}{8}$	$10\frac{2}{12}$	$12\frac{2}{10}$	$15\frac{1}{4}$	161	$10\frac{3}{16}$	$13\frac{7}{12}$	$16\frac{3}{10}$	$20\frac{3}{8}$
121	$7\frac{11}{16}$	$10\frac{3}{12}$	$12\frac{3}{10}$	$15\frac{3}{8}$	162	$10\frac{1}{4}$	$13\frac{8}{12}$	$16\frac{4}{10}$	$20\frac{1}{2}$
122	$7\frac{3}{4}$	$10\frac{4}{12}$	$12\frac{4}{10}$	$15\frac{1}{2}$	163	$10\frac{5}{16}$	$13\frac{9}{12}$	$16\frac{5}{10}$	$20\frac{5}{8}$
123	$7\frac{13}{16}$	$10\frac{5}{12}$	$12\frac{5}{10}$	$15\frac{5}{8}$	164	$10\frac{3}{8}$	$13\frac{10}{12}$	$16\frac{6}{10}$	$20\frac{3}{4}$
124	$7\frac{7}{8}$	$10\frac{6}{12}$	$12\frac{6}{10}$	$15\frac{3}{4}$	165	$10\frac{7}{16}$	$13\frac{11}{12}$	$16\frac{7}{10}$	$20\frac{7}{8}$
125	$7\frac{15}{16}$	$10\frac{7}{12}$	$12\frac{7}{10}$	$15\frac{7}{8}$	166	$10\frac{1}{2}$	14	$16\frac{8}{10}$	21
126	8	$10\frac{8}{12}$	$12\frac{8}{10}$	16	167	$10\frac{9}{16}$	$14\frac{1}{12}$	$16\frac{9}{10}$	$21\frac{1}{8}$
127	$8\frac{1}{16}$	$10\frac{9}{12}$	$12\frac{9}{10}$	$16\frac{1}{8}$	168	$10\frac{5}{8}$	$14\frac{2}{12}$	17	$21\frac{1}{4}$
128	$8\frac{1}{8}$	$10\frac{10}{12}$	13	$16\frac{1}{4}$	169	$10\frac{11}{16}$	$14\frac{3}{12}$	$17\frac{1}{10}$	$21\frac{3}{8}$
129	$8\frac{3}{16}$	$10\frac{11}{12}$	$13\frac{1}{10}$	$16\frac{3}{8}$	170	$10\frac{3}{4}$	$14\frac{4}{12}$	$17\frac{2}{10}$	$21\frac{1}{2}$
130	$8\frac{1}{4}$	11	$13\frac{2}{10}$	$16\frac{1}{2}$	171	$10\frac{13}{16}$	$14\frac{5}{12}$	$17\frac{3}{10}$	$21\frac{5}{8}$
131	$8\frac{5}{16}$	$11\frac{1}{12}$	$13\frac{3}{10}$	$16\frac{5}{8}$	172	$10\frac{7}{8}$	$14\frac{6}{12}$	$17\frac{4}{10}$	$21\frac{3}{4}$
132	$8\frac{3}{8}$	$11\frac{2}{12}$	$13\frac{4}{10}$	$16\frac{3}{4}$	173	$10\frac{15}{16}$	$14\frac{7}{12}$	$17\frac{5}{10}$	$21\frac{7}{8}$

GEAR TOOTH CUTTERS

Brown & Sharpe and other makes of standard cutters are made in sets of eight cutters for each pitch and are to be used as follows:

No. 1 will cut wheels from 135 teeth to a rack.
No. 2 will cut wheels from 55 teeth to 134 teeth.
No. 3 will cut wheels from 35 teeth to 54 teeth.
No. 4 will cut wheels from 26 teeth to 34 teeth.
No. 5 will cut wheels from 21 teeth to 25 teeth.
No. 6 will cut wheels from 17 teeth to 20 teeth.
No. 7 will cut wheels from 14 teeth to 16 teeth.
No. 8 will cut wheels from 12 teeth to 13 teeth.

TABLE SHOWING DEPTH OF SPACE AND THICKNESS OF TOOTH IN SPUR WHEELS, WHEN CUT WITH THESE CUTTERS.

Pitch of Cutter	Depth to be cut in Gear Inches	Thickness of Tooth at Pitch Line. Inches	Pitch of Cutter	Depth to be cut in Gear Inches	Thickness of Tooth at Pitch Line. Inches
$1\frac{1}{4}$	1.726	1.257	11	.196	.143
$1\frac{1}{2}$	1.438	1.047	12	.180	.131
$1\frac{3}{4}$	1.233	.898	14	.154	.112
2	1.078	.785	16	.135	.098
$2\frac{1}{4}$	.958	.697	18	.120	.087
$2\frac{1}{2}$	.863	.628	20	.108	.079
$2\frac{3}{4}$	.784	.570	22	.098	.071
3	.719	.523	24	.090	.065
$3\frac{1}{2}$	.616	.448	26	.083	.060
4	.539	.393	28	.077	.056
5	.431	.314	30	.072	.052
6	.359	.262	32	.067	.049
7	.308	.224	36	.060	.044
8	.270	.196	40	.054	.039
9	.240	.175	48	.045	.033
10	.216	.157			

BLOCK INDEXING IN CUTTING GEAR TEETH

Block or intermittent indexing is a method to increase the output of gear cutters by allowing the feed and cutting speed to be increased without unduly heating the work. This is done by jumping from the tooth just cut to a tooth far enough away to escape the local heating and on the following rounds to cut the intermediate teeth. While the indexing takes a trifle more time, the heat is distributed so that faster cutting can be done without heating and dulling the cutter.

The following table gives the indexing of gears from 25 to 200 teeth and is worked out for the Brown & Sharpe gear cutter but can be modified to suit other machines.

outside diameter would be $\frac{18}{8}$, the number of teeth being always two less than there are eighths — *when it is* 8 *pitch* — in the outside diameter.

The distance from the pitch line to the bottom of the teeth is the same as to the top, excepting the clearance, which varies from $\frac{1}{8}$ of the pitch to $\frac{1}{16}$ of the thickness of the tooth at the pitch line. This latter is used by Brown & Sharpe and many others, but the clearance being provided for in the cutters the two gears would be laid out to mesh together just $\frac{1}{8}$.

These rules apply to all pitches, so that the outside diameter of a 5-pitch gear with 24 teeth would be $\frac{26}{5}$; if a 3-pitch gear with 40 teeth it would be $\frac{42}{3}$. Again, if a blank be $4\frac{1}{6}$ ($\frac{25}{6}$) in diameter, and cut 6 pitch, it should contain 23 teeth.

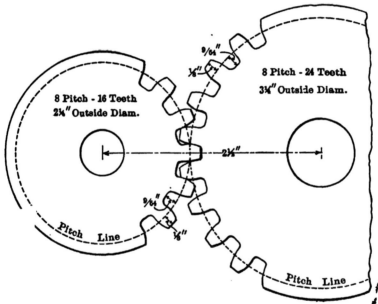

8 Pitch - 16 Teeth
2¼" Outside Diam.

8 Pitch - 24 Teeth
3¼" Outside Diam.

9/64"

⅛"

2½"

9/64"

⅛"

Pitch Line

Pitch Line

Fig. 4. — Laying out a Pair of Gears

ACTUAL SIZE OF DIAMETRAL PITCHES

It is not always easy to judge or imagine just how large a given pitch is when measured by the diametral system. To make it easy to see just what any pitch looks like the actual sizes of twelve diametral pitches are given on the following page, ranging from 20 to 4 teeth per inch of diameter on the pitch line, so that a good idea of the size of any of these teeth can be had at a glance.

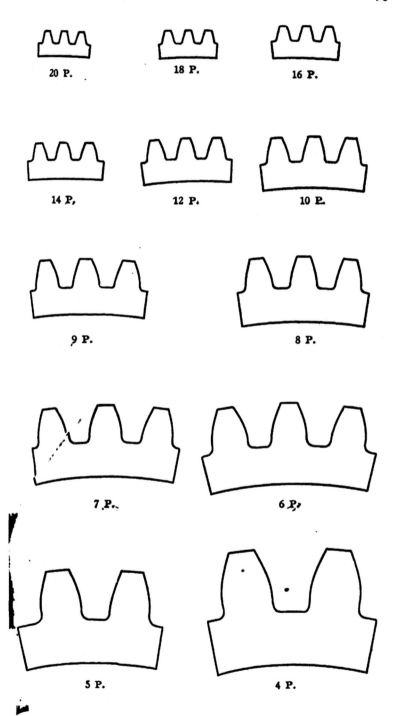

20 P. 18 P. 16 P.

14 P. 12 P. 10 P.

9 P. 8 P.

7 P. 6 P.

5 P. 4 P.

LAYING OUT SINGLE CURVE TOOTH

A VERY simple method of laying out a standard tooth is shown in Fig. 5, and is known as the single curve method. Having calculated the various proportions of the tooth by rules already given, draw the pitch, outside, working depth and clearance or whole depth circles as shown. With a radius one half the pitch radius draw the semicircle from the center to the pitch circle. Take one quarter the pitch radius and with one leg at top of pitch circle strike arc cutting the semicircle. This is the center for the first tooth curve and locates the base circle for all the tooth arcs. Lay off the tooth thickness and space distances around the pitch circle and draw the tooth curves through these points with the tooth curve radius already found. The fillets in the tooth corners may be taken as one seventh of the space between the tops of the teeth.

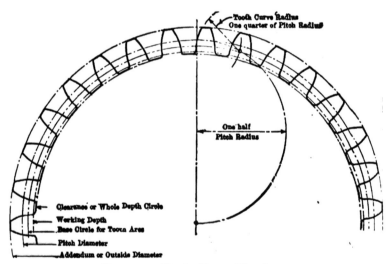

Tooth Curve Radius
One quarter of Pitch Radius

One half
Pitch Radius

Clearance or Whole Depth Circle
Working Depth
Base Circle for Tooth Arcs
Pitch Diameter
Addendum or Outside Diameter

FIG. 5. — Single Curve Tooth

PRESSURE ANGLES

WE next come to pressure angles of gear teeth, which means the angle at which one tooth presses against the other and can best be shown by the pinion and rack, Figs. 6 and 7.

The standard tooth has a $14\frac{1}{2}$ degree pressure angle, probably because it was so easy for the millwright to lay it out as he could obtain the angle without a protractor by using the method shown for laying out a thread tool (see Fig. 14). As the sides of an involute rack tooth are straight, and at the pressure angle from the perpendicular, draw the line of pressure at $14\frac{1}{2}$ degrees from the pitch line. The base circle of the tooth arcs can be found by drawing a line from the center of the gear to the line of pressure and at right angle to it as shown, or by the first method, and working from this the tooth curve can be drawn by either the single-curve method or, as is more usual, by stopping the curve from two or more points on this same circle.

The difference between the 14½- and 20-degree pressure angles can be seen by comparing Figs. 6 and 7. Not only is the tooth shorter, but the base is broader. The base circle for the tooth arcs is found in the same way as before.

This form of tooth is largely used in automobile transmission and similar work. William Sellers & Co. use a 20 degree pressure angle with a tooth of standard length.

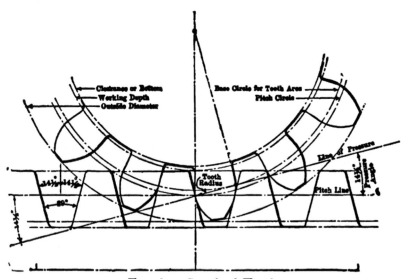

FIG. 6. — Standard Tooth

STUB-TOOTH GEARS

ANY tooth shorter than the regular standard length is called a "stub" tooth, but like the bastard thread there have been many kinds. In 1899 the Fellows Gear Shaper Company introduced a short tooth with a 20-degree pressure angle instead of the usual 14½-degree. This gives a broader flank to the tooth and makes a stronger gear, especially for small pinions where strength is needed. While the Fellows tooth is shorter than the standard tooth there is no fixed relation between them, as, on account of the tooth depth graduations of the gear shaper, it was thought best to give the new tooth depth in the same scale which is shown in the following table. This means that if the pitch is 4 it has the depth of a 5-pitch standard tooth divided as shown. The clearance is one-quarter the addendum or ledendum.

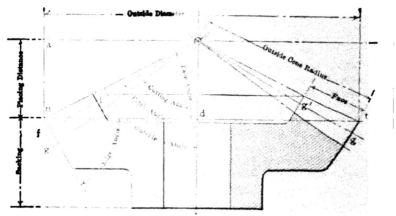

FIG. 11. — Bevel Gear Parts

BEVEL GEARS

BEVEL Gears are used to transmit power when shafts are not parallel. They can be made for any angle, but are more often at right angles than any other. Right angle bevel gears are often called miter gears. The teeth are or should be radial so that they are lower at the outer end. The names of the various parts are shown in Fig. 11.

LAYING OUT BEVEL GEAR BLANKS

IN laying out bevel gears, first decide upon the pitch, and draw the center lines $B B$ and $C C$, intersecting at right angles at A as shown in Fig. 12. Then draw the lines $D D$ to $E E$ the same distance each

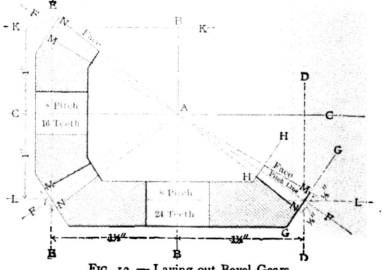

FIG. 12. — Laying out Bevel Gears

side of *B B* and parallel to it; the distance from *D D* to *E E* being as many eighths of an inch — if it be 8 pitch — as there are to be teeth in the gear. In the example the number of teeth is 24; therefore the distance from *D D* to *E E* will be ²⁴⁄₈, or 1½ inches each side of *B B*. *K K* and *L L* are similarly drawn, but there being only 16 teeth in the small gear, the distance from *K K* to *L L* will be ¹⁶⁄₈, or 1 inch each side of *C C*. Then through the intersections of *D D* and *L L*, *E E* and *L L*, and *E E* and *K K*, draw the diagonals *F A*. These are the pitch lines. Through the same point draw lines as *G G* at right angles to the pitch lines, forming the backs of the teeth. On these lines lay off ⅛ of an inch each side of the pitch lines, and draw *M A* and *N A*, forming the faces and bottoms of the teeth. The lines *H H* are drawn parallel to *G G*, the distance between them being the width of the face.

The face of the larger gear should be turned to the lines *M A*, and the small gear to *N A*. For other pitches the same rules apply. If 4 pitch, use 4ths instead of 8ths; if 3 pitch, 3ds, and so on.

Bevel gears should always be turned to the exact diameters and angles of the drawings and the teeth cut at the correct angle.

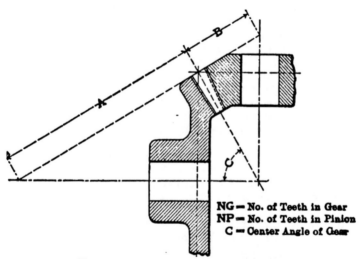

NG = No. of Teeth in Gear
NP = No. of Teeth in Pinion
C = Center Angle of Gear

FIG. 13. — Finding the Cutter to Use

CUTTERS FOR BEVEL GEARS

LAY out the bevel gears and draw lines *A* and *B* at right angles to the center angle line. Extend this to the center lines and measure *A* and *B*. The distance *A* = the radius of a spur gear of the same pitch, and finding the number of teeth in such a gear we have the right cutter for the bevel gear in question. Calling the gears 8 pitch and the distance *A* = 4 inches. Then 2 × 4 × 8 = 64 teeth, so that a No. 2 cutter is the one to use. For the pinion, if *R* is 2 inches, then 2 × 2 × 8 = 32 or a No. 4 cutter is the on use.

USING THE BEVEL GEAR TABLE

TAKE a pair of bevel gears 24 and 72 teeth, 8 diametral pitch. Divide the pinion by the gear — 24 ÷ 72 = .3333. This is the tangent of the center angles. Look in the seven columns under center angles for the nearest number to this. The nearest is .3346 in the center column, as all these are decimals to four places. Follow this out to the left and find 18 in the center angle column. As the .3346 is in the column marked .50 the center angle of the pinion is 18.50 degrees. Looking to the right under center angles for gears find 71 and add the .50 making the gear angle 71.50 degrees. This gives

> Center angle of pinion 18.5 degrees.
> Center angle of gear 71.5 degrees.

In the first column opposite 18 is 36. Divide this by the number of teeth in the pinion, 24, and get 1.5 degrees. This is the angle increase for this pair of gears, and is the amount to be added to the center angle to get the face angle and to be deducted to get the cut angles. This gives

> Pinion center angle 18.5 + 1.5 = 20 degrees face angle.
> Pinion center angle 18.5 − 1.5 = 17 degrees cut angle.
> Gear center angle 71.5 + 1.5 = 73 degrees face angle.
> Gear center angle 71.5 − 1.5 = 70 degrees cut angle.

For the outside diameter go to the column of diameter increase and in line with 18 find 1.90. Divide this by the pitch, 8, and get .237, which is the diameter increase for the pinion. Follow the same line to the right and find .65 for the gear increase. Divide this by the pitch, .8, and get .081 for gear increase. This gives

> Pinion, 24 teeth, 8 pitch = 3 inches + .237 = 3.237 inches outside diameter.
> Gear, 72 teeth, 8 pitch = 9 inches + .081 = 9.081 inches outside diameter.

Another way of selecting the cutter is to divide the number of teeth in the gear by the cosine of the center angle C and the answer is the number of teeth in a spur gear from which to select the cutter. For the pinion the process is the same except the number of teeth in the pinion is divided by the sine of the center angle. Formula

Tangent of $C = \dfrac{NG}{NP}$.

Number of teeth to use in selecting cutter for gear $= \dfrac{NG}{Cos\ C}$.

Number of teeth to use in selecting cutter for pinion $= \dfrac{NP}{Sin\ C}$.

Any pair of gears can be figured out in the same way, bearing in mind that when finding the center angle for the gear, to read the parts of a degree from the decimals at the bottom, and that for the pinion they are at the top. In the example worked out the tangent came in the center column so that it made no difference. If, however, the tangent had been .3476 we read the pinion angle at the top, 19.17 degrees and the gear angle at the bottom, 70.83. By noting that the sum of the two angles is 90 degrees, we can be sure we are right.

BEVEL GEAR TABLE

SHAFT ANGLES 90°

Angle Increase; Divide by Teeth in Pinion	Diameter Increase; Divide by Pitch for Pinion	Center Angle Degrees for Pinion	Center Angle Hundredth Degrees Left-hand Column read here							Center Angle Degrees for Gear	Diameter Increase; Divide by Pitch for Gear
			0	.17	.33	.50	.67	.83	1.00		
1	2.00	0	.0000	.0029	.0058	.0087	.0116	.0145	.0175	89	.03
2	2.00	1	.0175	.0204	.0233	.0262	.0291	.0320	.0349	88	.07
4	2.00	2	.0349	.0378	.0407	.0437	.0466	.0495	.0524	87	.10
6	2.00	3	.0524	.0553	.0582	.0612	.0641	.0670	.0699	86	.14
8	1.99	4	.0699	.0729	.0758	.0787	.0816	.0846	.0875	85	.17
10	1.99	5	.0875	.0904	.0934	.0963	.0992	.1022	.1051	84	.21
12	1.99	6	.1051	.1080	.1110	.1139	.1169	.1198	.1228	83	.24
14	1.98	7	.1228	.1257	.1278	.1317	.1346	.1376	.1405	82	.28
16	1.98	8	.1405	.1435	.1465	.1495	.1524	.1554	.1584	81	.31
18	1.98	9	.1584	.1614	.1644	.1673	.1703	.1733	.1763	80	.34
20	1.97	10	.1763	.1793	.1823	.1853	.1883	.1914	.1944	79	.38
22	1.96	11	.1944	.1974	.2004	.2035	.2065	.2095	.2126	78	.41
24	1.96	12	.2126	.2156	.2186	.2217	.2247	.2278	.2309	77	.45
26	1.95	13	.2309	.2339	.2370	.2401	.2432	.2462	.2493	76	.48
28	1.94	14	.2493	.2524	.2555	.2586	.2617	.2648	.2679	75	.51
30	1.93	15	.2679	.2711	.2742	.2773	.2805	.2836	.2867	74	.55
32	1.92	16	.2867	.2899	.2931	.2962	.2994	.3026	.3057	73	.58
34	1.91	17	.3057	.3089	.3121	.3153	.3185	.3217	.3249	72	.62
36	1.90	18	.3249	.3281	.3314	.3346	.3378	.3411	.3443	71	.65
37	1.89	19	.3443	.3476	.3508	.3541	.3574	.3607	.3640	70	.68
39	1.88	20	.3640	.3673	.3706	.3739	.3772	.3805	.3839	69	.71
41	1.86	21	.3839	.3872	.3906	.3939	.3973	.4006	.4040	68	.75
43	1.85	22	.4040	.4074	.4108	.4142	.4176	.4210	.4245	67	.78
45	1.84	23	.4245	.4279	.4314	.4348	.4383	.4417	.4452	66	.81
47	1.82	24	.4452	.4487	.4522	.4557	.4592	.4628	.4663	65	.84
49	1.81	25	.4663	.4699	.4734	.4770	.4806	.4841	.4877	64	.88
50	1.79	26	.4877	.4913	.4950	.4986	.5022	.5059	.5095	63	.91
52	1.78	27	.5095	.5132	.5169	.5206	.5243	.5280	.5317	62	.93
54	1.76	28	.5317	.5354	.5392	.5430	.5467	.5505	.5543	61	.97
56	1.74	29	.5543	.5581	.5619	.5658	.5696	.5735	.5774	60	1.00
57	1.73	30	.5774	.5812	.5851	.5890	.5930	.5969	.6009	59	1.03
59	1.71	31	.6009	.6048	.6088	.6128	.6168	.6208	.6249	58	1.05
61	1.69	32	.6249	.6289	.6330	.6371	.6412	.6453	.6494	57	1.08
63	1.67	33	.6494	.6536	.6577	.6619	.6661	.6703	.6745	56	1.11
64	1.65	34	.6745	.6787	.6830	.6873	.6916	.6959	.7002	55	1.14
66	1.63	35	.7002	.7046	.7089	.7133	.7177	.7221	.7265	54	1.17
68	1.61	36	.7265	.7310	.7355	.7400	.7445	.7490	.7536	53	1.20
69	1.59	37	.7536	.7581	.7627	.7673	.7720	.7766	.7813	52	1.23
71	1.57	38	.7813	.7860	.7907	.7954	.8002	.8050	.8098	51	1.25
72	1.55	39	.8098	.8146	.8195	.8243	.8292	.8342	.8391	50	1.28
73	1.53	40	.8391	.8441	.8491	.8541	.8591	.8642	.8693	49	1.31
75	1.51	41	.8693	.8744	.8796	.8847	.8899	.8952	.9004	48	1.33
77	1.48	42	.9004	.9057	.9110	.9163	.9217	.9271	.9325	47	1.36
79	1.46	43	.9325	.9380	.9435	.9490	.9545	.9601	.9657	46	1.39
80	1.43	44	.9657	.9713	.9770	.9827	.9884	.9942	1.0000	45	1.41
81	1.41	45	1.0000	1.0058	1.0117	1.0176	1.0235	1.0295	1.0355	44	1.43
			1.00	.83	.67	.50	.33	.17	0		

Right Hand Column read here

A TABLE FOR DIMENSIONS FOR MITER GEARS

THE accompanying table is of service in determining the principal dimensions of miter gears (center angle 45 degrees), the number of teeth and the pitch being known. The table covers most of the possible numbers of teeth from 12 to 60, inclusive, and pitches from 2 to 10, inclusive, omitting 9.

The arrangement and use of the table needs no explanation.

A TABLE OF DIMENSIONS FOR MITER GEARS
MITER GEARS, CENTER ANGLE 45 DEGREES

No. of Teeth	Outside Diameters 2 P.	3 P.	4 P.	5 P.	6 P.	7 P.	8 P.	10 P.	Face Angle	Cut Angle
12	6.71	4.48	3.35	2.68	2.24	1.92	1.68	1.34	38°17'	37°14'
13	7.21	4.80	3.60	2.88	2.40	2.06	1.81	1.44	38°48'	37°40'
14	7.71	5.14	3.85	3.08	2.57	2.20	1.93	1.54	39°13'	38°20'
15	8.21	5.46	4.10	3.28	2.73	2.35	2.06	1.64	39°37'	38°46'
16	8.71	5.80	4.35	3.48	2.90	2.49	2.18	1.74	39°57'	39°00'
17	9.21	6.14	4.60	3.68	3.07	2.63	2.31	1.84	40°15'	39°30'
18	9.71	6.48	4.85	3.88	3.24	2.77	2.43	1.94	40°31'	39°48'
19	10.21	6.80	5.10	4.08	3.40	2.92	2.56	2.04	40°45'	40°04'
20	10.71	7.14	5.35	4.28	3.57	3.06	2.68	2.14	40°57'	40°19'
21	11.21	7.46	5.60	4.48	3.73	3.20	2.81	2.24	41°09'	40°32'
22	11.71	7.80	5.85	4.68	3.90	3.35	2.93	2.34	41°19'	40°45'
23	12.21	8.14	6.10	4.88	4.07	3.49	3.06	2.44	41°20'	40°56'
24	12.71	8.48	6.35	5.08	4.24	3.63	3.18	2.54	41°38'	41°00'
25	13.21	8.80	6.60	5.28	4.40	3.77	3.31	2.64	41°46'	41°15'
26	13.71	9.14	6.85	5.48	4.57	3.92	3.43	2.74	41°53'	41°24'
27	14.21	9.46	7.10	5.68	4.73	4.06	3.56	2.84	42°00'	41°32'
28	14.71	9.80	7.35	5.88	4.90	4.20	3.68	2.94	42°06'	41°39'
29	15.21	10.14	7.60	6.08	5.07	4.35	3.81	3.04	42°13'	41°46'
30	15.71	10.48	7.85	6.28	5.24	4.49	3.93	3.14	42°18'	41°53'
31	16.21	10.80	8.10	6.48	5.40	4.63	4.06	3.24	42°23'	41°59'
32	16.71	11.14	8.35	6.68	5.57	4.77	4.18	3.34	42°28'	42°04'
33	17.21	11.46	8.60	6.88	5.73	4.92	4.31	3.44	42°33'	42°10'
34	17.71	11.80	8.85	7.08	5.90	5.06	4.43	3.54	42°37'	42°15'
35	18.21	12.14	9.10	7.28	6.07	5.20	4.56	3.64	42°41'	42°19'
36	18.71	12.48	9.35	7.48	6.24	5.35	4.68	3.74	42°45'	42°24'
37	19.21	12.80	9.60	7.68	6.40	5.49	4.81	3.84	42°40'	42°28'
38	19.71	13.14	9.85	7.88	6.57	5.63	4.93	3.94	42°52'	42°32'
40	20.71	13.80	10.35	8.28	6.90	5.92	5.18	4.14	42°59'	42°39'
42	21.71	14.48	10.85	8.68	7.24	6.20	5.43	4.34	43°04'	42°46'
44	22.71	15.14	11.35	9.08	7.57	6.49	5.68	4.54	43°10'	42°52'
46	23.71	15.80	11.85	9.48	7.90	6.77	5.93	4.74	43°14'	42°58'
48	24.71	16.48	12.35	9.88	8.24	7.06	6.18	4.94	43°18'	43°06'
50	25.71	17.14	12.85	10.28	8.57	7.35	6.43	5.14	43°23'	43°18'
54	27.71	18.48	13.85	11.08	9.24	7.92	6.93	5.54	43°29'	43°31'
58	29.71	19.80	14.85	11.88	9.91	8.49	7.44	5.94	43°44'	43°44'

Pitch Diameters (2 P. through 10 P.) are given by the number of teeth divided by the pitch.

SPIRAL GEARS

THE term spiral gear is usually applied to gears having angular teeth and which do not have their shafts or axis in parallel lines, and usually at right angles. Spiral gears take the place of bevel gears and give a smoother action as well as allowing greater speed ratios in a given space. When gears with angular or skew teeth run on parallel shafts they are usually called helical gears.

In considering speed ratios for spiral gears the driving gear can be taken as a worm having as many threads as there are teeth and the driven as the worm wheel with its number of teeth, so that one revolution of the driver will turn a point on the pitch circle of the driven gear as many inches as the product of the lead of the driver and the number of its teeth. Divide this by the circumference of the pitch circle of the driven gear to get the revolutions of the driven.

When the spiral angles are 45 degrees, the speed ratio depends entirely on the number of teeth as in bevel gears, but for other angles of spiral the following formula will be useful:

Let R_1 = Revs. of Driver.

R_2 = Revs. of Driven.

D_1 = Pitch diameter of Driver.

D_2 = Pitch diameter of Driven.

N_1 = Number of teeth in Driver.

N_2 = Number of teeth in Driven.

Then with shafts at 90 degrees we have

$$\frac{R_2}{R_1} = \frac{D_1}{D_2} \times \text{cotangent of helix angle of driver with its axis.}$$

$$R_2 = \frac{D_1 \times R_1}{D_2} \times \text{cotangent of helix angle of driver with its axis.}$$

$$D_1 = \frac{\dfrac{R_2 \times D_2}{R_1}}{\text{cotangent of helix}} \qquad D_2 = \frac{R_1 \times D_1}{R_2} \times \text{cotangent of helix.}$$

Or in the form of rules we have:

Having	To Find	Rule
Same Diam. of both Gears Speed of Driving Gear Cotangent of Tooth Angle of Driver	Speed of Driven Gear	Multiply speed of driver by cotangent of tooth angle of driver with axis.
Driven Gear the Largest Speed of Driving Gear Cotangent of Driver Angle Diameter of Driving Gear Diameter of Driven Gear	Speed of Driven Gear	Divide diameter of driver by diameter of driven. Multiply by speed of driver and by cotangent of driver tooth angle with axis.

Speed ratio of Gears or Revolutions of Driver and Driven Cotangent of Tooth Angle of driver Diameter of Driven	Diameter of Driver	Divide speed of driven by speed of driver and multiply by diameter of driven. Divide by cotangent of tooth angle of driver with axis.
Speed ratio or Revolution of Driver and Driven Cotangent of Tooth Angle of Driver Diameter of Driver	Diameter of Driven	Divide speed of driver by speed of driven and multiply by diameter and cotangent of tooth angle of driver and by diameter of driver with axis.

Other formulas for spiral gears are:

Lead = Distance the spiral advances in one turn.

Angle = The angle formed on the pitch surface by sides of teeth and a line parallel with the axis of the gear.

45° Angle Spiral = Two spirals are cut 45° when the pitch diameters and teeth are in the same ratio.

FORMULÆ FOR 45° SPIRALS

Diametral Pitch = Number teeth ÷ pitch diameter.

= (Number teeth + 1.414) ÷ outside diameter.

Normal Pitch = Diametral pitch ÷ .707.

Number Teeth = Pitch diameter × diametral pitch.

= Outside diameter × diametral pitch − 1.414.

Lead = Pitch diameter × 3.1416.

Cutter = Select spur gear cutter to nearest standard normal pitch for *three times* as many teeth as in spiral. Cut to get the tooth parts figured from the correct normal pitch.

Figure normal tooth parts the same as for a spur gear.

FORMULÆ FOR ANY ANGLE — AXIS AT RIGHT ANGLES

Diametral Pitch = Number of teeth ÷ pitch diameter.

= Number of teeth + 2 cosine angle ÷ outside diameter.

= Number teeth ÷ (pitch diameter × cosine of angle.)

= Diametral pitch ÷ cosine of angle.

Number of teeth = Pitch diameter × diametral pitch.

= Outside diameter × diametral pitch − 2 cosine spiral angle.

Lead = Pitch diameter × 3.1416 ÷ tangent of angle.

Cutter — Divide the number of teeth in the spiral gear by the cube of the cosine of the angle. Select a spur gear cutter for the standard normal pitch which is nearest to this. Then cut as before.

SPIRAL GEAR TABLE

WHILE it is better in every case to understand the principles involved before using a table as this tends to prevent errors, they can be used with good results by simply following directions carefully. The subject of spiral gears is so much more complicated than other gears that many will prefer to depend entirely on the tables.

This table gives the circular pitch and addendum or the diametral pitch and lead of spirals for one diametral pitch and with teeth having angles of from 1 to 89 degrees to 45 and 45 degrees. For other pitches divide the addendum given and the spiral number by the required pitch and multiply the results by the required number of teeth. This will give the pitch diameter and lead of spiral for each wheel. For the outside diameter add two diametral pitches as in spur gearing.

Suppose we want a pair of spiral gears with 10 and 80 degree angles, 8 diametral pitch cutter, with 16 teeth in the small gear, having 10-degree angle and 10 teeth in the large gear with its 80-degree angle.

Find the 10-degree angle of spiral and in the third column find 1.0154. Divide by pitch, 8, and get .1269. Multiply this by number of teeth — .1269 × 16 = 2.030 = pitch diameter. Add 2 pitches — two $\frac{1}{8}$ = $\frac{1}{4}$ and 2.030 + .25 = 2.28 inches outside diameter.

The lead of spiral for 10 degrees for small wheel is 18.092. Divide by pitch = 18.092 ÷ 8 = 2.2615. Multiply by number of teeth, 2.2615 × 16 = 36.18, the lead of spiral, which means that it makes one turn in 36.18 inches.

For the other gear with its 80-degree angle, find the addendum, 5.7587. Divide by pitch, 8, = .7198. Multiply by number of teeth, 10 = 7.198. Add two pitches, or .25, gives 7.448 as outside diameter.

The lead of spiral is 3.1901. Dividing by pitch, 8 = .3988. Multiplying by number of teeth = 3.988 the lead of spiral.

When racks are to mesh with spiral gears, divide the number in the circular pitch columns for the given angle by the required diametral pitch to get the corresponding circular pitch.

If we want to make a rack to mesh with a 40-degree spiral gear of 8 pitch: Look for circular pitch opposite 40 and find 4.101. Dividing by 8 gives .512 as the circular pitch for this angle. The greater the angle the greater the circular or linear pitch, as can be seen by trying an 80-degree angle. Here the circular pitch is 2.261 inches.

SPIRAL GEAR TABLE

SHAFT ANGLES 90° FOR ONE DIAMETRAL PITCH

Angle of Spiral Degrees	To obtain the circular pitch for one tooth, divide by the required diametral pitch	To obtain the pitch diameter, divide by the required diametral pitch and multiply the quotient by the required number of teeth	To obtain the lead of spiral, divide by the required diametral pitch and multiply quotient by the required number of teeth		divide by the pitch diameter, diametral pitch and multiply the quotient by the required number of teeth	To obtain the circular pitch for one tooth, divide by the required diametral pitch.	Angle of Spiral in Degrees
	Circular Pitch	One Tooth or Addend.	Lead of Spirals		One Tooth or Addendum	Circular Pitch	
Small Wheel	Small Wheel	Small Wheel	Small Wheel	Large Wheel	Large Wheel	Large Wheel	Large Wheel
1	3.1419	1.0001	180.05	3.1420	57.298	180.01	89
2	3.1435	1.0006	90.020	3.1435	28.653	90.016	88
3	3.1457	1.0013	60.032	3.1458	19.107	60.026	87
4	3.1491	1.0024	45.038	3.1492	14.335	45.035	86
5	3.1535	1.0038	37.077	3.1527	11.473	36.044	85
6	3.1589	1.0055	30.056	3.1589	9.5667	30.055	84
7	3.1652	1.0075	25.728	3.1651	8.2055	25.778	83
8	3.1724	1.0098	22.573	3.1724	7.1852	22.573	82
9	3.1806	1.0124	20.082	3.1807	6.3924	20.082	81
10	3.1900	1.0154	18.092	3.1901	5.7587	18.092	80
11	3.2003	1.0187	16.464	3.2003	5.2408	16.464	79
12	3.2145	1.0232	15.076	3.2105	4.8097	15.104	78
13	3.2242	1.0263	13.966	3.2294	4.4454	13.988	77
14	3.2377	1.0306	12.986	3.2378	4.1335	12.986	76
15	3.2522	1.0352	12.138	3.2524	3.8637	12.138	75
16	3.2679	1.0402	11.393	3.2678	3.6279	11.397	74
17	3.2848	1.0456	10.417	3.2821	3.4203	10.745	73
18	3.3116	1.0514	10.192	3.3032	3.2360	10.166	72
19	3.3225	1.0576	9.6494	3.3225	3.0715	9.6494	71
20	3.3430	1.0641	9.1848	3.3433	2.9238	9.1854	70
21	3.3650	1.0711	8.7662	3.3652	2.7904	8.7663	69
22	3.3882	1.0785	8.3862	3.3833	2.6694	8.3862	68
23	3.4127	1.0863	8.0399	3.4129	2.5593	8.0403	67
24	3.4451	1.0946	7.7379	3.4391	2.4585	7.7242	66
25	3.4661	1.1033	7.4332	3.4663	2.3662	7.4336	65
26	3.4953	1.1126	7.1664	3.4952	2.2811	7.1663	64
27	3.5258	1.1223	6.9198	3.5257	2.2026	6.9197	63
28	3.5579	1.1325	6.6912	3.5575	2.1300	6.6916	62
29	3.5918	1.1433	6.4799	3.5919	2.0626	6.4799	61
30	3.6276	1.1547	6.2778	3.6277	2.0000	6.2832	60
31	3.6650	1.1666	6.0979	3.6652	1.9416	6.0997	59
32	3.7043	1.1791	5.9282	3.7044	1.8870	5.9282	58
33	3.7457	1.1923	5.7710	3.7459	1.8360	5.7680	57
34	3.7894	1.2062	5.6181	3.7826	1.7882	5.6178	56
35	3.8349	1.2207	5.4754	3.8351	1.7434	5.4770	55
36	3.8830	1.2360	5.3431	3.8834	1.7013	5.3448	54
37	3.9336	1.2521	5.2201	3.9261	1.6616	5.2200	53
38	3.9867	1.2690	5.1028	3.9921	1.6242	5.1026	52
39	4.0482	1.2867	4.9866	4.0416	1.5890	4.9920	51
40	4.1010	1.3054	4.8873	4.1012	1.5557	4.8874	50
41	4.1626	1.3250	4.7885	4.1540	1.5242	4.7884	49
42	4.2273	1.3456	4.6949	4.2272	1.4944	4.6948	48
43	4.2956	1.3673	4.6065	4.2956	1.4662	4.6062	47
44	4.3671	1.3901	4.5223	4.3675	1.4395	4.5225	46
5	4.4428	1.4142	4.4428	4.4428	1.4142	4.4428	45

THREADS OF WORMS

Worms are cut with threads having a total angle of 29 degrees, similar to the Acme thread. Some use the same proportions as for the Acme, but most use a deeper thread such as the Brown & Sharpe, which is .6866 deep instead of .51 for a one-inch pitch as in the Acme. It is not easy to cut odd fractional pitches in most lathes, so regular pitches are cut and the circular pitch of the worm wheel is allowed to come in fractional measurements for pitch diameters and center distances. Having determined on the reduction as 40 to 1, the relative proportions can be considered as follows:

Assume a thread of 4 to the inch for the worm or a lead of ¼ inch. Then as the reduction of 40 to 1 there must be 40 teeth in the worm gear, ¼ inch from center to center of teeth or 10 inches in circumference on the pitch line or 3.18 inches. If a reduction of 20 to 1 is wanted we can use the same gear but cut a double thread of 2 per inch, which will give the same distance between teeth, but the worm gear will be moved two teeth every revolution of the worm.

Some of the commonly used proportions are:

$$\text{Pitch diam. of worm gear} = \frac{\text{No. of teeth} \times \text{pitch in inches}}{3.1416}.$$

$$\text{Diametral pitch} = \frac{3.1416}{\text{Linear pitch}}.$$

$$\text{Throat diam. of worm gear} = \text{Pitch diam.} + \frac{2}{\text{Diam. Pitch}}.$$

$$\text{Outside diam.} = \text{Pitch diam.} + \frac{4}{\text{Diam. Pitch}}.$$

Whole depth of tooth of worm or worm gear = .6866 × linear pitch.

Width at top of tooth of worm = .335 × linear pitch.
Width of bottom of tooth of worm = .31 × linear pitch.
Outside diam. of worm — single thread = 4 × linear pitch.
Outside diam. of worm — double thread = 5 × linear pitch.
Outside diam. of worm — triple thread = 6 × linear pitch.
Face of worm gear = ½ to ¼ outside diameter of worm.

Width of Face

A common practice for determining the width of face or thickness of worm wheels is shown in Fig. 15. Draw the diameter of the worm and lay off 60 degrees as shown; this gives the width of working face, the sides being made straight from the bottom of the teeth. Others make the face equal to ¼ the outside diameter of worm, but ½ the diameter of the worm is more common.

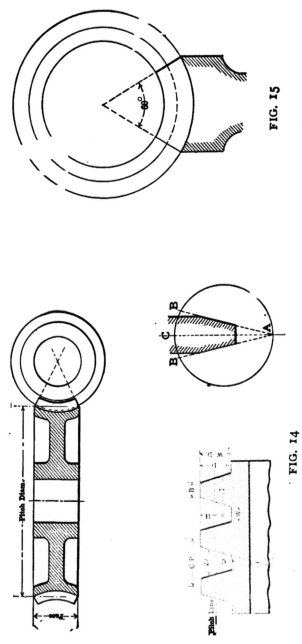

FIG. 15

FIG. 14

To Find the Pitch Diameter. (Given pitch and number of teeth.)

Rule.—Multiply the number of teeth by the circular pitch and divide by 3.1416.

Example.—Number of teeth 96; Circular pitch, ¼. Then 96 × ¼ = 24. 24 ÷ 3.1416 = 7.639 = pitch diameter.

Note.—Diameter of Hob equals diameter of Worm plus twice the clearance "C."

TABLE OF PROPORTIONS OF WORM THREADS TO RUN IN WORM WHEELS

C.P. Circular Pitch, Inches	pi. Threads per Inch ($\text{pi.}=\frac{1}{\text{C.P.}}$)	D.P. Diametrical Pitch ($\text{D.P.}=\frac{\pi}{\text{C.P.}}$)	H. Tooth above Pitch-Line ($\text{H.}=\frac{1}{\text{D.P.}}$)	D. Working Depth of Tooth ($\text{D.}=2\times\frac{1}{\text{D.P.}}$)	C. Clearance ($\text{C.}=\frac{\text{T}}{10}$)	S. Depth of Space below Pitch Line ($\text{S.}=\text{H.}+\text{C.}$)	W.D. Whole Depth of Tooth ($\text{W.D.}=\text{D.}+\text{C.}$)	T. Thickn. of Tooth on Pitch Line ($\text{T.}=\frac{\text{C.P.}}{2}$)	W. Width of Thread Tool at End ($\text{W.}=.31\times\text{C.P.}$)	B. Width of Thread at Top ($\text{B.}=.335\times\text{C.P.}$)
2	½	1.5708	.6366	1.2732	.1000	.7366	1.3732	1.0000	.6200	.6708
1¾	4/7	1.7952	.5570	1.1141	.0875	.6445	1.2016	.8750	.5425	.5862
1½	⅔	2.0944	.4775	.9549	.0750	.5525	1.0299	.7500	.4650	.5025
1¼	⅘	2.5133	.3979	.7958	.0625	.4604	.8583	.6250	.3875	.4187
1	1	3.1416	.3183	.6366	.0500	.3683	.6866	.5000	.3100	.3350
¾	1⅓	4.1888	.2387	.4775	.0375	.2762	.5150	.3750	.2325	.2512
⅔	1½	4.7124	.2122	.4244	.0333	.2455	.4577	.3333	.2066	.2233
½	2	6.2832	.1592	.3183	.0250	.1842	.3433	.2500	.1550	.1675
⅖	2½	7.8540	.1273	.2546	.0200	.1473	.2746	.2000	.1240	.1340
⅓	3	9.4248	.1061	.2122	.0166	.1227	.2288	.1666	.1033	.1117
2/7	3½	10.9956	.0909	.1819	.0143	.1052	.1962	.1429	.0886	.0957
¼	4	12.5664	.0796	.1591	.0125	.0921	.1716	.1250	.0775	.0838
2/9	4½	14.1372	.0707	.1415	.0111	.0818	.1526	.1111	.0689	.0744
⅕	5	15.7080	.0637	.1273	.0100	.0737	.1373	.1000	.0620	.0670
⅙	6	18.8496	.0531	.1061	.0083	.0614	.1144	.0833	.0517	.0558
1/7	7	21.9911	.0455	.0910	.0071	.0526	.0981	.0714	.0443	.0479
⅛	8	25.1327	.0398	.0796	.0062	.0460	.0858	.0625	.0388	.0419
1/9	9	28.2743	.0354	.0708	.0055	.0409	.0763	.0555	.0344	.0372
1/10	10	31.4159	.0318	.0636	.0050	.0368	.0687	.0500	.0310	.0335
1/12	12	37.6992	.0265	.0530	.0042	.0307	.0572	.0416	.0258	.0279
1/14	14	43.9824	.0227	.0454	.0036	.0263	.0490	.0357	.0221	.0239
1/16	16	50.2655	.0199	.0398	.0031	.0230	.0429	.0312	.0194	.0209
1/18	18	56.5488	.0176	.0352	.0025	.0201	.0377	.0255	.0172	.0186

Note. — The above table refers to single threads only. For Multiple threads, divide the sizes given in the table for the same pitch, by 2 for double, 3 for triple, 4 for quadruple threads, etc.

To find the Pitch Diameter (given pitch and number of teeth).

Rule. — Multiply the number of teeth by the circular pitch and divide by 3.1416.

Example. — Number of teeth, 96; Circular pitch, ¼. Then $96 \times \frac{1}{4} = 24$. $24 \div 3.1416 = 7.639 =$ pitch diameter. — *Note.* — Diameter of Hob equals diameter of Worm plus twice the clearance "C."

MILLING AND MILLING CUTTERS

MILLING MACHINE FEEDS AND SPEEDS

THE determining of the proper feeds of milling cutters in the past was usually a matter of guesswork, or experience, as a good many would term it, no absolute rule of any kind having ever been established.

On the other hand, the speeds of milling cutters have been fairly well established. A good average for the surface speed of an ordinary tool-steel cutter, when used in a milling machine, is 20, 40, 60 feet respectively for steel, cast iron, and brass. For the air-hardening tool-steel cutter, the surface speed may be increased from 50 to 75 per cent. The surface speed of an ordinary tool-steel cutter, when used in a standard gear cutting machine, is generally 40, 60 and 80 feet respectively for steel, cast iron, and brass. This increase in surface speed over the regular tool-steel milling cutter is due to the fact that the conditions of the standard gear-cutting machine have been provided to adapt it for just one kind of work where the overhang of parts is reduced to a minimum. A guide for determining the proper feed of milling cutters is found in ascertaining the thickness of the chip per tooth of the cutter.

Thickness of Chip

Taking, for example, an average size milling cutter working in cast iron, say 2½ inches diameter, 3 inches long, with eighteen teeth, which is quite commonly used, and it will be found that the thickness of the chip per tooth is quite small, resulting in .0018 inch, with a table feed of 2 inches per minute. This is entirely too slow. Now, comparing this cut of .0018 inch with a lathe tool cut, it will be seen that such a chip in a milling cutter is much smaller and is far more injurious to the cutter than a heavier feed, since the cutting edge of a tool will hold up longer in cutting into the metal instead of scraping it. A standard 6-pitch gear cutter, cutting cast iron at the rate of 5½ inches per minute, averages a chip per tooth of only .0067 inch. The gear cutter is here mentioned only to demonstrate the possibilities of other classes of milling cutters.

A cutter is very seldom ruined by the feed, but is generally ruined by overspeeding it. For instance, with a cutter of thirty teeth with a table feed of .300 inch per revolution, the chip per tooth will then only be $\frac{.300}{30T}$ = .010 inch thick — still quite a light cut when comparing it with a lathe tool chip. Hence in many cases of milling, if the feeds are guided by the thickness of chip per tooth, a much faster feed would be used, since it is evident that the heaviest feeds, comparatively, give only a thin chip per tooth.

The Question of Power

The question of power in milling is to-day quite an important subject, and a great deal depends on the form and shape of cutter used.   Roughing cutters should invariably be coarse-toothed, and, if possible, never have more than two or three teeth in contact with the metal.

Below are tabulated the results of a test of power consumed by two identical roughing cutters, Figs. 1 and 2, with the exception that one contained thirty teeth and the other fifteen teeth:

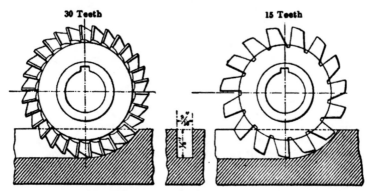

FIG. 1. — 5 Teeth in Contact   FIG. 2. — 3 Teeth in Contact

COMPARISON OF FINE AND COARSE PITCH CUTTERS

Test in Cast Iron

	Fig. 1 Cutter	Fig. 2 Cutter
Diameter of cutter	4″	4″
No. of teeth in cutter	30	15
Width of cutter	$\frac{9}{16}$″	$\frac{9}{16}$″
Depth of cut	$\frac{7}{8}$″	$\frac{7}{8}$″
No. of teeth in contact with metal	5	3
Volts	110	110
Amperes	13.5	10.5
Feed of table	.134″	.134″
Rev. of cutter p. m.	40	40
Thickness of chip per tooth	.0044	.0088

The machine on which this test was made was a No. 3 Cincinnati motor-driven, positive-feed miller connected with a Weston volt ammeter, so that very accurate readings of the power required fr the two cutters could be taken.   Observe that the 30-tooth cut

25

required 13.5 amperes, while the 15-tooth cutter required only 10.5 amperes on the same cut, a saving of about 22 per cent. This is plainly obvious, since the 30-tooth cutter had constantly 5 teeth in contact with the iron, while the 15-tooth cutter had only 3 teeth in contact, proving that the more bites a cutter must take the more power is required, irrespective of the fact that the 15-tooth cutter took a larger chip per tooth than did the 30-tooth cutter.

This clearly demonstrates the advantage of a coarse tooth cutter for heavy milling, not only through its economic use of power, but also because of its greater durability. A cutter of this kind will also permit of a heavier feed, especially on deep cuts, since ample space exists between the teeth for the escape of the chips, which would be impossible with a fine tooth cutter.

For finishing cuts a finer tooth cutter should be used, which will greatly lessen the feed marks for a smooth milled surface. This is quite readily permissible, since these cutters have no great amount of metal to remove, and the space between the teeth is ample for the escape of the chips.

The points enumerated above need not be absolutely adhered to for ordinary milling, but if the very best results are aimed at with a view of increased production, it would be well to bear them in mind, especially when constructing roughing and finishing gangs.

Clearance Angle, Lubrication, etc.

Great care should be exercised when sharpening cutters to give them the proper angle for clearances, the life of the cutter being thereby materially lengthened. A 5-degree clearance for finishing and 7-degree clearance for roughing produces very good results.

Whenever feasible, as small a cutter as possible should be used. They are not only less costly, but require less power for driving when milling.

When milling steel, a heavy flow of oil on a milling cutter, forced by means of an oil pump, is just as essential as the great volume of oil which is used on automatic screw machine tools, which would not hold up one-half hour if not so flushed. The life of a milling cutter amply lubricated will be materially prolonged and it will be capable of standing a much heavier feed.

The foregoing remarks apply to practice in general, but on jobs where conditions are favorable still better results can be obtained. For example, work has been milled with an end mill 3 inches diameter, 20 teeth, 2½-inch face, at a table feed of 15 inches per minute. This mill was mounted on the nose of a horizontal spindle miller where there is no overhang, and ran at a surface speed of 40 feet per minute. The milling machine table in this case advanced .300 inch per turn of the cutter, which gave a chip per tooth of .015 inch. This same piece was then finished with a table feed of 25 inches per minute with the cutter rotating at a surface speed of 65 feet per minute, producing a surface so smooth that the cutter marks could be easily polished off by means of emery cloth. After milling thirty pieces, ½ inches wide, 12 inches long, the cutter was still in good condition.

MILLING HEART-SHAPED CAMS

ONE method of producing heart-shaped cams is as follows:

Lay out the curve of the cam roughly, as in Fig. 3. Drill and remove the outside stock, being sure to leave sufficient stock to overcome errors in laying out. Put the cam on the nut arbor and tighten securely. If the roll of the cam is ⅛ radius, select a milling cutter having the same radius, as the roll of the cam must come to the lowest point, which it would not be able to do if a cutter of a smaller radius than that of the roll were used. It would also make a difference to the other points on the curve of the cam, which is not quite so apparent at first glance.

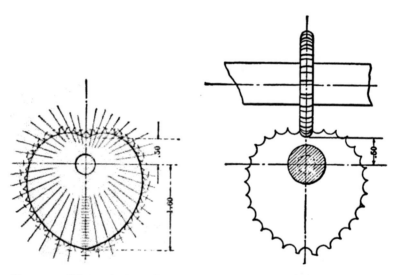

FIG. 3.—Method of Laying Out Cam

FIG. 4.—Position of Cam and Cutter when Commencing to Mill

Selecting an Index

THE next operation is to place the cam between centers on the milling machine, having the cutter in line with the vertical radius of the cam, at its lowest point. Next choose an index circle which will give a division of the cam such that the rise of each division will be in thousandths of an inch, if possible. For this cam take a circle which will give 200 divisions. As this will make 100 divisions on a side, the rise of each division will be 0.011 of an inch. Now raise the table to the required hight, starting at the lowest point of the cam, and mill across, as in Fig. 4.

Moving the Table

For the other cuts lower the table 0.011 each time, and revolve the cam one division until the highest point of the cam is reached, then raise the table 0.011 for each division of the cam.

When the cam comes from the milling machine there will be found to be small grooves left between the cuts. These may be easily removed by smoothing off with a file without impairing the accuracy of the cam.

Most screw-machine cams can be made in this manner, and they will be found to be more accurate than if laid out and filed to the line, and also much easier to make after one has become accustomed to the method.

MILLING CAMS BY GEARING UP THE DIVIDING HEAD

By the method here shown, cams of any rise may be milled with the gears regularly furnished with the milling-machine.

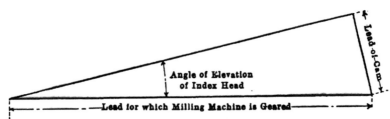

FIG. 5. — Diagram for Angle of Index Head

With the head set vertically the lead of the cam would be the same as the lead for which the machine is geared, while with the head horizontal and the milling spindle also, a concentric arc, or rest, would be milled on the cam, regardless of how the machine was geared. By inclining the head and milling spindle, we can produce any lead on the cam less than that for which the machine is geared.

The method of finding the inclination at which to set the index head is shown in Fig. 5, and is simply the solution of a plain right-angled triangle, in which the hypothenuse represents the lead of the machine, and one of the other sides represents the lead we wish to produce on the cam. By dividing the latter by the former we get the sine of the angle of inclination.

Take for illustration a plate cam having $\frac{1}{8}$-inch rise in 300 degrees.

$$\frac{360}{300} \times \frac{1}{8} = 0.15,$$

which is the lead we want on the cam, while the slowest lead for which the B & S. machine can be geared is 0.67

$$\frac{0.15}{0.67} = 0.224.$$

Consulting a table of sines, we find 0.224 approximates closely the sine of 13 degrees, which is the angle at which to set the head, and if the milling spindle is also set at the same angle, the edge of the cam will be parallel with the shaft on which it is to run. Fig. 6 shows a milling-machine set for this job.

When a cam has several lobes of different leads, we gear the machine up for a lead somewhat longer than the longest one called for in that cam, and then all the different lobes can be milled with the one setting of gears, by simply altering the inclination of head and milling spindle for each different lead on the cam.

If the diameter of the cam and the inclination of the head will admit, it is better to mill on the under side of the cam, as that brings the mill and the table nearer together and thus increases rigidity, besides enabling us to easily see any lines that may be laid out on the flat face of the cam. Also the chips do not accumulate on the work.

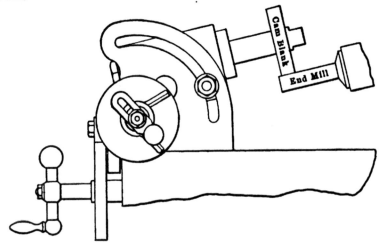

FIG. 6. — Dividing Head Set for Cam Milling

The work is fed against the cutter by turning the index crank, and on coming back for another cut we turn the handle of the milling-machine table. As a result the work will recede from the cutter before the cam blank commences to turn, owing to back lash in the gears, thus preventing the cutter from dragging over the work while running back.

In this way we use to advantage what is ordinarily considered a defect in machine construction.

The milling-machine, when used as shown in Fig. 6, will be found to be more rigid than when the head is set in the vertical position, and the cams will work more smoothly on account of the shearing action of the cutter.

One possible objection to the method here advocated is the necessity of using, in some cases, an end mill of extra length of tooth. In practise, an end mill ½-inch diameter and with a 3½-inch length of tooth is not unusual; but the results in both speed and quality w˙ be found entirely satisfactory.

TABLE OF PITCHES AND APPROXIMATE ANGLES FOR CUTTING SPIRALS ON THE UNIVERSAL MILLING MACHINE

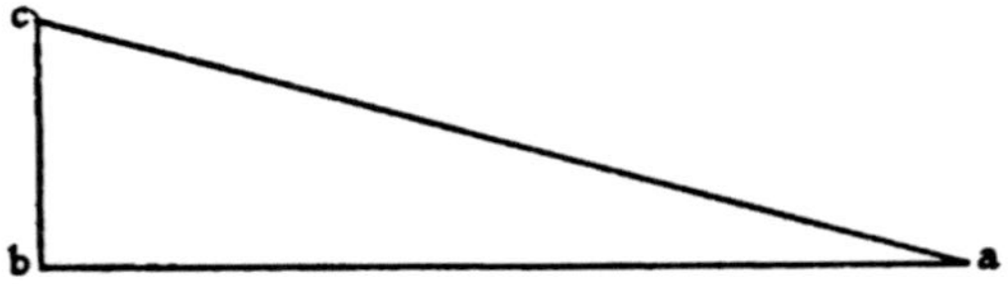

To find the angle for cutters of a larger diameter than given in the table, make a drawing as shown in the diagram, the angle b being a right angle. Let b c equal the circumference. Let a b equal the pitch. Connect c a by a line and measure the angle a with a protractor, or divide the circumference by the lead and the quotient will be the tangent of the angle. Find the angle in a table of tangents.

Diameter of Mill, Cutter, or Drill to be Cut

Inches

Values Given Under Diameters are Angles in Degrees

Gear on Worm	First Gear on Stud	Second Gear on Stud	Gear on Screw	Pitch in Inches to one Turn	3/16	1/4	3/8	1/2	5/8	3/4	7/8	1	1 1/4	1 1/2	1 3/4	2	2 1/4	2 1/2	3	3 1/4	3 1/2	3 3/4	4
24	64	24	72	1.25	17½	32¼																	
24	64	28	72	1.46	14¼	28	38½																
24	64	30	72	1.56	14¼	26¼	37																
24	64	32	72	1.67	12½	25	34½	43½															

																								45	44¼	44¼	44¼
																						45	42½	42¼	42		
																				44	43	42¼	40	39¼	39½		
																			43¾	41¼	40¼	39½	37½	37¼	6¼	34	
																	44¼	43¼	43	38½	3⅛	6¼	34½	34¼	34		
															44¼	43¼	41	40¼	39½	37	35¼	34	33¼	31¼	31¼	31	
													42	41½	40¼	39¾	37	36¾	36	33¼	30¾	30	28½	28	27¼		
											44¼	41½	41	37¾	3¼	36¼	35¼	33	32¼	32	28	27	26¾	24¼	24¼	24¼	
									43¼	41¼	39	36¼	36	33	32¼	31¼	30¼	28¼	7½	25¼	24	23	22½	21	20¼	20¼	
							44¼	40¼	39	37	35	33	30	27	26¾	25¼	25¼	23¼	9¼	18¼	18¼	7½	17	16¾			
					44¼	43¼	40¼	36¼	35¼	33¼	31¼	9½	27¼	26¼	24¼	23¼	23	22½	9¼	9¼	9½	18¼	17	16	14¾	14¼	
			43¼	40¾	39	36	32¼	31¼	29½	7¼	25¼	23¾	23¼	21	20¾	20	19¾	18	7¾	17¼	15¾	14¾	14	13¼	12¾	12¼	12¾
45	43¼	41¼	38	35¼	34	31¼	28	26¾	25¼	23¾	22	20¾	20	17½	17¼	16¾	16¼	15	14¾	14¾	13¾	12½	11¼	11¼	10¾	10¾	10¼
39	37	35	32	9¼	28¼	25¾	23	22	20¾	9½	7½	16¾	16	14¾	14¼	13¾	13¼	12½	12	11¾	10¾	10	9¾	9¼	8¾	8¾	8¼
31	29½	27¼	25	23	21¾	19¾	7½	7½	15¾	14¼	13¾	12¼	12	11	10¾	10¼	10	9¼	9	8¼	8	7¾	7	6¾	6¾	6¼	
21¾	20¾	9¼	17	15¼	15	13¾	11¾	11¾	10¾	9¾	9	8¼	8	7¾	7	6¾	6¾	6¼	6¼	5¾	5¼	5	4¾	4¼	4¼	4¼	
11¼	10¾	10¼	9¼	8¼	8	7¾	6¼	6	5¾	5¼	5	4¾	4	4	3¾	3½	3¾	3¼	3¼	3	2¾	2¾	2¾	2¾	2	2	2
1.94	2.08	2.22	2.50	2.78	2.92	3.24	3.70	3.89	4.17	4.53	4.86	5.33	5.44	6.12	6.22	6.48	6.67	7.29	7.41	7.62	8.33	9.00	9.33	9.52	10.29	10.37	10.50
72	72	72	72	72	72	72	72	64	72	72	72	72	64	72	72	56	64	72	56	64	72	56	56	72	64		
28	40	28	48	28	56	28	32	24	48	28	56	32	28	28	32	40	28	40	40	32	40	24	48	40	32	56	56
64	64	56	64	56	64	48	48	48	72	48	64	40	40	40	40	48	48	48	48	48	48	32	30	48	40	48	40
32	24	32	40	24	40	24	40	40	56	40	40	48	56	56	56	56	64	56	64	64	48	72	56	64	72	64	48

TABLE OF PITCHES AND APPROXIMATE ANGLES FOR CUTTING SPIRALS ON THE UNIVERSAL MILLING MACHINE

(Continued)

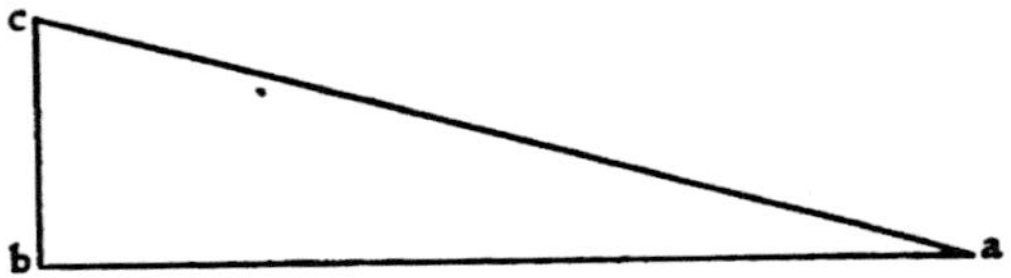

To find the angle for cutters of a larger diameter than given in the table, make a drawing as shn in the diagram; the angle *b* being a right angle. Let *b c* equal the circumference. Let *a b* qal the pitch. Connect *c a* by a line, and measure the angle *a* with a protractor; or divide the me by the lead and the quotient will be he tangent of the angle. Find the angle in a ble of tangents.

Diameter of Mill, Cutter, or Drill to be Cut — Inches — Values Given Under Diameters are Angles in Degrees

Gear on Worm	First Gear on Stud	Second Gear on Stud	Gear on Screw	Pitch in Inches to one Turn	1/8	1/4	3/8	1/2	5/8	3/4	7/8	1	1 1/4	1 1/2	1 3/4	2	2 1/4	2 1/2	2 3/4	3	3 1/4	3 1/2	3 3/4	4
64	40	48	72	10.67	2	4	6 1/4	8 1/4	10 1/4	12 1/4	14 1/4	16 1/2	20 1/4	24	27 1/4	30 1/2	33 1/2	36 1/2	39	41 1/2	43 3/4			
56	32	40	64	10.94	2	4	6	8 1/4	10 1/4	12	14	16 1/4	20	23 1/2	26 3/4	30	33	35 3/4	38 1/4	40 3/4	43			
54	32	40	72	11.11	2	4	6	8	10	11 3/4	13 3/4	16	19 3/4	23	26 1/2	29 1/2	32 1/2	35 1/4	38	40 1/4	42 1/2	44 3/4		
66	32	48	72	11.66	1 3/4	3 3/4	5 3/4	7 1/2	9 1/2	11 1/4	13 1/4	15 1/4	18 3/4	22	25 1/4	28 1/2	31 1/4	34	36 1/2	39	41 1/4	43 1/2		

12.00	13.12	13.33	13.71	15.24	15.56	15.75	16.87	17.14	18.75	19.29	19.59	19.69	21.43	22.50	23.33	26.25	26.67	28.00	30.86	31.50	36.00	41.14	45.00	48.00	51.43	60.00	68.57
48	64	72	56	72	72	40	64	56	48	56	56	64	56	64	48	64	48	40	40	40	40	32	40	32	32	28	
32	48	48	48	48	48	56	72	48	40	48	48	56	40	56	56	56	56	56	48	56	64	64	56	64	64	64	
40	32	28	40	28	32	32	32	32	32	28	32	24	28	32	24	28	32	28	32	28	28	24	28	24	24		
72	56	56	64	64	64	56	72	64	72	72	64	72	72	64	72	64	72	72	72	72	72	72	72	72	72	72	2

TABLE FOR PLAIN AND DIFFERENTIAL INDEXING, ON BROWN & SHARPE MILLING-MACHINES

DIFFERENTIAL INDEXING is simpler than the compound method for obtaining divisions that are prime numbers. It differs from the compound method in that the movement of the spiral head spindle in relation to the index crank is positively made by gearing. The index plate is geared to the spindle, as in Fig. 7, thus giving a differential motion that allows the indexing to be made with one circle of holes and the index crank to be turned in one direction, the same as in plain indexing. This enables spacings to be made that cannot be obtained with an index plate locked with a stop pin in the usual way.

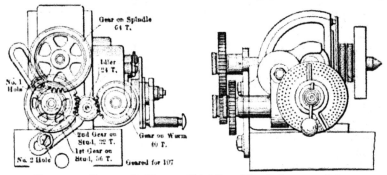

FIG. 7.—Brown & Sharpe Dividing Head Arranged for
Differential Indexing

EXAMPLE: REQUIRED TO INDEX FOR 107 DIVISIONS

If we use the plate having 20 holes and move 8 holes per division, 100 moves will be required to rotate the worm 40 turns, which in turn rotates the spindle once. If we made 107 moves, with the index plate fixed, we would obtain $107 \times \dfrac{8}{20} = 42.8$ revolutions of the worm, which is 2.8 in excess of what is required. Therefore the index plate must be geared so that it will move back 2.8 turns, while the spindle is revolving once; that is, the ratio of the gearing must be 2.8 to 1.

$$\frac{2.8}{1} = \frac{2.8}{2} \times \frac{2}{1}$$

$$\frac{2.8}{2} \times \frac{20}{20} = \frac{56}{40} \qquad\qquad \frac{2}{1} \times \frac{32}{32} = \frac{64}{32}$$

Then $\dfrac{2.8}{1} = \dfrac{56}{40} \times \dfrac{64}{32}$ and the gears will be 64 and 56 for the spindle and first gear on stud, and 40 and 32 for the worm and second gear on stud, as shown in Fig. 7.

As compound gears are used, but one idler is required to cause the index plate to move in a direction opposite to that of the crank. For this purpose an idler having 24 teeth is employed.

The table on the following pages gives the B. & S. spiral head gears for plain and differential indexing of all numbers up to 370.

Number of Divisions	Index Circle	No. of Turns of Index	Gear on Worm	No. 1 Hole		Gear on Spindle	Idlers	
				First Gear on Stud	Second Gear on Stud		No. 1 Hole	No. 2 Hole
2	Any	20						
3	39	$13\frac{11}{13}$						
4	Any	10						
5	Any	8						
6	39	$6\frac{24}{39}$						
7	49	$5\frac{15}{49}$						
8	Any	5						
9	27	$4\frac{12}{27}$						
10	Any	4						
11	33	$3\frac{21}{33}$						
12	39	$3\frac{14}{39}$						
13	39	$3\frac{3}{39}$						
14	49	$2\frac{42}{49}$						
15	39	$2\frac{26}{39}$						
16	20	$2\frac{10}{20}$						
17	17	$2\frac{6}{17}$						
18	27	$2\frac{6}{27}$						
19	19	$2\frac{2}{19}$						
20	Any	2						
21	21	$1\frac{19}{21}$						
22	33	$1\frac{27}{33}$						
23	23	$1\frac{17}{23}$						
24	39	$1\frac{26}{39}$						
25	20	$1\frac{12}{20}$						
26	39	$1\frac{30}{39}$						
27	27	$1\frac{13}{27}$						
28	49	$1\frac{21}{49}$						
29	29	$1\frac{11}{29}$						
30	39	$1\frac{18}{39}$						
31	31	$1\frac{9}{31}$						
32	20	$1\frac{5}{20}$						
33	33	$1\frac{7}{33}$						
34	17	$1\frac{3}{17}$						
35	49	$1\frac{7}{49}$						
36	27	$1\frac{3}{27}$						
37	37	$1\frac{3}{37}$						
38	19	$1\frac{1}{19}$						
39	39	$1\frac{1}{39}$						
40	Any	1						
41	41	$\frac{40}{41}$						
42	21	$\frac{20}{21}$						

Number of Divisions	Index Cile	No. of Turns of Index	Gear on Worm	No. 1 Hole		Gear on Spindle	Idlers	
				First Gear on Stud	Second Gear on Stud		No. 1 Hole	No. 2 Hole
43	43	40/43						
44	33	30/33						
45	27	24/27						
46	23	20/23						
47	47	40/47						
48	18	15/18						
49	49	40/49						
50	20	16/20						
51	17	14/17	24			48	24	44
52	39	30/39						
53	49	35/49	56	40	24	72		
54	27	20/27						
55	33	24/33						
56	49	35/49						
57	21	15/21	56			40	24	44
58	29	20/29						
59	39	26/39	48			32	44	
60	39	26/39						
61	39	26/39	48			32	24	44
62	31	20/31						
63	39	26/39	24			48	24	44
64	16	10/16						
65	39	24/39						
66	33	20/33						
67	21	12/21	28			48	44	
68	17	10/17						
69	20	12/20	40			56	24	44
70	49	28/49						
71	18	10/18	72			40	24	
72	27	15/27						
73	21	12/21	28			48	24	44
74	37	20/37						
75	15	8/15						
76	19	10/19						
77	20	10/20	32			48	44	
78	39	20/39						
79	20	10/20	48			24	44	
80	20	10/20						
81	20	10/20	48			24	24	44
82	41	20/41						
83	20	10/20	32			48	24	44

Number of Divisions	Index Circle	No. of Turns of Index	Gear on Worm	No. 1 Hole		Gear on Spindle	Idlers	
				First Gear on Stud	Second Gear on Stud		No. 1 Hole	No. 2 Hole
84	21	10/21						
85	17	8/17						
86	43	20/43						
87	15		40			24	24	44
88	33	15/33						
89	18		72			32	44	
90	27	12/27						
91	39		24			48	24	44
92	23	10/23						
93	18		24			32	24	44
94	47	20/47						
95	19	8/19						
96	21	9/21	28			32	24	44
97	20		40			48	44	
98	49	20/49						
99	20		56	28	40	32		
100	20	8/20						
101	20		72	24	40	48		24
102	20		40			32	24	44
103	20		40			48	24	44
104	39							
105	21							
106	43		86	24	24	48		
107	20		40	56	32	64		24
108	27							
109	16		32			28	24	44
110	33							
111	39		24			72	32	
112	39		24			64	44	
113	39		24			56	44	
114	39		24			48	44	
115	23							
116	29							
117	39		24			24	56	
118	39		48			32	44	
119	39		72			24	44	
120	39							
121	39		72			24	24	44
122	39		48			32	24	44
123	39		24			24	24	44
124	31							

Number of Divisions	Index Circle	No. of Turns of Index	Gear on Worm	No. 1 Hole First Gear on Stud	No. 1 Hole Second Gear on Stud	Gear on Spindle	Idlers No. 1 Hole	Idlers No. 2 Hole
125	39		24			40	24	44
126	39		24			48	24	44
127	39		24			56	24	44
128	16							
129	39		24			72	24	44
130	39							
131	20		40			28	44	
132	33							
133	21		24			48	44	
134	21		28			48	44	
135	27							
136	17							
137	21		28			24	56	
138	21		56			32	44	
139	21		56	32	48	24		
140	49							
141	18		48			40	44	
142	21		56			32	24	44
143	21		28			24	24	44
144	18							
145	29							
146	21		28			48	24	44
147	21		24			48	24	44
148	37							
149	21		28			72	24	44
150	15							
151	20		32			72	44	
152	19							
153	20		32			56	44	
154	20		32			48	44	
155	31							
156	39							
157	20		32			24	56	
158	20		48			24	44	
159	20		64	32	56	28		
160	20							
161	20		64	32	56	28		24
162	20		48			24	24	44
163	20		32			24	24	44
164	41							
165	33							

Number of Divisions	Index Circle	No. of Turns of Index	Gear on Worm	No. 1 Hole		Gear on Spindle	Idlers	
				First Gear on Stud	Second Gear on Stud		No. 1 Hole	No. 2 Hole
166	20	$\frac{8}{20}$	32			48	24	44
167	20	$\frac{8}{20}$	32			56	24	44
168	21	$\frac{5}{21}$						
169	20	$\frac{8}{20}$	32		.	72	24	44
170	17	$\frac{4}{17}$						
171	21	$\frac{5}{21}$	56			40	24	44
172	43	$\frac{10}{43}$						
173	18	$1\frac{4}{18}$	72	56	32	64		
174	18	$1\frac{4}{18}$	24			32	56	
175	18	$1\frac{4}{18}$	72	40	32	64		
176	18	$1\frac{4}{18}$	72	24	24	64		
177	18	$1\frac{4}{18}$	72			48	24	
178	18	$1\frac{4}{18}$	72			32	44	
179	18	$1\frac{4}{18}$	72	24	48	32		
180	18	$1\frac{4}{18}$						
181	18	$1\frac{4}{18}$	72	24	48	32		24
182	18	$1\frac{4}{18}$	72			32	24	44
183	18	$1\frac{4}{18}$	48			32	24	44
184	23	$\frac{5}{23}$						
185	37	$\frac{8}{37}$						
186	18 ·	$1\frac{4}{18}$	48			64	24	44
187	18	$1\frac{4}{18}$	72	48	24	56		24
188	47	$\frac{10}{47}$						
189	18	$1\frac{4}{18}$	32			64	24	44
190	19	$1\frac{5}{19}$						
191	20	$2\frac{4}{20}$	40			72	24	
192	20	$2\frac{4}{20}$	40			64	44	
193	20	$2\frac{4}{20}$	40			56	44	
194	20	$2\frac{4}{20}$	40			48	44	
195	39	$1\frac{9}{39}$						
196	49	$1\frac{11}{49}$						
197	20	$2\frac{4}{20}$	40			24	56	
198	20	$2\frac{4}{20}$	56	28	40	32		
199	20	$2\frac{4}{20}$	100	40	64	32		
200	20	$2\frac{4}{20}$						
201	20	$2\frac{4}{20}$	72	24	40	24		24
202	20	$2\frac{4}{20}$	72	24	40	48		24
203	20	$2\frac{4}{20}$	40			24	24	44
204	20	$3\frac{4}{20}$	40			32	24	44
205	41	$\frac{4}{41}$						
206	20	$3\frac{4}{20}$	40			48	24	44

Number of Divisions	Index Circle	No. of Turns of Index	Gear on Worm	No. 1 Hole		Gear on Spindle	Idlers	
				First Gear on Stud	Second Gear on Stud		No. 1 Hole	No. 2 Hole
207	20	4/20	40			56	24	44
208	20	4/20	40			64	24	44
209	20	4/20	40			72	24	44
210	21	4/21	.					
211	16	8/16	64			28	44	
212	43	8/43	86	24	24	48		
213	27	6/27	72			40	44	
214	20	4/20	40	56	32	64		24
215	43	8/43						
216	27	6/27						
217	21	4/21	48			64	24	44
218	16	8/16	64			56	24	44
219	21	4/21	28			48	24	44
220	33	6/33						
221	17	3/17	24			24	56	
222	18	4/18	24			72	44	
223	43	8/43	86	48	24	64		24
224	18	4/18	24			64	44	
225	27	6/27	24			40	24	44
226	18	3/18	24			56	44	
227	49	8/49	56	64	28	72		
228	18	8/18	24			48	44	
229	18	8/18	24			44	48	
230	23	4/23						
231	18	3/18	32			48	44	
232	29	8/29						
233	18	8/18	48			56	44	
234	18	8/18	24			24	56	
235	47	4/47						
236	18	8/18	48			32	44	
237	18	8/18	48			24	44	
238	18	8/18	72			24	44	
239	18	8/18	72	24	64	32		
240	18	8/18						
241	18	8/18	72	24	64	32		24
242	18	8/18	72			24	24	44
243	18	8/18	64			32	24	44
244	18	8/18	48			32	24	44
245	49	8/49						
246	18	8/18	24			24	24	44
247	18	3/18	48			56	24	44

Number of Divisions	Index Circle	No. of Turns of Index	Gear on Worm	No. 1 Hole		Gear on Spindle	Idlers	
				First Gear on Stud	Second Gear on Stud		No. 1 Hole	No. 2 Hole
248	31	$\frac{5}{31}$						
249	18	$\frac{8}{18}$	32			48	24	44
250	18	$\frac{8}{18}$	24			40	24	44
251	18	$\frac{8}{18}$	48	44	32	64		24
252	18	$\frac{8}{18}$	24			48	24	44
253	33	$\frac{8}{33}$	24			40	56	
254	18	$\frac{8}{18}$	24			56	24	44
255	18	$\frac{8}{18}$	48	40	24	72		24
256	18	$\frac{8}{18}$	24			64	24	44
257	49	$\frac{8}{49}$	56	48	28	64		24
258	43	$\frac{8}{43}$	32			64	24	44
259	21	$\frac{6}{21}$	24			72	44	
260	39	$\frac{6}{39}$						
261	29	$\frac{4}{29}$	48	64	24	72		
262	20	$\frac{3}{20}$	40			28	44	
263	49	$\frac{8}{49}$	56	64	28	72		24
264	33	$\frac{5}{33}$						
265	21	$\frac{8}{21}$	56	40	24	72		
266	21	$\frac{8}{21}$	32			64	44	
267	27	$\frac{4}{27}$	72			32	44	
268	21	$\frac{8}{21}$	28			48	44	
269	20	$\frac{8}{20}$	64	32	40	28		24
270	27	$\frac{7}{27}$						
271	21	$\frac{2}{21}$	56			72	24	
272	21	$\frac{8}{21}$	56			64	24	
273	21	$\frac{8}{21}$	24			24	56	
274	21	$\frac{8}{21}$	56			48	44	
275	21	$\frac{8}{21}$	56			40	44	
276	21	$\frac{8}{21}$	56			32	44	
277	21	$\frac{2}{21}$	56			24	44	
278	21	$\frac{8}{21}$	56	32	48	24		
279	27	$\frac{7}{27}$	24			32	24	44
280	49	$\frac{7}{49}$						
281	21	$\frac{8}{21}$	72	24	56	24		24
282	43	$\frac{6}{43}$	86	24	24	56		
283	21	$\frac{8}{21}$	56			24	24	44
284	21	$\frac{8}{21}$	56			32	24	44
285	21	$\frac{8}{21}$	56			40	24	44
286	21	$\frac{8}{21}$	56			48	24	44
287	21	$\frac{8}{21}$	24			24	24	44
288	21	$\frac{8}{21}$	28			32	24	44

Number of Divisions	Index Circle	No. of Turns of Index	Gear on Worm	No. 1 Hole		Gear on Spindle	Idlers	
				First Gear on Stud	Second Gear on Stud		No. 1 Hole	No. 2 Hole
289	21	$\frac{8}{21}$	56			72	24	44
290	29	$\frac{4}{29}$						
291	15	$\frac{2}{15}$	40			48	44	
292	21	$\frac{3}{21}$	28			48	24	44
293	15	$\frac{2}{15}$	48	32	40	56		
294	21	$\frac{3}{21}$	24			48	24	44
295	15	$\frac{2}{15}$	48			32	44	
296	37	$\frac{5}{37}$						
297	33	$\frac{5}{33}$	28	48	24	56		
298	21	$\frac{3}{21}$	28			72	24	44
299	23	$\frac{2}{23}$	24			24	56	
300	15	$\frac{2}{15}$						
301	43	$\frac{6}{43}$	24			48	24	44
302	16	$\frac{2}{16}$	32			72	24	
303	15	$\frac{2}{15}$	72	24	40	48		24
304	16	$\frac{2}{16}$	24			48	44	
305	15	$\frac{2}{15}$	48			32	24	44
306	15	$\frac{2}{15}$	40			32	24	44
307	15	$\frac{2}{15}$	72	48	40	56		24
308	16	$\frac{2}{16}$	32			48	44	
309	15	$\frac{2}{15}$	40			48	24	44
310	31	$\frac{3}{31}$						
311	16	$\frac{2}{16}$	64	24	24	72		
312	39	$\frac{5}{39}$						
313	16	$\frac{2}{16}$	32			28	56	
314	16	$\frac{2}{16}$	32			24	56	
315	16	$\frac{2}{16}$	64			40	24	
316	16	$\frac{2}{16}$	64			32	44	
317	16	$\frac{2}{16}$	64			24	44	
318	16	$\frac{2}{16}$	56	28	48	24		
319	29	$\frac{4}{29}$	48	64	24	72		24
320	16	$\frac{2}{16}$						
321	16	$\frac{2}{16}$	72	24	64	24		24
322	23	$\frac{3}{23}$	32			64	24	44
323	16	$\frac{2}{16}$	64			24	24	44
324	16	$\frac{2}{16}$	64			32	24	44
325	16	$\frac{2}{16}$	64			40	24	44
326	16	$\frac{2}{16}$	32			24	24	44
327	16	$\frac{2}{16}$	32			28	24	44
328	41	$\frac{5}{41}$						
329	16	$\frac{2}{16}$	64	24	24	72		24

Number of Divisions	Index Circle	No. of Turns of Index	Gear on Worm	No. 1 Hole		Gear on Spindle	Idlers	
				First Gear on Stud	Second Gear on Stud		No. 1 Hole	No. 2 Hole
330	33	$\frac{4}{33}$						
331	16		64	44	24	48		24
332	16		32			48	24	44
333	18		24			72	44	
334	16		32			56	24	44
335	33		72	48	44	40		24
336	16		32			64	24	44
337	43		86	40	32	56		
338	16		32			72	24	44
339	18		24			56	44	
340	17							
341	43		86	24	32	40		
342	18		32			64	44	
343	15		40	64	24	86		24
344	43							
345	18		24			40	56	
346	18		72	56	32	64		
347	43		86	24	32	40		24
348	18		24			32	56	
349	18		72	44	24	48		
350	18		72	40	32	64		
351	18		24			24	56	
352	18		72	24	24	64		
353	18		72			56	24	
354	18		72			48	24	
355	18		72			40	24	
356	18		72			32	24	
357	18		72			24	44	
358	18		72	32	48	24		
359	43		86	48	32	100		24
360	18							
361	19		32		..	64	44	
362	18		72	28	56	32		24
363	18		72			24	24	44
364	18		72			32	24	44
365	20		32	48	24	56		
366	18		48			32	24	44
367	18		72			56	24	24
368	18		72	24	24	64		24
369	41		32	56	28	64		
370	37							

MILLING CUTTER, REAMER AND TAP FLUTES

THE following tables give the number of teeth or flutes suitable for milling in various types of cutters, reamers, taps, etc., and also show the forms of fluting cutters used.

END MILLS

STRAIGHT TEETH		SPIRAL TEETH	
Dia. Mill	No. Teeth	Dia. Mill	No. Teeth
$\frac{1}{4}$	6	$\frac{1}{4}$ to $\frac{1}{2}$	8
$\frac{5}{16}$ to $\frac{9}{16}$	8	$\frac{9}{16}$ to $\frac{15}{16}$	10
$\frac{5}{8}$ to 1	10	1 to $1\frac{1}{4}$	12
$1\frac{1}{8}$ to $1\frac{3}{8}$	12	$1\frac{3}{8}$ to $1\frac{1}{2}$	14
$1\frac{1}{2}$	14		

SHELL END MILLS; STRAIGHT OR SPIRAL TEETH		INSERTED TOOTH CUTTERS, P. & W. FORM	
Dia. Mill	No. Teeth	Dia. Cutter	No. Blades
		4	10
		5	12
		6	16
		7	18
$1\frac{1}{4}$ to $1\frac{9}{16}$	16	8	20
$1\frac{5}{8}$ to $2\frac{9}{16}$	18	10	24
$2\frac{5}{8}$ to 3	20	12	28

METAL SLITTING CUTTERS

Thickness	Pitch	Thickness	Pitch
$\frac{1}{32}$	$\frac{3}{16}$	$\frac{5}{32}$	$\frac{5}{16}$
$\frac{1}{16}$	$\frac{1}{4}$	$\frac{3}{16}$	$\frac{11}{32}$
$\frac{3}{32}$	$\frac{9}{32}$	$\frac{1}{4}$	$\frac{3}{8}$
$\frac{1}{8}$	$\frac{5}{16}$		

SCREW SLOTTING CUTTERS

Cutters thinner than $\frac{1}{32}$ cut $\frac{3}{64}$ pitch.
Cutters $\frac{1}{32}$ to $\frac{3}{64}$ thick   cut $\frac{1}{16}$ pitch.
Cutters over $\frac{3}{64}$ thick   cut $\frac{3}{32}$ pitch.

Plain Milling Cutters	
Dia. of Cutter	No. of Teeth
2 to 2¾	18
3 to 3¾	20
4 to 5	22
5¼ to 6	24
6¼ to 8¾	26
9 to 9¼	28
9½ to 10	30
10¼ to 11	32

Fluting Cutters

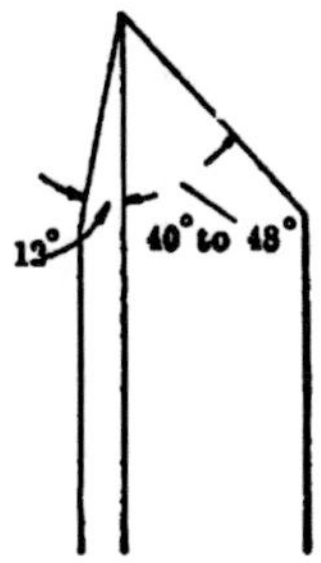

Form of Cutter for Milling Teeth in Plain Milling Cutters.

Plain cutters of ¾-inch face and over are generally made with spiral teeth. The 12-degree angle on side of fluting cutter gives ample clearance for cutting spiral grooves with the 12-degree face set on the center line of the work.

Side or Straddle Mills	
Dia. of Cutter	No. of Teeth
2	16
2½	20
2¾ to 3½	24
3¾ to 4½	26
5 to 5¾	28
6 to 6¾	30
7 to 7¾	32
8 to 8¾	34

Angular Cutter for Milling Teeth in Straddle Milling Cutters.

For milling teeth on periphery of straddle mills use angular cutter with 60 degree angle at *A;* for milling teeth on sides of cutters use 70°, 75° or 80° cutter according to number of teeth in cutter to be milled.

CORNER ROUNDING CUTTERS

Dia. Cutter	Rad. Circle	No. Teeth
2	$\frac{1}{16}$ to $\frac{1}{8}$	14
$2\frac{1}{4}$	$\frac{3}{32}$ to $\frac{1}{4}$	12
$2\frac{3}{4}$ to $3\frac{1}{2}$	$\frac{5}{16}$ to $\frac{5}{8}$	10
$3\frac{3}{4}$	$\frac{11}{16}$ to $\frac{3}{4}$	8

CONCAVE AND CONVEX CUTTERS

Dia. Cutter	Dia. Circle	No. Teeth
2	$\frac{1}{8}$ to $\frac{1}{4}$	12
$2\frac{1}{4}$ to $3\frac{1}{4}$	$\frac{3}{8}$ to $1\frac{1}{8}$	10
$3\frac{1}{2}$ to 4	$1\frac{1}{4}$ to 2	8

ANGULAR CUTTERS

Dia. of Cutter	No. of Teeth
$2\frac{1}{2}$	18
$2\frac{3}{4}$	20
3	22

DOUBLE ANGLE CUTTERS

Dia. of Cutter	No. of Teeth
$2\frac{1}{2}$ to 3	22

CUTTERS FOR SPIRAL MILLS

Dia. of Cutters	No. of Teeth
$2\frac{1}{2}$	18
$2\frac{3}{4}$	20
3	22

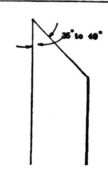

35° to 40°

Angular Cutter for Milling Teeth in Corner Rounding, Concave and Convex Cutters.

B   A

Cutter for Milling Teeth in Angular Cutters, Double Angle Cutters and Cutters for Spiral Mills. To cut Teeth on side A of this Cutter use 60° Cutter. To cut side B use 70°–75° Cutter.

HAND TAPS	
Dia. of Tap	No. Flutes
$\frac{8}{16}$ to $1\frac{3}{4}$	4
$1\frac{7}{8}$ to $2\frac{3}{4}$	6
3 to 4	8
3 to 4 mm.	3
5 to 44 mm.	4
46 to 50 mm.	6

MACHINE SCREW TAPS	
Dia. of Tap	No. Flutes
No. 1 to 7	3
No. 8 to 30	4

TAPPER TAPS	
Dia. of Tap	No. Flutes
$\frac{1}{4}$ to $1\frac{5}{8}$	4
$1\frac{3}{4}$ to 2	5
$2\frac{1}{4}$	6

MACHINE OR NUT TAPS	
Dia. of Tap	No. Flutes
$\frac{8}{16}$ to $\frac{5}{16}$	4
$\frac{3}{8}$ to $2\frac{1}{4}$	5
$2\frac{3}{8}$ to 3	6
$3\frac{1}{4}$ to 4	7

SCREW MACHINE TAPS	
Dia. of Tap	No. Flutes
$\frac{1}{4}$ to $1\frac{1}{2}$	4
$1\frac{9}{16}$ to 2	6

TAP FLUTING CUTTER

With
3 Flutes in Tap, B = $\frac{3}{8}$ Dia. Tap
4 Flutes in Tap, B = $\frac{1}{2}$ Dia. Tap
5 Flutes in Tap, B = $\frac{11}{16}$ Dia. Tap
6 Flutes in Tap, B = $\frac{11}{16}$ Dia. Tap
7 Flutes in Tap, B = $\frac{9}{32}$ Dia. Tap
8 Flutes in Tap, B = $\frac{1}{4}$ Dia. Tap

With
3 Flutes in Tap, E = $\frac{9}{32}$ Dia. Tap
4 Flutes in Tap, E = $\frac{1}{4}$ Dia. Tap
5 Flutes in Tap, E = $\frac{11}{64}$ Dia. Tap
6 Flutes in Tap, E = $\frac{11}{64}$ Dia. Tap
7 Flutes in Tap, E = $\frac{5}{32}$ Dia. Tap
8 Flutes in Tap, E = $\frac{9}{64}$ Dia. Tap

In milling taps with the convex cutter the cutter must be central with the tap.

TAPER PIPE TAPS	
Dia. of Tap	No. of Flutes
$\frac{1}{8}$ to $\frac{1}{2}$	4
$\frac{3}{4}$	4 or 5
1 to 1$\frac{1}{2}$	5
2	7
2$\frac{1}{2}$	8
3	9
3$\frac{1}{2}$ to 4	11

STRAIGHT PIPE TAPS

Dia. of Tap	No. of Flutes
$\frac{1}{4}$ to $\frac{1}{2}$	4
$\frac{3}{4}$ to 1$\frac{1}{4}$	5
1$\frac{1}{2}$	6
2	7
2$\frac{1}{2}$	8
3	9
3$\frac{1}{2}$	10

TAP FLUTING CUTTERS

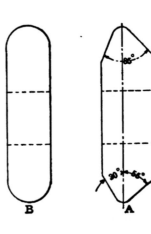

B   A

Cutter A is a regular tap fluting cutter that may be used if preferred for fluting any kind of tap in place of convex cutter B.

PIPE HOBS

Dia. Hob	No. Flutes	Dia. Hob	No. Flutes	Dia. Hob	No. Flutes
$\frac{1}{8}$ to $\frac{1}{4}$	6	1	12	3	28
$\frac{3}{8}$	8	1$\frac{1}{4}$ to 1$\frac{1}{2}$	16	3$\frac{1}{2}$	34
$\frac{1}{2}$	9	2	20	4	36
$\frac{3}{4}$	10	2$\frac{1}{2}$	24	4$\frac{1}{2}$	46

HOB FLUTING CUTTER

SELLERS HOBS

Dia. Hobs	No. Flutes	Dia. Hob	No. Flutes
$\frac{1}{4}$ to $\frac{7}{16}$	6	1$\frac{5}{8}$ to 2$\frac{1}{4}$	12
$\frac{1}{2}$ to $\frac{7}{8}$	8	2$\frac{5}{8}$ to 2$\frac{7}{8}$	14
1$\frac{1}{16}$ to 1$\frac{1}{2}$	10	3 to 4	16

HOB TAPS

Dia. Hob	No. Flutes
$\frac{1}{4}$ to $\frac{1}{2}$	6
$\frac{9}{16}$ to 1$\frac{1}{16}$	8
$\frac{7}{8}$ to 1$\frac{1}{2}$	10
1$\frac{5}{8}$ to 2	12

In fluting hobs leave land $\frac{3}{16}$ inch wide on top.

Shell Reamers

Dia. of Reamer	No. of Flutes	Dia. of Reamer	No. of Flutes	Dia. of Reamer	No. of Flutes
$\frac{1}{4}$ to $\frac{3}{8}$	6	$\frac{21}{32}$ to $1\frac{1}{2}$	10	$2\frac{9}{32}$ to $2\frac{3}{4}$	14
$\frac{13}{32}$ to $\frac{5}{8}$	8	$1\frac{17}{32}$ to $2\frac{1}{4}$	12	$2\frac{25}{32}$ to 4	16

Cutters for Fluting Reamers

Dia. of Reamer	R = Radius of Corner	Dia. of Reamer	R = Radius of Corner
$\frac{1}{8}$ to $\frac{3}{16}$	0	$1\frac{1}{8}$ to $1\frac{1}{2}$	$\frac{1}{16}$
$\frac{1}{4}$ to $\frac{3}{8}$	$\frac{1}{64}$	$1\frac{9}{16}$ to $2\frac{1}{4}$	$\frac{5}{64}$
$\frac{1}{2}$ to $\frac{5}{8}$	$\frac{1}{32}$	$2\frac{5}{16}$ to 3	$\frac{3}{16}$
$\frac{3}{4}$ to 1	$\frac{3}{64}$		

Dia. of Reamer	A = Am't Cutting Edge is Ahead of Center	Dia. of Reamer	A = Am't Cutting Edge is Ahead of Center
$\frac{1}{4}$	.011	$1\frac{1}{2}$	.066
$\frac{3}{8}$	.016	$1\frac{3}{4}$	.076
$\frac{1}{2}$	.022	2	.087
$\frac{5}{8}$	.027	$2\frac{1}{4}$	.098
$\frac{3}{4}$	.033	$2\frac{1}{2}$	.109
$\frac{7}{8}$	.038	$2\frac{3}{4}$	.120
1	.044	3	.131
$1\frac{1}{4}$	.055		

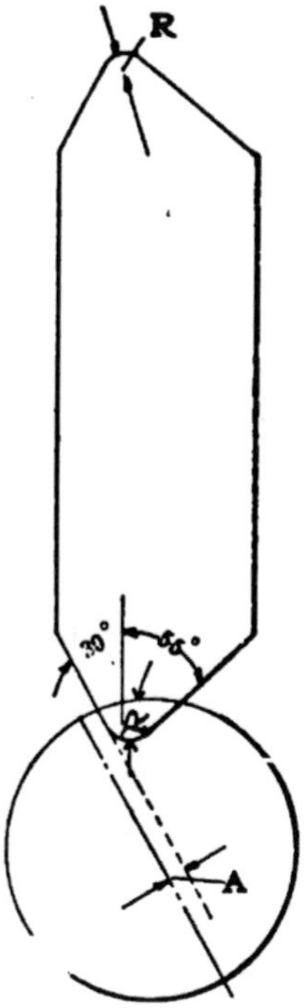

The type of cutter shown may be used for all classes of reamers except rose reamers.

Rose Chucking Reamers

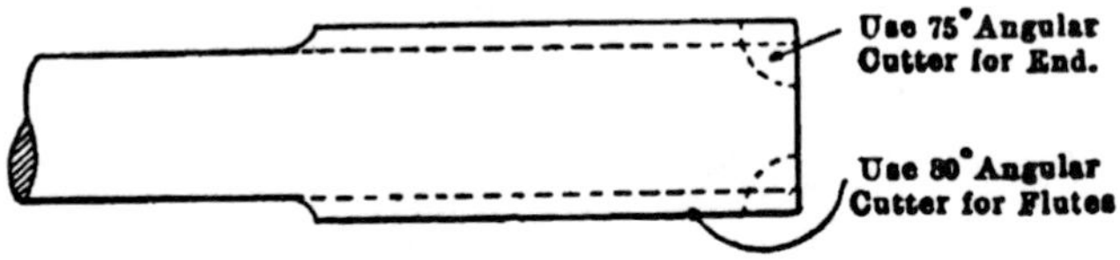
Use 75° Angular Cutter for End.
Use 80° Angular Cutter for Flutes

Depth of Groove = $\frac{1}{8}$ to $\frac{1}{6}$ Dia.

Dia. of Reamer	No. of End Cuts	No. of Flutes	Dia. of Reamer	No. of End Cuts	No. of Flutes
$\frac{1}{8}$ to $\frac{1}{2}$	6	3	$1\frac{3}{4}$ to 2	12	6
$\frac{5}{8}$ to 1	8	4	$2\frac{1}{4}$ to $2\frac{1}{2}$	14	7
$1\frac{1}{8}$ to $1\frac{1}{2}$	10	5	$2\frac{3}{4}$ to 3	16	8

Taper Reamers

Morse Taper		B. & S. Taper		Jarno Taper	
No. of Taper	No. of Flutes	No. of Taper	No. of Flutes	No. of Taper	No. of Flutes
0 to 1	6	1 to 5	6	2	4
2 to 4	8	6 to 10	8	3 to 4	6
5	10	11 to 12	10	5 to 10	8
6	14	13	12	11 to 15	10
7	16	14 to 16	14	16 to 18	12
		16 to 18	16	19 to 20	14

Taper Pin Reamers		Locomotive Taper Reamers	
No. of Reamer	No. of Flutes	Dia. of Reamer	No. of Flutes
0000 to 00	4	$\frac{1}{4}$ to $\frac{1}{2}$	6
0 to 7	6	$\frac{9}{16}$ to $1\frac{1}{4}$	8
8 to 10	8	$1\frac{5}{16}$ to $1\frac{3}{4}$	10
11 to 14	10	$1\frac{13}{16}$ to 2	12

CENTER REAMERS

DIAMETER OF STRADDLE MILL FOR FLUTING (3 FLUTES)

Size of Reamer	Outside Dia. of Cutter	Size of Reamer	Outside Dia. of Cutter
$\frac{1}{4}''$ cut	$2\frac{1}{2}$	$\frac{3}{4}''$ cut	$3\frac{1}{2}$
$\frac{3}{8}''$ cut	$2\frac{3}{4}$	$\frac{7}{8}''$ cut	$3\frac{3}{4}$
$\frac{1}{2}''$ cut	3	$1''$ cut	4
$\frac{5}{8}''$ cut	$3\frac{1}{4}$		

CUTTER KEYWAYS

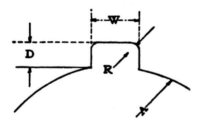

SQUARE KEYWAY

Dia. Hole, A	$\frac{1}{8}-\frac{9}{16}$	$\frac{5}{8}-\frac{7}{8}$	$\frac{15}{16}-1\frac{1}{8}$	$1\frac{3}{16}-1\frac{3}{8}$	$1\frac{7}{16}-1\frac{3}{4}$	$1\frac{13}{16}-2$	$2\frac{1}{16}-2\frac{1}{2}$	$2\frac{9}{16}-3$
Width Key, W	$\frac{3}{32}$	$\frac{1}{8}$	$\frac{5}{32}$	$\frac{3}{16}$	$\frac{1}{4}$	$\frac{5}{16}$	$\frac{3}{8}$	$\frac{7}{16}$
Depth, D	$\frac{3}{64}$	$\frac{1}{16}$	$\frac{5}{64}$	$\frac{3}{32}$	$\frac{1}{4}$	$\frac{5}{32}$	$\frac{3}{16}$	$\frac{7}{16}$
Radius, R ...	.020	.030	.035	.040	.050	.060	.060	.060

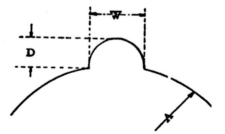

HALF-ROUND KEYWAY

Dia. Hole, A .	$\frac{1}{8}-\frac{5}{8}$	$\frac{11}{16}-\frac{13}{16}$	$\frac{7}{8}-1\frac{1}{16}$	$1\frac{1}{4}-1\frac{7}{16}$	$1\frac{1}{2}-2$	$2\frac{1}{16}-2\frac{7}{16}$	$2\frac{1}{2}-3$
Width, W	$\frac{1}{8}$	$\frac{3}{16}$	$\frac{1}{4}$	$\frac{5}{16}$	$\frac{3}{8}$	$\frac{7}{16}$	$\frac{1}{2}$
Depth, D	$\frac{1}{16}$	$\frac{3}{32}$	$\frac{1}{8}$	$\frac{3}{32}$	$\frac{3}{16}$	$\frac{7}{32}$	$\frac{1}{4}$

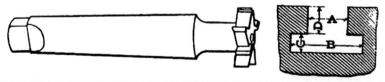

Width of Slot A Inches	Diameter of Neck of Cutter Inches	Width of Slot B Inches	Depth C Inches	Extreme Limit D Inches
1/4	7/32	1/2	5/32	5/16
5/16	9/32	5/8	3/32	3/8
3/8	11/32	3/4	3/32	7/16
7/16	3/8	7/8	7/32	9/16
1/2	7/16	1	3/32	3/4
5/8	1/2	1 1/16	9/32	7/8
3/4	9/16	1 3/16	3/8	1
7/8	11/16	1 1/4	13/32	1 1/16
1	29/32	1 3/8	15/16	1 1/16

These cutters are made 1/32 inch larger in diameter and 1/64 inch greater in thickness than the figures given, to allow for sharpening.

LARGEST SQUARES THAT CAN BE MILLED ON ROUND STOCK

Diam. of Stock	Decimal Equivalent	Size of Square	Nearest Fraction	Diam. of Stock	Decimal Equivalent	Size of Square	Nearest Fraction
1/8	.125	.088	3/32 −	1 9/16	1.5625	1.105	1 7/64 −
3/16	.1875	.133	1/8 +	1 5/8	1.625	1.149	1 5/32 −
1/4	.250	.177	11/64 −	1 11/16	1.6875	1.193	1 3/16 +
5/16	.3125	.221	7/32 +	1 3/4	1.750	1.237	1 15/64 +
3/8	.375	.265	17/64 −	1 13/16	1.8125	1.282	1 9/32
7/16	.4375	.309	5/16 −	1 7/8	1.875	1.326	1 21/64 −
1/2	.500	.354	23/64 −	1 15/16	1.9375	1.370	1 3/8 −
9/16	.5625	.398	25/64 +	2	2.000	1.414	1 13/32 +
5/8	.625	.442	7/16 +	2 1/16	2.0625	1.458	1 29/64 +
11/16	.6875	.486	31/64 +	2 1/8	2.125	1.502	1 1/2 +
3/4	.750	.530	17/32 −	2 3/16	2.1875	1.547	1 35/64
13/16	.8125	.574	37/64 −	2 1/4	2.250	1.591	1 19/32 −
7/8	.875	.619	5/8 −	2 5/16	2.3125	1.635	1 41/64 −
15/16	.9375	.663	21/32 +	2 3/8	2.375	1.679	1 43/64 +
1	1.000	.707	45/64 +	2 7/16	2.4375	1.723	1 23/32 +
1 1/16	1.0625	.755	3/4 +	2 1/2	2.500	1.768	1 49/64 +
1 1/8	1.125	.795	51/64 −	2 9/16	2.5625	1.813	1 13/16
1 3/16	1.1875	.840	27/32 −	2 5/8	2.625	1.856	1 55/64 −
1 1/4	1.250	.884	57/64 −	2 11/16	2.6875	1.900	1 29/32 −
1 5/16	1.3125	.928	59/64 +	2 3/4	2.750	1.944	1 15/16 −
1 3/8	1.375	.972	31/32 +	2 13/16	2.8125	1.989	1 63/64 +
1 7/16	1.4375	1.016	1 1/64 +	2 7/8	2.875	2.033	2 1/32 +
1 1/2	1.500	1.061	1 1/16 −	2 15/16	2.9375	2.077	2 5/64 −
				3	3.000	2.121	2 1/8 −

Side of Largest Square = Dia. of Stock × .707

TABLE OF DIVISIONS CORRESPONDING TO GIVEN CIRCUMFERENTIAL DISTANCES

This table gives approximate number of divisions and distances
part on circumference, corresponding to a known diameter of work.
t is useful in milling-machine work in cutting mills, saws, ratchets,
tc.

Dia. of Work	DISTANCE ON CIRCUMFERENCE														
	1/32″	1/16″	1/8″	3/16″	1/4″	5/16″	3/8″	7/16″	1/2″	9/16″	5/8″	11/16″	3/4″	7/8″	1″
1/4	25	12	6												
5/16	31	16	8												
3/8	38	19	9	6											
7/16	44	22	11	7	5										
1/2	50	25	13	8	6										
5/8	63	31	16	10	8	6									
3/4	75	38	19	13	9	8	6								
7/8	88	44	22	15	11	9	7	6							
1	100	50	25	17	13	10	8	7	6						
1/4	126	63	31	21	16	13	10	9	8	7	6				
1/2	150	75	38	25	19	15	13	11	10	8	7				
3/4	170	88	44	29	22	18	15	13	11	10	9	8	7	6	
2	200	100	50	34	25	20	17	14	12	11	10	9	8	7	6
1/4	226	113	56	38	28	23	19	16	14	13	11	10	9	8	7
1/2	251	125	63	42	31	25	21	18	16	14	12	11	10	9	8
3/4	277	138	69	46	35	28	23	20	17	15	14	13	12	10	9
3	302	151	75	50	38	30	25	22	19	17	15	14	13	11	9
1/4	327	163	82	54	41	33	27	23	20	18	16	15	14	12	10
1/2	352	176	88	59	44	35	30	25	22	20	18	16	15	13	11
3/4	378	189	94	63	47	38	31	27	24	21	19	17	16	14	12
4	402	201	100	67	50	40	34	29	25	22	20	18	17	15	13
1/4	428	214	107	71	53	43	36	31	27	24	21	19	18	15	13
1/2	454	227	114	76	57	45	38	32	28	25	23	21	19	16	14
3/4	478	239	119	79	60	48	40	34	30	27	24	22	20	17	15
5	503	252	126	84	63	50	42	36	31	28	25	23	21	18	16
1/4	528	264	132	88	66	53	44	38	33	29	26	24	22	19	16
1/2	554	277	138	92	69	55	46	40	35	31	27	25	23	20	17
3/4	570	289	145	96	73	58	48	41	36	32	28	26	24	21	18
6	604	302	151	101	76	61	50	44	38	34	30	27	25	22	19

For example: A straddle mill, say, 5 inches in diameter, is to be
cut with teeth $\frac{7}{16}$ apart. Without a table of this kind the workman
will have to go to the trouble of multiplying the diameter by 3.1416
and then divide by $\frac{7}{16}$ to find the number of teeth to set up for. In
the table, under $\frac{7}{16}$ and opposite 5, he can find at once the number
of divisions, as 36. Where the table shows an odd number of teeth,
one more or less can, of course, be taken if it is important to have
even number of teeth.

GRINDING AND LAPPING

GRINDING WHEELS AND GRINDING

The Commercial Abrasives

EMERY, corundum, carborundum, and alundum are the ordinary commercial abrasive materials. They vary in hardness, though it does not follow that the hardest grit is the best for cutting purposes; the shape and form of fracture of the particles must also be taken into consideration. We may imagine a wheel made up from diamonds, the hardest substance in nature, and whose individual kernels were of spherical form; it is quite obvious that it would be of little service as a cutting agent; on the other hand, if these kernels were crystalline or conchoidal in form it would probably be the ideal grinding wheel.

Emery is a form of corundum found with a variable percentage of impurity; it is of a tough consistency and breaks with a conchoidal fracture.

Corundum is an oxide of aluminum of a somewhat variable purity according to the neighborhood in which it is mined; its fracture is conchoidal and generally crystalline.

Carborundum is a silicide of carbon and is a product of the electric furnace; it breaks with a sharp crystalline fracture.

Alundum is an artificial product, being a fused oxide of aluminum. It is of uniform quality with about 98 per cent. of purity. It breaks with a sharp, conchoidal crystalline fracture and has all the toughness of emery.

Grading of Wheels

OF the many firms engaged in the manufacture of grinding wheels there are probably no two which have a similar method of grading or designating the hardness of their wheels. The Norton Company, which is probably the oldest in the field, uses the letter method, which may be said to be the simplest. That is, they take *M* for their medium-hard wheel and the letters before M denote in regular alphabetical progression the progressively softer wheels. Moreover they use a + mark for denoting wheels which vary in temper from the standards. Thus a wheel may be harder than the standard K, and still be not so hard as the standard L; in this case it is known as K+. The Carborundum Company adopts a somewhat similar method of grading, the difference being that although M denotes its medium-hard wheel the letters before M denote the progressively harder grades. Various other American companies use the letter method of grading to some extent, but all have individual ideas as to what degree of hardness should constitute an M or medium-grade wheel. Then there are firms both in America and on the continent of Europe which discard the letter method of grading or else use it in conjunction with numbers or fractions of numbers such as 2H, 1½M and so on.

The selection of suitable wheels for machine grinding may be said to be governed by the following points, namely, the texture of the material to be ground, the arc of wheel contact with work and the quality of finish required. The first and last of these points can for convenience' sake be taken in conjunction. The quality of surface finish is dependent on the condition of the wheel face and depth of cut rather than on the fineness of the grit in the wheel. A wheel of so fine a grit as 100 will give an indifferent finish if it is not turned true and smooth.

It may be assumed that for all general purposes the aim in view is to procure a wheel which will fulfil two conditions, that is, that it shall first remove stock rapidly and at the same time give a decent finish. Wheels made from a combination of grit of different sizes are the best for this purpose, as may be seen from the following explanation. Coarse wheels of an even number of grit will remove stock faster than will fine wheels of an even number, because their depth of cut or penetration is greater. They, however, fail in giving a high surface finish except in grinding very hard material, because they are not compact enough.

The Combination Grit Wheel

WITH the combination wheel the conditions are different and it seems better at removing stock than does the coarse, even grit wheel. It may be safe to assume from this that something of a *grindstone action* takes place, that is, that the finer particles of grit become detached from the bond and both roll and cut in their imprisoned condition between the larger particles. For finishing purposes this wheel has all the compactness and smooth face of a wheel which was made solely from its finest number of grit; and for roughing, it enables a depth of cut to be got which is within the capacity of its largest kernels.

With regard to the texture or hardness of material ground it may be taken as a general rule that the harder the material is, the softer the bond of wheel should be, and that cast iron and hardened steel bear some relation to each other as far as grinding wheels are concerned, for the same wheel is usually suitable for both materials.

Too large an assortment of wheels is likely to lead to confusion and we may take the Norton plain cylindrical grinding machine as being a case in point of a limited assortment of wheels; at the same time it will be a starting point to illustrate choice of wheels under various grinding conditions. In this machine four different grade wheels, all of 24 combination grit, are found sufficient for all classes of material that it is ordinarily required to grind. These include high- and low-carbon steels, cast iron, chilled iron, and bronze or composition metals. These wheels are graded J, K, L, and M.

Hard Wheels

ONE of the greatest advantages accruing from grinding is that it ignores the non-homogeneity of material and that it machines work with the lightest known method of tool pressure, thus avoiding all deflections and distortions of material which are a natural resul the more severe machining processes. Yet these objects are too often defeated by the desire for hard and long-lived wheels. A wheel that is too hard or whose bond will not crumble sufficiently under the pressure of cut will displace the work and give rise to many unforeseen troubles. It is also a prolific cause of vibration which is antagonistic to good and accurate work. The advantage claimed for it, that it gives a better surface finish, is a deceptive one, for it mostly obtains this finish at the expense of accuracy. Quality of finish, that is, accurate finish, is merely a question of arranging of work speed, condition of wheel face and depth of cut. In the machine mentioned the suitability of wheels to materials and conditions is found to be as follows, the wheels being in each case of a combination of alundum grit:

For hard chilled iron and large diameters of cast iron and
    hardened steel .... 24 J

For medium chilled iron and medium diameters of cast iron
    and hardened steel and bronze........................ 24 K

For all grades of steel which are not hardened and for bronze 24 L

For very low carbon machine steels .... 24 M

The table given may, speaking generally, be what would be chosen in the way of wheels for the materials given, and in actual practice they soon give evidence as to whether they are suitable. It may be gathered from the table that diameter of work is a factor in the choice of a wheel. This refers to area of wheel contact and is governed by what is shown in the table when broad differences of diameter occur; for instance, it might be necessary to use the K wheel for a large diameter of high carbon steel if the L wheel was evidently too hard.

Speed and Efficient Cutting

The efficient cutting of a wheel depends very much on the speed of the work, and an absence of knowledge in this respect may often lead to a suitable wheel's rejection. Revolving the wheel at the speed recommended by the maker is the first necessity, and if it is found unsuitable after experimenting with various speeds it should be changed for a softer or harder one as the conditions indicate. Starting from the point that a wheel is desired that shall remove the maximum amount of stock with the minimum amount of wear on the wheel, the indications and method of procedure may be as follows; only it must be understood that this refers to cases where an ample supply of water is being delivered at the *grinding point.*

If, after trying all reasonable work speeds, a wheel should burn the work, or refuse to cut without excessive pressure, or persistently glaze the surface of the work, it is too hard for that particular work and material and may be safely rejected. If, after trying all reasonably reduced work speeds, a wheel should lose its size quickly and show all signs of rapid wear, it is too soft for that particular work and material and may be rejected. These indications refer to all ordinary cases and it may be gathered that the most economical wheel is that which acts in such a manner as to be a medium between the two cases. There is still another point to bear in mind with regard to the size of the grit in the wheel, but which refers more especially

to very hard materials such as chilled iron. Either a coarse or combination wheel may go on cutting efficiently in roughing cuts because
pressure is exerted, but may begin to glaze when this pressure is
much relieved as in finishing cuts. A careful microscopic scrutiny
of a wheel that displays this tendency would seem to lead to the following assumption:

When a Wheel is Sharp

THE wheel face when newly trued with the diamond tool, which
is necessary to obtain an accurate finish, shows a promiscuous arrangement of particles, some of which present points and others
present a broader face with a rough and granular surface. When
the wheel is presented to the hard surface of the work the high points
of this granular face and the sharp contour of the kernels will go on
cutting until they are dulled and worn down, after which their face

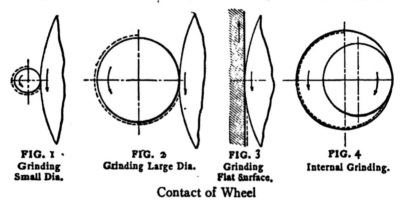

FIG. 1 ·   FIG. 2   FIG. 3   FIG. 4
Grinding   Grinding Large Dia.   Grinding   Internal Grinding.
Small Dia.   Flat Surface.

Contact of Wheel

area is too great to enter the surface without undue pressure. When
the wheel has reached this condition the microscope shows these
broader-faced kernels polished to a metallic luster, which bears out
the explanation tendered and also makes the remedy quite apparent. This is to use a wheel of very fine grit for finishing purposes
in these cases or else keep the coarser wheel in condition by repeated
dressings with the diamond tool.

Wheel Contact

REFERENCE to Figs. 1 to 4 will show what actual practice requires
in the choice of a wheel so far as the question of wheel contact is
concerned. A wheel is shown in contact with four different varieties of work, all of which we will suppose to be of the same material, the depth of cut, much exaggerated, being the same in each case.
In the first case it is a shaft of small diameter, and the wheel contact
being the smallest the harder grade of wheel would be suitable, comparatively speaking. Assuming that this wheel was found to be
suitable it would probably require a softer wheel for the next case,
which is a shaft of larger diameter, and the wheel contact propor-
tionately greater. To continue the comparison still further, the thir᷒

case shows the wheel engaged in grinding a flat surface, and the fourth is a wheel grinding internally. In each case practice demands that the wheel shall be progressively softer in bond or grade and is some proof of a consistency in the action of grinding wheels.

The Contact Area of a Wheel

THE most probable explanation of this may be that as the contact area increases more work is required from each individual kernel of grit and it the sooner becomes dulled; this requires that the bond must be more friable both to allow it to escape easily and to minimize the pressure required to make the wheel cut as the cutting area becomes greater. Following on this reasoning we are able to choose a list of wheels which would be suitable for almost all purposes, and which would be as follows if of Norton grade:

For plain cylindrical grinding
        J K L M
For grinding plane surfaces
        H I J K
For internal grinding
        F H I J

This collection of wheels would be suitable for almost any type of grinding machines, though when the wheels are exceptionally narrow a grade or one-half grade higher might be possible; it would, of course, be a matter for a little trial and experiment. The wheels for external cylindrical work may preferably be combination wheels, but for plane surface and internal work they are better made of single grit, about 36 or 46. The great contact area of wheel in these two classes of work is liable to generate much heat so that an open and porous wheel is preferable.

Wheel Pressure and Wear

As the wheel is a disk built up from a numerous assortment of minute cutting tools which are held in position by a more or less friable bond, in using it we must bring it to bear on the work with a pressure that shall not be so great as to tear these minute tools from their setting until their cutting efficiency is exhausted, for if we do so we are wasting the wheel. To gage the exact amount of the pressure required is a matter of judgment and experience, though where automatic feeds are provided on a machine the right amount of pressure or feed is soon determined. It will also be readily understood that a regular automatic feed is more reliable for the purpose than a possibly erratic hand one. The automatic feed may be set to give a certain depth of cut at each pass of the wheel, and its amount of wear noted; if this wear be found excessive the depth of cut may be reduced. It must not be here forgotten that work speed also enters into this consideration and that a high work speed will tend to wear the wheel excessively; inversely a reduced work speed will reduce the amount of wear. Having these points in mind the right combination of depth of cut and work speed is soon arrived at, and an approximate judgment attained for the future.

Grinding Allowances

THE amount of stock left for removal by the grinding wheel and the method of preparing the work have both much bearing on the economic use of grinding wheels, and heavy and unnoticed losses often occur through want of a few precautionary measures. The necessary amount of stock to leave on a piece of work as a grinding allowance depends firstly on the type of machine employed, the class of labor engaged in preparing it, and whether it has to be hardened or otherwise.

In powerful machines, which will remove stock rapidly, the grinding allowance may be anything from $\frac{1}{32}$ to $\frac{1}{64}$ inch. There are many cases of an especial character when the grinding allowance stated may be exceeded to advantage so long as discretion is used. Straight shafts may often be ground direct from the black bar of raw material $\frac{1}{16}$ inch above finished size, or when shafts of this character must have large reduction on the ends they can be roughly reduced in the turret lathe while in their black state and finished outright more economically in the grinding machine. Very hard qualities of steels or chilled rolls are other cases where it is often more economical to use the grinding machine without any previous machining process, and though there may be sometimes an alarming waste of abrasive material its cost is as nothing compared with other savings that are made.

Grinding allowances for hardened work are usually larger than for soft work, to allow for possible distortion; so that individual experience alone can determine the amount to be left. It is sufficient to say that the allowances on case-hardened or carbonized work should not be excessive; otherwise the hardened surface may be ground away.

Grinding Hardened Work

As far as the actual grinding of hardened work goes, it is indispensable that the whole portion of a piece that is to be ground should be roughed over previous to the final finishing; if it is at all possible to allow some little time to elapse between the two operations so much the better, more especially if it has bent in hardening and been afterward straightened; this will allow of the development of any strain that may be present. Both for special and standard work in a factory a table of grinding allowances can be compiled as a result of experience and posted in a conspicuous position. If this be done and trouble taken to see that it is adhered to, it will save much trouble and be a means of avoiding much unnecessary expense.

It is necessary to slightly undercut the corners of shoulders so as to preserve the corner of the grinder's wheel intact. A piece of work should never be prepared in such a manner as to form a radius on the corner of the wheel, for to get the wheel face flat again means much waste of wheel and wear of diamond. Where fillets or radii are necessary they are better got out with a tool, for even if they are to be ground they must be turned good to allow the wheel to conform to their shape. The only excusable reason for grinding a round corner is when the work is hardened or in some special case where the expense incurred is warranted.

case shows the wheel engaged in grinding a flat surface, and the fourth is a wheel grinding internally. In each case practice demands that the wheel shall be progressively softer in bond or grade and is some proof of a consistency in the action of grinding wheels.

The Contact Area of a Wheel

THE most probable explanation of this may be that as the contact area increases more work is required from each individual kernel of grit and it the sooner becomes dulled; this requires that the bond must be more friable both to allow it to escape easily and to minimize the pressure required to make the wheel cut as the cutting area becomes greater. Following on this reasoning we are able to choose a list of wheels which would be suitable for almost all purposes, and which would be as follows if of Norton grade:

For plain cylindrical grinding
J K L M
For grinding plane surfaces
H I J K
For internal grinding
F H I J

This collection of wheels would be suitable for almost any type of grinding machines, though when the wheels are exceptionally narrow a grade or one-half grade higher might be possible, it

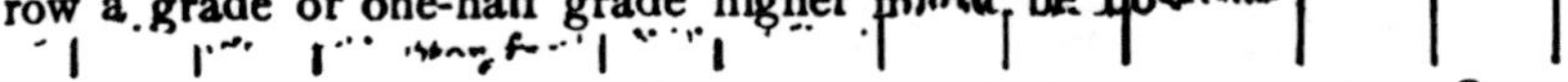

FIG. 5   FIG. 6   FIG. 7   FIG. 8

Methods of Setting Diamonds

The Use of Diamonds

HERE it is perhaps well to give the question of diamonds some little consideration as they are sometimes a very expensive item. A diamond is a very essential part of a grinding machine's equipment, for in its absence a good and highly finished grade of work is an impossibility. It is perhaps unnecessary to state that they should be the hardest rough stones procurable, and that the larger they are the cheaper they are in the end. With regard to their size: This is a known proportionate element in their price per carat, but a large stone allows of a more secure hold in its setting and so the danger of losing it is reduced. As a further precaution against this danger the diamond tool should always be held by mechanical means when using it except in cases which are unavoidable; this may be in cases where profile shapes have to be turned on the wheel face. An attempt to turn by hand a perfectly flat face on a wheel, which is necessary for finishing, must of a necessity end in failure.

As a means of preservation of the diamond a full stream of water should be run on it when in use and many light chips are preferable to a few heavy ones. The main thing is to watch that it does not get unduly heated, for this is disastrous to it. Where large quantities of material have to be removed from a wheel the ordinary wheel dresser may be employed to reduce the bulk of the stock, and the diamond only used for finishing to shape.

Grinding Allowances

THE amount of stock left for removal by the grinding wheel and the method of preparing the work have both much bearing on the economic use of grinding wheels, and heavy and unnoticed losses often occur through want of a few precautionary measures. The necessary amount of stock to leave on a piece of work as a grinding allowance depends firstly on the type of machine employed, the class of labor engaged in preparing it, and whether it has to be hardened or otherwise.

In powerful machines, which will remove stock rapidly, the grinding allowance may be anything from $\frac{1}{32}$ to $\frac{1}{64}$ inch. There are many cases of an especial character when the grinding allowance stated may be exceeded to advantage so long as discretion is used. Straight shafts may often be ground direct from the black bar of raw material $\frac{1}{16}$ inch above finished size, or when shafts of this character must have large reduction on the ends they can be roughly reduced in the turret lathe while in their black state and finished outright more economically in the grinding machine. Very hard qualities of steels or chilled rolls are other cases where it is often more economical to use the grinding machine without any previous machining process, and though there may be sometimes an alarming waste of abrasive material its cost is as nothing compared with other savings that are made.

The same allowances for hardened work are usually larger than various diameters of emery wheels, to cause them to run at the peripheral rates of 4000, 5000 and 6000 feet per minute. Ordinarily a speed of 5000 feet is employed, though sometimes the speed is somewhat lower or higher for certain cases.

Diam. Wheel	Rev. Per Minute for Surface Speed of 4,000 ft.	Rev. Per Minute for Surface Speed of 5,000 ft.	Rev. Per Minute for Surface Speed of 6,000 ft.
1 inch.	15,279	19,099	22,918
2 "	7,639	9,549	11,456
3 "	5,093	6,366	7,639
4 "	3,820	4,775	5,730
5 "	3,056	3,820	4,584
6 "	2,546	3,183	3,820
7 "	2,183	2,728	3,274
8 "	1,910	2,387	2,865
10 "	1,528	1,910	2,292
12 "	1,273	1,592	1,910
14 "	1,091	1,364	1,637
16 "	955	1,194	1,432
18 "	849	1,061	1,273
20 "	764	955	1,146
22 "	694	868	1,042
24 "	637	796	955
30 "	509	637	764
36 "	424	531	637

GRADING ABRASIVE WHEELS

THE Norton Company uses 26 grade marks, the Carborundum Company 19, while the Safety Emery Wheel Company uses 40. The following table is a comparison between the grade designations of the Norton Company and the Carborundum Company. Intermediate letters between the grade designations indicate relative degrees of hardness between them; the Norton Company manufacturing four degrees of each designation, while the Carborundum Company manufactures three.

Norton Co.	Grade Designation	Carborundum Co.
A...................	Extremely or Very Soft	 V
B..................		.... U
C.............		.... T
D........		
E.....	Soft	.... S
F........		.... R
G...........		.... Q
H..............		
I...................	Medium Soft	.... P
J.............		.... O
K.........		
L.........		.... N
M ...	Medium	.... M
N		 L
O............		
P		.... K
Q	Medium Hard	.... J
R		 I
S		.... H
T		
U	Hard	.... G
V.........		.... F
W		.... E
X		
Y	Extremely or Very Hard	.... D
Z		

The Safety Emery Wheel Company's grade list is an arbitrary one with the following designations:

C. Extra Soft
A. Soft

P. Medium
O. Hard
E. Extra Hard

H. Very Soft
M. Medium Soft

I. Medium Hard
N. Very Hard
D. Special Extra Hard

Intermediate figures between those designated as soft, medium
oft, etc., indicate so many degrees harder or softer, *e.g.*, A¼ is one
egree harder than soft. A¾ is three degrees harder than soft or
ne degree softer than medium soft.

Numbers and Grades of Abrasive Wheels

In the following table for the selection of grades will be found a
omparison of the grading used by the Norton Company, and that
)f the Carborundum Company:

Class of Work	Norton Co.		Carborundum Co.	
	Number Usually Furnished	Grade Usually Furnished	Number Usually Furnished	Grade Usually Furnished
Large Cast Iron and Steel Castings	16 to 20	Q to R	16 to 24	G to H
Small Cast Iron and Steel Castings	20 to 30	P to Q	20 to 30	G to H
Large Malleable Iron Castings ..	16 to 20	Q to R	16 to 24	G to H
Small Malleable Iron Castings ..	20 to 30	P to Q	20 to 30	H to I
Chilled Iron Castings	16 to 20	Q to R	16 to 24	H
Wrought Iron	16 to 30	P to Q	16 to 24	F to H
Brass Castings.................	16 to 30	O to P	20 to 36	H to I
Bronze Castings	16 to 30	P to Q	20 to 30	I
Rough Work in General........	16 to 30	P to Q	20 to 30	H
General Machine Shop Use	30 to 46	O to P	24 to 36	G to J
Lathe and Planer Tools	30 to 46	N to O	30 to 36	I to J
Small Tools	36 to 100	N to P	50 to 80	I to J
Wood-working Tools	36 to 60	M to N	40 to 60	L to M
Twist Drills (Hand Grinding)...	36 to 60	M to N	60	I to J
Twist Drills (Special Machines).	46 to 60	K to M	50	L to O
Reamers, Taps, Milling Cutters, etc. (Hand Grind)...........	46 to 100	N to P	50 to 80	K to N
Reamers, Taps, Milling Cutters, etc. (Spec. Mach.)...........	46 to 60	H to K	50 to 60	L to M
Edging and Jointing Agricultural Implements	16 to 30	Q to R	141 to 24	G to I
Grinding Plow Points..........	16 to 30	P to Q	20 to 24	H
Surfacing Plow Bodies	20 to 30	N to O	16 to 20	G
Stove Mounting...............	20 to 36	P to Q	24 to 30	G
Finishing Edges of Stoves	30 to 46	O to P	24 to 30	G
Drop Forgings................	20 to 30	P to Q	24 to 36	G to I
Gumming and Sharpening Saws.	36 to 60	M to N	403—603	J to L
Planing Mill and Paper Cutting Knives	30 to 46	J to K	202—60 to 80	M to R
Car Wheel Grinding...........	20 to 30	O to P	16 to 24	H

LAPPING

LAPPING may be defined as the process of finishing the surface of a piece of work by means of another piece of material, called a lap, the surface of which is charged with an abrasive.

Laps are roughly divided into three general classes. First, those where the form of the lap makes a line contact with the work, and the work is, if cylindrical, revolved to develop the cylindrical form, or, if straight, in one direction, is moved back and forth under the lap. Second, those which are used for straight surfaces with a full contact on the lap, and third, those which are used for male and female cylindrical surfaces with a full contact on the lap. In all cases the material from which the lap is made must be softer than the work. If this is not so, the abrasive will charge the work and cut the lap, instead of the lap cutting the work.

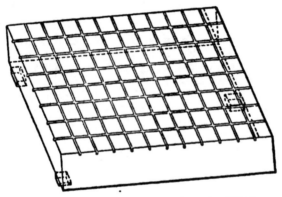

FIG. 9.—A Lapping Plate for Flat Work

The first class is used in the place of emery wheels, either where the work is too small to use an ordinary wheel or where a form is to be ground on the work and an emery wheel will not keep its shape. They are usually made of machinery steel and the abrasive used is crushed diamond rolled into the surface. In rolling in the diamond dust the sharp corners of the particles cause them to bed securely into the surface of the lap, and if a good quality of diamond is used, a lap will grind all day without recharging. Oil is used to lubricate the work and carry away the dust from the grinding. If a diamond lap is run dry the particles of diamond tear and raise "burs" in the work, which strip the lap very quickly. The speed should be about two-thirds that for an emery wheel of the same size; for if it is excessive, the lap will wear smooth and glaze instead of cutting. This kind of lap is used mainly in watch and clock shops, and shops making watch tools, sub-press dies, and similar work.

Lapping Flat Surfaces

IN lapping flat surfaces, which are usually on hardened steel, a cast-iron plate is used as a lap and emery as an abrasive. In order that the plate may stay reasonably straight, it should either be quite thick, or else ribbed sufficiently to make it rigid, and in any case it should be supported on three feet, the same as a surface place. For rough work or " blocking down," as it is called, the lap works better if scored with narrow grooves, about ⅛ inch apart, both lengthways and crossways, thus dividing the plate into small squares, as in Fig. 9. The emery is sprinkled loosely on the block, wet with lard oil and the work rubbed on it; care is taken to press hardest on the highest spots. The emery and oil get in the grooves, and are continually rolling in and out, getting between the plate and the work and are crushed into the cast iron, thus charging it thoroughly in a short time. About No. 100 or No. 120 emery is best for this purpose.

After blocking down, or if the work has first been ground on a surface grinder, the process is different. A plain plate is used with the best quality of flour of emery as an abrasive, as the least lump or coarseness will scratch the work so that it will be very hard to get the scratches out. Instead of oil, benzine is used as a lubricant and the lap should be cleaned off and fresh benzine and emery applied as often as it becomes sticky. The work should be tried from time to time with a straight-edge and care taken not to let the emery run in and out from under the work, as this will cause the edges to abrade more than the center, and will especially mar the corners. After getting a good surface, the plate and work should be cleaned perfectly dry, and then rubbed. The charging in the plate will cut just enough to remove whatever emery may have become charged in the work, will take away the dull surface and leave it as smooth as glass and as accurate as it is possible to produce.

Laps for Holes

IN lapping holes various kinds of laps are used, according to the accuracy required, and the conditions under which the work is done. The simplest is a piece of wood turned cylindrical with a longitudinal groove or split in which the edge of a piece of emery cloth is inserted. This cloth is wound around the wood until it fills the hole in the work. This is only fit for smoothing or enlarging rough holes and usually leaves them more out of round and bell-mouthed than they were at first. Another lap used for the same purpose — and which produces better results — is 'made by turning a piece of copper, brass, or cast iron to fit the hole and splitting it longitudinally for some distance from the end. Loose emery is sprinkled over it, with lard oil for a lubricant, and a taper wedge is driven into the end for adjustment as the lap wears.

For lapping common drill bushings, cam rolls, etc., in large quantities, where a little bell-mouthing can be allowed, and yet a reasonably good hole is required, a great many shops use adjustable copper laps made with more care than the above. One way of making them is to split the lap nearly the whole length, but leaving both ends solid. One side is drilled and tapped for spreading screws

for adjustment.   Either one screw half-way down the split may be
used or two screws dividing the split into thirds.   Another and
better means of adjustment is to drill a small longitudinal hole a
little over half the length of the lap, enlarge it for half its length,
and tap the large end for some distance.   This is done before splitting.   Into this hole a long screw with a taper point is fitted so that
when tightened it tries to force itself into a small hole, thus spreading the lap.

For nice work there is nothing better than a lead lap.   Lead charges
easily, holds the emery firmly and does not scratch or score the work.
It is easy to fit to the work and holds its shape well for light cuts.
Under hard usage, however, it wears easily.   For this reason, while
laps for a single hole or a special job are sometimes cast on straight
arbors, where much lapping is done it is customary to mold the laps
to taper arbors with means for a slight adjustment.   After any extensive adjustment the lap will be out of true and must be turned off.
All of these laps, as shown in Fig. 10, are to be held by one end in

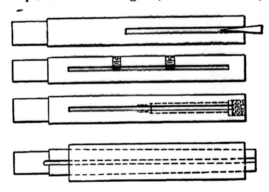

FIG. 10. — Laps for Holes

a lathe chuck, and the work run back and forth on them by hand,
or by means of a clamp held in the hand.   If a clamp is used care
should be taken not to spring the work.

How to Do Good Lapping

THERE are several points which must be taken into consideration
in order to get good results in lapping holes.   The most important
is that the lap shall always fill the hole.   If this condition is not com-
plied with the weight of the work and the impossibility of holding
it exactly right will cause it to lap out of round, or if it is out of round
at the start the lap will be free to follow the original surface.   If the
lap fits, it will bear hardest on the high spots and lap them off.   Next
in importance to getting a round hole is to have it straight.   To
attain this end the lap should be a little longer than the work, so
that it will lap the whole length of the hole at once, and not have a
tendency to follow any curvature there may be in it.   What is known
as bell-mouthing, or lapping large on the ends, is hard to prevent,
'specially if the emery is sprinkled on the lap and the work shoved

on it while it is running. The best way to avoid this condition when using cast-iron or copper laps, which do not charge easily, is to put the emery in the slot, near the center of the lap, and after the work is shoved on squirt oil in the slot to float the emery. Then, when the lathe is started the emery will carry around and gradually work out to the ends, lapping as it goes. Where lead is used the emery can be put on where it is desired to have the lap cut and rolled in with a flat strip of iron. It will not come out easily, so will not spread to any extent, and it is possible with a lap charged in this manner to avoid cutting the ends of the hole at all. The work should always be kept in motion back and forth to avoid lumping of the emery and cuttings which will score grooves in the work.

Ring Gage and Other Work

RING gages are lapped with a lead lap. They are first ground straight and smooth to within .0005 inch of size, and then, when lapped, are cooled as well as cleaned, before trying the plug, by

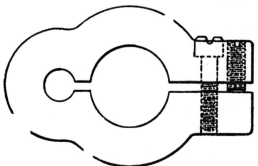

FIG. 11. — A Lap for Plugs

placing them in a pail of benzine for a long enough time to bring them down to the temperature of the room. Some shops leave a thin collar projecting from each side around the hole, so that, if there is any bell-mouthing, it will be in these collars, which are ground off after the lapping is done.

Other metals are lapped in this same manner, except that the abrasive is different. Cast iron is lapped with emery, but charges to some extent. This charging can be taken out without changing the work materially by rubbing it by hand with flour of emery cloth. In lapping bronze or brass, crocus and Vienna lime are used. Crocus is used with a cast-iron or lead lap, and the charging is removed by running the work for a few seconds on a hardwood stick which fits the hole. Unslaked Vienna lime, freshly crushed, is used with a lead or hardwood lap, and does not charge. It does a nice job, but is very slow, and is only used in watch factories.

For lapping plug gages, pistons, and other cylindrical articles, a cast-iron lap is usually used, split and fitted with a closing and a spreading screw, as shown in Fig. 11. Sometimes, where a very fine finish is required, or where the work is not hardened, the hole is made larger than the work, and a lead ring cast into it.

DIAMOND POWDER IN THE MACHINE SHOP

THE diamond used for this purpose, costing 85 cents per carat, is an inferior grade of diamond, not so hard as the black diamond used for drills and truing emery wheels, and not of a clear and perfect structure to permit it to enter the gem class. Many are a mixed black and white, others yellow and some pink; many are clear but flaky. Then there is the small débris from diamond cutting, which is reduced to powder and sells somewhat cheaper; but some find it more economical to use the above and powder it themselves, as the débris from diamond cutting is of a flaky nature, and does not charge into the lap so well.

Assuming there is 25 carats to reduce to powder, proceed as follows:

Into a mortar, as shown at Fig. 12, place about 5 carats, using an 8-ounce hammer to crush it. It takes from 3 to 4 minutes' steady pounding to reduce it to a good average. Scrape the powder free from the bottom and the sides and empty into one-half pint of oil. The oil used is the best olive oil obtainable, and is held in a cup-shaped receptacle that will hold a pint and one half. The 25 carats being reduced to powder, and in the oil, stir it until thoroughly mixed, and allow to stand 5 minutes; then pour off to another dish. The diamond that remains in the dish is coarse and should be washed in benzine and allowed to dry, and should be repounded, unless extremely coarse diamond is desired. In that case label it No. 0. Now stir that which has been poured from No. 0, and allow to stand 10 minutes. Then pour off into another dish. The residue will be No. 1. Repeat the operation, following the table below.

The settlings can be put into small dishes for convenient use, enough oil staying with the diamond to give it the consistency of paste. The dishes can be obtained from a jewelers' supply house.

TABLE FOR SETTLING DIAMOND

To obtain No. 0 — 5 minutes.
To obtain No. 1 — 10 minutes.
To obtain No. 2 — 30 minutes.
To obtain No. 3 — 1 hour.
To obtain No. 4 — 2 hours.
To obtain No. 5 — 10 hours .
To obtain No. 6 — until oil is clear.

Diamond is seldom hammered; it is generally rolled into the metal. For instance, several pieces of wire of various diameters charged with diamond may be desired for use in die work. Place the wire and a small portion of the diamond between two hardened surfaces, and under pressure roll back and forth until thoroughly charged. No. 2 diamond in this case is generally used. Or one can form the metal any desired shape and apply diamond and use a roll, as Fig. 14, to force the diamond into the metal. This is then a file which will work hard steel, but the moment this diamond file, or lap, is crowded it is stripped of the diamond, and is consequently of no ·e. It is to be used with comparatively light pressure.

Diamond Laps

COPPER is the best metal. It takes the diamond readily, and
etains it longer than other metals; brass next, then bessemer
teel. The latter is used when it is wished to preserve a form that
s often used.

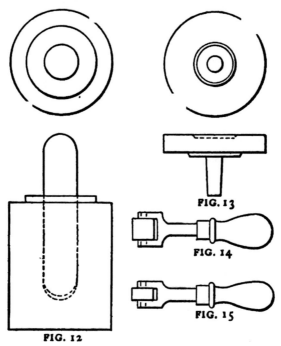

FIG. 13

FIG. 14

FIG. 15

FIG. 12

Diamond Laps and Tools

For sharpening small, flat drills, say 0.008 to 0.100, a copper lap
mounted on a taper shank, as in Fig. 13, and charged on the face with
No. 2 diamond, using pressure on the roll, makes a most satisfactory
method of sharpening drills. The diamond lasts for a long time if
properly used, and there is no danger of drawing the temper on the
drill. It is much quicker than any other method of sharpening.

To charge the lap use the roll, Fig. 15, supported on a T rest
pressing firmly against the lap, being careful to have the roll on the
center; otherwise instead of charging the lap it will be grinding the
roll. The diamond may be spread either on the lap or the roll, and
the first charging usually takes twice the amount of diamond that
subsequent charging takes. To avoid loss of diamond, wash the
lap in a dish of benzine kept exclusively for that purpose. This can
be reclaimed by burning the metal with acids, and the diamond can
be resettled.

For the grinding of taper holes in hard spindles or for position work in hard plates, where holes are too small to allow the use of emery wheels, No. 1 diamond does the work beautifully. Or if it is wished to grind sapphire centers or plugs as stops, etc., a bessemer lap made in the form of a wheel and charged with diamond on the diameter does the work nicely.

Nos. 5 and 6 diamond are used on boxwood laps, mounted on taper plugs or chucks, and the diamond smeared on with the finger. The lap is run at high speed and used for fine and slow cutting which also gives a high polish.

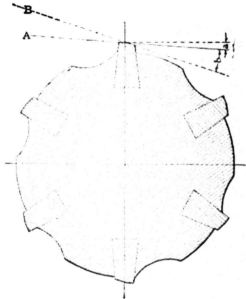

FIG. 16. — Cross-section of
Hand Reamer

REAMER AND CUTTER GRINDING

Reamer Clearances

AFTER constant experimenting for a period of more than a year, the Cincinnati Milling Machine Company succeeded in establishing tables for four styles of reamers for obtaining what they consider to be the best clearances, the object being to grind clearances on reamers which would ream the greatest number of smooth holes with a minimum amount of wear. The four styles of reamers are as follows: Hand reamers for steel, hand reamers for gray iron and bronze, chucking reamers for gray iron and bronze, chucking reamers for steel. The company uses adjustable blade reamers almost exclusively, all of which are ground in the toolroom on their universal cutter and tool ·inder.

Fig. 16 is a cross-section of a hand reamer.   Two clearance lines,
and *B*, are ground on the blades, *a* being the cutting clearance
d *b* the second clearance called for in the table.   The object of
*r*ing the adjustment for the second clearance so minutely is to pro-
de a proper width of land, which equals .025 inch on all hand reamers
r gray iron or bronze, and 0.005 inch on hand reamers for steel.

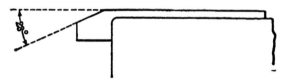

FIG. 17. — Chucking Reamer
Blade for Gray Iron and Bronze

*Chucking reamers for gray iron and bronze have, in this system, 23-
*degree beveled ends as shown in Fig. 17, and are provided with two
*clearances along the blades, for which the settings are given in Table
.   The beveled ends have only one clearance which is equal to the
*econd clearance given in Table 3.   Fig. 18 shows a chucking reamer
or reaming steel.   In these reamers the blades are circular ground
o the exact size of hole to be reamed and without clearance, the 45-
*degree beveled ends only having clearance as given in Table 4.   On
*all reamers of this style the blades are ground from .015 to .020 inch
*below size half of their length toward the shank end.

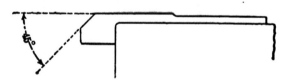

FIG. 18. — Chucking Reamer
Blade for Steel

In grinding the clearances for the various kinds of reamers as
given in Tables 1, 2, and 3, the tooth rest is held stationary on the
emery wheel head of the grinder, while in grinding the 45-degree
beveled ends on the chucking reamers for steel, the tooth rest is
supported from the grinder table and travels with the work.   The
front end of the hand reamer blades are tapered about 0.004 per
inch.   The back ends of the blades are also slightly tapered to prevent injuring the holes when backing the reamer out.

REAMER CLEARANCE

Ground with Cup Wheel 3″ dia.—Tooth Rest to be Set Central with Emery Wheel Spindle. Set Work holding Centers above Emery Wheel Center by Amount given below in Tables No. 1-2 and 3

Set Tooth Rest Below Work Holding Centers Amount given Below in Table No. 4

Size of Reamer	TABLE 1 Hand Reamer for Steel Cut'g Clearance Land .006 Wide		TABLE 2 Hand Reamer for Cast Iron and Bronze Cut'g Clearance Land .025 Wide		TABLE 3 Chucking Reamer for Cast Iron and Bronze Cut'g Clearance Land .025 Wide		TABLE 4 Chucking Reamers for Steel Circular Ground	
	For Cutting Clearance	For Second Clearance	For Cutting Clearance	For Second Clearance	For Cutting Clearance	For Second Clearance	Angle on End of Blade	For Cutting Clearance on Angle
1/2″	.012	.052	.032	.072	.040	.080	45 Degrees	.080
9/16″	.012	.057	.032	.072	.040	.080	45 "	.080
5/8″	.012	.062	.032	.072	.040	.090	45 "	.090
11/16″	.012	.067	.035	.095	.040	.100	45 "	.100
3/4″	.012	.072	.035	.095	.040	.100	45 "	.100
13/16″	.012	.077	.037	.095	.045	.125	45 "	.125
7/8″	.012	.082	.040	.120	.045	.125	45 "	.125
15/16″	.012	.087	.040	.120	.045	.125	45 "	.125
I	.012	.092	.040	.120	.045	.125	45 "	.125
1 1/16″	.012	.097	.040	.120	.045	.125	45 "	.125
1 1/8″	.012	.102	.040	.120	.045	.125	45 "	.125
1 3/16″	.012	.106	.042	.122	.045	.125	45 "	.125
1 1/4″	.012	.112	.045	.145	.050	.160	45 "	.160
1 5/16″	.012	.118	.045	.145	.050	.160	45 "	.160
1 3/8″	.012	.122	.045	.145	.050	.160	45 "	.175
1 7/16″	.012	.127	.045	.145	.055	.175	45 "	.175
1 1/2″	.012	.132	.048	.168	.055	.175	45 "	.175
1 9/16″	.012	.137	.050	.170	.055	.175	45 "	.175
1 5/8″	.012	.142	.050	.170	.060	.200	45 "	.200
1 11/16″	.012	.147	.050	.170	.060	.200	45 "	.200
1 3/4″	.012	.152	.052	.192	.060	.200	45 "	.200
1 13/16″	.012	.157	.052	.192	.060	.200	45 "	.200
1 7/8″	.012	.162	.056	.196	.060	.200	45 "	.200
1 15/16″	.012	.167	.056	.196	.064	.200	45 "	.200
2	.012	.172	.056	.216	.064	.224	45 "	.225
2 1/16″	.012	.172	.056	.216	.064	.224	45 "	.225
2 1/8″	.012	.172	.059	.219	.064	.224	45 "	.225
2 3/16″	.012	.172	.059	.219	.064	.224	45 "	.225
2 1/4″	.012	.172	.063	.223	.064	.224	45 "	.225
2 5/16″	.012	.172	.063	.223	.064	.224	45 "	.225
2 3/8″	.012	.172	.063	.223	.068	.228	45 "	.230
2 7/16″	.012	.172	.063	.223	.068	.228	45 "	.230
2 1/2″	.012	.172	.065	.225	.072	.232	45 "	.230
2 9/16″	.012	.172	.065	.225	.072	.232	45 "	.230
2 5/8″	.012	.172	.065	.225	.075	.235	45 "	.235
2 11/16″	.012	.172	.065	.225	.075	.235	45 "	.235
2 3/4″	.012	.172	.065	.225	.077	.237	45 "	.240

Mount Tooth Rest on Emery Wheel Head

Mount Tooth Rest on Table of Machine

	REAMER CLEARANCE Ground with Cup Wheel 3" dia.—Tooth Rest to be Set Central with Emery Wheel Spindle. Set Work holding Centers above Emery Wheel Center by Amount given below in Tables No. 1-2 and 3						Set Tooth Rest Below Work Holding Centers. Amount given Below in Table No. 4	
	TABLE 1 Hand Reamer for Steel Cut'g Clearance Land .006 Wide		TABLE 2 Hand Reamer for Cast Iron and Bronze Cut'g Clearance Land .025 Wide		TABLE 3 Chucking Reamer for Cast Iron and Bronze Cut'g Clearance Land .025 Wide		TABLE 4 Chucking Reamers for Steel Circular Ground	
Size of Reamer	For Cutting Clearance	For Second Clearance	For Cutting Clearance	For Second Clearance	For Cutting Clearance	For Second Clearance	Angle on End of Blade	For Cutting Clearance on Angle
2 1/8"	.012	.172	.065	.225	.077	.237	45 Degrees	.240
2 1/4"	.012	.172	.070	.230	.080	.240	45 "	.240
2 3/8"	.012	.172	.070	.230	.080	.240	45 "	.240
3	.012	.172	.072	.232	.080	.240	45 "	.240
3 1/16"	.012	.172	.072	.232	.080	.240	45 "	.240
3 1/8"	.012	.172	.075	.235	.083	.240	45 "	.240
3 3/16"	.012	.172	.075	.235	.083	.243	45 "	.240
3 1/4"	.012	.172	.078	.238	.083	.243	45 "	.245
3 5/16"	.012	.172	.078	.238	.087	.243	45 "	.245
3 3/8"	.012	.172	.081	.241	.087	.247	45 "	.245
3 7/16"	.012	.172	.081	.241	.090	.247	45 "	.245
3 1/2"	.012	.172	.084	.244	.090	.250	45 "	.250
3 9/16"	.012	.172	.084	.244	.090	.250	45 "	.250
3 5/8"	.012	.172	.087	.247	.093	.253	45 "	.250
3 11/16"	.012	.172	.087	.247	.093	.253	45 "	.250
3 3/4"	.012	.172	.090	.250	.097	.257	45 "	.255
3 7/8"	.012	.172	.090	.250	.097	.257	45 "	.255
3 7/8"	.012	.172	.093	.253	.100	.260	45 "	.255
3 15/16"	.012	.172	.093	.253	.100	.260	45 "	.255
4	.012	.172	.096	.256	.104	.264	45 "	.260
4 1/16"	.012	.172	.096	.256	.104	.264	45 "	.260
4 1/8"	.012	.172	.096	.256	.104	.264	45 "	.260
4 3/16"	.012	.172	.096	.256	.106	.266	45 "	.260
4 1/4"	.012	.172	.096	.256	.106	.266	45 "	.265
4 5/16"	.012	.172	.096	.256	.106	.266	45 "	.265
4 3/8"	.012	.172	.096	.256	.108	.268	45 "	.265
4 7/16"	.012	.172	.096	.256	.108	.268	45 "	.265
4 1/2"	.012	.172	.100	.260	.108	.268	45 "	.265
4 9/16"	.012	.172	.100	.260	.108	.268	45 "	.265
4 5/8"	.012	.172	.100	.260	.110	.270	45 "	.270
4 11/16"	.012	.172	.100	.260	.110	.270	45 "	.270
4 3/4"	.012	.172	.104	.264	.114	.274	45 "	.275
4 7/8"	.012	.172	.104	.264	.114	.274	45 "	.275
4 7/8"	.012	.172	.106	.266	.116	.276	45 "	.275
4 15/16"	.012	.172	.106	.266	.116	.276	45 "	.275
5	.012	.172	.110	.270	.118	.278	45 "	.275
5 1/8"	.012	.172	.118	278				
	Mount Tooth Rest on Emery Wheel Head						Mount Tooth Rest on Table of Machine	

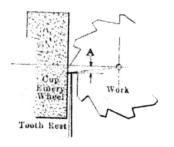

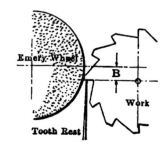

CUP WHEEL CLEARANCE TABLE	DISK WHEEL CLEARANCE TABLE

For setting tooth rest to obtain 5° or 7° clearance when grinding peripheral teeth of milling cutters with cup-shaped wheel. Tooth rest is set below work centers as at *A*, the distance being found in the table below.

Giving distance *B* for setting work centers and tooth rest below center of wheel spindle to obtain 5° or 7° clearance with wheels of different diameters when grinding with periphery of disk wheel.

Dia. Cutter Inches	For 5° Clearance A =	For 7° Clearance A =	Dia. of Emery Wheel Inches	For 5° Clearance B =	For 7° Clearance B =
$\frac{1}{4}$	.011	.015	2	.0937	.125
$\frac{3}{8}$	.015	.022	$2\frac{1}{4}$	.099	.141
$\frac{1}{2}$	.022	.030	$2\frac{1}{2}$	.110	.156
$\frac{5}{8}$	.028	.037	$2\frac{3}{4}$	.125	.172
$\frac{3}{4}$	.033	.045	3	.132	.187
$\frac{7}{8}$	.037	.052	$3\frac{1}{4}$	.143	.203
I	.044	.060	$3\frac{1}{2}$	.154	.219
$1\frac{1}{4}$	.055	.075	$3\frac{3}{4}$	.165	.234
$1\frac{1}{2}$	.066	.090	4	.176	.250
$1\frac{3}{4}$	.077	.105	$4\frac{1}{4}$	.187	.265
2	.088	.120	$4\frac{1}{2}$	.198	.281
$2\frac{1}{4}$	.099	.135	$4\frac{3}{4}$	.209	.297
$2\frac{1}{2}$	.110	.150	5	.220	.312
$2\frac{3}{4}$	.121	.165	$5\frac{1}{4}$	.231	.328
3	.132	.180	$5\frac{1}{2}$	.242	.344
$3\frac{1}{2}$	.154	.210	$5\frac{3}{4}$	.253	.359
4	.176	.240	6	.264	.375
$4\frac{1}{2}$	.198	.270	$6\frac{1}{4}$	.275	.390
5	.220	.300	$6\frac{1}{2}$	.286	.406
$5\frac{1}{2}$	.242	.330	$6\frac{3}{4}$	.297	.421
6	.264	.360	7	.308	.437

SCREW MACHINE TOOLS, SPEEDS
AND FEEDS

BOX TOOLS AND CUTTERS

THE general principles of two types of box tools using respectively
ngent and radial cutters are represented in Figs. 1 and 2. The
rmer type is generally used for roughing and the latter for finishing.
he tangent cutter in the type of box tool shown in Fig. 1 lies in a
ot formed parallel to the bottom of the box but at an angle, usually

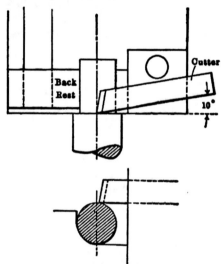

FIG. 1. — Roughing Box Tool with
Tangent Cutter

ten degrees, with the front of the box, thus giving the desired rake
at the cutting point. Finishing cutters of the type in Fig. 2 are
straight on the end, located square with the work and ordinarily
ground as indicated to give 7 to 10 degrees front clearance for steel
and 5 to 8 degrees for brass.

The tangent cutter is sharpened by grinding on the end, and com-
pensation for the grinding away of the metal is made by adjusting
the cutter forward, whereas in the radial type of cutter in Fig. 2,
frequent sharpening cannot be done without resulting in lowering
the cutting edge of the tool below the center of the work, unless a
substantial part of the tool be sacrificed. The radial tool, however,
is easily ground accurately on face a, which is the edge governing the
finish; while the corresponding face on the tangent tool is rather
difficult to grind so as to produce as smooth work.

169

The sizes of steel recommended for box-tool cutters are as follows: For box tools used for stock diameters up to $\frac{1}{16}$ inch, $\frac{3}{16}$ inch square; up to $\frac{1}{4}$ inch diameter, $\frac{7}{32}$ inch square; up to $\frac{1}{2}$ inch diameter, $\frac{1}{4}$ inch square; up to $\frac{3}{4}$ inch diameter, $\frac{5}{16}$ inch square; up to 1 inch diameter, $\frac{3}{8}$ inch square; up to $1\frac{1}{2}$ inches diameter, $\frac{1}{2}$ inch square.

HOLLOW MILLS

THE teeth of hollow mills should be radial or ahead of the center. With the cutting edge ahead of the center, as in Fig. 3, the chips so produced are caused to move outward away from the work and prevented from disfiguring it. With the cutting edge below the center, rough turning will result. With the cutting edge greatly above the center, chattering occurs. About one tenth of the cutting diameter is a good average amount to cut the teeth ahead of the center.

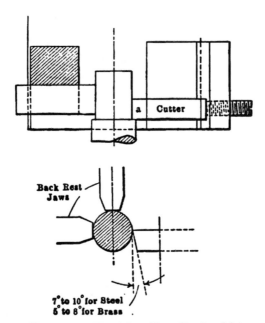

FIG. 2. — Finishing Box Tool with
Radial Cutter

When the chips produced from any turning or boring cut curl nicely, it is indicative of a free cutting action; but these chips are very troublesome on the automatic screw machine. In making hollow mills for the automatic, part or all of the rake to the cutting edge is generally sacrificed.

The table under the hollow mill in Fig. 3 gives proportions of mills from $\frac{1}{16}$ to $\frac{3}{4}$ diameter, showing the amount to cut the teeth ahead of the center, the taper of the hole, etc.

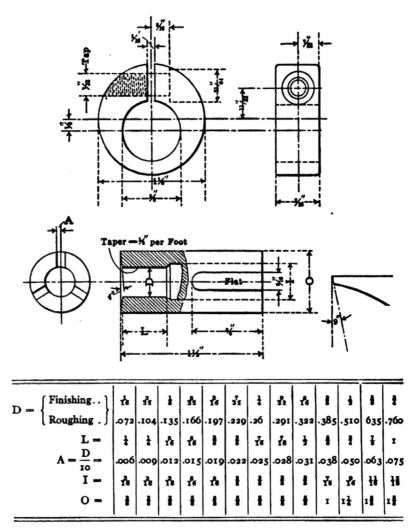

FIG. 3.—Hollow Mill Dimensions

D =	Finishing..	1/16	3/32	1/8	5/32	3/16	7/32	1/4	9/32	5/16	3/8	1/2	5/8	3/4
	Roughing .	.072	.104	.135	.166	.197	.229	.26	.291	.322	.385	.510	635	.760
	L =	1/4	1/4	5/16	5/16	3/8	3/8	7/16	7/16	1/2	5/8	3/4	7/8	1
A = $\frac{D}{10}$ =		.006	.009	.012	.015	.019	.022	.025	.028	.031	.038	.050	.063	.075
	I =	1/16	1/16	1/16	1/16	1/16	1/8	1/8	1/8	1/8	7/16	7/16	11/16	11/16
	O =	1/2	1/2	1/2	5/8	5/8	3/4	3/4	3/4	7/8	1	1 1/4	1 1/8	1 1/8

DIES AND TAPS

IT is good practice in making spring screw dies to either hob out the thread with a hob tap 0.005 to 0.015 inch over-size, according to size, and in use to spring the prongs to proper cutting size by a clamping ring, or to tap the die out from the rear with a hob tap tapering from $\frac{3}{16}$ inch to $\frac{1}{4}$ inch per foot, leaving the front end about 0.002 inch over cutting size, and in this case also to use a clamping ring. Both of these schemes are for the purpose of obtaining back clearance and are effective. Of the two the use of the taper hob is to be preferred.

Spring Die Sizes

The table of dimensions for spring screw dies, Fig. 4, should prove of service, particularly for steel. For brass the cutting edge is radial, thus eliminating dimension *A*. The width of land at bottom of thread is usually made about 1/4 outside diameter of cut, the milling between flutes being 70 degrees, leaving 50 degrees for the prong in the case of three-flute dies.

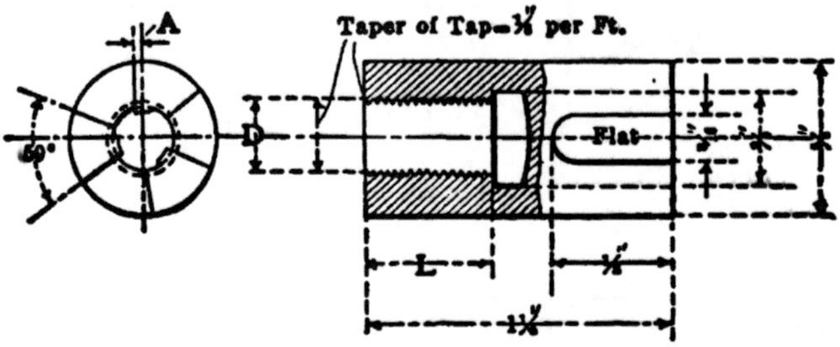

SMALL SIZES OF DIES
(Over all Dimensions Given in Sketch)

D =	1/16	1/8	3/16	1/4	5/16	No. 3	4	6	8	10	12
Threads P. I. =	64	40	32	20	18	56	40	32	24-32	24-32	24-32
$A = \frac{D}{10}$	.003	.012	.019	.025	.031	0.10	.011	.014	.016	0.19	.021
L =	3/16	9/32	3/8	1/2	1/2	7/32	1/4	9/32	5/16	11/32	3/8

SIZES 3/8 TO 1 INCH

D =	3/8 to 1/2	1/2 to 3/4	3/4 to 1
Th' s P.I. =	Std.	Std.	Std.
A =	D ÷ 10	D ÷ 10	D ÷ 10
L =	3/4	1"	1 1/2
O.S. Dia.	1"	1 1/8	1 5/8
Length	2"	2 1/4"	2 1/2"

FIG. 4.—Spring Die Dimensions

Sizing Work for Threading

In boring holes previously to tapping they should be somewhat larger than the theoretical diameter at bottom of thread, as the crowding action of the tap will cause the metal to flow some and compensate for this. Where no allowance is made, frequent tap breakage is liable to occur and torn threads in the work also. On external work it is for the same reasons advisable to turn the work undersize and the following table gives good average allowances for both internal and external work.

ALLOWANCES FOR THREADING IN THE SCREW MACHINE

Threads per Inch	External Work Turn Undersize	Internal Work Increase Over Theoretical Bottom of Thread
28	0.002	0.004
24	0.002	0.0045
22	0.0025	0.005
20	0.0025	0.0055
16	0.003	0.006
14	0.003	0.0065
13	0.0035	0.007
12	0.0035	0.007
11	0.0035	0.0075
10	0.004	0.008
9	0.004	0.0085
8	0.0045	0.009
7	0.0045	0.0095
6	0.005	0.010

Tap Length and Number of Lands

The number of teeth in taps and the width of land should be regulated by the diameter and pitch of work as well as the nature of the material being cut. On fine threads, where a drunken thread is to be insured against, more teeth are required than on a coarser pitch of the same diameter. A good average number of teeth on taps for United States standard threads is given in the following table. With

Outside Dia.	No. of Flutes	Width of Land
$\frac{3}{16}$	4	$\frac{3}{64}$
$\frac{1}{4}$	4	$\frac{1}{16}$
$\frac{5}{16}$	4	$\frac{5}{64}$
$\frac{3}{8}$	4	$\frac{3}{32}$
$\frac{7}{16}$	4	$\frac{7}{64}$
$\frac{1}{2}$	4	$\frac{1}{8}$
$\frac{9}{16}$	4	$\frac{5}{32}$
$\frac{5}{8}$	4	$\frac{3}{16}$
$\frac{3}{4}$	4	$\frac{7}{32}$
$\frac{7}{8}$	4	$\frac{1}{4}$
1	4	$\frac{1}{4}$
1 $\frac{1}{4}$	4	$\frac{5}{16}$

too few teeth and too short land very little support is afforded and this may cause chattering; too much land in contact causes heat due to excessive friction, welding of chips and torn threads.

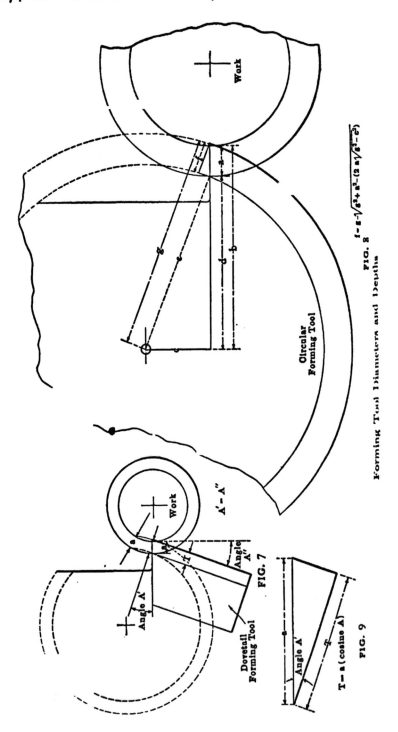

Forming Tool Diameters and Depths

FORMING TOOLS

THE two types of forming cutters commonly used in the screw machine are shown in Figs. 5 and 6. The circular forming cutter in Fig. 5 is usually cut away from ⅛ to 1/16 inch below center to give suitable cutting clearance and the center of the tool post on which it is mounted is a corresponding amount above the center of the machine, so that the cutting edge of the circular tool is brought on

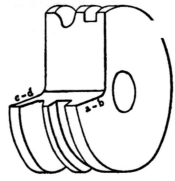

FIG. 5. — Circular Forming Tool

the center line of the work. The relative clearance ordinarily obtained by circular cutters and dovetail tools of the type shown in Fig. 6, is indicated in Fig. 7. It is obvious that with a given material the larger the diameter of the work the greater the angle of clearance required. Clearance angles are seldom less than 7 degrees or over 12 degrees.

The diameter of circular forming tools is an important matter for consideration. A small diameter has a more pronounced change of

FIG. 6. — Dovetail Forming Tool

clearance angle than a large diameter. In fact, when of an exceedingly large diameter the circular tool approaches in cutting action the dovetail type of tool which is usually provided with about 10 degrees clearance. Circular tools usually range from about 1¼ to 3 inches diameter, depending upon the size of machine in which they are used.

Getting the Tool Diameters at Different Points

In order to make a circular or a dovetail type of tool so that the contour of its cutting edge is such as to produce correct work, the amount a circular tool is cut below center, as at c in Fig. 8, and the clearance angle of a dovetail tool as at A', Fig. 7 must be known. Thus, referring to Fig. 8, the forming tool shown cuts two different diameters on the work, the step between being represented by dimension a. To find depth f to which the forming tool must be finished on the center line to give the correct depth of cut a in the work (the

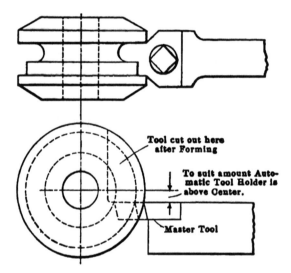

FIG. 10. — Finishing a Circular Tool

cutter being milled below center an amount represented by c) the following formula may be applied:

$$f = g - \sqrt{g^2 + a^2 - (2a \sqrt{g^2 - c^2})}.$$

Suppose the depth of cut in the work represented by a to be 0.157; the radius g of the forming cutter 1 inch; the distance c which the forming tool is milled below center, $\frac{3}{16}$ inch. Applying the above formula to find f and substituting the values just given for the letters in the formula we have $f = 1 - \sqrt{1 + .0231 - (.304 \sqrt{1 - .0351})}$

$$= 1 - \sqrt{1 + .0231 - (.304 \times .9823)}$$

$$= 1 - \sqrt{.724485} = 1 - .8512 = .1488$$

Then $f = .1488$

Dovetail Tool Depths

If a similar piece of work is to be formed with a dovetail type of cutter, the distance T, Figs. 7 and 9, to which it is necessary to plane the tool shoulder in order that it may cut depth a correctly in the work, is found by the formula: $T = a$ (cosine A'). As 10 degrees is the customary clearance on this form of tool, the cosine of this angle, which is .98481, may be considered a constant, making reference to a table of cosines unnecessary as a rule. Assuming the same depth for a as in the previous case, that is .152 inch, and multiplying by .98481, gives .1496 inch as the depth of T to which the tool must be planed.

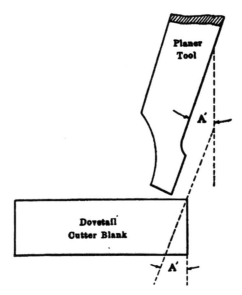

Planer Tool

A'

Dovetail Cutter Blank

A'

FIG. 11. — Finishing a Dovetail Forming Tool

While it frequently is necessary or advisable to determine by calculation the dimension computed in the preceding examples, in the majority of cases when making a cutter with a master tool of the same outline as the model, the correct form in the circular cutter is obtained automatically by dropping the master tool to the same distance below the lathe center as the circular cutter is to be milled off center and then feeding it in to finish the cutter. This procedure, shown in Fig. 10, assures the correct shape at all points being produced on the exact working plane of the cutter. Similarly, in finishing a dovetail cutter in the planer or shaper, the master tool may be set as in Fig. 11 at the same angle with the cutter (usually 10 degre͏ as the latter will afterward be applied to the work.

CIRCULAR TOOL FOR CONICAL POINTS

WHEN a circular cutter is to be made for forming a conical surface on a piece as in Fig. 12, a master tool of the exact angle required on the work may be used for finishing the cutter in the same way as the tool in Fig. 9 is applied; that is, the master is to be dropped below center the amount the cutter center is to be above the work center when in operation. The distance is represented by A in Fig. 12. Another method, which avoids the necessity of making a master tool, is to set the compound rest of the lathe to the exact angle required (in this case 30 degrees with the center line) and with a horizontal

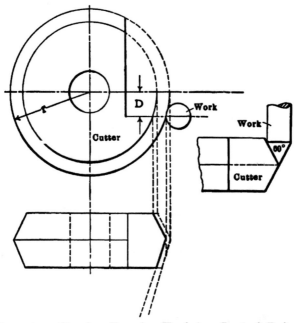

FIG. 12. — Circular Forming Tool for Conical Points

cutting tool set at distance D below center, turn off one side of the cutter blank and then set the compound rest around the other way and face off the other side. If desired a similar method may be followed for grinding the forming cutter after hardening. The arbor carrying the cutter should be located either above or below the grinding wheel a distance equal to $D \dfrac{(R + r)}{r}$, where D equals the depth the cutter is milled below center, r the radius of the cutter, and R the radius of the emery wheel. Assuming D to be .187 ($\frac{3}{16}$) inch; R, 2.5 inches; and r, 1 inch, the vertical distance between centers of forming tool and grinding wheel centers would equal .187 $\dfrac{(1 + 2.5)}{1}$

= .187 (3.5) = .6562 ($\frac{21}{32}$) inch.

TABLE 1. CUTTING SPEEDS AND FEEDS FOR SCREW STOCK

1/32 Inch Chip				1/16 Inch Chip				1/8 Inch Chip				1/4 Inch Chip			
Dia. of Stock	Feet Surface Speed	Rev. per Min.	Feed per Rev.	Dia. of Stock	Feet Surface Speed	Rev. per Min.	Feed per Rev.	Dia. of Stock	Feet Surface Speed	Rev. per Min.	Feed per Rev.	Dia. of Stock	Feet Surface Speed	Rev. per Min.	Feed per Rev.
1/8	80	2445	.002	1/4	60	916	.0035	3/8	55	560	.004	3/4	50	254	.004
3/16	70	1426	.003	3/8	60	611	.004	1/2	55	420	.005	1	50	191	.005
1/4	70	1069	.004	1/2	60	458	.005	3/4	55	280	.006	1 1/4	45	137	.005
3/8	70	713	.005	3/4	55	280	.006	1	50	191	.007	1 1/2	45	114	.006
1/2	60	458	.006	1	55	210	.007	1 1/4	50	152	.007	1 3/4	45	98	.006
3/4	60	305	.007	1 1/4	55	168	.007	1 1/2	45	114	.007	2	40	76	.006
1	60	229	.008	1 1/2	50	127	.008	1 3/4	45	98	.007	2 1/4	40	68	.007
1 1/4	60	183	.008	1 3/4	50	109	.008	2	40	76	.008	2 1/2	40	61	.007

TABLE 2. CUTTING SPEEDS AND FEEDS FOR BRASS

1/32 Inch Chip				1/16 Inch Chip				1/8 Inch Chip				1/4 Inch Chip			
Dia. of Stock	Feet Surface Speed	Rev. per Min.	Feed per Rev.	Dia. of Stock	Feet Surface Speed	Rev. per Min.	Feed per Rev.	Dia. of Stock	Feet Surface Speed	Rev. per Min.	Feed per Rev.	Dia. of Stock	Feet Surface Speed	Rev. per Min.	Feed per Rev.
1/8	180	5500	.003	1/4	180	2748	.004	3/8	165	1680	.004	3/4	150	762	.005
3/16	180	3668	.004	3/8	180	1833	.005	1/2	165	1260	.006	1	150	573	.006
1/4	180	2748	.005	1/2	180	1374	.0065	3/4	165	840	.007	1 1/4	135	411	.007
3/8	180	1833	.006	3/4	165	840	.0075	1	150	573	.008	1 1/2	135	342	.008
1/2	180	1374	.008	1	165	630	.0085	1 1/4	150	456	.009	1 3/4	135	294	.008
3/4	180	915	.010	1 1/4	165	504	.010	1 1/2	135	342	.010	2	120	228	.009
1	180	687	.011	1 1/2	150	381	.012	1 3/4	135	294	.010	2 1/4	120	204	.009
1 1/4	180	549	.012	1 3/4	150	327	.012	2	120	228	.011	2 1/2	120	183	.010

SPEEDS AND FEEDS FOR SCREW MACHINE WORK

THE accompanying tables of speeds and feeds for different types of tools used on materials commonly worked in the automatic screw machine have been compiled from data accumulated and thoroughly tested during extended experience in this class of work.

It is, of course, impossible, where a series of tools is used on an automatic machine, to select speeds theoretically correct for every tool carried by the turret and cross slide. A compromise is necessary and therefore speeds are selected which will fall within the range suitable for the different tools.

Speeds and Feeds for Turning

Tables 1 and 2, page 179, cover turning speeds and feeds for bright-drawn stock (screw stock) and brass, with various depths of chip (that is, stock removed on a side) from $\frac{1}{32}$ up to $\frac{1}{4}$ inch. These feeds and speeds and depths of cut are figured more especially for such tools

TABLE 3. SPEEDS AND FEEDS FOR FINISH BOX TOOL

Finished Diameter of Work	SCREW STOCK			BRASS ROD			Amount advisable to remove on a side
	Feet Surface Speed	Rev. per Min.	Feed per Rev.	Feet Surface Speed	Rev. per Min.	Feed per Rev.	
$\frac{1}{8}$	80	2445	.0045	180	5500	.0045	.0025
$\frac{3}{16}$	70	1426	.0055	180	3668	.0055	.0025
$\frac{1}{4}$	65	993	.0075	180	2750	.0075	.0045 .
$\frac{1}{2}$	60	458	.011	180	1375	.011	.006
$\frac{3}{4}$	60	305	.012	180	917	.012	.006
1	60	229	.012	175	668	.012	.0065
$1\frac{1}{2}$	55	140	.014	170	433	.014	.007
2	50	95	.014	170	325	.014.	.008

as roughing boxes where the cut, though frequently heavy, is taken by a single cutting tool. For a $\frac{3}{16}$-inch chip the feeds for various diameters of stock are practically midway between those tabulated for $\frac{1}{8}$-and $\frac{1}{4}$-inch chips. The feed per revolution for $\frac{1}{4}$ chip taken on diameters $1\frac{1}{4}$ inch and larger is the same as given for $\frac{1}{4}$ inch chip, the speed also being the same for corresponding diameters. Where hollow mills are used on steel and the work is divided among three or more cutting edges the feed per revolution for a given depth of chip is about 25 per cent coarser than given for box tools; with both classes of tools the feeds are, of course, increased as the diameter of the stock increases, the peripheral speeds being reduced as the feeds grow coarser. The speeds and feeds for finishing box tools as used on screw stock and brass are given in Table 3, the last column indicating the amount of stock which, generally speaking, it is advisable to remove in order to produce a good surface.

Forming-tool Speeds and Feeds

Speeds and feeds for forming tools are given in Tables 4 and 5. It will be seen that after a work diameter of about 1/4 inch has been reached, a tool about 1/8-inch wide is adapted to take the coarsest feed, tools from this width up to approximately $\frac{3}{16}$ (such as commonly employed for cutting-off purposes) admitting of heavier crowding as a rule than either the narrower or wider tools.

TABLE 4. SPEEDS FOR FORMING

Dia. of Work	SCREW STOCK		BRASS ROD		Dia. of Work	SCREW STOCK		BRASS ROD	
	Feet Surface Speed	Rev. per Min.	Feet Surface Speed	Rev. per Mid.		Feet Surface Speed	Rev. per Min.	Feet Surface Speed	Rev. per Min.
1/8	75	2292	200	6112	5/8	60	360	175	1050
3/16	75	1528	200	4074	3/4	60	305	175	882
1/4	70	1069	185	2827	1	60	229	175	667
3/8	65	662	185	1885	1½	60	153	170	432
1/2	65	497	185	1414	2	50	96	170	324

TABLE 5. FEEDS FOR FORMING TOOLS

Width of Form	SMALLEST DIAMETER OF FORM							
	1/16	1/8	3/16	1/4	3/8	1/2	3/4	1½
1/16	.0007	.0008	.001	.0012	.0012	.0012	.0012	.0012
1/8	.0005	.0008	.001	.0012	.0015	.0020	.0025	.0025
3/16		.0007	.001	.001	.0015	.0015	.0018	.0018
1/4			.0009	.001	.001	.0012	.0015	.0015
3/8			.0008	.0009	.001	.001	.0015	.0015
1/2				.0008	.0009	.001	.0011	.0012
3/4					.0008	.0009	.001	.0012
1					.0007	.0009	.001	.0012
1½						.0007	.0009	.0011
2							.0007	.001

Drilling Speeds and Feeds

Drilling speeds and feeds are given in Table 6. While these speeds are based on much higher peripheral velocities than drill-makers as a rule recommend for general purposes, it should be noted that conditions for drilling in the automatic, on the ordinary run of work, are usually ideal so far as lubrication, steadiness of feed, etc., are concerned, and it is possible where the holes drilled are comparatively shallow and the drill has ample opportunity for cooling during operation of the other tools, to maintain speeds that would be considered too high to be attempted in general shop practice.

TABLE 6. DRILLING FEEDS AND SPEEDS

Dia. of Drill	SCREW STOCK		BRASS ROD		Dia. of Drill	SCREW STOCK		BRASS ROD	
	Feed per Drill	R.P.M. at 60 Ft. Peripheral Speed	Feed per Rev.	R.P.M. at 175 Ft. Peripheral Speed		Feed per Rev.	R.P.M. at 55 Ft. Peripheral Speed	Feed per Rev.	R.P.M. at 165 Ft. Peripheral Speed
$\frac{1}{16}$	.0013	3667	.0017	10696	$\frac{1}{2}$	.005	420	.0065	1260
$\frac{5}{64}$	.0016	2093	.002	8555	$\frac{9}{16}$	.0057	373	.0074	1120
$\frac{3}{32}$	.0018	2445	.0023	7130	$\frac{5}{8}$	.0059	336	.0077	1008
$\frac{1}{8}$	.0025	1833	.0033	5348	$\frac{11}{16}$	.006	305	.0078	917
$\frac{5}{32}$	.003	1421	.0039	4144	$\frac{3}{4}$	.0065	280	.0084	84
$\frac{3}{16}$	.004	1222	.0052	3565	$\frac{7}{8}$	.0075	240	.0097	702
$\frac{7}{32}$	.004	1048	.0052	3050			50 Ft.		150 Ft.
$\frac{1}{4}$	.0045	916	.0058	2674	1	.0085	191	.0110	573
$\frac{9}{32}$	.0045	815	.0058	2377	$1\frac{1}{4}$	.0095	152	.0123	458
$\frac{5}{16}$	.0045	733	.0058	2139	$1\frac{1}{2}$	.011	127	.0143	382
$\frac{3}{8}$	.0045	611	.0061	1783	$1\frac{3}{4}$	.013	109	.0169	327
$\frac{7}{16}$	.005	524	.0065	1528	2	.014	96	.0182	294

Speeds and Feeds for Reaming

Table 7 is made up of speed and feed data for reamers. In this table the feed for different classes of material has been considered as constant for any given diameter of reamer, although it is probable that with certain materials, especially on brass alloys, etc., the feed per revolution might be increased somewhat, to advantage, over the rates given. These feeds have been tabulated, however, as presenting highly satisfactory practice in reaming the materials ted.

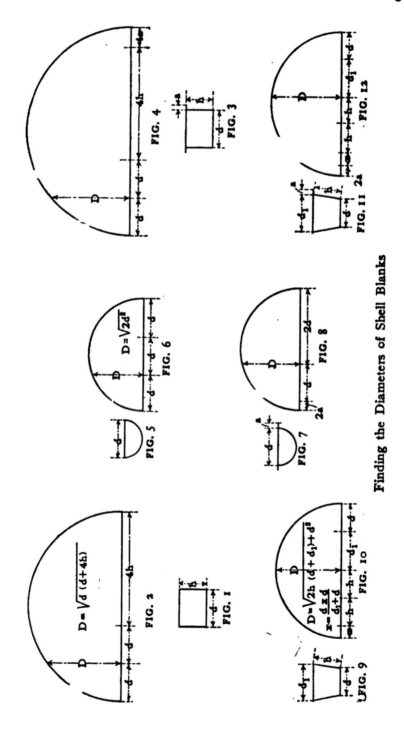

Finding the Diameters of Shell Blanks

PUNCH PRESS TOOLS

METHOD OF FINDING THE DIAMETERS OF SHELL BLANKS

THIS method for the finding of diameters of shell blanks, applies also to some other shapes which frequently occur in practice.

The method is based upon the surface of the shell in comparison with the area of the blank and should therefore be used only when light material is to be considered. In case of the flanged shapes the width of the flange should be small in proportion to the diameter.

Fig. 1 shows a cylindrical shell of the diameter d and the depth h. To find the diameter of the blank, lay down the diameter d of the shell twice on a horizontal line, Fig. 2, add to this a distance equal to four times the depth h of the shell and describe a semicircle of which the total distance is the diameter. The vertical line D from the intersecting point with the circle to the horizontal line gives the desired blank diameter. Line D is to be drawn at a distance d from the end of the horizontal.

If the shell has a flange as in Fig. 3, add four times the width of this flange to the horizontal line and proceed as above; see Fig. 4.

In the case of a hemisphere, Fig. 5, lay down the diameter three times on the horizontal line and draw the vertical line at the distance d from the end, as in Fig. 6.

If the hemisphere has a flange as in Fig. 7, add a distance equal to twice the width of the flange to the horizontal line, as in Fig. 8. In any case, the length of the vertical line D gives the desired diameter of blank.

If a shell with tapering sides, Fig. 9, has to be drawn, multiply first the bottom diameter by itself and divide the product by the sum of the two diameters d_1 and d in order to obtain the length x. Otherwise proceed as shown in Fig. 10.

If the taper shell has a flange of the width a, Fig. 11, add to the base line of the diagram twice this width, as shown in Fig. 12.

PUNCH AND DIE ALLOWANCE FOR ACCURATE WORK

IN the blanking, perforating and forming of flat stock in the power press for parts of adding machines, typewriters, etc., it is generally desired to make two different kinds of cuts with the dies used. First, leave the outside of the blank of a semi-smooth finish, with sharp

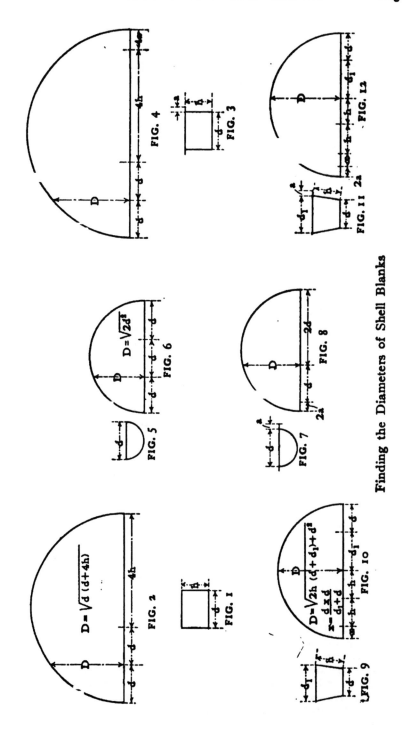

Finding the Diameters of Shell Blanks

corners, free from burrs, and with the least amount of rounding on the cutting side. Second, to leave the holes and slots that are perforated in the parts as smooth and straight as possible, and true to size. The table given is the result of considerable experimenting on this class of work, and has stood the test of years of use since it was compiled.

The die always governs the size of the work passing through it. The punch governs the size of the work that it passes through. In

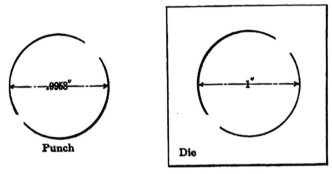

FIG. 13. — Blanking Tools

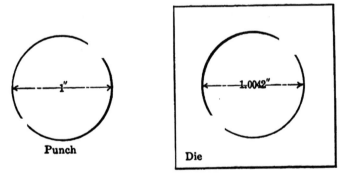

FIG. 14. — Perforating Tools

blanking work the die is made to the size of the work wanted and the punch smaller. In perforating work the punch is made to the size of the work wanted and the die larger than the punch. The clearance between the die and punch governs the results obtained.

Figs. 13 and 14 show the application of the table in determining the clearance for blanking or perforating hard rolled steel .060 inch thick. The clearance given in the table for this thickness of metal is .0042, and Fig. 13 shows that for blanking to exactly 1 inch diameter this amount is deducted from the diameter of the punch, while for perforating the same amount is added, as in Fig. 14, to the diameter of the die.   For a sliding fit make punch and die .00025 to .0005 inch larger; and for a driving fit make punch and die .0005 to .0015 inch smaller.

TABLE OF ALLOWANCES FOR PUNCH AND DIE FOR DIFFERENT THICKNESS AND MATERIALS

Thickness of Stock Inch	Clearance for Brass and Soft Steel Inch	Clearance for Medium Rolled Steel Inch	Clearance for Hard Rolled Steel Inch
.010	.0005	.0006	.0007
.020	.001	.0012	.0014
.030	.0015	.0018	.0021
.040	.002	.0024	.0028
.050	.0025	.003	.0035
.060	.003	.0036	.0042
.070	.0035	.0042	.0049
.080	.004	.0048	.0056
.090	.0045	.0054	.0063
.100	.005	.006	.007
.110	.0055	.0066	.0077
.120	.006	.0072	.0084
.130	.0065	.0078	.0091
.140	.007	.0084	.0098
.150	.0075	.009	.0105
.160	.008	.0096	.0112
.170	.0085	.0102	.0119
.180	.009	.0108	.0126
.190	.0095	.0114	.0133
.200	.010	.012	.014

CLEARANCE FOR PUNCHES AND DIES FOR BOILER WORK

THE practice of the Baldwin Locomotive Works on sizes up to $1\frac{1}{4}$ inches is to make the punch $\frac{1}{64}$ inch below nominal size and the die $\frac{1}{64}$ inch above size, which gives $\frac{1}{32}$ inch clearance.   Above $1\frac{1}{4}$ inches the punches are made to nominal size and the dies $\frac{1}{32}$ inch large, which allows the same clearance as before.   The taper on dies below $1\frac{1}{4}$ inches is 1 inch in 12; on sizes above $1\frac{1}{4}$ inches it is half this or $\frac{1}{2}$ inch in 12 inches.

LUBRICANTS FOR PRESS TOOLS

ALTHOUGH there are some shops in which no lubricant is used when working sheet metal, and where good results are obtained, it is best to use a lubricant on all classes of sheet-metal work.

For all cutting dies on brass and steel a heavy animal oil is best. Pure lard oil is very satisfactory, although expensive.

When punching copper, or German silver, a thin coating of lard oil or sperm oil should be spread over the sheets or strips before punching. A good way to do this evenly is to coat one sheet thickly and then feed it through a pair of rolls, after which a number of other sheets may be run through the rolls and thus coated evenly. For drawn work this method of coating the sheets from which the shells are to be drawn will be found to be the best, as the coating of oil on the stock will be very thin and it will not be found necessary to clean the shells afterward, the oil having disappeared during the blanking and drawing process. When oil is applied with a pad or brush the coating will be so thick that it will be necessary to clean the article produced.

In drawing steel shells a mixture of equal parts of oil and black lead is very useful and while it may be used warm it does not affect the work as much as the speed of the drawing press does; the thicker the stock the slower must be the speed of the punch. A heavy grease with a small proportion of white lead mixed in with it is also recommended for this purpose.

If the drawing die is very smooth and hard at the corner of the "draw," or edge of the die, the liability of clogging will be reduced to a minimum. Often it will help to give to the die a lateral polish by taking a strip of emery cloth and changing the grain of the polish from circular to the same direction as the drawing.

For drawing brass or copper a clean soap water is considered most satisfactory. One of the largest brass firms in this country uses a preparation made by putting 15 pounds of Fuller's soap in a barrel of hot water, and boiling until all the lumps are dissolved. This is used as hot as possible. If the work is allowed to lie in the water until a slime has formed on the shell it will draw all the better. A soap that is strong in resin or potash will not give good results.

In drawing zinc the water should be hot, or the percentage of broken shells will be large.

Aluminum is an easy metal to draw, but it hardens up very quickly. For lubricants lard oil, melted Russian tallow and vaseline are all good. The lubricant should be applied to both sides of the metal.

U. S. STANDARD BOLTS AND NUTS

THE U. S. Standard for bolts, nuts, etc., called also Sellers' Standard, Franklin Institute Standard, and American Standard, was recommended in 1864 by the Franklin Institute for general adoption by engineers. The distance between parallel sides of the bolt head and nut for a rough bolt is one and one-half diameters of the bolt plus one-eighth of an inch. The thickness of the head in this system for a rough bolt is equal to one-half the distance between its parallel sides. The thickness of the nut is equal to the diameter of the bolt. The thickness of the head for a finished bolt is equal to the thickness of the nut. The distance between the parallel sides of a bolt head and nut and the thickness of the nut is one sixteenth of an inch less for finished work than for rough.

STRENGTH OF U. S. STANDARD BOLTS FROM ¼ TO 3″ DIAMETER

Bolt		Areas		Tensile Strength			Shearing Strength			
							Full Bolt		Bottom of Thread	
Diameter of Bolt	No. of Threads per Inch	Full Bolt	Bottom of Thread	At 10,000 lbs. Sq. In.	At 12,500 lbs. Sq. In.	At 17,500 lbs. Sq. In.	At 7,500 lbs. per Sq. In.	At 10,000 lbs. per Sq. In.	At 7,500 lbs. per Sq. In.	At 10,000 lbs. per Sq. In.
¼	20	.049	.027	270	340	470	380	490	200	270
5/16	18	.077	.045	450	570	790	580	770	340	450
⅜	16	.110	.068	680	850	1,190	830	1,100	510	680
7/16	14	.150	.093	930	1,170	1,630	1,130	1,500	700	930
½	13	.196	.126	1,260	1,570	2,200	1,470	1,960	940	1,260
9/16	12	.248	.162	1,620	2,030	2,840	1,860	2,480	1,220	1,620
⅝	11	.307	.202	2,020	2,520	3,530	2,300	3,070	1,510	2,020
¾	10	.442	.302	3,020	3,770	5,290	3,310	4,420	2,270	3,020
⅞	9	.601	.419	4,190	5,240	7,340	4,510	6,010	3,150	4,190
1	8	.785	.551	5,510	6,890	9,640	5,890	7,850	4,130	5,510
1⅛	7	.994	.693	6,930	8,660	12,130	7,450	9,940	5,200	6,930
1¼	7	1.227	.890	8,890	11,120	15,570	9,200	12,270	6,670	8,900
1⅜	6	1.485	1.054	10,540	13,180	18,450	11,140	14,850	7,910	10,540
1½	6	1.767	1.294	12,940	16,170	22,640	13,250	17,670	9,700	12,940
1⅝	5½	2.074	1.515	15,150	18,940	26,510	15,550	20,740	11,360	15,150
1¾	5	2.405	1.745	17,450	21,800	30,520	18,040	24,050	13,080	17,440
1⅞	5	2.761	2.049	20,490	25,610	35,860	20,710	27,610	15,370	20,490
2	4½	3.142	2.300	23,000	28,750	40,250	23,560	31,420	17,250	23,000
2¼	4½	3.976	3.021	30,210	37,770	52,870	29,820	39,760	22,660	30,210
2½	4	4.909	3.716	37,160	46,450	65,040	36,820	49,090	27,870	37,160
2¾	4	5.940	4.620	46,200	57,750	80,840	44,580	59,400	34,650	46,200
3	3½	7.069	5.428	54,280	67,850	94,990	53,020	70,690	40,710	54,280

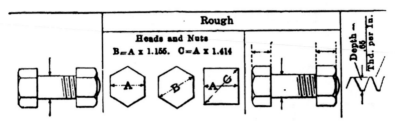

U. S. STANDARD BOLTS AND NUTS
ROUGH

Dia. of Bolt	Threads per Inch	Across Flats	Across Corners		THICKNESS		Depth of Thread
					Head	Nut	
1/4	20	1/2	37/64	25/32	1/4	1/4	.0325
5/16	18	19/32	11/16	27/32	19/64	5/16	.0361
3/8	16	11/16	51/64	31/32	11/64	3/8	.0406
7/16	14	25/32	29/32	1 7/64	21/64	7/16	.0464
1/2	13	7/8	1 1/64	1 1/4	7/16	1/2	.0500
9/16	12	31/32	1 1/8	1 3/8	31/64	9/16	.0542
5/8	11	1 1/16	1 15/64	1 1/2	17/32	5/8	.0590
3/4	10	1 1/4	1 29/64	1 25/32	5/8	3/4	.0650
7/8	9	1 7/16	1 43/64	2 1/32	23/32	7/8	.0722
1	8	1 5/8	1 7/8	2 19/64	13/16	1	.0812
1 1/8	7	1 13/16	2 3/32	2 9/16	29/32	1 1/8	.0928
1 1/4	7	2	2 5/16	2 41/64	1	1 1/4	.0928
1 3/8	6	2 3/16	2 17/32	3 3/32	1 3/32	1 3/8	.1083
1 1/2	6	2 3/8	2 3/4	3 25/64	1 11/64	1 1/2	.1083
1 3/4	5	2 3/4	3 1/8	3 57/64	1 3/8	1 3/4	.1300
2	4 1/2	3 1/8	3 39/64	4 27/64	1 9/16	2	.1444
2 1/4	4 1/2	3 1/2	4 3/64	4 31/64	1 3/4	2 1/4	.1444
2 1/2	4	3 7/8	4 31/64	5 21/64	1 15/16	2 1/2	.1625
2 3/4	4	4 1/4	4 49/64	6 1/64	2 1/8	2 3/4	.1625
3	3 1/2	4 5/8	5 11/32	6 17/32	2 5/16	3	.1857

NOTE. — U. S. Government Standard Bolts and Nuts are made to above U. S. or Sellers' Standard Rough Dimensions. The sizes of finished bolt heads and nuts are the same as the sizes of the rough ones, that is for finished work the forgings must be larger than for rough, thus the same wrench may be used on both black and finished heads and nuts.

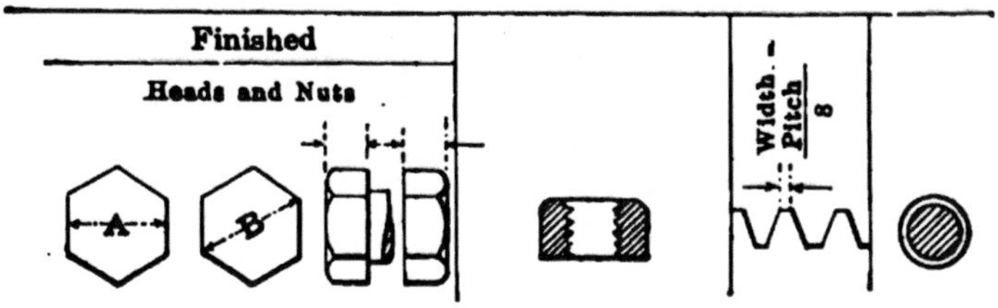

U. S Standard Bolts and Nuts

FINISHED HEADS AND NUTS

Dia. of Bolt	Across Flats	Across Corners	Thickness	Exact Size of Hole	Tap Drill Used	Width of Flat	Area at Root of Thread	Safe Strain in lbs. Iron at 50,000 lbs. per Sq. In. Factor of Safety = 5
$\frac{1}{4}$	$\frac{7}{16}$	$\frac{1}{2}$	$\frac{3}{16}$	.185	.191	.0063	.0260	260
$\frac{5}{16}$	$\frac{17}{32}$	$\frac{39}{64}$	$\frac{1}{4}$	.2408	.246	.0069	.0452	452
$\frac{3}{8}$	$\frac{5}{8}$	$\frac{23}{32}$	$\frac{5}{16}$	.2938	$\frac{19}{64}$	.0078	.0677	677
$\frac{7}{16}$	$\frac{23}{32}$	$\frac{53}{64}$	$\frac{3}{8}$	.3447	$\frac{23}{64}$	.0089	.0932	932
$\frac{1}{2}$	$\frac{13}{16}$	$\frac{15}{16}$	$\frac{7}{16}$	.4001	$\frac{13}{32}$	.0096	.1257	1257
$\frac{9}{16}$	$\frac{29}{32}$	$1\frac{1}{64}$	$\frac{1}{2}$	.4542	$\frac{15}{32}$	.0104	.1620	1620
$\frac{5}{8}$	1	$1\frac{5}{32}$	$\frac{9}{16}$	.5069	$\frac{33}{64}$	.0114	.2018	2018
$\frac{3}{4}$	$1\frac{3}{16}$	$1\frac{21}{64}$	$\frac{11}{16}$	.6201	$\frac{5}{8}$	.0124	.3020	3020
$\frac{7}{8}$	$1\frac{3}{8}$	$1\frac{19}{32}$	$\frac{13}{16}$	.7307	$\frac{47}{64}$	.0139	.4194	4194
1	$1\frac{9}{16}$	$1\frac{13}{16}$	$\frac{15}{16}$	.8376	$\frac{27}{32}$	.0156	.5509	5509
$1\frac{1}{8}$	$1\frac{3}{4}$	$2\frac{1}{64}$	$1\frac{1}{16}$	.9394	$\frac{61}{64}$	.0179	.6930	6930
$1\frac{1}{4}$	$1\frac{15}{16}$	$2\frac{15}{64}$	$1\frac{3}{16}$	1.0644	$1\frac{5}{64}$	.0179	.8890	8890
$1\frac{3}{8}$	$2\frac{1}{8}$	$2\frac{29}{64}$	$1\frac{5}{16}$	1.1585	$1\frac{11}{64}$	.0208	1.054	10540
$1\frac{1}{2}$	$2\frac{5}{16}$	$2\frac{43}{64}$	$1\frac{7}{16}$	1.2835	$1\frac{19}{64}$	.0208	1.293	12930
$1\frac{3}{4}$	$2\frac{11}{16}$	$3\frac{7}{64}$	$1\frac{11}{16}$	1.4902	$1\frac{33}{64}$	.0250	1.744	17440
2	$3\frac{1}{16}$	$3\frac{17}{32}$	$1\frac{15}{16}$	1.7113	$1\frac{23}{32}$	.0278	2.3	23000
$2\frac{1}{4}$	$3\frac{7}{16}$	$3\frac{31}{32}$	$2\frac{3}{16}$	1.9613	$1\frac{31}{32}$	.0278	3.021	30210
$2\frac{1}{2}$	$3\frac{13}{16}$	$4\frac{13}{32}$	$2\frac{7}{16}$	2.1752	$2\frac{3}{16}$	.0313	3.714	37140
$2\frac{3}{4}$	$4\frac{3}{16}$	$4\frac{27}{32}$	$2\frac{11}{16}$	2.4252	$2\frac{7}{16}$	.0313	4.618	46180
3	$4\frac{9}{16}$	$5\frac{9}{32}$	$2\frac{15}{16}$	2.6288	$2\frac{41}{64}$	.0357	5.427	54270

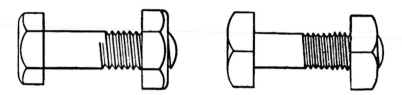

MACHINE BOLTS WITH MANUFACTURERS STD. HEADS.

Dia. of Bolt	No. of Threads per Inch	Hex. and Square Heads				Hex. and Square Nuts			
		Across Flats Hex. and Sq.	Across Corners Hex. Head	Across Corners Square Head	Thickness Hex. and Sq.	Across Flats Hex. and Sq.	Across Corners Hex. Nut	Across Corners Square Nut	Thickness Hex. and Sq.
$\frac{1}{4}$	20	$\frac{3}{8}$	$\frac{7}{16}$	$\frac{17}{32}$	$\frac{3}{16}$	$\frac{7}{16}$	$\frac{1}{2}$	$\frac{5}{8}$	$\frac{3}{16}$
$1\frac{5}{16}$	18	$\frac{1}{2}$	$\frac{9}{16}$	$\frac{11}{16}$	$\frac{1}{4}$	$\frac{1}{2}$	$\frac{9}{16}$	$\frac{3}{4}$	$\frac{1}{4}$
$\frac{3}{8}$	16	$\frac{9}{16}$	$\frac{21}{32}$	$\frac{51}{64}$	$\frac{5}{16}$	$\frac{5}{8}$	$\frac{23}{32}$	$\frac{57}{64}$	$1\frac{5}{16}$
$\frac{7}{16}$	14	$\frac{21}{32}$	$\frac{3}{4}$	$1\frac{5}{8}$	$\frac{3}{8}$	$\frac{23}{32}$	$\frac{53}{64}$	1	$\frac{3}{8}$
$\frac{1}{2}$	13	$\frac{3}{4}$	$\frac{55}{64}$	$1\frac{1}{16}$	$\frac{7}{16}$	$\frac{13}{16}$	$\frac{15}{16}$	$1\frac{5}{32}$	$\frac{7}{16}$
$\frac{9}{16}$	12	$\frac{27}{32}$	$\frac{31}{32}$	$1\frac{3}{16}$	$\frac{1}{2}$	$\frac{29}{32}$	$1\frac{1}{64}$	$1\frac{9}{32}$	$\frac{1}{2}$
$\frac{5}{8}$	11	$\frac{15}{16}$	$1\frac{5}{64}$	$1\frac{21}{64}$	$\frac{17}{32}$	1	$1\frac{5}{32}$	$1\frac{27}{64}$	$\frac{9}{16}$
$\frac{3}{4}$	10	$1\frac{1}{8}$	$1\frac{19}{64}$	$1\frac{19}{32}$	$\frac{5}{8}$	$1\frac{1}{16}$	$1\frac{21}{64}$	$1\frac{31}{32}$	$\frac{11}{16}$
$\frac{7}{8}$	9	$1\frac{5}{16}$	$1\frac{33}{64}$	$1\frac{55}{64}$	$\frac{3}{4}$	$1\frac{3}{8}$	$1\frac{19}{32}$	$1\frac{13}{16}$	$\frac{13}{16}$
1	8	$1\frac{1}{2}$	$1\frac{23}{32}$	$2\frac{1}{8}$	$\frac{7}{8}$	$1\frac{9}{16}$	$1\frac{13}{16}$	$2\frac{3}{16}$	$\frac{15}{16}$
$1\frac{1}{8}$	7	$1\frac{11}{16}$	$1\frac{61}{64}$	$2\frac{21}{64}$	1	$1\frac{11}{16}$	$2\frac{3}{32}$	$2\frac{9}{16}$	$1\frac{1}{8}$
$1\frac{1}{4}$	7	$1\frac{7}{8}$	$2\frac{11}{64}$	$2\frac{41}{64}$	$1\frac{1}{8}$	2	$2\frac{5}{16}$	$2\frac{53}{64}$	$1\frac{1}{4}$
$1\frac{3}{8}$	6	$2\frac{1}{16}$	$2\frac{3}{8}$	$2\frac{19}{32}$	$1\frac{1}{4}$	$2\frac{3}{16}$	$2\frac{17}{32}$	$3\frac{3}{32}$	$1\frac{3}{8}$
$1\frac{1}{2}$	6	$2\frac{1}{4}$	$2\frac{19}{32}$	$3\frac{3}{16}$	$1\frac{3}{8}$	$2\frac{3}{8}$	$2\frac{3}{4}$	$3\frac{21}{64}$	$1\frac{1}{2}$
$1\frac{5}{8}$	$5\frac{1}{2}$	$2\frac{7}{16}$	$2\frac{13}{16}$	$3\frac{7}{16}$	$1\frac{1}{2}$	$2\frac{9}{16}$	$2\frac{31}{32}$	$3\frac{5}{8}$	$1\frac{5}{8}$
$1\frac{3}{4}$	5	$2\frac{5}{8}$	$3\frac{1}{32}$	$3\frac{23}{32}$	$1\frac{5}{8}$	$2\frac{3}{4}$	$3\frac{3}{16}$	$3\frac{27}{64}$	$1\frac{3}{4}$
$1\frac{7}{8}$	5	$2\frac{13}{16}$	$3\frac{5}{16}$	$3\frac{61}{64}$	$1\frac{3}{4}$	$2\frac{15}{16}$	$3\frac{13}{32}$	$4\frac{5}{32}$	$1\frac{7}{8}$
2	$4\frac{1}{2}$	3	$3\frac{15}{32}$	$4\frac{1}{4}$	$1\frac{7}{8}$	$3\frac{1}{8}$	$3\frac{39}{64}$	$4\frac{27}{64}$	2

NOTE.—Nuts supplied by different makers for manufacturers standard bolts vary somewhat as regards thickness. The above nut sizes are Hoopes and Townsend Standard.

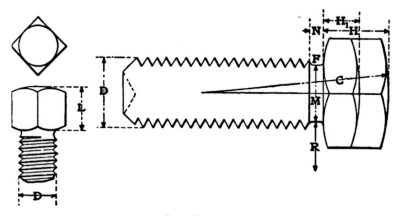

SET SCREWS

			Hartford Machine Screw Co. Standard						
Dia. of Screw	No. of Threads per Inch	Length of Head	Length of Head	Length of Low Head	Radius of Crown	Dia. of Neck	Radius of Neck	Fillet — Neck	Length of Neck
D		L	H	H_1	C	M	R	F	N
$\frac{1}{4}$	20	$\frac{1}{4}$	$\frac{7}{32}$	$\frac{1}{8}$	$\frac{5}{8}$	$\frac{7}{32}$	$\frac{1}{4}$	.019	.075
$\frac{5}{16}$	18	$\frac{5}{16}$	$\frac{35}{128}$	$\frac{5}{32}$	$\frac{25}{32}$	$\frac{35}{128}$	$\frac{5}{16}$	.021	.083
$\frac{3}{8}$	16	$\frac{3}{8}$	$\frac{21}{64}$	$\frac{3}{16}$	$\frac{15}{16}$	$\frac{21}{64}$	$\frac{3}{8}$	.023	.094
$\frac{7}{16}$	14	$\frac{7}{16}$	$\frac{49}{128}$	$\frac{7}{32}$	$1\frac{3}{32}$	$\frac{49}{128}$	$\frac{7}{16}$	.027	.107
$\frac{1}{2}$	13	$\frac{1}{2}$	$\frac{7}{16}$	$\frac{1}{4}$	$1\frac{1}{4}$	$\frac{7}{16}$	$\frac{1}{2}$	.031	.125
$\frac{9}{16}$	12	$\frac{9}{16}$	$\frac{63}{128}$	$\frac{9}{32}$	$1\frac{13}{32}$	$\frac{63}{128}$	$\frac{9}{16}$	.032	.125
$\frac{5}{8}$	11	$\frac{5}{8}$	$\frac{35}{64}$	$\frac{5}{16}$	$1\frac{9}{16}$	$\frac{35}{64}$	$\frac{5}{8}$	.034	.130
$\frac{3}{4}$	10	$\frac{3}{4}$	$\frac{21}{32}$	$\frac{3}{8}$	$1\frac{7}{8}$	$\frac{21}{32}$	$\frac{3}{4}$	.037	.150
$\frac{7}{8}$	9	$\frac{7}{8}$	$\frac{49}{64}$	$\frac{7}{16}$	$2\frac{3}{16}$	$\frac{49}{64}$	$\frac{7}{8}$	.041	.166
1	8	1	$\frac{7}{8}$	$\frac{1}{2}$	$2\frac{1}{2}$	$\frac{7}{8}$	1	.047	.187

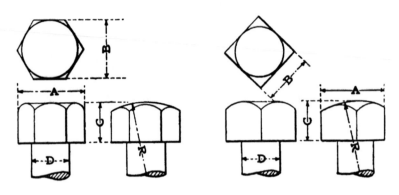

		Hexagon Head Cap Screws				Square Head Cap Screws			
Dia. of Screw	No. of Threads per In.	Distance Across Corners	Distance Across Flats	Thickness of Head	Radius of Head	Distance Across Corners	Distance Across Flats	Thickness of Head	Radius of Head
D		A	B	C	R	A	B	C	R
$\frac{1}{4}$	20	$\frac{1}{2}$	$\frac{7}{16}$	$\frac{1}{4}$	$\frac{11}{32}$	$\frac{17}{32}$	$\frac{3}{8}$	$\frac{1}{4}$	$\frac{9}{16}$
$\frac{5}{16}$	18	$\frac{37}{64}$	$\frac{1}{2}$	$\frac{5}{16}$	$\frac{3}{4}$	$\frac{5}{8}$	$\frac{7}{16}$	$\frac{5}{16}$	$\frac{21}{32}$
$\frac{3}{8}$	16	$\frac{21}{32}$	$\frac{9}{16}$	$\frac{3}{8}$	$\frac{27}{32}$	$\frac{45}{64}$	$\frac{1}{2}$	$\frac{3}{8}$	$\frac{3}{4}$
$\frac{7}{16}$	14	$\frac{23}{32}$	$\frac{5}{8}$	$\frac{7}{16}$	$1\frac{5}{16}$	$\frac{51}{64}$	$\frac{9}{16}$	$\frac{7}{16}$	$\frac{27}{32}$
$\frac{1}{2}$	13	$\frac{7}{8}$	$\frac{3}{4}$	$\frac{1}{2}$	$1\frac{1}{8}$	$\frac{57}{64}$	$\frac{5}{8}$	$\frac{1}{2}$	$1\frac{5}{16}$
$\frac{9}{16}$	12	$1\frac{1}{16}$	$\frac{13}{16}$	$\frac{9}{16}$	$1\frac{7}{32}$	$\frac{51}{32}$	$\frac{11}{16}$	$\frac{9}{16}$	$1\frac{13}{32}$
$\frac{5}{8}$	11	$1\frac{1}{64}$	$\frac{7}{8}$	$\frac{5}{8}$	$1\frac{5}{16}$	$1\frac{1}{16}$	$\frac{3}{4}$	$\frac{5}{8}$	$1\frac{1}{8}$
$\frac{3}{4}$	10	$1\frac{5}{32}$	1	$\frac{3}{4}$	$1\frac{1}{2}$	$1\frac{15}{64}$	$\frac{7}{8}$	$\frac{3}{4}$	$1\frac{5}{16}$
$\frac{7}{8}$	9	$1\frac{19}{64}$	$1\frac{1}{8}$	$\frac{7}{8}$	$1\frac{11}{16}$	$1\frac{13}{32}$	$1\frac{1}{8}$	$\frac{7}{8}$	$1\frac{11}{16}$
1	8	$1\frac{7}{16}$	$1\frac{1}{4}$	1	$1\frac{7}{8}$	$1\frac{41}{64}$	$1\frac{1}{4}$	1	$1\frac{7}{8}$
$1\frac{1}{8}$	7	$1\frac{19}{32}$	$1\frac{3}{8}$	$1\frac{1}{8}$	$2\frac{1}{16}$	$1\frac{13}{16}$	$1\frac{3}{8}$	$1\frac{1}{8}$	$2\frac{1}{16}$
$1\frac{1}{4}$	7	$1\frac{47}{64}$	$1\frac{1}{2}$	$1\frac{1}{4}$	$2\frac{1}{4}$	$2\frac{1}{8}$	$1\frac{1}{2}$	$1\frac{1}{4}$	$2\frac{1}{4}$

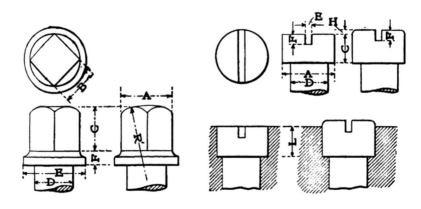

		Collar Head Screws						Fillister Head Cap Screws (P. & .W. St'd)						
Dia. of Screw	No. of Threads per Inch	Distance Across Corners	Distance Across Flats	Length of Head	Dia. of Collar	Thickness of Collar	Radius of Head	Dia. of Head	Length of Flat Head	Hight of Round Corner	Length of Round Head	Width of Slot	Depth of Slot	Depth of Counterbore
D		A	B	C	E	F	R	A	C	H	$\frac{C+}{H}$	E	F	L
$\frac{1}{8}$	40	$\frac{11}{64}$	$\frac{1}{8}$	$\frac{9}{64}$	$\frac{1}{4}$	$\frac{1}{16}$	$\frac{1}{4}$	$\frac{11}{64}$	$\frac{3}{32}$	$\frac{1}{32}$	$\frac{1}{8}$	.025	$\frac{1}{32}$	$\frac{3}{32}$
$\frac{3}{16}$	32	$\frac{17}{64}$	$\frac{3}{16}$	$\frac{7}{32}$	$\frac{11}{32}$	$\frac{5}{64}$	$\frac{3}{8}$	$\frac{1}{4}$	$\frac{1}{8}$	$\frac{1}{32}$	$\frac{5}{32}$	.039	$\frac{3}{64}$	$\frac{1}{8}$
$\frac{1}{4}$	20	$\frac{11}{32}$	$\frac{1}{4}$	$\frac{9}{32}$	$\frac{7}{16}$	$\frac{3}{32}$	$\frac{1}{2}$	$\frac{11}{32}$	$\frac{5}{32}$	$\frac{3}{32}$	$\frac{7}{16}$	.058	$\frac{5}{64}$	$\frac{5}{32}$
$\frac{5}{16}$	18	$\frac{7}{16}$	$\frac{5}{16}$	$\frac{11}{32}$	$\frac{1}{2}$	$\frac{1}{8}$	$\frac{11}{16}$	$\frac{7}{16}$	$\frac{3}{16}$	$\frac{8}{64}$	$\frac{13}{64}$	.071	$\frac{5}{64}$	$\frac{3}{16}$
$\frac{3}{8}$	16	$\frac{17}{32}$	$\frac{3}{8}$	$\frac{13}{32}$	$\frac{5}{8}$	$\frac{5}{32}$	$\frac{13}{16}$	$\frac{1}{2}$	$\frac{1}{4}$	$\frac{9}{64}$	$\frac{19}{64}$	.086	$\frac{7}{64}$	$\frac{1}{4}$
$\frac{7}{16}$	14	$\frac{5}{8}$	$\frac{7}{16}$	$\frac{15}{32}$	$\frac{11}{16}$	$\frac{3}{16}$	$\frac{15}{16}$	$\frac{9}{16}$	$\frac{1}{4}$	$\frac{9}{64}$	$\frac{19}{64}$	.099	$\frac{7}{64}$	$\frac{1}{4}$
$\frac{1}{2}$	13	$\frac{45}{64}$	$\frac{1}{2}$	$\frac{9}{16}$	$\frac{13}{16}$	$\frac{13}{64}$	$1\frac{1}{16}$	$\frac{11}{16}$	$\frac{5}{16}$	$\frac{9}{64}$	$\frac{23}{64}$	.112	$\frac{1}{8}$	$\frac{5}{16}$
$\frac{9}{16}$	12	$\frac{51}{64}$	$\frac{9}{16}$	$\frac{5}{8}$	$\frac{15}{16}$	$\frac{7}{32}$	$1\frac{3}{16}$	$\frac{3}{4}$	$\frac{3}{8}$	$\frac{9}{64}$	$\frac{27}{64}$	.133	$\frac{9}{64}$	$\frac{3}{8}$
$\frac{5}{8}$	11	$\frac{57}{64}$	$\frac{5}{8}$	$\frac{11}{16}$	1	$\frac{1}{4}$	$1\frac{5}{16}$	$\frac{7}{8}$	$\frac{7}{16}$	$\frac{5}{16}$	$\frac{1}{2}$	.133	$\frac{5}{32}$	$1\frac{7}{8}$
$\frac{3}{4}$	10	$1\frac{1}{16}$	$\frac{3}{4}$	$\frac{13}{16}$	$1\frac{1}{4}$	$\frac{5}{16}$	$1\frac{9}{16}$	$1\frac{1}{16}$	$\frac{1}{2}$	$\frac{5}{16}$	$\frac{9}{16}$	.133	$\frac{3}{16}$	$\frac{1}{2}$
$\frac{7}{8}$	9							$1\frac{3}{16}$	$\frac{9}{16}$	$\frac{5}{16}$	$\frac{5}{8}$	.133	$\frac{7}{32}$	$\frac{9}{16}$
1	8							$1\frac{3}{8}$	$\frac{5}{8}$	$\frac{7}{16}$	$\frac{11}{16}$	.165	$\frac{1}{4}$	$\frac{5}{8}$

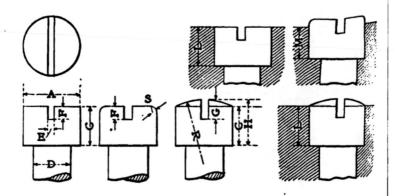

FLAT, ROUND AND OVAL FILLISTER HEAD CAP SCREWS

Dia. of Screw	No. of Threads per Inch	Dia. of Head	Length of Flat and Round Head	Radius of Corner	Length of Oval Head	Radius of Crown	Width of Slot	Depth of Slot	Depth of Slot Oval Head	Depth of Counter Bore, Flat and Oval Head	Depth of Counterbore Round Head
D		A	C	S	H	R	E	F	G	L	M
$\frac{1}{8}$	40	$\frac{3}{16}$	$\frac{1}{8}$	$\frac{1}{32}$	$\frac{9}{64}$	$\frac{1}{4}$	.032	$\frac{1}{16}$	$\frac{5}{64}$	$\frac{1}{8}$	$\frac{3}{32}$
$\frac{3}{16}$	32	$\frac{1}{4}$	$\frac{3}{16}$	$\frac{1}{32}$	$\frac{7}{32}$	$\frac{5}{16}$	.040	$\frac{1}{16}$	$\frac{3}{32}$	$\frac{3}{16}$	$\frac{5}{32}$
$\frac{1}{4}$	20	$\frac{3}{8}$	$\frac{1}{4}$	$\frac{1}{32}$	$\frac{9}{32}$	$\frac{1}{2}$	.064	$\frac{1}{16}$	$\frac{3}{32}$	$\frac{1}{4}$	$\frac{3}{32}$
$\frac{5}{16}$	18	$\frac{7}{16}$	$\frac{5}{16}$	$\frac{1}{16}$	$\frac{23}{64}$	$\frac{5}{8}$	.072	$\frac{5}{64}$	$\frac{1}{8}$	$\frac{5}{16}$	$\frac{1}{4}$
$\frac{3}{8}$	16	$\frac{9}{16}$	$\frac{3}{8}$	$\frac{1}{16}$	$\frac{15}{32}$	$\frac{3}{4}$	.091	$\frac{3}{32}$	$\frac{9}{64}$	$\frac{3}{8}$	$\frac{5}{16}$
$\frac{7}{16}$	14	$\frac{5}{8}$	$\frac{7}{16}$	$\frac{1}{16}$	$\frac{1}{2}$	$\frac{7}{8}$	.102	$\frac{7}{64}$	$\frac{11}{64}$	$\frac{7}{16}$	$\frac{3}{8}$
$\frac{1}{2}$	13	$\frac{3}{4}$	$\frac{1}{2}$	$\frac{1}{16}$	$\frac{9}{16}$	$1\frac{1}{16}$	.114	$\frac{1}{8}$	$\frac{3}{16}$	$\frac{1}{2}$	$\frac{7}{16}$
$\frac{9}{16}$	12	$\frac{13}{16}$	$\frac{9}{16}$	$\frac{1}{16}$	$\frac{41}{64}$	$1\frac{1}{8}$	.114	$\frac{9}{64}$	$\frac{7}{32}$	$\frac{9}{16}$	$\frac{1}{2}$
$\frac{5}{8}$	11	$\frac{7}{8}$	$\frac{5}{8}$	$\frac{1}{16}$	$\frac{45}{64}$	$1\frac{1}{4}$	.128	$\frac{5}{32}$	$\frac{15}{64}$	$\frac{5}{8}$	$\frac{9}{16}$
$\frac{3}{4}$	10	1	$\frac{3}{4}$	$\frac{1}{16}$	$\frac{27}{32}$	$1\frac{1}{2}$	.133	$\frac{3}{16}$	$\frac{9}{32}$	$\frac{3}{4}$	$\frac{11}{16}$
$\frac{7}{8}$	9	$1\frac{1}{8}$	$\frac{7}{8}$	$\frac{1}{16}$	$\frac{63}{64}$	$1\frac{5}{8}$	.133	$\frac{7}{32}$	$\frac{21}{64}$	$\frac{7}{8}$	$\frac{11}{16}$
1	8	$1\frac{1}{4}$	1	$\frac{3}{32}$	$1\frac{1}{8}$	$1\frac{3}{4}$	.165	$\frac{1}{4}$	$\frac{3}{8}$	1	$\frac{13}{16}$

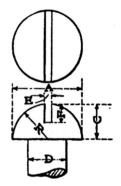
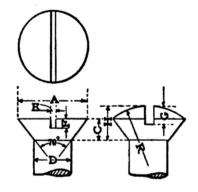

		Button Head Cap Screws					Flat and Oval Countersunk Head Cap Screws						
Dia. of Screw	No. of Threads per Inch	Dia. of Head	Length of Head	Radius of Head	Width of Slot	Depth of Slot	Dia. of Head	Length of Flat Head	Radius of Crown	Length of Oval Head	Width of Slot	Depth of Slot Flat Head	Depth of Slot Oval Head
D		A	C	R	E	F	A	C	R	H	E	F	G
$\frac{1}{8}$	40	$\frac{7}{32}$	$\frac{7}{64}$	$\frac{7}{64}$	.035	$\frac{1}{16}$	$\frac{1}{4}$	$\frac{3}{32}$	$\frac{3}{16}$	$\frac{3}{64}$	.040	$\frac{3}{64}$	$\frac{3}{64}$
$\frac{5}{32}$	32	$\frac{9}{16}$	$\frac{5}{32}$	$\frac{9}{32}$	.051	$\frac{3}{32}$	$\frac{3}{8}$	$\frac{9}{64}$	$\frac{9}{32}$	$\frac{11}{64}$	.064	$\frac{3}{64}$	$\frac{3}{32}$
$\frac{1}{4}$	20	$\frac{7}{16}$	$\frac{7}{32}$	$\frac{7}{32}$	.072	$\frac{1}{8}$	$\frac{11}{32}$	$\frac{5}{32}$	$\frac{3}{8}$	$\frac{17}{64}$	.072	$\frac{1}{16}$	$\frac{1}{8}$
$\frac{5}{16}$	18	$\frac{9}{16}$	$\frac{9}{32}$	$\frac{9}{32}$	.091	$\frac{5}{32}$	$\frac{5}{8}$	$\frac{7}{32}$	$\frac{15}{16}$	$\frac{11}{32}$	.102	$\frac{3}{64}$	$\frac{3}{32}$
$\frac{3}{8}$	16	$\frac{5}{8}$	$\frac{5}{16}$	$\frac{5}{16}$	.102	$\frac{3}{16}$	$\frac{3}{4}$	$\frac{17}{64}$	$\frac{9}{16}$	$\frac{13}{32}$	.114	$\frac{3}{32}$	$\frac{3}{16}$
$\frac{7}{16}$	14	$\frac{3}{4}$	$\frac{3}{8}$	$\frac{3}{8}$	.114	$\frac{7}{32}$	$\frac{13}{16}$	$\frac{17}{64}$	$\frac{21}{32}$	$\frac{15}{32}$	.114	$\frac{3}{32}$	$\frac{7}{32}$
$\frac{1}{2}$	13	$\frac{13}{16}$	$\frac{13}{32}$	$\frac{13}{32}$	.114	$\frac{1}{4}$	$\frac{7}{8}$	$\frac{21}{64}$	$\frac{3}{4}$	$\frac{17}{32}$	.128	$\frac{9}{32}$	$\frac{1}{4}$
$\frac{9}{16}$	12	$\frac{15}{16}$	$\frac{15}{32}$	$\frac{15}{32}$	.114	$\frac{9}{32}$	1	$\frac{5}{16}$	$\frac{27}{32}$	$\frac{15}{32}$	.133	$\frac{7}{64}$	$\frac{9}{32}$
$\frac{5}{8}$	11	1	$\frac{1}{2}$	$\frac{1}{2}$	.133	$\frac{5}{16}$	$1\frac{1}{8}$	$\frac{23}{64}$	$1\frac{5}{16}$	$\frac{23}{64}$	.133	$\frac{1}{8}$	$\frac{5}{16}$
$\frac{3}{4}$	10	$1\frac{1}{4}$	$\frac{5}{8}$	$\frac{5}{8}$	.133	$\frac{3}{8}$	$1\frac{3}{8}$	$\frac{7}{16}$	$1\frac{1}{8}$	$\frac{21}{32}$	.133	$\frac{5}{32}$	$\frac{3}{8}$

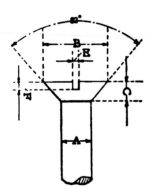 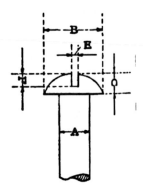

MACHINE SCREWS. AMERICAN SCREW COMPANY

No.	A	FLAT HEAD				ROUND HEAD			
		B	C	E	F	B	C	E	F
2	.0842	.1631	.0454	.030	.0151	.1544	.0672	.030	.0403
3	.0973	.1894	.0530	.032	.0177	.1786	.0746	.032	.0448
4	.1105	.2158	.0605	.034	.0202	.2028	.0820	.034	.0492
5	.1236	.2421	.0681	.036	.0227	.2270	.0894	.036	.0536
6	.1368	.2684	.0757	.039	.0252	.2512	.0968	.039	.0580
7	.1500	.2947	.0832	.041	.0277	.2754	.1042	.041	.0625
8	.1631	.3210	.0908	.043	.0303	.2996	.1116	.043	.0670
9	.1763	.3474	.0984	.045	.0328	.3238	.1190	.045	.0714
10	.1894	.3737	.1059	.048	.0353	.3480	.1264	.048	.0758
12	.2158	.4263	.1210	.052	.0403	.3922	.1412	.052	.0847
14	.2421	.4790	.1362	.057	.0454	.4364	.1560	.057	.0936
16	.2684	.5316	.1513	.061	.0504	.4806	.1708	.061	.1024
18	.2947	.5842	.1665	.066	.0555	.5248	.1856	.066	.1114
20	.3210	.6368	.1816	.070	.0605	.5690	.2004	.070	.1202
22	.3474	.6895	.1967	.075	.0656	.6106	.2152	.075	.1291
24	.3737	.7421	.2118	.079	.0706	.6522	.2300	.079	.1380
26	.4000	.7421	.1967	.084	.0656	.6938	2448	.084	.1469
28	.4263	.7948	.2118	.088	.0706	.7354	.2596	.088	.1558
30	.4526	.8474	.2270	.093	.0757	.7770	.2744	.093	.1646

Dimensions given are maximum, the necessary working variations being below them.

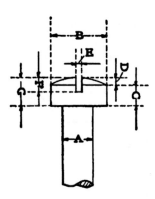

MACHINE SCREWS. AMERICAN SCREW COMPANY

No.	A	FILLISTER HEAD					G
		B	C	D	E	F	
2	.0842	.1350	.0549	.0126	.030	.0338	.0675
3	.0973	.1561	.0634	.0146	.032	.0390	.0780
4	.1105	.1772	.0720	.0166	.034	.0443	.0886
5	.1236	.1984	.0806	.0186	.036	.0496	.0992
6	.1368	.2195	.0892	.0205	.039	.0549	.1097
7	.1500	.2406	.0978	.0225	.041	.0602	.1203
8	.1631	.2617	.1063	.0245	.043	.0654	.1308
9	.1763	.2828	.1149	.0265	.045	.0707	.1414
10	.1894	.3040	.1235	.0285	.048	.0760	.1520
12	.2158	.3462	.1407	.0324	.052	.0866	.1731
14	.2421	.3884	.1578	.0364	.057	.0971	.1942
16	.2684	.4307	.1750	.0403	.061	.1077	.2153
18	.2947	.4729	.1921	.0443	.066	.1182	.2364
20	.3210	.5152	.2093	.0483	.070	.1288	.2576
22	.3474	.5574	.2267	.0520	.075	.1384	.2787
24	.3737	.5996	.2436	.0562	.079	.1499	.2998
26	.4000	.6419	.2608	.0601	.084	.1605	.3209
28	.4263	.6841	.2779	.0641	.088	.1710	.3420
30	.4526	.7264	.2951	.0681	.093	.1816	.3632

AMERICAN SCREW COMPANY. STANDARD THREADS PER INCH

No.	2	3	4	5	6	7	8	9	10	12
Threads per Inch	48 56 64	48 56	32, 36 40		30 32 36	30 32	30 32 36	24, 30, 32		20 24

No.		14	16	18	20	22	24	26	28	30
Threads per Inch		18 20 24	16, 18, 20		16, 18		14 16 18	14, 16		

A.S.M.E. STANDARD PROPORTIONS OF MACHINE SCREWS

THE diagram and tables herewith show the proportions of machine screws as recommended by the committee of the American Society of Mechanical Engineers on Standard Proportions for Machine Screws, the report of this committee being adopted by the Society at its spring meeting, 1907.

The included angle is 60 degrees, and the flat at top and bottom of thread is one eighth of the pitch for the basic or standard diameter. There is a uniform increment of 0.013 inch, between all sizes from 0.06 to 0.19 (numbers 0 to 10 in the tables which follow) and of 0.026 inch in the remaining sizes. This change has been made in the interest of simplicity and because the resulting pitch diameters are more nearly in accord with the pitch diameters of screws in present use.

The pitches are a function of the diameter as expressed by the formula

$$\text{Threads per inch} = \frac{6.5}{D + 0.02},$$

with the results given approximately, so as to avoid the use of fractional threads.

The diagram shows the various sizes for both 16 and 72 threads per inch, and shows, among other things, the allowable difference in the flat surface, between the maximum tap and the minimum screw, this variation being from one-eighth to one sixteenth.

The minimum tap conforms to the basic standard in all respects, except diameter. The difference between the minimum tap and the maximum screw provides an allowance for error in pitch and for wear of tap in service.

The form of tap thread shown is recommended as being stronger and more serviceable than the so-called V-thread, but as some believe a strict adherence to the form shown might add to the cost of small taps, they have decided that taps having the correct angle and pitch diameter are permissible even with the V-thread. This will allow a large proportion of the taps now in stock to be utilized.

The tables given by the committee were combined into the present compact form by the Corbin Screw Corporation.

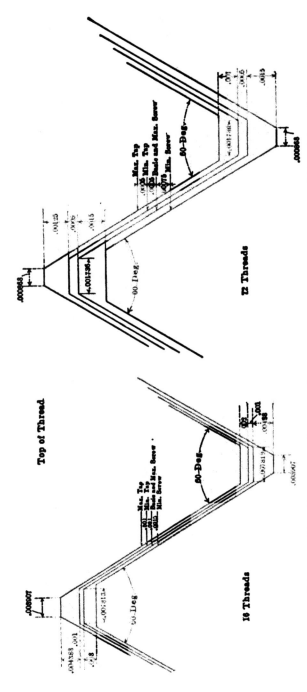

Diagram Showing Form of Basic Maximum and Minimum Screw and Tap Threads

A. S. M. E. Machine Screw Standard

A. S. M. E. Standard Machine Screws

Size		Outside Diameters			Pitch Diameters			Root Diameters		
No.	Out. Dia. and Thds. P. I.	Minimum	Maximum	Difference	Minimum	Maximum	Difference	Minimum	Maximum	Difference
0	.060–80	.0572	.060	.0028	.0505	.0519	.0014	.0410	.0438	.0028
1	.073–72	.070	.073	.003	.0625	.064	.0015	.052	.055	.0030
2	.086–64	.0828	.086	.0032	.0743	.0759	.0016	.0624	.0657	.0033
3	.099–56	.0955	.099	.0035	.0857	.0874	.0017	.0721	.0758	.0037
4	.112–48	.1082	.112	.0038	.0966	.0985	.0019	.0807	.0849	.0042
5	.125–44	.1210	.125	.0040	.1082	.1102	.0020	.0910	.0955	.0045
6	.138–40	.1338	.138	.0042	.1197	.1218	.0021	.1007	.1055	.0048
7	.151–36	.1466	.151	.0044	.1308	.1330	.0022	.1097	.1149	.0052
8	.164–36	.1596	.164	.0044	.1438	.146	.0022	.1227	.1279	.0052
9	.177–32	.1723	.177	.0047	.1544	.1567	.0023	.1307	.1364	.0057
10	.190–30	.1852	.190	.0048	.166	.1684	.0024	.1407	.1467	.0060
12	.216–28	.2111	.216	.0049	.1904	.1928	.0024	.1633	.1696	.0063
14	.242–24	.2368	.242	.0052	.2123	.2149	.0026	.1808	.1879	.0071
16	.268–22	.2626	.268	.0054	.2358	.2385	.0027	.2014	.209	.0076
18	.294–20	.2884	.294	.0056	.2587	.2615	.0028	.2208	.229	.0082
20	.320–20	.3144	.320	.0056	.2847	.2875	.0028	.2468	.255	.0082
22	.346–18	.3402	.346	.0058	.3070	.3099	.0029	.2649	.2738	.0089
24	.372–16	.366	.372	.0060	.3284	.3314	.0030	.281	.2908	.0098
26	.398–16	.392	.398	.0060	.3544	.3574	.0030	.307	.3168	.0098
28	.424–14	.4178	.424	.0062	.3745	.3776	.0031	.3204	.3312	.0108
30	.450–14	.4438	.450	.0062	.4005	.4036	.0031	.3464	.3572	.0108

TAPS FOR A. S. M. E. STANDARD MACHINE SCREWS

No.	Size — Out. Dia. and Thds. P. In.	Outside Diameters			Pitch Diameters			Root Diameters			Tap Drill Diameters
		Minimum	Maximum	Difference	Minimum	Maximum	Difference	Minimum	Maximum	Difference	
0	.060–80	.0609	.0632	.0023	.0528	.0538	.001	.0447	.0466	.0019	.0465
1	.073–72	.074	.0765	.0025	.065	.066	.001	.056	.058	.0020	.0595
2	.086–64	.0871	.0898	.0027	.0770	.0781	.0011	.0668	.0689	.0021	.070
3	.099–56	.1002	.1033	.0031	.0886	.0897	.0011	.077	.0793	.0023	.0785
4	.112–48	.1133	.1168	.0035	.0998	.101	.0012	.0862	.0887	.0025	.089
5	.125–44	.1263	.1301	.0038	.1116	.1129	.0013	.0968	.0995	.0027	.0995
6	.138–40	.1394	.1435	.0041	.1232	.1246	.0014	.1069	.1097	.0028	.110
7	.151–36	.1525	.1569	.0044	.1345	.1359	.0014	.1164	.1193	.0029	.120
8	.164–36	.1655	.1699	.0044	.1475	.1489	.0014	.1294	.1323	.0029	.136
9	.177–32	.1786	.1835	.0049	.1583	.1598	.0015	.138	.1411	.0031	.1405
10	.190–30	.1916	.1968	.0052	.170	.1716	.0016	.1483	.1515	.0032	.152
12	.216–28	.2176	.2232	.0056	.1944	.1961	.0017	.1712	.1745	.0033	.173
14	.242–24	.2438	.250	.0062	.2167	.2184	.0017	.1897	.1932	.0035	.1935
16	.268–22	.2698	.2765	.0067	.2403	.2421	.0018	.2108	.2144	.0036	.213
18	.294–20	.2959	.3031	.0072	.2634	.2652	.0018	.2309	.2346	.0037	.234
20	.320–20	.3219	.3291	.0072	.2894	.2912	.0018	.2569	.2606	.0037	.261
22	.346–18	.3479	.3559	.0080	.3118	.3138	.0020	.2757	.2796	.0039	.281
24	.372–16	.374	.3828	.0088	.3334	.3354	.0020	.2928	.2968	.0040	.2968
26	.398–16	.400	.4088	.0088	.3594	.3614	.0020	.3188	.3228	.0040	.323
28	.424–14	.4261	.4359	.0098	.3797	.3818	.0021	.3333	.3374	.0041	.339
30	.450–14	.4521	.4619	.0098	.4057	.4078	.0021	.3593	.3634	.0041	.368

A. S. M. E. SPECIAL MACHINE SCREWS

No.	Out. Dia. and Thds. P.I.	Outside Diameters Minimum	Maximum	Difference	Pitch Diameters Minimum	Maximum	Difference	Root Diameters Minimum	Maximum	Difference
1	073–64	.0698	.073	.0032	.0613	.0629	.0016	.0494	.0527	.0033
2	086–56	.0825	.086	.0035	.0727	.0744	.0017	.0591	.0628	.0037
3	099–48	.0952	.099	.0038	.0836	.0855	.0019	.0677	.0719	.0042
4	112–40	.1078	.112	.0042	.0937	.0958	.0021	.0747	.0795	.0048
	36	.1076	.112	.0044	.0918	.094	.0022	.0707	.0759	.0052
5	125–40	.1208	.125	.0042	.1067	.1088	.0021	.0877	.0925	.0048
	36	.1206	.125	.0044	.1048	.107	.0022	.0837	.0889	.0052
6	138–36	.1336	.138	.0044	.1178	.120	.0022	.0967	.1019	.0052
	32	.1333	.138	.0047	.1154	.1177	.0023	.0917	.0974	.0057
7	151–32	.1463	.151	.0047	.1284	.1307	.0023	.1047	.1104	.0057
	30	.1462	.151	.0048	.1270	.1294	.0024	.1017	.1077	.60
8	164–32	.1593	.164	.0047	.1414	.1437	.0023	.1177	.1234	.0057
	30	.1592	.164	.0048	.1400	.1424	.0024	.1147	.1207	.0060
9	177–30	.1722	.177	.0048	.1529	.1553	.0024	.1277	.1337	.0060
	24	.1718	.177	.0052	.1473	.1499	.0026	.1158	.1229	.91
10	190–32	.1853	.190	.0047	.1674	.1697	.0023	.1437	.1494	.0057
	24	.1848	.190	.0052	.1603	.1629	.0026	.1288	.1359	.91
12	216–24	.2108	.216	.0052	.1863	.1889	.0026	.1548	.1619	.91
14	242–20	.2364	.242	.0056	.2067	.2095	.0028	.1688	.1770	.0082
16	268–20	.2624	.268	.0056	.2327	.2355	.0028	.1948	.203	.0082
18	294–18	.2882	.294	.1058	.255	.2579	.0029	.2129	.2218	.0089
20	320–18	.3142	.320	.0058	.281	.2839	.0029	.2389	.2478	.0089
22	346–16	.340	.346	.0060	.3024	.3054	.0030	.255	.2648	.0098
24	372–18	.3662	.372	.0058	.333	.3359	.0029	.2909	.2998	.0089
26	398–14	.3918	.398	.0062	.3485	.3516	.0031	.2944	.3052	.0108
28	424–16	.418	.424	.0060	.3804	.3834	.0030	.333	.3428	.0098
	450–16	.444	.450	.0060	.4064	.4094	.0030	.359	.3688	.0098

No.	Out. Dia. and Thds. P. In.	Outside Diameters			Pitch Diameters			Root Diameters			Tap Drill Diameters
		Minimum	Maximum	Difference	Minimum	Maximum	Difference	Minimum	Maximum	Difference	
1	.073–64	.0741	.0768	.0027	.064	.0651	.0011	.0538	.0559	.0021	.055
2	.086–56	.0872	.0903	.0031	.0756	.0767	.0011	.064	.0663	.0023	.067
3	.099–48	.1003	.1038	.0035	.0868	.088	.0012	.0732	.0757	.0025	.076
4	.112–40	.1134	.1175	.0041	.0972	.0986	.0014	.0809	.0837	.0028	.082
	36	.1135	.1179	.0044	.0955	.0969	.0014	.0774	.0803	.0029	.081
5	.125–40	.1264	.1305	.0041	.1102	.1116	.0014	.0939	.0967	.0028	.098
	36	.1265	.1309	.0044	.1085	.1099	.0014	.0904	.0933	.0029	.0935
6	.138–36	.1395	.1439	.0044	.1215	.1229	.0014	.1034	.1063	.0029	.1065
	32	.1396	.1445	.0049	.1193	.1208	.0015	.099	.1021	.0031	.1015
7	.151–32	.1526	.1575	.0049	.1323	.1338	.0015	.112	.1151	.0031	.116
	30	.1526	.1578	.0052	.131	.1326	.0016	.1093	.1125	.0032	.113
8	.164–32	.1656	.1705	.0049	.1453	.1468	.0015	.125	.1281	.0031	.1285
	30	.1656	.1708	.0052	.144	.1456	.0016	.1223	.1255	.0032	.1285
9	.177–30	.1786	.1838	.0052	.1569	.1585	.0016	.1353	.1385	.0032	.1405
	24	.1788	.185	.0062	.1517	.1534	.0017	.1247	.1282	.0035	.1285
10	.190–32	.1916	.1965	.0049	.1713	.1728	.0015	.151	.1541	.0031	.154
	24	.1918	.198	.0062	.1647	.1664	.0017	.1377	.1412	.0035	.1405
12	.216–24	.2178	.224	.0062	.1907	.1924	.0017	.1637	.1672	.0035	.166
14	.242–20	.2439	.2511	.0072	.2114	.2132	.0018	.1789	.1826	.0037	.182
16	.268–20	.2699	.2771	.0072	.2374	.2392	.0018	.2049	.2086	.0037	.209
18	.294–18	.2959	.3039	.008	.2598	.2618	.002	.2237	.2276	.0039	.228
20	.320–18	.3219	.3299	.008	.2858	.2878	.002	.2497	.2536	.0039	.257
22	.346–16	.348	.3568	.0088	.3074	.3094	.002	.2668	.2708	.004	.272
24	.372–18	.3739	.3819	.008	.3378	.3398	.002	.3017	.3056	.0039	.3125
26	.398–14	.4001	.4099	.0098	.3537	.3558	.0021	.3073	.3114	.0041	.316
28	.424–16	.426	.4348	.0088	.3854	.3874	.002	.3448	.3488	.004	.348
30	.450–16	.452	.4608	.0088	.4114	.4134	.002	.3708	.3748	.004	.377

PROPORTIONS OF MACHINE SCREW HEADS
A. S. M. E. Standard

THE four standard heads are given herewith. These proportions are based on and include the diameter of the screw, diameter of the head, thickness of head, width and depth of slot, radius for round and fillister heads, and included angle of the flat-head screw.

OVAL FILLISTER HEAD MACHINE SCREWS. A. S. M. E. STANDARD

OVAL FILLISTER HEAD SCREWS

A = Diameter of Body.
B = $1.64A$ − .009 = Diam. of Head and Rad. for Oval
C = $0.66A$ − .002 = Hight of Side
D = .173A + .015
E = $\frac{1}{2}F$ = Depth of Slot
F = .134B + C = Hight of Head.

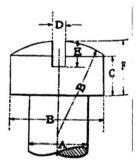

A	B	C	D	E	F
.060	.0894	.0376	.025	.025	.0496
.073	.1107	.0461	.028	.030	.0609
.086	.132	.0548	.030	.036	.0725
.099	.153	.0633	.032	.042	.0838
.112	.1747	.0719	.034	.048	.0953
.125	.196	.0805	.037	.053	.1068
.138	.217	.089	.039	.059	.1180
.151	.2386	.0976	.041	.065	.1296
.164	.2599	.1062	.043	.071	.1410
.177	.2813	.1148	.046	.076	.1524
.190	.3026	.1234	.048	.082	.1639
.216	.3452	.1405	.052	.093	.1868
.242	.3879	.1577	.057	.105	.2097
.268	.4305	.1748	.061	.116	.2325
.294	.4731	.192	.066	.128	.2554
.320	.5158	.2092	.070	.140	.2783
.346	.5584	.2263	.075	.150	.3011
.372	.601	.2435	.079	.162	.3240
.398	.6437	.2606	.084	.173	.3460
.424	.6863	.2778	.088	.185	.3608
.450	.727	.295	.093	.201	.4024

FLAT FILLISTER HEAD MACHINE SCREWS. A. S. M. E. STANDARD

FLAT FILLISTER HEAD SCREWS

A = Diameter of Body
B = $1.64A - .009$ = Diam. of Head.
C = $0.66A - .002$ = Hight of Head
D = $0.173A + .015$ = Width of Slot
E = $\frac{1}{2}C$ = Depth of Slot

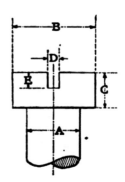

A	B	C	D	E
.060	.0894	.0376	.025	.019
.073	.1107	.0461	.028	.023
.086	.132	.0548	.030	.027
.099	.153	.0633	.032	.032
.112	.1747	.0719	.034	.036
.125	.196	.0805	.037	.040
.138	.217	.0890	.039	.044
.151	.2386	.0976	.041	.049
.164	.2599	.1062	.043	.053
.177	.2813	.1148	.046	.057
.190	.3026	.1234	.048	.062
.216	.3452	.1405	.052	.070
.242	.3879	.1577	.057	.079
.268	.4305	.1748	.061	.087
.294	.4731	.1920	.066	.096
.320	.5158	.2092	.070	.104
.346	.5584	.2263	.075	.113
.372	.601	.2435	.079	.122
.398	.6437	.2606	.084	.130
.424	.6863	.2778	.088	.139
.450	.727	.295	.093	.147

FLAT HEAD MACHINE SCREWS. A. S. M. E. STANDARD

FLAT HEAD SCREWS

A = Diameter of Body

$B = 2A - .008$ = Diam. of Head

$C = \dfrac{A - .008}{1.739}$ = Depth of Head

$D = .173A + .015$ = Width of Slot

$E = \frac{1}{3}C$ = Depth of Slot

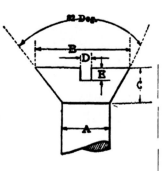

A	B	C	D	E
.060	.112	.029	.025	.010
.073	.138	.037	.028	.012
.086	.164	.045	.030	.015
.099	.190	.052	.032	.017
.112	.216	.060	.034	.020
.125	.242	.067	.037	.022
.138	.262	.075	.039	.025
.151	.294	.082	.041	.027
.164	.320	.090	.043	.030
.177	.346	.097	.046	.032
.190	.372	.105	.048	.035
.216	.424	.120	.052	.040
.242	.472	.135	.057	.045
.268	.528	.150	.061	.050
.294	.580	.164	.066	.055
.320	.632	.179	.070	.060
.346	.682	.194	.075	.065
.372	.732	.209	.079	.070
.398	.788	.224	.084	.075
.424	.840	.239	.088	.080
.450	.892	.254	.093	.085

Round Head Machine Screws. A. S. M. E. Standard

ROUND HEAD SCREWS

A = Diameter of Body
B = 1.85A − .005 = Diam. of Head
C = .7A = Hight of Head
D = .173A + .015 = Width of Slot
E = ½C + .01 = Depth of Slot

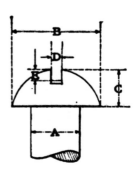

A	B	C	D	E
.060	.106	.042	.025	.031
.073	.130	.051	.028	.035
.086	.154	.060	.030	.040
.099	.178	.069	.032	.044
.112	.202	.078	.034	.049
.125	.226	.087	.037	.053
.138	.250	.096	.039	.058
.151	.274	.105	.041	.062
.164	.298	.114	.043	.067
.177	.322	.123	.046	.071
.190	.346	.133	.048	.076
.216	.394	.151	.052	.085
.242	.443	.169	.057	.094
.268	.491	.187	.061	.103
.294	.539	.205	.066	.112
.320	.587	.224	.070	.122
.346	.635	.242	.075	.131
.372	.683	.260	.079	.140
.398	.731	.278	.084	.149
.424	.779	.296	.088	.158
.450	.827	.315	.093	.167

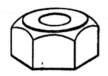

HOT PRESSED AND COLD PUNCHED NUTS

U. S. STANDARD HOT PRESSED AND COLD PUNCHED NUTS				COLD PUNCHED CHECK AND JAM NUTS			
HEXAGON AND SQUARE				HEXAGON			
Dia. Bolt	Across Flats	Thickness	Dia. Hole	Dia. Bolt	Across Flats	Thickness	Dia. Hole
$\frac{1}{4}$	$\frac{1}{2}$	$\frac{1}{4}$	$\frac{5}{16}$	$\frac{1}{4}$	$\frac{1}{2}$	$\frac{3}{16}$	$\frac{13}{64}$
$\frac{5}{16}$	$\frac{19}{32}$	$\frac{5}{16}$	$\frac{1}{4}$	$\frac{5}{16}$	$\frac{19}{32}$	$\frac{7}{32}$	$\frac{1}{4}$
$\frac{3}{8}$	$\frac{11}{16}$	$\frac{3}{8}$	$\frac{19}{64}$	$\frac{3}{8}$	$\frac{11}{16}$	$\frac{1}{4}$	$\frac{19}{64}$
$\frac{7}{16}$	$\frac{25}{32}$	$\frac{7}{16}$	$\frac{23}{64}$	$\frac{7}{16}$	$\frac{25}{32}$	$\frac{5}{16}$	$\frac{13}{32}$
$\frac{1}{2}$	$\frac{7}{8}$	$\frac{1}{2}$	$\frac{15}{32}$	$\frac{1}{2}$	$\frac{7}{8}$	$\frac{5}{16}$	$\frac{15}{32}$
$\frac{9}{16}$	$\frac{31}{32}$	$\frac{9}{16}$	$\frac{29}{64}$	$\frac{9}{16}$	$\frac{31}{32}$	$\frac{11}{32}$	$\frac{19}{64}$
$\frac{5}{8}$	$1\frac{1}{16}$	$\frac{5}{8}$	$\frac{33}{64}$	$\frac{5}{8}$	$1\frac{1}{16}$	$\frac{3}{8}$	$\frac{5}{8}$
$\frac{3}{4}$	$1\frac{1}{4}$	$\frac{3}{4}$	$\frac{5}{8}$	$\frac{3}{4}$	$1\frac{1}{4}$	$\frac{7}{16}$	$\frac{41}{64}$
$\frac{7}{8}$	$1\frac{7}{16}$	$\frac{7}{8}$	$\frac{47}{64}$	$\frac{7}{8}$	$1\frac{7}{16}$	$\frac{1}{2}$	$\frac{9}{16}$
1	$1\frac{5}{8}$	1	$\frac{27}{32}$	1	$1\frac{5}{8}$	$\frac{9}{16}$	$\frac{11}{16}$
$1\frac{1}{8}$	$1\frac{13}{16}$	$1\frac{1}{8}$	$\frac{15}{16}$	$1\frac{1}{8}$	$1\frac{13}{16}$	$\frac{5}{8}$	$\frac{15}{16}$
$1\frac{1}{4}$	2	$1\frac{1}{4}$	$1\frac{1}{16}$	$1\frac{1}{4}$	2	$\frac{3}{4}$	$1\frac{1}{16}$
$1\frac{3}{8}$	$2\frac{3}{16}$	$1\frac{3}{8}$	$1\frac{5}{32}$	$1\frac{3}{8}$	$2\frac{3}{16}$	$\frac{13}{16}$	$1\frac{5}{32}$
$1\frac{1}{2}$	$2\frac{3}{8}$	$1\frac{1}{2}$	$1\frac{9}{32}$	$1\frac{1}{2}$	$2\frac{3}{8}$	$\frac{7}{8}$	$1\frac{9}{32}$
$1\frac{5}{8}$	$2\frac{9}{16}$	$1\frac{5}{8}$	$1\frac{25}{32}$	$1\frac{5}{8}$	$2\frac{9}{16}$	$\frac{15}{16}$	$1\frac{11}{32}$
$1\frac{3}{4}$	$2\frac{3}{4}$	$1\frac{3}{4}$	$1\frac{1}{2}$	$1\frac{3}{4}$	$2\frac{3}{4}$	1	$1\frac{1}{2}$
$1\frac{7}{8}$	$2\frac{15}{16}$	$1\frac{7}{8}$	$1\frac{5}{8}$	$1\frac{7}{8}$	$2\frac{15}{16}$	$1\frac{1}{16}$	$1\frac{5}{8}$
2	$3\frac{1}{8}$	2	$1\frac{23}{32}$	2	$3\frac{1}{8}$	$1\frac{1}{8}$	$1\frac{23}{32}$
$2\frac{1}{8}$	$3\frac{5}{16}$	$2\frac{1}{8}$	$1\frac{13}{16}$				
$2\frac{1}{4}$	$3\frac{1}{2}$	$2\frac{1}{4}$	$1\frac{15}{16}$				
$2\frac{3}{8}$	$3\frac{11}{16}$	$2\frac{3}{8}$	$2\frac{1}{16}$				
$2\frac{1}{2}$	$3\frac{7}{8}$	$2\frac{1}{2}$	$2\frac{11}{64}$				
$2\frac{3}{4}$	$4\frac{1}{4}$	$2\frac{3}{4}$	$2\frac{27}{64}$				
3	$4\frac{5}{8}$	3	$2\frac{5}{8}$				

Finished case-hardened and semi-finished nuts are made to the above dimensions. Semi-finished nuts are tapped and faced true on the bottom.

 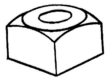

HOT PRESSED NUTS

HOT PRESSED NUTS, MANUFACTURERS STANDARD				HOT PRESSED AND FORGED NUTS, MANUFACTURERS STANDARD			
HEXAGON				SQUARE			
Dia. Bolt	Across Flats	Thick-ness	Dia. Hole	Dia. Bolt	Across Flats	Thick-ness	Dia. Hole
$\frac{1}{4}$	$\frac{1}{2}$	$\frac{1}{4}$	$\frac{7}{32}$	$\frac{1}{4}$	$\frac{1}{2}$	$\frac{1}{4}$	$\frac{7}{32}$
$\frac{5}{16}$	$\frac{5}{8}$	$\frac{5}{16}$	$\frac{9}{32}$	$\frac{5}{16}$	$\frac{5}{8}$	$\frac{5}{16}$	$\frac{9}{32}$
$\frac{3}{8}$	$\frac{3}{4}$	$\frac{3}{8}$	$\frac{11}{32}$	$\frac{3}{8}$	$\frac{3}{4}$	$\frac{3}{8}$	$\frac{11}{32}$
$\frac{7}{16}$	$\frac{7}{8}$	$\frac{7}{16}$	$\frac{25}{64}$	$\frac{7}{16}$	$\frac{7}{8}$	$\frac{7}{16}$	$\frac{25}{64}$
$\frac{1}{2}$	1	$\frac{1}{2}$	$\frac{7}{16}$	$\frac{1}{2}$	1	$\frac{1}{2}$	$\frac{7}{16}$
$\frac{9}{16}$	$1\frac{1}{8}$	$\frac{9}{16}$	$\frac{1}{2}$	$\frac{9}{16}$	$1\frac{1}{8}$	$\frac{9}{16}$	$\frac{1}{2}$
$\frac{5}{8}$	$1\frac{1}{4}$	$\frac{5}{8}$	$\frac{9}{16}$	$\frac{5}{8}$	$1\frac{1}{4}$	$\frac{5}{8}$	$\frac{9}{16}$
$\frac{3}{4}$	$1\frac{3}{8}$	$\frac{3}{4}$	$\frac{21}{32}$	$\frac{3}{4}$	$1\frac{1}{2}$	$\frac{3}{4}$	$\frac{21}{32}$
$\frac{7}{8}$	$1\frac{5}{8}$	$\frac{7}{8}$	$\frac{49}{64}$	$\frac{7}{8}$	$1\frac{3}{4}$	$\frac{7}{8}$	$\frac{49}{64}$
1	$1\frac{3}{4}$	1	$\frac{7}{8}$	1	2	1	$\frac{7}{8}$
$1\frac{1}{8}$	2	$1\frac{1}{8}$	$\frac{31}{32}$	$1\frac{1}{8}$	$2\frac{1}{4}$	$1\frac{1}{8}$	$\frac{31}{32}$
$1\frac{1}{4}$	$2\frac{1}{4}$	$1\frac{3}{8}$	$1\frac{3}{32}$	$1\frac{1}{4}$	$2\frac{1}{2}$	$1\frac{1}{4}$	$1\frac{3}{32}$
$1\frac{3}{8}$	$2\frac{1}{2}$	$1\frac{1}{2}$	$1\frac{3}{16}$	$1\frac{3}{8}$	$2\frac{3}{4}$	$1\frac{3}{8}$	$1\frac{3}{16}$
$1\frac{1}{2}$	$2\frac{3}{4}$	$1\frac{5}{8}$	$1\frac{5}{16}$	$1\frac{1}{2}$	3	$1\frac{1}{2}$	$1\frac{5}{16}$
$1\frac{5}{8}$	3	$1\frac{3}{4}$	$1\frac{7}{16}$	$1\frac{5}{8}$	$3\frac{1}{4}$	$1\frac{5}{8}$	$1\frac{7}{16}$
$1\frac{3}{4}$	$3\frac{1}{4}$	$1\frac{7}{8}$	$1\frac{9}{16}$	$1\frac{3}{4}$	$3\frac{1}{2}$	$1\frac{3}{4}$	$1\frac{9}{16}$
$1\frac{7}{8}$	$3\frac{1}{2}$	2	$1\frac{11}{16}$	$1\frac{7}{8}$	$3\frac{3}{4}$	$1\frac{7}{8}$	$1\frac{11}{16}$
2	$3\frac{1}{2}$	2	$1\frac{13}{16}$	2	4	2	$1\frac{13}{16}$
$2\frac{1}{8}$	$3\frac{3}{4}$	$2\frac{1}{8}$	$1\frac{7}{8}$	$2\frac{1}{8}$	4	$2\frac{1}{8}$	$1\frac{7}{8}$
$2\frac{1}{4}$	$3\frac{3}{4}$	$2\frac{1}{4}$	2	$2\frac{1}{4}$	$4\frac{1}{4}$	$2\frac{1}{4}$	2
$2\frac{3}{8}$	4	$2\frac{3}{8}$	$2\frac{1}{8}$	$2\frac{3}{8}$	$4\frac{1}{4}$	$2\frac{3}{8}$	$2\frac{1}{8}$
$2\frac{1}{2}$	$4\frac{1}{4}$	$2\frac{1}{2}$	$2\frac{1}{4}$	$2\frac{1}{2}$	$4\frac{1}{2}$	$2\frac{1}{2}$	$2\frac{1}{4}$
$2\frac{3}{4}$	$4\frac{1}{2}$	$2\frac{3}{4}$	$2\frac{7}{16}$	$2\frac{3}{4}$	$4\frac{3}{4}$	$2\frac{3}{4}$	$2\frac{7}{16}$
3	$4\frac{3}{4}$	3	$2\frac{11}{16}$	3	5	3	$2\frac{11}{16}$
$3\frac{1}{4}$	5	$3\frac{1}{4}$	$2\frac{15}{16}$	$3\frac{1}{4}$	$5\frac{1}{2}$	$3\frac{1}{4}$	$2\frac{15}{16}$
$3\frac{1}{2}$	$5\frac{1}{4}$	$3\frac{1}{2}$	$3\frac{1}{8}$	$3\frac{1}{2}$	6	$3\frac{1}{2}$	$3\frac{1}{8}$

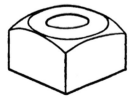

Cold Punched Nuts, Manufacturers Standard

Hexagon				Square			
Dia. Bolt	Across Flats	Thickness	Dia. Hole	Dia. Bolt	Across Flats	Thickness	Dia. Hole
$\frac{1}{4}$	$\frac{1}{2}$	$\frac{1}{4}$	$\frac{7}{32}$	$\frac{1}{4}$	$\frac{1}{2}$	$\frac{1}{4}$	$\frac{7}{32}$
$\frac{5}{16}$	$\frac{5}{8}$	$\frac{5}{16}$	$\frac{9}{32}$	$\frac{5}{16}$	$\frac{5}{8}$	$\frac{5}{16}$	$\frac{9}{32}$
$\frac{3}{8}$	$\frac{3}{4}$	$\frac{3}{8}$	$\frac{11}{32}$	$\frac{3}{8}$	$\frac{3}{4}$	$\frac{3}{8}$	$\frac{11}{32}$
$\frac{7}{16}$	$\frac{7}{8}$	$\frac{7}{16}$	$\frac{13}{32}$	$\frac{7}{16}$	$\frac{7}{8}$	$\frac{7}{16}$	$\frac{13}{32}$
$\frac{1}{2}$	$\frac{7}{8}$	$\frac{1}{2}$	$\frac{7}{16}$	$\frac{1}{2}$	$\frac{7}{8}$	$\frac{1}{2}$	$\frac{7}{16}$
$\frac{9}{16}$	1	$\frac{9}{16}$	$\frac{7}{16}$	$\frac{9}{16}$	1	$\frac{9}{16}$	$\frac{1}{2}$
$\frac{5}{8}$	1	$\frac{9}{16}$	$\frac{9}{16}$	$\frac{5}{8}$	$1\frac{1}{8}$	$\frac{9}{16}$	$\frac{9}{16}$
$\frac{11}{16}$	$1\frac{1}{8}$	$\frac{9}{16}$	$\frac{9}{16}$		$1\frac{1}{8}$		$\frac{9}{16}$
$\frac{3}{4}$	$1\frac{1}{8}$	$\frac{5}{8}$	$\frac{9}{16}$	$\frac{3}{4}$	$1\frac{1}{4}$	$\frac{5}{8}$	$\frac{5}{8}$
$\frac{13}{16}$	$1\frac{1}{4}$	$\frac{5}{8}$	$\frac{11}{16}$		$1\frac{3}{8}$		$\frac{11}{16}$
$\frac{7}{8}$	$1\frac{1}{4}$	$\frac{3}{4}$	$\frac{3}{4}$	$\frac{7}{8}$	$1\frac{1}{2}$	$\frac{3}{4}$	$\frac{3}{4}$
$\frac{15}{16}$	$1\frac{3}{8}$	$\frac{3}{4}$	$\frac{13}{16}$		$1\frac{5}{8}$		$\frac{13}{16}$
1	$1\frac{3}{8}$	$\frac{7}{8}$	$\frac{13}{16}$	1	$1\frac{3}{4}$	$\frac{7}{8}$	$\frac{13}{16}$
$1\frac{1}{16}$	$1\frac{1}{2}$	$\frac{7}{8}$	$\frac{7}{8}$		$1\frac{3}{4}$		$\frac{7}{8}$
$1\frac{1}{8}$	$1\frac{1}{2}$	$\frac{15}{16}$	$\frac{15}{16}$	1	$1\frac{3}{4}$	1	$\frac{15}{16}$
$1\frac{3}{16}$	$1\frac{5}{8}$	1	$\frac{15}{16}$	1	2	1	$\frac{15}{16}$
$1\frac{1}{4}$	$1\frac{5}{8}$	$\frac{7}{8}$	1	$1\frac{1}{8}$	2	$1\frac{1}{8}$	1
$1\frac{5}{16}$	$1\frac{3}{4}$	1	1	$1\frac{1}{8}$	$2\frac{1}{4}$	$1\frac{1}{8}$	$1\frac{1}{16}$
$1\frac{3}{8}$	$1\frac{3}{4}$	1	$1\frac{1}{16}$	$1\frac{1}{4}$	$2\frac{1}{4}$	$1\frac{1}{4}$	$1\frac{1}{16}$
1	$1\frac{3}{4}$	1	$\frac{7}{8}$	$1\frac{1}{4}$	$2\frac{1}{2}$	$1\frac{1}{4}$	$1\frac{1}{16}$
1	$1\frac{3}{4}$	$1\frac{1}{8}$	$\frac{7}{8}$	$1\frac{3}{8}$	$2\frac{3}{4}$	$1\frac{3}{8}$	$1\frac{1}{16}$
$1\frac{1}{8}$	2	$1\frac{1}{4}$	$\frac{15}{16}$	$1\frac{1}{2}$	3	$1\frac{1}{2}$	$1\frac{1}{16}$
$1\frac{1}{4}$	$2\frac{1}{4}$	$1\frac{3}{8}$	$1\frac{1}{16}$	$1\frac{5}{8}$	$3\frac{1}{4}$	$1\frac{5}{8}$	$1\frac{1}{16}$
$1\frac{3}{8}$	$2\frac{1}{2}$	$1\frac{1}{2}$	$1\frac{3}{16}$	$1\frac{3}{4}$	$3\frac{1}{2}$	$1\frac{3}{4}$	$1\frac{3}{16}$
$1\frac{1}{2}$	$2\frac{3}{4}$	$1\frac{5}{8}$	$1\frac{5}{16}$	$1\frac{7}{8}$	$3\frac{3}{4}$	$1\frac{7}{8}$	$1\frac{5}{16}$
$1\frac{5}{8}$	3	$1\frac{3}{4}$	$1\frac{7}{16}$	2	4	2	$1\frac{7}{16}$
$1\frac{3}{4}$	$3\frac{1}{4}$	$1\frac{7}{8}$	$1\frac{9}{16}$				
$1\frac{7}{8}$	$3\frac{1}{2}$	2	$1\frac{11}{16}$				
2	$3\frac{1}{2}$	2	$1\frac{13}{16}$				
2	$3\frac{1}{2}$	$2\frac{1}{8}$	$1\frac{13}{16}$				

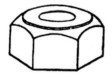 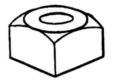

HOT PRESSED NUTS. MANUFACTURERS STANDARD NARROW GAGE SIZES

HEXAGON				SQUARE			
Dia. Bolt	Across Flats	Thickness	Dia. Hole	Dia. Bolt	Across Flats	Thickness	Dia. Hole
				$\frac{1}{8}$	$\frac{11}{32}$	$\frac{5}{32}$	$\frac{5}{32}$
				$\frac{3}{16}$	$\frac{11}{32}$	$\frac{5}{16}$	$\frac{6}{32}$
				$\frac{1}{4}$	$\frac{11}{32}$	$\frac{7}{32}$	$\frac{7}{32}$
				$\frac{5}{16}$	$\frac{9}{16}$	$\frac{9}{32}$	$\frac{9}{32}$
$\frac{3}{8}$	$\frac{11}{16}$	$\frac{3}{8}$	$\frac{21}{64}$	$\frac{3}{8}$	$\frac{11}{16}$	$\frac{3}{8}$	$\frac{21}{64}$
$\frac{7}{16}$	$\frac{25}{32}$	$\frac{7}{16}$	$\frac{25}{64}$	$\frac{7}{16}$	$\frac{25}{32}$	$\frac{7}{16}$	$\frac{25}{64}$
$\frac{1}{2}$	$\frac{7}{8}$	$\frac{1}{2}$	$\frac{7}{16}$	$\frac{1}{2}$	$\frac{7}{8}$	$\frac{1}{2}$	$\frac{7}{16}$
$\frac{9}{16}$	1	$\frac{9}{16}$	$\frac{1}{2}$	$\frac{9}{16}$	1	$\frac{9}{16}$	$\frac{1}{2}$
$\frac{5}{8}$	$1\frac{1}{8}$	$\frac{5}{8}$	$\frac{9}{16}$	$\frac{5}{8}$	$1\frac{1}{8}$	$\frac{5}{8}$	$\frac{9}{16}$
$\frac{3}{4}$	$1\frac{1}{4}$	$\frac{3}{4}$	$\frac{21}{32}$	$\frac{3}{4}$	$1\frac{3}{8}$	$\frac{3}{4}$	$\frac{21}{32}$
$\frac{7}{8}$	$1\frac{1}{2}$	$\frac{7}{8}$	$\frac{49}{64}$	$\frac{7}{8}$	$1\frac{5}{8}$	$\frac{7}{8}$	$\frac{49}{64}$
1	$1\frac{5}{8}$	1	$\frac{7}{8}$	1	$1\frac{3}{4}$	1	$\frac{7}{8}$
$1\frac{1}{8}$	2	$1\frac{1}{8}$	$\frac{31}{32}$	$1\frac{1}{8}$	2	$1\frac{1}{8}$	$\frac{31}{32}$
$1\frac{1}{4}$	$2\frac{1}{4}$	$1\frac{1}{4}$	$1\frac{3}{32}$	$1\frac{1}{4}$	$2\frac{1}{4}$	$1\frac{1}{4}$	$1\frac{3}{32}$
$1\frac{3}{8}$	$2\frac{1}{2}$	$1\frac{1}{2}$	$1\frac{3}{16}$	$1\frac{3}{8}$	$2\frac{1}{2}$	$1\frac{1}{2}$	$1\frac{3}{16}$
$1\frac{1}{2}$	$2\frac{1}{2}$	$1\frac{1}{2}$	$1\frac{3}{16}$	$1\frac{1}{2}$	$2\frac{1}{2}$	$1\frac{1}{2}$	$1\frac{3}{16}$

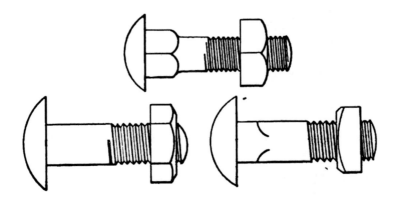

BUTTON HEAD MACHINE, CARRIAGE AND LOOM BOLTS

Diameter Bolt	$\frac{1}{4}$	$\frac{5}{16}$	$\frac{3}{8}$	$\frac{7}{16}$	$\frac{1}{2}$	$\frac{9}{16}$	$\frac{5}{8}$	$\frac{3}{4}$	$\frac{7}{8}$	1
Diameter Head	$\frac{1}{2}$	$\frac{5}{8}$	$\frac{3}{4}$	$\frac{7}{8}$	1	$1\frac{1}{8}$	$1\frac{1}{4}$	$1\frac{1}{2}$	$1\frac{3}{4}$	2
Thickness of Head	$\frac{1}{8}$	$\frac{5}{32}$	$\frac{3}{16}$	$\frac{7}{32}$	$\frac{1}{4}$	$\frac{9}{32}$	$\frac{5}{16}$	$\frac{3}{8}$	$\frac{7}{16}$	$\frac{1}{2}$

LENGTHS OF BOLTS

SQUARE HEAD, Hexagon Head, Button Head, Round Head and Cone Head bolts are measured under the head. Countersunk Head bolts, bolt ends and rods are measured over all.

LENGTHS OF THREADS CUT ON BOLTS

Length of Bolts	$\frac{1}{4}$ & $\frac{5}{16}$	$\frac{3}{8}$ & $\frac{7}{16}$	$\frac{1}{2}$ & $\frac{9}{16}$	$\frac{5}{8}$	$\frac{3}{4}$	$\frac{7}{8}$	1	$1\frac{1}{8}$	$1\frac{1}{4}$
1 to 1$\frac{1}{2}$"	$\frac{3}{4}$	$\frac{7}{8}$	1	$1\frac{1}{4}$	..	..	..	..	..
1$\frac{5}{8}$ to 2 "	$\frac{7}{8}$	1	1	$1\frac{1}{4}$	$1\frac{1}{2}$	$1\frac{1}{4}$	..	..	..
2$\frac{1}{8}$ to 2$\frac{1}{2}$"	1	1	1	$1\frac{1}{4}$	$1\frac{1}{2}$	$1\frac{1}{4}$	$1\frac{3}{4}$	..	..
2$\frac{5}{8}$ to 3 "	1	1	1	$1\frac{1}{4}$	$1\frac{1}{2}$	$1\frac{3}{4}$	2	$2\frac{1}{4}$	..
3$\frac{1}{8}$ to 4 "	1	$1\frac{1}{4}$	$1\frac{1}{4}$	$1\frac{1}{2}$	$1\frac{1}{2}$	$1\frac{3}{4}$	2	$2\frac{1}{2}$	$2\frac{1}{4}$
4$\frac{1}{8}$ to 8 "	1	$1\frac{1}{4}$	$1\frac{1}{4}$	$1\frac{1}{2}$	$1\frac{3}{4}$	2	$2\frac{1}{4}$	$2\frac{3}{4}$	3
8$\frac{1}{8}$ to 12 "	1	$1\frac{1}{4}$	$1\frac{1}{2}$	$1\frac{3}{4}$	2	$2\frac{1}{4}$	$2\frac{1}{2}$	3	$3\frac{1}{4}$
12$\frac{1}{8}$ to 20 "	1	$1\frac{1}{2}$	2	2	2	$2\frac{1}{2}$	3	$3\frac{1}{2}$	$3\frac{1}{2}$

Bolts longer than 20 inches and larger than 1$\frac{1}{4}$ inch in diameter are usually threaded about 3 times the diameter of the rod.

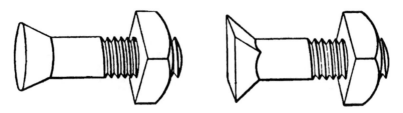

ROUND AND SQUARE COUNTERSUNK HEAD BOLTS

Diameter Bolt	$\frac{1}{4}$	$\frac{5}{16}$	$\frac{3}{8}$	$\frac{7}{16}$	$\frac{1}{2}$	$\frac{9}{16}$	$\frac{5}{8}$	$\frac{3}{4}$	$\frac{7}{8}$	1
Diameter Round Head	$\frac{1}{2}$	$\frac{19}{32}$	$\frac{11}{16}$	$\frac{25}{32}$	$\frac{7}{8}$	$\frac{31}{32}$	$1\frac{1}{16}$	$1\frac{1}{4}$	$1\frac{7}{16}$	$1\frac{5}{8}$
Distance across Flats Square Head	$\frac{1}{2}$	$\frac{19}{32}$	$\frac{11}{16}$	$\frac{25}{32}$	$\frac{7}{8}$	$\frac{31}{32}$	$1\frac{1}{16}$	$1\frac{1}{4}$	$1\frac{7}{16}$	$1\frac{5}{8}$
Thickness Square and Round Heads	$\frac{5}{16}$	$\frac{5}{16}$	$\frac{7}{32}$	$\frac{1}{4}$	$\frac{1}{4}$	$\frac{9}{32}$	$\frac{5}{16}$	$\frac{11}{32}$	$\frac{13}{32}$	$\frac{7}{16}$

TAP BOLTS

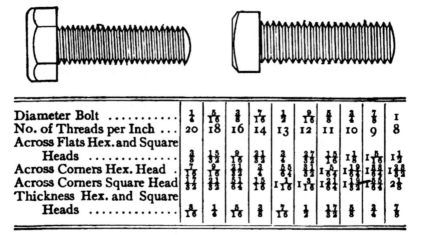

Diameter Bolt	$\frac{1}{4}$	$\frac{5}{16}$	$\frac{3}{8}$	$\frac{7}{16}$	$\frac{1}{2}$	$\frac{9}{16}$	$\frac{5}{8}$	$\frac{3}{4}$	$\frac{7}{8}$	1
No. of Threads per Inch	20	18	16	14	13	12	11	10	9	8
Across Flats Hex. and Square Heads	$\frac{3}{8}$	$\frac{15}{32}$	$\frac{9}{16}$	$\frac{21}{32}$	$\frac{3}{4}$	$\frac{27}{32}$	$\frac{15}{16}$	$1\frac{1}{8}$	$1\frac{5}{16}$	$1\frac{1}{2}$
Across Corners Hex. Head	$\frac{7}{16}$	$\frac{9}{16}$	$\frac{21}{32}$	$\frac{3}{4}$	$\frac{61}{64}$	$\frac{31}{32}$	$1\frac{5}{64}$	$1\frac{19}{64}$	$1\frac{33}{64}$	$1\frac{47}{64}$
Across Corners Square Head	$\frac{17}{32}$	$\frac{21}{32}$	$\frac{51}{64}$	$1\frac{1}{8}$	$1\frac{1}{16}$	$1\frac{3}{16}$	$1\frac{21}{64}$	$1\frac{37}{32}$	$1\frac{53}{64}$	$2\frac{1}{8}$
Thickness Hex. and Square Heads	$\frac{5}{16}$	$\frac{1}{4}$	$\frac{5}{16}$	$\frac{3}{8}$	$\frac{7}{16}$	$\frac{1}{2}$	$\frac{17}{32}$	$\frac{5}{8}$	$\frac{3}{4}$	$\frac{7}{8}$

STOVE BOLT DIAMETERS AND THREADS

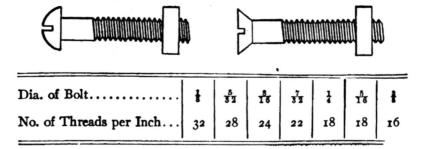

Dia. of Bolt	$\frac{1}{8}$	$\frac{5}{32}$	$\frac{3}{16}$	$\frac{7}{32}$	$\frac{1}{4}$	$\frac{5}{16}$	$\frac{3}{8}$
No. of Threads per Inch	32	28	24	22	18	18	16

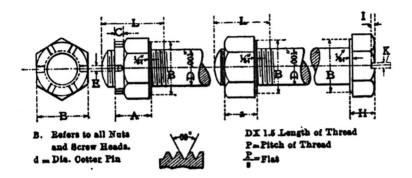

B. Refers to all Nuts
and Screw Heads.
d = Dia. Cotter Pin

DX 1.5 Length of Thread
P = Pitch of Thread
$\frac{P}{8}$ = Flat

AUTOMOBILE SCREW AND NUT STANDARDS ADOPTED BY THE A.L.A.M.

D	$\frac{1}{4}$	$\frac{5}{16}$	$\frac{3}{8}$	$\frac{7}{16}$	$\frac{1}{2}$	$\frac{9}{16}$	$\frac{5}{8}$	$\frac{11}{16}$	$\frac{3}{4}$	$\frac{7}{8}$	1
P	28	24	24	20	20	18	18	16	16,	14	14
A	$\frac{9}{32}$	$\frac{21}{64}$	$\frac{11}{32}$	$\frac{22}{64}$	$\frac{7}{16}$	$\frac{12}{64}$	$\frac{11}{32}$	$\frac{42}{64}$	$\frac{18}{16}$	$\frac{21}{32}$	1
a	$\frac{7}{32}$	$\frac{17}{64}$	$\frac{21}{64}$	$\frac{3}{8}$	$\frac{7}{16}$	$\frac{11}{64}$	$\frac{11}{64}$	$\frac{11}{32}$	$\frac{21}{32}$	$\frac{11}{64}$	$\frac{7}{8}$
B	$\frac{3}{8}$	$\frac{1}{2}$	$\frac{9}{16}$	$\frac{11}{16}$	$\frac{3}{4}$	$\frac{7}{8}$	$\frac{15}{16}$	1	$1\frac{1}{8}$	$1\frac{1}{4}$	$1\frac{7}{16}$
C	$\frac{3}{32}$	$\frac{3}{32}$	$\frac{1}{8}$	$\frac{1}{8}$	$\frac{3}{16}$	$\frac{3}{16}$	$\frac{1}{4}$	$\frac{1}{4}$	$\frac{1}{4}$	$\frac{1}{4}$	$\frac{1}{4}$
E	$\frac{5}{64}$	$\frac{5}{64}$	$\frac{1}{8}$	$\frac{1}{8}$	$\frac{1}{8}$	$\frac{5}{32}$	$\frac{5}{32}$	$\frac{5}{32}$	$\frac{5}{32}$	$\frac{5}{32}$	$\frac{5}{32}$
H	$\frac{5}{16}$	$\frac{15}{64}$	$\frac{9}{32}$	$\frac{21}{64}$	$\frac{3}{8}$	$\frac{27}{64}$	$\frac{15}{32}$	$\frac{33}{64}$	$\frac{9}{16}$	$\frac{21}{32}$	$\frac{3}{4}$
I	$\frac{3}{32}$	$\frac{7}{64}$	$\frac{1}{8}$	$\frac{1}{8}$	$\frac{1}{8}$	$\frac{1}{8}$	$\frac{1}{8}$	$\frac{1}{8}$	$\frac{1}{8}$	$\frac{1}{8}$	$\frac{1}{8}$
K	$\frac{1}{16}$	$\frac{1}{16}$	$\frac{3}{32}$	$\frac{3}{32}$	$\frac{3}{32}$	$\frac{3}{32}$	$\frac{3}{32}$	$\frac{3}{32}$	$\frac{3}{32}$	$\frac{3}{32}$	$\frac{3}{32}$
d	$\frac{1}{16}$	$\frac{1}{16}$	$\frac{3}{32}$	$\frac{3}{32}$	$\frac{3}{32}$	$\frac{1}{8}$	$\frac{1}{8}$	$\frac{1}{8}$	$\frac{1}{8}$	$\frac{1}{8}$	$\frac{1}{8}$
L	$\frac{3}{8}$	$\frac{15}{32}$	$\frac{9}{16}$	$\frac{21}{32}$	$\frac{3}{4}$	$\frac{27}{32}$	$\frac{15}{16}$	$1\frac{1}{32}$	$1\frac{1}{8}$	$1\frac{5}{16}$	$1\frac{1}{2}$

PLANER NUTS

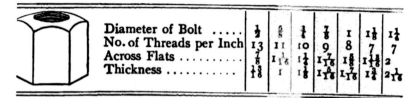

Diameter of Bolt	$\frac{1}{2}$	$\frac{5}{8}$	$\frac{3}{4}$	$\frac{7}{8}$	1	$1\frac{1}{8}$	$1\frac{1}{4}$
No. of Threads per Inch	13	11	10	9	8	7	7
Across Flats	$\frac{7}{8}$	$1\frac{1}{16}$	$1\frac{1}{2}$	$1\frac{7}{16}$	$1\frac{5}{8}$	$1\frac{13}{16}$	2
Thickness	$1\frac{3}{16}$	1	$1\frac{1}{8}$	$1\frac{1}{16}$	$1\frac{7}{16}$	$1\frac{3}{4}$	$2\frac{1}{16}$

COUPLING BOLTS

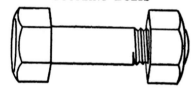

Diameter of Bolt	$\frac{1}{2}$	$\frac{5}{8}$	$\frac{3}{4}$	$\frac{7}{8}$	1	$1\frac{1}{8}$	$1\frac{1}{4}$
No. of Threads per Inch	13	11	10	9	8	7	7
Short Diameter of Head	$\frac{7}{8}$	$1\frac{1}{16}$	$1\frac{1}{4}$	$1\frac{7}{16}$	$1\frac{5}{8}$	$1\frac{13}{16}$	2
Length of Head	$\frac{1}{2}$	$\frac{5}{8}$	$\frac{3}{4}$	$\frac{7}{8}$	1	$1\frac{1}{8}$	$1\frac{1}{4}$
Thickness of Nut	$\frac{1}{2}$	$\frac{5}{8}$	$\frac{3}{4}$	$\frac{7}{8}$	1	$1\frac{1}{8}$	$1\frac{1}{4}$
Short Diameter of Nut	$\frac{7}{8}$	$1\frac{1}{16}$	$1\frac{1}{4}$	$1\frac{7}{16}$	$1\frac{5}{8}$	$1\frac{13}{16}$	2

PLANER HEAD BOLTS AND NUTS

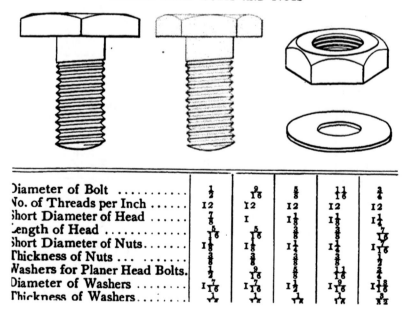

Diameter of Bolt	$\frac{1}{2}$	$\frac{9}{16}$	$\frac{5}{8}$	$\frac{11}{16}$	$\frac{3}{4}$
No. of Threads per Inch	12	12	12	12	12
Short Diameter of Head	$\frac{7}{8}$	1	$1\frac{1}{8}$	$1\frac{1}{8}$	$1\frac{1}{4}$
Length of Head	$\frac{5}{16}$	$\frac{5}{16}$	$\frac{3}{8}$	$\frac{3}{8}$	$\frac{7}{16}$
Short Diameter of Nuts.......	$1\frac{1}{8}$	$1\frac{1}{8}$	$1\frac{1}{4}$	$1\frac{1}{4}$	$1\frac{7}{16}$
Thickness of Nuts	$\frac{3}{8}$	$\frac{3}{8}$	$\frac{3}{8}$	$\frac{3}{8}$	$\frac{1}{2}$
Washers for Planer Head Bolts.	$\frac{1}{2}$	$\frac{9}{16}$	$\frac{5}{8}$	$\frac{11}{16}$	$\frac{3}{4}$
Diameter of Washers	$1\frac{7}{16}$	$1\frac{7}{16}$	$1\frac{1}{2}$	$1\frac{9}{16}$	$1\frac{11}{16}$
Thickness of Washers........	$\frac{1}{7}$	$\frac{1}{7}$	$\frac{1}{7}$	$\frac{1}{7}$	$\frac{3}{32}$

Depths to Drill and Tap for Studs

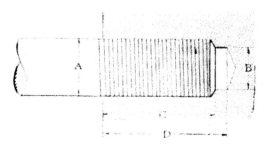

Dia. of Stud	A									
Dia. of Drill	B									
Depth of Thread	C									
Depth of Drill	D									

Bolt Heads for Standard T-Slots

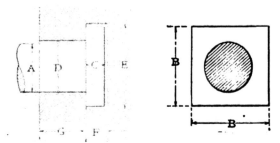

Dia. of Bolt	Width of Head	Thickness of Head	Width of Slot	Width of Slot	Depth of Slot	Maximum Depth with St'd Cutter
A	B	C	D	E	F	G
$\frac{1}{4}$	$\frac{9}{16}$	$\frac{1}{8}$	$\frac{5}{16}$	$\frac{5}{8}$	$\frac{5}{32}$	$\frac{3}{8}$
$\frac{5}{16}$	$\frac{5}{8}$	$\frac{3}{16}$	$\frac{3}{8}$	$\frac{11}{16}$	$\frac{7}{32}$	$\frac{7}{16}$
$\frac{3}{8}$	$\frac{3}{4}$	$\frac{3}{16}$	$\frac{7}{16}$	$\frac{13}{16}$	$\frac{7}{32}$	$\frac{9}{16}$
$\frac{7}{16}$	$\frac{7}{8}$	$\frac{1}{4}$	$\frac{1}{2}$	$\frac{15}{16}$	$\frac{9}{32}$	$\frac{3}{4}$
$\frac{1}{2}$	$1\frac{1}{8}$	$\frac{3}{8}$	$\frac{5}{8}$	$1\frac{3}{16}$	$\frac{13}{32}$	
$\frac{5}{8}$	$1\frac{1}{4}$	$\frac{1}{2}$	$\frac{3}{4}$	$1\frac{5}{16}$	$\frac{15}{32}$	1
$\frac{3}{4}$	$1\frac{9}{16}$	$\frac{5}{8}$		$1\frac{8}{8}$	$\frac{11}{16}$	$1\frac{1}{16}$
$\frac{7}{8}$	$1\frac{13}{16}$	$\frac{3}{4}$	1	$1\frac{7}{8}$	$\frac{15}{16}$	$1\frac{3}{16}$

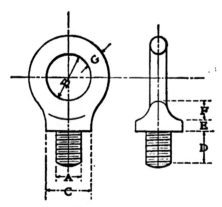

EYE BOLTS

A	B	C	D	E	F	G
3/8	2	3/4	7/8	3/16	3/8	5/16
1/2	2 1/8	1	1	1/4	1/2	3/8
5/8	2 1/4	1 1/2	1 1/8	5/16	5/8	7/16
3/4	2 3/8	1 7/16	1 1/4	5/16	11/16	9/16
7/8	2 1/2	1 11/16	1 3/8	3/8	3/4	5/8
1	2 3/4	1 7/8	1 1/2	7/16	7/8	3/4
1 1/8	2 7/8	2 3/8	1 5/8	1/2	1	13/16
1 1/4	3	2 5/8	1 3/4	1/2	1 1/8	7/8
1 3/8	3 1/8	2 5/8	1 7/8	9/16	1 1/16	1
1 1/2	3 1/4	2 3/4	2	5/8	1 1/4	1 1/16
1 5/8	3 3/8	3	2 1/8	11/16	1 3/8	1 1/8
1 3/4	3 1/2	3 1/4	2 1/4	3/4	1 1/2	1 1/4
1 7/8	3 5/8	3 1/2	2 3/8	13/16	1 5/8	1 5/16
2	3 3/4	3 3/4	2 1/2	7/8	1 3/4	1 3/8

SPRING COTTERS

No. of Wire Gage	13	12	11	10	9	8	7	6
Dia. Inches	3/32	7/64	1/8	9/64	5/32	11/64	3/16	13/64
Lengths. Inches	1/2 to 2	1/2 to 2	1/2 to 2 1/2	1/2 to 2 1/2	1/2 to 2 1/2	1/2 to 2 1/2	3/4 to 3	3/4 to 3

No. of Wire Gage	5	4	1				
Dia. Inches	5/32	1/4	1/16	3/8	7/16	1/2	5/8
Lengths. Inches	1 to 3	1 to 4	1 to 4	1 1/2 to 4	1 3/4 to 5	2 to 6	3 to 6

Regular Lengths vary by 1/4 inch up to 4 inches and by 1 inch from 4 to 6 inches. Lengths are measured under the eye.

Round and Square Washers

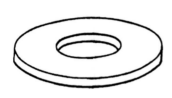

U. S. Standard Washers

Size of Bolt	Size of Hole	Outside Diameter	Thickness Wire Gage No.
$\frac{3}{16}$	$\frac{1}{4}$	$\frac{9}{16}$	18 ($\frac{9}{64}$)
$\frac{1}{4}$	$\frac{5}{16}$	$\frac{3}{4}$	16 ($\frac{1}{16}$)
$\frac{5}{16}$	$\frac{3}{8}$	$\frac{7}{8}$	16 ($\frac{1}{16}$)
$\frac{3}{8}$	$\frac{7}{16}$	1	14 ($\frac{5}{64}$)
$\frac{7}{16}$	$\frac{1}{2}$	$1\frac{1}{4}$	14 ($\frac{5}{64}$)
$\frac{1}{2}$	$\frac{9}{16}$	$1\frac{3}{8}$	12 ($\frac{3}{32}$)
$\frac{9}{16}$	$\frac{5}{8}$	$1\frac{1}{2}$	12 ($\frac{3}{32}$)
$\frac{5}{8}$	$\frac{11}{16}$	$1\frac{3}{4}$	10 ($\frac{1}{8}$)
$\frac{3}{4}$	$\frac{13}{16}$	2	10 ($\frac{1}{8}$)
$\frac{7}{8}$	$\frac{15}{16}$	$2\frac{1}{4}$	9 ($\frac{5}{32}$)
1	$1\frac{1}{16}$	$2\frac{1}{2}$	9 ($\frac{5}{32}$)
$1\frac{1}{8}$	$1\frac{1}{4}$	$2\frac{3}{4}$	9 ($\frac{5}{32}$)
$1\frac{1}{4}$	$1\frac{3}{8}$	3	9 ($\frac{5}{32}$)
$1\frac{3}{8}$	$1\frac{1}{2}$	$3\frac{1}{4}$	8 ($\frac{11}{64}$)
$1\frac{1}{2}$	$1\frac{5}{8}$	$3\frac{1}{2}$	8 ($\frac{11}{64}$)
$1\frac{5}{8}$	$1\frac{3}{4}$	$3\frac{3}{4}$	8 ($\frac{11}{64}$)
$1\frac{3}{4}$	$1\frac{7}{8}$	4	8 ($\frac{11}{64}$)
$1\frac{7}{8}$	2	$4\frac{1}{4}$	8 ($\frac{11}{64}$)
2	$2\frac{1}{8}$	$4\frac{1}{2}$	8 ($\frac{11}{64}$)
$2\frac{1}{4}$	$2\frac{3}{8}$	$4\frac{3}{4}$	6 ($\frac{7}{16}$)
$2\frac{1}{2}$	$2\frac{5}{8}$	5	5 ($\frac{7}{32}$)

Narrow Gage Washers

Size of Bolt	Size of Hole	Outside Diameter	Thickness Wire Gage No.
$\frac{1}{4}$	$\frac{5}{16}$	$\frac{5}{8}$	16 ($\frac{1}{16}$)
$\frac{5}{16}$	$\frac{3}{8}$	$\frac{3}{4}$	16 ($\frac{1}{16}$)
$\frac{3}{8}$	$\frac{7}{16}$	$\frac{7}{8}$	16 ($\frac{1}{16}$)
$\frac{7}{16}$	$\frac{1}{2}$	$1\frac{1}{8}$	14 ($\frac{5}{64}$)
$\frac{1}{2}$	$\frac{9}{16}$	$1\frac{1}{4}$	12 ($\frac{3}{32}$)
$\frac{9}{16}$	$\frac{5}{8}$	$1\frac{3}{8}$	12 ($\frac{3}{32}$)
$\frac{5}{8}$	$\frac{11}{16}$	$1\frac{1}{2}$	10 ($\frac{1}{8}$)
$\frac{3}{4}$	$\frac{13}{16}$	$1\frac{3}{4}$	10 ($\frac{1}{8}$)
$\frac{7}{8}$	$\frac{15}{16}$	2	9 ($\frac{5}{32}$)
1	$1\frac{1}{16}$	$2\frac{1}{4}$	9 ($\frac{5}{32}$)
$1\frac{1}{8}$	$1\frac{1}{4}$	$2\frac{1}{2}$	9 ($\frac{5}{32}$)
$1\frac{1}{4}$	$1\frac{3}{8}$	$2\frac{3}{4}$	9 ($\frac{5}{32}$)
$1\frac{3}{8}$	$1\frac{1}{2}$	3	8 ($\frac{11}{64}$)
$1\frac{1}{2}$	$1\frac{5}{8}$	$3\frac{1}{4}$	8 ($\frac{11}{64}$)

Square Washers
Standard Sizes

Size of Bolt	Size of Hole	Width	Thickness
$\frac{3}{8}$	$\frac{7}{16}$	$1\frac{1}{2}$	$\frac{1}{8}$
$\frac{7}{16}$	$\frac{1}{2}$	$1\frac{3}{4}$	
$\frac{1}{2}$	$\frac{9}{16}$	2	
$\frac{5}{8}$	$\frac{11}{16}$	$2\frac{1}{4}$	
$\frac{3}{4}$	$\frac{13}{16}$	$2\frac{1}{2}$	$\frac{3}{16}$
$\frac{7}{8}$	$\frac{15}{16}$	3	
1	$1\frac{1}{32}$	$3\frac{1}{2}$	
$1\frac{1}{8}$	$1\frac{1}{4}$	4	
$1\frac{1}{4}$	$1\frac{3}{8}$	$4\frac{1}{2}$	
$1\frac{3}{8}$	$1\frac{1}{2}$	5	
$1\frac{1}{2}$	$1\frac{5}{8}$	6	
$1\frac{3}{4}$	$1\frac{7}{8}$	6	
2	$2\frac{1}{8}$	6	

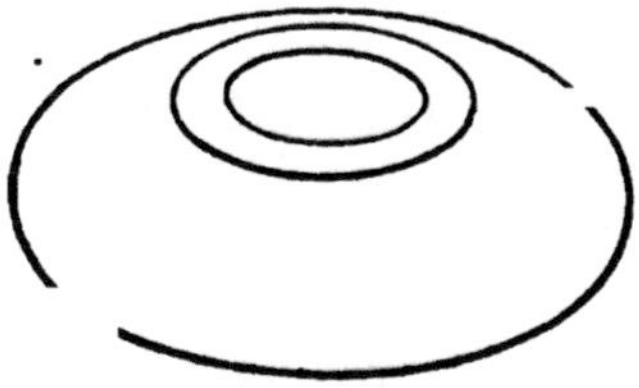

Cast-Iron Washers

Size of Bolt	Outside Diameter	Thickness	Size of Bolt	Outside Diameter	Thickness
3/8	1 1/2	5/16	1 1/8	4 1/2	1
1/2	2	3/8	1 1/4	5	1 1/8
5/8	2 1/2	1/2	1 3/8	5 1/2	1 1/4
3/4	3	5/8	1 1/2	6	1 3/8
7/8	3 1/2	3/4	1 3/4	7	1 1/2
1	4	7/8	2	7 1/2	1 5/8

Riveting Washers

Size of Rivet	Size of Hole	Outside Diameter	Thickness Wire Gage	Size of Rivet	Size of Hole	Outside Diameter	Thickness Wire Gag
7 (.180)	3/16	1/2	18	3/8	13/32	1 1/4	12
6 (.203)	7/32	9/16	18	7/16	15/32	1	14
5 (.220)	15/64	5/8	18	7/16	15/32	1 1/4	12
1/4	17/64	5/8	16	1/2	17/32	1 1/8	12
1/4	17/64	3/4	16	1/2	17/32	1 1/4	12
5/16	11/32	3/4	14	1/2	17/32	1 1/2	11
5/16	11/32	7/8	14	5/8	21/32	1 1/2	11
3/8	13/32	7/8	14	5/8	21/32	1 3/4	10
3/8	13/32	1	14				

Machine and Wood Screw Gage

No. of Screw Gage	Size of Number in Decimals	No. of Screw Gage	Size of Number in Decimals	No. of Screw Gage	Size of Number in Decimals	No. of Screw Gage	Size of Number in Decimals
000	.03152	12	.21576	25	.38684	38	.55792
00	.04468	13	.22892	26	.40000	39	.57108
0	.05784	14	.24208	27	.41316	40	.58424
1	.07100	15	.25524	28	.42632	41	.59740
2	.08416	16	.26840	29	.43948	42	.61056
3	.09732	17	.28156	30	.45264	43	.62372
4	.11048	18	.29472	31	.46580	44	.63688
5	.12364	19	.30788	32	.47896	45	.65004
6	.13680	20	.32104	33	.49212	46	.66320
7	.14996	21	.33420	34	.50528	47	.67636
8	.16312	22	.34736	35	.51844	48	.68952
9	.17628	23	.36052	36	.53160	49	.70268
10	.18944	24	.37368	37	.54476	50	.71584
11	.20260	The difference between consecutive sizes is .01316"					

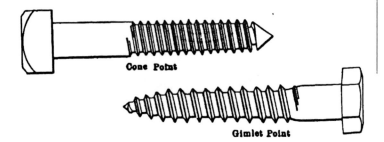

Cone Point

Gimlet Point

Coach and Lag Screws

Diameter Screw	$\frac{1}{4}$	$\frac{5}{16}$	$\frac{3}{8}$	$\frac{7}{16}$	$\frac{1}{2}$	$\frac{9}{16}$	$\frac{5}{8}$	$\frac{3}{4}$	$\frac{7}{8}$	1
No. of Threads per Inch	10	$9\frac{1}{2}$	7	7	6	5	5	$4\frac{1}{2}$	$4\frac{1}{2}$	3
Across Flats Hex. and Square Heads	$\frac{3}{8}$	$\frac{15}{32}$	$\frac{9}{16}$	$\frac{21}{32}$	$\frac{3}{4}$	$\frac{27}{32}$	$\frac{15}{16}$	$1\frac{1}{8}$	$1\frac{5}{16}$	$1\frac{1}{2}$
Thickness Hex. and Square Heads	$\frac{3}{16}$	$\frac{1}{4}$	$\frac{5}{16}$	$\frac{3}{8}$	$\frac{7}{16}$	$\frac{1}{2}$	$\frac{17}{32}$	$\frac{5}{8}$	$\frac{3}{4}$	$\frac{7}{8}$

Lengths of Threads on Coach and Lag Screws
of all Diameters

Length of Screw	Length of Thread	Length of Screw	Length of Thread
$1\frac{1}{2}''$	To Head	$5''$	$4''$
$2\frac{1}{4}''$	$1\frac{1}{2}''$	$5\frac{1}{2}''$	$4''$
$2\frac{1}{2}''$	$2''$	$6''$	$4\frac{1}{2}''$
$3''$	$2\frac{1}{4}''$	$7''$	$5''$
$3\frac{1}{2}''$	$2\frac{1}{2}''$	$8''$	$6''$
$4''$	$3''$	$9''$	$6''$
$4\frac{1}{2}''$	$3\frac{1}{2}''$	10 to $12''$	$7''$

Lag-Screw Test
(Screws Drawn Out of Yellow Pine)
Test by Hoopes and Townsend

Diameter Screw	$\frac{1}{2}$ in.	$\frac{5}{8}$ in.	$\frac{3}{4}$ in.	$\frac{7}{8}$ in.	1 in.
Depth in Wood	$3\frac{1}{2}$ in.	4 in.	4 in.	5 in.	6 in.
Force in Pounds	4,960	6,000	7,685	11,500	12,620

WOOD SCREWS

WOOD screws range in size from No. 0 to No. 30, by the American Screw Company's gage and in lengths from ¼ inch to 6 inches. The increase in length is by eights of an inch up to 1 inch, then by quarters of an inch up to 3 inches and by half inches up to 5 inches. As a rule the threaded portion is about seven tenths of the total length. The included angle of the flat head is 82 degrees. The table below gives the body and head diameters, and the threads per inch as generally cut, although there is no fixed standard as to number of threads which is universally adhered to by all wood-screw manufacturers.

WOOD-SCREW DIMENSIONS

(ANGLE OF FLAT HEAD = 82 DEGREES)

No. of Screw Gage	Diameter of Screw	Diameter of Head	Threads per Inch	No. of Screw Gage	Diameter of Screw	Diameter of Head	Threads per Inch
0	.05784 (1/16 −)	.110 (7/64 +)	32	16	.26840 (17/64 +)	.526 (17/32 −)	9
1	.07100 (5/64 −)	.136 (9/64 −)	28	17	.28156 (9/32)	.552 (35/64 +)	9
2	.08416 (5/64 +)	.162 (5/32 +)	26	18	.29472 (19/64 −)	.578 (37/64)	8
3	.09732 (3/32 +)	.188 (3/16)	24	19	.30788 (5/16 −)	.604 (39/64 −)	8
4	.11048 (7/64 +)	.214 (7/32 −)	22	20	.32104 (21/64 −)	.630 (5/8 +)	8
5	.12364 (1/8 −)	.240 (15/64 +)	20	21	.33420 (21/64 +)	.656 (21/32)	8
6	.13680 (9/64 −)	.266 (17/64 +)	18	22	.34736 (11/32 +)	.682 (11/16 −)	7
7	.14996 (5/32 −)	.292 (19/64 −)	16	23	.36052 (23/64)	.708 (45/64 +)	7
8	.16312 (5/32 +)	.318 (5/16 +)	15	24	.37368 (3/8 −)	.734 (47/64)	7
9	.17628 (11/64 +)	.344 (11/32 +)	14	25	.38684 (25/64 −)	.760 (49/64 −)	7
10	.18944 (3/16 +)	.370 (3/8 −)	13	26	.40000 (13/32 −)	.786 (25/32 +)	6
11	.20260 (13/64 +)	.396 (25/64 +)	12	27	.41316 (13/32 +)	.812 (13/16)	6
12	.21576 (7/32 −)	.422 (27/64 −)	11	28	.42632 (27/64 +)	.838 (27/32 −)	6
13	.22892 (15/64 −)	.448 (29/64 −)	11	29	.43948 (7/16 +)	.864 (55/64 +)	6
14	.24208 (1/4 −)	.474 (15/32 +)	10	30	.45264 (29/64)	.890 (57/64)	6
15	.25524 (1/4 +)	.500 (1/2)	10				

BOILER RIVET HEADS

Size of Rivet	Button Heads		Cone Heads		Countersunk Heads	
	Dia.	Thickness	Dia.	Thickness	Dia.	Thickness
$\frac{1}{2}$	$\frac{7}{8}$	$\frac{3}{8}$	$\frac{7}{8}$	$\frac{15}{32}$	$\frac{7}{8}$	$\frac{1}{4}$
$\frac{9}{16}$	$\frac{15}{16}$	$\frac{3}{8}$	$\frac{31}{32}$	$\frac{31}{64}$	$\frac{15}{16}$	$\frac{9}{32}$
$\frac{5}{8}$	$1\frac{1}{16}$	$\frac{7}{16}$	$1\frac{1}{16}$	$\frac{9}{16}$	$1\frac{1}{8}$	$\frac{9}{32}$
$\frac{11}{16}$	$1\frac{1}{8}$	$\frac{1}{2}$	$1\frac{1}{8}$	$\frac{39}{64}$	$1\frac{1}{16}$	$\frac{5}{16}$
$\frac{3}{4}$	$1\frac{1}{4}$	$\frac{9}{16}$	$1\frac{1}{4}$	$\frac{21}{32}$	$1\frac{1}{4}$	$\frac{3}{8}$
$\frac{7}{8}$	$1\frac{7}{16}$	$\frac{3}{4}$	$1\frac{7}{16}$	$\frac{3}{4}$	$1\frac{3}{8}$	$\frac{7}{16}$
1	$1\frac{5}{8}$	$\frac{3}{4}$	$1\frac{5}{8}$	$\frac{27}{32}$	$1\frac{5}{8}$	$\frac{1}{2}$
$1\frac{1}{8}$	$1\frac{3}{4}$	$1\frac{3}{16}$	$1\frac{27}{32}$	$\frac{15}{16}$	$1\frac{7}{8}$	$\frac{9}{16}$
$1\frac{1}{4}$	2	1	$2\frac{1}{16}$	$1\frac{1}{32}$	$2\frac{1}{8}$	$\frac{5}{8}$

TANK RIVET HEADS

Size of Rivet	Button Heads		Flat Heads		Countersunk Heads	
	Dia.	Thickness	Dia.	Thickness	Dia.	Thickness
$\frac{1}{8}$	$\frac{9}{32}$	$\frac{5}{64}$	$\frac{3}{32}$	$\frac{3}{64}$	$\frac{1}{4}$	$\frac{1}{16}$
$\frac{5}{16}$	$\frac{13}{32}$	$\frac{8}{64}$	$\frac{3}{8}$	$\frac{1}{16}$	$\frac{11}{32}$	$\frac{3}{32}$ full
$\frac{1}{4}$	$\frac{1}{32}$	$\frac{5}{32}$	$\frac{17}{32}$	$\frac{7}{32}$	$\frac{15}{32}$	$\frac{3}{32}$ "
$\frac{5}{16}$	$\frac{9}{16}$	$\frac{3}{16}$	$\frac{21}{32}$	$\frac{7}{64}$	$\frac{5}{16}$	$\frac{1}{16}$
$\frac{3}{8}$ sct.	$\frac{11}{16}$	$\frac{1}{4}$	$\frac{25}{32}$	$\frac{5}{32}$	$\frac{5}{8}$	$\frac{7}{32}$
$\frac{3}{8}$ ex.	$\frac{25}{32}$	$\frac{5}{16}$	$\frac{25}{32}$	$\frac{11}{64}$	$\frac{11}{16}$	$\frac{1}{4}$
$\frac{7}{16}$	$\frac{13}{16}$	$\frac{5}{8}$	$\frac{13}{16}$	$\frac{11}{64}$	$\frac{3}{4}$	$\frac{9}{32}$
$\frac{1}{2}$	$\frac{7}{8}$	$\frac{3}{8}$	$\frac{7}{8}$	$\frac{5}{32}$	$\frac{7}{8}$	$\frac{5}{16}$

CALIPERING AND FITTING

THE VERNIER AND HOW TO READ IT

This method of measuring or of dividing known distances into very small parts is credited to the invention of Pierre Vernier in 1631. The principle is shown in Figs. 1 to 3 and its application in Figs. 4 and 5. In Figs. 1 and 2 both distances 0–1 are the same but they are divided into different divisions. Calling 0 — 1 = 1 inch then in Fig. 1 it is clear that moving the lower seal one division will divide

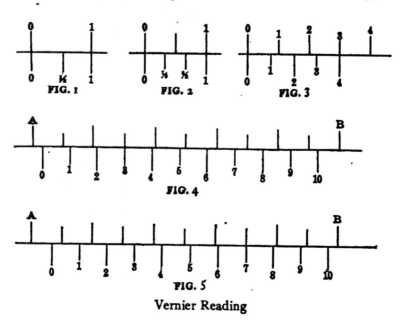

FIG. 1   FIG. 2   FIG. 3

FIG. 4

FIG. 5

Vernier Reading

the upper one in half. In Fig. 2 the upper scale is divided in half and the lower one in thirds. If the lower scale is moved either way until ⅓ or ⅔ comes under the end line, it has moved ⅓ of an inch but if either of these are moved to the center line then it is only moved ½ of this amount or ⅙

Figure 3 shows the usual application of the principle except that it is divided in four parts instead of ten. Here both the scales have four parts but on the lower scale the four parts just equal three parts of the upper scale. It is evident that if we move the lower scale so that 0 goes to 1 and 4 goes to 4 that it will be moved ¼ the length of the distance 0 — 4 on the upper scale. If this distance was 1 inch, each division on the upper scale equals ¼ inch and moving the lower scale so that the line 1 just matches the line next to 0 on the upper scale gives ¼ of one of these divisions or 1/16 of an inch.

225

Figures 4 and 5 show the usual application in which the lower or vernier scale is divided into 10 parts which equals 9 parts of the upper scale. The same division holds good, however, and when the lower scale is moved so that the first division of the vernier just matches the first line of the scale, it has been moved just one tenth of a division. In Fig. 4 the third lines match so that it has moved $\frac{3}{10}$ and in Fig. 5, $\frac{1}{10}$ of a division. So if A B is one inch then each division is $\frac{1}{10}$ of an inch and each line of the vernier is $\frac{1}{10}$ of that or $\frac{1}{100}$ of an inch.

To find the reading of any vernier, divide one division of the upper or large scale by the number of divisions in the small scale. So if we had a vernier with 16 divisions in each, the large scale being 1 inch long, then the movement of one division is $\frac{1}{16}$ of $\frac{1}{16}$ or $\frac{1}{256}$ of an inch.

READING THE MICROMETER

THE commercial micrometer consists of a frame, the anvil or fixed measuring point, the spindle which has a thread cut 40 to the inch on the portion inside the sleeve or barrel and the thimble which goes outside the sleeve and turns the spindle. One turn of the

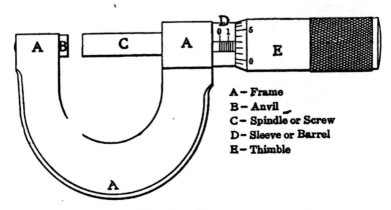

A – Frame
B – Anvil
C – Spindle or Screw
D – Sleeve or Barrel
E – Thimble

FIG. 6. — Micrometer

screw moves the spindle $\frac{1}{40}$ or .025 of an inch and the marks on the sleeve show the number of turns the screw is moved. Every fourth graduation is marked 1, 2, 3, etc., representing tenths of an inch or as each mark is .025 the first four means .025 × 4 = .100, the third means .025 × 4 × 3 = .300.

The thimble has a beveled edge divided into 25 parts and numbered 0, 5, 10, 15, 20 and to 0 again. Each of these mean $\frac{1}{25}$ of a turn or $\frac{1}{25}$ of $\frac{1}{40} = \frac{1}{1000}$ of an inch. To read, multiply the marks on the barrel by 25 and add the graduations on the edge of the thimble. In the cut there are 7 marks on the sleeve and 3 on the thimble so we say 7 × 25 = 175, plus 3 = 178 or .178.

In shop practice it is common to read them without any multiplying by using mental addition. Beginning at the largest number shown on the sleeve and calling it hundreds and add 25 for each mark, we say in the case show 100 and 25, 50, 75 and then add the numbers shown on the thimble 3, making .178 in all. If it showed 4 and one mark, with the thimble showing 8 marks, the reading would be 400 + 25 + 8 = 433 thousandths or .433.

THE TEN-THOUSANDTH MICROMETER

THIS adds a vernier to the micrometer sleeve or barrel as shown in Fig. 7, which is read the same as any vernier as has been explained. First note the thousandths as in the ordinary micrometer and then look at the line on the sleeve which just matches a line on

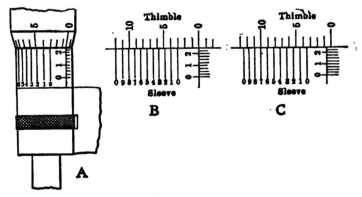

FIG. 7. — Micrometer Graduations

the thimble. If the two zero lines match two lines on the thimble, the measurement is in even thousandths as at B which reads .250. At C the seventh line matches a line on the thimble so the reading is .2507 inch.

MEASURING THREE-FLUTED TOOLS WITH THE MICROMETER

THE sketch, Fig. 8 on page 228, shows a V-block or gage for measuring three-fluted drills, counterbores, etc.

The angle being 60 degrees, the distances A, B, and C are equal. Consequently to determine the correct diameter of the piece to be measured, apply the gage as indicated in the sketch and deduct one third of the total measurement.

The use of this gage has a decided advantage over the old way of soldering on a piece of metal opposite a tooth or boring out a ring to fit to.

Using a standard 60-degree triangle for setting and a few different sizes of standard cylindrical plug gages for testing, the V-block may be easily and very accurately made.

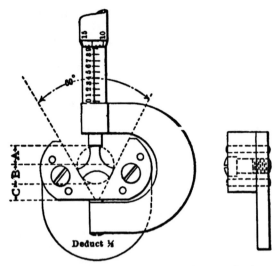

FIG. 8. — Measuring Three-Fluted Tools

PRESS AND RUNNING FITS

Parallel Press, Drive and Close Fits

TABLE 1, page 229, gives the practice of the C. W. Hunt Company, New York, for press, drive and close or hand fits for parallel shafts ranging between one and ten inches in diameter. In accordance with general practice, the holes for all parallel fits are made standard, except for unavoidable variation due to the wear of the reamer, the variation from standard diameter for the various kinds of fits being made in the shaft. This variation is, however, not positive, but is made between limits of accuracy or tolerance. Taking the case of a press fit on a two-inch shaft, for example, it will be seen that the hole — that is, the reamer — is kept between the correct size and 0.002 inch below size, while the shaft must be between 0.002 and 0.003 inch over size. For a drive or hand fit the limits for the hole are the same as for a press fit, while the shaft in the former case must be between 0.001 and 0.002 large and in the latter between 0.001 and 0.002 small.

Parallel Running Fits

Table 2, page 230, gives in the same way the allowances made by the same concern for parallel running fits of three grades of closeness. The variations allowed in the holes are not materially different from those of the preceding table, but the shafts are, of course, below instead of above the nominal size.

In all cases the tables apply to steel shafts and cast-iron wheels or other members. In the right-hand columns of the tables the formulas from which the allowances are calculated are given, and from which the range of tables may be extended.

TABLE 1. LIMITS TO DIAMETERS OF PARALLEL SHAFTS AND BUSHINGS (SHAFTS CHANGING)

Diameters		1 in.	2 in.	3 in.	4 in.	5 in.	Formula:
Press Fit	Shaft {	+.001 +.002	+.002 +.003	+.003 +.004	+.004 +.005	+.005 +.006	+(.001 d +.000) +(.001 d +.001)
Drive Fit	Shaft {	+.0005 +.0015	+.001 +.002	+.0015 +.0025	+.002 +.003	+.0025 +.0035	+(.0005 d +.000) +(.0005 d +.001)
Hand Fit	Shaft {	−.001 −.002	−.001 −.002	−.001 −.002	−.002 −.003	−.002 −.003	....
All Fits	Hole {	+.000 −.002	+.000 −.002	+.000 −.002	+.000 −.003	+.000 −.003	....

Diameters		6 in.	7 in.	8 in.	9 in.	10 in.	Formula:
Press Fit	Shaft {	+.006 +.007	+.007 +.008	+.008 +.009	+.009 +.010	+.010 +.011	+(.001 d +.000) +(.001 d +.001)
Drive Fit	Shaft {	+.003 +.004	+.0035 +.0045	+.004 +.005	+.0045 +.0055	+.005 +.006	+(.0005 d +.000) +(.0005 d +.001)
Hand Fit	Shaft {	−.002 −.003	−.003 −.004	−.003 −.004	−.003 −.004	−.003 −.004	....
All Fits	Hole {	+.000 −.003	+.000 −.004	+.000 −.004	+.000 −.004	+.000 −.004	....

TABLE 2. LIMITS TO DIAMETERS OF PARALLEL JOURNALS AND BEARINGS (JOURNALS CHANGING)

Diameters		1 in.	2 in.	3 in.	4 in.	5 in.	Formula:
Close Fit	Shaft {	−.003 −.005	−.004 −.006	−.005 −.007	−.006 −.008	−.007 −.009	−(.001 d+.02) −(.001 d+.004)
Free Fit	Shaft {	−.008 −.011	−.009 −.012	−.010 −.013	−.011 −.014	−.012 −.015	−(.001 d+.07) −(.001 d+.010)
Loose Fit	Shaft {	−.023 −.028	−.026 −.031	−.029 −.034	−.032 −.037	−.035 −.040	−(.003 d+.020) −(.003 d+.025)
All Fits	Hole {	+.000 −.002	+.000 −.002	+.000 −.002	+.000 −.002	+.000 −.003	......

Diameters		6 in.	7 in.	8 in.	9 in.	10 in.	Formula:
Close Fit	Shaft {	−.008 −.010	−.009 −.011	−.010 −.012	−.011 −.013	−.012 −.014	−(.01 d+.02) −(.01 d+.04)
Free Fit	Shaft {	−.013 −.016	−.014 −.017	−.015 −.018	−.016 −.019	−.017 −.020	−(.01 d+.07) −(.001 d+.010)
Loose Fit	Shaft {	−.038 −.043	−.041 −.044	−.044 −.049	−.047 −.052	−.050 −.055	−(.003 d+.020) −(.03 d+.025)
All Fits	Hole {	+.000 −.003	+.000 −.03	+.00 −.004	+.000 −.004	+.000 −.004	......

Shrink Fits

Table 3 gives the practice of the General Electric Company, Schenectady, New York, in regard to shrink fits, the same allowances also being made for press fits on heavy work such as couplings, etc.

TABLE 3. ALLOWANCES FOR SHRINK FITS

Dia. In.	Allowance	Dia. In.	Allowance	Dia. In.	Allowance
1	.001	20	.008	42	.0143
2	.0015	22	.0088	44	.015
3	.0020	24	.0093	46	.0155
4	.0028	26	.0098	48	.016
6	.0035	28	.0105	60	.020
8	.0045	30	.011	72	.024
10	.0053	32	.0115	84	.027
12	.0058	34	.012	96	.030
14	.0065	36	.0128	108	.033
16	.007	38	.0133	120	.0355
18	.0075	40	.0138	132	.038
				144	.040

LIMITS FOR GAGES

THE Newall Engineering Company, when developing their system of limit gages, investigated the practice of the leading English, Continental and American engineering concerns relative to allowances for different kinds of fits and prepared a table which is the average of all the data received, every point included being covered by the practice of some prominent establishment. The limits and allowances thus arrived at for shop gages are given in Table 4, which is self-explanatory.

TABLE 4. LIMITS AND ALLOWANCES IN SHOP GAGES FOR DIFFERENT KINDS OF FITS

Nominal Diameters	½″	1″	2″	3″	4″	5″	6″
Over size	.00025	.00050	.00075	.00100	.00100	.00100	.00150
Under size	.00025	.00025	.00025	.00050	.00050	.00050	.00050
Margin	.00050	.00075	.00100	.00150	.00150	.00200	.00200

Limits in Plug Gages for Standard Holes

(Table Continued on Page 232)

TABLE 4 *Continued.* — LIMITS IN SHOP GAGES

Allowances — over Standard — for Force Fits

Nominal Diameters	½	1"	2"	3"	4"	5"	6"
Mean	.00075	.00175	.00350	.00525	.00700	.00900	.01100
High	.00100	.00200	.00400	.00600	.00800	.01000	.01200
Low	.00050	.00150	.00300	.00450	.00600	.00800	.01000
Margin	.00050	.00050	.00100	.00150	.00200	.00200	.00200

Allowances — over Standard — for Driving Fits

Nominal Diameters	½"	1"	2"	3"	4"	5"	6"
Mean	.000375	.000875	.00125	.00200	.00250	.00300	.00350
High	.00050	.00100	.00150	.00250	.00300	.00350	.00400
Low	.00025	.00075	.00100	.00150	.00200	.00250	.00300
Margin	.00025	.00025	.00050	.00100	.00100	.00100	.00100

Allowances — Below Standard — for Push or Keying Fits

Nominal Diameters	½"	1"	2"	3"	4"	5"	6"
High	.00025	.00050	.00100	.00150	.00150	.00200	.00200
Low	.00075	.00100	.00150	.00200	.00250	.00250	.00250
Margin	.00050	.00050	.00050	.00050	.00050	.00050	.00050

Clearances for Running Fits

Class of Gage	Diameters	½"	1	2"	3"	4"	5"	6"
X.	Mean	.00150	.00200	.00260	.00320	.00380	.00440	.00500
	High	.00100	.00125	.00175	.00200	.00250	.00300	.00350
	Low	.00200	.00275	.00350	.00425	.00500	.00575	.00650
	Margin	.00100	.00150	.00175	.00225	.00250	.00275	.00300
Y.	Mean	.00100	.00150	.00190	.00230	.00270	.00310	.00350
	High	.00075	.00100	.00125	.00150	.00200	.00225	.00250
	Low	.00125	.00200	.00250	.00300	.00350	.00400	.00450
	Margin	.00050	.00100	.00125	.00150	.00150	.00175	.00200
Z.	Mean	.000625	.00100	.00120	.00140	.00160	.00180	.00200
	High	.00050	.00075	.00075	.00100	.00100	.00125	.00125
	Low	.00075	.00125	.00150	.00200	.00225	.00250	.00275
	Margin	.00025	.00050	.00075	.00100	.00125	.00125	.00150

Class X is suitable for engine and other work requiring easy fits.
Class Y is suitable for high speeds and good average machine work.
Class Z is suitable for fine tool work.

MAKING ALLOWANCES WITH THE CALIPERS FOR RUNNING, SHRINK, AND PRESS FITS

ONE of the familiar devices of the machinist consists in giving the inside calipers a certain amount of side play, when it is desirable to obtain a measure minutely less than the full diameter of the hole, as in making a loose or running fit, or a sliding fit as of a plunger in a cylinder. Thus in Fig, 9, A is the diameter of the bore, B the caliper setting and C the side play permitted the caliper in the hole.

In the table below is given a list of the reduced dimensions for different amounts of side play of the calipers in a 12-inch hole. From this, the dimensions may be obtained for holes of other diameters by division. Where in the table the side play is 2 inches, if we divide the items by 4 we have the side play and the reduced dimension for a 3-inch hole, or 0.5 inch and 2.9894 inches respectively.

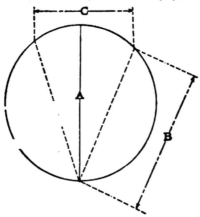

FIG. 9. — Side Play of the Calipers

TABLE OF REDUCED DIAMETERS INDICATED BY INSIDE CALIPERS
FOR DIFFERENT AMOUNTS OF SIDE PLAY IN A 12-INCH HOLE

Side play

0.1	11.9999
0.2	11.9991
0.4	11.9983
0.6	11.9962
0.8	11.9933
1.0	11.9895
1.2	11.9849
1.4	11.9795
1.6	11.9730
1.8	11.9660
2.0	11.9579
2.2	11.9490
2.4	11.9391
2.5	11.9339
3.0	11.9044

Axial Inclination of the Calipers in Measuring for Shrink or Press Fits

In the case worked out on page 233, it was desired to produce a hole slightly larger than the piece to go into it, or a piece slightly smaller than the hole. In operations where a hole is wanted somewhat smaller than the piece to be shrunk or pressed into it, a similar plan of measuring can be employed, and a table giving the tightness can be computed. The sketch, Fig. 10, will serve to make the meaning clear. The distance A is the diameter of a hole and line a is the length of a gage the exact size of the piece to be pressed or shrunk into the hole. The distance b is the amount the gage lacks of assuming a position square or at a right angle to the axis of the hole.

It is an easy matter to make a table as suggested. It is only necessary to find the different lengths for the hypotenuse a for the right-angle triangle of which A is the constant base and b the perpendicular, taking b at different lengths from $\frac{1}{8}$ inch to 2 inches. Assuming the diameter to be 12 inches, then the lengths indicated for different inclinations in the direction of the axis will be as given in the following table.

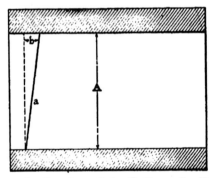

FIG. 10. — Inclination of the Calipers for Press Fits

TABLE FOR AXIAL INCLINATION OF CALIPERS IN ALLOWING FOR
SHRINK OR FORCE FITS IN A 12-INCH HOLE

Inclination of
calipers

$\frac{1}{8}$ inch	12.00065
$\frac{1}{4}$ inch	12.00260
$\frac{3}{8}$ inch	12.00580
$\frac{1}{2}$ inch	12.01040
$\frac{5}{8}$ inch	12.01626
$\frac{3}{4}$ inch	12.02340
$\frac{7}{8}$ inch	12.03180
1 inch	12.04159
$1\frac{1}{4}$ inches	12.06490
$1\frac{1}{2}$ inches	12.09338
$1\frac{3}{4}$ inches	12.12689
2 inches	12.16550

Side Play of Calipers in Boring Holes Larger than a Piece of Known Diameter

The following is an approximate rule for obtaining the variation in the size of a hole corresponding to a given amount of side play in the calipers. The rule has the merit of extreme simplicity and can be applied equally well to all diameters except the very smallest. In most cases the calculation is so simple that it can be done mentally without having recourse to pencil or paper.

The Calculation

Let A in Fig. 11 = side play of calipers or end measuring rod *in sixteenths of an inch.*

        B = dimensions to which calipers are set, or length of measuring rod *in inches.*

        C = difference between diameter of hole and length of B *in thousandths of an inch.*

Then $C = \dfrac{A^2}{2B}$, within a very small limit.

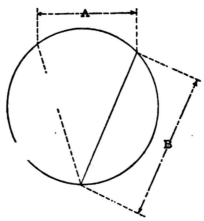

FIG. 11. — Caliper Side Play

Example: A standard end measuring rod, $5\frac{1}{2}$ inches long, has $\frac{3}{8}$ inch of side play in a hole. What is the size of the hole? In this case $A = 6$ and $B = 5\frac{1}{2}$. Apply the above formula:

$$C = \frac{6 \times 6}{11} = \frac{36}{11} = 3.27 \text{ thousandths of an inch, or } 0.00327 \text{ inch.}$$

The diameter of the hole, therefore, is $5\frac{1}{2} + 0.00327$ or 5.50327.

The method will be found to be correct within a limit of about 0.0002 inch if the amount of side play is not more than one eighth of the diameter of the hole for holes up to 6 inches diameter; within 0.0005 inch for holes from 6 inches up to 12 inches; and within 0.001 for holes from 12 inches up to 24 inches.

Allowing for Running and Driving Fits

This rule has been found to be useful for boring holes of large diameters in which allowances have to be made for running or driving fits, as only a single measuring rod for each nominal size is required. The rods should be of standard length, or a known amount less than standard, the allowances being obtained by varying the amount of side play when boring. The rule is also capable of determining limits, as the maximum and minimum amount of side play allowable can be specified. The measuring rods should be tapered at each end and the points slightly rounded. For accurate work, the body of the rod should be encased in some non-conducting material to nullify the effect of the heat of the hand.

In comparing this method with that described on page 233, it should be remembered that the conditions are reversed — that is to say, the first method is for setting calipers to a given dimension *smaller* than a hole of known diameter, whereas the method now described is for boring a hole a given amount *larger* than a gage of known length.

In measuring the side play it is sufficient to take it to the nearest sixteenth of an inch, and if anything like accuracy is required it should be measured- not guessed at.

DIMENSIONS OF KEYS AND KEY-SEATS

THE following rules and table on page 237, as prepared by Baker Bros., Toledo, Ohio, give dimension of keys and key-seats.

The width of the key should equal one fourth the diameter of the shaft.

The thickness of the key should equal one sixth the diameter of the shaft.

The depth in the hub for a straight key-seat should be one half the thickness of the key.

The depth in the hub at the large end, for a taper key-seat, should be three fifths the thickness of the key.

The taper for all key-seats should be $\frac{1}{8}$ inch in 1 foot of length.

The depth to be cut in the hub for taper key-seats, at the large end, is greater than those cut straight, for the reason that unless this is done the depth in the hub at the small end will not be sufficient, especially in long key-seats.

The depths of key-seats in the table are given in thousandths of an inch and measured from the edge of the key-seat, and not from the center. In this manner the exact depth of key-seat can be measured at any time after it is cut.

For extra long key-seats the depth cut in the hub may be slightly increased, but for the average work the table will be found correct.

DIMENSIONS OF KEYS AND KEY-SEATS.   (BAKER BROS.)

Size of Hole	Decimal Equivalent	Preferred Width of Key-Seat	Nearest Size of Cutter	Preferred Thickness of Key	Nearest Fractional Thickness	Depth to be Cut in Hub for Straight Key	Depth at Large End for Taper Key
1	1.	.25	1/4	.166	3/16	.093	.112
1 1/16	1.062	.265	1/4	.177	3/16	.093	.112
1 1/8	1.125	.281	1/4	.187	3/16	.093	.112
1 3/16	1.187	.296	5/16	.198	13/64	.109	.131
1 1/4	1.25	.312	5/16	.208	13/64	.109	.131
1 5/16	1.312	.328	5/16	.219	7/32	.109	.131
1 3/8	1.375	.343	3/8	.229	1/4	.125	.15
1 7/16	1.437	.359	3/8	.239	1/4	.125	.15
1 1/2	1.5	.375	3/8	.25	1/4	.125	.15
1 9/16	1.562	.39	3/8	.26	1/4	.125	.15
1 5/8	1.625	.406	7/16	.271	9/32	.141	.168
1 11/16	1.687	.421	7/16	.281	9/32	.141	.168
1 3/4	1.75	.437	7/16	.292	9/32	.141	.168
1 13/16	1.812	.453	7/16	.302	9/32	.141	.168
1 7/8	1.875	.468	1/2	.312	5/16	.171	.206
1 15/16	1.937	.484	1/2	.323	5/16	.171	.206
2	2.	.5	1/2	.333	11/32	.171	.206
2 1/16	2.062	.515	1/2	.344	11/32	.171	.206
2 1/8	2.125	.531	1/2	.354	11/32	.171	.206
2 3/16	2.187	.547	1/2	.364	3/8	.171	.206
2 1/4	2.25	.563	1/2	.375	3/8	.171	.206
2 5/16	2.312	.578	1/2	.385	3/8	.171	.206
2 3/8	2.375	.593	5/8	.396	13/32	.218	.262
2 7/16	2.437	.609	5/8	.406	13/32	.218	.262
2 1/2	2.5	.625	5/8	.416	7/16	.218	.262
2 9/16	2.562	.641	5/8	.427	7/16	.218	.262
2 5/8	2.625	.656	5/8	.437	7/16	.218	.262
2 11/16	2.687	.672	5/8	.448	7/16	.218	.262
2 3/4	2.75	.687	5/8	.458	15/32	.218	.262
2 13/16	2.812	.703	5/8	.469	15/32	.218	.262
2 7/8	2.875	.719	3/4	.479	1/2	.25	.3
2 15/16	2.937	.734	3/4	.49	1/2	.25	.3
3	3.	.75	3/4	.5	1/2	.25	.3
3 1/8	3.125	.781	3/4	.521	1/2	.25	.3
3 3/16	3.187	.797	3/4	.531	17/32	.25	.3
3 1/4	3.25	.812	3/4	.542	17/32	.25	.3
3 3/8	3.375	.844	7/8	.562	9/16	.312	.375
3 7/16	3.437	.859	7/8	.573	9/16	.312	.375
3 1/2	3.5	.875	7/8	.583	9/16	.312	.375
3 5/8	3.625	.906	7/8	.604	19/32	.312	.375
3 11/16	3.687	.923	7/8	.614	5/8	.312	.375
3 3/4	3.75	.937	7/8	.625	5/8	.312	.375
3 7/8	3.875	.969	1	.646	11/16	.343	.412
3 15/16	3.937	.984	1	.656	11/16	.343	.412
4	4.	1.	1	.666	11/16	.343	.412

DIMENSIONS OF STRAIGHT KEYS

ANOTHER system of keys used by a good many manufacturers is given in the table following, the sizes of shafts ranging by sixteenths from $\frac{1}{16}$ inch to 4 inches and by eighths from 4 to 6 inches. The keys are square until the $1\frac{1}{4}$ inch shaft is reached, when the thickness of the key becomes $\frac{1}{16}$ less than the width. With the $4\frac{1}{2}$ size the thickness of the key becomes $\frac{1}{4}$ inch less than the width and this difference is constant up to the $5\frac{1}{4}$ shaft when the width exceeds the thickness by $\frac{5}{16}$ inch, this difference in the two dimensions continuing throughout the remainder of the table.

DIMENSIONS OF STRAIGHT KEYS

Dia. of Shaft	Width of Key	Thickness of Key	Dia. of Shaft	Width of Key	Thickness of Key	Dia. of Shaft	Width of Key	Thickness of Key	Dia. of Shaft	Width of Key	Thickness of Key	Dia. of Shaft	Width of Key	Thickness of Key
0			$1\frac{1}{4}$	$\frac{5}{16}$	$\frac{5}{16}$	$2\frac{1}{2}$	$\frac{1}{2}$	$\frac{7}{16}$	$3\frac{3}{4}$	$\frac{7}{8}$	$\frac{11}{16}$	6	$1\frac{7}{16}$	$1\frac{1}{16}$
$\frac{1}{16}$			$1\frac{5}{16}$	$\frac{5}{16}$	$\frac{5}{16}$	$2\frac{9}{16}$	$\frac{9}{16}$	$\frac{7}{16}$	$3\frac{13}{16}$	$\frac{7}{8}$	$\frac{11}{16}$			
$\frac{1}{8}$			$1\frac{3}{8}$	$\frac{5}{16}$	$\frac{5}{16}$	$2\frac{5}{8}$	$\frac{9}{16}$	$\frac{9}{16}$	$3\frac{7}{8}$	$\frac{7}{8}$	$\frac{11}{16}$			
$\frac{3}{16}$			$1\frac{7}{16}$	$\frac{5}{16}$	$\frac{5}{16}$	$2\frac{11}{16}$	$\frac{9}{16}$	$\frac{9}{16}$	$3\frac{15}{16}$	$\frac{15}{16}$	$\frac{11}{16}$			
$\frac{1}{4}$			$1\frac{1}{2}$	$\frac{3}{8}$	$\frac{3}{8}$	$2\frac{3}{4}$	$\frac{9}{16}$	$\frac{9}{16}$	4	$\frac{15}{16}$	$\frac{11}{16}$			
$\frac{5}{16}$	$\frac{3}{32}$	$\frac{3}{32}$	$1\frac{9}{16}$	$\frac{3}{8}$	$\frac{3}{8}$	$2\frac{13}{16}$	$\frac{9}{16}$	$\frac{9}{16}$	$4\frac{1}{8}$	$\frac{15}{16}$	$\frac{3}{4}$			
$\frac{3}{8}$	$\frac{3}{32}$	$\frac{3}{32}$	$1\frac{5}{8}$	$\frac{3}{8}$	$\frac{3}{8}$	$2\frac{7}{8}$	$\frac{5}{8}$	$\frac{9}{16}$	$4\frac{1}{4}$	1	$\frac{3}{4}$			
$\frac{7}{16}$	$\frac{1}{8}$	$\frac{1}{8}$	$1\frac{11}{16}$	$\frac{3}{8}$	$\frac{3}{8}$	$2\frac{15}{16}$	$\frac{5}{8}$	$\frac{9}{16}$	$4\frac{3}{8}$	1	$\frac{3}{4}$			
$\frac{1}{2}$	$\frac{1}{8}$	$\frac{1}{8}$	$1\frac{3}{4}$	$\frac{7}{16}$	$\frac{3}{8}$	3	$\frac{5}{8}$	$\frac{9}{16}$	$4\frac{1}{2}$	1	$\frac{3}{4}$			
$\frac{9}{16}$	$\frac{5}{32}$	$\frac{5}{32}$	$1\frac{13}{16}$	$\frac{7}{16}$	$\frac{3}{8}$	$3\frac{1}{16}$	$\frac{5}{8}$	$\frac{9}{16}$	$4\frac{5}{8}$	1	$\frac{3}{4}$			
$\frac{5}{8}$	$\frac{5}{32}$	$\frac{5}{32}$	$1\frac{7}{8}$	$\frac{7}{16}$	$\frac{3}{8}$	$3\frac{1}{8}$	$\frac{5}{8}$	$\frac{5}{8}$	$4\frac{3}{4}$	$1\frac{1}{16}$	$\frac{13}{16}$			
$\frac{11}{16}$	$\frac{3}{16}$	$\frac{3}{16}$	$1\frac{15}{16}$	$\frac{7}{16}$	$\frac{3}{8}$	$3\frac{3}{16}$	$\frac{11}{16}$	$\frac{5}{8}$	$4\frac{7}{8}$	$1\frac{1}{16}$	$\frac{13}{16}$			
$\frac{3}{4}$	$\frac{3}{16}$	$\frac{3}{16}$	2	$\frac{7}{16}$	$\frac{3}{8}$	$3\frac{1}{4}$	$\frac{11}{16}$	$\frac{5}{8}$	5	$1\frac{1}{8}$	$\frac{7}{8}$			
$\frac{13}{16}$	$\frac{3}{16}$	$\frac{3}{16}$	$2\frac{1}{16}$	$\frac{1}{2}$	$\frac{7}{16}$	$3\frac{5}{16}$	$\frac{11}{16}$	$\frac{5}{8}$	$5\frac{1}{8}$	$1\frac{1}{8}$	$\frac{7}{8}$			
$\frac{7}{8}$	$\frac{1}{4}$	$\frac{1}{4}$	$2\frac{1}{8}$	$\frac{1}{2}$	$\frac{7}{16}$	$3\frac{3}{8}$	$\frac{3}{4}$	$\frac{11}{16}$	$5\frac{1}{4}$	$1\frac{3}{16}$	$\frac{7}{8}$			
$\frac{15}{16}$	$\frac{1}{4}$	$\frac{1}{4}$	$2\frac{3}{16}$	$\frac{1}{2}$	$\frac{7}{16}$	$3\frac{7}{16}$	$\frac{3}{4}$	$\frac{11}{16}$	$5\frac{3}{8}$	$1\frac{3}{16}$	$\frac{7}{8}$			
1	$\frac{1}{4}$	$\frac{1}{4}$	$2\frac{1}{4}$	$\frac{1}{2}$	$\frac{7}{16}$	$3\frac{1}{2}$	$\frac{3}{4}$	$\frac{11}{16}$	$5\frac{1}{2}$	$1\frac{1}{4}$	$\frac{15}{16}$			
$1\frac{1}{16}$	$\frac{1}{4}$	$\frac{1}{4}$	$2\frac{5}{16}$	$\frac{9}{16}$	$\frac{1}{2}$	$3\frac{9}{16}$	$\frac{3}{4}$	$\frac{11}{16}$	$5\frac{5}{8}$	$1\frac{1}{4}$	$\frac{15}{16}$			
$1\frac{1}{8}$	$\frac{1}{4}$	$\frac{1}{4}$	$2\frac{3}{8}$	$\frac{9}{16}$	$\frac{1}{2}$	$3\frac{5}{8}$	$\frac{3}{4}$	$\frac{11}{16}$	$5\frac{3}{4}$	$1\frac{5}{16}$	1			
$1\frac{3}{16}$	$\frac{1}{4}$	$\frac{1}{4}$	$2\frac{7}{16}$	$\frac{9}{16}$	$\frac{1}{2}$	$3\frac{11}{16}$	$\frac{7}{8}$	$\frac{11}{16}$	$5\frac{7}{8}$	$1\frac{5}{16}$	1			

SQUARE FEATHER KEYS AND STRAIGHT KEY SIZES

THE tables on page 239 give the sizes of square feather keys and regular straight keys in accordance with the practice of Jones & Laughlin, Pittsburg. For taper keys, this concern and many other use a $\frac{1}{8}$-inch per foot taper.

SQUARE FEATHER KEY SIZES

Dia. of Shaft	Size of Key	Dia. of Shaft	Size of Key
1 to $1\frac{1}{8}$	$\frac{1}{4} \times \frac{1}{4}$	$3\frac{3}{16}$ to $3\frac{3}{8}$	$\frac{11}{16} \times \frac{11}{16}$
$1\frac{3}{16}$ to $1\frac{3}{8}$	$\frac{5}{16} \times \frac{5}{16}$	$3\frac{7}{16}$ to $3\frac{5}{8}$	$\frac{7}{8} \times \frac{7}{8}$
$1\frac{7}{16}$ to $1\frac{5}{8}$	$\frac{3}{8} \times \frac{3}{8}$	$3\frac{11}{16}$ to $3\frac{7}{8}$	$\frac{15}{16} \times \frac{15}{16}$
$1\frac{11}{16}$ to $1\frac{7}{8}$	$\frac{7}{16} \times \frac{7}{16}$	$3\frac{15}{16}$ to $4\frac{1}{8}$	1×1
$1\frac{15}{16}$ to $2\frac{1}{8}$	$\frac{1}{2} \times \frac{1}{2}$	$4\frac{3}{16}$ to $4\frac{1}{8}$	$1\frac{1}{16} \times 1\frac{1}{16}$
$2\frac{3}{16}$ to $2\frac{3}{8}$	$\frac{9}{16} \times \frac{9}{16}$	$4\frac{7}{16}$ to $4\frac{3}{4}$	$1\frac{1}{8} \times 1\frac{1}{8}$
$2\frac{7}{16}$ to $2\frac{5}{8}$	$\frac{5}{8} \times \frac{5}{8}$	$4\frac{13}{16}$ to $5\frac{1}{4}$	$1\frac{1}{4} \times 1\frac{1}{4}$
$2\frac{11}{16}$ to $2\frac{7}{8}$	$\frac{11}{16} \times \frac{11}{16}$	$5\frac{5}{16}$ to $5\frac{3}{4}$	$1\frac{3}{8} \times 1\frac{3}{8}$
$2\frac{15}{16}$ to $3\frac{1}{8}$	$\frac{3}{4} \times \frac{3}{4}$	$5\frac{13}{16}$ to $6\frac{1}{4}$	$1\frac{1}{2} \times 1\frac{1}{2}$

STRAIGHT KEY SIZES

Dia. of Shaft	Size of Key	Dia. of Shaft	Size of Key
1 to $1\frac{1}{8}$	$\frac{1}{4} \times \frac{3}{16}$	$3\frac{3}{16}$ to $3\frac{1}{4}$	$\frac{11}{16} \times \frac{17}{32}$
$1\frac{3}{16}$ to $1\frac{3}{8}$	$\frac{5}{16} \times \frac{7}{32}$	$3\frac{5}{16}$ to $3\frac{5}{8}$	$\frac{7}{8} \times \frac{19}{32}$
$1\frac{7}{16}$ to $1\frac{5}{8}$	$\frac{3}{8} \times \frac{1}{4}$	$3\frac{11}{16}$ to $3\frac{7}{8}$	$\frac{15}{16} \times \frac{5}{8}$
$1\frac{11}{16}$ to $1\frac{7}{8}$	$\frac{7}{16} \times \frac{9}{32}$	$3\frac{15}{16}$ to $4\frac{1}{8}$	$1 \times \frac{11}{16}$
$1\frac{15}{16}$ to $2\frac{1}{8}$	$\frac{1}{2} \times \frac{11}{32}$	$4\frac{3}{16}$ to $4\frac{1}{8}$	$1\frac{1}{16} \times \frac{11}{16}$
$2\frac{3}{16}$ to $2\frac{5}{8}$	$\frac{5}{8} \times \frac{11}{16}$	$4\frac{7}{16}$ to $4\frac{3}{4}$	$1\frac{1}{8} \times \frac{3}{4}$
$2\frac{11}{16}$ to $2\frac{7}{8}$	$1\frac{1}{16} \times \frac{11}{16}$	$4\frac{13}{16}$ to $5\frac{1}{4}$	$1\frac{1}{4} \times \frac{7}{8}$
$2\frac{15}{16}$ to $3\frac{1}{8}$	$\frac{3}{4} \times \frac{1}{2}$	$5\frac{5}{16}$ to $5\frac{3}{4}$	$1\frac{1}{2} \times 1$

THE BARTH KEY

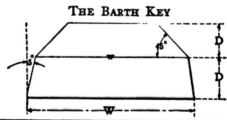

No. of Key	w	W	D
1	$\frac{1}{4}$	.132	$\frac{5}{32}$
2	$\frac{5}{32}$	.165	$\frac{3}{64}$
3	$\frac{3}{16}$	.199	$\frac{5}{16}$
4	$\frac{1}{4}$	.264	$\frac{5}{64}$
5	$\frac{5}{16}$	.329	$\frac{6}{32}$

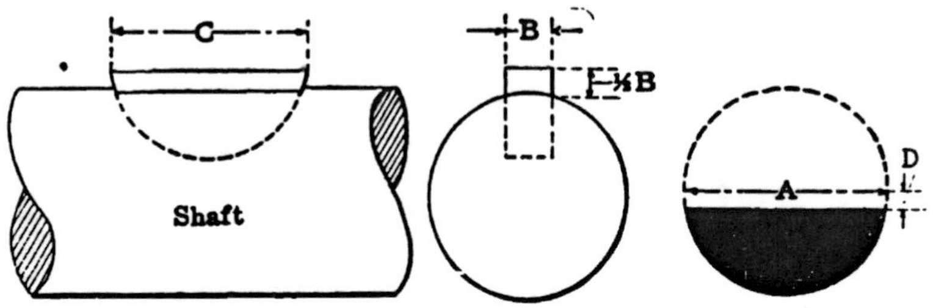

Whitney Keys and Cutters. Nos. 1 to 26
(Woodruff's Patent)

No. of Key and Cutter	Dia. of Cutter	Thickness of Key and Cutter	Length of Key	Key Cut Below Center	No. of Key and Cutter	Dia. of Cutter	Thickness of Key and Cutter	Length of Key	Key Cut Below Center
	A	B	C	D		A	B	C	D
1	1/2	1/16	1/2	3/64	16	1 1/8	3/16	1 1/8	5/64
2	1/2	3/32	1/2	3/64	17	1 1/8	7/32	1 1/8	5/64
3	1/2	1/8	1/2	3/64	18	1 1/8	1/4	1 1/8	5/64
4	5/8	3/32	5/8	1/16	C	1 1/8	5/16	1 1/8	5/64
5	5/8	1/8	5/8	1/16	19	1 1/4	3/16	1 1/4	5/64
6	5/8	5/32	5/8	1/16	20	1 1/4	7/32	1 1/4	5/64
7	3/4	1/8	3/4	1/16	21	1 1/4	1/4	1 1/4	5/64
8	3/4	5/32	3/4	1/16	D	1 1/4	5/16	1 1/4	5/64
9	3/4	3/16	3/4	1/16	E	1 1/4	3/8	1 1/4	5/64
10	7/8	5/32	7/8	1/16	22	1 3/8	1/4	1 3/8	3/32
11	7/8	3/16	7/8	1/16	23	1 3/8	5/16	1 3/8	3/32
12	7/8	7/32	7/8	1/16	F	1 3/8	3/8	1 3/8	3/32
A	7/8	1/4	7/8	1/16	24	1 1/2	1/4	1 1/2	7/64
13	1	3/16	1	1/16	25	1 1/2	5/16	1 1/2	7/64
14	1	7/32	1	1/16	G	1 1/2	3/8	1 1/2	7/64
15	1	1/4	1	1/16					
B	1	5/16	1	1/16					

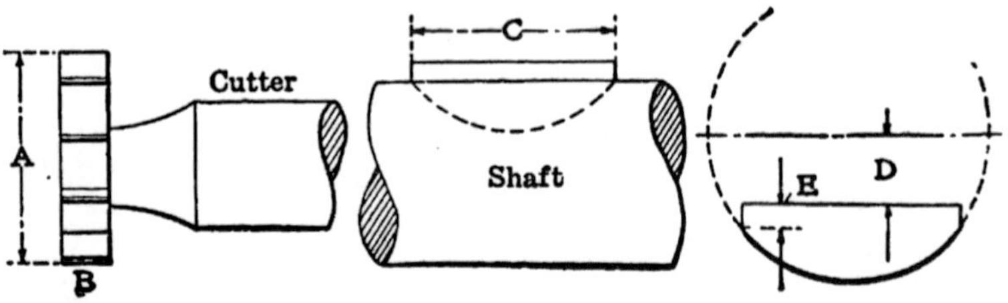

Note: Refer to table at top of page 241 for values of dimension E.

WHITNEY KEYS AND CUTTERS. NOS. 26 TO 36

No. of Key and Cutter	Dia. of Cutter	Thickness of Key and Cutter	Length of Key	Key Cut Below Center	Flat at End of Key	No. of Key and Cutter	Dia. of Cutter	Thickness of Key and Cutter	Length of Key	Key Cut Below Center	Flat at End of Key
	A	B	C	D	E		A	B	C	D	E
26	$2\frac{1}{8}$	$\frac{3}{16}$	$1\frac{23}{32}$	$1\frac{17}{32}$	$\frac{3}{32}$	30	$3\frac{1}{2}$	$\frac{3}{8}$	$2\frac{7}{8}$	$1\frac{13}{16}$	$\frac{7}{16}$
27	$2\frac{1}{8}$	$\frac{1}{4}$	$1\frac{23}{32}$	$1\frac{17}{32}$	$\frac{3}{32}$	31	$3\frac{1}{2}$	$\frac{7}{16}$	$2\frac{7}{8}$	$1\frac{13}{16}$	$\frac{7}{16}$
28	$2\frac{1}{8}$	$\frac{5}{16}$	$1\frac{23}{32}$	$1\frac{17}{32}$	$\frac{3}{32}$	32	$3\frac{1}{2}$	$\frac{1}{2}$	$2\frac{7}{8}$	$1\frac{13}{16}$	$\frac{7}{16}$
29	$2\frac{1}{8}$	$\frac{3}{8}$	$1\frac{23}{32}$	$1\frac{17}{32}$	$\frac{3}{32}$	33	$3\frac{1}{2}$	$\frac{9}{16}$	$2\frac{7}{8}$	$1\frac{13}{16}$	$\frac{7}{16}$
R	$2\frac{1}{4}$	$\frac{1}{4}$	$2\frac{5}{16}$	$\frac{5}{8}$	$\frac{1}{8}$	34	$3\frac{1}{2}$	$\frac{5}{8}$	$2\frac{7}{8}$	$1\frac{13}{16}$	$\frac{7}{16}$
S	$2\frac{1}{4}$	$\frac{5}{16}$	$2\frac{5}{16}$	$\frac{5}{8}$	$\frac{1}{8}$	35	$3\frac{1}{2}$	$\frac{11}{16}$	$2\frac{7}{8}$	$1\frac{13}{16}$	$\frac{7}{16}$
T	$2\frac{1}{4}$	$\frac{3}{8}$	$2\frac{5}{16}$	$\frac{5}{8}$	$\frac{1}{8}$	36	$3\frac{1}{2}$	$\frac{3}{4}$	$2\frac{7}{8}$	$1\frac{13}{16}$	$\frac{7}{16}$
U	$2\frac{1}{4}$	$\frac{7}{16}$	$2\frac{5}{16}$	$\frac{5}{8}$	$\frac{1}{8}$						
V	$2\frac{1}{4}$	$\frac{1}{2}$	$2\frac{5}{16}$	$\frac{5}{8}$	$\frac{1}{8}$						

PROPORTIONS OF KEY HEADS
(STANDARD GAGE STEEL CO.)

A	B	C	D	A	B	C	D
$\frac{1}{8}$	$\frac{1}{8}$	$\frac{1}{2}$	$\frac{7}{32}$	$1\frac{5}{8}$	$1\frac{5}{8}$	$2\frac{3}{4}$	$1\frac{7}{8}$
$\frac{3}{16}$	$\frac{3}{16}$	$\frac{5}{8}$	$\frac{9}{32}$	$1\frac{11}{16}$	$1\frac{11}{16}$	$2\frac{7}{8}$	$1\frac{15}{16}$
$\frac{1}{4}$	$\frac{1}{4}$	$\frac{3}{4}$	$\frac{11}{32}$	$1\frac{3}{4}$	$1\frac{3}{4}$	3	2
$\frac{5}{16}$	$\frac{5}{16}$	$\frac{13}{16}$	$\frac{13}{32}$	$1\frac{13}{16}$	$1\frac{13}{16}$	$3\frac{1}{8}$	$2\frac{1}{16}$
$\frac{3}{8}$	$\frac{3}{8}$	$\frac{7}{8}$	$\frac{15}{32}$	$1\frac{7}{8}$	$1\frac{7}{8}$	$3\frac{1}{4}$	$2\frac{1}{8}$
$\frac{7}{16}$	$\frac{7}{16}$	$\frac{15}{16}$	$\frac{17}{32}$	$1\frac{15}{16}$	$1\frac{15}{16}$	$3\frac{5}{8}$	$2\frac{3}{16}$
$\frac{1}{2}$	$\frac{1}{2}$	$\frac{7}{8}$	$\frac{19}{32}$	2	2	$3\frac{3}{4}$	$2\frac{1}{4}$
$\frac{9}{16}$	$\frac{9}{16}$	1	$\frac{21}{32}$	$2\frac{1}{16}$	$2\frac{1}{16}$	$3\frac{7}{8}$	$2\frac{7}{16}$
$\frac{5}{8}$	$\frac{5}{8}$	$1\frac{1}{8}$	$\frac{23}{32}$	$2\frac{1}{8}$	$2\frac{1}{8}$	4	$2\frac{1}{2}$
$\frac{11}{16}$	$\frac{11}{16}$	$1\frac{1}{16}$	$\frac{25}{32}$	$2\frac{3}{16}$	$2\frac{3}{16}$	$4\frac{1}{8}$	$2\frac{9}{16}$
$\frac{3}{4}$	$\frac{3}{4}$	$1\frac{1}{4}$	$\frac{7}{8}$	$2\frac{1}{4}$	$2\frac{1}{4}$	$4\frac{1}{4}$	$2\frac{5}{8}$
$\frac{13}{16}$	$\frac{13}{16}$	$1\frac{5}{16}$	$\frac{15}{16}$	$2\frac{5}{16}$	$2\frac{5}{16}$	$4\frac{3}{8}$	$2\frac{11}{16}$
$\frac{7}{8}$	$\frac{7}{8}$	$1\frac{3}{8}$	1	$2\frac{3}{8}$	$2\frac{3}{8}$	$4\frac{1}{2}$	$2\frac{3}{4}$
$\frac{15}{16}$	$\frac{15}{16}$	$1\frac{7}{16}$	$1\frac{1}{16}$	$2\frac{7}{16}$	$2\frac{7}{16}$	$4\frac{5}{8}$	$2\frac{13}{16}$
1	1	$1\frac{1}{2}$	$1\frac{1}{8}$	$2\frac{1}{2}$	$2\frac{1}{2}$	$4\frac{3}{4}$	$2\frac{7}{8}$
$1\frac{1}{16}$	$1\frac{1}{16}$	$1\frac{9}{16}$	$1\frac{1}{16}$	$2\frac{9}{16}$	$2\frac{9}{16}$	$4\frac{7}{8}$	$2\frac{15}{16}$
$1\frac{1}{8}$	$1\frac{1}{8}$	$1\frac{5}{8}$	$1\frac{5}{16}$	$2\frac{5}{8}$	$2\frac{5}{8}$	5	3
$1\frac{3}{16}$	$1\frac{3}{16}$	$1\frac{11}{16}$	$1\frac{7}{16}$	$2\frac{11}{16}$	$2\frac{11}{16}$	5	$3\frac{1}{16}$
$1\frac{1}{4}$	$1\frac{1}{4}$	2	$1\frac{7}{16}$	$2\frac{3}{4}$	$2\frac{3}{4}$	$5\frac{1}{4}$	$3\frac{1}{8}$
$1\frac{5}{16}$	$1\frac{5}{16}$	$2\frac{1}{8}$	$1\frac{1}{2}$	$2\frac{13}{16}$	$2\frac{13}{16}$	$5\frac{3}{8}$	$3\frac{1}{16}$
$1\frac{3}{8}$	$1\frac{3}{8}$	$2\frac{1}{4}$	$1\frac{9}{16}$	$2\frac{7}{8}$	$2\frac{7}{8}$	$5\frac{1}{2}$	$3\frac{1}{4}$
$1\frac{7}{16}$	$1\frac{7}{16}$	$2\frac{3}{8}$	$1\frac{5}{8}$	$2\frac{15}{16}$	$2\frac{15}{16}$	$5\frac{3}{4}$	$3\frac{7}{16}$
$1\frac{1}{2}$	$1\frac{1}{2}$	$2\frac{1}{2}$	$1\frac{11}{16}$	3	3	$5\frac{7}{8}$	$3\frac{3}{4}$
$1\frac{9}{16}$	$1\frac{9}{16}$	$2\frac{5}{8}$	$1\frac{13}{16}$				

TABLE FOR FINDING TOTAL KEYWAY DEPTH

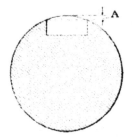

In the column marked "Size of Shaft" find the number representing the size; then to the right find the column representing the keyway to be cut and the decimal there is the distance A, which added to the depth of the keyway will give the total depth from the point where the cutter first begins to cut.

Size of Shaft	⅛ Keyway	3/16 Keyway	¼ Keyway	5/16 Keyway	⅜ Keyway
½	0.0325	....	....	....	....
9/16	0.0289	....	....	....	....
⅝	0.0254	0.0413	....	....	....
11/16	0.0236	0.0379	....	....	....
¾	0.022	0.0346	0.0511	....	....
13/16	0.0198	0.0314	0.0465	....	....
⅞	0.0177	0.0283	0.042	0.0583	....
15/16	0.0164	0.0264	0.0392	0.0544	....
1	0.0152	0.0246	0.0365	0.0506	0.067
1 1/16	0.0143	0.0228	0.0342	0.0476	0.0625
1⅛	0.0136	0.021	0.0319	0.0446	0.0581
1 3/16	0.0131	0.0204	0.0304	0.0421	0.0551
1¼	0.0127	0.0198	0.029	0.0397	0.0522
1 5/16	0.0123	0.0191	0.0279	0.038	0.0499
1⅜	0.012	0.0185	0.0268	0.0364	0.0477
1 7/16	0.0114	0.0174	0.0254	0.0346	0.0453
1½	0.011	0.0164	0.024	0.0328	0.0429
1 9/16	0.0107	0.0158	0.0231	0.0309	0.0412
1⅝	0.0105	0.0153	0.0221	0.0291	0.0395
1 11/16	0.0102	0.0147	0.0214	0.0282	0.0383
1¾	0.0099	0.0142	0.0207	0.0274	0.0371
1 13/16	0.0095	0.0136	0.0198	0.0265	0.0355
1⅞	0.0093	0.013	0.019	0.0257	0.0339
1 15/16	0.009	0.0127	0.0184	0.025	0.0328
2	0.0088	0.0124	0.0179	0.0243	0.0317
2 1/16	0.0083	0.0117	0.0173	0.0236	0.0308
2⅛	0.0078	0.0111	0.0168	0.0229	0.0299
2 3/16	0.0073	0.0109	0.0163	0.0222	0.0291
2¼	0.007	0.0107	0.0159	0.0216	0.0282

TABLE FOR FINDING TOTAL KEYWAY DEPTH

In the column marked "Size of Shaft" find the number representing the size; then to the right find the column representing the keyway to be cut and the decimal there is the distance A, which added to the depth of the keyway will give the total depth from the point where the cutter first begins to cut.

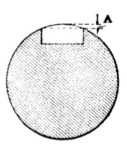

Size of Shaft	¼ Keyway	₃⁄₁₆ Keyway	⅜ Keyway	₇⁄₁₆ Keyway	½ Keyway
2 5/16	0.0068	0.0104	0.0155	0.0209	0.0274
2 3/8	0.0066	0.0102	0.0152	0.0202	0.0267
2 7/16	0.0064	0.01	0.0149	0.0198	0.026
2 1/2	0.0063	0.0098	0.0146	0.0194	0.0253
2 9/16	0.0061	0.0094	0.0142	0.0189	0.0247
2 5/8	0.006	0.009	0.0139	0.0185	0.0242
2 11/16	0.0059	0.0089	0.0136	0.018	0.0236
2 3/4	0.0058	0.0088	0.0133	0.0176	0.023
2 13/16	0.0057	0.0086	0.0129	0.0172	0.0226
2 7/8	0.0056	0.0084	0.0126	0.0168	0.022
2 15/16	0.0054	0.0083	0.0122	0.0164	0.0216
3	0.0053	0.0081	0.0119	0.0161	0.0211
3 1/16	0.0052	0.008	0.0116	0.0158	0.0207
3 1/8	0.0051	0.0078	0.0114	0.0155	0.0202
3 3/16	0.005	0.0076	0.0112	0.0152	0.0198
3 1/4	0.0049	0.0075	0.011	0.0149	0.0194
3 5/16	0.0048	0.0074	0.0108	0.0146	0.0191
3 3/8	0.0047	0.0072	0.0106	0.0143	0.0187
3 7/16	0.0046	0.0071	0.0104	0.014	0.0184
3 1/2	0.0045	0.007	0.0102	0.0138	0.018
3 9/16	0.0044	0.0069	0.0101	0.0135	0.0188
3 5/8	0.0043	0.0067	0.01	0.0133	0.0174
3 11/16	0.0042	0.0066	0.0099	0.0131	0.0171
3 3/4	0.0042	0.0065	0.0098	0.0128	0.0168
3 13/16	0.0041	0.0064	0.0097	0.0126	0.0166
3 7/8	0.0041	0.0063	0.0096	0.0124	0.0163
3 15/16	0.0041	0.0062	0.0095	0.0123	0.0161
4	0.004	0.0061	0.0094	0.0121	0.016

TAPERS FOR KEYS, ETC., FROM $\frac{1}{16}$ TO 1 INCH PER FOOT. AMOUNT OF TAPER FOR LENGTHS VARYING BY $\frac{1}{2}$ INCH

	LENGTH									
1	1½	2	2½	3	3½	4	4½	5	5½	6
.0052	.0078	.0104	.0130	.0156	.0182	.0208	.0234	.0260	.0286	
		.0208	.0260	.0312	.0364	.0416	.0468	.0520	.0572	
.0156	.0234	.0312	.0390	.0468	.0546	.0625	.0703	.0781	.0859	
.0208		.0416	.0520	.0625	.0729	.0833	.0937	.1041	.1145	
		.0520	.0650	.0781	.0911	.1041	.1171	.1302	.1432	
		.0625	.0780	.0937	.1092	.1250	.1406	.1562	.1718	
		.0729	.0911		.1275	.1458	.1640	.1822	.2004	
.0416	.0624		.1041	.1250	.1457	.1666	.1874	.2083	.2291	
.0468	.0702		.1171	.1406	.1641	.1875	.2109	.2343	.2577	
.0520	.0780	.1041		.1562	.1823	.2083	.2343	.2604	.2864	
.0572	.0858	.1145		.1718	.2004	.2291	.2577	.2864	.3150	
.0625	.0938	.1250		.1875	.2188	.2500	.2813	.3125	.3438	
.0677	.1016	.1354		.2031	.2370	.2708	.3047	.3385	.3724	
.0729	.1093	.1458		.2187	.2552	.2916	.3281	.3645	.4010	
.0781	.1172	.1562		.2343	.2734	.3125	.3515	.3906	.4296	
.0833	.1250	.1666		.2500	.2916	.3333	.3749	.4166	.4582	

	LENGTH									
6½	7	7½	8	8½	9	9½	10	10½	11	11½
.0338	.0364	.0390	.0416	.0442	.0468	.0494	.0520	.0546	.0572	
.0677	.0729	.0781	.0833	.0885	.0937	.0989	.1041	.1093	.1145	
.1015	.1093	.1171	.1250	.1328	.1406	.1484	.1562	.1640	.1718	
.1354	.1458	.1562	.1666	.1770	.1875	.1979	.2083	.2187	.2291	
.1692	.1822	.1952	.2083	.2213	.2343	.2473	.2604	.2734	.2864	
.2031	.2187	.2343	.2500	.2656	.2812	.2968	.3125	.3281	.3437	
.2369	.2552	.2734	.2916	.3099	.3281	.3463	.3645	.3827	.4010	
.2708	.2916	.3124	.3333	.3541	.3750	.3958	.4166	.4374	.4583	
.3046	.3281	.3515	.3750	.3984	.4218	.4452	.4687	.4921	.5156	
.3385	.3645	.3905	.4166	.4426	.4687	.4947	.5208	.5468	.5729	
.3723	.4010	.4296	.4583	.4869	.5156	.5442	.5729	.6015	.6302	
.4063	.4375	.4688	.5000	.5312	.5625	.5938	.6250	.6563	.6875	
.4401	.4739	.5078	.5416	.5755	.6093	.6432	.6770	.7109	.7447	
.4739	.5104	.5468	.5833	.6197	.6562	.6926	.7291	.7655	.8020	
.5078	.5468	.5859	.6250	.6641	.7031	.7422	.7812	.8203	.8593	
.5416	.5833	.6249	.6666	.7083	.7500	.7917	.8333	.8750	.9166	

TAPERS AND DOVETAILS

MEASURING TAPERS

An Accurate Taper Gage

THE gage illustrated in Fig. 1 is an exceedingly accurate device for the gaging of tapers.

It is evident that if two round disks of unequal diameter are placed on a surface plate a certain distance apart, two straight-edges touching these two disks will represent a certain taper. It is also evident

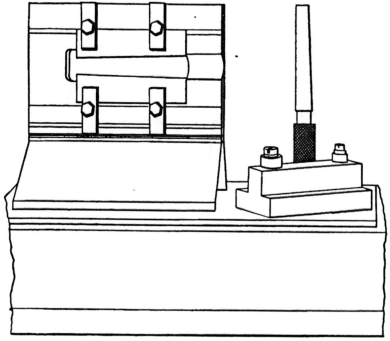

FIG. 1. — Accurate Taper Gage

that with the measuring instruments now in use it is a simple matter to measure accurately the diameters of the two disks, and the distance these disks are apart. These three dimensions accurately and positively determine the taper represented by the straight-edges touching the rolls. If a record is made of these three dimensions these conditions can be reproduced at any time, thus making it possible to duplicate a taper piece even though the part may not at the time be accessible.

The formulas on the following pages may be of service in connection with a gage of this character:

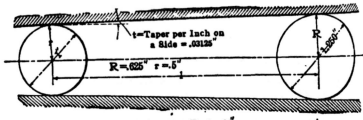

t = Taper per Inch on a Side = .03125"

R = .625"   r = .5"

Taper per Foot = ¼"

FIG. 2

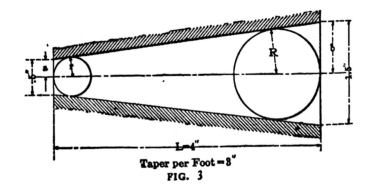

L = 4"

Taper per Foot = 3"

FIG. 3

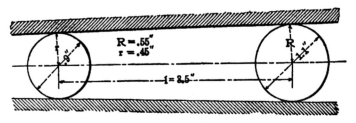

R = .55"
r = .45"

l = 3.5"

FIG. 4

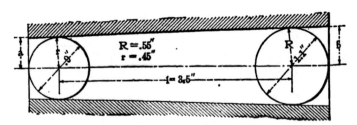

R = .55"
r = .45"

l = 3.5"

FIG. 5

Applications of Taper Gage

Formulas for Use in Connection with Taper Gage

To find Center Distance (l), refer to Fig. 2.

$$l = \frac{R - r}{t} \sqrt{1 + t^2}$$

To find Disk Diameters, refer to Fig. 3.

$$r = \frac{a}{L}\left[\sqrt{L^2 + (b - a)^2} + (b - a)\right]$$

Dia. Small Disk $= 2r$

$$R = \frac{b}{L}\left[\sqrt{L^2 + (b - a)^2} - (b - a)\right]$$

Dia. Large Disk $= 2R$

To find Taper Per Foot (T), refer to Fig. 4.

$$T = 24\left(\frac{R - r}{\sqrt{t^2 - (R - r)^2}}\right)$$

To find Width of Opening at Ends, refer to Fig. 5.

$$a = r\left[\frac{l}{\sqrt{t^2 - (R - r)^2}} - \sqrt{\left(\frac{l}{\sqrt{t^2 - (R - r)^2}}\right)^2 - 1}\right]$$

Width of Opening at Small End $= 2a$.

$$b = R\left[\frac{l}{\sqrt{t^2 - (R - r)^2}} + \sqrt{\left(\frac{l}{\sqrt{t^2 - (R - r)^2}}\right)^2 - 1}\right]$$

Width of Opening at Large End $= 2b$.

Applications of Formulas

To Find Center Distance Between Disks

Suppose there are two disks as shown in Fig. 2, whose diameters are respectively $1\frac{1}{4}$ and 1 inch. It is desired to construct a taper of $\frac{3}{4}$ to the foot and the center distance l between disks must be determined in order that the gage jaws when touching both disks shall give that taper.

Let $R =$ radius of large disk, or 0.625 inch.
$r =$ radius of small disk, or 0.500 inch.
$t =$ taper per inch on side, or

$$\frac{0.750}{24} = 0.03125 \text{ inch.}$$

Then

$$l = \frac{R - r}{t} \sqrt{1 + t^2} =$$

$$\frac{0.125}{0.03125} \sqrt{1.000976} = 4 \times 1.0005 = 4.002 \text{ inches.}$$

To Find Disk Diameters

Suppose the gage jaws are to be set as in Fig. 3 for a three-inch per foot taper whose length is to be four inches. The small end is to be exactly $\frac{1}{2}$ inch and the large end for this taper will, therefore, be $1\frac{1}{2}$ inches. What diameter must the disks be made so that when the jaws are in contact with them and the distance L over the disks measures 4 inches, the taper will be exactly three inches per foot? Here a represents $\frac{1}{2}$ the width of opening at the small end, and b one half the width of opening at the large end. The radius of the small disk may be found by the formula:

$$r = \frac{a}{L} \left\{ \sqrt{L^2 + (b - a)^2} + (b + a) \right\}.$$

Then

$$r = \frac{0.250}{4} \left(\sqrt{16 + 0.25} + 0.5 \right)$$

$$= 0.0625 \, (4.0311 + 0.5) = 0.2832.$$

Diameter small disk = 0.2832 inch $\times$ 2 = 0.5664 inch. For the large disk:

$$R = \frac{b}{L} \left\{ \sqrt{L^2 + (b - a)^2} - (b - a) \right\}.$$

Then

$$R = \frac{0.75}{4} \left(\sqrt{16 + 0.25} - 0.5 \right)$$

$$= 0.1875 \, (4.0311 - 0.5) = 0.6621.$$

Diameter large disk = 0.6621 inch $\times$ 2 = 1.3242 inches.

To Find Taper Per Foot

In duplicating a taper the gage jaws may be set to the model and by placing between the jaws a pair of disks whose diameters are known the taper per foot may be readily found. For example, the jaws in Fig. 4 are set to a certain model, two disks 0.9 and 1.1 inch diameter are placed between them and the distance over the disks measured, from which dimension l (which is 3.5 inches) is readily found by subtracting half the diameters of the disks. Here l represents the center distance as in Fig. 2. To determine the taper per foot which may be represented by T, the formula is:

$$T = 24 \left(\frac{R - r}{\sqrt{l^2 - (R - r)^2}} \right).$$

Then

$$T = 24\left(\frac{0.1}{\sqrt{12.25 - 0.01}}\right) = 24\left(\frac{0.1}{3.4985}\right) = 0.684$$

Taper per foot = 0.684 inch.

To Find Width of Opening at Ends

If, with the ends of the gage jaws flush with a line tangent to the disk peripheries as in Fig. 5, it is required to find the width of the opening at the small end where a represents one half that width, the following formula may be applied, the disks being as in the last example 0.9 and 1.1 inch diameter respectively, and the center distance 3.5 inches:

$$a = r\left\{\frac{l}{\sqrt{l^2 - (R - r)^2}} - \sqrt{\left(\frac{l}{\sqrt{l^2 - (R - r)^2}}\right)^2 - 1}\right\}$$

Then

$$a = 0.45\left\{\frac{3.5}{\sqrt{12.24}} - \sqrt{\left(\frac{3.5}{\sqrt{12.24}}\right) - 1}\right\}$$

$$= 0.45\left(1.00043 - \sqrt{0.00086}\right) = 0.45(1.00043 - 0.0293) = 0.437.$$

0.437 inch × 2 = 0.874 inch width of opening at small end of gage.

Similarly the width of opening at the large end of the gage may be found as follows where b = half the width at large end.

$$b = R\left\{\frac{l}{\sqrt{l^2 - (R - r)^2}} + \sqrt{\left(\frac{l}{\sqrt{l^2 - (R - r)^2}}\right)^2 - 1}\right\}$$

Then

$$b = 0.55\left\{\frac{3.5}{\sqrt{12.24}} + \sqrt{\left(\frac{3.5}{\sqrt{12.24}}\right)^2 - 1}\right\}$$

$$= 0.55\left(1.00043 + \sqrt{0.00086}\right) = 0.55(1.00043 + 0.0293) = 0.5663$$

0.5663 inch × 2 = 1.1326 inches = width of gage opening at large end.

The formulas for a and b appear a little complicated; actually, however, they are simple enough. The expression

$$\frac{l}{\sqrt{l^2 - (R - r)^2}}$$

which appears twice in each formula is readily given its numerical value and upon this substitution the appearance is generally simplified.

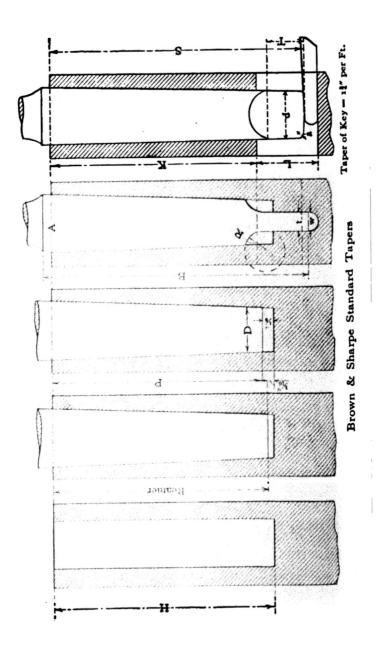

Brown & Sharpe Standard Tapers

BROWN & SHARPE TAPERS

Number of Taper	Dia. of Plug at Small End (D)	Dia. at End of Socket for length P. (A)	Standard Plug Depth (P)	Whole Length of Shank (B)	Depth of Hole (H)	End of Socket to Keyway (K)	Length of Keyway (L)	Width of Key-way (W)	Length of Tongue (T)	Dia. of Tongue (d)	Thickness of Tongue (t)	Radius of Mill for Tongue (R)	Radius of Tongue (a)	Shank Depth (S)	Taper per Foot	Taper per Inch
1	.20	.2391	1 1/16	1 1/4	1 1/16	1	1/2	.135	3/16	.170	5/32	3/16	.030	1 3/16	.500	.0416
2	.25	.2995	1 3/16	1 3/8	1 3/16	1	1/2	.166	1/4	.220	3/32	3/16	.030	1 3/16	.500	.0416
3	.312	.3952	2	1 5/8	1 9/16	1 1/8	9/16	.197	5/16	.280	1/8	5/16	.040	1 1/4	.500	.0416
4	.35	.4020	1 3/16	1 3/4	1 11/16	1 1/4	5/8	.228	5/16	.320	1/8	5/16	.050	1 5/16	.500	.0416
5	.45	.5229	1 7/16	2	1 15/16	1 1/2	11/16	.260	3/8	.420	3/16	3/8	.060	2 1/16	.500	.0416
6	.50	.5989	2	2 3/8	2 5/16	1 3/4	13/16	.291	7/16	.460	3/16	3/8	.060	2 3/16	.500	.0416
7	.60	.7250	3	3 1/8	3	2 1/8	15/16	.322	1/2	.560	7/32	7/16	.070	3 1/16	.500	.0416
8	.75	.8985	3 1/16	3 3/4	3 5/8	2 5/8	1 1/16	.353	9/16	.710	1/4	1/2	.080	3 7/16	.500	.0416
9	.90	1.0667	4	4 1/4	4 1/16	3 1/16	1 3/16	.385	5/8	.860	9/32	9/16	.100	4 1/16	.500	.0416
9	.90	1.0770	4 1/4	4 3/8	4 3/16	3 3/16	1 3/16	.385	11/16	.860	9/32	9/16	.100	4 5/8	.5161	.043
10	1.0446	1.2596	5	5	4 13/16	3 15/16	1 7/16	.447	3/4	1.010	5/16	11/16	.110	5 1/4	.5161	.043
10	1.0446	1.2891	5 1/16	6 1/16	5 7/16	4 3/16	1 7/16	.447	7/8	1.010	5/16	11/16	.110	6 1/8	.500	.0416
11	1.25	1.5312	6 1/4	6 7/8	6 11/16	5 1/16	1 11/16	.510	15/16	1.210	3/8	3/4	.130	7 1/16	.500	.0416
12	1.50	1.7968	7 1/4	7 13/16	7 9/16	6 1/16	1 15/16	.510	1 1/16	1.460	7/16	7/8	.150	7 11/16	.500	.0416
13	1.75	2.0729	8 1/4	8 7/8	8 9/16	7 1/16	2 3/16	.572	1 3/16	1.710	1/2	1	.170	8 1/16	.500	.0416
14	2.	2.3437	8 1/4	9 1/2	8 15/16	7 15/16	2 7/16	.572	1 5/16	1.960	9/16	1 1/8	.190	9 3/16	.500	.0416
15	2.25	2.6145	9 1/4	10	9 13/16	8 3/16	2 11/16	.635	1 7/16	2.210	5/8	1 1/4	.210	9 3/4	.500	.0416
16	2.50	2.8855	9 1/4	10 1/4	9 15/16	9	2 15/16	.635	1 9/16	2.450	11/16	1	.230	10 1/4	.500	.0416

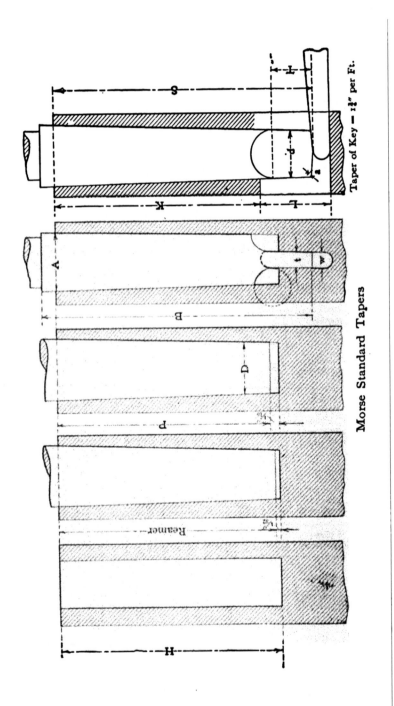

Morse Standard Tapers

		No. of Key	0	1	2	3	4	5	6	7
		Taper per Inch	.05208	.05	.05016	.05016	.05191	.0525	.05216	.05208
		Taper per Foot	.625	.600	.602	.602	.623	.630	.626	.625
S		Shank Depth	$2\frac{7}{32}$	$2\frac{3}{8}$	$2\frac{5}{8}$	$3\frac{9}{16}$	$4\frac{5}{8}$	$5\frac{1}{4}$	8	$11\frac{1}{4}$
a		Radius of Tongue	.04	.05	.06	.08	.10	.12	.15	.18
R		Radius of Mill for Tongue	$\frac{5}{32}$	$\frac{3}{16}$	$\frac{1}{4}$	$\frac{9}{32}$	$\frac{5}{16}$	$\frac{3}{8}$	$\frac{1}{2}$	$\frac{3}{4}$
t		Thickness of Tongue	$\frac{5}{32}$	$\frac{13}{64}$	$\frac{1}{4}$	$\frac{5}{16}$	$\frac{15}{32}$	$\frac{5}{8}$	$\frac{3}{4}$	$1\frac{1}{8}$
d		Dia. of Tongue	.235	.343	$\frac{17}{32}$	$\frac{3}{4}$	$\frac{31}{32}$	$1\frac{13}{32}$	2	$2\frac{7}{16}$
T		Length of Tongue	$\frac{1}{4}$	$\frac{3}{8}$	$\frac{7}{16}$	$\frac{9}{16}$	$\frac{5}{8}$	$\frac{3}{4}$	$1\frac{1}{8}$	$1\frac{3}{8}$
W		Width of Key-way	.160	.213	.26	.322	.478	.635	.76	1.135
L		Length of Key-way	$\frac{9}{16}$	$\frac{3}{4}$	$\frac{7}{8}$	$1\frac{3}{16}$	$1\frac{1}{4}$	$1\frac{3}{8}$	$1\frac{3}{4}$	$2\frac{5}{8}$
K		End of Socket to Keyway	$1\frac{15}{16}$	$2\frac{1}{16}$	$2\frac{1}{2}$	$3\frac{1}{16}$	$3\frac{7}{8}$	$4\frac{15}{16}$	7	$9\frac{1}{4}$
H		Depth of Hole	$2\frac{1}{32}$	$2\frac{1}{16}$	$2\frac{5}{8}$	$3\frac{1}{4}$	$4\frac{1}{8}$	$5\frac{1}{4}$	$7\frac{3}{8}$	$10\frac{3}{8}$
B		Whole Length of Shank	$2\frac{11}{32}$	$2\frac{9}{16}$	$3\frac{1}{16}$	$3\frac{3}{4}$	$4\frac{7}{8}$	6	$8\frac{5}{16}$	$11\frac{5}{8}$
P		Standard Plug Depth	2	$2\frac{1}{8}$	$2\frac{9}{16}$	$3\frac{3}{16}$	$4\frac{1}{16}$	$5\frac{3}{16}$	$7\frac{1}{4}$	10
A		Dia. at End of Socket	.356	.475	.7	.938	1.231	1.748	2.494	3.27
D		Dia. of Plug at Small End	.252	.369	.572	.778	1.02	1.475	2.116	2.75
		No. of Taper	0	1	2	3	4	5	6	7

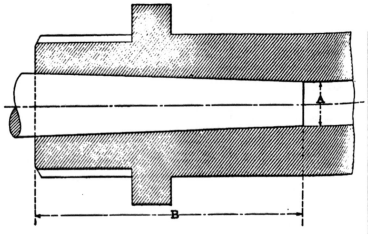

THE REED TAPER

THE F. E. Reed Company, Worcester, Mass., uses in its lathe spindles the 1 in 20 taper (0.6 per foot) which the Jarno system is based on. The diameters of the Reed tapers, however, differ from the Jarno, and the lengths in most cases are somewhat less. The dimensions are given in the table below.

F. E. REED LATHE CENTER TAPERS

TAPER PER FOOT = 0.6 INCH.   TAPER PER INCH = 0.05 INCH

Size of Lathe	Dia. of Small End of Taper	Length of Taper	Size of Lathe	Dia. of Small End of Taper	Length of Taper
	A	B		A	B
12″	$\frac{9}{16}$	$3\frac{5}{8}$	20″	$1\frac{1}{2}$	$5\frac{1}{16}$
14″	$\frac{15}{16}$	$4\frac{1}{8}$	22″	$1\frac{1}{2}$	$5\frac{5}{16}$
16″	$1\frac{1}{4}$	$4\frac{1}{4}$	24″	$1\frac{5}{8}$	$5\frac{1}{4}$
Special 16″	$1\frac{3}{8}$	$4\frac{3}{8}$	27″	$1\frac{3}{4}$	$5\frac{1}{4}$
18″	$1\frac{1}{4}$	$5\frac{1}{16}$	30″	2	$5\frac{3}{4}$

THE JARNO TAPER

WHILE the majority of American tool builders use the Brown & Sharpe taper in their milling-machine spindles and the Morse taper in their lathes, a number of firms, among them the Pratt & Whitney Company, Hartford, Conn., and the Norton Grinding Company, Worcester, Mass., have adopted the "Jarno" taper, the proportions of which are given in the accompanying table. In this system the taper of which is 0.6 inch per foot or 1 in 20, the number of the taper is the key by which all the dimensions are immediately deter

nined without the necessity even of referring to the table. That is,
he number of the taper is the number of tenths of an inch in diam-
·ter at the small end, the number of eighths of an inch at the large
·nd, and the number of halves of an inch in length or depth. For
·xample: the No. 6 taper is six eighths (⅜) inch diameter at large
·nd, six tenths (₁⁶₀) diameter at the small end and six halves (3
nches) in length. Similarly, the No. 16 taper is ¹⁶⁄₈, or 2 inches
liameter at the large end; ¹⁶⁄₈ or 1.6 inches at the small end; ¹⁶⁄₂ or
; inches in length.

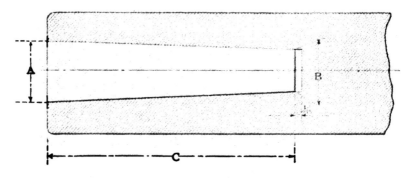

JARNO TAPERS

TAPER PER FOOT = 0.6 INCH.   TAPER PER INCH = 0.05 INCH.

$$\text{Dia. Large End} = \frac{\text{No. of Taper}}{8}$$

$$\text{Dia. Small End} = \frac{\text{No. of Taper}}{10}$$

$$\text{Length of Taper} = \frac{\text{No. of Taper}}{2}$$

No. of Taper	Dia. Large End A	Dia. Small End B	Length of Taper C	No. of Taper	Dia. Large End A	Dia. Small End B	Length of Taper C
1	.125	.10	.5	11	1.375	1.10	5.5
2	.250	.20	1.	12	1.500	1.20	6.0
3	.375	.30	1.5	13	1.625	1.30	6.5
4	.500	.40	2.0	14	1.750	1.40	7.0
5	.625	.50	2.5	15	1.875	1.50	7.5
6	.750	.60	3.0	16	2.000	1.60	8.0
7	.875	.70	3.5	17	2.125	1.70	8.5
8	1.000	.80	4.0	18	2.250	1.80	9.0
9	1.125	.90	4.5	19	2.375	1.90	9.5
10	1.250	1.00	5.0	20	2.500	2.00	10.0

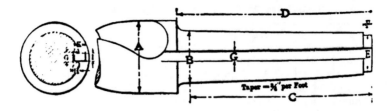

SELLERS TAPERS

Dia. of Drill	Dia. of Shank at Gage Point B	Length of Shank from Point B	Length of Drill Body from Point B	Total Length of Drill	Length of Shank Over All	Dia. at Reduced Portion of Shank	Length of Reduced Portion of Shank	Approximate Pitch of Spiral Grooves	Width of Spline in Shank	Depth of Spline in Shank	Width of Key	Hight of Key
A	B	C			D	E	F		G	H	I	K
$\frac{1}{4}$	$\frac{1}{2}$	$2\frac{1}{4}$	$4\frac{1}{4}$	$6\frac{1}{2}$	$2\frac{5}{16}$	$1\frac{3}{4}$	$\frac{3}{16}$	3.70	$\frac{3}{8}$	$\frac{5}{64}$		
$\frac{5}{16}$	"	"	$4\frac{1}{2}$	$6\frac{1}{2}$	"	"	"	3.70	"	"		
$\frac{3}{8}$	"	"	$4\frac{3}{4}$	7	"	"	"	3.70	"	"		
$\frac{7}{16}$	"	"	$5\frac{1}{2}$	$7\frac{1}{2}$	"	"	"	3.70	"	"		
$\frac{1}{2}$	"	"	$5\frac{3}{4}$	8	"	"	"	5.32	"	"		
$\frac{9}{16}$	$1\frac{1}{16}$	$2\frac{3}{4}$	$6\frac{1}{4}$	8	$2\frac{7}{8}$	$\frac{1}{2}$	$\frac{1}{4}$	5.32	$\frac{1}{2}$	$\frac{3}{32}$		
$\frac{5}{8}$	"	"	$6\frac{1}{2}$	$9\frac{1}{2}$	"	"	"	5.32	"	"		
$\frac{11}{16}$	"	"	$6\frac{3}{4}$	$9\frac{1}{2}$	"	"	"	6.24	"	"		
$\frac{3}{4}$	"	"	$7\frac{1}{4}$	10	"	"	"	6.24	"	"		
$\frac{13}{16}$	"	"	$7\frac{1}{4}$	10	"	"	"	6.24	"	"		
$\frac{7}{8}$	$\frac{7}{8}$	$3\frac{1}{2}$	8	$11\frac{1}{2}$	$3\frac{11}{16}$	$\frac{5}{8}$	$\frac{5}{16}$	7.28	"	"		
$\frac{15}{16}$	"	"	8	$11\frac{1}{2}$	"	"	"	7.28	"	"		
1	"	"	$8\frac{1}{2}$	12	"	"	"	9.50	"	"		
$1\frac{1}{16}$	"	"	$8\frac{1}{2}$	12	"	"	"	9.50	"	"		
$1\frac{1}{8}$	$1\frac{1}{8}$	$4\frac{1}{2}$	9	$13\frac{1}{2}$	$4\frac{3}{4}$	$\frac{25}{32}$	$\frac{3}{8}$	"	$\frac{3}{16}$	$\frac{1}{8}$		
$1\frac{3}{16}$	"	"	9	$13\frac{1}{2}$	"	"	"	"	"	"		
$1\frac{1}{4}$	"	"	9	$13\frac{1}{2}$	"	"	"	"	"	"		
$1\frac{5}{16}$	"	"	9	$13\frac{1}{2}$	"	"	"	"	"	"		
$1\frac{3}{8}$	"	"	$9\frac{1}{2}$	14	"	"	"	"	"	"		
$1\frac{7}{16}$	"	"	$9\frac{1}{2}$	14	"	"	"	"	"	"		
$1\frac{1}{2}$	"	"	10	$14\frac{1}{2}$	"	"	"	"	"	"		
$1\frac{9}{16}$	"	"	10	$14\frac{1}{2}$	"	"	"	"	"	"		
$1\frac{5}{8}$	$1\frac{5}{8}$	$6\frac{1}{2}$	10	$16\frac{1}{2}$	$6\frac{3}{4}$	$1\frac{1}{8}$	$\frac{7}{16}$	13.72	$\frac{1}{4}$	$\frac{3}{16}$		
$1\frac{11}{16}$	"	"	10	$16\frac{1}{2}$	"	"	"	"	"	"		
$1\frac{3}{4}$	"	"	$10\frac{1}{2}$	17	"	"	"	"	"	"		
$1\frac{13}{16}$	"	"	$10\frac{1}{2}$	17	"	"	"	"	"	"		
$1\frac{7}{8}$	"	"	11	$17\frac{1}{2}$	"	"	"	"	"	"		
$1\frac{15}{16}$	"	"	11	$17\frac{1}{2}$	"	"	"	"	"	"		
2	"	"	$11\frac{1}{2}$	18	"	"	"	"	"	"		

THE SELLERS' TAPER

THE system of tapers used by William Sellers & Company, Inc., of Philadelphia, Pa., in lathes, drilling and boring machines, is given in the preceding table. The taper is ¾ inch per foot and each size of taper is splined as shown for a key the dimensions of which are included in the table. The pitch of the spiral for the drills used by the company is also included.

TAPER PINS AND REAMERS

TAPER REAMERS AND PINS

(PRATT & WHITNEY CO.)

Taper = ¼ inch per foot or .0208 inch per inch

Size. No.	Dia. of Small End of Reamer	Dia. of Large End of Reamer	Length of Flute	Total Length of Reamer	Size Drill for Reamer	Longest Limit Length of Pin	Dia. of Large End of Pin	Approx. Fractional Size at Large End of Pin
0	0.135″	.162″	1 3/16″	2″	28	1″	.156″	5/32″
1	.146″	.179″	1 9/16″	2 3/8″	25	1 1/4″	.172″	11/64″
2	.162″	.200″	1 13/16″	2 11/16″	19	1 1/2″	.193″	3/16″
3	.183″	.226″	2 1/16″	3″	12	1 3/4″	.219″	7/32″
4	.208″	.257″	2 1/4″	3 7/16″	3	2″	.250″	1/4″
5	.240″	.300″	2 7/8″	4 7/8″	1/4	2 1/4″	.289″	19/64″
6	.279″	.354″	3 5/8″	5″	5/16	3 1/4″	.341″	11/32″
7	.331″	.423″	4 1/16″	6 1/16″	11/32	3 3/4″	.409″	13/32″
8	.398″	.507″	5 1/8″	7 1/16″	13/32	4 1/2″	.492″	1/2″
9	.482″	.609″	6 1/8″	8 1/8″	31/64	5 1/4″	.591″	19/32″
10	.581″	.727″	7″	9 1/2″	19/32	6″	.706″	23/32″
11	.706″	.878″	8 1/4″	11 1/4″	23/32	7 1/4″	.857″	55/64″
12	.842″	1.050″	10″	13 3/8″	53/64	8 1/2″	1.013″	1 1/64″
13	1.009″	1.259″	12″	16″	1 3/16 r	10 3/4″	1.233″	1 1/4″

These reamer sizes are so proportioned that each overlaps the size smaller about ⅛ inch

Tapers from $\frac{1}{16}$ to $1\frac{1}{4}$ Inch per Foot
Amount of Taper for Lengths up to 24 Inches

Length of Tapered Portion	Taper per Foot									
	$\frac{1}{16}$	$\frac{3}{32}$	$\frac{1}{8}$	$\frac{1}{4}$	$\frac{3}{8}$	$\frac{1}{2}$	$\frac{5}{8}$	$\frac{3}{4}$	1	$1\frac{1}{4}$
$\frac{1}{32}$	.0002	.0002	.0003	.0007	.0010	.0013	.0016	.0020	.0026	.0033
$\frac{1}{16}$	.0003	.0005	.0007	.0013	.0020	.0026	.0033	.0039	.0052	.0065
$\frac{1}{8}$	.0007	.0010	.0013	.0026	.0039	.0052	.0065	.0078	.0104	.0130
$\frac{3}{16}$	.0010	.0015	.0020	.0039	.0059	.0078	.0098	.0117	.0156	.0195
$\frac{1}{4}$	.0013	.0020	.0026	.0052	.0078	.0104	.0130	.0136	.0208	.0260
$\frac{5}{16}$	.0016	.0024	.0033	.0065	.0098	.0130	.0163	.0195	.0260	.0326
$\frac{3}{8}$	.0020	.0029	.0039	.0078	.0117	.0156	.0195	.0234	.0312	.0391
$\frac{7}{16}$	.0023	.0034	.0046	.0091	.0137	.0182	.0228	.0273	.0365	.0456
$\frac{1}{2}$	.0026	.0039	.0052	.0104	.0156	.0208	.0260	.0312	0.417	.0521
$\frac{9}{16}$	.0029	.0044	.0059	.0117	.0176	.0234	.0293	.0352	.0469	.0586
$\frac{5}{8}$	.0033	.0049	.0065	.0130	.0195	.0260	.0326	.0391	.0521	.0651
$\frac{11}{16}$	.0036	.0054	.0072	.0143	.0215	.0286	.0358	.0430	.0573	.0710
$\frac{3}{4}$	.0039	.0059	.0078	.0156	.0234	.0312	.0391	.0469	.0625	.0781
$\frac{13}{16}$	.0042	.0063	.0085	.0169	.0254	.0330	.0423	.0508	.0677	.0846
$\frac{7}{8}$	.0046	.0068	.0091	.0182	.0273	.0365	.0456	.0547	.0729	.0911
$\frac{15}{16}$	.0049	.0073	.0098	.0195	.0293	.0391	.0488	.0586	.0781	.0977
1	.0052	.0078	.0104	.0208	.0312	.0417	.0521	.0625	.0833	.1042
2	.0104	.0156	.0208	.0417	.0625	.0833	.1042	.125	.1667	.2083
3	.0156	.0234	.0312	.0625	.0937	.1250	.1562	.1875	.250	.3125
4	.0208	.0312	.0417	.0833	.125	.1667	.2083	.250	.3333	.4167
5	.0260	.0391	.0521	.1042	.1562	.2083	.2604	.3125	.4167	.5208
6	.0312	.0469	.0625	.125	.1875	.250	.3125	.375	.500	.625
7	.0365	.0547	.0729	.1458	.2187	.2917	.3646	.4375	.5833	.729
8	.0417	.0625	.0833	.1667	.250	.3333	.4167	.500	.6667	.833
9	.0469	.0703	.0937	.1875	.2812	.375	.4687	.5625	.750	.9375
10	.0521	.0781	.1042	.2083	.3125	.4167	.5208	.625	.8333	1.0417
11	.0573	.0859	.1146	.2292	.3437	.4583	.5729	.6875	.9167	1.1458
12	.0625	.0937	.125	.250	.375	.500	.625	.750	1.000	1.250
13	.0677	.1016	.1354	.2708	.4062	.5417	.6771	.8125	1.0833	1.3542
14	.0729	.1094	.1458	.2917	.4375	.5833	.7292	.875	1.1667	1.4583
15	.0781	.1172	.1562	.3125	.4687	.625	.7812	.9375	1.250	1.5625
16	.0833	.125	.1667	.3333	.500	.6667	.8333	1.000	1.3333	1.6667
17	.0885	.1328	.1771	.3542	.5312	.7083	.8854	1.0625	1.4167	1.7708
18	.0937	.1406	.1875	.3750	.5625	.750	.9375	1.125	1.500	1.875
19	.0990	.1484	.1979	.3958	.5937	.7917	.9896	1.1875	1.5833	1.9791
20	.1042	.1562	.2083	.4167	.625	.8333	1.0417	1.250	1.6667	2.0833
21	.1094	.1641	.2187	.4375	.6562	.875	1.0937	1.3125	1.750	2.1875
22	.1146	.1719	.2292	.4583	.6875	.9167	1.1458	1.375	1.8333	2.2917
23	.1198	.1797	.2396	.4792	.7187	.9583	1.1970	1.4375	1.9167	2.3958
24	.125	.1875	.250	.500	.750	1.000	1.250	1.500	2.000	2.500

TAPERS PER FOOT IN INCHES AND CORRESPONDING ANGLES

Taper per Foot	Included Angle			Angle with Center Line			Taper per Foot	Included Angle			Angle with Center Line		
	Deg.	Min.	Sec.	Deg.	Min.	Sec.		Deg.	Min.	Sec.	Deg.	Min.	Sec.
1/64	0	4	28	0	2	14	1	4	46	18	2	23	9
1/32	0	8	58	0	4	29	1 1/8	5	21	44	2	40	52
1/16	0	17	54	0	8	57	1 1/4	5	57	48	2	58	54
3/32	0	26	52	0	13	26	1 3/8	6	33	26	3	16	43
1/8	0	35	48	0	17	54	1 1/2	7	9	10	3	34	35
5/32	0	44	44	0	22	22	1 5/8	7	44	48	3	52	24
3/16	0	53	44	0	26	52	1 3/4	8	20	26	4	10	13
7/32	1	2	34	0	31	17	1 7/8	8	56	2	4	28	1
1/4	1	11	36	0	35	48	2	9	31	36	4	45	48
9/32	1	20	30	0	40	15	2 1/4	10	42	42	5	21	21
5/16	1	29	30	0	44	45	2 1/2	11	53	36	5	56	48
11/32	1	38	22	0	49	11	2 3/4	13	4	24	6	32	12
3/8	1	47	24	0	53	42	3	14	15	0	7	7	30
13/32	1	56	24	0	58	12	3 1/4	15	25	24	7	42	42
7/16	2	5	18	1	2	39	3 1/2	16	35	40	8	17	50
15/32	2	14	16	1	7	8	3 3/4	17	45	40	8	52	50
1/2	2	23	10	1	11	35	4	18	55	28	9	27	44
17/32	2	32	4	1	16	2	4 1/4	20	5	2	10	2	31
9/16	2	41	4	1	20	32	4 1/2	21	14	2	10	37	1
19/32	2	50	2	1	25	1	4 3/4	22	23	22	11	11	41
5/8	2	59	42	1	29	51	5	23	32	12	11	46	6
21/32	3	7	56	1	33	58	5 1/4	24	40	42	12	20	21
11/16	3	16	54	1	38	27	5 1/2	25	48	48	12	54	24
23/32	3	25	50	1	42	55	5 3/4	26	56	46	13	28	23
3/4	3	34	44	1	47	22	6	28	4	2	14	2	1
25/32	3	43	44	1	51	52	6 1/4	29	11	34	14	35	47
13/16	3	52	38	1	56	19	6 1/2	30	18	26	15	9	13
27/32	4	1	36	2	0	48	6 3/4	31	25	2	15	42	31
7/8	4	10	32	2	5	16	7	32	31	12	16	15	36
29/32	4	19	34	2	9	47	7 1/4	33	36	40	16	48	20
15/16	4	28	24	2	14	12	7 1/2	34	42	30	17	21	15
31/32	4	37	20	2	18	40	7 3/4	35	47	32	17	53	46
							8	36	52	12	18	26	6

TABLE FOR COMPUTING TAPERS

The Tabulated Quantities = Twice the Tangent of Half the Angle.

Deg.	0′	10′	20′	30′	40′	50′	60′
0	.00000	.00290	.00582	.00872	.01164	.01454	.01746
1	.01746	.02036	.02326	.02618	.02910	.03200	.03492
2	.03492	.03782	.04072	.04364	.04656	.04946	.05238
3	.05238	.05528	.05820	.06110	.06402	.06692	.06984
4	.06984	.07276	.07566	.07858	.08150	.08440	.08732
5	.08732	.09024	.09316	.09606	.09898	.10190	.10482
6	.10482	.10774	.11066	.11356	.11648	.11940	.12232
7	.12232	.12524	.12816	.13108	.13400	.13694	.13986
8	.13986	.14278	.14570	.14862	.15156	.15448	.15740
9	.15740	.16034	.16326	.16618	.16912	.17204	.17498
10	.17498	.17790	.18084	.18378	.18670	.18964	.19258
11	.19258	.19552	.19846	.20138	.20432	.20726	.21020
12	.21020	.21314	.21610	.21904	.22198	.22492	.22788
13	.22788	.23082	.23376	.23672	.23966	.24262	.24556
14	.24556	.24852	.25148	.25444	.25738	.26034	.26330
15	.26330	.26626	.26922	.27218	.27516	.27812	.28108
16	.28108	.28404	.28702	.28998	.29296	.29592	.29890
17	.29890	.30188	.30486	.30782	.31080	.31378	.31676
18	.31676	.31976	.32274	.32572	.32870	.33170	.33468
19	.33468	.33768	.34066	.34366	.34666	.34966	.35266
20	.35266	.35566	.35866	.36166	.36466	.36768	.37068
21	.37068	.37368	.37670	.37972	.38272	.38574	.38876
22	.38876	.39178	.39480	.39782	.40084	.40388	.40690
23	.40690	.40994	.41296	.41600	.41904	.42208	.42512
24	.42512	.42816	.43120	.43424	.43728	.44034	.44338
25	.44338	.44644	.44950	.45256	.45562	.45868	.46174
26	.46174	.46480	.46786	.47094	.47400	.47708	.48016
27	.48016	.48324	.48632	.48940	.49248	.49556	.49866
28	.49866	.50174	.50484	.50794	.51004	.51414	.51724
29	.51724	.52034	.52344	.52656	.52966	.53278	.53590
30	.53590	.53902	.54214	.54526	.54838	.55152	.55464
31	.55464	.55778	.56092	.56406	.56720	.57034	.57350
32	.57350	.57664	.57980	.58294	.58610	.58926	.59242
33	.59242	.59560	.59876	.60194	.60510	.60828	.61146
34	.61146	.61464	.61782	.62102	.62420	.62740	.63060
35	.63060	.63380	.63700	.64020	.64342	.64662	.64984
36	.64984	.65306	.65628	.65950	.66272	.66596	.66920
37	.66920	.67242	.67566	.67890	.68216	.68540	.68866
38	.68866	.69192	.69516	.69844	.70170	.70496	.70824
39	.70824	.71152	.71480	.71808	.72136	.72464	.72794
40	.72794	.73124	.73454	.73784	.74114	.74446	.74776
41	.74776	.75108	.75440	.75774	.76106	.76440	.76772
42	.76772	.77106	.77442	.77776	.78110	.78446	.78782
43	.78782	.79118	.79454	.79792	.80130	.80468	.80806
44	.80806	.81144	.81482	.81822	.82162	.82502	.82842
45	.82842	.83184	.83526	.83866	.84210	.84552	.84804

TABLE FOR COMPUTING TAPERS

he Tabulated Quantities = Twice the Tangent of Half the Angle.

eg.	0′	10′	20′	30′	40′	50′	60′
16	.84894	.85238	.85582	.85926	.86272	.86616	.86962
17	.86962	.87308	.87656	.88002	.88350	.88698	.89046
18	.89046	.89394	.89744	.90094	.90444	.90794	.91146
19	.91146	.91496	.91848	.92202	.92554	.92908	.93262
50	.93262	.93616	.93970	.94326	.94682	.95038	.95396
51	.95396	.95752	.96110	.96468	.96828	.97186	.97546
52	.97546	.97906	.98268	.98630	.98990	.99354	.99716
53	.99716	1.00080	1.00444	1.00808	1.01174	1.01538	1.01906
54	1.01906	1.02272	1.02638	1.03006	1.03376	1.03744	1.04114
55	1.04114	1.04484	1.04854	1.05226	1.05596	1.05970	1.06342
56	1.06342	1.06716	1.07090	1.07464	1.07840	1.08214	1.08592
57	1.08592	1.08968	1.09346	1.09724	1.10102	1.10482	1.10862
58	1.10862	1.11242	1.11624	1.12006	1.12388	1.12770	1.13154
59	1.13154	1.13538	1.13924	1.14310	1.14696	1.15082	1.15470
60	1.15470	1.15858	1.16248	1.16636	1.17026	1.17418	1.17810
61	1.17810	1.18202	1.18594	1.18988	1.19382	1.19776	1.20172
62	1.20172	1.20568	1.20966	1.21362	1.21762	1.22160	1.22560
63	1.22560	1.22960	1.23362	1.23764	1.24166	1.24570	1.24974
64	1.24974	1.25378	1.25784	1.26190	1.26598	1.27006	1.27414
65	1.27414	1.27824	1.28234	1.28644	1.29056	1.29468	1.29882
66	1.29882	1.30296	1.30710	1.31126	1.31542	1.31960	1.32378
67	1.32378	1.32796	1.33216	1.33636	1.34056	1.34478	1.34902
68	1.34902	1.35326	1.35750	1.36176	1.36602	1.37028	1.37456
69	1.37456	1.37984	1.38314	1.38744	1.39176	1.39608	1.40042
70	1.40042	1.40476	1.40910	1.41346	1.41782	1.42220	1.42658
71	1.42658	1.43098	1.43538	1.43980	1.44422	1.44864	1.45308
72	1.45308	1.45754	1.46200	1.46646	1.47094	1.47542	1.47992
73	1.47992	1.48442	1.48894	1.49348	1.49800	1.50256	1.50710
74	1.50710	1.51168	1.51624	1.52084	1.52544	1.53004	1.53466
75	1.53466	1.53928	1.54392	1.54856	1.55322	1.55790	1.56258
76	1.56258	1.56726	1.57196	1.57668	1.58140	1.58612	1.59088
77	1.59088	1.59562	1.60040	1.60516	1.60996	1.61476	1.61966
78	1.61956	1.62440	1.62922	1.63406	1.63892	1.64380	1.64868
79	1.64868	1.65356	1.65846	1.66338	1.66830	1.67324	1.67820
80	1.67820	1.68316	1.68814	1.69312	1.69812	1.70314	1.70816
81	1.70816	1.71320	1.71824	1.72332	1.72836	1.73348	1.73858
82	1.73858	1.74368	1.74882	1.75396	1.75910	1.76428	1.76946
83	1.76946	1.77464	1.77984	1.78506	1.79030	1.79554	1.80080
84	1.80080	1.80608	1.81138	1.81668	1.82198	1.82732	1.83266
85	1.83266	1.83802	1.84340	1.84878	1.85418	1.85960	1.86504
86	1.86504	1.87048	1.87594	1.88142	1.88690	1.89240	1.89792
87	1.89792	1.90346	1.90902	1.91458	1.92016	1.92576	1.93138
88	1.93138	1.93700	1.94266	1.94832	1.95400	1.95968	1.96540
89	1.96540	1.97112	1.97686	1.98262	1.98840	1.99420	2.00000
90	2.						

Refer to page 262 for explanation of table.

TABLE FOR USE IN COMPUTING TAPERS

In the table on pages 260 and 261 the quantities when expressed in inches represent the taper per inch corresponding to various angles advancing by 10 minutes from 10 minutes to 90 degrees. If an angle is given as, say, 27½ degrees and it is desired to find the corresponding taper in inches, the amount, 0.4894 may be taken directly from the table. This is the taper per inch of length measured as in Fig. 6, along the axis. The taper in inches per foot of length is found by multiplying

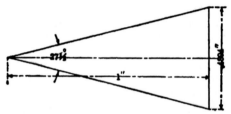

FIG. 6. — Taper per Inch and Corresponding Angle

the tabulated quantity by 12, and in this particular case would be 0.4894″ × 12 = 5.8728″. Where the included angle is not found directly in the table, the taper per inch is found as follows: Assume that the angle in question is 12¼ degrees, then the nearest angles in the table are 12° 10′ and 12° 20′, the respective quantities tabulated under these angles being 0.21314 and 0.21610. The difference between the two is 0.00296, and as 12¼° is half way between 12° 10′ and 12° 20′ one half of 0.00296, or 0.00148 is added to 0.21314, giving 0.21462″ as the taper of a piece 1 inch in length and of an included angle of 12¼ degree. The taper per foot equals 0.21362″ × 12 = 2.5634″.

TABLE FOR DIMENSIONING DOVETAIL SLIDES AND GIBS

THE table on page 263 is figured for machine-tool work, so as to enable one to tell at a glance the amount to be added or subtracted in dimensioning dovetail slides and their gibs, for the usual angles up to 60 degrees. The column for 45-degree dovetails is omitted, as A and B would, of course, be alike for this angle.

In the application of the table, assuming a base with even dimensions, as in the sketch Fig. 7, to obtain the dimensions x and y of the slide Fig. 8, allowing for the gib which may be assumed to be ¼ inch thick, the perpendicular depth of the dovetail being ⅝ inch, and the angle 60 degrees, look under column A for ⅝ inch and it will be found opposite this that B is 0.360 inch, which subtracted from 2 inches gives 1.640 inches, the dimension x. To find y first get the dimension 1.640 inches, then under the column for 60-d gibs (where C is ¼ inch), D is found to be 0.289 inch, which is add to 1.640, giving 1.929 inches.

In practice this dimension is usually made a little larger, say t the nearest 64th, to allow for fitting the gib.

TABLE FOR DIMENSIONING DOVETAIL SLIDES AND GIBS

A	B	B	B	C	D	D	D	D
$\frac{1}{32}''$	.018″	.022″	.027″	$\frac{1}{8}''$	.144″	.152″	.163″	.176″
$\frac{1}{16}''$	.036″	.044″	.053″	$\frac{3}{16}''$	.216″	.228″	.244″	.264″
$\frac{1}{8}''$	.072″	.087″	.105″	$\frac{1}{4}''$	.289″	.305″	.326″	.353″
$\frac{1}{4}''$	.144″	.175″	.210″	$\frac{5}{16}''$	.361″	.381″	.407″	.442″
$\frac{3}{8}''$	.216″	.262″	.314″	$\frac{3}{8}''$	.433″	.457″	.489″	.530″
$\frac{1}{2}''$	.288″	.350″	.420″	$\frac{1}{2}''$	.577″	.610″	.652″	.707″
$\frac{5}{8}''$	.360″	.437″	.525″	$\frac{5}{8}''$	.721″	.762″	.815″	.883″
$\frac{3}{4}''$	.433″	.525″	.629″	$\frac{3}{4}''$	.866″	.915″	.979″	1.060″
$\frac{7}{8}''$	.505″	.612″	.734″	$\frac{7}{8}''$	1.010″	1.067″	1.142″	1.237″
1″	.577″	.700″	.839″	1″	1.154″	1.220″	1.305″	1.414″
$1\frac{1}{8}''$	.649″	.787″	.944″					
$1\frac{1}{4}''$	.721″	.875″	1.049″					
$1\frac{3}{8}''$	.794″	.962″	1.153″					
$1\frac{1}{2}''$	.866″	1.050″	1.259″					
$1\frac{3}{4}''$	1.010″	1.225″	1.469″					
2″	1.154″	1.400″	1.677″					
$2\frac{1}{4}''$	1.298″	1.575″	1.888″					
$2\frac{1}{2}''$	1.442″	1.750″	2.097″					
$2\frac{3}{4}''$	1.588″	1.925″	2.307″					
3″	1.732″	2.100″	2.517″					
$3\frac{1}{2}''$	2.020″	2.450″	2.937″					
4″	2.308″	2.800″	3.356″					
$4\frac{1}{2}''$	2.598″	3.150″	3.776″					
5″	2.885″	3.501″	4.195″					

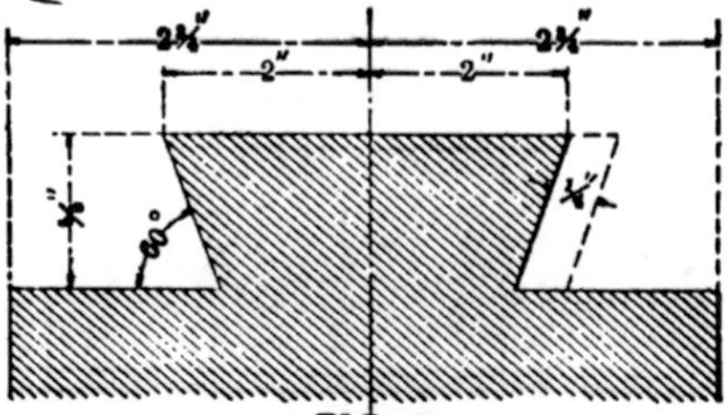

FIG. 7

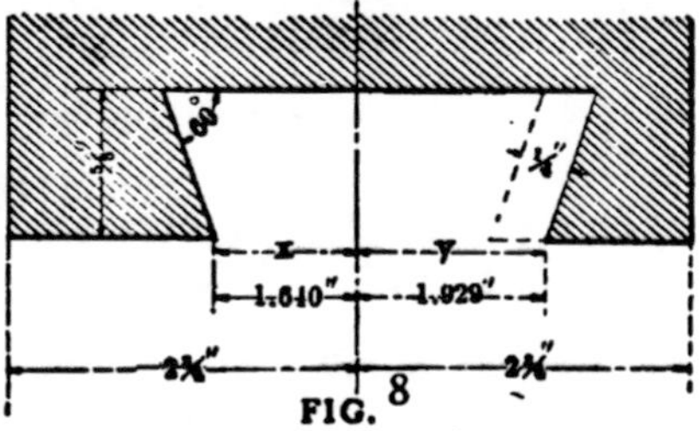

FIG. 8

MEASURING EXTERNAL AND INTERNAL DOVETAILS

THE accompanying table of constants is for use with the plug method of sizing dovetail gages, etc. The constants are calculated for the plugs and angles most in use; and to use them a knowledge of arithmetic is all that is required. The formulas by which they were obtained are added for the convenience of those who may have an unusual angle to make.

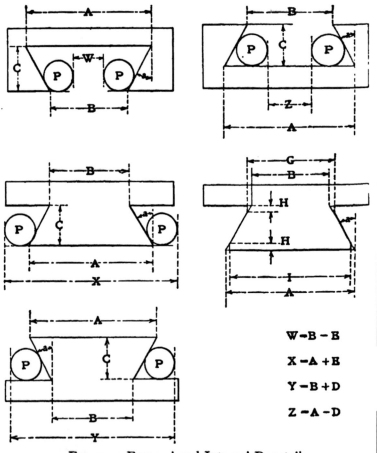

$$W = B - E$$

$$X = A + E$$

$$Y = B + D$$

$$Z = A - D$$

FIG. 9. — External and Internal Dovetails

As an example of the use of the table, suppose that Z, Fig. 9, is the dimension wanted, and that the dimension A and the angle a are known. A glance at the formulas above shows that $Z = A - D$. Then the constant D corresponding to the size of plug and the angle used is subtracted from A and the remainder equals Z. For instance, if $A = 4''$, the plug used = $\frac{3}{4}''$, and the angle = 30 degrees, then $Z = A - D = 4'' - 1.0245'' = 2.9755''$.

If A is not known but B and C are given, according to the formula below the table $A = B + C F$. Then if $B = 3.134''$, $C = \frac{3}{4}''$, and the angle is 30 degrees, as before, $A = B + C F = 3.134'' + (.75'' \times 1.1547) = 4''$, whence Z can be found, as already shown.

If the corners of the dovetail are flat, as shown in Fig. 9 at I and G, and the dimensions I and H and the angles are known, it will be found from the formulas below the table that A also $= I + H F$; so that if $I = 3.8557''$, $H = \frac{1}{8}''$, and the angle $= 30$ degrees, then $A = I + H F = 3.8557'' + (.125'' \times 1.1547) = 4''$, from which Z is found as before.

CONSTANTS FOR DOVETAILS

Plug		60°	55°	50°	45°	40°	35°	30°
$\frac{1}{4}''$	D	1.1830	1.0429	.9368	.8535	.7861	.7302	.6830
	E	.3170	.3233	.3410	.3536	.3666	.3802	.3943
$\frac{3}{8}''$	D	1.7745	1.5643	1.4053	1.2803	1.1792	1.0954	1.0245
	E	.4755	.4932	.5115	.5303	.5499	.5702	.5915
$\frac{1}{2}''$	D	2.3660	2.0858	1.8730	1.7070	1.5722	1.4604	1.3660
	E	.6340	.6576	.6820	.7072	.7332	.7603	.7886
$\frac{3}{4}''$	D	3.5490	3.1286	2.8106	2.5606	2.3534	2.1903	2.0490
	E	.9510	.9864	1.0230	1.0606	1.0998	1.1404	1.1830
	F	3.4641	2.8563	2.3836	2	1.6782	1.4004	1.1547

$$A = B + CF = I + HF$$

$$B = A - CF = G - HF$$

$$E = P\left(\cot. \frac{90 + a}{2}\right) + P$$

$$D = P\left(\cot. \frac{90 - a}{2}\right) + P$$

$$F = 2 \tan a$$

STANDARD JIG PARTS

Drill Bushings

WHEN drilling and reaming operations are to be performed in the same jig, two slip bushings, one for the drill and the other for the reamer, should be used; if the jig is to be used for a large number of parts, the hole for the bushings should in turn be bushed with a steel lining to prevent wearing. The soft cast-iron will wear rapidly if this is not done, and the jig will soon have to be rebored and rebushed.

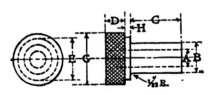

FIG. 1
Loose Bushings

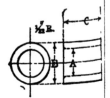

FIG. 2
Fixed Bushings

	LOOSE BUSHINGS						FIXED BUSHINGS		
A	**B**	**C**	**D**	**E**	**G**	**H**	**A**	**B**	**C**
No. 52	$\frac{1}{4}$	$\frac{9}{16}$	$\frac{1}{4}$	$\frac{7}{16}$	$\frac{9}{16}$	$\frac{1}{16}$	52	$\frac{1}{4}$	$\frac{9}{16}$
No. 30	$\frac{5}{16}$	$\frac{5}{8}$	$\frac{1}{4}$	$\frac{1}{2}$	$\frac{5}{8}$	$\frac{1}{16}$	30	$\frac{5}{16}$	$\frac{9}{16}$
No. 12	$\frac{3}{8}$	$\frac{5}{8}$	$\frac{5}{16}$	$\frac{9}{16}$	$\frac{5}{8}$	$\frac{1}{16}$	12	$\frac{3}{8}$	$\frac{5}{8}$
$\frac{1}{4}$	$\frac{7}{16}$	$\frac{11}{16}$	$\frac{5}{16}$	$\frac{9}{16}$	$\frac{11}{16}$	$\frac{1}{16}$	$\frac{1}{4}$	$\frac{7}{16}$	$\frac{11}{16}$
$\frac{5}{16}$	$\frac{9}{16}$	$\frac{3}{4}$	$\frac{3}{8}$	$\frac{11}{16}$	$\frac{3}{4}$	$\frac{1}{16}$	$\frac{5}{16}$	$\frac{9}{16}$	$\frac{3}{4}$
$\frac{3}{16}$	$\frac{11}{16}$	$\frac{13}{16}$	$\frac{3}{8}$	$\frac{7}{8}$	1	$\frac{1}{16}$	$\frac{3}{16}$	$\frac{5}{8}$	$\frac{13}{16}$
$\frac{9}{16}$	$\frac{3}{4}$	$\frac{7}{8}$	$\frac{7}{16}$	$\frac{15}{16}$	$1\frac{1}{16}$	$\frac{1}{16}$	$\frac{9}{16}$	$\frac{3}{4}$	$\frac{11}{16}$
$\frac{11}{16}$	$\frac{13}{16}$	$\frac{15}{16}$	$\frac{1}{2}$	$1\frac{1}{16}$	$1\frac{1}{8}$	$\frac{1}{16}$	$\frac{11}{16}$	$\frac{7}{8}$	$\frac{15}{16}$
$\frac{11}{16}$	$1\frac{1}{16}$	1	$\frac{9}{16}$	$1\frac{1}{8}$	$1\frac{7}{16}$	$\frac{1}{16}$	$\frac{11}{16}$	1	1
$\frac{13}{16}$	$1\frac{1}{8}$	$1\frac{1}{16}$	$\frac{9}{16}$	$1\frac{5}{16}$	$1\frac{1}{2}$	$\frac{1}{16}$	$\frac{13}{16}$	$1\frac{1}{16}$	$1\frac{1}{8}$
$\frac{7}{8}$	$1\frac{1}{8}$	$1\frac{1}{8}$	$\frac{5}{8}$	$1\frac{7}{16}$	$1\frac{5}{8}$	$\frac{1}{16}$	$\frac{7}{8}$	$1\frac{1}{8}$	$1\frac{1}{8}$
$\frac{15}{16}$	$1\frac{1}{16}$	$1\frac{3}{16}$	$\frac{3}{4}$	$1\frac{1}{2}$	$1\frac{11}{16}$	$\frac{1}{16}$	$\frac{15}{16}$	$1\frac{3}{16}$	1
1	$1\frac{1}{4}$	$1\frac{1}{4}$	$\frac{3}{4}$	$1\frac{9}{16}$	$1\frac{3}{4}$	$\frac{1}{16}$	1	$1\frac{1}{4}$	1

Three different styles of bushings with their dimensions are shown in Figs. 1, 2 and 3. These can be blanked out in quantities and finished to required sizes as needed, and should be made of tool steel. Allowances should be made in the blanks for grinding and

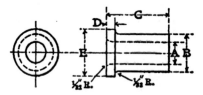

A	B	C	D	E	A	B	C	D	E
$\frac{1}{4}$	$\frac{7}{16}$	$\frac{11}{16}$	$\frac{3}{32}$	$\frac{9}{16}$	$\frac{11}{16}$	$1\frac{1}{8}$	$1\frac{1}{16}$	$\frac{1}{8}$	$1\frac{1}{8}$
$\frac{5}{16}$	$\frac{1}{2}$	$\frac{3}{4}$	$\frac{3}{32}$	$\frac{5}{8}$	$\frac{3}{4}$	1	$1\frac{1}{8}$	$\frac{1}{8}$	$1\frac{1}{16}$
$\frac{3}{8}$	$\frac{9}{16}$	$\frac{13}{16}$	$\frac{3}{32}$	$\frac{11}{16}$	$\frac{13}{16}$	$1\frac{1}{16}$	$1\frac{3}{16}$	$\frac{1}{8}$	$1\frac{1}{4}$
$\frac{7}{16}$	$\frac{5}{8}$	$\frac{7}{16}$	$\frac{3}{32}$	$\frac{3}{4}$	$\frac{7}{8}$	$1\frac{1}{8}$	$1\frac{1}{4}$	$\frac{5}{32}$	$1\frac{5}{16}$
$\frac{1}{2}$	$\frac{11}{16}$	$\frac{7}{8}$	$\frac{3}{32}$	$\frac{13}{16}$	$\frac{15}{16}$	$1\frac{3}{16}$	$1\frac{5}{16}$	$\frac{5}{32}$	$1\frac{3}{8}$
$\frac{9}{16}$	$\frac{3}{4}$	$\frac{15}{16}$	$\frac{1}{8}$	$\frac{7}{8}$	1	$1\frac{1}{4}$	$1\frac{5}{16}$	$\frac{5}{32}$	$1\frac{7}{16}$
$\frac{5}{8}$	$\frac{7}{8}$	1	$\frac{1}{8}$	$1\frac{1}{16}$					

FIG. 3. — Fixed Bushings

lapping after hardening. Fig. 1 shows a slip bushing; Fig. 2 a stationary bushing, and Fig. 3 a stationary bushing where tools with stop collars are to be used. Such bushings as shown in Figs. 2 and 3 are also used for linings for slip bushings.

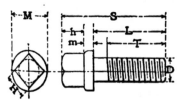

FIG. 4
Collar-Head Jig Screws.

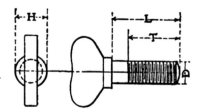

FIG. 5
Winged Jig Screws

D	Thrd.	L	T	h	m	H	M	S	D	Thrd.	H	L	T
$\frac{1}{4}$	20	1	$\frac{3}{8}$	$\frac{1}{4}$	$\frac{3}{32}$	$\frac{1}{4}$	$\frac{7}{16}$	$1\frac{11}{32}$	$\frac{1}{4}$	20	$\frac{3}{8}$	1	$\frac{3}{8}$
$\frac{5}{16}$	18	1	$\frac{3}{4}$	$\frac{5}{16}$	$\frac{1}{8}$	$\frac{5}{16}$	$\frac{1}{2}$	$1\frac{7}{16}$	$\frac{5}{16}$	18	$\frac{7}{16}$	1	$\frac{3}{4}$
$\frac{3}{8}$	16	$1\frac{1}{2}$	1	$\frac{3}{8}$	$\frac{3}{8}$	$\frac{5}{8}$	$\frac{3}{4}$	2	$\frac{3}{8}$	16	$\frac{1}{2}$	$1\frac{1}{2}$	1
$\frac{7}{16}$	14	2	$1\frac{1}{2}$	$\frac{7}{16}$	$\frac{3}{8}$	$\frac{7}{16}$	$1\frac{1}{16}$	$2\frac{5}{8}$					
$\frac{1}{2}$	12	2	$1\frac{1}{4}$	$\frac{1}{2}$	$\frac{3}{8}$	$\frac{1}{2}$	$1\frac{1}{4}$	$2\frac{1}{2}$					

Binding Screws

Binding-screws should be made in various sizes and with threads to conform to the standard taps with which the shop is provided. When drills of a very large size are used, a screw with a square or hexagon head is best, as the work requires firm clamping. If the drills used are small, a winged screw will be sufficient and more convenient, as it will require less time to manipulate. Some good screws for clamping straps are shown in Figs. 4 and 5. Of course the screws can be made of any length desired.

When the work is to be held against the seat or a stop by means of a set-screw, such screws as shown in Figs. 6 and 7 will be found very useful. If, however, the work is very light, a wing screw can be used.

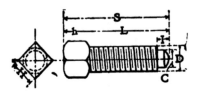
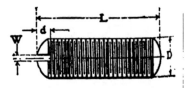

D	Thrd.	L	h	I	C	H	S	D	Thrd.	L	W	d
$\frac{1}{4}$	20	1	$\frac{1}{4}$	$\frac{1}{8}$	$\frac{5}{32}$	$\frac{1}{4}$	$1\frac{1}{4}$	$\frac{1}{4}$	20	$\frac{3}{4}$	.040	$\frac{11}{64}$
$\frac{5}{16}$	18	1	$\frac{5}{16}$	$\frac{1}{8}$	$\frac{5}{64}$	$\frac{5}{16}$	$1\frac{5}{16}$	$\frac{5}{16}$	18	1	.057	$\frac{11}{64}$
$\frac{3}{8}$	16	$1\frac{1}{4}$	$\frac{3}{8}$	$\frac{1}{8}$	$\frac{8}{8}$	$\frac{3}{8}$	$1\frac{3}{8}$	$\frac{3}{8}$	16	1	$\frac{1}{16}$	$\frac{1}{8}$
$\frac{7}{16}$	14	$1\frac{1}{2}$	$\frac{7}{16}$	$\frac{7}{16}$	$\frac{11}{16}$	$\frac{7}{16}$	$1\frac{11}{16}$	$\frac{7}{16}$	14	$1\frac{1}{2}$	$\frac{5}{64}$	$\frac{1}{8}$
$\frac{1}{2}$	13	$1\frac{1}{2}$	$\frac{1}{2}$	$\frac{7}{16}$	$\frac{11}{32}$	$\frac{1}{2}$	2	$\frac{1}{2}$	13	$1\frac{1}{2}$	$\frac{5}{64}$	$\frac{1}{8}$

FIG. 6
Square-Head Jig Screws

FIG. 7
Headless Jig Screws

Supporting Screws

Figs. 8 and 9 show screws that are useful in supporting work against the thrust of drills when the work is of such a nature that it cannot be supported otherwise.

Locking Screws

A convenient hinge-cover locking screw is shown in Fig. 10. This screw, when used, should be adjusted so that only a quarter turn will be needed to clamp or release the cover, which should be slotted to admit the head of the screw.

The different sizes of the styles of screws shown are not only used with drilling jigs, but are equally useful with other jigs and fixtures. These screws should be made of screw stock and case-hardened.

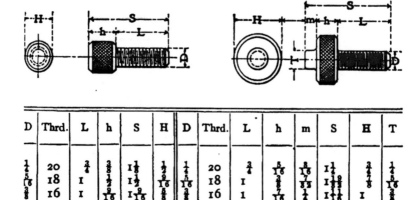

D	Thrd.	L	h	S	H	D	Thrd.	L	h	m	S	H	T
$\frac{1}{4}$	20	$\frac{3}{4}$	$\frac{3}{8}$	$1\frac{1}{8}$	$\frac{1}{2}$	$\frac{1}{4}$	20	$\frac{3}{4}$	$\frac{5}{16}$	$\frac{5}{16}$	$1\frac{1}{4}$	$\frac{3}{8}$	$\frac{1}{4}$
$\frac{5}{16}$	18	1	$\frac{1}{2}$	$1\frac{1}{4}$	$\frac{5}{16}$	$\frac{5}{16}$	18	1	$\frac{3}{8}$	$\frac{7}{32}$	$1\frac{9}{32}$	$\frac{7}{16}$	$\frac{5}{16}$
$\frac{3}{8}$	16	1	$\frac{9}{16}$	$1\frac{11}{16}$	$\frac{3}{8}$	$\frac{3}{8}$	16	1	$\frac{7}{16}$	$\frac{1}{2}$	$1\frac{11}{16}$	1	$\frac{3}{4}$

FIG. 8
Nurled-Head Jig Screws

FIG. 9
Nurled-Head Jig Screws

Strap Dimensions

A convenient strap to use with these jigs is shown in Fig. 11. The straps should be made of bessemer steel and case-hardened after finishing. The slot G can be located in the proper position and made of such dimensions as to allow the strap to be slipped back out of the way when work is being placed in and taken from the jig.

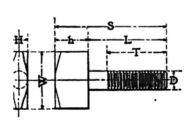

IG. 10. — Locking Jig Screws

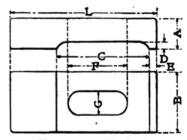

FIG. 11. — Jig Straps

D	Thrd.	H	h	L	S	T	W	A	B	C	D	E	L
$\frac{5}{16}$	18	$\frac{5}{16}$	$\frac{5}{8}$	$1\frac{1}{2}$	$2\frac{1}{2}$	$1\frac{1}{8}$	$1\frac{1}{16}$	$\frac{1}{2}$	1	$1\frac{5}{8}$	$\frac{1}{8}$	$\frac{1}{8}$	$2\frac{1}{2}$
	16	$\frac{3}{8}$	$\frac{11}{16}$	$1\frac{5}{8}$	$2\frac{5}{16}$	$1\frac{1}{8}$	$1\frac{8}{16}$	$\frac{1}{2}$	1	$1\frac{5}{8}$	$\frac{1}{8}$	$\frac{1}{8}$	3
	14	$\frac{7}{16}$	$\frac{3}{4}$	$1\frac{3}{4}$	$2\frac{5}{8}$	$1\frac{1}{4}$	$1\frac{5}{16}$	$\frac{5}{8}$	$1\frac{1}{4}$	2	$\frac{1}{8}$	$\frac{3}{16}$	$3\frac{1}{2}$
	13	$\frac{1}{2}$	$\frac{3}{4}$	$1\frac{7}{8}$	$2\frac{5}{8}$	$1\frac{1}{4}$	$1\frac{1}{16}$	$\frac{5}{8}$	$1\frac{1}{4}$	2	$\frac{1}{8}$	$\frac{3}{16}$	4

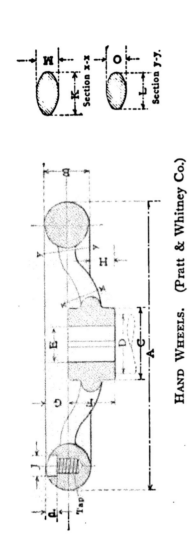

Section x-x.

Section y-y.

HAND WHEELS. (Pratt & Whitney Co.)

A	B	C	D	E	F	G	H	J	K	L	M	O	P	Tap	Spline	Handle No.	No. of Arms
3 in.														—28	×	8	3
4														—28	×	8	3
4½														—24	×	0	3
5														—24	×	1	3
6														—24	×	1	3
7														—24	×	2	3
8														—24	×	3	3
9														—24	×	4	3
10														—24	×	4	3
11														—20	×	5	3
12														—16	×	5	3

HANDLES FOR HAND-WHEELS

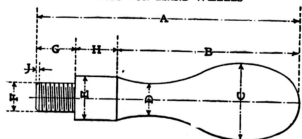

No.	A	B	C	D	E	F	G	H	J	P. I.
∞	$1\frac{15}{16}$	$1\frac{5}{16}$	$\frac{17}{32}$	$\frac{1}{4}$	$\frac{3}{8}$	$\frac{5}{16}$	$\frac{3}{8}$	$\frac{1}{4}$	$\frac{1}{32}$	28
0	$2\frac{1}{16}$	$1\frac{15}{16}$	$\frac{15}{16}$	$1\frac{1}{8}$	$\frac{15}{32}$	$\frac{5}{16}$	$\frac{3}{8}$	$\frac{9}{16}$	$\frac{1}{32}$	28
1	$2\frac{5}{16}$	2	$\frac{15}{16}$	$1\frac{1}{8}$	$\frac{15}{32}$	$\frac{5}{16}$	$\frac{3}{8}$	$\frac{9}{16}$	$\frac{1}{32}$	24
2	$3\frac{7}{16}$	$2\frac{3}{8}$	1	$\frac{7}{8}$	$\frac{19}{32}$	$\frac{3}{8}$	$\frac{1}{2}$	$\frac{9}{16}$	$\frac{1}{32}$	24
3	$3\frac{15}{16}$	$2\frac{15}{32}$	$1\frac{1}{8}$	$\frac{15}{32}$	$\frac{19}{32}$	$\frac{3}{8}$	$\frac{1}{2}$	$\frac{5}{8}$	$\frac{1}{32}$	24
4	$4\frac{1}{2}$	$3\frac{1}{8}$	$1\frac{1}{4}$	$\frac{1}{2}$	$\frac{19}{32}$	$\frac{3}{8}$	$\frac{1}{2}$	$\frac{5}{8}$	$\frac{1}{32}$	24
5	$4\frac{1}{2}$	$3\frac{7}{16}$	$1\frac{5}{16}$	$\frac{9}{16}$	$\frac{21}{32}$	$\frac{7}{16}$	$\frac{1}{2}$	$\frac{11}{16}$	$\frac{1}{32}$	20
6	$5\frac{1}{16}$	$3\frac{5}{8}$	$1\frac{7}{16}$	$\frac{5}{8}$	$\frac{3}{4}$	$\frac{1}{2}$	$\frac{11}{16}$	$\frac{7}{8}$	$\frac{1}{32}$	16

KNOBS

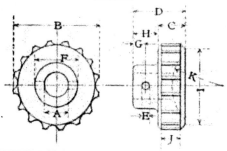

A	B	C	D	E	F	G	H	I	J	K
$\frac{3}{16}$	1	$\frac{3}{16}$	$\frac{3}{16}$	$\frac{5}{32}$	$\frac{7}{16}$	$\frac{1}{8}$	$\frac{1}{4}$	$1\frac{1}{16}$	$\frac{5}{32}$	
$\frac{1}{4}$	$1\frac{3}{16}$	$\frac{5}{16}$	$\frac{11}{16}$	$\frac{5}{32}$	$\frac{9}{16}$	$\frac{5}{32}$	$\frac{1}{4}$	$1\frac{1}{16}$		
$\frac{5}{16}$	$1\frac{3}{16}$	$\frac{5}{16}$	$\frac{11}{16}$	$\frac{1}{4}$	$\frac{9}{16}$	$\frac{5}{32}$	$\frac{5}{16}$	$1\frac{1}{16}$		
$\frac{3}{8}$	$1\frac{1}{2}$	$\frac{7}{16}$	$\frac{7}{8}$	$\frac{1}{4}$	$\frac{11}{16}$	$\frac{5}{32}$	$\frac{7}{16}$	$1\frac{1}{2}$		
$\frac{7}{16}$	$1\frac{1}{2}$	$\frac{7}{16}$	$\frac{7}{8}$	$\frac{1}{4}$	$\frac{11}{16}$	$\frac{5}{32}$	$\frac{7}{16}$	$1\frac{1}{2}$		
$\frac{1}{2}$	$1\frac{3}{4}$	$\frac{9}{16}$	$1\frac{1}{16}$	$\frac{5}{32}$	$\frac{3}{4}$	$\frac{1}{4}$	$\frac{1}{2}$	$1\frac{5}{8}$	$\frac{7}{16}$	
$\frac{9}{16}$	$1\frac{3}{4}$	$\frac{9}{16}$	$1\frac{1}{16}$	$\frac{3}{16}$	$\frac{3}{4}$	$\frac{1}{4}$	$\frac{1}{2}$	$1\frac{5}{8}$	$\frac{7}{16}$	
$\frac{5}{8}$	$2\frac{1}{4}$	$\frac{11}{16}$	$1\frac{1}{4}$	$\frac{3}{16}$	$1\frac{1}{16}$	$\frac{9}{32}$	$\frac{9}{16}$	$2\frac{1}{16}$		
$\frac{11}{16}$	$2\frac{1}{4}$	$\frac{11}{16}$	$1\frac{1}{4}$	$\frac{3}{16}$	$1\frac{1}{16}$	$\frac{9}{32}$	$\frac{9}{16}$	$2\frac{1}{16}$		
$\frac{3}{4}$	$2\frac{1}{4}$	$\frac{11}{16}$	$1\frac{1}{4}$	$\frac{3}{16}$	$1\frac{1}{16}$	$\frac{9}{32}$	$\frac{9}{16}$	$2\frac{1}{16}$		Cup Out to Suit

BALL HANDLES
(Pratt & Whitney Co.)

Center Ball

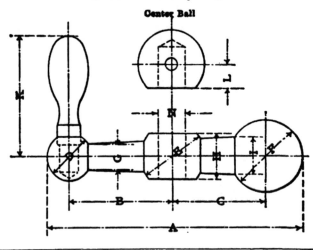

A	B	C	D	E	F	G	H	I	K	L	N
3	$1\frac{9}{16}$	1	$\frac{5}{8}$	$\frac{7}{8}$	1	$\frac{3}{8}$	$\frac{9}{16}$	$\frac{1}{2}$	$1\frac{7}{8}$	$\frac{9}{32}$	$\frac{1}{8}$
4	$1\frac{21}{32}$	$1\frac{7}{16}$	$\frac{15}{16}$	1	$1\frac{1}{8}$	$\frac{7}{16}$	$\frac{11}{16}$	$\frac{9}{16}$	$1\frac{15}{16}$	$\frac{11}{32}$	$\frac{1}{16}$
$4\frac{1}{2}$	$1\frac{7}{8}$	$1\frac{5}{8}$	$\frac{3}{4}$	$1\frac{1}{16}$	$1\frac{1}{4}$	$\frac{7}{16}$	$\frac{3}{4}$	$\frac{5}{8}$	$2\frac{1}{8}$	$\frac{3}{8}$	$\frac{1}{8}$
5	$2\frac{1}{16}$	$1\frac{13}{16}$	$\frac{3}{4}$	$1\frac{5}{16}$	$1\frac{3}{8}$	$\frac{1}{2}$	$\frac{13}{16}$	$\frac{11}{16}$	$2\frac{11}{16}$	$\frac{13}{32}$	$\frac{1}{8}$
$5\frac{1}{2}$	$2\frac{9}{32}$	$2\frac{1}{32}$	$\frac{15}{16}$	$1\frac{1}{4}$	$1\frac{7}{8}$	$\frac{1}{2}$	$\frac{7}{8}$	$\frac{13}{16}$	$2\frac{1}{4}$	$\frac{7}{16}$	$\frac{9}{16}$
6	$2\frac{1}{2}$	$2\frac{7}{32}$	1	$1\frac{5}{16}$	$1\frac{9}{16}$	$\frac{1}{2}$	$\frac{15}{16}$	$\frac{3}{4}$	$3\frac{1}{8}$	$\frac{15}{32}$	$\frac{9}{16}$
7	$2\frac{31}{32}$	$2\frac{21}{32}$	$1\frac{1}{16}$	$1\frac{7}{16}$	$1\frac{13}{16}$	$\frac{9}{16}$	1	$\frac{13}{16}$	$3\frac{1}{16}$	$\frac{1}{2}$	$\frac{5}{8}$
8	$3\frac{7}{16}$	$3\frac{5}{32}$	$1\frac{1}{8}$	$1\frac{1}{2}$	$1\frac{15}{16}$	$\frac{9}{16}$	$1\frac{1}{16}$	$\frac{7}{8}$	$3\frac{5}{8}$	$\frac{17}{32}$	

BINDER HANDLES

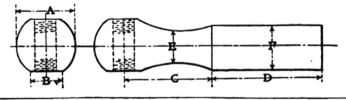

A	B		C		D		E		F		Dia. of Tap
1	$\frac{3}{4}$		$1\frac{5}{8}$		$1\frac{5}{8}$		$\frac{7}{16}$		$\frac{3}{4}$		$\frac{7}{16}$
$1\frac{1}{16}$		$\frac{13}{16}$		$1\frac{3}{4}$	2		$\frac{1}{2}$		$\frac{13}{16}$		$\frac{1}{2}$
$1\frac{1}{4}$	$\frac{7}{8}$		$2\frac{1}{8}$		$2\frac{3}{8}$		$\frac{9}{16}$		$\frac{15}{16}$		$\frac{9}{16}-\frac{3}{4}$
	$1\frac{1}{2}$		$1\frac{1}{8}$		$2\frac{5}{8}$	$2\frac{3}{4}$		$\frac{5}{8}$		$1\frac{1}{16}$	$\frac{5}{8}-1\frac{1}{4}$

Single End Ball Handles. (Walcott & Wood)

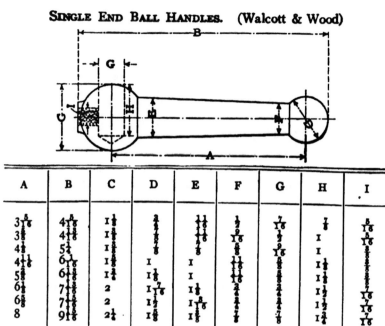

A	B	C	D	E	F	G	H	I
$3\frac{5}{16}$	$4\frac{5}{8}$	$1\frac{1}{8}$	$\frac{7}{8}$	$1\frac{1}{16}$	$\frac{7}{8}$	$\frac{7}{16}$	$\frac{7}{8}$	$\frac{5}{16}$
$3\frac{3}{8}$	$4\frac{1}{16}$	$1\frac{1}{8}$	$\frac{7}{8}$	$1\frac{1}{16}$	$\frac{9}{16}$	$\frac{1}{2}$	1	$\frac{5}{16}$
$4\frac{1}{2}$	$5\frac{7}{16}$	$1\frac{3}{8}$	$\frac{7}{8}$	$\frac{7}{8}$	$\frac{3}{4}$	$\frac{9}{16}$	1	$\frac{5}{16}$
$4\frac{11}{16}$	$6\frac{1}{16}$	$1\frac{3}{8}$	1	1	$1\frac{1}{16}$	$\frac{3}{4}$	$1\frac{1}{8}$	$\frac{3}{8}$
$5\frac{3}{8}$	$6\frac{5}{8}$	$1\frac{1}{4}$	$1\frac{1}{8}$	1	$1\frac{1}{8}$	$\frac{3}{4}$	$1\frac{1}{8}$	$\frac{3}{8}$
$6\frac{1}{4}$	$7\frac{1}{16}$	2	$1\frac{7}{16}$	$1\frac{1}{8}$	$1\frac{1}{4}$	$\frac{3}{4}$	$1\frac{1}{2}$	$\frac{7}{16}$
$6\frac{5}{8}$	$7\frac{5}{8}$	2	$1\frac{1}{2}$	$1\frac{5}{16}$	$1\frac{3}{8}$	$\frac{3}{4}$	$1\frac{1}{2}$	$\frac{7}{16}$
8	$9\frac{1}{16}$	$2\frac{1}{4}$	$1\frac{5}{8}$	$1\frac{5}{8}$	$1\frac{1}{2}$	$\frac{7}{8}$	$1\frac{3}{4}$	$\frac{7}{16}$

Ball Lever Handles

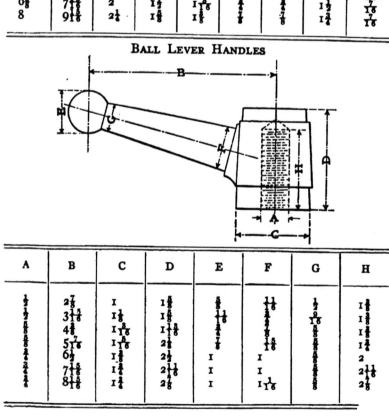

A	B	C	D	E	F	G	H
$\frac{7}{16}$	$2\frac{7}{8}$	1	$1\frac{5}{8}$	$\frac{5}{8}$	$1\frac{1}{16}$	$\frac{1}{2}$	$1\frac{3}{8}$
$\frac{1}{2}$	$3\frac{5}{16}$	$1\frac{1}{8}$	$1\frac{5}{8}$	$1\frac{1}{16}$	$1\frac{1}{16}$	$\frac{9}{16}$	$1\frac{3}{8}$
$\frac{5}{8}$	$4\frac{5}{8}$	$1\frac{5}{16}$	$1\frac{3}{8}$	$\frac{3}{4}$	$1\frac{3}{8}$	$\frac{5}{8}$	$1\frac{3}{4}$
$\frac{3}{4}$	$5\frac{7}{16}$	$1\frac{5}{16}$	$2\frac{1}{2}$	$\frac{7}{8}$	$1\frac{3}{8}$	$\frac{3}{4}$	$1\frac{3}{4}$
$\frac{7}{8}$	$6\frac{1}{2}$	$1\frac{3}{8}$	$2\frac{1}{2}$	1	1	$\frac{7}{8}$	2
$\frac{3}{4}$	$7\frac{5}{16}$	$1\frac{3}{8}$	$2\frac{11}{16}$	1	1	$\frac{7}{8}$	$2\frac{11}{16}$
$\frac{3}{4}$	$8\frac{5}{16}$	$1\frac{1}{4}$	$2\frac{5}{8}$	1	$1\frac{1}{16}$	$\frac{3}{4}$	$2\frac{5}{8}$

WING NUTS

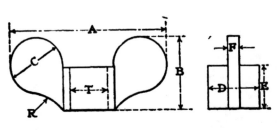

A	B	C	D	E	F	R	T
13/16	11/16	3/32	7/32	5/32	3/32	9/32	1/8
7/8	3/4			1/4	1/8		5/32
15/16	13/16	5/16		1/4		3/32	
1 1/16	7/8	3/8	7/16	3/8	1/8	5/16	1/4
1 1/2	1	1/2	1/2	5/16	5/32	7/16	5/16
1 5/8	1 1/8	9/16	1/2	1/2	9/32	7/16	3/8
1 13/16	7/8	5/8	5/8	9/16	5/32	1/2	7/16
2 1/16	1 1/16	3/4	3/4	11/16	5/32	3/4	1/2

MACHINE HANDLES

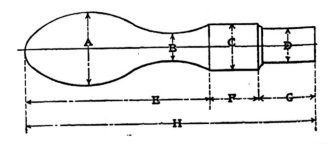

A	B	C	D	E	F	G	H
1 1/8		11/32		15/32	5/16	1 15/16	7/16	1/2	2 7/8
	1 3/8	7/16	1 3/8	11/32	5/16	2	5/8	11/16	3 1/8
			11/32	1/2	5/16	2 3/8	5/8	3/4	3 3/4
	1 3/16	1 1/16	19/32	5/16	2 11/16	11/16	11/16	4 1/16
1 1/4		1/2	19/32	3/8	3 1/8	3/4	11/8	4 7/8
	1 5/16	5/8	9/16	21/32	3/8	3 7/16	7/8	13/16	5 1/8
1 7/16		3/4	7/16	3 7/8	13/16	1 1/8	5 11/16

THUMB NUTS

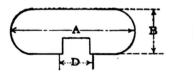

D	A	B	C	Mill
$\frac{3}{16}$	$\frac{7}{8}$	$\frac{1}{4}$	$\frac{3}{32}$	$\frac{5}{16}$
$\frac{1}{4}$		$\frac{5}{16}$	$\frac{1}{8}$	$\frac{5}{16}$
$\frac{5}{16}$	$1\frac{1}{4}$	$\frac{3}{8}$	$\frac{5}{32}$	$\frac{7}{16}$
$\frac{3}{8}$	$1\frac{1}{2}$	$\frac{7}{16}$	$\frac{5}{32}$	$\frac{7}{16}$
$\frac{7}{16}$	$1\frac{3}{4}$	$\frac{1}{2}$	$\frac{3}{16}$	$\frac{1}{2}$
$\frac{1}{2}$		$\frac{9}{16}$	$\frac{3}{16}$	$\frac{1}{2}$
$\frac{9}{16}$	$2\frac{1}{4}$	$\frac{5}{8}$	$\frac{3}{16}$	$\frac{5}{8}$
$\frac{5}{8}$	$2\frac{1}{2}$	$\frac{11}{16}$	$\frac{7}{32}$	$\frac{5}{8}$

HOOK BOLTS

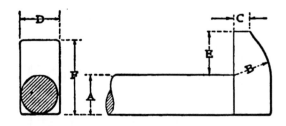

Dia. of Bolt	Thickness of Head	Thickness at End	Width of Head	Off Set of Head	Length of Head
A	B	C	D	E	F
$\frac{3}{8}$	$\frac{7}{16}$	$\frac{3}{16}$	$\frac{3}{8}$	$\frac{7}{16}$	$\frac{13}{16}$
$\frac{1}{2}$	$\frac{9}{16}$	$\frac{1}{4}$	$\frac{1}{2}$	$\frac{9}{16}$	$1\frac{1}{16}$
$\frac{5}{8}$	$\frac{11}{16}$	$\frac{5}{16}$	$\frac{5}{8}$	$\frac{11}{16}$	$1\frac{5}{16}$
$\frac{3}{4}$	$\frac{13}{16}$	$\frac{3}{8}$	$\frac{3}{4}$	$\frac{13}{16}$	$1\frac{9}{16}$
$\frac{7}{8}$	$\frac{15}{16}$	$\frac{7}{16}$	$\frac{7}{8}$	$\frac{15}{16}$	$1\frac{13}{16}$
1	$1\frac{1}{16}$	$\frac{1}{2}$	1	$1\frac{1}{16}$	$2\frac{1}{16}$

DIMENSIONS OF STANDARD PLUG AND RING GAGES

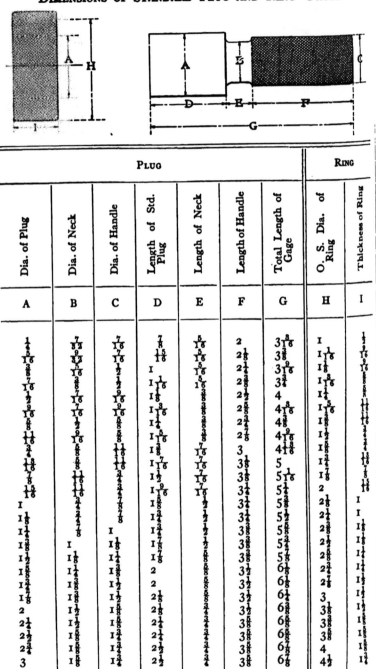

	PLUG						RING	
Dia. of Plug	Dia. of Neck	Dia. of Handle	Length of Std. Plug	Length of Neck	Length of Handle	Total Length of Gage	O. S. Dia. of Ring	Thickness of Ring
A	B	C	D	E	F	G	H	I
$\frac{1}{4}$	$\frac{7}{32}$	$\frac{7}{16}$	$\frac{7}{8}$	$\frac{5}{8}$	2	$3\frac{8}{16}$	1	$1\frac{7}{16}$
$\frac{5}{16}$	$\frac{9}{32}$	$\frac{7}{16}$	$1\frac{5}{16}$	$\frac{5}{8}$	$2\frac{1}{8}$	$3\frac{8}{16}$	$1\frac{1}{16}$	$1\frac{9}{16}$
$\frac{3}{8}$	$\frac{5}{16}$	$\frac{7}{16}$	1	$\frac{5}{8}$	$2\frac{1}{4}$	$3\frac{9}{16}$	$1\frac{3}{16}$	$1\frac{9}{16}$
$\frac{7}{16}$	$\frac{3}{8}$	$\frac{9}{16}$	$1\frac{3}{16}$	$\frac{5}{8}$	$2\frac{3}{8}$	$3\frac{1}{4}$	$1\frac{1}{4}$	$1\frac{1}{2}$
$\frac{1}{2}$	$\frac{7}{16}$	$\frac{9}{16}$	$1\frac{3}{8}$	$\frac{3}{4}$	$2\frac{1}{2}$	4	$1\frac{5}{16}$	$1\frac{11}{16}$
$\frac{5}{8}$	$\frac{9}{16}$	$\frac{9}{16}$	$1\frac{1}{4}$	$\frac{3}{4}$	$2\frac{5}{8}$	$4\frac{8}{16}$	$1\frac{3}{8}$	$1\frac{11}{16}$
$\frac{3}{4}$	$\frac{5}{8}$	$\frac{5}{8}$	$1\frac{1}{4}$	$\frac{3}{4}$	$2\frac{3}{4}$	$4\frac{8}{16}$	$1\frac{1}{2}$	$1\frac{3}{4}$
$\frac{11}{16}$	$\frac{9}{16}$	$\frac{5}{8}$	$1\frac{3}{8}$	$\frac{3}{4}$	$2\frac{7}{8}$	$4\frac{9}{16}$	$1\frac{5}{8}$	$1\frac{3}{4}$
$\frac{3}{4}$	$\frac{11}{16}$	$\frac{11}{16}$	$1\frac{5}{8}$	$\frac{7}{16}$	3	$4\frac{1}{2}$	$1\frac{5}{8}$	$1\frac{13}{16}$
$\frac{13}{16}$	$\frac{11}{16}$	$\frac{3}{4}$	$1\frac{7}{8}$	$\frac{7}{16}$	$3\frac{1}{8}$	5	$1\frac{3}{4}$	$1\frac{13}{16}$
$\frac{7}{8}$	$\frac{11}{16}$	$\frac{3}{4}$	$1\frac{3}{4}$	$\frac{7}{16}$	$3\frac{1}{8}$	$5\frac{1}{16}$	$1\frac{7}{8}$	$1\frac{13}{16}$
$\frac{15}{16}$	$\frac{3}{4}$	$\frac{13}{16}$	$1\frac{9}{16}$	$\frac{7}{16}$	$3\frac{1}{4}$	$5\frac{1}{4}$	2	$1\frac{15}{16}$
1	$\frac{13}{16}$	$\frac{13}{16}$	$1\frac{7}{8}$	$\frac{1}{2}$	$3\frac{1}{4}$	$5\frac{3}{8}$	$2\frac{1}{8}$	1
$1\frac{1}{8}$	$\frac{7}{8}$	$\frac{7}{8}$	$1\frac{3}{4}$	$\frac{1}{2}$	$3\frac{3}{8}$	$5\frac{1}{2}$	$2\frac{1}{4}$	1
$1\frac{1}{4}$	1	1	$1\frac{7}{8}$	$\frac{1}{2}$	$3\frac{5}{8}$	$5\frac{5}{8}$	$2\frac{3}{8}$	$1\frac{1}{8}$
$1\frac{3}{8}$	1	$1\frac{1}{8}$	$1\frac{7}{8}$	$\frac{1}{2}$	$3\frac{5}{8}$	$5\frac{5}{8}$	$2\frac{1}{2}$	$1\frac{1}{8}$
$1\frac{1}{2}$	$1\frac{1}{8}$	$1\frac{1}{8}$	2	$\frac{1}{2}$	$3\frac{5}{8}$	$5\frac{3}{4}$	$2\frac{5}{8}$	$1\frac{1}{4}$
$1\frac{5}{8}$	$1\frac{1}{4}$	$1\frac{1}{8}$	2	$\frac{1}{2}$	$3\frac{1}{2}$	$6\frac{1}{8}$	$2\frac{3}{4}$	$1\frac{1}{4}$
$1\frac{3}{4}$	$1\frac{1}{2}$	$1\frac{1}{4}$	$2\frac{1}{8}$	$\frac{1}{2}$	$3\frac{1}{2}$	$6\frac{1}{8}$	$2\frac{7}{8}$	$1\frac{1}{2}$
2	$1\frac{1}{2}$	$1\frac{3}{8}$	$2\frac{1}{8}$	$\frac{1}{2}$	$3\frac{1}{2}$	$6\frac{1}{4}$	3	$1\frac{1}{2}$
$2\frac{1}{4}$	$1\frac{1}{2}$	$1\frac{3}{8}$	$2\frac{1}{4}$	$\frac{1}{2}$	$3\frac{5}{8}$	$6\frac{5}{8}$	$3\frac{1}{8}$	$1\frac{5}{8}$
$2\frac{1}{2}$	$1\frac{5}{8}$	$1\frac{1}{4}$	$2\frac{1}{4}$	$\frac{1}{2}$	$3\frac{5}{8}$	$6\frac{5}{8}$	$3\frac{3}{8}$	$1\frac{5}{8}$
$2\frac{3}{4}$	$1\frac{5}{8}$	$1\frac{1}{4}$	$2\frac{1}{2}$	$\frac{1}{2}$	$3\frac{5}{8}$	$6\frac{5}{8}$	4	$1\frac{5}{8}$
3	$1\frac{5}{8}$	$1\frac{1}{4}$	$2\frac{1}{2}$	$\frac{1}{2}$	$3\frac{5}{8}$	$6\frac{7}{8}$	$4\frac{1}{2}$	$1\frac{3}{4}$

COUNTERBORES WITH INSERTED PILOTS

Diam. A	K	L	M	N	O
3/8	1 7/32	1/4	1 15/32	5/16	From 3/16" to 1 1/2" in 32nds.
7/16	"	"	"	"	
1/2					
9/16	1 27/32	3/8	2 7/32	5/16	
5/8	"	"	"	"	
11/16	"	"	"	"	From 1/2" to 3 1/2" in 32nds.
3/4	"	"	"	"	
13/16	"	"	"	"	
7/8	"	"	"	"	
15/16	"	"	"	"	
1	"	"	"	"	
1 1/16	2 11/32	1/2	2 27/32	1/2	From 1" to 1 1/2" in 32nds.
1 1/8	"	"	"	"	
1 3/16	"	"	"	"	
1 1/4	"	"	"	"	
1 5/16	"	"	"	"	
1 3/8	"	"	"	"	
1 7/16	"	"	"	"	
1 1/2	"	"	"	"	

3/8 to 1" Counterbores to have 4 Flutes
1 1/16 to 1 1/2" " " " 6 "

Diam. A	B	C	D	E	F	G	H	I	J
3/8	5/16	1 1/32	5/16	1	1 1/2	3 1/2	5/16	1 1/16	1 1/4
7/16	"	1 3/32	"	1 1/32	1 9/16	3 11/16	"	1 3/32	"
1/2	"	1 5/32	"	1 1/16	1 5/8	3 7/8	"	1 1/8	"
9/16	5/16	1 7/32	1/2	1 1/8	1 11/16	4 1/8	1/4	1 5/32	1 7/8
5/8	"	1 9/32	"	1 5/32	1 3/4	4 5/16	"	1 3/16	"
11/16	"	1 11/32	"	1 1/4	1 13/16	4 1/2	"	1 7/32	"
3/4	"	1 13/32	"	1 5/16	1 7/8	4 11/16	"	1 1/4	"
13/16	"	1 15/32	"	1 3/8	1 15/16	4 15/16	"	1 9/32	"
7/8	"	1 17/32	"	1 7/16	2	5 1/8	"	1 15/32	"
15/16	"	1 19/32	"	1 1/2	2 1/16	5 3/8	"	1 17/32	"
1	"	1 11/16	"	1 9/16	2 1/8	5 9/16	"	1 3/8	"
1 1/16	1/2	1	3/4	1 5/8	2 3/16	5 13/16	5/16	1 13/32	2 3/8
1 1/8	"	1 1/16	"	1 11/16	2 1/4	6	"	1 7/16	"
1 3/16	"	1 1/16	"	1 3/4	2 3/8	6 1/4	"	1 1/2	"
1 1/4	"	1 1/8	"	1 13/16	2 3/8	6 7/16	"	1 17/32	"
1 5/16	"	1 1/8	"	1 7/8	2 7/16	6 5/8	"	1 9/16	"
1 3/8	"	1 1/4	"	1 15/16	2 1/2	6 7/8	"	1 19/32	"
1 7/16	"	1 1/4	"	2	2 9/16	7 1/16	"	1 5/8	"
1 1/2	"	1 3/8	"	2 1/16	2 5/8	7 1/16	"	1 5/8	"

Two-Point Ball-Bearing Dimensions (Garford Co.)

Dia. Balls	0.25"			0.3125"			0.375"	
	X	Y	Z	X	Y	Z	X	Z
8		0.661	0.921	0.520	0.824	1.147		1.373
9		0.740	1.000		0.923	1.246		1.492
10	0.559	0.819	1.080	0.699	1.022	1.344	0.839	1.610
11	0.638	0.898	1.159	0.798	1.121	1.443	0.957	1.728
12	0.717	0.978	1.238	0.897	1.219	1.541	1.075	1.846
13	0.797	1.057	1.318	0.996	1.319	1.641	1.194	1.965
14	0.877	1.137	1.398	1.096	1.418	1.741	1.314	2.085
15	0.957	1.217	1.477	1.195	1.518	1.840	1.433	2.204
16	1.037	1.297	1.557	1.295	1.617	1.940	1.552	2.323
17	1.117	1.377	1.637	1.395	1.717	2.040	1.672	2.443
18	1.197	1.457	1.717	1.495	1.817	2.139	1.791	2.563
19	1.277	1.537	1.797	1.595	1.917	2.239	1.911	2.682
20	1.357	1.617	1.878	1.694	2.017	2.339	2.031	2.802
21	1.437	1.698	1.958	1.795	2.117	2.439	2.151	2.922
22	1.518	1.778	2.038	1.895	2.217	2.539	2.271	3.042
23	1.598	1.858	2.118	1.995	2.317	2.640	2.390	3.162
24	1.678	1.938	2.199	2.095	2.417	2.740	2.510	3.282
25	1.759	2.019	2.279	2.195	2.517	2.840	2.631	3.402
26	1.839	2.099	2.359	2.295	2.618	2.940	2.751	3.522
27	1.919	2.180	2.440	2.396	2.718	3.040	2.871	3.642
28	2.000	2.260	2.520	2.496	2.818	3.141	2.991	3.762
29	2.080	2.340	2.600	2.596	2.918	3.241	3.111	3.882
30	2.160	2.420	2.681	2.696	3.018	3.341	3.231	4.002

Diameters

Dia. of Balls	R	A	B	C	S	S-D
.250	.175	.050	.045	.022	.260	.010
.3125	.205	.049	.044	.021	.322	.010
.375	.240	.053	.047	.023	.386	.011
.4375	.275	.056	.051	.025	.449	.011
.500	.315	.065	.058	.029	.513	.013
.5625	.354	.073	.065	.032	.577	.015
.625	.394	.082	.073	.036	.641	.016
.6875	.433	.089	.080	.039	.705	.018
.750	.473	.098	.088	.043	.770	.020

D = Dia. of Ball.
R = Rad. of Race.
$A = R - \dfrac{D}{2}$
$B = A \cos. 26°$
$C = A \sin. 26°$
$S = 2(R - B)$.
$S - D$ = Clearance of Ball.
N = No. of Balls.
$Y = D + 0.003"$
$\sin. \dfrac{180°}{N}$

TWO-POINT BALL-BEARING DIMENSIONS

Dia. of Balls	0.4375″			0.50″			0.5625″			0.625″			0.6875″			0.750″		
No. of Balls	X	Y	Z	X	Y	Z	X	Y	Z	X	Y	Z	X	Y	Z	X	Y	Z
8	0.702	1.151	1.509	0.801	1.314	1.828	0.901	1.478	2.055	1.000	1.641	2.282	1.099	1.804	2.510	1.198	1.968	2.738
9	0.840	1.289	1.738	0.959	1.472	1.985	1.078	1.655	2.232	1.196	1.839	2.479	1.315	2.020	2.726	1.433	2.203	2.973
10	0.978	1.427	1.875	1.116	1.629	2.142	1.255	1.832	2.409	1.392	2.034	2.675	1.531	2.236	2.942	1.669	2.439	3.208
11	1.115	1.564	2.013	1.273	1.786	2.300	1.432	2.009	2.586	1.589	2.230	2.872	1.747	2.452	3.158	1.904	2.674	3.444
12	1.253	1.702	2.151	1.430	1.943	2.457	1.608	2.185	2.762	1.785	2.426	3.068	1.963	2.668	3.373	2.140	2.909	3.679
13	1.392	1.841	2.290	1.589	2.102	2.615	1.786	2.363	2.940	1.983	2.625	3.266	2.180	2.886	3.591	2.377	3.147	3.917
14	1.531	1.980	2.419	1.748	2.261	2.774	1.964	2.541	3.118	2.181	2.823	3.464	2.398	3.104	3.809	2.615	3.385	4.154
15	1.671	2.119	2.563	1.906	2.420	2.933	2.142	2.719	3.296	2.380	3.021	3.662	2.616	3.322	4.027	2.852	3.622	4.392
16	1.809	2.258	2.707	2.065	2.578	3.092	2.322	2.809	3.476	2.578	3.219	3.860	2.834	3.539	4.245	3.090	3.860	4.630
17	1.949	2.397	2.846	2.224	2.738	3.251	2.501	3.078	3.655	2.777	3.418	4.059	3.053	3.758	4.464	3.328	4.098	4.868
18	2.088	2.537	2.986	2.384	2.897	3.410	2.680	3.257	3.834	2.975	3.617	4.258	3.271	3.977	4.682	3.567	4.337	5.107
19	2.228	2.676	3.125	2.543	3.056	3.569	2.859	3.436	4.013	3.174	3.816	4.457	3.490	4.196	4.901	3.805	4.575	5.345
20	2.367	2.816	3.265	2.702	3.215	3.729	3.038	3.615	4.192	3.373	4.015	4.656	3.709	4.414	5.119	4.044	4.814	5.583
21	2.507	2.956	3.404	2.862	3.375	3.888	3.217	3.794	4.371	3.572	4.214	4.855	3.928	4.633	5.338	4.285	5.055	5.825
22	2.647	3.095	3.543	3.021	3.535	4.048	3.396	3.973	4.550	3.772	4.413	5.054	4.147	4.852	5.557	4.522	5.291	6.061
23	2.786	3.235	3.684	3.181	3.694	4.207	3.575	4.152	4.729	3.971	4.612	5.254	4.366	5.071	5.776	4.760	5.530	6.300
24	2.926	3.375	3.824	3.340	3.854	4.367	3.756	4.333	4.910	4.170	4.811	5.453	4.585	5.290	5.996	4.999	5.769	6.539
25	3.066	3.515	3.964	3.500	4.014	4.527	3.936	4.513	5.090	4.370	5.011	5.652	4.804	5.510	6.215	5.238	6.008	6.778
26	3.206	3.655	4.104	3.660	4.173	4.687	4.116	4.693	5.270	4.569	5.211	5.852	5.014	5.729	6.435	5.478	6.248	7.017
27	3.346	3.795	4.244	3.820	4.333	4.847	4.296	4.873	5.450	4.769	5.410	6.052	5.243	5.949	6.654	5.717	6.487	7.257
28	3.486	3.935	4.384	3.980	4.493	5.006	4.474	5.051	5.628	4.968	5.610	6.251	5.463	6.168	6.873	5.956	6.726	7.496
29	3.626	4.075	4.524	4.139	4.652	5.166	4.653	5.230	5.807	5.167	5.809	6.450	5.682	6.387	7.092	6.195	6.965	7.735
30	3.765	4.214	4.663	4.299	4.812	5.325	4.831	5.408	5.985	5.367	6.008	6.649	5.901	6.606	7.311	6.434	7.204	7.974

FOUR-POINT BALL-BEARING DIMENSIONS. (GARFORD CO.)

Diameters
D = Dia. of Balls.
N = No. of Balls.
Dia. of Ball Circle $Y = \dfrac{D + .003''}{\sin\frac{180°}{N}}$

Diameter of Balls	W	R	C
.25	.330	.08	.018
.3125	.415	.10	.023
.375	.505	.12	.028
.4375	.595	.14	.033
.50	.680	.16	.038

Dia. of Balls	.25"			.3125"			.375"			.4375"			.50"		
No. of Balls	X	Y	Z	X	Y	Z	X	Y	Z	X	Y	Z	X	Y	Z
8	.64	.665	.69	.80	.825	.85	.96	.985	1.01	1.13	1.155	1.18	1.29	1.315	1.34
9	.72	.745	.77	.90	.925	.95	1.08	1.105	1.13	1.27	1.295	1.32	1.45	1.475	1.49
10	.80	.825	.85	1.00	1.025	1.05	1.20	1.225	1.25	1.40	1.425	1.45	1.61	1.635	1.66
11	.88	.905	.93	1.10	1.125	1.15	1.32	1.345	1.37	1.54	1.565	1.59	1.76	1.785	1.81
12	.96	.985	1.01	1.20	1.225	1.25	1.44	1.465	1.49	1.68	1.705	1.73	1.92	1.945	1.97
13	1.04	1.065	1.09	1.30	1.325	1.35	1.56	1.585	1.61	1.82	1.845	1.87	2.08	2.105	2.13
14	1.12	1.145	1.17	1.40	1.425	1.45	1.68	1.705	1.73	1.96	1.985	2.10	2.24	2.265	2.29
15	1.20	1.225	1.25	1.50	1.525	1.55	1.80	1.825	1.85	2.10	2.125	2.15	2.40	2.425	2.45
16	1.28	1.305	1.33	1.60	1.625	1.65	1.92	1.945	1.97	2.24	2.265	2.29	2.56	2.585	2.61
17	1.36	1.385	1.41	1.70	1.725	1.75	2.04	2.065	2.09	2.38	2.405	2.43	2.72	2.745	2.77
18	1.44	1.465	1.49	1.80	1.825	1.85	2.16	2.185	2.21	2.51	2.535	2.56	2.87	2.895	2.92
19	1.52	1.545	1.57	1.90	1.925	1.95	2.28	2.305	2.33	2.65	2.675	2.70	3.03	3.055	3.08
20	1.60	1.625	1.65	2.00	2.025	2.05	2.40	2.425	2.45	2.79	2.815	2.84	3.19	3.215	3.24
21	1.68	1.705	1.73	2.10	2.125	2.15	2.52	2.545	2.57	2.93	2.955	2.98	3.35	3.375	3.40
22	1.76	1.785	1.81	2.20	2.225	2.25	2.64	2.665	2.69	3.07	3.095	3.12	3.51	3.535	3.56
23	1.84	1.865	1.89	2.30	2.325	2.35	2.76	2.785	2.81	3.21	3.235	3.26	3.67	3.695	3.72
24	1.92	1.945	1.97	2.40	2.425	2.45	2.88	2.905	2.93	3.35	3.375	3.40	3.83	3.855	3.88
25	2.00	2.025	2.05	2.50	2.525	2.55	3.00	3.025	3.05	3.49	3.515	3.54	3.99	4.015	4.04
26	2.08	2.105	2.13	2.60	2.625	2.65	3.12	3.145	3.17	3.63	3.655	3.68	4.15	4.175	4.20
27	2.16	2.185	2.21	2.70	2.725	2.75	3.24	3.265	3.29	3.77	3.795	3.82	4.31	4.335	4.36
28	2.24	2.265	2.29	2.80	2.825	2.85	3.36	3.385	3.41	3.91	3.935	3.96	4.47	4.495	4.52
29	2.32	2.345	2.37	2.90	2.925	2.95	3.48	3.505	3.59	4.05	4.075	4.10	4.63	4.655	4.68
30	2.40	2.425	2.45	3.00	3.025	3.05	3.60	3.625	3.65	4.19	4.215	4.24	4.79	4.815	4.84

Above Dimensions are Given to Nearest .005"

THE erection of a perpendicular by the construction of a triangle whose sides are respectively 3, 4 and 5 units in length is a familiar and handy device. The following table gives a greater range of choice in the shape or proportions of the triangle employed. The table is a list of all integral, or whole-number, right-angled triangles the units of whose least sides do not exceed 20.

Hight	Base	Hypot-enuse	Hight	Base	Hypot-enuse	Hight	Base	Hypot-enuse
3	4	5	12	16	20	17	144	195
5	12	13	12	35	37	18	24	30
6	8	10	13	84	85	18	80	82
7	24	25	14	48	50	19	180	181
8	15	17	15	20	25	20	21	29
9	12	15	15	36	39	20	48	52
9	40	41	15	112	113	20	99	101
10	24	26	16	30	34			
11	60	61	16	63	65			

TABLE OF CHORDS

To construct any angle from the table of chords, page 282: Let the required angle be 36° 38′; the nearest angles in the table are 36° 30′ and 36° 40′, and the chords are respectively 0.6263 and 0.6291, the difference 0.0028 corresponding to an angular difference of 10′. To find the amount which must be added to 0.6263 (the chord corresponding to 36° 30′) in order to obtain the chord for a 36° 38′ arc, multiply **0.0028** by $\frac{8}{10}$ = 0.00224. 0.6263 + 0.00224 = 0.62854. Then, if the radius is 1″ and the angle 36° 38′ the chord will be 0.62854″.

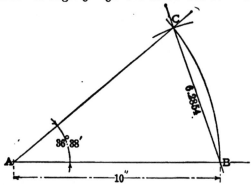

In laying out an angle as in the accompanying illustration a base line *A B* can be drawn, say 10 inches long, then with a radius *A B* and center *A*, arc *B C* can be struck. Multiply chord 0.62854 inch by 10 giving 6.2854 inches, as the radius of an arc to be struck from center *B* and cutting arc *B C* at *C*. Through point *C* draw a line *A C* and the angle *B A C* will equal 36° 38′.

Where the angle required is in even degrees or sixths of degrees (as 10′, 20′, etc.) the corresponding chord may be taken directly from the table. A 10 to 1 layout is particularly convenient as the multiplication of the tabulated chords by 10 is readily performed mentally.

TABLE OF CHORDS

THE TABULATED QUANTITIES = TWICE THE SINE OF HALF THE ARC

Deg.	0'	10'	20'	30'	40'	50'	60'
0	.0000	.0029	.0058	.0087	.0116	.0145	.0174
1	.0174	.0204	.0233	.0262	.0291	.0320	.0349
2	.0349	.0378	.0407	.0436	.0465	.0494	.0523
3	.0523	.0553	.0582	.0611	.0640	.0669	.0698
4	.0698	.0727	.0756	.0785	.0814	.0843	.0872
5	.0872	.0901	.0930	.0959	.0988	.1017	.1047
6	.1047	.1076	.1105	.1134	.1163	.1192	.1221
7	.1221	.1250	.1279	.1308	.1337	.1366	.1395
8	.1395	.1424	.1453	.1482	.1511	.1540	.1569
9	.1569	.1598	.1627	.1656	.1685	.1714	.1743
10	.1743	.1772	.1801	.1830	.1859	.1888	.1917
11	.1917	.1946	.1975	.2004	.2033	.2062	.2090
12	.2090	.2119	.2148	.2177	.2206	.2235	.2264
13	.2264	.2293	.2322	.2351	.2380	.2409	.2437
14	.2437	.2466	.2495	.2524	.2553	.2582	.2610
15	.2610	.2639	.2668	.2697	.2726	.2755	.2783
16	.2783	.2812	.2841	.2870	.2899	.2927	.2956
17	.2956	.2985	.3014	.3042	.3071	.3100	.3129
18	.3129	.3157	.3186	.3215	.3243	.3272	.3301
19	.3301	.3330	.3358	.3387	.3416	.3444	.3473
20	.3473	.3502	.3530	.3559	.3587	.3616	.3645
21	.3645	.3673	.3702	.3730	.3759	.3788	.3816
22	.3816	.3845	.3873	.3902	.3930	.3959	.3987
23	.3987	.4016	.4044	.4073	.4101	.4130	.4158
24	.4158	.4187	.4215	.4243	.4272	.4300	.4329
25	.4329	.4357	.4385	.4414	.4442	.4471	.4499
26	.4499	.4527	.4556	.4584	.4612	.4641	.4669
27	.4669	.4697	.4725	.4754	.4782	.4810	.4838
28	.4838	.4867	.4895	.4923	.4951	.4979	.5008
29	.5008	.5036	.5064	.5092	.5120	.5148	.5176
30	.5176	.5204	.5232	.5261	.5289	.5317	.5345
31	.5345	.5373	.5401	.5429	.5457	.5485	.5513
32	.5513	.5541	.5569	.5596	.5624	.5652	.5680
33	.5680	.5708	.5736	.5764	.5792	.5820	.5847
34	.5847	.5875	.5903	.5931	.5959	.5986	.6014
35	.6014	.6042	.6069	.6097	.6125	.6153	.6180
36	.6180	.6208	.6236	.6263	.6291	.6318	.6346
37	.6346	.6374	.6401	.6429	.6456	.6484	.6511
38	.6511	.6539	.6566	.6594	.6621	.6649	.6676
39	.6676	.6703	.6731	.6758	.6786	.6813	.6840
40	.6840	.6868	.6895	.6922	.6950	.6977	.7004
41	.7004	.7031	.7059	.7086	.7113	.7140	.7167
42	.7167	.7194	.7222	.7249	.7276	.7303	.7330
43	.7330	.7357	.7384	.7411	.7438	.7465	.7492
44	.7492	.7519	.7546	.7573	.7600	.7627	.7654
45	.7654	.7680	.7707	.7734	.7761	.7788	.7815

TABLE OF CHORDS

HE TABULATED QUANTITIES = TWICE THE SINE OF HALF THE ARC

Deg.	o′	10′	20′	30′	40′	50′	60′
46	.7815	.7841	.7868	.7895	.7921	.7948	.7975
47	.7975	.8001	.8028	.8055	.8081	.8108	.8135
48	.8135	.8161	.8188	.8214	.8241	.8267	.8294
49	.8294	.8320	.8347	.8373	.8400	.8426	.8452
50	.8452	.8479	.8505	.8531	.8558	.8584	.8610
51	.8610	.8636	.8663	.8689	.8715	.8741	.8767
52	.8767	.8793	.8820	.8846	.8872	.8898	.8924
53	.8924	.8950	.8976	.9002	.9028	.9054	.9080
54	.9080	.9106	.9132	.9157	.9183	.9209	.9235
55	.9235	.9261	.9286	.9312	.9338	.9364	.9389
56	.9389	.9415	.9441	.9466	.9492	.9518	.9543
57	.9543	.9569	.9594	.9620	.9645	.9671	.9696
58	.9696	.9722	.9747	.9772	.9798	.9823	.9848
59	.9848	.9874	.9899	.9924	.9949	.9975	1.0000
60	1.0000	1.0025	1.0050	1.0075	1.0100	1.0126	1.0151
61	1.0151	1.0176	1.0201	1.0226	1.0251	1.0276	1.0301
62	1.0301	1.0326	1.0350	1.0375	1.0400	1.0425	1.0450
63	1.0450	1.0475	1.0500	1.0524	1.0550	1.0574	1.0598
64	1.0598	1.0623	1.0648	1.0672	1.0697	1.0721	1.0746
65	1.0746	1.0770	1.0795	1.0819	1.0844	1.0868	1.0893
66	1.0893	1.0917	1.0941	1.0966	1.0990	1.1014	1.1039
67	1.1039	1.1063	1.1087	1.1111	1.1135	1.1159	1.1184
68	1.1184	1.1208	1.1232	1.1256	1.1280	1.1304	1.1328
69	1.1328	1.1352	1.1376	1.1400	1.1424	1.1448	1.1471
70	1.1471	1.1495	1.1519	1.1543	1.1567	1.1590	1.1614
71	1.1614	1.1638	1.1661	1.1685	1.1708	1.1732	1.1756
72	1.1756	1.1780	1.1803	1.1826	1.1850	1.1873	1.1896
73	1.1896	1.1920	1.1943	1.1966	1.1990	1.2013	1.2036
74	1.2036	1.2059	1.2083	1.2106	1.2129	1.2152	1.2175
75	1.2175	1.2198	1.2221	1.2244	1.2267	1.2290	1.2313
76	1.2313	1.2336	1.2360	1.2382	1.2405	1.2427	1.2450
77	1.2450	1.2473	1.2496	1.2518	1.2541	1.2564	1.2586
78	1.2586	1.2609	1.2631	1.2654	1.2677	1.2699	1.2721
79	1.2721	1.2744	1.2766	1.2789	1.2811	1.2833	1.2856
80	1.2856	1.2878	1.2900	1.2922	1.2945	1.2967	1.2989
81	1.2989	1.3011	1.3033	1.3055	1.3077	1.3099	1.3121
82	1.3121	1.3143	1.3165	1.3187	1.3209	1.3231	1.3252
83	1.3252	1.3274	1.3296	1.3318	1.3340	1.3361	1.3383
84	1.3383	1.3404	1.3426	1.3447	1.3469	1.3490	1.3512
85	1.3512	1.3533	1.3555	1.3576	1.3597	1.3619	1.3640
86	1.3640	1.3661	1.3682	1.3704	1.3725	1.3746	1.3767
87	1.3767	1.3788	1.3809	1.3030	1.3851	1.3872	1.3893
88	1.3893	1.3914	1.3935	1.3956	1.3977	1.3997	1.4018
89	1.4018	1.4039	1.4060	1.4080	1.4101	1.4121	1.4142
90	1.4142						

TABLE FOR SPACING HOLES IN CIRCLES

No. of Divisions in Circle	Deg. of Arc	Length of Chord Dia. 1	Length of Chord Dia. 2	Length of Chord Dia. 3	Length of Chord Dia. 4	Length of Chord Dia. 5	Length of Chord Dia. 6
3	120	.866	1.732	2.598	3.464	4.330	5.196
4	90	.707	1.414	2.121	2.828	3.536	4.243
5	72	.588	1.176	1.763	2.351	2.938	3.527
6	60	.500	1.000	1.500	2.000	2.500	3.000
7	51°–25′	.434	.868	1.302	1.736	2.170	2.604
8	45	.383	.765	1.148	1.531	1.913	2.296
9	40	.342	.684	1.026	1.368	1.710	2.052
10	36	.309	.618	.927	1.236	1.545	1.854
11	32°–43′	.282	.564	.845	1.127	1.409	1.691
12	30	.259	.518	.776	1.035	1.294	1.553
13	27°–41′	.239	.479	.718	.958	1.197	1.436
14	25°–42′	.222	.445	.667	.890	1.112	1.334
15	24	.208	.416	.624	.832	1.040	1.247
16	22°–30′	.195	.390	.585	.780	.975	1.171
17	21°–11′	.184	.367	.551	.735	.918	1.102
18	20	.174	.347	.521	.695	.868	1.042
19	18°–57′	.164	.329	.493	.658	.822	.987
20	18	.156	.318	.469	.626	.782	.937
21	17°– 8′	.149	.298	.447	.596	.745	.894
22	16°–22′	.142	.286	.427	.569	.712	.855
23	15°–39′	.136	.273	.409	.545	.681	.818
24	15	.130	.261	.392	.522	.653	.783
25	14°–24′	.125	.251	.375	.501	.627	.752
26	13°–51′	.120	.241	.361	.482	.602	.723
27	13°–20′	.116	.232	.348	.464	.580	.697
28	12°–51′	.112	.224	.336	.448	.560	.672
29	12°–25′	.108	.216	.324	.432	.540	.648
30	12	.104	.209	.314	.418	.522	.627
31	11°–37′	.101	.202	.303	.404	.505	.606
32	11°–15′	.098	.196	.294	.393	.491	.589

TABLE FOR SPACING HOLES IN CIRCLES

No. of Divisions in Circle	Deg. of Arc	Length of Chord Dia. 7	Length of Chord Dia. 8	Length of Chord Dia. 9	Length of Chord Dia. 10	Length of Chord Dia. 11	Length of Chord Dia. 12
3	120	6.062	6.928	7.794	8.660	9.526	10.392
4	90	4.950	5.657	6.364	7.081	7.778	8.485
5	72	4.115	4.702	5.290	5.878	6.465	7.053
6	60	3.500	4.000	4.500	5.000	5.500	6.000
7	51°-25'	3.037	3.471	3.905	4.339	4.773	5.207
8	45	2.679	3.061	3.444	3.827	4.210	4.592
9	40	2.394	2.736	3.078	3.420	3.762	4.104
10	36	2.163	2.472	2.781	3.090	3.399	3.708
11	32°-43'	1.973	2.254	2.536	2.818	3.100	3.381
12	30	1.812	2.069	2.329	2.588	2.847	3.106
13	27°-41'	1.676	1.915	2.154	2.394	2.633	2.873
14	25°-42'	1.557	1.779	2.000	2.224	2.446	2.669
15	24	1.455	1.663	1.871	2.079	2.287	2.495
16	22°-30'	1.366	1.561	1.756	1.951	2.146	2.341
17	21°-11'	1.286	1.469	1.653	1.837	2.020	2.204
18	20	1.216	1.389	1.563	1.737	1.910	2.084
19	18°-57'	1.151	1.316	1.480	1.645	1.809	1.974
20	18	1.095	1.251	1.408	1.564	1.721	1.877
21	17°- 8'	1.043	1.192	1.341	1.489	1.639	1.788
22	16°-22'	.996	1.139	1.281	1.423	1.566	1.708
23	15°-39'	.954	1.092	1.227	1.363	1.499	1.635
24	15	.914	1.044	1.175	1.305	1.436	1.566
25	14°-24'	.877	1.003	1.128	1.253	1.379	1.504
26	13°-51'	.843	.963	1.084	1.204	1.325	1.445
27	13°-20'	.813	.929	1.045	1.161	1.277	1.393
28	12°-51'	.784	.896	1.008	1.121	1.233	1.345
29	12°-25'	.756	.864	.972	1.080	1.188	1.296
30	12	.732	.836	.941	1.045	1.150	1.254
31	11°-37'	.707	.808	.910	1.011	1.112	1.213
32	11°-15'	.687	.785	.883	.982	1.080	1.178

TABLE FOR SPACING HOLES IN CIRCLES

THE table on pages 284 and 285 will be found of service when it is desired to space any number of holes up to and including 32, in a circle. The number of divisions or holes desired will be found in the first column, the corresponding angle included at the center being given as a convenience in the second column. The remaining column heads cover various diameters of circles from 1 to 12 inches, and under these different heads and opposite the required number of holes will be found the lengths of chords or distances between hole centers for the given circle diameter.

Thus, if it is required to space off 18 holes in an 8-inch circle, by following down the first column until 18 is reached and then reading directly to the right, in the column headed "Length of Chord-Dia. 8," will be found the distance 1.389 as the chord length for that number of divisions and diameter of circle. Or, suppose a circle of 12 inches diameter is to be spaced off for a series of 27 holes to be drilled at equal distances apart: Opposite 27 found in the first column, and under the heading, "Dia. 12," will be found the chord 1.393 as the length to which the dividers may be set directly for laying off the series of holes.

If it is desired to lay off a series of holes in a circle of some diameter not given in the table, say 10 holes in an 11½-inch circle, subtract the chord for 10 holes in an 11-inch circle, or, 3.399 from the chord in the "Dia. 12" column, or 3.708, and add half the difference (.154) to 3.399, giving 3.553 as the chord or center distance between holes. Or, if 24 holes are to be equally spaced in a 20-inch circle, all that is necessary in order to find the chord, or center distance, is to find opposite 24, and in the column headed, "Dia. 10," the quantity 1.305 and multiply this by 2, giving a length of 2.610 inches as the center distance.

TABLE OF SIDES, ANGLES AND SINES

THE table on pages 287 to 291 is carried out for a much higher number of sides or spaces than are included in the preceding table and will be found useful in many cases not covered by that table. It was originally computed for finding the thicknesses of commutator bars and also for calculating the chord for spacing slots in armature punchings. In using this table the diameter of the circle is, of course, multiplied by the sine opposite the desired number of holes or sides.

Assuming for illustration that a series of 51 holes are to be equally spaced about a circle having a diameter of 17 inches, opposite 51 in the column headed "No. of Sides," find the quantity .06156 in the column headed "Sine," and multiply this quantity by 17. The product 1.0456 is the length of the chord or the required distance between centers of the holes for this circle. Or, if 40 equidistant points are to be spaced about a circle 16 inches diameter, opposite the number of sides, 40, will be found the quantity .078459 which multiplied by 16 gives 1.255 inch as the distance between centers.

MULTIPLY DIAMETER BY SINE TO GET LENGTH OF SIDE
(Angle given is half of angle subtended at center)

No. Sides	Angle Deg. Min. Sec.	Sine	No. Sides	Angle Deg. Min. Sec.	Sine
3	60	.8660254	52	3–27–41.53	.0603784
4	45	.7071067	53	3–23–46.41	.0592405
5	36	.5877852	54	3–20	.0581448
6	30	.5000000	55	3–16–21.81	.0570887
7	25–42–51.42	.4338837	56	3–12–51.42	.0560704
8	22–30	.3826834	57	3– 9–28.42	.0550877
9	20–	.3420201	58	3– 6–12.41	.0541388
10	18–	.3090170	59	3– 3– 3.05	.0532221
11	16–21–49.09	.2817325	60	3–	.0523360
12	15–	.2588190	61	2–57– 2.95	.0514787
13	13–50–46.15	.2393157	62	2–54–11.61	.0506491
14	12–51–25.71	.2225208	63	2–51–25.71	.0498458
15	12	.2079116	64	2–48– 45	.0490676
16	11–15	.1950903	65	2–46– 9.23	.0483133
17	10–35–17.64	.1837495	66	2–43–38.18	.0475819
18	10–	.1736481	67	2–41–11.64	.0468722
19	9–28–25.26	.1645945	68	2–38–49.41	.0461834
20	9–	.1564344	69	2–36–31.30	.0455145
21	8–34–17.14	.1490422	70	2–34–17.14	.0448648
22	8–10–54.54	.1423148	71	2–32– 6.76	.0442333
23	7–49–33.91	.1361666	72	2–30	.0436194
24	7–30–	.1305262	73	2–27–56.71	.0430222
25	7–12–	.1253332	74	2–25–56.75	.0424411
26	6–55–23.07	.1205366	75	2–24–	.0418757
27	6–40	.1160929	76	2–22– 6.31	.0413249
28	6–25–42.85	.1119644	77	2–20–15.58	.0407885
29	6–12–24.82	.1081189	78	2–18–27.69	.0402659
30	6–	.1045284	79	2–16–42.53	.0397565
31	5–48–23.22	.1011683	80	2–15–	.0392598
32	5–37–30	.0980171	81	2–13–20	.0387753
33	5–27–16.36	.0950560	82	2–11–42.45	.0383027
34	5–17–38.82	.0922683	83	2–10– 7.22	.0378414
35	5– 8–34.28	.0896392	84	2– 8–34.28	.0373911
36	5–	.0871557	85	2– 7– 3.54	.0369515
37	4–51–53.51	.0848058	86	2– 5–34.88	.0365220
38	4–44–12.63	.0825793	87	2– 4– 8.27	.0361023
39	4–36–55.38	.0804665	88	2– 2–43.63	.0356923
40	4–30–	.0784591	89	2– 1–20.89	.0352914
41	4–23–24.87	.0765492	90	2–	.0348995
42	4–17– 8.57	.0747301	91	1–58–40.87	.0345160
43	4–11– 9.76	.0729952	92	1–57–23.47	.0341410
44	4– 5–27.27	.0713391	93	1–56– 7.74	.0337741
45	4	.0697565	94	1–54–53.61	.0334149
46	3–54–46.95	.0682423	95	1–53–41.05	.0330633
47	3–49–47.23	.0667926	96	1–52–30.	.0327190
48	3–45–	.0654031	97	1–51–20.41	.0323818
49	3–40–24.49	.0640702	98	1–50–12.24	.0320515
50	3–36–	.0627905	99	1–49– 5.45	.0317279
51	3–31–45.88	.0615609	100	1–48–	.031410

MULTIPLY DIAMETER BY SINE TO GET LENGTH OF SIDE
(Angle given is half of angle subtended at center)

No. Sides	Angle Deg. Min Sec.	Sine	No. Sides	Angle Deg. Min. Sec.	Sine
101	1–46–55.84	.0310998	151	1–11–31.39	.0208037
102	1–45–52.94	.0307950	152	1–11– 3.15	.0206668
103	1–44–51.26	.0304961	153	1–10–35.29	.0205318
104	1–43–50.76	.0302029	154	1–10– 7.79	.0203985
105	1–42–51.42	.0299154	155	1– 9–40.64	.0202669
106	1–41–53.20	.0296332	156	1– 9–13.84	.0201370
107	1–40–56.07	.0293564	157	1– 8–47.38	.0200087
108	1–40–	.0290847	158	1– 8–21.26	.0198821
109	1–39– 4.95	.0288179	159	1– 7–55.47	.0197571
110	1–38–10.90	.0285560	160	1– 7–30	.0196336
111	1–37–17.83	.0282488	161	1– 7– 4.84	.0195117
112	1–36–25.71	.0280462	162	1– 6–40	.0193913
113	1–35–34.51	.0277981	163	1– 6–15.46	.0192723
114	1–34–44.21	.0275543	164	1– 5–51.21	.0191548
115	1–33–54.78	.0273147	165	1– 5–27.27	.0190387
116	1–33– 6.20	.0270793	166	1– 5– 3.61	.0189241
117	1–32–18.46	.0268479	167	1– 4–40.23	.0188107
118	1–31–31.52	.0266204	168	1– 4–17.14	.0186988
119	1–30–45.38	.0263968	169	1– 3–54.31	.0185881
120	1–30–	.0261769	170	1– 3–31.76	.0184788
121	1–29–15.37	.0259606	171	1– 3– 9.47	.0183708
122	1–28–31.47	.0257478	172	1– 2–47.44	.0182640
123	1–27–48.29	.0255386	173	1– 2–25.66	.0181584
124	1–27– 5.80	.0253326	174	1– 2– 4.13	.0180541
125	1–26–24	.0251300	175	1– 1–42.85	.0179509
126	1–25–42.85	.0249306	176	1– 1–21.81	.0178489
127	1–25– 2.36	.0247344	177	1– 1– 1.01	.0177481
128	1–24–22.50	.0245412	178	1– 0–40.44	.0176484
129	1–23–43.25	.0243509	179	1– 0–20.11	.0175498
130	1–23– 4.61	.0241637	180	1– –	.0174524
131	1–22–26.56	.0239793	181	–59–40.11	.0173559
132	1–21–49.09	.0237976	182	–59–20.43	.0172605
133	1–21–12.18	.0236188	183	–59– 0.98	.0171663
134	1–20–35.82	.0234425	184	–58–41.73	.0170730
135	1–20–	.0232689	185	–58–22.70	.0169807
136	1–19–24.70	.0230978	186	–58– 3.87	.0168894
137	1–18–49.92	.0229292	187	–57–45.24	.0167991
138	1–18–15.65	.0227631	188	–57–26.30	.0167097
139	1–17–41.87	.0225994	189	–57– 8.57	.0166214
140	1–17– 8.57	.0224380	190	–56–50.52	.0165339
141	1–16–35.74	.0222789	191	–56–32.67	.0164473
142	1–16– 3.38	.0221220	192	–56–15	.0163617
143	1–15–31.46	.0219673	193	–55–57.51	.0162769
144	1–15–	.0218148	194	–55–40.20	.0161930
145	1–14–28.96	.0216644	195	–55–23.07	.0161100
146	1–13–58.35	.0215160	196	–55– 6.12	.0160278
147	1–13–28.16	.0213697	197	–54–49.34	.0159464
148	1–12–58.37	.0212253	198	–54–32.72	.0158659
149	1–12–28.99	.0210829	199	–54–16.28	.0157862
150	1–12–	.0209424	200	–54–	.0157073

MULTIPLY DIAMETER BY SINE TO GET LENGTH OF SIDE
(Angle given is half of angle subtended at center)

s	Angle Min. Sec.	Sine	No. Sides	Angle Min. Sec.	Sine
	53–43.88	.0156244	251	43– 1.67	.0125160
	53–27.92	.0155518	252	42–51.43	.0124663
.	53–12.12	.0154752	253	42–41.26	.0124171
.	52–56.47	.0153993	254	42–31.18	.0123682
!	52–40.97	.0153242	255	42–21.18	.0123197
)	52–25.63	.0152498	256	42–11.25	.0122715
'	52–10.44	.0151764	257	42– 1.40	.0122238
;	51–55.38	.0151033	258	41–51.63	.0121764
)	51–40.48	.0150310	259	41–41.93	.0121294
)	51–25.71	.0149595	260	41–32.31	.0120827
[	51–11.09	.0148886	261	41–22.76	.0120364
2	50–56.60	.0148183	262	41–13.28	.0119905
3	50–42.25	.0147487	263	41– 3.88	.0119449
4	50–28.04	.0146798	264	40–54.54	.0118997
5	50–13.96	.0146115	265	40–45.28	.0118548
6	50–	.0145439	266	40–36.09	.0118102
7	49–46.17	.0144769	267	40–26.96	.0117660
8	49–32.48	.0144104	268	40–17.91	.0117221
9	49–18.91	.0143446	269	40– 8.93	.0116786
0	49– 5.46	.0142794	270	40–	.0116353
1	48–52.13	.0142148	271	39–51.14	.0115923
2	48–38.92	.0141508	272	39–42.35	.0115497
!3	48–25.83	.0140874	273	39–33.63	.0115074
!4	48–12.86	.0140245	274	39–24.96	.0114654
!5	48–	.0139622	275	39–16.36	.0114237
!6	47–47.26	.0139004	276	39– 7.83	.0113823
!7	47–34.63	.0138392	277	38–59.35	.0113412
!8	47–22.11	.0137785	278	38–50.94	.0113004
29	47– 9.69	.0137183	279	38–42.58	.0112599
30	46–57.39	.0136587	280	38–34.28	.0112197
31	46–45.19	.0135995	281	38–26.05	.0111798
32	46–33.10	.0135409	282	38–17.87	.0111401
33	46–21.11	.0134828	283	38– 9.75	.0111008
34	46– 9.23	.0134252	284	38– 1.69	.0110617
35	45–57.45	.0133681	285	37–53.68	.0110229
36	45–45.76	.0133115	286	17–45.73	.0109844
37	45–34.18	.0132553	287	37–37.84	.0109461
38	45–22.69	.0131996	288	37–30	.0109081
39	45–11.29	.0131444	289	37–22.21	.0108704
40	45–	.0130896	290	37–14.48	.0108329
41	44–48.80	.0130353	291	37– 6.80	.0107957
42	44–37.68	.0129814	292	36–59.18	.0107587
43	44–26.67	.0129280	293	36–51.60	.0107220
44	44–15.74	.0128750	294	36–44.08	.0106855
45	44– 4.90	.0128225	295	36–36.61	.0106493
246	43–54.15	.0127704	296	36–29.19	.0106133
247	43–43.48	.0127187	297	36–21.82	.0105776
248	43–32.40	.0126674	298	36–14.50	.0105421
249	43–22.41	.0126165	299	36– 7.22	.0105068
250	43–12	.0125661	300	36–	.0104718

MULTIPLY DIAMETER BY SINE TO GET LENGTH OF SIDE
(Angle given is half of angle subtended at center)

No. Sides	Angle Min. Sec.	Sine	No. Sides	Angle Min. Sec.	Sine
301	35–52.82	.0104370	351	30–46.15	.0089502
302	15–45.69	.0104024	352	30–40.91	.0089248
303	35–38.61	.0103681	353	30–35.69	.0088996
304	35–31.58	.0103340	354	30–30.51	.0088744
305	35–24.59	.0103001	355	30–25.35	.0088494
306	35–17.65	.0102665	356	30–20.22	.0088245
307	35–10.75	.0102330	357	30–15.12	.0087998
308	35– 3.90	.0101998	358	30–10.05	.0087753
309	34–57.09	.0101668	359	30– 5.01	.0087508
310	34–50.32	.0101340	360	30–	.0087265
311	34–43.60	.0101014	361	29–55.01	.0087023
312	34–36.92	.0100690	362	29–50.05	.0086783
313	34–30.29	.0100368	363	29–45.12	.0086544
314	34–23.69	.0100049	364	29–40.22	.0086306
315	34–17.14	.0099731	365	29–35.34	.0086070
316	34–10.63	.0099415	366	29–30.49	.0085835
317	34– 4.16	.0099102	367	29–25.67	.0085601
318	33–57.74	.0098791	368	29–20.87	.0085368
319	33–51.35	.0098482	369	29–16.10	.0085137
320	33–45	.0098174	370	29–11.35	.0084907
321	33–38.69	.0097868	371	29– 6.63	.0084678
322	33–32.42	.0097564	372	29– 1.94	.0084451
323	33–26.19	.0097261	373	28–57.27	.0084224
324	33–20	.0096961	374	28–52.62	.0083999
325	33–13.85	.0096663	375	28–48	.0083775
326	33– 7.73	.0096367	376	28–43.40	.0083552
327	33– 1.65	.0096072	377	28–38.83	.0083331
328	32–55.61	.0095779	378	28–34.28	.0083110
329	32–49.60	.0095488	379	28–29.76	.0082891
330	32–43.64	.0095198	380	28–25.26	.0082673
331	32–37.70	.0094911	381	28–20.78	.0082456
332	32–31.81	.0094625	382	28–16.33	.0082240
333	32–25.95	.0094341	383	28–11.91	.0082025
334	32–20.12	.0094059	384	28– 7.50	.0081812
335	32–14.33	.0093778	385	28– 3.12	.0081599
336	32– 8.57	.0093499	386	27–58.76	.0081387
337	32– 2.85	.0093221	387	27–54.42	.0081177
338	31–57.16	.0092945	388	27–50.10	.0080968
339	31–51.50	.0092671	389	27–45.81	.0080760
340	31–45.88	.0092398	390	27–41.54	.0080553
341	31–40.29	.0092127	391	27–37.29	.0080347
342	31–34.74	.0091858	392	27–33.06	.0080142
343	31–29.21	.0091590	393	27–28.85	.0079938
344	31–23.72	.0091324	394	27–24.67	.0079735
345	31–18.26	.0091059	395	27–20.51	.0079533
346	31–12.83	.0090796	396	27–16.36	.0079332
347	31– 7.44	.0090534	397	27–12.24	.0079132
348	31– 2.07	.0090274	398	27– 8.14	.0078934
349	30–56.73	.0090016	399	27– 4.06	.0078736
50	30–51.43	.0089758	400	27–	.0078534

MULTIPLY DIAMETER BY SINE TO GET LENGTH OF SIDE
(Angle given is half of angle subtended at center)

No. Sides	Angle Min. Sec.	Sine	No. Sides	Angle Min. Sec.	Sine
401	26–55.96	.0078343	451	23–56.81	.0069658
402	26–51.94	.0078148	452	23–53.63	.0069504
403	26–47.94	.0077954	453	23–50.46	.0069351
404	26–43.96	.0077761	454	23–47.31	.0069198
405	26–40	.0077569	455	23–44.17	.0069046
406	26–36.06	.0077378	456	23–41.05	.0068894
407	26–32.14	.0077188	457	23–37.94	.0068744
408	26–28.23	.0076999	458	23–34.84	.0068594
409	26–24.35	.0076811	459	23–31.76	.0068444
410	26–20.49	.0076623	460	23–28.69	.0068295
411	26–16.64	.0076437	461	23–25.64	.0068147
412	26–12.82	.0076251	462	23–22.60	.0067999
413	26– 9.01	.0076067	463	23–19.57	.0067852
414	26– 5.22	.0075883	464	23–16.55	.0067706
415	26– 1.45	.0075700	465	23–13.55	.0067561
416	25–57.70	.0075518	466	23–10.56	.0067416
417	25–53.96	.0075337	467	23– 7.58	.0067272
418	25–50.24	.0075157	468	23– 4.61	.0067128
419	25–46.54	.0074977	469	23– 1.66	.0066985
420	25–42.86	.0074799	470	22–58.72	.0066842
421	25–39.19	.0074621	471	22–55.79	.0066700
422	25–35.54	.0074444	472	22–52.88	.0066559
423	25–31.91	.0074268	473	22–49.98	.0066418
424	25–28.30	.0074093	474	22–47.09	.0066278
425	25–24.70	.0073919	475	22–44.21	.0066138
426	25–21.12	.0073745	476	22–41.34	.0065999
427	25–17.56	.0073573	477	22–38.49	.0065861
428	25–14.02	.0073401	478	22–35.65	.0065723
429	25–10.49	.0073230	479	22–32.82	.0065585
430	25– 6.98	.0073059	480	22–30	.0065449
431	25– 3.48	.0072890	481	22–27.20	.0065313
432	25–	.0072721	482	22–24.40	.0065178
433	24–56.54	.0072553	483	22–21.61	.0065043
434	24–53.09	.0072386	484	22–18.84	.0064909
435	24–49.66	.0072220	485	22–16.08	.0064775
436	24–46.24	.0072054	486	22–13.33	.0064641
437	24–42.84	.0071889	487	22–10.59	.0064509
438	24–39.45	.0071725	488	22– 7.87	.0064377
439	24–36.08	.0071562	489	22– 5.16	.0064245
440	24–32.73	.0071399	490	22– 2.45	.0064114
441	24–29.39	.0071237	491	21–59.75	.0063983
442	24–26.06	.0071076	492	21–57.07	.0063853
443	24–22.75	.0070916	493	21–54.40	.0063723
444	24–19.46	.0070756	494	21–51.74	.0063594
445	24–16.18	.0070597	495	21–49.09	.0063466
446	24–12.91	.0070439	496	21–46.45	.0063338
447	24– 9.66	.0070281	497	21–43.82	.0063211
448	24– 6.43	.0070124	498	21–41.20	.0063084
449	24– 3.21	.0069968	499	21–38.59	.0062957
450	24–	.0069813	500	21–36	.0062831

ACTUAL CUTTING SPEED OF PLANERS IN FEET PER MINUTE

Forward Cutting Speed in Feet per Minute	Return Speed						
	2 to 1	3 to 1	4 to 1	5 to 1	6 to 1	7 to 1	8 to 1
20	13.3	15.	16	16.66	17.14	17.5	17.76
25	16.6	18.75	20	20.83	21.42	21.87	22.26
30	20.	22.5	24	25.	25.71	26.25	26.56
35	23.3	26.25	28	29.16	30.	30.62	31.04
40	26.6	30.	32	33.33	34.28	35.	35.52
45	30.	33.75	36	37.5	38.56	39.37	40.
50	33.3	37.5	40	41.66	42.84	43.75	44.48
55	36.6	41.25	44	45.83	47.12	48.12	48.95
60	40.	45.	48	50.	51.42	52.50	53.43
65	43.3	48.75	52	54.16	55.70	56.87	57.91
70	46.6	52.5	56	58.33	60.	61.25	62.3
75	50.	56.25	60	62.5	64.28	66.62	66.71

The table shows very clearly that a slight increase in forward cutting speed is better than a high return speed. A 25-foot forward speed at 4 to 1 return is much better than an 8 to 1 return with 20-feet forward speed.

ALLOWANCES FOR BOLT HEADS AND UPSETS

STOCK ALLOWED FOR STANDARD UPSETS BY ACME MACHINERY CO

1/2	in. Upset to	5/8	in. Length of Upset,	3	in. Stock required,	1 3/4	in.
5/8	" "	3/4	" " "	3 1/2	" " "	1 5/8	"
3/4	" "	7/8	" " "	3 1/2	" " "	1 1/2	"
7/8	" "	1 1/8	" " "	4	" " "	2 3/4	"
1	" "	1 1/4	" " "	4	" " "	2 3/8	"
1 1/8	" "	1 3/8	" " "	4 1/2	" " "	2 1/4	"
1 1/4	" "	1 1/2	" " "	4	" " "	2 1/2	"
1 3/8	" "	1 5/8	" " "	5	" " "	2 1/8	"
1 1/2	" "	1 3/4	" " "	5	" " "	2	"
1 1/2	" "	1 7/8	" " "	5	" " "	2 7/8	"
1 5/8	" "	1 7/8	" " "	5 1/2	" " "	1 7/8	"
1 3/4	" "	2	" " "	5	" " "	1 3/4	"
1 7/8	" "	2 1/8	" " "	6	" " "	2	"
2	" "	2 3/8	" " "	6	" " "	2 1/2	"
2	" "	2 1/2	" " "	6	" " "	3 1/2	"
2 1/4	" "	2 5/8	" " "	6 1/2	" " "	2 1/2	"
2 1/2	" "	2 7/8	" " "	6 1/2	" " "	2 1/4	"
2 3/4	" "	3 1/8	" " "	7	" " "	2	"
[illegible]	" "	3 1/2	" " "	7	" " "	2 5/8	"

STOCK REQUIRED TO MAKE SQUARE AND HEXAGON BOLT HEADS
MANUFACTURERS' STANDARD SIZES

Diameter	Distance Across Flats Square	Distance Across Flats Hexagon	Thickness of Head	Stock for Square Head	Stock for Hexagon Head	Diameter
$\frac{1}{4}$	$\frac{3}{8}$	$\frac{7}{16}$	$\frac{3}{16}$	$\frac{17}{32}$	$\frac{15}{32}$	$\frac{1}{4}$
$\frac{3}{8}$	$\frac{9}{16}$	$\frac{5}{8}$	$\frac{9}{32}$	$\frac{25}{32}$	$\frac{11}{16}$	$\frac{3}{8}$
$\frac{1}{2}$	$\frac{3}{4}$	$\frac{13}{16}$	$\frac{3}{8}$	$1\frac{1}{16}$	$1\frac{1}{32}$	$\frac{1}{2}$
$\frac{5}{8}$	$\frac{15}{16}$	1	$\frac{15}{32}$	$1\frac{11}{32}$	$1\frac{5}{32}$	$\frac{5}{8}$
$\frac{3}{4}$	$1\frac{1}{8}$	$1\frac{3}{16}$	$\frac{9}{16}$	$1\frac{5}{8}$	$1\frac{9}{32}$	$\frac{3}{4}$
$\frac{7}{8}$	$1\frac{5}{16}$	$1\frac{3}{8}$	$\frac{21}{32}$	$1\frac{7}{8}$	$1\frac{1}{2}$	$\frac{7}{8}$
1	$1\frac{1}{2}$	$1\frac{9}{16}$	$\frac{3}{4}$	$2\frac{5}{32}$	$2\frac{7}{32}$	1
$1\frac{1}{8}$	$1\frac{11}{16}$	$1\frac{3}{4}$	$\frac{27}{32}$	$2\frac{15}{32}$	$2\frac{5}{32}$	$1\frac{1}{8}$
$1\frac{1}{4}$	$1\frac{7}{8}$	$1\frac{15}{16}$	$\frac{15}{16}$	$2\frac{3}{16}$	$2\frac{7}{16}$	$1\frac{1}{4}$
$1\frac{3}{8}$	$2\frac{1}{16}$	$2\frac{1}{8}$	$1\frac{1}{16}$	$3\frac{5}{32}$	$2\frac{27}{32}$	$1\frac{3}{8}$
$1\frac{1}{2}$	$2\frac{1}{4}$	$2\frac{5}{16}$	$1\frac{1}{8}$	$3\frac{11}{32}$	3	$1\frac{1}{2}$
$1\frac{5}{8}$	$2\frac{7}{16}$	$2\frac{1}{2}$	$1\frac{1}{4}$	$3\frac{21}{32}$	$3\frac{1}{4}$	$1\frac{5}{8}$
$1\frac{3}{4}$	$2\frac{5}{8}$	$2\frac{11}{16}$	$1\frac{5}{16}$	$3\frac{3}{4}$	$3\frac{15}{32}$	$1\frac{3}{4}$
$1\frac{7}{8}$	$2\frac{13}{16}$	$2\frac{7}{8}$	$1\frac{7}{16}$	$4\frac{3}{32}$	$3\frac{23}{32}$	$1\frac{7}{8}$
2	3	$3\frac{1}{16}$	$1\frac{1}{2}$	$4\frac{5}{16}$	4	2
$2\frac{1}{4}$	$3\frac{3}{8}$	$3\frac{7}{16}$	$1\frac{3}{4}$	5	$4\frac{9}{16}$	$2\frac{1}{4}$
$2\frac{1}{2}$	$3\frac{3}{4}$	$3\frac{13}{16}$	$1\frac{7}{8}$	$5\frac{3}{8}$	$4\frac{11}{16}$	$2\frac{1}{2}$
$2\frac{3}{4}$	$4\frac{1}{8}$	$4\frac{3}{16}$	$2\frac{1}{16}$	$5\frac{23}{32}$	$5\frac{3}{8}$	$2\frac{3}{4}$
3	$4\frac{1}{2}$	$4\frac{9}{16}$	$2\frac{1}{4}$	$6\frac{7}{16}$	$5\frac{11}{16}$	3

UNITED STATES STANDARD SIZES

Diameter	Distance Across Flats Square and Hexagon	Distance Across Corners Hexagon	Thickness of Head	Stock for Square Head	Stock for Hexagon Head	Diameter
$\frac{1}{4}$	$\frac{1}{2}$	$\frac{9}{16}$	$\frac{1}{4}$	$1\frac{9}{32}$	$1\frac{3}{32}$	$\frac{1}{4}$
$\frac{3}{8}$	$\frac{11}{16}$	$\frac{25}{32}$	$\frac{11}{32}$	$1\frac{1}{2}$	$1\frac{1}{4}$	$\frac{3}{8}$
$\frac{1}{2}$	$\frac{7}{8}$	1	$\frac{7}{16}$	$1\frac{23}{32}$	$1\frac{1}{2}$	$\frac{1}{2}$
$\frac{5}{8}$	$1\frac{1}{16}$	$1\frac{7}{32}$	$\frac{17}{32}$	2	$1\frac{11}{16}$	$\frac{5}{8}$
$\frac{3}{4}$	$1\frac{1}{4}$	$1\frac{7}{16}$	$\frac{5}{8}$	$2\frac{5}{32}$	$1\frac{7}{8}$	$\frac{3}{4}$
$\frac{7}{8}$	$1\frac{7}{16}$	$1\frac{21}{32}$	$\frac{21}{32}$	$2\frac{1}{2}$	$2\frac{5}{32}$	$\frac{7}{8}$
1	$1\frac{5}{8}$	$1\frac{7}{8}$	$\frac{13}{16}$	$2\frac{3}{4}$	$2\frac{3}{8}$	1
$1\frac{1}{8}$	$1\frac{13}{16}$	$2\frac{3}{32}$	$\frac{29}{32}$	3	$2\frac{19}{32}$	$1\frac{1}{8}$
$1\frac{1}{4}$	2	$2\frac{5}{16}$	1	$3\frac{1}{4}$	$2\frac{27}{32}$	$1\frac{1}{4}$
$1\frac{3}{8}$	$2\frac{3}{16}$	$2\frac{1}{2}$	$1\frac{1}{32}$	$3\frac{1}{16}$	3	$1\frac{3}{8}$
$1\frac{1}{2}$	$2\frac{3}{8}$	$2\frac{3}{4}$	$1\frac{1}{16}$	$3\frac{1}{16}$	$3\frac{1}{16}$	$1\frac{1}{2}$
$1\frac{5}{8}$	$2\frac{9}{16}$	$2\frac{15}{16}$	$1\frac{1}{32}$	$4\frac{1}{16}$	$3\frac{1}{4}$	$1\frac{5}{8}$
$1\frac{3}{4}$	$2\frac{3}{4}$	$3\frac{3}{16}$	$1\frac{1}{8}$	$4\frac{1}{4}$	$3\frac{1}{4}$	$1\frac{3}{4}$
$1\frac{7}{8}$	$2\frac{15}{16}$	$3\frac{3}{32}$	$1\frac{13}{32}$	$4\frac{1}{2}$	4	$1\frac{7}{8}$
2	$3\frac{1}{8}$	$3\frac{5}{8}$	$1\frac{1}{16}$	$4\frac{3}{4}$	$4\frac{1}{4}$	2
$2\frac{1}{4}$	$3\frac{1}{2}$	$4\frac{1}{16}$	$1\frac{1}{4}$	$5\frac{1}{16}$	$4\frac{25}{32}$	$2\frac{1}{4}$
$2\frac{1}{2}$	$3\frac{7}{8}$	$4\frac{1}{2}$	$1\frac{15}{16}$	$5\frac{1}{8}$	$5\frac{5}{8}$	$2\frac{1}{2}$
$2\frac{3}{4}$	$4\frac{1}{4}$	$4\frac{7}{8}$	$2\frac{1}{8}$	$6\frac{7}{16}$	$5\frac{1}{2}$	$2\frac{3}{4}$
3	$4\frac{5}{8}$	$5\frac{11}{32}$	$2\frac{5}{16}$	7	$6\frac{7}{16}$	3

QUICK WAY OF ESTIMATING LUMBER FOR A PATTERN

MULTIPLY length, breadth, and thickness in inches together and this by 7, pointing off three places.

Board 8 inches wide, 18 inches long, 1 inch thick. $8 \times 18 \times 1 \times 7 = 1.008$ square feet. This is .008 too much, but near enough for most work. Board $1\frac{1}{2} \times 10 \times 36 = 540 \times 7 = 3.780$. The correct answer is 3.75.

TABLE GIVING PROPORTIONATE WEIGHT OF CASTINGS TO WEIGHT OF WOOD PATTERNS

A Pattern Weighing One Pound Made of (Less weight of Core Prints)	Cast Iron	Brass	Copper	Bronze	Bell Metal	Zinc
Pine or Fir..	16	18.8	19.7	19.3	17	15.5
Oak........	9	10.1	10.4	10.3	10.9	8.6
Beech	9.7	10.9	11.4	11.3	11.9	9.1
Linden	13.4	15.1	16.7	15.5	16.3	12.9
Pear........	10.2	11.5	11.9	11.8	12.4	9.8
Birch.......	10.6	11.9	12.3	12.3	12.9	10.2
Alder.......	12.8	14.3	14.9	14.7	15.5	12.2
Mahogany ..	11.7	13.2	13.7	13.5	14.2	11.2
Brass.......	0.85	0.95	0.99	0.98	1.0	0.81

DEGREES OBTAINED BY OPENING A TWO-FOOT RULE

Degrees	Inches	Degrees	Inches	Degrees	Inches
1	.21	15	3.12	55	11.08
2	.422	20	4.17	60	12
3	.633	25	5.21	65	12.89
4	.837	30	6.21	70	13.76
5	1.04	35	7.20	75	14.61
7.5	1.57	40	8.21	80	15.43
10	2.09	45	9.20	85	16.21
14.5	3.015	50	10.12	90	16.97

Open a two-foot rule until open ends are distance apart given in table when degrees given in table can be scribed. Same results can be had with two 12-inch steel scales placed together at one end.

WEIGHT OF FILLETS

To facilitate the calculations of the weights of the different parts of a machine from the drawings, the accompanying table of areas or volumes of fillets having radii from $\frac{1}{16}$ to 3 inches can be used. It has been calculated for fillets connecting sides that are at right angles to each other.

TABLE OF AREAS OR VOLUMES OF FILLETS

Radius of Fillet in Inches	Area or Volume of Fillet in Sq. or Cubic Inches	Radius of Fillet in Inches	Area or Volume of Fillet in Sq. or Cubic Inches
$\frac{1}{16}$	.0008	$1\frac{9}{16}$	.5240
$\frac{1}{8}$	.0033	$1\frac{5}{8}$	.5667
$\frac{3}{16}$	.0075	$1\frac{11}{16}$	.6119
$\frac{1}{4}$	.0134	$1\frac{3}{4}$	.6572
$\frac{5}{16}$	.0209	$1\frac{13}{16}$	.7050
$\frac{3}{8}$	.0302	$1\frac{7}{8}$	.7543
$\frac{7}{16}$	.0410	$1\frac{15}{16}$	.8056
$\frac{1}{2}$	.0537	2	.8584
$\frac{9}{16}$	.0678	$2\frac{1}{16}$	.9129
$\frac{5}{8}$	.0838	$2\frac{1}{8}$	.9690
$\frac{11}{16}$	.1013	$2\frac{3}{16}$	1.0269
$\frac{3}{4}$	.1207	$2\frac{1}{4}$	1.0864
$\frac{13}{16}$	.1417	$2\frac{5}{16}$	1.1475
$\frac{7}{8}$	.1643	$2\frac{3}{8}$	1.2105
$\frac{15}{16}$	.1886	$2\frac{7}{16}$	1.2749
1	.2146	$2\frac{1}{2}$	1.3413
$1\frac{1}{16}$	.2423	$2\frac{9}{16}$	1.4086
$1\frac{1}{8}$	.2716	$2\frac{5}{8}$	1.4787
$1\frac{3}{16}$	.3026	$2\frac{11}{16}$	1.5500
$1\frac{1}{4}$	.3353	$2\frac{3}{4}$	1.6229
$1\frac{5}{16}$	.3697	$2\frac{13}{16}$	1.6869
$1\frac{3}{8}$	.4057	$2\frac{7}{8}$	1.7739
$1\frac{7}{16}$	.4434	$2\frac{15}{16}$	1.8518
$1\frac{1}{2}$	.4829	3	1.9314

To find the volume of a fillet by this table when the radius and length are given, multiply the value in the table opposite the given radius by the length of the fillet in inches, and this result multiplied by the weight of a cubic inch of the material will give the weight of the fillet.

WIRE GAGES AND STOCK WEIGHTS

TWIST DRILL AND STEEL WIRE GAGE SIZES

THE Twist Drill and Steel Wire Gage is used for measuring the sizes of twist drills and steel drill rods. Rod sizes by this gage should not be confused with Stubs' Steel Wire Gage sizes. The difference between the sizes of corresponding numbers in the two gages ranges from about .0005 to .004 inch, the Stubs sizes being the smaller except in the cases of a few numbers where the systems coincide exactly.

TWIST DRILL AND STEEL WIRE GAGE SIZES

No. of Gage	Dia. in Inches	No. of Gage	Dia. in Inches	No. of Gage	Dia. in Inches	No. of Gage	Dia. in Inches
1	.2280	21	.1590	41	.0960	61	.0390
2	.2210	22	.1570	42	.0935	62	.0380
3	.2130	23	.1540	43	.0890	63	.0370
4	.2090	24	.1520	44	.0860	64	.0360
5	.2055	25	.1495	45	.0820	65	.0350
6	.2040	26	.1470	46	.0810	66	.0330
7	.2010	27	.1440	47	.0785	67	.0320
8	.1990	28	.1405	48	.0760	68	.0310
9	.1960	29	.1360	49	.0730	69	.02925
10	.1935	30	.1285	50	.0700	70	.0280
11	.1910	31	.1200	51	.0670	71	.0260
12	.1890	32	.1160	52	.0635	72	.0250
13	.1850	33	.1130	53	.0595	73	.0240
14	.1820	34	.1110	54	.0550	74	.0225
15	.1800	35	.1100	55	.0520	75	.0210
16	.1770	36	.1065	56	.0465	76	.0200
17	.1730	37	.1040	57	.0430	77	.0180
18	.1695	38	.1015	58	.0420	78	.0160
19	.1660	39	.0995	59	.0410	79	.0145
20	.1610	40	.0980	60	.0400	80	.0135

STUBS' GAGES

IN using Stubs' Gages, the difference between the Stubs Iron Wire Gage and the Stubs Steel Wire Gage should be kept in mind. The Stubs Iron Wire Gage is the one commonly known as the English Standard Wire, or Birmingham Gage, and designates the Stubs soft wire sizes. The Stubs Steel Wire Gage is used in measuring drawn steel wire or drill rods of Stubs' make and is also used by many American makers of drill rods.

DIMENSIONS IN DECIMAL PARTS OF AN INCH

Number of Gage	American or Brown & Sharpe	Birmingham or Stubs' Iron Wire	Washburn & Moen Mfg. Co.	Trenton Iron Co.	Stubs' Steel Wire	Imperial Wire Gage	U. S. Standard for Plate
00000	....	......	......	......	......	.464	.46875
0000	....	......	......	.450	......	.432	.4375
000	.46	.454	.3938	.400	......	.400	.40625
00	.40964	.425	.3625	.360	......	.372	.375
0	.3648	.380	.3310	.330	......	.348	.34375
0	.32486	.340	.3065	.305	......	.324	.3125
1	.2893	.300	.2830	.285	.227	.300	.28125
2	.25763	.284	.2625	.265	.219	.276	.265625
3	.22942	.259	.2437	.245	.212	.252	.25
4	.20431	.238	.2253	.225	.207	.232	.234375
5	.18194	.220	.2070	.205	.204	.212	.21875
6	.16202	.203	.1920	.190	.201	.192	.203125
7	.14428	.180	.1770	.175	.199	.176	.1875
8	.12849	.165	.1620	.160	.197	.160	.171875
9	.11443	.148	.1483	.145	.194	.144	.15625
10	.10189	.134	.1350	.130	.191	.128	.140625
11	.090742	.120	.1205	.1175	.188	.116	.125
12	.080808	.109	.1055	.105	.185	.104	.109375
13	.071961	.095	.0915	.0925	.182	.092	.09375
14	.064084	.083	.0800	.080	.180	.080	.078125
15	.057068	.072	.0720	.070	.178	.072	.0703125
16	.05082	.065	.0625	.061	.175	.064	.0625
17	.045257	.058	.0540	.0525	.172	.056	.05625
18	.040303	.049	.0475	.045	.168	.048	.05
19	.03589	.042	.0410	.040	.164	.040	.04375
20	.031961	.035	.0348	.035	.161	.036	.0375
21	.028462	.032	.03175	.031	.157	.032	.034375
22	.025347	.028	.0286	.028	.155	.028	.03125
23	.022571	.025	.0258	.025	.153	.024	.028125
24	.0201	.022	.0230	.0225	.151	.022	.025
25	.0179	.020	.0204	.020	.148	.020	.021875
26	.01594	.018	.0181	.018	.146	.018	.01875
27	.014195	.016	.0173	.017	.143	.0164	.0171875
28	.012641	.014	.0162	.016	.139	.0149	.015625
29	.011257	.013	.0150	.015	.134	.0136	.0140625
30	.010025	.012	.0140	.014	.127	.0124	.0125
31	.008928	.010	.0132	.013	.120	.0116	.0109375
32	.00795	.009	.0128	.012	.115	.0108	.01015625
33	.00708	.008	.0118	.011	.112	.0100	.009375
34	.006304	.007	.0104	.010	.110	.0092	.00859375
35	.005614	.005	.0095	.0095	.108	.0084	.0078125
36	.005	.004	.0090	.009	.106	.0076	.00703125
37	.004453			.0085	.103	.0068	.006640625
38	.003965			.008	.101	.0060	.00625
39	.003531			.0075	.099	.0052	

WIRE AND DRILL SIZES ARRANGED CONSECUTIVELY

Dia. of Wire	American or B. & S.	B'ham or Stubs' Wire	Stubs' Steel Wire	Twist Drill and Steel Wire
		Gage Number		
.00314	40			
.00353	39			
.00397	38			
.004		36		
.0045	37			
.005	36	35		
.0056	35			
.0063	34			
.007		34		
.0071	33			
.008	32	33		
.0089	31			
.009		32		
.010	30	31		
.0113	29			
.012		30		
.0126	28			
.013		29	80	
.0135				80
.014		28	79	
.0142	27			
.0145				79
.015			78	
.0159	26			
.016		27	77	78
.0179	25			
.018		26	76	77
.020		25	75	76
.0201	24			
.021				75
.022		24	74	
.0225				74
.0226	23			
.023			73	
.024			72	73
.025		23		72
.0253	22			
.026			71	71
.027			70	
.028		22		70
.0285	21			
.029			69	
.0293				69
.030			68	
.031			67	68
.032	20	21	66	67
.033			65	66
.035		20	64	65
.0359	19			
.036			63	64
.037			62	63
.038			61	62
.039			60	61
.040			59	60
.0403	18			

Dia. of Wire	American or B. & S.	B'ham or Stubs' Wire	Stubs' Steel Wire	Twist Drill or Steel Wire
		Gage Number		
.041			58	
.042		19	57	
.043				
.045			56	
.0453	17			
.0465				
.049		18		
.050			55	
.0508	16			
.052				55
.055			54	54
.0571	15			
.058		17	53	
.0595				53
.063			52	
.0635				52
.0641	14			
.065		16	51	
.066				51
.067			50	
.069				50
.070			49	
.072	13	15		49
.073			48	
.075				48
.076			47	
.077				47
.0785			46	
.079				46
.0808	12		45	
.081				45
.082		14		
.083			44	
.085				44
.086			43	
.088				43
.089				
.0907	11		42	
.092				42
.0935		13	41	
.095				41
.096			40	
.097				40
.098				
.099			39	
.0995				39
.101			38	
.1015				38
.1019	10			
.103			37	
.104				37
.106			36	
.1065				36
.108			35	

WIRE AND DRILL SIZES ARRANGED CONSECUTIVELY

Dia. of Wire	American or B. & S.	B'ham or Stubs' Wire	Stubs' Steel Wire	Twist Drill and Steel Wire
	Gage Number			
.109		12		
.110			34	35
.111				34
.112			33	
.113				33
.1144	9			
.115			32	
.116				32
.120		11	31	31
.127			30	
.1285	8			30
.134		10	29	
.136				29
.139			28	
.1405				28
.143			27	
.144				27
.1443	7			
.146			26	
.147				26
.148		9	25	
.1495				25
.151			24	
.152				24
.153			23	
.154				23
.155			22	
.157			21	22
.159				21
.161			20	20
.162	6			
.164			19	
.165		8		
.166				19
.168			18	
.1695				18
.172			17	
.173				17
.175			16	
.177				16
.178			15	
.180		7	14	15
.1819	5			
.182			13	14
.185			12	13
.188			11	
.189				12
.191			10	11
.1935				10
.194			9	
.196				9
.197			8	
.199			7	8

Dia. of Wire	American or B. & S.	B'ham or Stubs' Wire	Stubs' Steel Wire	Twist Drill and Steel Wire
	Gage Number			
.203		6		
.204			5	6
.2043	4			
.2055				5
.207			4	
.209				4
.212			3	
.213				3
.219			2	
.220		5		
.221				2
.227			1	
.228				1
.2294	3			
.234			A	
.238		4	B	
.242			C	
.246			D	
.250			E	
.257			F	
.2576	2			
.259		3		
.261			G	
.266			H	
.272			I	
.277			J	
.281			K	
.284		2		
.2893	1			
.290			L	
.295			M	
.300		1		
.302			N	
.316			O	
.323			P	
.3249	0			
.332			Q	
.339			R	
.340		0		
.348			S	
.358			T	
.3648	00			
.368			U	
.377			V	
.380		00		
.386			W	
.397			X	
.404			Y	
.4096	000			
.413			Z	
.425		000		
.454		0000		
.460	0000			

STUBS' STEEL WIRE SIZES AND WEIGHTS

As stated in the explanatory note regarding Stubs' Gages at the bottom of page 296 the Stubs steel wire gage is used for measuring drawn steel wire and drill rods of Stubs' make and is also used by various drill rod makers in America.

STUBS' STEEL WIRE SIZES, AND WEIGHT IN POUNDS PER LINEAR FOOT

Letter and No. of Gage	Dia. in Inches	Weight per Foot	No. of Wire Gage	Dia. in Inches	Weight per Foot	No. of Wire Gage	Dia. in Inches	Weight per Foot
Z	.413	.456	10	.191	.098	46	.079	.017
Y	.404	.437	11	.188	.095	47	.077	.016
X	.397	.422	12	.185	.092	48	.075	.015
W	.386	.399	13	.182	.089	49	.072	.014
V	.377	.380	14	.180	.087	50	.069	.013
U	.368	.362	15	.178	.085	51	.066	.012
T	.358	.335	16	.175	.082	52	.063	.011
S	.348	.324	17	.172	.079	53	.058	.009
R	.339	.307	18	.168	.075	54	.055	.008
Q	.332	.295	19	.164	.072	55	.050	.007
P	.323	.280	20	.161	.069	56	.045	.006
O	.316	.267	21	.157	.066	57	.042	.0047
N	.302	.244	22	.155	.064	58	.041	.0045
M	.295	233	23	.153	.063	59	.040	.0042
L	.290	.225	24	.151	.061	60	.039	.0040
K	.281	.211	25	.148	.059	61	.038	.0039
J	.277	.205	26	.146	.057	62	.037	.0037
I	.272	.192	27	.143	.055	63	.036	.0035
H	.266	.189	28	.139	.052	64	.035	.0033
G	.261	.182	29	.134	.048	65	.033	.0029
F	.257	.177	30	.127	.043	66	.032	.0027
E	.250	.167	31	.120	.039	67	.031	.0020
D	.246	.162	32	.115	.035	68	.030	.0024
C	.242	.159	33	.112	.034	69	.029	.0022
B	.238	.152	34	.110	.032	70	.027	.0020
A	.234	.146	35	.108	.031	71	.026	.0018
1	.227	.138	36	.106	.030	72	.024	.0015
2	.219	.128	37	.103	.028	73	.023	.0014
3	.212	.120	38	.101	.027	74	.022	.0013
4	.207	.115	39	.099	.026	75	.020	.0011
5	.204	.111	40	.097	.025	76	.018	.0009
6	.201	.108	41	.095	.024	77	.016	.0007
7	.199	.106	42	.092	.023	78	.015	.0006
8	.197	.104	43	.088	.020	79	.014	.0005
9	.194	.101	44	.085	.019	80	.013	.0004
			45	.081	.018			

MUSIC WIRE SIZES

No. of Gage	Washburn & Moen	Webster & Horsefall	No. of Gage	Washburn & Moen	Webster & Horsefall	No. of Gage	Washburn & Moen	Webster & Horsefall
8–0	.0083		6	.0215	.016	19	.0414	.043
7–0	.0087		7	.023	.018	20	.0434	.045
6–0	.0095		8	.0243	.020	21	.046	.047
5–0	.010		9	.0256	.022	22	.0483	.052
4–0	.011	.006	10	.027	.024	23	.051	.055
3–0	.012	.007	11	.0284	.026	24	.055	.059
2–0	.0133	.008	12	.0296	.029	25	.0586	.061
1–0	.0144	.009	13	.0314	.031	26	.0626	.065
1	.0156	.010	14	.0326	.033	27	.0658	.070
2	.0166	.011	15	.0345	.035	28	.072	.072
3	.0178	.012	16	.036	.037	29	.076	.077
4	.0188	.013	17	.0377	.039	30	.080	.083
5	.0202	.014	18	.0395	.041			

WEIGHTS OF SHEET STEEL AND IRON
UNITED STATES STANDARD GAGE
(Adopted by U. S. Government, July 1, 1893)

Number of Gage	App. Thickness	WEIGHT PER SQ. FOOT		No. of Gage	App. Thickness	WEIGHT PER SQ. FOOT	
		Steel	Iron			Steel	Iron
0000000	.5	20.320	20.00	17	.05625	2.286	2.25
000000	.46875	19.050	18.75	18	.05	2.032	2.
00000	.4375	17.780	17.50	19	.04375	1.778	1.75
0000	.40625	16.510	16.25	20	.0375	1.524	1.50
000	.375	15.240	15.00	21	.03437	1.397	1.375
00	.34375	13.970	13.75	22	.03125	1.270	1.25
0	.3125	12.700	12.50	23	.02812	1.143	1.125
1	.28125	11.430	11.25	24	.025	1.016	1.
2	.26562	10.795	10.625	25	.02187	.889	.875
3	.25	10.160	10.00	26	.01875	.762	.75
4	.23437	9.525	9.375	27	.01718	.698	.687
5	.21875	8.890	8.75	28	.01562	.635	.623
6	.20312	8.255	8.125	29	.01406	.571	.562
7	.1875	7.620	7.5	30	.0125	.508	.5
8	.17187	6.985	6.875	31	.01093	.694	.437
9	.15625	6.350	6.25	32	.01015	.413	.406
10	.14062	5.715	5.625	33	.00937	.381	.375
11	.125	5.080	5.00	34	.00859	.349	.343
12	.10937	4.445	4.375	35	.00781	.317	.312
13	.09375	3.810	3.75	36	.00703	.285	.281
14	.07812	3.175	3.125	37	.00664	.271	.265
15	.07031	2.857	2.812	38	.00625	.254	.25
16	.0625	2.540	2.50				

Weight of 1 cubic foot is assumed to be 487.7 lbs. for steel plates and 480 lbs. for iron plates.

WEIGHTS OF STEEL, WROUGHT IRON, BRASS AND COPPER PLATES

AMERICAN OR BROWN & SHARPE GAGE

No. of Gage	Thickness in Inches	WEIGHT IN LBS. PER SQUARE FOOT			
		Steel	Iron	Brass	Copper
0000	.46	18.77	18.40	19.688	20.838
000	.4096	16.71	16.38	17.533	18.557
00	.3648	14.88	14.59	15.613	16.525
0	.3249	13.26	13.00	13.904	14.716
1	.2893	11.80	11.57	12.382	13.105
2	.2576	10.51	10.30	11.027	11.670
3	.2294	9.39	9.18	9.819	10.392
4	.2043	8.34	8.17	8.745	9.255
5	.1819	7.42	7.28	7.788	8.242
6	.1620	6.61	6.48	6.935	7.340
7	.1443	5.89	5.77	6.175	6.536
8	.1285	5.24	5.14	5.499	5.821
9	.1144	4.67	4.58	4.898	5.183
10	.1019	4.16	4.08	4.361	4.616
11	.0908	3.70	3.63	3.884	4.110
12	.0808	3.30	3.23	3.458	3.660
13	.0720	2.94	2.88	3.080	3.260
14	.0641	2.62	2.56	2.743	2.903
15	.0571	2.33	2.28	2.442	2.585
16	.0508	2.07	2.03	2.175	2.302
17	.0453	1.85	1.81	1.937	2.050
18	.0403	1.64	1.61	1.725	1.825
19	.0359	1.46	1.44	1.536	1.626
20	.0320	1.31	1.28	1.367	1.448
21	.0285	1.16	1.14	1.218	1.289
22	.0253	1.03	1.01	1.085	1.148
23	.0226	.922	.904	.966	1.023
24	.0201	.820	.804	.860	.910
25	.0179	.730	.716	.766	.811
26	.0159	.649	.636	.682	.722
27	.0142	.579	.568	.608	.643
28	.0126	.514	.504	.541	.573
29	.0113	.461	.452	.482	.510
30	.0100	.408	.400	.429	.454
31	.0089	.363	.356	.382	.404
32	.0080	.326	.320	.340	.360
33	.0071	.290	.284	.303	.321
34	.0063	.257	.252	.269	.286
35	.0056	.228	.224	.240	.254
36	.0050	.190	.188	.214	.226
37	.0045	.169	.167	.191	.202
38	.0040	.151	.149	.170	.180
39	.0035	.134	.132	.151	.160
40	.0031	.119	.118	.135	.142

Weights of Steel, Wrought Iron, Brass and Copper Plates
Birmingham or Stubs' Gage

No. of Gage	Thickness in Inches	Weight in Lbs. per Square Foot			
		Steel	Iron	Brass	Copper
0000	.454	18.52	18.16	19.431	20.556
000	.425	17.34	17.00	18.190	19.253
00	.380	15.30	15.20	16.264	17.214
0	.340	13.87	13.60	14.552	15.402
1	.300	12.24	12.00	12.840	13.590
2	.284	11.59	11.36	12.155	12.865
3	.259	10.57	10.36	11.085	11.733
4	.238	9.71	9.52	10.186	10.781
5	.220	8.98	8.80	9.416	9.966
6	.203	8.28	8.12	8.689	9.196
7	.180	7.34	7.20	7.704	8.154
8	.165	6.73	6.60	7.062	7.475
9	.148	6.04	5.92	6.334	6.704
10	.134	5.47	5.36	5.735	6.070
11	.120	4.90	4.80	5.137	5.436
12	.109	4.45	4.36	4.667	4.938
13	.095	3.88	3.80	4.066	4.303
14	.083	3.39	3.32	3.552	3.769
15	.072	2.94	2.88	3.081	3.262
16	.065	2.65	2.60	2.782	2.945
17	.058	2.37	2.32	2.482	2.627
18	.049	2.00	1.96	2.097	2.220
19	.042	1.71	1.68	1.797	1.902
20	.035	1.43	1.40	1.498	1.585
21	.032	1.31	1.28	1.369	1.450
22	.028	1.14	1.12	1.198	1.270
23	.025	1.02	1.00	1.070	1.132
24	.022	.898	.88	.941	.997
25	.020	.816	.80	.856	.906
26	.018	.734	.72	.770	.815
27	.016	.653	.64	.685	.725
28	.014	.571	.56	.599	.634
29	.013	.530	.52	.556	.589
30	.012	.490	.48	.514	.544
31	.010	.408	.40	.428	.453
32	.009	.367	.36	.385	.408
33	.008	.326	.32	.342	.362
34	.007	.286	.28	.2996	.317
35	.005	.204	.20	.214	.227
36	.004	.163	.16	.171	.181

WEIGHTS OF STEEL, IRON, BRASS AND COPPER WIRE

AMERICAN OR BROWN & SHARPE GAGE

No. of Gage	Dia. in Inches	WEIGHT IN LBS. PER 1000 LINEAR FEET			
		Steel	Iron	Brass	Copper
0000	.4600	566.03	560.74	605.18	640.51
000	.4096	448.88	444.68	479.91	507.95
00	.3648	355.99	352.66	380.67	402.83
0	.3247	282.30	279.67	301.82	319.45
1	.2893	223.89	221.79	239.35	253.34
2	.2576	177.55	175.89	189.82	200.91
3	.2294	140.80	139.48	150.52	159.32
4	.2043	111.66	110.62	119.38	126.35
5	.1819	88.548	87.720	94.666	100.20
6	.1620	70.221	69.565	75.075	79.462
7	.1447	55.685	55.165	59.545	63.013
8	.1285	44.164	43.751	47.219	49.976
9	.1144	35.026	34.699	37.437	39.636
10	.1019	27.772	27.512	29.687	31.426
11	.0907	22.026	21.820	23.549	24.924
12	.0808	17.468	17.304	18.676	19.766
13	.0720	13.851	13.722	14.809	15.674
14	.0641	10.989	10.886	11.746	12.435
15	.0571	8.712	8.631	9.315	9.859
16	.0508	6.909	6.845	7.587	7.819
17	.0453	5.478	5.427	5.857	6.199
18	.0403	4.344	4.304	4.645	4.916
19	.0359	3.445	3.413	3.684	3.899
20	.0320	2.734	2.708	2.920	3.094
21	.0285	2.167	2.147	2.317	2.452
22	.0253	1.719	1.703	1.838	1.945
23	.0226	1.363	1.350	1.457	1.542
24	.0201	1.081	1.071	1.155	1.223
25	.0179	.8571	.8491	.9163	.9699
26	.0159	.6797	.6734	.7267	.7692
27	.0142	.5391	.5340	.5763	.6099
28	.0126	.4275	.4235	.4570	.4837
29	.0113	.3389	.3358	.3624	.3835
30	.0100	.2688	.2663	.2874	.3042
31	.0089	.2132	.2113	.2280	.2413
32	.0080	.1691	.1675	.1808	.1913
33	.0071	.1341	.1328	.1434	.1517
34	.0063	.1063	.1053	.1137	.1204
35	.0056	.0844	.0836	.0901	.0956
36	.0050	.0668	.0662	.0715	.0757
37	.0045	.0530	.0525	.0567	.0600
38	.0040	.0420	.0416	.0449	.0475
39	.0035	.0333	.0330	.0356	.0375
40	.0031	.0264	.0262	.0282	.0299

WEIGHTS OF IRON, BRASS, AND COPPER WIRE

BIRMINGHAM OR STUBS' GAGE

No. of Gage	Dia. in Inches	WEIGHT IN LBS. PER 1000 LINEAR FEET		
		Iron	Brass	Copper
0000	.454	546.21	589.29	623.2
000	.425	478.65	516.41	546.1
00	.380	382.66	412.84	436.6
0	.340	306.34	330.50	349.5
1	.300	238.50	257.31	272.1
2	.284	213.74	230.60	243.9
3	.259	177.77	191.79	202.8
4	.238	150.11	161.95	171.3
5	.220	128.26	138.37	146.3
6	.203	109.20	117.82	124.6
7	.180	85.86	92.63	97.96
8	.165	72.14	77.83	82.31
9	.148	58.05	62.62	66.23
10	.134	47.58	51.34	54.29
11	.120	38.16	41.17	43.54
12	.109	31.49	33.97	35.92
13	.095	23.92	25.80	27.29
14	.083	18.26	19.70	20.83
15	.072	13.73	14.82	15.67
16	.065	11.19	12.08	12.77
17	.058	8.92	9.62	10.17
18	.049	6.36	6.86	7.259
19	.042	4.67	5.04	5.333
20	.035	3.25	3.52	3.704
21	.032	2.71	2.93	3.096
22	.028	2.08	2.24	2.370
23	.025	1.66	1.79	1.890
24	.022	1.28	1.39	1.463
25	.020	1.06	1.14	1.209
26	.018	.863	.926	.979
27	.016	.680	.732	.774
28	.014	.529	.560	.592
29	.013	.438	.483	.511
30	.012	.382	.412	.435
31	.010	.266	.286	.302
32	.009	.212	.232	.244
33	.008	.167	.183	.193
34	.007	.133	.140	.148
35	.005	.066	.071	.075
36	.004	.046	.048	.052

WEIGHTS OF STEEL AND IRON BARS PER LINEAR FOOT

Dia. or Distance Across Flats	STEEL Weight per Foot				IRON Weight per Foot	
	Round	Square	Hexagon	Octagon	Round	S
1/16	.010	.013	.012	.011	.010	.013
1/8	.042	.053	.046	.044	.041	.054
3/16	.094	.119	.103	.099	.092	.117
1/4	.167	.212	.185	.177	.164	.208
5/16	.261	.333	.288	.277	.256	.326
3/8	.375	.478	.414	.398	.368	.469
7/16	.511	.651	.564	.542	.501	.638
1/2	.667	.850	.737	.708	.654	.833
9/16	.845	1.076	.932	.896	.828	1.055
5/8	1.043	1.328	1.151	1.107	1.023	1.302
11/16	1.262	1.608	1.393	1.331	1.237	1.576
3/4	1.502	1.913	1.658	1.584	1.473	1.875
13/16	1.763	2.245	1.944	1.860	1.728	2.201
7/8	2.044	2.603	2.256	2.156	2.004	2.552
15/16	2.347	2.989	2.591	2.482	2.301	2.930
1	2.670	3.400	2.947	2.817	2.618	3.333
1 1/16	3.014	3.838	3.327	3.182	2.955	3.763
1 1/8	3.379	4.303	3.730	3.568	3.313	4.210
1 3/16	3.766	4.795	4.156	3.977	3.692	4.701
1 1/4	4.173	5.312	4.605	4.407	4.091	5.208
1 5/16	4.600	5.857	5.077	4.858	4.510	5.742
1 3/8	5.049	6.428	5.571	5.331	4.950	6.302
1 7/16	5.518	7.026	6.091	5.827	5.410	6.888
1 1/2	6.008	7.650	6.631	6.344	5.890	7.500
1 9/16	6.520	8.301	7.195	6.905	6.392	8.138
1 5/8	7.051	8.978	7.776	7.446	6.913	8.802
1 11/16	7.604	9.682	8.392	8.027	7.455	9.492
1 3/4	8.178	10.41	9.025	8.635	8.018	10.21
1 13/16	8.773	11.17	9.682	9.264	8.601	10.95
1 7/8	9.388	11.95	10.36	9.918	9.204	11.72
1 15/16	10.02	12.76	11.06	10.58	9.828	12.51
2	10.68	13.60	11.79	11.28	10.47	13.33
2 1/8	12.06	15.35	13.31	12.71	11.82	15.05
2 1/4	13.52	17.22	14.92	14.24	13.25	16.88
2 3/8	15.07	19.18	16.62	15.88	14.77	18.80
2 1/2	16.69	21.25	18.42	17.65	16.36	20.83
2 5/8	18.40	23.43	20.31	19.45	18.04	22.97
2 3/4	20.20	25.71	22.29	21.28	19.80	25.21
2 7/8	22.07	28.10	24.36	23.28	21.64	27.55
3	24.03	30.60	26.53	25.36	23.56	30.00
3 1/8	26.08	33.20	28.78	27.50	25.57	32.55
3 1/4	28.20	35.92	31.10	29.28	27.65	35.21
3 3/8	30.42	38.78	33.57	32.10	29.82	37.97
3 1/2	32.71	41.65	36.10	34.56	32.07	40.83
3 5/8	35.09	44.68	38.73	37.05	34.40	43.80
3 3/4	37.56	47.82	41.45	39.68	36.82	46.88
3 7/8	40.10	51.05	44.26	42.35	39.31	50.05
4	42.73	54.40	47.16	45.12	41.89	53.33

EIGHTS OF BRASS, COPPER AND ALUMINUM BARS PER LINEAR FOOT

Dia. or Dis-tance cross flats	BRASS Weight per Foot			COPPER Weight per Foot		ALUMINUM Weight per Foot	
	Round	Square	Hexagon	Round	Square	Round	Square
1/16	.011	.014	.013	.012	.015	.003	.004
1/8	.045	.055	.048	.047	.060	.014	.018
3/16	.100	.125	.108	.106	.135	.032	.041
1/4	.175	.225	.194	.189	.241	.057	.072
5/16	.275	.350	.301	.296	.377	.089	.114
3/8	.395	.510	.436	.426	.542	.128	.163
7/16	.540	.690	.592	.579	.737	.174	.222
1/2	.710	.905	.773	.757	.964	.227	.290
9/16	.900	1.15	.978	.958	1.22	.288	.367
5/8	1.10	1.40	1.24	1.18	1.51	.356	.453
11/16	1.35	1.72	1.45	1.43	1.82	.430	.548
3/4	1.66	2.05	1.73	1.70	2.17	.516	.652
13/16	1.85	2.40	2.03	2.00	2.54	.601	.766
7/8	2.15	2.75	2.36	2.32	2.95	.697	.888
15/16	2.48	3.15	2.71	2.66	3.39	.800	1.02
1	2.85	3.65	3.10	3.03	3.86	.911	1.16
1 1/16	3.20	4.08	3.49	3.42	4.35	1.03	1.31
1 1/8	3.57	4.55	3.91	3.81	4.88	1.15	1.47
1 3/16	3.97	5.08	4.38	4.27	5.44	1.28	1.64
1 1/4	4.41	5.65	4.82	4.72	6.01	1.42	1.81
1 5/16	4.86	6.22	5.33	5.21	6.63	1.57	2.00
1 3/8	5.35	6.81	5.76	5.72	7.24	1.72	2.19
1 7/16	5.86	7.45	6.38	6.26	7.97	1.88	2.40
1 1/2	6.37	8.13	6.92	6.81	8.67	2.05	2.61
1 9/16	6.92	8.83	7.54	7.39	9.41	2.22	2.83
1 5/8	7.48	9.55	8.15	7.99	10.18	2.41	3.06
1 11/16	8.05	10.27	8.80	8.45	10.73	2.59	3.30
1 3/4	8.65	11.00	9.47	9.27	11.80	2.79	3.55
1 13/16	9.29	11.82	10.15	9.76	12.43	2.99	3.81
1 7/8	9.95	12.68	10.86	10.64	13.55	3.20	4.08
1 15/16	10.58	13.50	11.68	11.11	14.15	3.41	4.35
2	11.25	14.35	12.36	12.11	15.42	3.64	4.64
2 1/8	12.78	16.27	13.92	13.67	17.42	4.11	5.24
2 1/4	14.32	18.24	15.72	15.33	19.51	4.61	5.87
2 3/8	15.96	20.32	17.52	17.08	21.74	5.14	6.54
2 1/2	17.68	22.53	19.44	18.92	24.09	5.69	7.25
2 5/8	19.50	24.83	21.24	20.86	26.56	6.27	7.99
2 3/4	21.40	27.25	23.40	22.89	29.05	6.89	8.53
2 7/8	23.39	29.78	25.82	25.02	31.86	7.52	9.58
3	25.47	32.43	27.84	27.24	34.69	8.20	10.44
3 1/4	30.45	38.77	32.76	31.97	40.71	9.62	12.25
3 1/2	35.31	44.96	37.80	37.08	47.22	11.16	14.21
3 3/4	40.07	51.01	43.56	42.11	53.61	12.81	16.31
4	46.12	58.73	49.44	48.43	61.67	14.56	18.56

WEIGHTS OF FLAT SIZES OF STEEL IN POUNDS PER LINEAR FOOT

	½	⅝	¾	⅞	1	1⅛	1¼	1⅜	1½	1¾	2	2¼	2½	2¾	3	3½	4	5	6
⅛	.213	.266	.320	.372	.426	.479	.530	.585	.640	.745	.850	.955	1.07	1.18	1.28	1.49	1.70	2.13	2.56
3/16	.319	.399	.480	.558	.639	.718	.790	.878	.960	1.12	1.28	1.43	1.60	1.76	1.92	2.24	2.55	3.20	3.83
¼	.425	.533	.640	.743	.852	.958	1.06	1.17	1.28	1.49	1.70	1.91	2.13	2.34	2.56	2.98	3.40	4.26	5.11
5/16	.531	.665	.800	.929	1.06	1.20	1.33	1.46	1.60	1.86	2.13	2.39	2.66	2.92	3.19	3.72	4.25	5.32	6.38
⅜	.638	.798	.960	1.12	1.28	1.43	1.59	1.75	1.91	2.23	2.55	2.87	3.20	3.51	3.83	4.46	5.10	6.40	7.66
7/16	.744	.931	1.12	1.30	1.49	1.67	1.86	2.05	2.23	2.60	2.98	3.35	3.72	4.09	4.46	5.21	5.95	7.44	8.92
½		1.07	1.28	1.49	1.70	1.91	2.13	2.34	2.55	2.98	3.40	3.83	4.26	4.68	5.10	5.96	6.80	8.52	10.20
9/16		1.20	1.44	1.67	1.91	2.15	2.39	2.63	2.87	3.35	3.83	4.30	4.78	5.26	5.74	6.69	7.65	9.56	11.50
⅝			1.60	1.86	2.12	2.39	2.66	2.92	3.19	3.72	4.26	4.79	5.32	5.86	6.39	7.44	8.52	10.64	12.78
11/16			1.76	2.04	2.34	2.63	2.92	3.22	3.51	4.09	4.68	5.26	5.84	6.43	7.01	8.18	9.35	11.70	14.00
¾				2.23	2.55	2.86	3.19	3.50	3.83	4.46	5.10	5.74	6.40	7.02	7.65	8.92	10.20	12.80	15.30
13/16				2.41	2.76	3.11	3.45	3.80	4.14	4.83	5.53	6.22	6.91	7.60	8.29	9.67	11.10	13.80	16.60
⅞					2.98	3.34	3.72	4.09	4.46	5.21	5.96	6.70	7.46	8.19	8.94	10.42	11.92	14.92	17.88
15/16					3.19	3.59	3.98	4.38	4.78	5.58	6.38	7.17	7.97	8.77	9.56	11.20	12.80	15.90	19.10
1						3.82	4.25	4.68	5.10	5.96	6.80	7.66	8.52	9.36	10.20	11.92	13.60	17.04	20.40
1⅛							4.78	5.27	5.74	6.71	7.65	8.61	9.59	10.54	11.48	13.41	15.30	19.17	22.95
1¼								5.85	6.38	7.45	8.50	9.57	10.65	11.71	12.76	14.90	17.00	21.30	25.61
1½								7.02	7.67	8.94	10.20	11.49	12.78	14.04	15.30	17.88	20.40	25.56	30.

No. of Gage	Thickness in Inches	⅛	³⁄₁₆	¼	⁵⁄₁₆	⅜	⁷⁄₁₆	½	⁹⁄₁₆	⅝	¾	⅞	1	1¼	1½	1¾	2	2¼	2½
1	.300												2.42	3.28	4.10	5.03	5.88	6.75	7.62
2	.284												2.35	3.16	4.03	4.80	5.57	6.45	7.26
3	.259											1.85	2.22	2.98	3.72	4.48	5.23	5.96	6.72
4	.238											1.76	2.10	2.79	3.49	4.18	4.82	5.51	6.25
5	.220											1.68	1.99	2.60	3.26	3.89	4.53	5.14	5.80
6	.203									.99	1.29	1.58	1.88	2.46	3.04	3.60	4.20	4.80	5.39
7	.180							.667	.808	.93	1.17	1.44	1.71	2.23	2.77	3.28	3.78	4.33	4.83
8	.165						.52	.64	.760	.88	1.117	1.35	1.59	2.07	2.54	3.02	3.50	3.98	4.46
9	.148				.28	.39	.50	.60	.717	.82	1.031	1.25	1.46	1.89	2.32	2.75	3.17	3.60	4.03
10	.134				.277	.37	.470	.567	.667	.76	.955	1.16	1.35	1.74	2.12	2.51	2.90	3.28	3.67
11	.120			.18	.267	.354	.440	.527	.614	.70	.875	1.05	1.22	1.57	1.91	2.27	2.61	2.96	3.31
12	.109			.177	.256	.335	.414	.493	.571	.65	.808	.965	1.12	1.43	1.76	2.07	2.39	2.70	3.01
13	.095			.170	.239	.308	.376	.445	.514	.58	.717	.855	1.00	1.27	1.55	1.82	2.09	2.37	2.65
14	.083			.160	.220	.280	.340	.400	.460	.52	.640	.760	.88	1.12	1.36	1.61	1.84	2.08	2.32
15	.072		.096	.148	.201	.252	.304	.356	.409	.46	.564	.667	.77	.99	1.19	1.40	1.61	1.81	2.02
16	.065		.092	.138	.186	.233	.280	.327	.374	.42	.515	.609	.70	.89	1.07	1.26	1.45	1.64	1.82
17	.058	.045	.087	.128	.169	.213	.254	.296	.338	.38	.463	.548	.64	.80	.97	1.14	1.30	1.48	1.64
18	.049	.043	.078	.113	.150	.185	.220	.256	.291	.33	.397	.467	.54	.67	.82	.96	1.10	1.24	1.39
19	.042	.040	.070	.101	.130	.162	.192	.222	.253	.28	.343	.404	.46	.58	.71	.83	.95	1.07	1.19
20	.035	.036	.062	.087	.113	.138	.163	.188	.213	.24	.289	.339	.39	.49	.59	.69	.80	.893	1.00
21	.032	.034	.057	.081	.104	.127	.150	.173	.196	.22	.266	.312	.357	.450	.542	.635	.727	.820	.913
22	.028	.031	.051	.072	.093	.112	.133	.153	.173	.19	.233	.275	.315	.396	.477	.556	.638	.718	.799
23	.025	.029	.047	.065	.082	.102	.119	.137	.155	.17	.209	.245	.281	.354	.426	.497	.571	.641	.714
24	.022	.026	.042	.058	.074	.090	.107	.122	.137	.15	.186	.218	.249	.312	.376	.438	.502	.566	.629
25	.020	.024	.039	.052	.068	.081	.096	.110	.126	.140	.169	.197	.226	.285	.342	.399	.457	.516	.573

For Weights of Seamless Copper Tubing, add 5 per cent. to the weights above.

BELTS AND SHAFTING

BELT FASTENINGS

THE best fastening for a belt is the cement splice. It is far beyond any form of lacing, belt hooks, riveting, or any other method of joining together the ends of a belt. The cement joint is easily applied to leather and to rubber belts, but to make a good cement splice in a canvas belt requires more time and apparatus than is usually at hand. Good glue makes a fine cement for leather belts, and fish glue is less affected by moisture than the other. Many of the liquid glues are fish glue treated with acid so as not to gelatinize when cold. A little bichromate of potash added to ordinary hot glue just before it is used will render it insoluble in water.

There are many styles of belt hooks in use, some of the more common kind being shown in Figs. 1, 2, 3 and 4. Fig. 2 is practically a double rivet, Fig. 3 a malleable iron fastening, although similar hooks have been made of pressed steel, and Fig. 4 is the Blake stud which has the advantage of not weakening the belt but makes a hump on the outside where the ends turn up. Fig. 5 is the Bristol hook of stamped steel which is driven in the points turned over on the other side. Fig. 6 is the Jackson belt lacing and is applied by a hand machine which screws a spiral wire across the ends of a belt. These are then flattened and a rawhide pin or a heavy soft cord used as a hinge joint between them. These joints are probably equal to 90 per cent. of the belt strength.

Lacing Belts

Belts fastened by lacing are weakened according to the amount of material punched out in making the holes to receive the lacing. It is preferable to lace with a small lacing put many times through small holes. Such a joint is stronger than a few pieces of wide lacing through a number of large holes. Figs. 7 and 9 illustrate two forms of belt lacing, the latter being far preferable to the other. The lacing shown by Fig. 7 is in a double leather belt 5 inches wide. The width makes no difference as the strength is figured in percentage of the total width. There are four holes in this piece of belt, each hole $\frac{3}{8}$ inch in diameter. The aggregate width thus cut out of the belt is $4 \times \frac{3}{8}$ inch $= \frac{12}{8} = 1\frac{1}{2}$ inches. Then $1.5 \div 5 = 0.30$, or 30 per cent. of the belt has been cut away — nearly one third of the total strength.

In Fig. 9 a different method is followed. Instead of there being a few large holes, there are more smaller ones — one fourth more, in fact. There are five holes, each $\frac{3}{16}$ inch in diameter, making a total of $\frac{15}{16}$ inch, or $0.9375 \div 5 = 18\frac{3}{4}$ per cent., leaving $81\frac{1}{4}$ per cent. of the total belt strength, against 70 per cent. in the belt with large holes. A first-class double leather belt will tear in two under

ı strain of about 500 pounds to each lace hole, the strain being
ıpplied in the holes by means of lacings.

The belt shown by Fig. 9 has 81¼ per cent. of 1.875 square inches
of section, = 1.525 square inches left after cutting out the five holes.
This amount is good for 3000 × 1.875 = 5625 pounds breaking
strain, and as the lacing will tear out under 2500 pounds, it will be
seen that we cannot afford to use lacings if the full power of the
leather is to be utilized. This, under a factor of safety of 5, would
be 1125 pounds to the square inch, or 1125 × 1.525 = 1715 pounds

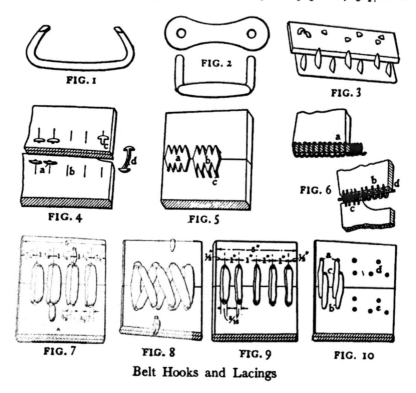

FIG. 1

FIG. 2

FIG. 3

FIG. 4

FIG. 5

FIG. 6

FIG. 7

FIG. 8

FIG. 9

FIG. 10

Belt Hooks and Lacings

working strain for the belt, or 1715 ÷ 5 = 343.5 pounds to each
lace. This, too, is too much, as it is less than a factor of safety of 2.

The belt to carry 40 pounds working tension to the inch of width
must also carry about 40 pounds standing tension, making a strain
of 80 pounds to the inch, or 80 × 5 = 400 pounds. This is a better
showing, and gives a factor of safety of 2500 ÷ 400 = 6¼. Still, we
are wasting a belt of 5625 pounds ultimate strength in order to get
from it 400 pounds working strain. This means a factor of safety
of over 14 in the body of the belt but of only 6¼ at the lacing, which
shows the advantage of a cement splice.

Fig. 10 shows a method sometimes used to relieve the lace-holes
of some of the strain. Double rows of holes are punched as at *a b*,
and the lacing distributed among them. As far as helping the

strength of the belt is concerned, this does nothing, for all the stress put upon the belt by the lacing at *c* must be carried by the belt section at *a*; therefore this way of punching holes does not increase the section strength. Neither does staggering the holes as shown at *d* and *e*. The form of hole-punching shown at *a b c* is desirable for another reason. It distributes the lacing very nicely and does not make such a lump to thump when it passes over the pulleys.

ALINING SHAFTING BY A STEEL WIRE

A STEEL wire is often used for alining shafting by stretching it parallel with the direction of the shaft and measuring from the shaft to the wire in a horizontal direction. This steel wire can also be used for leveling or alining in a direction at right angles to the other.

SAGS OF A STEEL ALINING WIRE FOR SHAFTING

Distance in Feet, from Reel to Point of Measurement

Distance in Feet, from Reel to Point of Support	10	20	30	40	50	60	70	80	90	100	110	120	130
280													
270													
260													
250													
240													
230													
220													
210													
200													
190													
180													
170													
160													
150													
140													
130													
120													
110													
100													
90													
80													
70													
60													
50													
40													
30													

Sag of the Wire in Inches

by making vertical measurements, if it is stretched under established conditions and if the sags at the points of measurement are known. The accompanying table gives the sags in inches from a truly level line passing through the points of support of the wire, at successive points beginning 10 feet from the reel and spaced 10 feet apart for a No. 17 Birmingham gage high grade piano wire, stretched with a weight of 60 pounds, wound on a reel of a minimum diameter of three and one-half inches and for total distances between the reel and point of support of the wire varying by increments of 10 feet from 40 to 280 feet. Thus a wire of any convenient length, of the kind indicated, can be selected, so long as this length is a multiple of 10 feet and between the limits specified, and the table gives the sags from a truly level line at points 10 feet apart for its entire length when it is stretched under the conditions designated. These sags

SAGS OF A STEEL ALINING WIRE FOR SHAFTING

Distance in Feet, from Reel to Point of Measurement

Sag of the Wire in Inches

being known, direct measurements can be made to level or aline a shaft by vertical measurements.

The method was originally developed for alining the propeller shafts of vessels, but it is equally serviceable for semi-flexible shafting, as factory line shafts.

SPEEDS OF PULLEYS AND GEARS

THE fact that the circumference of a pulley or gear is always 3.1416 or 3⅐ times the diameter makes it easy to figure speeds by considering only the diameter of both driver and driven pulleys. Belting from one 6-inch pulley to another gives the same speed to both; but if the driving pulley is 16 inches and the driven pulley only 4 inches it is clear that the small pulley will turn 4 times for every turn of the large pulley. If this is reversed and the small pulley is the driver, the large pulley will only make one turn for every four of the small pulley. The same rule applies to gears if the *pitch diameter* and not the outside diameter is taken. The following rules have been arranged for convenience in finding any desired information about pulley or gear speeds.

HAVING	TO FIND	RULE
Diam. of Driving Pulley Diam. of Driven Pulley Speed of Driving Pulley	Speed of Driven Pulley	Multiply Diam. of Driving Pulley by its Speed and divide by Diam. of Driven Pulley.
Diam. of Driving Pulley Speed of Driving Pulley Speed of Driven Pulley	Diam. of Driven Pulley	Multiply Diam. of Driving Pulley by its Speed and Divide by Speed of Driven Pulley.
Diam. of Driving Pulley Diam. of Driven Pulley Speed of Driven Pulley	Speed of Driving Pulley	Multiply Diam. of Driven Pulley by its Speed and Divide by Diam. of Driving Pulley.
Diam. of Driven Pulley Speed of Driven Pulley Speed of Driving Pulley	Diam. of Driving Pulley	Multiply Diam. of Driven Pulley by its Speed and Divide by Speed of Driving Pulley.

These rules apply equally well to a number of pulleys belts together or to a train of gears if *all* the driving and *all* the driven pulley diameters and speeds are grouped together.

STEEL AND OTHER METALS

HEAT TREATMENT OF STEEL

ΓHE theory of the heat treatment of steel rests upon the influence
the rate of cooling on certain molecular changes in structure occur-
g at different temperatures in the solid state. These changes are
two classes, critical and progressive; the former occur periodically
:ween certain narrow temperature limits, while the latter proceed
ιdually with the rise in temperature, each change producing alter-
ons in the physical characteristics. By controlling the rate of
ɔling, these changes can be given a permanent set, and the physical
ιaracteristics can thus be made different from those in the metal
its normal state.

The highest temperature that it is safe to submit a steel to for
at-treating is governed by the chemical composition of the steel.
ιre carbon steel should be raised to about 1650 degrees Fahr.,
ιile some of the high-grade alloy steels may safely be raised to 1750
grees Fahr., and the high-speed steels may be raised to just below
ɔ melting point. It is necessary to raise the metal to these points
that the active cooling process will have the desired effect of check-
ჳ the crystallization of the structure.

Methods of Heating

Furnaces using solid fuel such as coal, coke, charcoal, etc., are the
ɔst numerous and have been used the longest. These furnaces
nsist of a grate to place the fuel on, an arch to reflect the heat and
plate to put the pieces on. The plate should be so arranged that
e flames will not strike the pieces to be heated, and for that reason
me use cast-iron or clay retorts which are open on the side toward
e doors of the furnace.

Liquid fuel furnaces, which have open fires and which use liquid
.els, are not very numerous at present, but their use is increasing,
ving to the ease with which the fire is handled and the cleanliness
ἱ compared with a coal, coke or charcoal fire.

Crude oil and kerosene are the fuels generally used in these fur-
ιces, owing to their cheapness and the fact that they can be easily
ɔtained. These fuels are usually stored in a tank near the furnaces
ιd are pumped to them or flow by force of gravity.

Heating in Liquids

Furnaces using liquid for heating have a receptacle to hold the
quid, which is heated by coal, oil, gas or any other economical
ιeans; the liquid is kept at the highest temperature to which the
iece should be heated. The piece should be heated slowly in an
rdinary furnace to about 800 degrees, after which it should b

315

immersed in the liquid bath and kept there long enough to attain the temperature of the bath and then removed to be annealed or hardened.

The bath usually consists of lead, although antimony, cyanate of potassium, chloride of barium, a mixture of chloride of barium and chloride of potassium in the proportion of 3 to 2, mercury, common salt and metallic salts have been successfully used.

This method gives good results, as no portion of the piece to be treated can reach a temperature above that of the liquid bath; a pyrometer attachment will indicate exactly when the piece has arrived at that temperature, and its surface cannot be acted upon chemically. The bath can be maintained easily at the proper temperature and the entire process is under perfect control.

When lead is used it is liable to stick to the steel unless it is pure and retard the cooling of the spots where it adheres. Impurities, such as sulphur, are liable to be absorbed by the steel and thus affect its chemical composition. With high temperatures lead and cyanate of potassium throw off poisonous vapors which make them prohibitive, and even at comparatively low temperatures these vapors are detrimental to the health of the workmen in the hardening room. The metallic salts, however, do not give off these poisonous vapors, and are much better to use for this purpose, but many times the fumes are unbearable.

Gas as Fuel

Furnaces using gaseous fuel are very numerous and are so constructed that they can use either natural gas, artificial gas, or producer gas. They are very easy to regulate and if well built are capable of maintaining a constant temperature within a wide range.

In first cost this style of furnace is greater than that of the solid fuel furnaces, but where natural or producer gas is used the cost of operating is so much less that the saving soon pays for the cost of installation. Illuminating gas, however, is more expensive than the solid fuels and is only used where high-grade work demands the best results from heat treatment.

COOLING THE STEEL

COOLING apparatus is divided into two classes — baths for hardening and the different appliances for annealing.

The baths for quenching are composed of a large variety of materials. Some of the more commonly used are as follows, being arranged according to their intensity on 0.85 per cent. carbon steel: Mercury; water with sulphuric acid added; nitrate of potassium; sal ammoniac; common salt; carbonate of lime; carbonate of magnesia; pure water; water containing soap, sugar, dextrine or alcohol; sweet milk; various oils; beef suet; tallow; wax. These baths, however, do not act under all conditions with the same relative intensity, as their conductivity and viscosity vary greatly with the temperature.

With the exception of the oils and some of the greases, the quenching effect increases as the temperature of the bath lowers. Sperm and linseed oils, however, at all temperatures between 32 and 250 rees Fahr., act about the same as distilled water at 160 degrees

The baths for hardening which give the best results are those in which some means are provided for keeping the liquid at an even temperature. Where but few pieces are to be quenched, or a considerable time elapses between the quenching of pieces, the bath will retain an atmospheric temperature from its own natural radiation. Where a bath is in continuous use, for quenching a large number of pieces throughout the day, some means must be provided to keep the temperature of the bath at a low even temperature. The hot pieces from the heating furnace will raise the temperature of the bath many degrees, and the last piece quenched will not be nearly as hard as the first.

Annealing

The appliances for annealing are as numerous as the baths for quenching, and where a few years ago the ashes from the forge were all that were considered necessary for properly annealing a piece of steel, to-day many special preparations are being manufactured and sold for this purpose.

The more common materials used for annealing are powdered charcoal, charred bone, charred leather, slacked lime, sawdust, sand, fire clay, magnesia or refractory earth. The piece to be annealed is usually packed in a cast-iron box, using some of these materials or combinations of them for the packing, the whole is then heated in a furnace to the proper temperature and set aside, with the cover left on, to cool gradually to the atmospheric temperature.

For certain grades of steel these materials give good results; but for all kinds of steels and for all grades of annealing the slow-cooling furnace no doubt gives the best satisfaction, as the temperature can be easily raised to the right point, kept there as long as necessary, and then regulated to cool down as slowly as is desired. The gas, oil or electric furnaces are the easiest to handle and regulate.

The Hardening Bath

.In hardening steels the influence of the bath depends upon its temperature, its mass and its nature; or to express this in another way, upon its specific heat, its conductivity, its volatility and its viscosity. With other things equal, the lower the temperature of the bath, the quicker will the metal cool and the more pronounced will be the hardening effect. Thus water at 60 degrees will make steel harder than water at 150 degrees, and when the bath is in constant use the first piece quenched will be harder than the tenth or twentieth, owing to the rise in temperature of the bath. Therefore if uniform results are to be obtained in using a water bath, it must either be of a very large volume or kept cool by some mechanical means. In other words, the bath must be maintained at a constant temperature.

The mass of the bath can be made large so no great rise in temperature is made by the continuous cooling of pieces, or it can be made small and its rise in temperature used for hardening tools that are to remain fairly soft, as, if this temperature is properly regulated, the tool will not have to be re-heated and tempered later, and cracks and fissures are not as liable to occur.

Another way of arriving at the same results would be to use the double bath for quenching, that is, to have one bath of some product similar to salt which fuses at 575 degrees Fahr. Quench the piece in that until it has reached its temperature, after which it can be quenched in a cold bath or cooled in the air.

HIGH-SPEED STEELS

THESE steels are made by alloying tungsten and chromium or molybdenum and chromium with steel. These compositions completely revolutionize the points of transformation. Chromium, which has a tendency to raise the critical temperature, when added to a tungsten steel, in the proportions of 1 or 2 per cent., reduces the critical temperature to below that of the atmosphere. Tungsten and molybdenum prolong the critical range of temperatures of the steel on slow cooling so that it begins at about 1300 degrees Fahr. and spreads out all the way down to 600 degrees.

These steels are heated to 1850 degrees for the molybdenum and 2200 degrees for the tungsten, and cooled moderately fast, usually in an air blast, to give them the property known as "red-hardness." This treatment prevents the critical changes altogether and preserves the steel in what is known as the austenitic condition. The austenitic condition is one of hardness and toughness.

One rule which has given good results in heat-treating these high-speed steels is to heat slowly to 1500 degrees Fahr., then heat fast to 2200 degrees; after which cool rapidly in an air blast to 1550 degrees; then cool either rapidly or slowly to the temperature of the air.

CASE-HARDENING

CASE-HARDENING, carbonizing, or, as it is called in Europe, "cementation," is largely used so that the outer shell can be made hard enough to resist wear and the core of the piece can be left soft enough to withstand the shock strains to which it is subjected.

Several methods different from the old established one of packing the metal in a box filled with some carbonizing material, and then subjecting it to heat, have been devised in the last few years. Among them might be mentioned the Harveyizing process which is especially applicable to armor plate. The Harveyizing process uses a bed of charcoal over the work, the plates being pressed up against it in a pit or furnace and gas turned on so that the steel will be heated through the charcoal, thus allowing the carbon to soak in from the top.

Factors Governing Carbonizing

The result of the carbonizing operation is determined by five factors, which are as follows: First, the nature of the steel; second, the nature of the carbonizing material; third, the temperature of the carbonizing furnace; fourth, the time the piece is submitted to the carbonizing process; fifth, the heat treatment which follows carbonizing.

The nature of the steel has no influence on the speed of penetration of the carbon, but has an influence on the final result of the operation.

If steel is used that has a carbon content up to 0.56 per cent., the rate of penetration in carbonizing is constant; but the higher the carbon content is, in the core, the more brittle it becomes by prolonged annealing after carbonizing. Therefore it is necessary that the carbon content should be low in the core, and for this reason a preference is given to steels containing from 0.12 to 0.15 per cent. of carbon for carbonizing or case-hardening purposes.

TABLE I. — PENETRATION OF CARBON PER HOUR WITH
DIFFERENT ALLOYS

Component of Alloys	Speed of Penetration per Hr. in Inches
0.5 per cent. manganese	0.043
1.0 per cent. manganese	0.047
1.0 per cent. chromium	0.039
2.0 per cent. chromium	0.043
2.0 per cent. nickel	0.028
5.0 per cent. nickel	0.020
0.5 per cent. tungsten	0.035
1.0 per cent. tungsten	0.036
2.0 per cent. tungsten	0.047
0.5 per cent. silicon	0.024
1.0 per cent. silicon	0.020
2.0 per cent. silicon	0.016
5.0 per cent. silicon	0.000
1.0 per cent. titanium	0.032
2.0 per cent. titanium	0.028
1.0 per cent. molybdenum	0.036
2.0 per cent. molybdenum	0.043
1.0 per cent. aluminum	0.016
3.0 per cent. aluminum	0.008

The rate of penetration for ordinary carbonizing steel under the same conditions would have been 0.035 inch. Thus it will be seen that manganese, chromium, tungsten and molybdenum increase the rate of penetration. These seem to exist in the state of a double carbide and release a part of the cementite iron.

Nickel, silicon, titanium and aluminum retard the rate of penetration — 5 per cent. of silicon reducing it to zero — and these exist in the state of solution in the iron.

The Carbonizing Materials

The nature of the carbonizing materials has an influence on the speed of penetration and it is very essential that the materials be of a known chemical composition as this is the only way to obtain like results on the same steel at all times.

These materials or cements are manufactured in many special and patented preparations. The following materials are used and compounded in these preparations, but many of them give as good results when used alone as when compounded with others in varying percentages: Powdered bone; wood charcoal; charred sugar; charred leather; cyanide of potassium; ferro-cyanide of potassium; bichromate of potassium; animal black, acid cleaned. Prussiate of potash, anthracite, mixture of barium carbonate, graphite, petroleum gas, acetylene, horn, etc.

How Wood Charcoal Acts

Wood charcoal is very largely used in carbonizing steels, but the value of this material varies with the wood used, the method employed in making the charcoal, and other factors. Used alone it gives the normal rate of penetration for the first hour, but after that the rate gradually decreases until at eight hours it gives the lowest rate of penetration of any of the carbonizing materials. The best wood charcoal is that made from hickory.

Powdered charcoal and bone give good results as a carbonizing material and are successfully used in carbonizing nickel-chrome steel by packing in a cast-iron pot and keeping at a temperature of about 2000 degrees Fahr. for four hours, and then cooling slowly before taking out of the pot or uncovering.

TABLE 2

Temperature in Degrees Fahrenheit	Materials Used and Rate of Penetration in Inches			
	Charcoal 60 per cent. + 40 per cent. of Carbonate of Borium	Ferro-cyanide 66 per cent. + 34 per cent. of Bichromate	Ferro-cyanide Alone	Powdered Wood Charcoal Alone
1300	....	....	....	....
1475	0.020	0.033	0.020	0.020
1650	0.088	0.069	0.079	0.048
1825	0.128	0.128	0.128	0.098
2000	0.177	0.177	0.198	0.138

The speed of penetration caused by the action of different cements at different temperatures for the same time, i.e., eight hours, is best shown by Table 2.

The nature of the carbonizing material has a very pronounced effect on the rate of carbonization, or the percentage of the carbon content in the surface layer of the piece, or both.

Another Test of Penetration

At the same temperature, i.e., 1825 degrees Fahr., for different lengths of time and with different cements, the rate of penetration obtained was according to Table 3.

Eighty per cent. charcoal + 20 per cent. carbonate of barium 40 per cent. charcoal + 60 per cent. carbonate of barium, ferro-cyanide alone and 66 per cent. ferro-cyanide + 34 per cent.

bichromate were used with practically the same results for eight hours' time.

TABLE 3

Length of Time in Hours	Materials Used and Rate of Penetration in Inches				
	Carbon 60 per cent. + 40 per cent. of Carbonate	Ferro-cyanide 66 per cent. + 34 per cent. of Bichromate	Powdered Wood Charcoal Alone	Charcoal and Carbonate of Potassium	Unwashed Animal Black
1	0.031	0.033	0.028	0.059	0.035
2	0.039	0.037	0.053	0.078	0.059
4	0.047	0.049	0.063	0.094	0.088
6	0.078	0.074	0.072	0.011	0.106
8	0.118	0.128	0.098	0.138	0.128

Another set of tests was carried out for a longer period of time, with other materials and at a uniform temperature of 1650 degrees Fahr., with the results given in Table 4.

TABLE 4

Length of Time in Hours	Materials Used and Rate of Penetration in Inches		
	Charred Leather	Ground Wood Charcoal	Barium Carbonate and Wood Charcoal
2	0.045	0.028	0.055
4	0.062	0.042	0.087
8	0.080	0.062	0.111
12	0.110	0.070	0.125

Carbonizing with Gas

In the use of hydrocarbons, or gases, a fresh supply can be kept flowing into the carbonizing receptacle and the time greatly reduced for deep penetration with an appreciable reduction of time for the shallow penetrations.

Effect of Composition and Hardening

The constitution of a given steel is not the same in the hardened as in the normal state, owing to the carbon not being in the same state. In the annealed or normal steel it is in a free state, while in a hardened steel it is in a state of solution which we may call martensite; and this contains more or less carbon according to the original carbon content of the steel. The composition, and therefore th mechanical properties, depend principally upon the carbon content, the mechanical properties being changed slowly and gradually by an increase in carbon.

This is best shown by Table 5 in which it will be seen that the tensile strength and elastic limit gradually increased with the increase in the percentage of carbon, both in the annealed and hardened state,

TABLE 5. — EFFECT OF COMPOSITION AND HARDENING ON THE STRENGTH

	Case Harden- ing Steel	Very Low Carbon	Low Carbon	Medium Carbon	High Carbon	Very High Carbon
Carbon	0.10	0.14	0.23	0.52	0.60	0.71
Silicon	0.09	0.05	0.15	0.18	0.10	0.17
Manganese	0.19	0.33	0.45	0.35	0.40	0.38
Phosphorus	0.016	0.023	0.091	0.021	0.035	0.03
Sulphur	0.025	0.052	0.062	0.043	0.025	0.00

MECHANICAL PROPERTIES WHEN ANNEALED

Tensile Strength (in pounds per square inch)	60,300	61,500	66,500	97,800	116,400	130,7x
Elastic Limit (in pounds per square inch)	36,300	35,200	41,200	52,600	66,500	75,8x
Elongation (percent- age in 4 inches...	29	27	26	20	14	

MECHANICAL PROPERTIES WHEN HARDENED

Tensile Strength (in pounds per square inch)	66,400	73,100	99,400	132,100	153,400	180,10
Elastic Limit (in pounds per square inch)	40,300	39,600	54,000	81,400	102,100	105,30
Elongation (percent- age in 4 inches)..	24	22	14	9	4	

while the elongation gradually decreased. These tests were made with a bar ½ inch in diameter and 4 inches in length. It will also be seen that there was considerable change in the steels which were so low in carbon to be made so hard that they could not be filed. The reduction in elongation when the test bars were heated and quenched show that the metal was harder than when in the annealed state.

A hardening process that will produce a steel that is as homogeneous as possible is always sought for in practice. This is easily obtained in a high-carbon steel and especially if it contains 0.85 per cent. carbon, by passing the recalescent point before quenching. The desired homogeneity is not so easily obtained, however, in the low-carbon steels as they have several points of transformation. If these are quenched at a point a little above the lowest point of transformation the carbon will pass into solution, but the solution is not homogeneous. To obtain this result it is necessary that the quenching be done from a little above the highest point of transformation. This is higher in the low- than in the high-carbon steels. In practice this calls for a quenching of the low-carbon steels at about 1650 degrees Fahr., while a high-carbon steel should be quenched at about 1450 degrees.

FAHRENHEIT AND CENTIGRADE THERMOMETER SCALES

F	C	F	C	F	C	F	C	F	C
− 40	− 40.	70	21.1	185	85.	950	510.	2100	1149.
− 35	− 37.2	75	23.9	190	87.8	1000	537.8	2150	1176.5
− 30	− 34.4	80	26.7	195	90.6	1050	565.5	2200	1204.
− 25	− 31.7	85	29.4	200	93.3	1100	593.	2250	1232.
− 20	− 28.9	90	32.2	205	96.1	1150	621.	2300	1260.
− 15	− 26.1	95	35.	210	98.9	1200	648.5	2350	1287.5
− 10	− 23.3	100	37.8	212	100.	1250	676.5	2400	1315.5
− 5	− 20.6	105	40.6	215	101.7	1300	704.	2450	1343.
0	− 17.8	110	43.3	225	107.2	1350	732.	2500	1371.
+ 5	− 15.	115	46.1	250	121.2	1400	760.	2550	1399.
10	− 12.2	120	48.9	300	148.9	1450	788.	2600	1426.5
15	− 9.4	125	51.7	350	176.7	1500	816.	2650	1455.
20	− 6.7	130	54.4	400	204.4	1550	844.	2700	1483.
25	− 3.9	135	57.2	450	232.2	1600	872.	2750	1510.
30	− 1.1	140	60.	500	260.	1650	899.	2800	1537.5
32	0	145	62.8	550	287.8	1700	926.	2850	1565.
35	+ 1.7	150	65.6	600	315.6	1750	954.	2900	1593.
40	4.4	155	68.3	650	343.3	1800	982.	2950	1621.
45	7.2	160	71.1	700	371.1	1850	1010.	3000	1648.5
50	10.	165	73.9	750	398.9	1900	1038.	3050	1676.
55	12.8	170	76.7	800	426.7	1950	1065.5	3100	1705.
60	15.6	175	79.4	850	454.4	2000	1093.	3150	1732.
65	18.3	180	82.2	900	482.2	2050	1121.	3200	1760

To convert Fahrenheit into Centigrade: Subtract 32 from Fahrenheit, divide remainder by 9 and multiply by 5.
Example: 212 Fahr.

$$\frac{32}{180} \qquad 180 \div 9 = 20. \quad 20 \times 5 = 100.$$

Ans. 212 Fahr. = 100 Cent.
Centigrade to Fahrenheit: Divide by 5, multiply by 9 and add 32.
Example: 260 Cent. ÷ 5 = 52. 52 × 9 = 468 + 32 = 500 Fahr.
Ans. 260 Cent. = 500 Fahr

PROPERTIES OF METALS

Metal	Melting Point	Wt. per Cu. In.	Wt. per Cu. Ft.	Tensile Strength	Specific Gravity	Chemical Symbol
Aluminum	1157	.0924	159.63	20,000	2.56	Al.
Antimony	1130	.2424	418.86		6.71	Sb.
Bismuth	505	.354	611.76		9.83	Bi.
Brass, cast	1692	.3029	523.2	24,000	8.393	
Bronze	1692	.319	550.	36,000	8.83	
Chromium	3500	.2457	429.49		6.8	Cr.
Cobalt	2732	.307	530.6		8.5	Co.
Copper	1929	.322	556.	36,000	8.9	Cu.
Gold	1965	.6979	1206.05	20,000	19.32	Au.
Iridium	3992	.8099	1400.		22.42	Ir.
Iron, cast	2700	.26	450.	16,500	7.21	Fe.
Iron, wrought......	2920	.278	480.13	50,000	7.7	Fe.
Lead	618	.41	710.	3,000	11.37	Pb.
Manganese	3452	.289	499.4		8.	Mn.
Mercury	− 39	.4909	848.35		13.59	Hg.
Nickel	2700	.3179	549.34		8.8	Ni.
Platinum..........	3227	.7769	1342.13		21.5	Pt.
Silver	1733	.3805	657.33	40,000	10.53	Ag.
Steel — cast	2450	.28	481.2	50,000	7.81	
Steel — rolled	2600	.2833	489.6	65,000	7.854	
Tin	445	.2634	455.08	4,600	7.29	Sn.
Tungsten	3600	.69	1192.31		19.10	W.
Vanadium.........	3230	.1987	343.34		5.50	V.
Zinc	779	.245	430.	7,500	6.86	Zn.

ALLOYS FOR COINAGE

	Gold	Copper	Silver	Other Constituents	Remarks
Gold coin...	91.66	8.33	—	—	British standard.
" " ...	90.0	10.0	—	—	"Latin Union" and American.
" " ...	1.33	82.73	15.93	—	Roman, Septimus Severus, 265 A.D.
" " ...	40.35	19.63	40.02	—	Early British B.C. 5
Silver coin...	0.1	7.1	92.5	Lead 0.2	Roman, B.C. 31, most same as British silver coin.
Silver coin...	—	7.5	92.5	—	British standard.

Composition of Bronzes (Navy Department)

hite Metal: | PARTS
- Tin 7.6
- Copper 2.3
- Zinc 83.3
- Antimony 3.8
- Lead 3.0

ird Bronze for Piston Rings:
- Tin 22.0
- Copper 78.0

arings — Wearing Surfaces, etc.:
- Copper 6
- Tin 1
- Zinc .... ¼

aval Brass:
- Copper 62.0
- Tin 1.0
- Zinc 37.0

azing Metal:
- Copper 85.0
- Zinc 15.0

ntifriction Metal:
- Copper — (best refined) 3.7
- Banca tin 88.8
- Regulus of antimony 7.5
    - Well fluxed with borax and rosin in mixing.

aring Metal — (Pennsylvania Railroad):
- Copper 77.0
- Tin 8.0
- Lead 15.0

BEARING METAL

In the Journal of the Franklin Institute G. H. Clamer states that
parts antimony and 87 parts lead make an excellent bearing
etal, these being exactly the proportions which give a homogeneous
ucture. For heavier duty tin should be added.

Bismuth Alloys (Fusible Metals)

	Bismuth	Lead	Tin	Cadmium	Melting Point
					C°
ewton's alloys......	50.0	31.25	18.75	—	95
ose's "	50.0	28.10	24.64	—	100
arcet's "	50.0	25.00	25.00	—	93
ood's "	50.0	24.00	14.00	12.00	66–71
ipowitz's "	50.0	27.00	13.00	10.00	60

ALLOYS

	Copper	Tin	Lead	Zinc	Nickel	Antimony	
Babbitt........	8.	92.				4.	Very hard.
Bell Metal.....	76.5	23.5					"Big Ben"
	74.8	25.2					Westminster.
Brass	63-72			27–34			Typical brass.
" wire	70.29	0.17		29.26			
Britannia......	1.46	90.62				7.81	Birmingham sheet.
Bronze	95	4.		1.			British coinage.
	80–90	12–18		7.		·	Heavy bearings.
German silver..	60			20.	20		Nickel varies.
Gun metal.....	91	9.					Cannons.
Mannheim Gold	80–88			20–12			
Muntz metal...	60–62			38–40			Ship sheating.
Packfong......	43.8			40.6			Chinese alloy.
Shot metal.....			99.6			15.6	Trace of arsenic.
Speculum.....	70.24	29.11		trace		·	Telescope mirror.
Type metal	2.0	10.	70			18.	
		3.2	82			14.	Stereotyping.
White metal....	6.	82.				12.	For bearings.

SHRINKAGE OF CASTINGS

Aluminum — pure	.2031 inch per f
" Nickel Alloy	.1875 " " "
" Special "	.1718 " " "
Iron, Small Cylinders	.0625 " " "
" Pipes	.125 " " "
" Girders and Beams	.100 " " "
" Large Cylinders, Contraction of Diameter at Top	.625 " " "
" Large Cylinders, Contraction of Diameter at Bottom	.083 " " "
" Large Cylinders, Contraction of Length .	.094 " " "
Brass — Thin	.167 " " "
" Thick	.150 " " "
Copper	.1875 " " "
Bismuth	.1563 " " "
Lead	.3125 " "
Zinc	.3125 " "

ALUMINUM

CAN be melted in ordinary plumbago crucibles the same as brass
nd will not absorb silicon or carbon to injure it unless overheated.
Melts at 1157 degrees Fahr. or 625 Cent. Becomes granular and
usily broken at about 1000 Fahr.

hrinkage of pure aluminum . . . . .2031" per foot
ickel Aluminum Casting Alloy . . . . . . .1875" " "
pecial Casting Alloy . . . . .1718" " "

The most used alloys have a strength of about 20,000 pounds to
quare inch at a weight of one third that of brass.

Iron or sand molds can be used and should be poured as cool as
will run to avoid blowholes.

Burnishing. — Use a bloodstone or steel burnisher, with mixture
melted vaseline and kerosene oil or two tablespoonfuls of ground
orax, dissolved in a quart of hot water and a few drops of ammonia
lded.

Frosting. — Clean with benzine. Dip in strong solution of caustic
da or potash, then in solution of undiluted nitric acid. Wash
noroughly in water and dry in hot sawdust.

Polishing. — Any good metal polish that will not scratch will
ean aluminum. One that is recommended is made of

Stearic Acid — One Part ⎫
Fuller's earth — one part ⎬ Grind fine and mix very well.
Rotten Stone — Six parts ⎭

Castings are cleaned with a brass scratch brush, run at a high
eed. Sand blasting is also used both alone and before scratch
rushing.

Spinning. — A high speed, about 4000 feet per minute, is best for
)inning. This means that for work 5 to 8 inches in diameter,
300 to 2600 revolutions per minute is good, while for smaller work
f 4 inches this would go up to 3200 r.p.m.

Turning. — Use a tool with shearing edge similar to a wood-
atting tool as they clear themselves better. Use kerosene or water
s a lubricant, or if a bright cut is wanted use benzine. For drawing
n a press use vaseline.

GENERAL REFERENCE TABLES

COMMON WEIGHTS AND MEASURES

LINEAR OR MEASURE OF LENGTH

12 inches = 1 foot.   3 feet = 1 yard.
5½ yards = 1 rod.   40 rods = furlong.
8 furlongs = 1 mile.

EQUIVALENT MEASURES

Inches	Feet	Yards	Rods	Furlongs	Mile
36 =	3 =	1			
198 =	16.5 =	5.5 =	1		
7920 =	660 =	220 =	40 =	1	
63,360 =	5280 =	1760 =	320 =	8 =	1

SQUARE MEASURE

144 square inches = 1 sq. foot.   30¼ square yards = 1 sq. rod.
9 square feet = 1 sq. yd.   160 square rods = 1 acre.
640 acres = 1 sq. mile.

EQUIVALENT MEASURE

Sq. Mi.	A.	Sq. Rd.	Sq. Yd.	Sq. Ft.	Sq. In.
1 =	640 =	102,400 =	3,097,600 =	27,878,400 =	4,014,489,600

CUBIC MEASURE

1728 cubic inches = 1 cubic foot.   128 cubic feet = 1 cord.
27 cubic feet = 1 cubic yard.   24¾ cubic feet = 1 perch.
1 cu. yd. = 27 cu. ft. = 46,656 cu. in.

WEIGHT — AVOIRDUPOIS

437.5 grains = 1 ounce.   100 pounds = 1 hundred weight.
16 ounces = 1 pound.   2000 pounds = 1 ton.
2240 pounds = 1 long ton.
1 ton = 20 cwt. = 2000 lbs. = 32,000 oz. = 14,000,000 gr.

WEIGHT — TROY

24 grains = 1 pennyweight.   20 pwt. = 1 ounce.
12 ounces = 1 pound.
1 lb. = 12 oz. = 240 pwt. = 5760 gr.

328

DRY MEASURE

2 pints = 1 quart.   8 quarts = 1 peck.
4 pecks = 1 bushel
1 bu. = 4 pk. = 32 qt. = 64 pt.
U. S. bushel = 2150.42 cu. in.   British = 2218.19 cu. in.

LIQUID MEASURE

4 gills = 1 pint.   4 quarts = 1 gallon.
2 pints = 1 quart. 31½ gallons = 1 barrel.
2 barrels or 63 gals = 1 hogshead.
1 hhd. = 2 bbl. = 63 gals. = 252 qt. = 504 pt. = 2016 gi.

The U. S. gallon contains 231 cu. in. = .134 cu. ft.
One cubic foot = 7.481 gallons.
One cubic foot weighs 62.425 lbs. at 39.2 deg. Fahr.
One gallon weighs 8.345 lbs.
For rough calculations 1 cu. ft. is called 7½ gallons and 1 gallon
s 8½ lbs.

ANGLES OR ARCS

10 seconds = 1 minute.   90 degrees = 1 rt. angle or quadrant
10 minutes = 1 degree.   360 degrees = 1 circle.
1 circle   = 360° = 21,600′ = 1,296,000″.

1 minute of arc on the earth's surface is 1 nautical mile = 1.17
imes a land mile or 6080 feet.

WATER CONVERSION FACTORS

U. S. gallons	× 8.33	= pounds.
U. S. gallons	× 0.13368	= cubic feet.
U. S. gallons	× 231	= cubic inches.
U. S. gallons	× 0.83	= English gallons.
U. S. gallons	× 3.78	= liters.
English gallons (Imperial)	× 10	= pounds.
English gallons (Imperial)	× 0.16	= cubic feet.
English gallons (Imperial)	× 277.274	= cubic inches.
English gallons (Imperial)	× 1.2	= U. S. gallons.
English gallons (Imperial)	× 4.537	= liters.
Cubic inches of water (39.1°)	× 0.036024	= pounds.
Cubic inches of water (39.1°)	× 0.004329	= U. S. gallons.
Cubic inches of water (39.1°)	× 0.003607	= English gallons.
Cubic inches of water (39.1°)	× 0.576384	= ounces.
Cubic feet (of water) (39.1°)	× 62.425	= pounds.
Cubic feet (of water) (39.1°)	× 7.48	= U. S. gallons.
Cubic feet (of water) (39.1°)	× 6.232	= English gallons.
Cubic feet (of water) (39.1°)	× 0.028	= tons.
Pounds of water	× 27.72	= cubic inches.
Pounds of water	× 0.01602	= cubic feet.
Pounds of water	× 0.083	= U. S. gallons.
Pounds of water	× 0.10	= English gallons.

CONVENIENT MULTIPLIERS

Inches	× 0.08333	= feet.	Sq. inches	× 0.00695	= Sq. feet.
Inches	× 0.02778	= yards.	Sq. inches	× 0.0007716	= Sq. yards
Inches	× 0.00001578	= miles.	Cu. inches	× 0.00058	= Cu. feet.
			Cu. inches	× 0.0000214	= Cu. yards.

Feet	× 0.3334	= yards.	Sq. feet	× 144	= Sq. inches.
Feet	× 0.00019	= miles.	Sq. feet	× 0.1112	= Sq. yards.

Yards	× 36	= inches.	Cu. feet	× 1728	= Cu. inches.
Yards	× 3	= feet.	Cu. feet	× 0.03704	= Cu. yards.
Yards	× 0.0005681	= miles.	Sq. yards	× 1296	= Sq. inches.

Miles	× 63360	= inches.	Sq. yards	× 9	= Sq. feet.
Miles	× 5280	= feet.	Cu. yards	× 46656	= Cu. inches.
Miles	× 1760	= yards.	Cu. yards	× 27	= Cu. feet.

Avoir. oz.	× 0.0625	= pounds.	Avoir. lbs.	× 0.0005	= tons.
Avoir. oz.	× 0.00003125	= tons.	Avoir. tons	× 32000	= ounces.
Avoir. lbs.	× 16	= ounces.	Avoir. tons	× 2000	= pounds.

THE METRIC SYSTEM

The Metric System is based on the Meter which was designed to be one ten-millionth ($\frac{1}{10000000}$) part of the earth's meridian quadrant, through Dunkirk and Formentera. Later investigations, however, have shown that the Meter exceeds one ten-millionth part by almost one part in 6400. The value of the Meter, as authorized by the U. S. Government, is 39.37 inches. The Metric system was legalized by the U. S. Government in 1866.

The three principal units are the Meter, the unit of length, the liter, the unit of capacity, and the gram, the unit of weight. Multiples of these are obtained by prefixing the Greek words: deka (10), hekto (100), and kilo (1000). Divisions are obtained by prefixing the Latin words: deci ($\frac{1}{10}$), centi ($\frac{1}{100}$), and milli ($\frac{1}{1000}$). Abbreviations of the multiples begin with a capital letter, and of the divisions with a small letter, as in the following tables:

MEASURES OF LENGTH

10 millimeters (mm)	= 1 centimeter	cm.
10 centimeters	= 1 decimeter	dm.
10 decimeters	= 1 meter	m.
10 meters	= 1 dekameter	Dm.
10 dekameters	= 1 hektometer	Hm.
10 hektometers	= 1 kilometer	Km.

MEASURES OF SURFACE (NOT LAND)

100 square millimeters (mm^2)	= 1 square centimeter	cm^2.
100 square centimeters	= 1 square decimeter	dm^2.
100 square decimeters	= 1 square meter	m^2.

MEASURES OF VOLUME

1000 cubic millimeters (mm^3)	= 1 cubic centimeter	cm^3
1000 cubic centimeters	= 1 cubic decimeter	dm^3
1000 cubic decimeters	= 1 cubic meter	m^3

MEASURES OF CAPACITY

ɔ milliliters (ml)	= 1 centiliter	cl.
ɔ centiliters	= 1 deciliter	dl.
ɔ deciliters	= 1 liter	l.
ɔ liters	= 1 dekaliter	Dl.
ɔ dekaliters	= 1 hektoliter	Hl.
ɔ hektoliters	= 1 kiloliter	Kl.

NOTE. — The liter is equal to the volume occupied by 1 cubic decimeter.

MEASURES OF WEIGHT

ɔ milligrams (mg)	= 1 centigram	cg.
ɔ centigrams	= 1 decigram	dg.
ɔ decigrams	= 1 gram	g.
ɔ grams	= 1 dekagram	Dg.
ɔ dekagrams	= 1 hektogram	Hg.
ɔ hektograms	= 1 kilogram	Kg.
ɔɔo kilograms	= 1 ton	T.

NOTE. — The gram is the weight of one cubic centimeter of pure distilled water t a temperature of 39.2° F., the kilogram is the weight of 1 liter of water; the ton is ɪe weight of 1 cubic meter of water.

METRIC AND ENGLISH CONVERSION TABLE

MEASURES OF LENGTH

meter = $\begin{cases} 39.37 \text{ inches.} \\ 3.28083 \text{ feet.} \\ 1.0936 \text{ yds.} \end{cases}$

centimeter = .3937 inch.

millimeter = $\begin{cases} .03937 \text{ inch, or} \\ \dfrac{1}{25} \text{ inch nearly.} \end{cases}$

kilometer = 0.62137 mile.

1 foot = .3048 meter.

1 inch = $\begin{cases} 2.54 \text{ centimeters.} \\ 25.4 \text{ millimeters.} \end{cases}$

MEASURES OF SURFACE

square meter = $\begin{cases} 10.764 \text{ square feet.} \\ 1.196 \text{ square yds.} \end{cases}$

square centimeter = .155 sq. in.

square millimeter = .00155 sq. in.

1 square yard = .836 square meter.

1 square foot = .0929 square meter.

1 square in. = $\begin{cases} 6.452 \text{ sq. centimeters:} \\ 645.2 \text{ sq. millimeters.} \end{cases}$

MEASURES OF VOLUME AND CAPACITY

ɪ cubic meter = $\begin{cases} 35.314 \text{ cubic feet.} \\ 1.308 \text{ cubic yards.} \\ 264.2 \text{ gallons (231} \\ \text{cubic inch).} \end{cases}$

ɪ cubic decimeter = $\begin{cases} 61.023 \text{ cubic in.} \\ .0353 \text{ cubic ft.} \end{cases}$

ɪ cubic centimeter = .061 cubic inch.

ɪ liter = $\begin{cases} 1 \text{ cubic decimeter.} \\ 61.023 \text{ cubic inches.} \\ .0353 \text{ cubic foot.} \\ 1.0567 \text{ quarts (U. S.)} \\ .2642 \text{ gallons (U. S.)} \\ 2.202 \text{ lbs. of water at } 62° \text{ F.} \end{cases}$

1 cubic yard = .7645 cubic meter.

1 cubic ft. = $\begin{cases} .02832 \text{ cubic meter.} \\ 28.317 \text{ cubic decimeters.} \\ 28.317 \text{ liters.} \end{cases}$

1 cubic inch = 16.387 cubic centimeters.

1 gallon (British) = 4.543 liters.

1 gallon (U. S.) = 3.785 liters.

MEASURES OF WEIGHT

1 gram = 15.432 grains.

1 kilogram = 2.2046 pounds.

ɪ metric ton = $\begin{cases} .9842 \text{ ton of 2240 lbs.} \\ 19.68 \text{ cwts.} \\ 2204.6 \text{ lbs.} \end{cases}$

1 grain = .0648 grams.

1 ounce avoirdupois = 28.35 grams.

1 pound = .4536 kilograms.

1 ton of 2240 lbs. = $\begin{cases} 1.016 \text{ metric tons.} \\ 1016 \text{ kilograms.} \end{cases}$

Miscellaneous Conversion Factors

1 kilogram per meter = .6720 pounds per foot.
1 gram per square millimeter = 1.422 pounds per square inch.
1 kilogram per square meter = 0.2084 pounds per square foot.
1 kilogram per cubic meter = .0624 pounds per cubic foot.
1 degree centigrade = 1.8 degrees Fahrenheit.
1 pound per foot = 1.488 kilograms per meter.
1 pound per square foot = 4.882 kilograms per square meter.
1 pound per cubic foot = 16.02 kilograms per cubic meter.
1 degree Fahrenheit = .5556 degrees centigrade.
1 Calorie (French Thermal Unit) = 3.968 B. T. U. (British Thermal Unit).

1 Horse Power = $\begin{cases} \text{33,000 foot pounds per minute.} \\ \text{746 Watts.} \end{cases}$

1 Watt (Unit of Electrical Power) = $\begin{cases} \text{.00134 Horse Power.} \\ \text{44.24 foot pounds per minute} \end{cases}$

1 Kilowatt = $\begin{cases} \text{1000 Watts.} \\ \text{1.34 Horse Power.} \\ \text{44240 foot pounds per minute.} \end{cases}$

Decimal Equivalents of Fractions of Millimeters. (Advancing by $\frac{1}{100}$ MM.)

mm.	Inches	mm.	Inches	mm.	Inches	mm.	Inches
$\frac{1}{100}$	.00039	$\frac{26}{100}$	.01024	$\frac{51}{100}$	.02008	$\frac{76}{100}$	.02992
$\frac{2}{100}$	.00079	$\frac{27}{100}$	.01063	$\frac{52}{100}$	.02047	$\frac{77}{100}$	.03032
$\frac{3}{100}$	.00118	$\frac{28}{100}$	.01102	$\frac{53}{100}$	.02087	$\frac{78}{100}$	.03071
$\frac{4}{100}$	.00157	$\frac{29}{100}$	.01142	$\frac{54}{100}$	.02126	$\frac{79}{100}$	.03110
$\frac{5}{100}$	.00197	$\frac{30}{100}$	.01181	$\frac{55}{100}$	.02165	$\frac{80}{100}$	.03150
$\frac{6}{100}$	.00236	$\frac{31}{100}$	.01220	$\frac{56}{100}$	.02205	$\frac{81}{100}$	.03189
$\frac{7}{100}$	.00276	$\frac{32}{100}$	.01260	$\frac{57}{100}$	.02244	$\frac{82}{100}$	.03228
$\frac{8}{100}$	.00315	$\frac{33}{100}$	.01299	$\frac{58}{100}$	.02283	$\frac{83}{100}$	.03268
$\frac{9}{100}$	.00354	$\frac{34}{100}$	.01339	$\frac{59}{100}$	.02323	$\frac{84}{100}$	.03307
$\frac{10}{100}$	.00394	$\frac{35}{100}$	.01378	$\frac{60}{100}$	.02362	$\frac{85}{100}$	.03346
$\frac{11}{100}$	.00433	$\frac{36}{100}$	.01417	$\frac{61}{100}$	.02402	$\frac{86}{100}$	.03386
$\frac{12}{100}$	.00472	$\frac{37}{100}$	.01457	$\frac{62}{100}$	.02441	$\frac{87}{100}$	.03425
$\frac{13}{100}$	.00512	$\frac{38}{100}$	.01496	$\frac{63}{100}$	.02480	$\frac{88}{100}$	.03465
$\frac{14}{100}$	.00551	$\frac{39}{100}$	.01535	$\frac{64}{100}$	.02520	$\frac{89}{100}$	.03504
$\frac{15}{100}$	.00591	$\frac{40}{100}$	.01575	$\frac{65}{100}$	.02559	$\frac{90}{100}$	.03543
$\frac{16}{100}$	.00630	$\frac{41}{100}$	.01614	$\frac{66}{100}$	.02598	$\frac{91}{100}$	.03583
$\frac{17}{100}$	.00669	$\frac{42}{100}$	.01654	$\frac{67}{100}$	.02638	$\frac{92}{100}$	.03622
$\frac{18}{100}$	.00709	$\frac{43}{100}$	.01693	$\frac{68}{100}$	.02677	$\frac{93}{100}$	.03661
$\frac{19}{100}$	.00748	$\frac{44}{100}$	.01732	$\frac{69}{100}$	.02717	$\frac{94}{100}$	.03701
$\frac{20}{100}$	.00787	$\frac{45}{100}$	.01772	$\frac{70}{100}$	.02756	$\frac{95}{100}$	.03740
$\frac{21}{100}$	.00827	$\frac{46}{100}$	.01811	$\frac{71}{100}$	.02795	$\frac{96}{100}$	.03780
$\frac{22}{100}$	.00866	$\frac{47}{100}$	.01850	$\frac{72}{100}$	.02835	$\frac{97}{100}$	.03819
$\frac{23}{100}$	.00906	$\frac{48}{100}$	.01890	$\frac{73}{100}$	.02874	$\frac{98}{100}$	.03858
$\frac{24}{100}$	.00945	$\frac{49}{100}$	.01929	$\frac{74}{100}$	.02913	$\frac{99}{100}$	.03898
$\frac{25}{100}$	.00984	$\frac{50}{100}$	.01969	$\frac{75}{100}$	.02953	1	.03937

ecimal Equivalents of Millimeters and Fractions of Mil. limeters. (Advancing by $\frac{1}{50}$ mm. and 1 mm.)

mm.	Inches	mm.	Inches	mm.	Inches	mm.	Inches
$\frac{1}{50}$ = .00079	$\frac{40}{50}$ = .03150	31 = 1.22047	71 = 2.79527
$\frac{2}{50}$ = .00157	$\frac{41}{50}$ = .03228	32 = 1.25984	72 = 2.83464
$\frac{3}{50}$ = .00236	$\frac{42}{50}$ = .03307	33 = 1.29921	73 = 2.87401
$\frac{4}{50}$ = .00315	$\frac{43}{50}$ = .03386	34 = 1.33858	74 = 2.91338
	$\frac{44}{50}$ = .03465	35 = 1.37795	75 = 2.95275
$\frac{5}{50}$ = .00394	$\frac{45}{50}$ = .03543	36 = 1.41732	76 = 2.99212
$\frac{6}{50}$ = .00472	$\frac{46}{50}$ = .03622	37 = 1.45669	77 = 3.03149
$\frac{7}{50}$ = .00551	$\frac{47}{50}$ = .03701	38 = 1.49606	78 = 3.07086
$\frac{8}{50}$ = .00630	$\frac{48}{50}$ = .03780	39 = 1.53543	79 = 3.11023
$\frac{9}{50}$ = .00709	$\frac{49}{50}$ = .03858	40 = 1.57480	80 = 3.14960
$\frac{10}{50}$ = .00787	1 = .03937	41 = 1.61417	81 = 3.18897
$\frac{11}{50}$ = .00866	2 = .07874	42 = 1.65354	82 = 3.22834
$\frac{12}{50}$ = .00945	3 = .11811	43 = 1.69291	83 = 3.26771
$\frac{13}{50}$ = .01024	4 = .15748	44 = 1.73228	84 = 3.30708
$\frac{14}{50}$ = .01102	5 = .19685	45 = 1.77165	85 = 3.34645
$\frac{15}{50}$ = .01181	6 = .23622	46 = 1.81102	86 = 3.38582
$\frac{16}{50}$ = .01260	7 = .27559	47 = 1.85039	87 = 3.42519
$\frac{17}{50}$ = .01339	8 = .31496	48 = 1.88976	88 = 3.46456
$\frac{18}{50}$ = .01417	9 = .35433	49 = 1.92913	89 = 3.50393
$\frac{19}{50}$ = .01496	10 = .39370	50 = 1.96850	90 = 3.54330
$\frac{20}{50}$ = .01575	11 = .43307	51 = 2.00787	91 = 3.58267
$\frac{21}{50}$ = .01654	12 = .47244	52 = 2.04724	92 = 3.62204
$\frac{22}{50}$ = .01732	13 = .51181	53 = 2.08661	93 = 3.66141
$\frac{23}{50}$ = .01811	14 = .55118	54 = 2.12598	94 = 3.70078
$\frac{24}{50}$ = .01890	15 = .59055	55 = 2.16535	95 = 3.74015
$\frac{25}{50}$ = .01969	16 = .62992	56 = 2.20472	96 = 3.77952
$\frac{26}{50}$ = .02047	17 = .66929	57 = 2.24409	97 = 3.81889
$\frac{27}{50}$ = .02126	18 = .70866	58 = 2.28346	98 = 3.85826
$\frac{28}{50}$ = .02205	19 = .74803	59 = 2.32283	99 = 3.89763
$\frac{29}{50}$ = .02283	20 = .78740	60 = 2.36220	100 = 3.93700
$\frac{30}{50}$ = .02362	21 = .82677	61 = 2.40157	
$\frac{31}{50}$ = .02441	22 = .86614	62 = 2.44094	
$\frac{32}{50}$ = .02520	23 = .90551	63 = 2.48031	
$\frac{33}{50}$ = .02598	24 = .94488	64 = 2.51968	
$\frac{34}{50}$ = .02677	25 = .98425	65 = 2.55905	
$\frac{35}{50}$ = .02756	26 = 1.02362	66 = 2.59842	
$\frac{36}{50}$ = .02835	27 = 1.06299	67 = 2.63779	
$\frac{37}{50}$ = .02913	28 = 1.10236	68 = 2.67716	
$\frac{38}{50}$ = .02992	29 = 1.14173	69 = 2.71653	
$\frac{39}{50}$ = .03071	30 = 1.18110	70 = 2.75590	

1 mm. = 0.03937 in.
1 cm. = 0.3937 "
1 dm. = 3.937 "
1 m. = 39.37 "

EQUIVALENTS OF ENGLISH INCHES IN MILLIMETERS (39.37 inches = 1 meter)

Inch	0	1/16	1/8	3/16	1/4	5/16	3/8	7/16	1/2	9/16	5/8	11/16	3/4	13/16	7/8	15/16	Inch
0	0.0	1.6	3.2	4.8	6.4	7.9	9.5	11.1	12.7	14.3	15.9	17.5	19.1	20.6	22.2	23.8	0
1	25.4	27.0	28.6	30.2	31.7	33.3	34.9	36.5	38.1	39.7	41.3	42.9	44.4	46.0	47.6	49.2	1
2	50.8	52.4	54.0	55.6	57.1	58.7	60.3	61.9	63.5	65.1	66.7	68.3	69.8	71.4	73.0	74.6	2
3	76.2	77.8	79.4	81.0	82.5	84.1	85.7	87.3	88.9	90.5	92.1	93.7	95.2	96.8	98.4	100.0	3
4	101.6	103.2	104.8	106.4	108.0	109.5	111.1	112.7	114.3	115.9	117.5	119.1	120.7	122.2	123.8	125.4	4
5	127.0	128.6	130.2	131.8	133.4	134.9	136.5	138.1	139.7	141.3	142.9	144.5	146.1	147.6	149.2	150.8	5
6	152.4	154.0	155.6	157.2	158.8	160.3	161.9	163.5	165.1	166.7	168.3	169.9	171.5	173.0	174.6	176.2	6
7	177.8	179.4	181.0	182.6	184.2	185.7	187.3	188.9	190.5	192.1	193.7	195.3	196.9	198.4	200.0	201.6	7
8	203.2	204.8	206.4	208.0	209.6	211.1	212.7	214.3	215.9	217.5	219.1	220.7	222.3	223.8	225.4	227.0	8
9	228.6	230.2	231.8	233.4	235.0	236.5	238.1	239.7	241.3	242.9	244.5	246.1	247.7	249.2	250.8	252.4	9
10	254.0	255.6	257.2	258.8	260.4	261.9	263.5	265.1	266.7	268.3	269.9	271.5	273.1	274.6	276.2	277.8	10
11	279.4	281.0	282.6	284.2	285.7	287.3	288.9	290.5	292.1	293.7	295.3	296.9	298.4	300.0	301.6	303.2	11
12	304.8	306.4	308.0	309.6	311.1	312.7	314.3	315.9	317.5	319.1	320.7	322.3	323.8	325.4	327.0	328.6	12
13	330.2	331.8	333.4	335.0	336.5	338.1	339.7	341.3	342.9	344.5	346.1	347.7	349.2	350.8	352.4	354.0	13
14	355.6	357.2	358.8	360.4	361.9	363.5	365.1	366.7	368.3	369.9	371.5	373.1	374.6	376.2	377.8	379.4	14
15	381.0	382.6	384.2	385.8	387.3	388.9	390.5	392.1	393.7	395.3	396.9	398.5	400.0	401.6	403.2	404.8	15
16	406.4	408.0	409.6	411.2	412.7	414.3	415.9	417.5	419.1	420.7	422.3	423.9	425.4	427.0	428.6	430.2	16
17	431.8	433.4	435.0	436.6	438.1	439.7	441.3	442.9	444.5	446.1	447.7	449.3	450.8	452.4	454.0	455.6	17
18	457.2	458.8	460.4	462.0	463.5	465.1	466.7	468.3	469.9	471.5	473.1	474.7	476.2	477.8	479.4	481.0	18
19	482.6	484.2	485.8	487.4	488.9	490.5	492.1	493.7	495.3	496.9	498.5	500.1	501.6	503.2	504.8	506.4	19
20	508.0	509.6	511.2	512.8	514.3	515.9	517.5	519.1	520.7	522.3	523.9	525.5	527.0	528.6	530.2	531.8	20
21	533.4	535.0	536.6	538.2	539.7	541.3	542.9	544.5	546.1	547.7	549.3	550.9	552.4	554.0	555.6	557.2	21
22	558.8	560.4	562.0	563.6	565.1	566.7	568.3	569.9	571.5	573.1	574.7	576.3	577.8	579.4	581.0	582.6	22
23	584.2	585.8	587.4	589.0	590.5	592.1	593.7	595.3	596.9	598.5	600.1	601.7	603.2	604.8	606.4	608.0	23

DECIMAL EQUIVALENTS OF FRACTIONS OF AN INCH. (ADVANCING BY 8THS, 16THS, 32NDS AND 64THS.)

8ths	32nds	64ths	64ths
$\frac{1}{8} = .125$	$\frac{1}{32} = .03125$	$\frac{1}{64} = .015625$	$\frac{33}{64} = .515625$
$\frac{1}{4} = .250$	$\frac{3}{32} = .09375$	$\frac{3}{64} = .046875$	$\frac{35}{64} = .546875$
$\frac{3}{8} = .375$	$\frac{5}{32} = .15625$	$\frac{5}{64} = .078125$	$\frac{37}{64} = .578125$
$\frac{1}{2} = .500$	$\frac{7}{32} = .21875$	$\frac{7}{64} = .109375$	$\frac{39}{64} = .609375$
$\frac{5}{8} = .625$	$\frac{9}{32} = .28125$	$\frac{9}{64} = .140625$	$\frac{41}{64} = .640625$
$\frac{3}{4} = .750$	$\frac{11}{32} = .34375$	$\frac{11}{64} = .171875$	$\frac{43}{64} = .671875$
$\frac{7}{8} = .875$	$\frac{13}{32} = .40625$	$\frac{13}{64} = .203125$	$\frac{45}{64} = .703125$
16ths.	$\frac{15}{32} = .46875$	$\frac{15}{64} = .234375$	$\frac{47}{64} = .734375$
$\frac{1}{16} = .0625$	$\frac{17}{32} = .53125$	$\frac{17}{64} = .265625$	$\frac{49}{64} = .765625$
$\frac{3}{16} = .1875$	$\frac{19}{32} = .59375$	$\frac{19}{64} = .296875$	$\frac{51}{64} = .796875$
$\frac{5}{16} = .3125$	$\frac{21}{32} = .65625$	$\frac{21}{64} = .328125$	$\frac{53}{64} = .828125$
$\frac{7}{16} = .4375$	$\frac{23}{32} = .71875$	$\frac{23}{64} = .359375$	$\frac{55}{64} = .859375$
$\frac{9}{16} = .5625$	$\frac{25}{32} = .78125$	$\frac{25}{64} = .390625$	$\frac{57}{64} = .890625$
$\frac{11}{16} = .6875$	$\frac{27}{32} = .84375$	$\frac{27}{64} = .421875$	$\frac{59}{64} = .921875$
$\frac{13}{16} = .8125$	$\frac{29}{32} = .90625$	$\frac{29}{64} = .453125$	$\frac{61}{64} = .953125$
$\frac{15}{16} = .9375$	$\frac{31}{32} = .96875$	$\frac{31}{64} = .484375$	$\frac{63}{64} = .984375$

DECIMAL EQUIVALENTS OF FRACTIONS OF AN INCH. (ADVANCING BY 64THS.)

$\frac{1}{64} = .015625$	$\frac{17}{64} = .265625$	$\frac{33}{64} = .515625$	$\frac{49}{64} = .765625$
$\frac{1}{32} = .03125$	$\frac{9}{32} = .28125$	$\frac{17}{32} = .53125$	$\frac{25}{32} = .78125$
$\frac{3}{64} = .046875$	$\frac{19}{64} = .296875$	$\frac{35}{64} = .546875$	$\frac{51}{64} = .796875$
$\frac{1}{16} = .0625$	$\frac{5}{16} = .3125$	$\frac{9}{16} = .5625$	$\frac{13}{16} = .8125$
$\frac{5}{64} = .078125$	$\frac{21}{64} = .328125$	$\frac{37}{64} = .578125$	$\frac{53}{64} = .828125$
$\frac{3}{32} = .09375$	$\frac{11}{32} = .34375$	$\frac{19}{32} = .59375$	$\frac{27}{32} = .84375$
$\frac{7}{64} = .109375$	$\frac{23}{64} = .359375$	$\frac{39}{64} = .609375$	$\frac{55}{64} = .859375$
$\frac{1}{8} = .125$	$\frac{3}{8} = .375$	$\frac{5}{8} = .625$	$\frac{7}{8} = .875$
$\frac{9}{64} = .140625$	$\frac{25}{64} = .390625$	$\frac{41}{64} = .640625$	$\frac{57}{64} = .890625$
$\frac{5}{32} = .15625$	$\frac{13}{32} = .40625$	$\frac{21}{32} = .65625$	$\frac{29}{32} = .90625$
$\frac{11}{64} = .171875$	$\frac{27}{64} = .421875$	$\frac{43}{64} = .671875$	$\frac{59}{64} = .921875$
$\frac{3}{16} = .1875$	$\frac{7}{16} = .4375$	$\frac{11}{16} = .6875$	$\frac{15}{16} = .9375$
$\frac{13}{64} = .203125$	$\frac{29}{64} = .453125$	$\frac{45}{64} = .703125$	$\frac{61}{64} = .953125$
$\frac{7}{32} = .21875$	$\frac{15}{32} = .46875$	$\frac{23}{32} = .71875$	$\frac{31}{32} = .96875$
$\frac{15}{64} = .234375$	$\frac{31}{64} = .484375$	$\frac{47}{64} = .734375$	$\frac{63}{64} = .984375$
$\frac{1}{4} = .25$	$\frac{1}{2} = .50$	$\frac{3}{4} = .75$	

DECIMAL EQUIVALENTS OF FRACTIONS BELOW ½

Decimal Equivalents	Fractional Parts of an Inch										Decimal Equivalents
	6	7	8	12	14	16	24	28	32	64	
.015625										1	.015625
.03125									1	2	.03125
.035714								1			.035714
.041667							1				.041667
.046875										3	.046875
.0625						1			2	4	.0625
.071429					1			2			.071429
.078125										5	.078125
.083333				1			2				.083333
.09375									3	6	.09375
.107143								3			.107143
.109375										7	.109375
.125			1			2	3		4	8	.125
.140625										9	.140625
.142857		1			2			4			.142857
.15625									5	10	.15625
.166666	1			2			4				.166666
.171875										11	.171875
.178571								5			.178571
.1875						3			6	12	.1875
.203125										13	.203125
.208333							5				.208333
.214286					3			6			.214286
.21875									7	14	.21875
.234375										15	.234375
.25			2	3		4	6	7	8	16	.25
.265625										17	.265625
.28125									9	18	.28125
.285714		2			4			8			.285714
.291666							7				.291666
.296875										19	.296875
.3125						5			10	20	.3125
.321429								9			.321429
.328125										21	.328125
.333333	2			4			8				.333333
.34375									11	22	.34375
.357143					5			10			.357143
.359375										23	.359375
.375			3			6	9		12	24	.375
.390625										25	.390625
.392857								11			.392857
.40625									13	26	.40625
.41666				5			10				.41666
.421875										27	.421875
.428571		3			6			12			.428571
.4375						7			14	28	.4375
.453125										29	.453125
.458333							11				.458333
.464286								13			.464286
.46875									15	30	.46875
.484375										31	.484375
.5	3		4	6	7	8	12	14	16	32	.5

DECIMAL EQUIVALENTS OF FRACTIONS BETWEEN ½″ AND 1″

Decimal Equivalents	Fractional Parts of an Inch										Decimal Equivalents
	6	7	8	12	14	16	24	28	32	64	
.515625										33	.515625
.53125									17	34	.53125
.535714								15			.535714
.541666							13				.541666
.546875										35	.546875
.5625						9			18	36	.5625
.571429		4			8			16			.571429
.578125										37	.578125
.583333				7			14				.583333
.59375									19	38	.59375
.607143								17			.607143
.609375										39	.609375
.625			5			10	15		20	40	.625
.640625										41	.640625
.642867					9			18			.642867
.65625									21	42	.65625
.666666	4			8			16				.666666
.671875										43	.671875
.678571								19			.678571
.6875						11			22	44	.6875
.703125										45	.703125
.708333							17				.708333
.714286		5			10			20			.714286
.71875									23	46	.71875
.734375										47	.734375
.75			6	9		12	18	21	24	48	.75
.765625										49	.765625
.78125									25	50	.78125
.785714					11			22			.785714
.791666							19				.791666
.796875										51	.796875
.8125						13			26	52	.8125
.821429								23			.821429
.828125										53	.828125
.833333	5			10			20				.833333
.84375									27	54	.84375
.857143		6			12			24			.857143
.859375										55	.859375
.875			7			14	21		28	56	.875
.890625										57	.890625
.892857								25			.892857
.90625									29	58	.90625
.916666				11			22				.916666
.921875										59	.921875
.928571					13			26			.928571
.9375						15			30	60	.9375
.953125										61	.953125
.958333							23				.958333
.964286								27			.964286
.96875									31	62	.96875
.984375										63	.984375

EQUIVALENTS OF INCHES AND FRACTIONS OF INCHES IN DECIMALS
OF A FOOT

In.	o In.	1 In.	2 In.	3 In.	4 In.	5 In.
		.0833	.1667	.2500	.3333	.4167
1/32	.0026	.0859	.1693	.2526	.3359	.4193
1/16	.0052	.0885	.1719	.2552	.3385	.4219
3/32	.0078	.0911	.1745	.2578	.3411	.4245
1/8	.0104	.0938	.1771	.2604	.3438	.4271
5/32	.0130	.0964	.1797	.2630	.3464	.4297
3/16	.0156	.0990	.1823	.2656	.3490	.4323
7/32	.0182	.1016	.1849	.2682	.3516	.4349
1/4	.0208	.1042	.1875	.2708	.3542	.4375
9/32	.0234	.1068	.1901	.2734	.3568	.4401
5/16	.0260	.1094	.1927	.2760	.3594	.4427
11/32	.0286	.1120	.1953	.2786	.3620	.4453
3/8	.0313	.1146	.1979	.2813	.3646	.4470
13/32	.0339	.1172	.2005	.2839	.3672	.4505
7/16	.0365	.1198	.2031	.2865	.3698	.4531
15/32	.0391	.1224	.2057	.2891	.3724	.4557
1/2	.0417	.1253	.2083	.2917	.3750	.4583
17/32	.0443	.1276	.2091	.2943	.3776	.4609
9/16	.0469	.1302	.2135	.2969	.3802	.4635
19/32	.0495	.1328	.2161	.2995	.3828	.4661
5/8	.0521	.1354	.2188	.3021	.3854	.4688
21/32	.0547	.1380	.2214	.3047	.3880	.4714
11/16	.0573	.1406	.2240	.3073	.3906	.4740
23/32	.0599	.1432	.2266	.3099	.3932	.4766
3/4	.0625	.1458	.2292	.3125	.3958	.4792
25/32	.0651	.1484	.2318	.3151	.3984	.4818
13/16	.0677	.1510	.2344	.3177	.4010	.4844
27/32	.0703	.1536	.2370	.3203	.4036	.4870
7/8	.0729	.1563	.2396	.3229	.4063	.4896
29/32	.0755	.1589	.2422	.3255	.4089	.4922
15/16	.0781	.1615	.2448	.3281	.4115	.4948
31/32	.0807	.1641	.2474	.3307	.4141	.4974

QUIVALENTS OF INCHES AND FRACTIONS OF INCHES IN DECIMALS OF A FOOT

n.	6 In.	7 In.	8 In.	9 In.	10 In.	11 In.
	.5000	.5833	.6667	.7500	.8333	.9167
1/32	.5026	.5859	.6693	.7526	.8359	.9193
1/16	.5052	.5885	.6719	.7552	.8385	.9219
3/32	.5078	.5911	.6745	.7578	.8411	.9245
1/8	.5104	.5938	.6771	.7604	.8438	.9271
5/32	.5130	.5964	.6797	.7630	.8464	.9297
3/16	.5156	.5990	.6823	.7656	.8490	.9323
7/32	.5182	.6016	.6849	.7682	.8516	.9349
1/4	.5208	.6042	.6875	.7708	.8542	.9375
9/32	.5234	.6068	.6901	.7734	.8568	.9401
5/16	.5260	.6094	.6927	.7760	.8594	.9427
11/32	.5286	.6120	.6953	.7786	.8620	.9453
3/8	.5313	.6146	.6979	.7813	.8646	.9479
13/32	.5339	.6172	.7005	.7839	.8672	.9505
7/16	.5365	.6198	.7031	.7865	.8698	.9531
15/32	.5391	.6224	.7057	.7891	.8724	.9557
1/2	.5417	.6250	.7083	.7917	.8750	.9583
17/32	.5443	.6276	.7109	.7943	.8776	.9609
9/16	.5469	.6302	.7135	.7969	.8802	.9635
19/32	.5495	.6328	.7161	.7995	.8828	.9661
5/8	.5521	.6354	.7188	.8021	.8854	.9688
21/32	.5547	.6380	.7214	.8047	.8880	.9714
11/16	.5573	.6406	.7240	.8073	.8906	.9740
23/32	.5599	.6432	.7266	.8099	.8932	.9766
3/4	.5625	.6458	.7292	.8125	.8958	.9792
25/32	.5651	.6484	.7318	.8151	.8984	.9818
13/16	.5677	.6510	.7344	.8177	.9010	.9844
27/32	.5703	.6536	.7370	.8203	.9036	.9870
7/8	.5729	.6563	.7396	.8229	.9063	.9896
29/32	.5755	.6589	.7422	.8255	.9089	.9922
15/16	.5781	.6615	.7448	.8281	.9115	.9948
31/32	.5807	.6641	.7474	.8307	.9141	.9974

DECIMAL EQUIVALENTS, SQUARES, SQUARE ROOTS, CUBES AND CUBE ROOTS OF FRACTIONS; CIRCUMFERENCES AND AREAS OF CIRCLES FROM 1/64 TO 1 INCH

Fraction	Dec. Equiv.	Square	Sq. Root	Cube	Cube Root	Circum. Circle	Area Circle
1/64	.015625	.000244	.1250	.000003815	.2500	.04909	.000192
1/32	.03125	.0009765	.1768	.00003052	.3150	.09818	.000767
3/64	.046875	.002197	.2165	.000103	.3606	.1473	.001726
1/16	.0625	.003906	.2500	.0002442	.3968	.1963	.003068
5/64	.078125	.006104	.2795	.0004768	.4275	.2455	.004794
3/32	.09375	.008789	.3062	.0008240	.4543	.2945	.006903
7/64	.109375	.01196	.3307	.001308	.4782	.3436	.009396
1/8	.1250	.01563	.3535	.001953	.5000	.3927	.01228
9/64	.140625	.01978	.3750	.002781	.5200	.4438	.01553
5/32	.15625	.02441	.3953	.003815	.5386	.4909	.01916
11/64	.171875	.02954	.4161	.005078	.5560	.5400	.02321
3/16	.1875	.03516	.4330	.006592	.5724	.5890	.02761
13/64	.203125	.04126	.4507	.008381	.5878	.6381	.03241
7/32	.21875	.04786	.4677	.01047	.6025	.6872	.03758
15/64	.234375	.05493	.4841	.01287	.6166	.7363	.04314
1/4	.2500	.0625	.5000	.01562	.6300	.7854	.04909
17/64	.265625	.07056	.5154	.01874	.6428	.8345	.05541
9/32	.28125	.07910	.5303	.02225	.6552	.8836	.06213
19/64	.296875	.08813	.5449	.02616	.6671	.9327	.06922
5/16	.3125	.09766	.5590	.03052	.6786	.9817	.07670
21/64	.328125	.1077	.5728	.03533	.6897	1.031	.08456
11/32	.34375	.1182	.5863	.04062	.7005	1.080	.09281
23/64	.359375	.12913	.5995	.04641	.7110	1.129	.1014
3/8	.3750	.1406	.6124	.05273	.7211	1.178	.1104
25/64	.390625	.1526	.6250	.05960	.7310	1.227	.1226
13/32	.40625	.1650	.6374	.06705	.7406	1.276	.1296
27/64	.421875	.17800	.6495	.07508	.7500	1.325	.1398
7/16	.4375	.1914	.6614	.08374	.7592	1.374	.1503
29/64	.453125	.2053	.6732	.09304	.7681	1.424	.1613
15/32	.46875	.2197	.6847	.1030	.7768	1.473	.1726
31/64	.484375	.2346	.6960	.1136	.7853	1.522	.1843
1/2	.5000	.2500	.7071	.1250	.7937	1.571	.1963

DECIMAL EQUIVALENTS, SQUARES, SQUARE ROOTS, CUBES, CUBE ROOTS OF FRACTIONS; CIRCUMFERENCES AND AREAS OF CIRCLES FROM $\frac{1}{64}$ TO 1 INCH.

Fraction	Dec. Equiv.	Square	Sq. Root	Cube	Cube Root	Circum. Circle	Area Circle
33/64	.515625	.2659	.7181	.1371	.8019	1.620	.2088
17/32	.53125	.2822	.7289	.1499	.8099	1.669	.2217
35/64	.546875	.2991	.7395	.1636	.8178	1.718	.2349
9/16	.5625	.3164	.7500	.1780	.8255	1.767	.2485
37/64	.578125	.3342	.7603	.1932	.8331	1.816	.2625
19/32	.59375	.3525	.7706	.2093	.8405	1.865	.2769
39/64	.609375	.3713	.7806	.2263	.8478	1.914	.2916
5/8	.6250	.3906	.7906	.2441	.8550	1.963	.3068
41/64	.640625	.4104	.8004	.2629	.8621	2.013	.3223
21/32	.65625	.4307	.8101	.2826	.8690	2.062	.3382
43/64	.671875	.4514	.8197	.3033	.8758	2.111	.3545
11/16	.6875	.4727	.8292	.3250	.8826	2.160	.3712
45/64	.703125	.4944	.8385	.3476	.8892	2.209	.3883
23/32	.71875	.5166	.8478	.3713	.8958	2.258	.4057
47/64	.734375	.5393	.8569	.3961	.9022	2.307	.4236
3/4	.7500	.5625	.8660	.4219	.9086	2.356	.4418
49/64	.765625	.5862	.8750	.4488	.9148	2.405	.4604
25/32	.78125	.6104	.8839	.4768	.9210	2.454	.4794
51/64	.796875	.6350	.8927	.5060	.9271	2.503	.4987
13/16	.8125	.6602	.9014	.5364	.9331	2.553	.5185
53/64	.828125	.6858	.9100	.5679	.9391	2.602	.5386
27/32	.84375	.7119	.9186	.6007	.9449	2.651	.5592
55/64	.859375	.7385	.9270	.6347	.9507	2.700	.5801
7/8	.8750	.7656	.9354	.6699	.9565	2.749	.6013
57/64	.890625	.7932	.9437	.7064	.9621	2.798	.6230
29/32	.90625	.8213	.9520	.7443	.9677	2.847	.6450
59/64	.921875	.8499	.9601	.7835	.9732	2.896	.6675
15/16	.9375	.8789	.9682	.8240	.9787	2.945	.6903
61/64	.953125	.9084	.9763	.8659	.9841	2.994	.7135
31/32	.96875	.9385	.9843	.9091	.9895	3.043	.7371
63/64	.984375	.9690	.9922	.9539	.9948	3.093	.7610
1	1	1	1	1	1	3.1416	.7854

Areas and Circumferences of Circles from 1 to 100

Dia.	Area	Circum.	Dia.	Area	Circum.	Dia.	Area	Circum.
1/32	0.00077	0.098175	2	3.1416	6.28319	5	19.635	15.7080
3/64	0.00173	0.147262	2 1/16	3.3410	6.47953	5 1/16	20.129	15.9043
1/16	0.00307	0.196350	2 1/8	3.5466	6.67588	5 1/8	20.629	16.1007
3/32	0.00690	0.294524	2 3/16	3.7583	6.87223	5 3/16	21.135	16.2970
1/8	0.01227	0.392699	2 1/4	3.9761	7.06858	5 1/4	21.648	16.4934
5/32	0.01917	0.490874	2 5/16	4.2000	7.26493	5 5/16	22.166	16.6897
3/16	0.02761	0.589049	2 3/8	4.4301	7.46128	5 3/8	22.691	16.8861
7/32	0.03758	0.687223	2 7/16	4.6664	7.65763	5 7/16	23.221	17.0824
1/4	0.04909	0.785398	2 1/2	4.9087	7.85398	5 1/2	23.758	17.2788
9/32	0.06213	0.883573	2 9/16	5.1572	8.05033	5 9/16	24.301	17.4751
5/16	0.07670	0.981748	2 5/8	5.4119	8.24668	5 5/8	24.850	17.6715
11/32	0.09281	1.07992	2 11/16	5.6727	8.44303	5 11/16	25.406	17.8678
3/8	0.11045	1.17810	2 3/4	5.9396	8.63938	5 3/4	25.967	18.0642
13/32	0.12962	1.27627	2 13/16	6.2126	8.83573	5 13/16	26.535	18.2605
7/16	0.15033	1.37445	2 7/8	6.4918	9.03208	5 7/8	27.109	18.4569
15/32	0.17257	1.47262	2 15/16	6.7771	9.22843	5 15/16	27.688	18.6532
1/2	0.19635	1.57080	3	7.0686	9.42478	6	28.274	18.8496
17/32	0.22166	1.66897	3 1/16	7.3662	9.62113	6 1/8	29.465	19.2423
9/16	0.24850	1.76715	3 1/8	7.6699	9.81748	6 1/4	30.680	19.6350
19/32	0.27688	1.86532	3 3/16	7.9798	10.0138	6 3/8	31.919	20.0277
5/8	0.30680	1.96350	3 1/4	8.2958	10.2102	6 1/2	33.183	20.4204
21/32	0.33824	2.06167	3 5/16	8.6179	10.4065	6 5/8	34.472	20.8131
11/16	0.37122	2.15984	3 3/8	8.9462	10.6029	6 3/4	35.785	21.2058
23/32	0.40574	2.25802	3 7/16	9.2806	10.7992	6 7/8	37.122	21.5984
3/4	0.44179	2.35619	3 1/2	9.6211	10.9956	7	38.485	21.9911
25/32	0.47937	2.45437	3 9/16	9.9678	11.1919	7 1/8	39.871	22.3838
13/16	0.51849	2.55254	3 5/8	10.321	11.3883	7 1/4	41.282	22.7765
27/32	0.55914	2.65072	3 11/16	10.680	11.5846	7 3/8	42.718	23.1692
7/8	0.60132	2.74889	3 3/4	11.045	11.7810	7 1/2	44.179	23.5619
29/32	0.64504	2.84707	3 13/16	11.416	11.9773	7 5/8	45.664	23.9546
15/16	0.69029	2.94524	3 7/8	11.793	12.1737	7 3/4	47.173	24.3473
31/32	0.73708	3.04342	3 15/16	12.177	12.3700	7 7/8	48.707	24.7400
1	0.78540	3.14159	4	12.566	12.5664	8	50.265	25.1327
1 1/16	0.88664	3.33794	4 1/16	12.962	12.7627	8 1/8	51.849	25.5224
1 1/8	0.99402	3.53429	4 1/8	13.364	12.9591	8 1/4	53.456	25.9181
1 3/16	1.1075	3.73064	4 3/16	13.772	13.1554	8 3/8	55.088	26.3108
1 1/4	1.2272	3.92699	4 1/4	14.186	13.3518	8 1/2	56.745	26.7035
1 5/16	1.3530	4.12334	4 5/16	14.607	13.5481	8 5/8	58.426	27.0962
1 3/8	1.4849	4.31969	4 3/8	15.033	13.7445	8 3/4	60.132	27.4880
1 7/16	1.6230	4.51604	4 7/16	15.466	13.9408	8 7/8	61.862	27.8810
1 1/2	1.7671	4.71239	4 1/2	15.904	14.1372	9	63.617	28.2743
1 9/16	1.9175	4.90874	4 9/16	16.349	14.3335	9 1/8	65.397	28.6670
1 5/8	2.0739	5.10509	4 5/8	16.800	14.5299	9 1/4	67.201	29.0597
1 11/16	2.2365	5.30144	4 11/16	17.257	14.7262	9 3/8	69.029	29.4524
1 3/4	2.4053	5.49779	4 3/4	17.721	14.9226	9 1/2	70.882	29.8451
1 13/16	2.5802	5.69414	4 13/16	18.190	15.1189	9 5/8	72.760	30.2378
1 7/8	2.7612	5.89049	4 7/8	18.665	15.3153	9 3/4	74.662	30.6305
1 15/16	2.9483	6.08684	4 15/16	19.147	15.5116	9 7/8	76.589	31.0232

AREAS AND CIRCUMFERENCES OF CIRCLES FROM 1 TO 100

Dia.	Area	Circum.	Dia.	Area	Circum.	Dia.	Area	Circum.
10	78.540	31.4159	16	201.06	50.2655	22	380.13	69.1150
1/8	80.516	31.8086	1/8	204.22	50.6582	1/8	384.46	69.5077
1/4	82.516	32.2013	1/4	207.39	51.0509	1/4	388.82	69.9004
3/8	84.541	32.5940	3/8	210.60	51.4436	3/8	393.20	70.2931
1/2	86.590	32.9867	1/2	213.82	51.8363	1/2	397.61	70.6858
5/8	88.664	33.3794	5/8	217.08	52.2290	5/8	402.04	71.0785
3/4	90.763	33.7721	3/4	220.35	52.6217	3/4	406.49	71.4712
7/8	92.886	34.1648	7/8	223.65	53.0144	7/8	410.97	71.8639
11	95.033	34.5575	17	226.98	53.4071	23	415.48	72.2566
1/8	97.205	34.9502	1/8	230.33	53.7998	1/8	420.00	72.6493
1/4	99.402	35.3429	1/4	233.71	54.1925	1/4	424.56	73.0420
3/8	101.62	35.7356	3/8	237.10	54.5852	3/8	429.13	73.4347
1/2	103.87	36.1283	1/2	240.53	54.9779	1/2	433.74	73.8274
5/8	106.14	36.5210	5/8	243.98	55.3706	5/8	438.36	74.2201
3/4	108.43	36.9137	3/4	247.45	55.7633	3/4	443.01	74.6128
7/8	110.75	37.3064	7/8	250.95	56.1560	7/8	447.69	75.0055
12	113.10	37.6991	18	254.47	56.5487	24	452.39	75.3982
1/8	115.47	38.0918	1/8	258.02	56.9414	1/8	457.11	75.7909
1/4	117.86	38.4845	1/4	261.59	57.3341	1/4	461.86	76.1836
3/8	120.28	38.8772	3/8	265.18	57.7268	3/8	466.64	76.5783
1/2	122.72	39.2699	1/2	268.80	58.1195	1/2	471.44	76.9690
5/8	125.19	39.6626	5/8	272.45	58.5122	5/8	476.26	77.3617
3/4	127.68	40.0553	3/4	276.12	58.9049	3/4	481.11	77.7544
7/8	130.19	40.4480	7/8	279.81	59.2976	7/8	485.98	78.1471
13	132.73	40.8407	19	283.53	59.6903	25	490.87	78.5398
1/8	135.30	41.2334	1/8	287.27	60.0830	1/8	495.79	78.9325
1/4	137.89	41.6261	1/4	291.04	60.4757	1/4	500.74	79.3252
3/8	140.50	42.0188	3/8	294.83	60.8684	3/8	505.71	79.7179
1/2	143.14	42.4115	1/2	298.65	61.2611	1/2	510.71	80.1105
5/8	145.80	42.8042	5/8	302.49	61.6538	5/8	515.72	80.5033
3/4	148.49	43.1969	3/4	306.35	62.0465	3/4	520.77	80.8960
7/8	151.20	43.5896	7/8	310.24	62.4392	7/8	525.84	81.2887
14	153.94	43.9823	20	314.16	62.8319	26	530.93	81.6814
1/8	156.70	44.3750	1/8	318.10	63.2246	1/8	536.05	82.0741
1/4	159.48	44.7677	1/4	322.06	63.6173	1/4	541.19	82.4668
3/8	162.30	45.1604	3/8	326.05	64.0100	3/8	546.35	82.8595
1/2	165.13	45.5531	1/2	330.06	64.4026	1/2	551.55	83.2522
5/8	167.99	45.9458	5/8	334.10	64.7953	5/8	556.76	83.6449
3/4	170.87	46.3385	3/4	338.16	65.1880	3/4	562.00	84.0376
7/8	173.78	46.7312	7/8	342.25	65.5807	7/8	567.27	84.4303
15	176.71	47.1239	21	346.36	65.9734	27	572.56	84.8230
1/8	179.67	47.5166	1/8	350.50	66.3661	1/8	577.87	85.2157
1/4	182.65	47.9093	1/4	354.66	66.7588	1/4	583.21	85.6084
3/8	185.66	48.3020	3/8	358.84	67.1515	3/8	588.57	86.0011
1/2	188.69	48.6947	1/2	363.05	67.5442	1/2	593.96	86.3938
5/8	191.75	49.0874	5/8	367.28	67.9369	5/8	599.37	86.786
3/4	194.83	49.4801	3/4	371.54	68.3296	3/4	604.81	87.17
7/8	197.93	49.8728	7/8	375.83	68.7223	7/8	610.27	87.57

GENERAL REFERENCE TABLES

Areas and Circumferences of Circles from 1 to 100

Area	Circum.	Dia.	Area	Circum.	Dia.	Area	Circum.
615.75	87.9646	34	907.02	106.814	40	1256.6	125.664
621.26	88.3573	⅛	914.61	107.207	⅛	1264.5	126.056
626.80	88.7500	¼	921.32	107.600	¼	1272.4	126.449
632.36	89.1427	⅜	928.06	107.992	⅜	1280.3	126.842
637.94	89.5354	½	934.82	108.385	½	1288.2	127.235
643.55	89.9281	⅝	941.61	108.788	⅝	1296.2	127.627
649.18	90.3208	¾	948.42	109.170	¾	1304.2	128.020
656.84	90.7135	⅞	955.25	109.563	⅞	1312.2	128.413
660.52	91.1062	35	962.11	109.956	41	1320.3	128.805
666.23	91.4989	⅛	969.00	110.348	⅛	1328.3	129.198
671.96	91.8916	¼	975.91	110.741	¼	1336.4	129.591
677.71	92.2843	⅜	982.84	111.134	⅜	1344.5	129.993
683.49	92.6770	½	989.80	111.527	½	1352.7	130.376
689.30	93.0697	⅝	996.78	111.919	⅝	1360.8	130.769
695.13	93.4624	¾	1003.8	112.312	¾	1369.0	131.161
700.98	93.8551	⅞	1010.8	112.705	⅞	1377.2	131.554
706.86	94.2478	36	1017.9	113.097	42	1385.4	131.947
712.76	94.6405	⅛	1025.0	113.490	⅛	1393.7	132.340
718.69	95.0332	¼	1032.1	113.883	¼	1402.0	132.732
724.64	95.4259	⅜	1039.2	114.275	⅜	1410.3	133.125
730.62	95.8186	½	1046.3	114.668	½	1418.6	133.518
736.62	96.2113	⅝	1053.5	115.061	⅝	1427.0	133.910
742.64	96.6040	¾	1060.7	115.454	¾	1435.4	134.303
748.69	96.9967	⅞	1068.0	115.846	⅞	1443.8	134.696
754.77	97.3894	37	1075.2	116.239	43	1452.2	135.088
760.87	97.7821	⅛	1082.5	116.632	⅛	1460.7	135.481
766.99	98.1748	¼	1089.8	117.024	¼	1469.1	135.874
773.14	98.5675	⅜	1097.1	117.417	⅜	1477.6	136.267
779.31	98.9602	½	1104.5	117.810	½	1486.2	136.659
785.51	99.3529	⅝	1111.8	118.202	⅝	1494.7	137.052
791.73	99.7456	¾	1119.2	118.596	¾	1503.3	137.445
797.98	100.138	⅞	1126.7	118.988	⅞	1511.9	137.837
804.25	100.531	38	1134.1	119.381	44	1520.5	138.230
810.54	100.924	⅛	1141.6	119.773	⅛	1529.2	138.623
816.86	101.316	¼	1149.1	120.166	¼	1537.9	139.015
823.21	101.709	⅜	1156.6	120.559	⅜	1546.6	139.408
829.58	102.102	½	1164.2	120.951	½	1555.3	139.801
835.97	102.494	⅝	1171.7	121.344	⅝	1564.0	140.194
842.39	102.887	¾	1179.3	121.737	¾	1572.8	140.586
848.83	103.280	⅞	1186.9	122.129	⅞	1581.6	140.979
855.30	103.673	39	1194.6	122.522	45	1590.4	141.372
861.79	104.065	⅛	1202.3	122.915	⅛	1599.3	141.764
868.31	104.458	¼	1210.0	123.308	¼	1608.2	142.157
874.85	104.851	⅜	1217.7	123.700	⅜	1617.0	142.550
881.41	105.243	½	1125.4	124.093	½	1626.0	142.942
888.00	105.636	⅝	1233.2	124.486	⅝	1634.9	143.335
894.62	106.029	¾	1241.0	124.878	¾	1643.9	143.728
901.26	106.421	⅞	1248.8	125.271	⅞	1652.9	144.121

AREAS AND CIRCUMFERENCES OF CIRCLES FROM 1 TO 100

Dia.	Area	Circum.	Dia.	Area	Circum.	Dia.	Area	Circum.
46	1661.9	144.513	52	2123.7	163.363	58	2642.1	182.212
1/8	1670.9	144.906	1/8	2133.9	163.756	1/8	2653.5	182.605
1/4	1680.0	145.299	1/4	2144.2	164.148	1/4	2664.9	182.998
3/8	1689.1	145.691	3/8	2154.5	164.541	3/8	2676.4	183.390
1/2	1698.2	146.084	1/2	2164.8	164.934	1/2	2687.8	183.783
5/8	1707.4	146.477	5/8	2175.1	165.326	5/8	2699.3	184.176
3/4	1716.5	146.869	3/4	2185.4	165.719	3/4	2710.9	184.569
7/8	1725.7	147.262	7/8	2195.8	166.112	7/8	2722.4	184.961
47	1734.9	147.655	53	2206.2	166.504	59	2734.0	185.354
1/8	1744.2	148.048	1/8	2216.6	166.897	1/8	2745.6	185.747
1/4	1753.5	148.440	1/4	2227.0	167.290	1/4	2757.2	186.139
3/8	1762.7	148.833	3/8	2237.5	167.683	3/8	2768.8	186.532
1/2	1772.1	149.226	1/2	2248.0	168.075	1/2	2780.5	186.925
5/8	1781.4	149.618	5/8	2258.5	168.468	5/8	2792.2	187.317
3/4	1790.8	150.011	3/4	2269.1	168.861	3/4	2803.9	187.710
7/8	1800.1	150.404	7/8	2279.6	169.253	7/8	2815.7	188.103
48	1809.6	150.796	54	2290.2	169.646	60	2827.4	188.496
1/8	1819.0	151.189	1/8	2300.8	170.039	1/8	2839.2	188.888
1/4	1828.5	151.582	1/4	2311.5	170.431	1/4	2851.0	189.281
3/8	1837.9	151.975	3/8	2322.1	170.824	3/8	2862.9	189.674
1/2	1847.5	152.367	1/2	2332.8	171.217	1/2	2874.8	190.066
5/8	1857.0	152.760	5/8	2343.5	171.609	5/8	2886.6	190.459
3/4	1866.5	153.153	3/4	2354.3	172.002	3/4	2898.6	190.852
7/8	1876.1	153.544	7/8	2365.0	172.395	7/8	2910.5	191.244
49	1885.7	153.938	55	2375.8	172.788	61	2922.5	191.637
1/8	1895.4	154.331	1/8	2386.6	173.180	1/8	2934.5	192.030
1/4	1905.0	154.723	1/4	2397.5	173.573	1/4	2946.5	192.423
3/8	1914.7	155.116	3/8	2408.3	173.966	3/8	2958.5	192.815
1/2	1924.2	155.509	1/2	2419.2	174.358	1/2	2970.6	193.208
5/8	1934.2	155.904	5/8	2430.1	174.751	5/8	2982.7	193.601
3/4	1943.9	156.294	3/4	2441.1	175.144	3/4	2994.8	193.993
7/8	1953.7	156.687	7/8	2452.0	175.536	7/8	3006.9	194.386
50	1963.5	157.080	56	2463.0	175.929	62	3019.1	194.779
1/8	1973.3	157.472	1/8	2474.0	176.322	1/8	3031.3	195.171
1/4	1983.2	157.865	1/4	2485.0	176.715	1/4	3043.5	195.564
3/8	1993.1	158.258	3/8	2496.1	177.107	3/8	3055.7	195.957
1/2	2003.0	158.650	1/2	2507.2	177.500	1/2	3068.0	196.350
5/8	2012.9	159.043	5/8	2518.3	177.893	5/8	3080.3	196.742
3/4	2022.8	159.436	3/4	2529.4	178.285	3/4	3092.6	197.135
7/8	2032.8	159.829	7/8	2540.6	178.678	7/8	3104.9	197.528
51	2042.8	160.221	57	2551.8	179.071	63	3117.2	197.920
1/8	2052.8	160.614	1/8	2563.0	179.463	1/8	3129.6	198.313
1/4	2062.9	161.007	1/4	2574.2	179.856	1/4	3142.0	198.706
3/8	2073.0	161.399	3/8	2585.4	180.249	3/8	3154.5	199.098
1/2	2083.1	161.792	1/2	2596.7	180.642	1/2	3166.9	199.491
5/8	2093.2	162.185	5/8	2608.0	181.034	5/8	3179.4	199.884
3/4	2103.3	162.577	3/4	2619.4	181.427	3/4	3191.9	200.277
7/8	2113.5	162.970	7/8	2630.7	181.820	7/8	3204.4	200.669

Areas and Circumferences of Circles from 1 to 100

Dia.	Area	Circum.	Dia.	Area	Circum.	Dia.	Area	Circum.
64	3217.0	201.062	70	3848.5	219.911	76	4536.5	238.76
1/8	3229.6	201.455	1/8	3862.2	220.304	1/8	4551.4	239.15
1/4	3242.2	201.847	1/4	3876.0	220.697	1/4	4566.4	239.54
3/8	3254.8	202.240	3/8	3889.8	221.090	3/8	4581.3	239.9
1/2	3267.5	202.633	1/2	3903.6	221.482	1/2	4596.3	240.33
5/8	3280.1	203.025	5/8	3917.5	221.875	5/8	4611.4	240.7
3/4	3292.8	203.418	3/4	3931.4	222.268	3/4	4626.4	241.11
7/8	3305.6	203.811	7/8	3945.3	222.660	7/8	4641.5	241.51
65	3318.3	204.204	71	3959.2	223.053	77	4656.6	241.90
1/8	3331.1	204.596	1/8	3973.1	223.446	1/8	4671.8	242.20
1/4	3343.9	204.989	1/4	3987.1	223.838	1/4	4686.9	242.68
3/8	3356.7	205.382	3/8	4001.1	224.231	3/8	4702.1	243.08
1/2	3369.6	205.774	1/2	4015.2	224.624	1/2	4717.3	243.47
5/8	3382.4	206.167	5/8	4029.2	225.017	5/8	4732.5	243.86
3/4	3395.3	206.560	3/4	4043.3	225.409	3/4	4747.8	244.25
7/8	3408.2	206.952	7/8	4057.4	225.802	7/8	4763.1	244.65
66	3421.2	207.345	72	4071.5	226.195	78	4778.4	245.04
1/8	3434.3	207.738	1/8	4085.7	226.587	1/8	4793.7	245.4
1/4	3447.2	208.131	1/4	4099.8	226.930	1/4	4809.0	245.83
3/8	3460.2	208.523	3/8	4114.0	227.373	3/8	4824.4	246.22
1/2	3473.2	208.916	1/2	4128.2	227.765	1/2	4839.8	246.61
5/8	3486.3	209.309	5/8	4142.5	228.158	5/8	4855.2	247.00
3/4	3499.4	209.701	3/4	4156.8	228.551	3/4	4870.7	247.40
7/8	3512.5	210.094	7/8	4171.1	228.944	7/8	4886.2	247.79
67	3525.7	210.487	73	4185.4	229.336	79	4901.7	248.18
1/8	3538.8	210.879	1/8	4199.7	229.729	1/8	4917.2	248.57
1/4	3552.0	211.272	1/4	4214.1	230.122	1/4	4932.7	248.97
3/8	3565.2	211.665	3/8	4228.5	230.514	3/8	4948.3	249.36
1/2	3578.5	212.058	1/2	4242.9	230.907	1/2	4963.9	249.75
5/8	3591.7	212.450	5/8	4257.4	231.300	5/8	4979.5	250.14
3/4	3605.0	212.843	3/4	4271.8	231.692	3/4	4995.2	250.54
7/8	3618.3	213.236	7/8	4286.3	232.085	7/8	5010.9	250.93
68	3631.7	213.628	74	4300.8	232.478	80	5026.5	251.37
1/8	3645.0	214.021	1/8	4315.4	232.871	1/8	5042.3	251.72
1/4	3658.4	214.414	1/4	4329.9	233.263	1/4	5058.0	252.11
3/8	3671.8	214.806	3/8	4344.5	233.656	3/8	5073.8	252.50
1/2	3685.3	215.199	1/2	4359.2	234.049	1/2	5089.6	252.88
5/8	3698.7	215.592	5/8	4373.8	234.441	5/8	5105.4	253.29
3/4	3712.2	215.984	3/4	4388.5	234.334	3/4	5121.2	253.68
7/8	3725.7	216.337	7/8	4403.1	235.227	7/8	5137.1	254.07
69	3739.3	216.770	75	4417.9	235.619	81	5153.0	254.46
1/8	3752.8	217.163	1/8	4432.6	236.012	1/8	5168.9	254.80
1/4	3766.4	217.555	1/4	4447.4	236.405	1/4	5184.9	255.25
3/8	3780.0	217.948	3/8	4462.2	236.798	3/8	5200.8	255.64
1/2	3793.7	218.341	1/2	4477.0	237.190	1/2	5216.8	256.04
5/8	3807.3	218.733	5/8	4491.8	237.583	5/8	5232.8	256.43
3/4	3821.0	219.126	3/4	4506.7	237.976	3/4	5248.9	256.82
7/8	3834.7	219.519	7/8	4521.5	238.368	7/8	5264.9	257.21

AREAS AND CIRCUMFERENCES OF CIRCLES FROM 1 TO 100

Dia.	Area	Circum.	Dia.	Area	Circum.	Dia.	Area	Circum.
82	5281.0	257.611	88	6082.1	276.460	94	6939.8	295.310
	5297.1	258.003		6099.4	276.853		6958.2	295.702
	5313.3	258.396		6116.7	277.246		6976.7	296.095
	5329.4	258.789		6134.1	277.638		6995.3	296.488
	5345.6	259.181		6151.4	278.031		7013.8	296.881
	5361.8	259.574		6168.8	278.424		7032.4	297.273
	5378.1	259.967		6186.2	278.816		7051.0	297.666
	5394.3	260.359		6203.7	279.209		7069.6	298.059
83	5410.6	260.752	89	6221.1	279.602	95	7088.2	298.451
	5426.9	261.145		6238.6	279.994		7106.9	298.844
	5443.3	261.538		6256.1	280.387		7125.6	299.237
	5459.6	261.930		6273.7	280.780		7144.3	299.629
	5476.0	262.323		6291.2	281.173		7163.0	300.022
	5492.4	262.716		6308.8	281.565		7181.8	300.415
	5508.8	263.103		6326.4	281.958		7200.6	300.807
	5525.3	263.501		6344.1	282.351		7219.4	301.200
84	5541.8	263.894	90	6361.7	282.743	96	7238.2	301.593
	5558.3	264.286		6379.4	283.136		7257.1	301.986
	5574.8	264.679		6397.1	283.529		7276.0	302.378
	5591.4	265.072		6414.9	283.921		7294.9	302.771
	5607.9	265.465		6432.6	284.314		7313.8	303.164
	5624.5	265.857		6450.4	284.707		7332.8	303.556
	5641.2	266.250		6468.2	285.100		7351.8	303.949
	5657.8	266.643		6486.0	285.492		7370.8	304.342
85	5674.5	267.035	91	6503.9	285.885	97	7389.8	304.734
	5691.2	267.428		6521.8	286.278		7408.9	305.127
	5707.9	267.821		6539.7	286.670		7428.0	305.520
	5724.7	268.213		6557.6	287.063		7447.1	305.913
	5741.5	268.606		6575.5	287.456		7466.2	306.305
	5758.3	268.999		6593.5	287.848		7485.3	306.698
	5775.1	269.392		6611.5	288.241		7504.5	307.091
	5791.9	269.784		6629.6	288.634		7523.7	307.483
86	5808.8	270.177	92	6647.6	289.027	98	7543.0	307.876
	5825.7	270.570		6665.7	289.419		7562.2	308.269
	5842.6	270.962		6683.8	289.812		7581.5	308.661
	5859.6	271.355		6701.9	290.205		7600.8	309.064
	5876.5	271.748		6720.1	290.597		7620.1	309.447
	5893.5	272.140		6738.2	290.990		7639.5	309.840
	5910.6	272.533		6756.4	291.383		7658.9	310.232
	5927.6	272.926		6774.7	291.775		7678.3	310.625
87	5944.7	273.319	93	6792.9	292.168	99	7697.7	311.018
	5961.8	273.711		6811.2	292.561		7717.1	311.410
	5978.9	274.104		6829.5	292.954		7736.6	311.803
	5996.0	274.497		6847.8	293.346		7756.1	312.196
	6013.2	274.889		6866.1	293.739		7775.6	312.588
	6030.4	275.282		6884.5	294.132		7795.2	312.981
	6047.6	275.675		6902.9	294.524		7814.8	313.374
	6064.9	276.067		6921.3	294.917		7834.4	313.764

SQUARES, CUBES, SQUARE ROOTS, CUBE ROOTS, CIRCUMFERENCES AND CIRCULAR AREAS OF NOS. FROM 1 TO 520

No.	Square	Cube	Sq. Root	Cube Root	Circle	
					Circum.	Area
1	1	1	1.0000	1.0000	3.142	0.785
2	4	8	1.4142	1.2599	6.283	3.141
3	9	27	1.7321	1.4422	9.425	7.068
4	16	64	2.0000	1.5874	12.566	12.566
5	25	125	2.2361	1.7100	15.708	19.635
6	36	216	2.4495	1.8171	18.850	28.274
7	49	343	2.6458	1.9129	21.991	38.484
8	64	512	2.8284	2.0000	25.133	50.265
9	81	729	3.0000	2.0801	28.274	63.617
10	100	1000	3.1623	2.1544	31.416	78.539
11	121	1331	3.3166	2.2240	34.558	95.033
12	144	1728	3.4641	2.2894	37.699	113.097
13	169	2197	3.6056	2.3513	40.841	132.732
14	196	2744	3.7417	2.4101	43.982	153.938
15	225	3375	3.8730	2.4662	47.124	176.715
16	256	4096	4.0000	2.5198	50.265	201.062
17	289	4913	4.1231	2.5713	53.407	226.980
18	324	5832	4.2426	2.6207	56.549	254.460
19	361	6859	4.3589	2.6684	59.690	283.520
20	400	8000	4.4721	2.7144	62.832	314.150
21	441	9261	4.5826	2.7589	65.973	346.361
22	484	10648	4.6904	2.8020	69.115	380.133
23	529	12167	4.7958	2.8439	72.257	415.470
24	576	13824	4.8990	2.8845	75.398	452.385
25	625	15625	5.0000	2.9240	78.540	490.874
26	676	17576	5.0990	2.9625	81.681	530.929
27	729	19683	5.1962	3.0000	84.823	572.555
28	784	21952	5.2915	3.0366	87.965	615.752
29	841	24389	5.3852	3.0723	91.106	660.520
30	900	27000	5.4772	3.1072	94.248	706.858
31	961	29791	5.5678	3.1414	90.389	754.768
32	1024	32768	5.6569	3.1748	100.531	804.248
33	1089	35937	5.7446	3.2075	103.673	855.220
34	1156	39304	5.8310	3.2396	106.814	907.920
35	1225	42875	5.9161	3.2711	109.956	962.113
36	1206	46656	6.0000	3.3019	113.097	1017.88
37	1369	50653	6.0828	3.3322	116.239	1075.21
38	1444	54872	6.1644	3.3620	119.381	1134.11
9	1521	59319	6.2450	3.3912	122.522	1194.59
0	1600	64000	6.3246	3.4200	125.660	1256.04

UARES, CUBES, SQUARE ROOTS, CUBE ROOTS, CIRCUMFERENCES AND CIRCULAR AREAS OF NOS. FROM 1 TO 520

No.	Square	Cube	Sq. Root	Cube Root	Circle	
					Circum.	Area
41	1681	68921	6.4031	3.4482	128.81	1320.25
42	1764	74088	6.4807	3.4760	131.95	1385.44
43	1849	79507	6.5574	3.5034	135.09	1452.20
44	1936	85184	6.6332	3.5303	138.23	1520.53
45	2025	91125	6.7082	3.5569	141.37	1590.43
46	2116	97336	6.7823	3.5830	144.51	1661.90
47	2209	103823	6.8557	3.6088	147.65	1734.94
48	2304	110592	6.9282	3.6342	150.80	1809.56
49	2401	117649	7.0000	3.6593	153.94	1885.74
50	2500	125000	7.0711	3.6840	157.08	1963.50
51	2601	132651	7.1414	3.7084	160.22	2042.82
52	2704	140608	7.2111	3.7325	163.36	2123.72
53	2809	148877	7.2801	3.7563	166.50	2206.18
54	2916	157464	7.3485	3.7798	169.65	2290.22
55	3025	166375	7.4162	3.8030	172.79	2375.83
56	3136	175616	7.4833	3.8259	175.93	2463.01
57	3249	185193	7.5498	3.8485	179.07	2551.76
58	3364	195112	7.6158	3.8709	182.21	2642.08
59	3481	205379	7.6811	3.8930	185.35	2733.97
60	3600	216000	7.7460	3.9149	188.50	2827.43
61	3721	226981	7.8102	3.9365	191.64	2922.47
62	3844	238328	7.8740	3.9579	194.78	3019.07
63	3969	250047	7.9373	3.9791	197.92	3117.25
64	4096	262144	8.0000	4.0000	201.06	3216.99
65	4225	274625	8.0623	4.0207	204.20	3318.31
66	4356	287496	8.1240	4.0412	207.35	3421.19
67	4489	300763	8.1854	4.0615	210.49	3525.65
68	4624	314432	8.2462	4.0817	213.63	3631.68
69	4761	328509	8.3066	4.1016	216.77	3739.28
70	4900	343000	8.3666	4.1213	219.91	3848.45
71	5041	357911	8.4261	4.1408	223.05	3959.19
72	5184	373248	8.4853	4.1602	226.19	4071.50
73	5329	389017	8.5440	4.1793	229.34	4185.39
74	5476	405224	8.6023	4.1983	232.48	4300.84
75	5625	421875	8.6603	4.2172	235.62	4417.86
76	5776	438976	8.7178	4.2358	238.76	4536.46
77	5929	456533	8.7750	4.2543	241.90	4656.63
78	6084	474552	8.8318	4.2727	245.04	4778.36
79	6241	493039	8.8882	4.2908	248.19	4901.67
80	6400	512000	8.9443	4.3089	251.33	5026.55

SQUARES, CUBES, SQUARE ROOTS, CUBE ROOTS, CIRCUMFERENCE AND CIRCULAR AREAS OF NOS. FROM I TO 520

No.	Square	Cube	Sq. Root	Cube Root	CIRCLE Circum.	CIRCLE Area
81	6561	531441	9.0000	4.3267	254.47	5153.00
82	6724	551368	9.0554	4.3445	257.61	5281.0?
83	6889	571787	9.1104	4.3621	260.75	5410.0?
84	7056	592704	9.1652	4.3795	263.89	554?.??
85	7225	614125	9.2195	4.3968	267.04	5674.??
86	7396	636056	9.2736	4.4140	270.18	5808.??
87	7569	658503	9.3274	4.4310	273.32	5944.0?
88	7744	681472	9.3808	4.4480	276.46	6082.1?
89	7921	704969	9.4340	4.4647	279.60	6221.1?
90	8100	729000	9.4868	4.4814	282.74	6361.7?
91	8281	753571	9.5394	4.4979	285.88	6503.??
92	8464	778688	9.5917	4.5144	289.03	6647.0?
93	8649	804357	9.6437	4.5307	292.17	6792.??
94	8836	830584	9.6954	4.5468	295.31	6939.??
95	9025	857375	9.7468	4.5629	298.45	7088.??
96	9216	884736	9.7980	4.5789	301.59	7238.??
97	9409	912673	9.8489	4.5947	304.73	7389.81
98	9604	941192	9.8995	4.6104	307.88	7542.0?
99	9801	970299	9.9499	4.6261	311.02	7667.6?
100	10000	1000000	10.0000	4.6416	314.16	7853.??
101	10201	1030301	10.0499	4.6570	317.30	8011.??
102	10404	1061208	10.0995	4.6723	320.44	8171.??
103	10609	1092727	10.1489	4.6875	323.58	8332.??
104	10816	1124864	10.1980	4.7027	326.73	8494.8?
105	11025	1157625	10.2470	4.7177	329.87	8659.01
106	11236	1191016	10.2956	4.7326	333.01	8824.7?
107	11449	1225043	10.3441	4.7475	336.15	8992.0?
108	11664	1259712	10.3923	4.7622	339.29	9160.8?
109	11881	1295029	10.4403	4.7769	342.43	9331.??
110	12100	1331000	10.4881	4.7914	345.58	9503.??
111	12321	1367631	10.5357	4.8059	348.72	9676.8?
112	12544	1404928	10.5830	4.8203	351.86	9852.0?
113	12769	1442897	10.6301	4.8346	355.00	10028.?
114	12996	1481544	10.6771	4.8488	358.14	10207.0
115	13225	1520875	10.7238	4.8629	361.28	10386.0
116	13456	1560896	10.7703	4.8770	364.42	10568.?
117	13689	1601613	10.8167	4.8910	367.57	10751.?
118	13924	1643032	10.8628	4.9049	370.71	10935.?
119	14161	1685159	10.9087	4.9187	373.85	11122.0
20	14400	1728000	10.9545	4.9324	376.99	11300.?

No.	Square	Cube	Sq. Root	Cube Root	CIRCLE	
					Circum.	Area
21	14641	1771561	11.0000	4.9461	380.13	11499.0
22	14884	1815848	11.0454	4.9597	383.27	11689.9
23	15129	1860867	11.0905	4.9732	386.42	11882.3
24	15376	1906624	11.1355	4.9866	389.56	12076.3
25	15625	1953125	11.1803	5.0000	392.70	12271.8
26	15876	2000376	11.2250	5.0133	395.84	12469.0
27	16129	2048383	11.2694	5.0265	398.98	12667.7
28	16384	2097152	11.3137	5.0397	402.12	12868.0
29	16641	2146689	11.3578	5.0528	405.27	13069.8
30	16900	2197000	11.4018	5.0658	408.41	13273.2
31	17161	2248091	11.4455	5.0788	411.55	13478.2
32	17424	2299968	11.4891	5.0916	414.69	13684.8
33	17689	2352637	11.5326	5.1045	417.83	13892.9
34	17956	2406104	11.5758	5.1172	420.97	14102.6
35	18225	2460375	11.6190	5.1299	424.12	14313.9
36	18496	2515456	11.6619	5.1426	427.26	14526.7
37	18769	2571353	11.7047	5.1551	430.40	14741.1
38	19044	2628072	11.7473	5.1676	433.54	14957.1
39	19321	2685619	11.7898	5.1801	436.68	15174.7
40	19600	2744000	11.8322	5.1925	439.82	15393.8
41	19881	2803221	11.8743	5.2048	442.96	15614.5
42	20164	2863288	11.9164	5.2171	446.11	15836.8
43	20449	2924207	11.9583	5.2293	449.25	16060.6
44	20736	2985984	12.0000	5.2415	452.39	16286.0
45	21025	3048625	12.0416	5.2536	455.53	16513.0
46	21316	3112136	12.0830	5.2656	458.67	16741.5
47	21609	3176523	12.1244	5.2776	461.81	16971.7
48	21904	3241792	12.1655	5.2896	464.96	17203.4
49	22201	3307949	12.2066	5.3015	468.10	17436.6
50	22500	3375000	12.2474	5.3133	471.24	17671.5
51	22801	3442951	12.2882	5.3251	474.38	17907.9
52	23104	3511808	12.3288	5.3368	477.52	18145.8
53	23409	3581577	12.3693	5.3485	480.66	18385.4
54	23716	3652264	12.4097	5.3601	483.81	18626.5
55	24025	3723875	12.4499	5.3717	486.95	18869.2
56	24336	3796416	12.4900	5.3832	490.09	19113.4
57	24649	3869893	12.5300	5.3947	493.23	19359.3
58	24964	3944312	12.5698	5.4061	496.37	19606.7
59	25281	4019679	12.6095	5.4175	499.51	19855.7
60	25600	4096000	12.6401	5.4288	502.65	20106.2

SQUARES, CUBES, SQUARE ROOTS, CUBE ROOTS, CIRCUMFERENCE
AND CIRCULAR AREAS OF NOS. FROM 1 TO 520

No.	Square	Cube	Sq. Root	Cube Root	CIRCLE	
					Circum.	Area
161	25921	4173281	12.6886	5.4401	505.80	20358.3
162	26244	4251528	12.7279	5.4514	508.94	20611.2
163	26569	4330747	12.7671	5.4626	512.08	20867.2
164	26896	4410944	12.8062	5.4737	515.22	21124.1
165	27225	4492125	12.8452	5.4848	518.36	21382.5
166	27556	4574296	12.8841	5.4959	521.50	21642.4
167	27889	4657463	12.9228	5.5069	524.65	21904.2
168	28224	4741632	12.9615	5.5178	527.79	22167.2
169	28561	4826809	13.0000	5.5288	530.93	22431.8
170	28900	4913000	13.0384	5.5397	534.07	22698.2
171	29241	5000211	13.0767	5.5505	537.21	22965.3
172	29584	5088448	13.1149	5.5613	540.35	23235.2
173	29929	5177717	13.1529.	5.5721	543.50	23506.2
174	30276	5268024	13.1909	5.5828	546.64	23778.7
175	30625	5359375	13.2288	5.5934	549.78	24052.2
176	30976	5451776	13.2665	5.6041	552.92	24328.5
177	31329	5545233	13.3041	5.6147	556.06	24605.7
178	31684	5639752	13.3417	5.6252	559.20	24884.6
179	32041	5735339	13.3791	5.6357	562.35	25164.0
180	32400	5832000	13.4164	5.6462	565.49	25446.0
181	32761	5929741	13.4536	5.6567	568.63	25730.4
182	33124	6028568	13.4907	5.6671	571.77	26015.5
183	33489	6128487	13.5277	5.6774	574.91	26302.2
184	33856	6229504	13.5647	5.6877	578.05	26590.4
185	34225	6331625	13.6015	5.6980	581.19	26880.3
186	34596	6434856	13.6382	5.7083	584.34	27171.6
187	34969	6539203	13.6748	5.7185	587.48	27464.6
188	35344	6644672	13.7113	5.7287	590.62	27750.1
189	35721	6751269	13.7477	5.7388	593.76	28055.2
190	36100	6859000	15.7840	5.7489	596.90	28352.0
191	36481	6967871	13.8203	5.7590	600.04	28652.1
192	36864	7077888	13.8564	5.7690	603.19	28952.0
193	37249	7189057	13.8924	5.7790	606.33	29255.3
194	37636	7301384	13.9284	5.7890	609.47	29559.2
195	38025	7414875	13.9642	5.7989	612.61	29864.8
196	38416	7529536	14.0000	5.8088	615.75	30171.9
197	38809	7645373	14.0357	5.8186	618.89	30480.5
198	39204	7762392	14.0712	5.8285	622.04	30790.7
199	39601	7880599	14.1067	5.8383	625.18	31102.0
200	40000	8000000	14.1421	5.8480	628.32	31415.9

SQUARES, CUBES, SQUARE ROOTS, CUBE ROOTS, CIRCUMFERENCES
AND CIRCULAR AREAS OF NOS. FROM 1 TO 520

No.	Square	Cube	Sq. Root	Cube Root	CIRCLE	
					Circum.	Area
201	40401	8120601	14.1774	5.8578	631.46	31730.9
202	40804	8242408	14.2127	5.8675	634.60	32047.4
203	41209	8365427	14.2478	5.8771	637.74	32365.5
204	41616	8489664	14.2829	5.8868	640.89	32685.1
205	42025	8615125	14.3178	5.8964	644.03	33006.4
206	42436	8741816	14.3527	5.9059	647.17	33329.2
207	42849	8869743	14.3875	5.9155	650.31	33653.5
208	43264	8998912	14.4222	5.9250	653.45	33979.5
209	43681	9129329	14.4568	5.9345	656.59	34307.0
210	44100	9261000	14.4914	5.9439	659.73	34636.1
211	44521	9393931	14.5258	5.9533	662.88	34966.7
212	44944	9528128	14.5602	5.9627	666.02	35298.9
213	45369	9663597	14.5945	5.9721	669.16	35632.7
214	45796	9800344	14.6287	5.9814	672.30	35968.1
215	46225	9938375	14.6629	5.9907.	675.44	36305.0
216	46656	10077696	14.6969	6.0000	678.58	36643.5
217	47089	10218313	14.7309	6.0092	681.73	36983.6
218	47524	10360232	14.7648	6.0185	684.87	37325.3
219	47961	10503459	14.7986	6.0277	688.01	37668.5
220	48400	10648000	14.8324	·6.0368	691.15	38013.3
221	48841	10793861	14.8661	6.0459	694.29	38359.6
222	49284	10941048	14.8997	6.0550	697.43	38707.6
223	49729	11089567	14.9332	6.0641	700.58	39057.1
224	50176	11239424	14.9666	6.0732	703.72	39408.1
225	50625	11390625	15.0000	6.0822	706.86	39760.8
226	51076	11543176	15.0333	6.0912	710.00	40115.0
227	51529	11697083	15.0665	6.1002	713.14	40470.8
228	51984	11852352	15.0997	6.1091	716.28	40828.1
229	52441	12008989	15.1327	6.1180	719.42	41187.1
230	52900	12167000	15.1658	6.1269	722.57	41547.6
231	53361	12326391	15.1987	6.1358	725.71	41909.6
232	53824	12487168	15.2315	6.1446	728.85	42273.3
233	54289	12649337	15.2643	6.1534	731.99	42638.5
234	54756	12812904	15.2971	6.1622	735.13	43005.3
235	55225	12977875	15.3297	6.1710	738.27	43373.6
236	55696	13144256	15.3623	6.1797	741.42	43743.5
237	56169	13312053	15.3948	6.1885	744.56	44115.0
238	56644	13481272	15.4272	6.1972	747.70	44488.1
239	57121	13651919	15.4596	6.2058	750.84	44862.7
240	57600	13824000	15.4919	6.2145	753.98	45238.9

SQUARES, CUBES, SQUARE ROOTS, CUBE ROOTS, CIRCUMFERENCES
AND CIRCULAR AREAS OF NOS. FROM 1 TO 520

No.	Square	Cube	Sq. Root	Cube Root	CIRCLE	
					Circum.	Area
241	58081	13997521	15.5242	6.2231	757.12	45616.7
242	58564	14172488	15.5563	6.2317	760.27	45996.2
243	59049	14348907	15.5885	6.2403	763.41	46377.2
244	59536	14526784	15.6205	6.2488	766.55	46759.5
245	60025	14706125	15.6525	6.2573	769.69	47143.5
246	60516	14886936	15.6844	6.2658	772.83	-47529.2
247	61009	15069223	15.7162	6.2743	775.97	47916.4
248	61504	15252992	15.7480	6.2828	779.12	48305.1
249	62001	15438249	15.7797	6.2912	782.26	48695.5
250	62500	15625000	15,8114	6.2996	785.40	49087.4
251	63001	15813251	15.8430	6.3080	788.54	49480.0
252	63504	16003008	15.8745	6.3164	791.68	49875.0
253	64009	16194277	15.9060	6.3247	794.82	50272.6
254	64516	16387064	15.9374	6.3330	797.96	50670.7
255	65025	16581375	15.9687	6.3413	801.11	51070.5
256	65536	16777216	16.0000	6.3496	804.25	51471.5
257	66049	16974593	16.0312	6.3579	807.39	51874.8
258	66564	17173512	16.0624	6.3661	810.53	52279.2
259	67081	17373979	16.0935	6.3743	813.67	52685.3
260	67600	17576000	16.1245	6.3825	816.81	53092.0
261	68121	17779581	16.1555	6.3907	819.96	53502.1
262	68644	17984728	16.1864	6.3988	823.10	53912.0
263	69169	18191447	16.2173	6.4070	826.24	54325.2
264	69696	18399744	16.2481	6.4151	829.38	54739.1
265	70225	18609625	16.2788	6.4232	832.52	55154.6
266	70756	18821096	16.3095	6.4312	835.66	55571.6
267	71289	19034163	16.3401	6.4393	838.81	55990.3
268	71824	19248832	16.3707	6.4473	841.95	56410.4
269	72361	19465109	16.4012	6.4553	845.09	56832.2
270	72900	19683000	16.4317	6.4633	848.23	57255.5
271	73441	19902511	16.4621	6.4713	851.37	57680.4
272	73984	20123648	16.4924	6.4792	854.51	58106.0
273	74529	20346417	16.5227	6.4872	857.66	58534.0
274	75076	20570824	16.5529	6.4951	860.80	58964.0
275	75625	20796875	16.5831	6.5030	863.94	59395.7
276	76176	21024576	16.6132	6.5108	867.08	59828.5
277	76729	21253933	16.6433	6.5187	870.22	60262.8
278	77284	21484952	16.6733	6.5265	873.36	60698.7
279	77841	21717639	16.7033	6.5343	876.50	61136.2
280	78400	21952000	16.7332	6.5421	879.65	61575.2

QUARES, CUBES, SQUARE ROOTS, CUBE ROOTS, CIRCUMFERENCES
AND CIRCULAR AREAS OF NOS. FROM 1 TO 520

No.	Square	Cube	Sq. Root	Cube Root	Circle	
					Circum.	Area
281	78961	22188041	16.7631	6.5499	882.79	62015.8
282	79524	22425768	16.7929	6.5577	885.93	62458.0
283	80089	22665187	16.8226	6.5654	889.07	62901.8
284	80656	22906304	16.8523	6.5731	892.21	63347.1
285	81225	23149125	16.8819	6.5808	895.35	63794.0
286	81796	23393656	16.9115	6.5885	898.50	64242.4
287	82369	23639903	16.9411	6.5962	901.64	64692.5
288	82944	23887872	16.9706	6.6039	904.78	65144.1
289	83521	24137569	17.0000	6.6115	907.92	65597.2
290	84100	24389000	17.0294	6.6191	911.06	66052.0
291	84681	24642171	17.0587	6.6267	914.20	66508.3
292	85264	24897088	17.0880	6.6343	917.35	66966.2
293	85849	25153757	17.1172	6.6419	920.49	67425.6
294	86436	25412184	17.1464	6.6494	923.63	67886.7
295	87025	25672375	17.1756	6.6569	926.77	68349.3
296	87616	25934336	17.2047	6.6644	929.91	68813.5
297	88209	26198073	17.2337	6.6719	933.05	69279.2
298	88804	26463592	17.2627	6.6794	936.19	69746.5
299	89401	26730899	17.2916	6.6869	939.34	70215.4
300	90000	27000000	17.3205	6.6943	942.48	70685.8
301	90601	27270901	17.3494	6.7018	945.62	71157.9
302	91204	27543608	17.3781	6.7092	948.76	71631.5
303	91809	27818127	17.4069	6.7166	951.90	72106.6
304	92416	28094464	17.4356	6.7240	955.04	72583.4
305	93025	28372625	17.4642	6.7313	958.19	73061.7
306	93636	28652616	17.4929	6.7387	961.33	73541.5
307	94249	28934443	17.5214	6.7460	964.47	74023.0
308	94864	29218112	17.5499	6.7533	967.61	74506.0
309	95481	29503629	17.5784	6.7606	970.75	74990.6
310	96100	29791000	17.6068	6.7679	973.89	75476.8
311	96721	30080231	17.6352	6.7752	977.04	75964.5
312	97344	30371328	17.6635	6.7824	980.18	76453.8
313	97969	30664297	17.6918	6.7897	983.32	76944.7
314	98596	30959144	17.7200	6.7969	986.46	77437.1
315	99225	31255875	17.7482	6.8041	989.60	77931.1
316	99856	31554496	17.7764	6.8113	992.74	78426.7
317	100489	31855013	17.8045	6.8185	995.88	78923.9
318	101124	32157432	17.8326	6.8256	999.03	79422.6
319	101761	32461759	17.8606	6.8328	1002.20	79922.9
320	102400	32768000	17.8885	6.8399	1005.30	80424.8

SQUARES, CUBES, SQUARE ROOTS, CUBE ROOTS, CIRCUMFERENCES
AND CIRCULAR AREAS OF NOS. FROM 1 TO 520

No.	Square	Cube	Sq. Root	Cube Root	Circle Circum.	Circle Area
321	103041	33076161	17.9165	6.8470	1008.5	80928.2
322	103684	33386248	17.9444	6.8541	1011.6	81433.2
323	104329	33698267	17.9722	6.8612	1014.7	81939.8
324	104976	34012224	18.0000	6.8683	1017.9	82448.2
325	105625	34328125	18.0278	6.8753	1021.0	82957.7
326	106276	34645976	18.0555	6.8824	1024.2	83469.2
327	106929	34965783	18.0831	6.8894	1027.3	83981.8
328	107584	35287552	18.1108	6.8964	1030.4	84496.3
329	108241	35611289	18.1384	6.9034	1033.6	85012.3
330	108900	35937000	18.1659	6.9104	1036.7	85529.9
331	109561	36264691	18.1934	6.9174	1039.9	86049.2
332	110224	36594368	18.2209	6.9244	1043.0	86569.7
333	110889	36926037	18.2483	6.9313	1046.2	87092.2
334	111556	37259704	18.2757	6.9382	1049.3	87615.7
335	112225	37595375	18.3030	6.9451	1052.4	88141.3
336	112896	37933056	18.3303	6.9521	1055.6	88668.3
337	113569	38272753	18.3576	6.9589	1058.7	89196.9
338	114244	38614472	18.3848	6.9658	1061.9	89727.0
339	114921	38958219	18.4120	6.9727	1065.0	90258.7
340	115600	39304000	18.4391	6.9795	1068.1	90792.0
341	116281	39651821	18.4662	6.9864	1071.3	91326.9
342	116964	40001688	18.4932	6.9932	1074.4	91863.3
343	117649	40353607	18.5203	7.0000	1077.6	92401.3
344	118336	40707584	18.5472	7.0068	1080.7	92940.9
345	119025	41063625	18.5742	7.0136	1083.8	93482.0
346	119716	41421736	18.6011	7.0203	1087.0	94024.7
347	120409	41781923	18.6279	7.0271	1090.1	94568.2
348	121104	42144192	18.6548	7.0338	1093.3	95114.9
349	121801	42508549	18.6815	7.0406	1096.4	95662.3
350	122500	42875000	18.7083	7.0473	1099.6	96211.3
351	123201	43243551	18.7350	7.0540	1102.7	96761.8
352	123904	43614208	18.7617	7.0607	1105.8	97314.0
353	124609	43986977	18.7883	7.0674	1109.0	97867.7
354	125316	44361864	18.8149	7.0740	1112.1	98423.0
355	126025	44738875	18.8414	7.0807	1115.3	98979.3
356	126736	45118016	18.8680	7.0873	1118.4	99538.2
357	127449	45499293	18.8944	7.0940	1121.5	100098
358	128164	45882712	18.9209	7.1006	1124.7	100660
359	128881	46268279	18.9473	7.1072	1127.8	101223
360	129600	46656000	18.9737	7.1138	1131.0	101788

QUARES, CUBES, SQUARE ROOTS, CUBE ROOTS, CIRCUMFERENCES
AND CIRCULAR AREAS OF NOS. FROM 1 TO 520

No.	Square	Cube	Sq. Root	Cube Root	CIRCLE	
					Circum.	Area
361	130321	47045881	19.0000	7.1204	1134.1	102354
362	131044	47437928	19.0263	7.1269	1137.3	102922
363	131769	47832147	19.0526	7.1335	1140.4	103491
364	132496	48228544	19.0788	7.1400	1143.5	104062
365	133225	48627125	19.1050	7.1466	1146.7	104635
366	133956	49027896	19.1311	7.1531	1149.8	105209
367	134689	49430863	19.1572	7.1596	1153.0	105785
368	135424	49836032	19.1833	7.1661	1156.1	106362
369	136161	50243409	19.2094	7.1726	1159.2	106941
370	136900	50653000	19.2354	7.1791	1162.4	107521
371	137641	51064811	19.2614	7.1855	1165.5	108103
372	138384	51478848	19.2873	7.1920	1168.7	108687
373	139129	51895117	19.3132	7.1984	1171.8	109272
374	139876	52313624	19.3391	7.2048	1175.0	109858
375	140625	52734375	19.3649	7.2112	1178.1	110447
376	141376	53157376	19.3907	7.2177	1181.2	111036
377	142129	53582633	19.4165	7.2240	1184.4	111628
378	142884	54010152	19.4422	7.2304	1187.5	112221
379	143641	54439939	19.4679	7.2368	1190.7	112815
380	144400	54872000	19.4936	7.2432	1193.8	113411
381	145161	55306341	19.5192	7.2495	1196.9	114009
382	145924	55742968	19.5448	7.2558	1200.1	114608
383	146689	56181887	19.5704	7.2622	1203.2	115209
384	147456	56623104	19.5959	7.2685	1206.4	115812
385	148225	57066625	19.6214	7.2748	1209.5	116416
386	148996	57512456	19.6469	7.2811	1212.7	117021
387	149769	57960603	19.6723	7.2874	1215.8	117628
388	150544	58411072	19.6977	7.2936	1218.9	118237
389	151321	58863869	19.7231	7.2999	1222.1	118847
390	152100	59319000	19.7484	7.3061	1225.2	119459
391	152881	59776471	19.7737	7.3124	1228.4	120072
392	153664	60236288	19.7990	7.3186	1231.5	120687
393	154449	60698457	19.8242	7.3248	1234.6	121304
394	155236	61162984	19.8494	7.3310	1237.8	121922
395	156025	61629875	19.8746	7.3372	1240.9	122542
396	156816	62099136	19.8997	7.3434	1244.1	123163
397	157609	62570773	19.9249	7.3496	1247.2	123786
398	158404	63044792	19.9499	7.3558	1250.4	124410
399	159201	63521199	19.9750	7.3619	1253.5	125036
400	160000	64000000	20.0000	7.3684	1256.6	125664

SQUARES, CUBES, SQUARE ROOTS, CUBE ROOTS. CIRCUMFERENCES, AND CIRCULAR AREAS OF NOS. FROM 1 TO 520

No.	Square	Cube	Sq. Root	Cube Root	Circle	
					Circum.	Area
401	160801	64481201	20.0250	7.3742	1259.8	126293
402	161604	64964808	20.0499	7.3803	1262.9	126923
403	162409	65450827	20.0749	7.3864	1266.1	127556
404	163216	65939264	20.0998	7.3925	1269.2	128190
405	164025	66430125	20.1246	7.3986	1272.3	128825
406	164836	66923416	20.1494	7.4047	1275.5	129462
407	165649	67419143	20.1742	7.4108	1278.6	130100
408	166464	67917312	20.1990	7.4169	1281.8	130741
409	167281	68417929	20.2237	7.4229	1284.9	131382
410	168100	68921000	20.2485	7.4290	1288.1	132025
411	168921	69426531	20.2731	7.4350	1291.2	132670
412	169744	69934528	20.2978	7.4410	1294.3	133317
413	170569	70444997	20.3224	7.4470	1297.5	133965
414	171396	70957944	20.3470	7.4530	1300.6	134614
415	172225	71473375	20.3715	7.4590	1303.8	135265
416	173056	71991296	20.3961	7.4650	1306.9	135918
417	173889	72511713	20.4206	7.4710	1310.0	136572
418	174724	73034632	20.4450	7.4770	1313.2	137228
419	175561	73560059	20.4695	7.4829	1316.3	137885
420	176400	74088000	20.4939	7.4889	1319.5	138544
421	177241	74618461	20.5183	7.4948	1322.6	139205
422	178084	75151448	20.5426	7.5007	1325.8	139867
423	178929	75686967	20.5670	7.5067	1328.9	140531
424	179776	76225024	20.5913	7.5126	1332.0	141196
425	180625	76765625	20.6155	7.5185	1335.2	141863
426	181476	77308776	20.6398	7.5244	1338.3	142531
427	182329	77854483	20.6640	7.5302	1341.5	143201
428	183184	78402752	20.6882	7.5361	1344.6	143872
429	184041	78953589	20.7123	7.5420	1347.7	144545
430	184900	79507000	20.7364	7.5478	1350.9	145220
431	185761	80062991	20.7605	7.5537	1354.0	145896
432	186624	80621568	20.7846	7.5595	1357.2	146574
433	187489	81182737	20.8087	7.5654	1360.3	147254
434	188356	81746504	20.8327	7.5712	1363.5	147934
435	189225	82312875	20.8567	7.5770	1366.6	148617
436	190096	82881856	20.8806	7.5828	1369.7	149301
437	190969	83453453	20.9045	7.5886	1372.9	149987
438	191844	84027672	20.9284	7.5944	1376.0	150674
439	192721	84604519	20.9523	7.6001	1379.2	151363
440	193600	85184000	20.9762	7.6059	1382.3	152053

SQUARES, CUBES, SQUARE ROOTS, CUBE ROOTS, CIRCUMFERENCES
AND CIRCULAR AREAS OF NOS. FROM 1 TO 520

No.	Square	Cube	Sq. Root	Cube Root	Circle	
					Circum.	Area
441	194481	85766121	21.0000	7.6117	1385.4	152745
442	195364	86350888	21.0238	7.6174	1388.6	153439
443	196249	86938307	21.0476	7.6232	1391.7	154134
444	197136	87528384	21.0713	7.6289	1394.9	154830
445	198025	88121125	21.0950	7.6346	1398.0	155528
446	198916	88716536	21.1187	7.6403	1401.2	156228
447	199809	89314623	21.1424	7.6460	1404.3	156930
448	200704	89915392	21.1660	7.6517	1407.4	157633
449	201601	90518849	21.1896	7.6574	1410.6	158337
450	202500	91125000	21.2132	7.6631	1413.7	159043
451	203401	91733851	21.2368	7.6688	1416.9	159751
452	204304	92345408	21.2603	7.6744	1420.0	160460
453	205209	92959677	21.2838	7.6801	1423.1	161171
454	206116	93576664	21.3073	7.6857	1426.3	161883
455	207025	94196375	21.3307	7.6914	1429.4	162597
456	207936	94818816	21.3542	7.6970	1432.6	163313
457	208849	95443993	21.3776	7.7026	1435.7	164030
458	209764	96071912	21.4009	7.7082	1438.9	164748
459	210681	96702579	21.4243	7.7138	1442.0	165468
460	211600	97336000	21.4476	7.7194	1445.1	166190
461	212521	97972181	21.4709	7.7250	1448.3	166914
462	213444	98611128	21.4942	7.7306	1451.4	167639
463	214369	99252847	21.5174	7.7362	1454.6	168365
464	215296	99897344	21.5407	7.7418	1457.7	169093
465	216225	100544625	21.5639	7.7473	1460.8	169823
466	217156	101194696	21.5870	7.7529	1464.0	170554
467	218089	101847563	21.6102	7.7584	1467.1	171287
468	219024	102503232	21.6333	7.7639	1470.3	172021
469	219961	103161709	21.6564	7.7695	1473.4	172757
470	220900	103823000	21.6795	7.7750	1476.5	173494
471	221841	104487111	21.7025	7.7805	1479.7	174234
472	222784	105154048	21.7256	7.7860	1482.8	174974
473	223729	105823817	21.7486	7.7915	1486.0	175716
474	224676	106496424	21.7715	7.7970	1489.1	176460
475	225625	107171875	21.7945	7.8025	1492.3	177205
476	226576	107850176	21.8174	7.8079	1495.4	17795
477	227529	108531333	21.8403	7.8134	1498.5	1787c
478	228484	109215352	21.8632	7.8188	1501.7	1794!
479	229441	109902239	21.8861	7.8243	1504.8	1802c
480	230400	110592000	21.9089	7.8297	1508.0	1809!

SQUARES, CUBES, SQUARE ROOTS, CUBE ROOTS, CIRCUMFERENCES AND CIRCULAR AREAS OF NOS. FROM 1 TO 520

No.	Square	Cube	Sq. Root	Cube Root	Circle	
					Circum.	Area
481	231361	111284641	21.9317	7.8352	1511.1	181711
482	232324	111980168	21.9545	7.8406	1514.3	182467
483	233289	112678587	21.9773	7.8460	1517.4	183225
484	234256	113379904	22.0000	7.8514	1520.5	183984
485	235225	114084125	22.0227 -	7.8568	1523.7	184745
486	236196	114791256	22.0454	7.8622	1526.8	185508
487	237169	115501303	22.0681	7.8676	1530.0	186272
488	238144	116214272	22.0907	7.8730	1533.1	187038
489	239121	116930169	22.1133	7.8784	1536.2	187805
490	240100	117649000	22.1359	7.8837	1539.4	188574
491	241081	118370771	22.1585	7.8891	1542.5	189345
492	242064	119095488	22.1811	7.8944	1545.7	190117
493	243049	119823157	22.2036	7.8998	1548.8	190890
494	244036	120553784	22.2261	7.9051	1551.9	191665
495	245025	121287375	22.2486	7.9105	1555.1	192442
496	246016	122023936	22.2711	7.9158	1558.2	193221
497	247009	122763473	22.2935	7.9211	1561.4	194000
498	248004	123505992	22.3159	7.9264	1564.5	194782
499	249001	124251499	22.3383	7.9317	1567.7	195565
500	250000	125000000	22.3607	7.9370	1570.8	196350
501	251001	125751501	22.3830	7.9423	1573.9	197136
502	252004	126506008	22.4054	7.9476	1577.1	197923
503	253009	127263527	22.4277	7.9528	1580.2	198713
504	254016	128024064	22.4499	7.9581	1583.4	199504
505	255025	128787625	22.4722	7.9634	1586.5	200296
506	256036	129554216	22.4944	7.9686	1589.7	201090
507	257049	130323843	22.5167	7.9739	1592.8	201886
508	258064	131096512	22.5389	7.9791	1595.9	202683
509	259081	131872229	22.5610	7.9843	1599.1	203482
510	260100	132651000	22.5832	7.9896	1602.2	204282
511	261121	133432831	22.6053	7.9948	1605.4	205084
512	262144	134217728	22.6274	8.0000	1608.5	205887
513	263169	135005697	22.6495	8.0052	1611.6	206692
514	264196	135796744	22.6716	8.0104	1614.8	207499
515	265225	136590875	22.6936	8.0156	1617.9	208307
516	266256	137388096	22.7156	8.0208	1621.1	209117
517	267289	138188413	22.7376	8.0260	1624.2	209928
518	268324	138991832	22.7596	8.0311	1627.3	210741
519	269361	139798359	22.7816	8.0363	1630.5	211556
520	270400	140608000	22.8035	8.0415	1633.6	212372

Cir-um.	Diameter	Cir-cum.	Diameter	Cir-cum.	Diameter	Cir-cum.	Diameter
1	.3183	51	16.2338	101	32.1493	151	48.0648
2	.6366	52	16.5521	102	32.4676	152	48.3831
3	.9549	53	16.8704	103	32.7859	153	48.7014
4	1.2732	54	17.1887	104	33.1042	154	49.0197
5	1.5915	55	17.5070	105	33.4225	155	49.3380
6	1.9099	56	17.8254	106	33.7408	156	49.6563
7	2.2282	57	18.1437	107	34.0592	157	49.9747
8	2.5465	58	18.4620	108	34.3775	158	50.2930
9	2.8648	59	18.7803	109	34.6958	159	50.6113
10	3.1831	60	19.0986	110	35.0141	160	50.9296
11	3.5014	61	19.4169	111	35.3324	161	51.2479
12	3.8197	62	19.7352	112	35.6507	162	51.5662
13	4.1380	63	20.0535	113	35.9690	163	51.8845
14	4.4563	64	20.3718	114	36.2873	164	52.2028
15	4.7746	65	20.6901	115	36.6056	165	52.5211
16	5.0930	66	21.0085	116	36.9239	166	52.8394
17	5.4113	67	21.3268	117	37.2423	167	53.1578
18	5.7296	68	21.6451	118	37.5606	168	53.4761
19	6.0479	69	21.9634	119	37.8789	169	53.7944
20	6.3662	70	22.2817	120	38.1972	170	54.1127
21	6.6845	71	22.6000	121	38.5155	171	54.4310
22	7.0028	72	22.9183	122	38.8338	172	54.7493
23	7.3211	73	23.2366	123	39.1521	173	55.0676
24	7.6394	74	23.5549	124	39.4704	174	55.3859
25	7.9577	75	23.8732	125	39.7887	175	55.7042
26	8.2761	76	24.1916	126	40.1070	176	56.0225
27	8.5944	77	24.5099	127	40.4254	177	56.3408
28	8.9127	78	24.8282	128	40.7437	178	56.6592
29	9.2310	79	25.1465	129	41.0620	179	56.9775
30	9.5493	80	25.4648	130	41.3803	180	57.2958
31	9.8676	81	25.7831	131	41.6986	181	57.6141
32	10.1859	82	26.1014	132	42.0169	182	57.9324
33	10.5042	83	26.4197	133	42.3352	183	58.2507
34	10.8225	84	26.7380	134	42.6535	184	58.5690
35	11.1408	85	27.0563	135	42.9718	185	58.8873
36	11.4592	86	27.3747	136	43.2901	186	59.2056
37	11.7775	87	27.6930	137	43.6085	187	59.5239
38	12.0958	88	28.0113	138	43.9268	188	59.8423
39	12.4141	89	28.3296	139	44.2451	189	60.1606
40	12.7324	90	28.6479	140	44.5634	190	60.4789
41	13.0507	91	28.9662	141	44.8817	191	60.7972
42	13.3690	92	29.2845	142	45.2000	192	61.1155
43	13.6873	93	29.6028	143	45.5183	193	61.4338
44	14.0056	94	29.9211	144	45.8366	194	61.7521
45	14.3239	95	30.2394	145	46.1549	195	62.0704
46	14.6423	96	30.3577	146	46.4732	196	62.3887
47	14.9606	97	30.8761	147	46.7916	197	62.7070
48	15.2789	98	31.1944	148	47.1099	198	63.0254
49	15.5972	99	31.5127	149	47.4282	199	63.343?
50	15.9155	100	31.8310	150	47.7465	200	63.662.

SHOP TRIGONOMETRY

THE laying out of angles is sometimes difficult by ordinary methods and a little knowledge of shop "trig" is very useful and much easier than as though we called it by its full name.

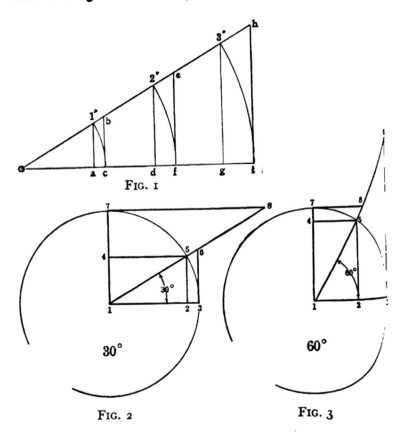

FIG. 1

FIG. 2   30°   FIG. 3   60°

It is really a system of constants or multipliers based on the fact that there are always fixed proportions between the sides and angles of triangles and other figures. Fig. 1 shows a 30-degree angle with 1, 2 and 3-inch arcs, 1 c, 2 f, and 3 i. It will be found that every similar measurement is in exact proportion to the radius. thus: is exactly twice the length of 1 a, and h i is just three times b c. So if we know the distance a c for a 1-inch radius for any angle, a similar distance as g i for the same angle, will be in exact proportion to the radius of the circle to one, which is the base. All these parts are named as shown in Figs. 2, 3, and 4.

45°

FIG. 4   FIG. 5

IS ALWAYS TAKEN EACH SIDE OF THE CENTER LI
SHOWN

ies 1-3 and 1-7 are called radius of the circle.
1-2 is called cosine of the angle.
4-5 is always the same as cosine of the angle.
the versed sine of the angle.
" co-versed sine of the angle.
sine   "   "   "
tangent   "   "   "
tangent   "   "   "
nt   "   "   "
ant   "   "   "

angle on one side of the
ed angle. In dealing with
center line and call the ang
verything is based on the 1
d as this base. Perhaps the
, the *tangent*, and the *secant*,
three of the figures. From t
the chord or the distance fro
tangent 3-6 is the distance

the horizontal radius to an extension of the radius at the angle given. The secant is the distance along the radius from the center to the tangent. From 2 to 3 is called the *versed sine*, and is the distance from the center of the chord to the outer circle.

The angle considered in this work is always less than 90 degrees, and the angle between the angle used and 90 degrees, or the angle which is necessary to complete this to 90 degrees is called the complementary angle. In the first case the complementary angle is 60 degrees, in the second case 45 degrees, and in the third case 75 degrees. The *co-sine* is the distance 4–5, the *co-tangent* is 7–8, the *co-secant* is 1–8, and the *co-versed sine* is 4–7 in all three examples. In the 45-degree angle it will be seen that the various parts are alike in both angles, as the sine is the same as the cosine, while the sine of the angle of 30 degrees is the same as the cosine of the angle of 60 degrees. These facts will be borne out by the tables and can be seen by studying the diagrams or by making any calculation and then proving it as near as may be on the drawing board.

All this is interesting, but unless it is useful it has no value to the practical man, so we will see where it can be used to advantage in saving time and labor.

Perhaps the easiest application is in finding the depth of a V thread without making any figures. The angle is 60 degrees or 30 degrees each side of the center line. The pitch is 1 inch so that each side is also an inch, and so the radius is an inch, the depth of the thread is the distance 1–2 or 4–5, and is the cosine of the angle. Looking in the table for the cosine of the angle of 30 degrees we find 0.86603, and as the radius is 1 this gives us the depth directly as 0.86603 inch. If the radius was 2 inches we would multiply by 2, or if it was ½ inch, divide by 2 and get the exact depth with almost no figuring. Suppose, on the other hand, that the thread was one inch deep and we want to find the length of one side, the angle remaining the same as before. In this case we have the depth which is the line 1–3, and we wish to find 1–6 which is the secant, so we look at the table again and find the secant of 30 degrees to be 1.1547 inches as the length of the side.

Suppose you have a square bar 2½ inches on each side, what is the distance across the corners? Looking at the second example we see that the side of the square bar is represented by line 1–3 and the corner distance by the secant 1–6 so we look for the secant of 45 degrees (because we know that half the 90 degree angle of a square bar must be 45 degrees) and find 1.4142 which would be the distance if the bar was one inch square, so we multiply 1.4142 by 2½ and get 3.5355 inches as the distance across the corners, and can know that this is closer than we can measure, and is not a guess by any means.

Reversing this we can find the side of a square that can be milled out of a round bar, such as the end of a reamer or tap. What square can we make on a 2-inch round reamer shank? The diameter of the bar is the radius as 1–5 and the angle 45 degrees as before, half the side of the square will be the sine 2–5, which the table shows to be 0.70711, and as this is half the chord which makes the flat across the bar, we multiply this by 2 and get 1.41422 inches as the distance across the flats for a reamer shank of this size.

Suppose we have a bar of $1\frac{1}{2} \times \frac{3}{4}$-inch steel and want to find the distance across the corners, and the angle it will make with the base. The $1\frac{1}{2}$-inch side is the radius, the diagonal is the secant, and the $\frac{3}{4}$-inch side is the tangent of the angle. Reducing these to a basis of one inch we have a bar 1 inch by $\frac{1}{2}$ inch and the $\frac{1}{2}$ inch is the tangent of the angle. Looking in the table we find this to be almost exactly the tangent of 26 degrees and 45 minutes. With this angle the secant or diagonal is 1.1198 for a radius of 1 inch and $1\frac{1}{2}$ times this gives 1.6797 as the distance across corners.

A very practical use for this kind of calculation is in spacing bolt holes or otherwise dividing a circle into any number of equal parts. It is easy enough to get the length of each arc of

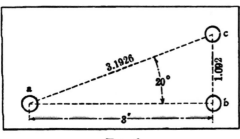

FIG. 6

the circumference by dividing 360 degrees by the number of divisions, but what we want is to find the chord or the distance from one point to the next in a straight line as a pair of dividers would step it off. First divide 360 by the number of divisions — say 9 — and get 40 degrees in each part. Fig. 5 shows this and we want the distance shown or the chord of the angle. This equals twice the sine of half the angle. Half the angle is 20 degrees and the sine for this is .342. Twice this or 0.684 is the chord of the 40-degree angle for every inch

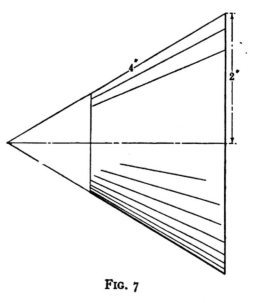

FIG. 7

of radius. If the circle is 14 inches in diameter the distance between the holes will be 7 times 0.684 or 4.788 inches. This is very quick and the most accurate method known.

Draftsmen often lay out jigs with the angles marked in degrees as in Fig. 6, overlooking the fact that the toolmaker has no convenient or accurate protractor for measuring the angle. Assume that a drawing shows three holes as a, b, and c, with b and c 20 degrees apart. The distance from a to b is 3 inches, what is the distance from b to c or from a to c?

As the known radius is from *a* to *b*, the distance *b c* is the tangent of the angle and the tangent for a one-inch radius is .36397, so for a 3-inch radius it is 3 × .36397 = 1.09191 inches from *b* to *c* and at right angles to it.

But we need not depend on the accuracy of the square or of the way we use it, as we can find the distance from *a* to *c* just as easily and just as accurately as we did *b c*. This distance is the secant and is 1.0642 for a one-inch radius. Multiplying this by 3 = 3.192 as the distance which can be accurately measured.

If the distance between *a* and *c* had been 3 inches, then *b c* would have been the sine and *a b* the cosine of the angle, both of which can be easily found from the tables.

It often happens that we want to find the angle of a roller or other piece of work as Fig. 7. Always work from the center line and continue the lines to complete the angle. Every triangle has the sides and they are called the side "opposite," "side adjacent," and "hypotenuse," the first being opposite the angle, the second the base line, and the third the slant line.

The following rules are very useful in this kind of work:

(1) Sine $= \dfrac{\text{Side Opp.}}{\text{Hypot.}}$ (6) Side Opp. = Hypot. × Sine.

(2) Cosine $= \dfrac{\text{Side Adj.}}{\text{Hypot.}}$ (7) Side Adj. = Hypot. × Cosine.

(3) Tangent $= \dfrac{\text{Side Opp.}}{\text{Side Adj.}}$ (8) Side Opp. = Side Adj. × Tangent.

(4) Co-Tangent $= \dfrac{\text{Side Adj.}}{\text{Side Opp.}}$ (9) Side Adj. = Co-Tan. × Side Opp.

(5) Hypot. $= \dfrac{\text{Side Opp.}}{\text{Sine.}}$ (10) Hypot. $= \dfrac{\text{Side Adj.}}{\text{Cosine.}}$

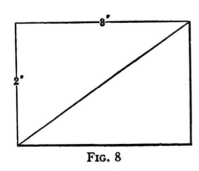

FIG. 8

If we have the dimensions shown in Fig. 7, the side opposite and the hypotenuse, we use formula No. 1, and dividing 2 by 4 we get ½ or .5 as the sine of the angle. The table shows this to be the sine of the angle of 30 degrees, consequently this is a 30-degree angle.

If we have the side opposite and the side adjacent we use formula No. 3, and find that ½ = ½ or .5 = the tangent of the angle. The table shows this to be the tangent of 26 degrees and 34 minutes.

Should it happen that we only knew the hypotenuse and the angle we use formula No. 6 and multiply 4 × .5 = 2, the side opposite. In the same way we can find the side adjacent by using formula No. 7. The cosine of 30 degrees in .866 and 4 × .866 = 3.404 inches as the side adjacent.

Having a bar of steel 2 by 3 inches, Fig. 8, what is the distance across the corners? Either formulas 3 or 4 will answer for this

Taking No. 4 we have 2 as the side opposite, 3 as the side adjacent. Dividing 3 by 2 gives 1.5. Looking under co-tangents for this we find 1.5003 after 33 degrees 41 minutes, which is nearly the correct angle. Then look for the secant of this and find 1.2017. Multiply this by 3 and get 3.6051 as the distance across the corners.

Complete tables of sines, tangents, secants, etc., will be on pages 371 to 405.

USING THE TABLE OF REGULAR POLYGONS

THE easiest way to lay out figures of this kind is to draw a circle and space it off, but it saves lots of time to know what spacing to use or how large a circle to draw to get a figure of the right size. Suppose we wish to lay out any regular figure, such as pentagon or five-sided figure, having sides 1½ inches long.

Number of Sides	Name of Figure	Diameter of Circle that will just enclose when side is 1	Diameter of circle that will just go inside when side is 1	Length of side where diameter of enclosure circle equals 1	Length of side where inside circle equals 1	Angle formed by lines drawn from center to corners	Angle formed by outer sides of figures	To find Area of Figure multiply side by itself and by number in this column
3	Triangle ..	1.1546	.5774	.866	1.732	120°	60°	.4330
4	Square	1.4142	1.	.7071	1.	90	90	1.
5	Pentagon..	1.7012	1.3764	.5878	.7265	72	108	1.7204
6	Hexagon ..	2.	1.732	.5	.5774	60	120	2.5980
7	Heptagon ..	2.3048	2.0766	.4338	.4815	51°-26'	128 ⁴⁄₇	3.6339
8	Octagon...	2.6132	2.4142	.3827	.4142	45	135	4.8284
9	Nonagon ..	2.9238	2.7474	.342	.3639	40	140	6.1818
10	Decagon ..	3.236	3.0776	.309	.3247	36	144	7.6942
11	Undecagon	3.5494	3.4056	.2817	.2936	32°-43	147 ⁸⁄₁₁	9.3656
12	Dodecagon	3.8638	3.732	.2588	.2679	30	150	11.1961

Table of Regular Polygons

Looking in the third column we find "Diameter of circle that will just enclose it," and opposite pentagon we find 1.7012 as the circle that will just enclose a pentagon having a side equal to 1. This may be 1 inch or 1 anything else, so as we are dealing in inches we call it inches. As the side of the pentagon is to be 1½ inches we multiply 1.7012 by 1½ and get 2.5518 as the diameter of circle to draw, and take half of this or the radius 1.2759 in the compass to draw the circle. Then with 1½ inches in the dividers we space round circle, and if the work has been carefully done it will just divide it into five equal parts. Connect these points by straight lines, and you have a pentagon with sides 1½ inches long.

If the pentagon is to go inside a circle of given diameter, say : inches, look under column 5 which gives "Length of side when diameter of enclosing circle equals 1," and find .5878. Multiply by 2 ½ this is for a 2-inch circle, and the side will be 2 × .5878 = 1.175c. Take this distance in the dividers and step around the 2-inch circle.

Assume that it is necessary to have a triangular end on a round shaft, how large must the shaft be to give a triangle 1.5 inches a a side?

Look in the table under column 3, and opposite triangle is 1.1546, meaning that where the side of a triangle is 1, the diameter of a circle that will just enclose it is 1.1546. As the side is 1.5, we have 1.5 × 1.1546 = 1.7318, the diameter of the shaft required. If the corners need not be sharp probably a shaft 1.625 would be ample.

Reversing this to find the size of a bearing that can be turned a a triangular bar of this size, look in column 4, which gives the largest circle that will go inside a triangle with a side equal to 1. This gives .5774. Multiply this by 1.5 = .8661.

A square taper reamer is to be used which must ream 1 inch at the small end and 1.5 at the back, what size must this be across the flats at both places?

Under column 5 find .7071 as the length of the side of a square when the diameter of the enclosing circle is 1, so this will be the side of the small end of the reamer and 1.5 × .7071 = 1.0606 is the side of the reamer at the large end.

FINDING THE RADIUS WITHOUT THE CENTER

It sometimes happens in measuring up a machine that we need to know the radius of curves when the center is not accessible. Three such cases are shown in Figs. 9, 10, and 11, the first two being a machine and the last a broken pulley. In Fig. 9 the rule is short enough to go in the curve while in Fig. 10 it has one end touching and the other across the sides. It makes no difference which is used so long as the distances are measured correctly, the short distance or *versed sine* being taken at the exact center of the chord and at right angles to it. It is easier figuring when the chord or the hight are even inches, so in measuring slip the rule until one or the other comes even; sometimes it is better to make the hight come 1 inch and let the chord go as it will, while at others the reverse may be true. The rule for finding the diameter is: Square half the chord, add to this the square of the hight, and divide the whole thing by the hight.

If the chord is 6 inches, as in Fig. 9, and the hight 1½ inches we have

$$\frac{\frac{1}{2} \text{ chord}^2 + \text{hight}^2}{\text{hight}} = \frac{3^2 + 1\frac{1}{2}^2}{1\frac{1}{2}} = \frac{9 + 2\frac{1}{4}}{1\frac{1}{2}} = \frac{11\frac{1}{4}}{1\frac{1}{2}} = 7\frac{1}{2} \text{ inches.}$$

Or as shown in Fig. 10 the chord is 10 inches and the hight 1 inch then the figures are

$$\frac{5^2 + 1^2}{1} = \frac{25 + 1}{1} = 26 \text{ inches.}$$

In Fig. 11 we have a piece of a broken pulley, and find the chord B to be 24 inches, and the hight A to be 2 inches. This becomes

$$\frac{12^2 + 2^2}{2} = \frac{144 + 4}{2} = \frac{148}{2} = 74,$$ so that the diameter of the pulley is 74 inches.

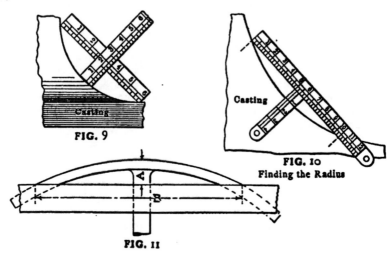

FIG. 9

FIG. 10
Finding the Radius

FIG. 11
Finding the Radius without Center

PROPERTIES OF REGULAR FIGURES

The Circle

A circle is a continuous curved line having every point at an equal distance from the center.

Its *perimeter* or circumference is always 3.14159265359 times the diameter, although 3.1416 is generally used and $3\frac{1}{7}$ is a very close approximation.

Area equals the diameter squared × .7854, or half the diameter squared × 3.1416, or half the diameter × half the circumference.

Diameter of a square having equal area = diameter of circle times .89 very nearly.

Triangle

Equilateral triangle is a regular figure having three equal sides and three equal angles of 60 degrees each.

The *side* equals .866 times the diameter of enclosing circle.

Distance from one side to opposite point equals the side times .866 or diameter of enclosing circle × .75 or inside circle × $1\frac{1}{2}$.

Diameter of enclosing circle equals the side times 1.1546 or $1\frac{1}{3}$ times distance from side to point or twice inside circle.

Diameter of inside circle equals side times .5774 or $\frac{1}{2}$ the enclosing circle.

The *area* equals one side multiplied by itself and by .433013.

Diameter of circle having equal area equals side of triangle times .73.

The Square

A square is a figure with four equal sides and four equal angles of 90 degrees.

Its *perimeter* or outside surface is four times the length of one side.

Area equals one side multiplied by the other which is the same as multiplying by itself or "squaring."

Diagonal or "long diameter," or "distance across corners," equals the side multiplied by 1.414.

Area of circle that will go inside the square equals one side multiplied by itself times .7854 or .7854 times the area of the square.

Area of circle that will just enclose the square equals diagonal multiplied by itself times .7854 or 1.27 times the area of the square.

Diameter of a circle having an equal area is 1.126 or practically 1⅛ times the side of the square.

The Hexagon

A hexagon is a regular figure with six equal sides and six equal angles of 120 degrees. It can be drawn inside a circle by spacing around with the radius of the circle.

The *side* equals half the diameter of enclosing circle.

The *long diameter* equals diameter of enclosing circle or twice the length of one side.

The *short diameter* equals the long diameter multiplied by .866 or 1.732 times one side.

The *area* equals one side multiplied by itself and by 2.5981.

The *area* of enclosing circle is one side multiplied by itself and by 3.1416.

The *area* of an inside circle is the short diameter multiplied by itself and by .7854.

Diameter of circle having equal area is practically .9 times long diameter.

The Octagon

An octagon is a regular figure with eight equal sides and eight equal angles of 135 degrees.

The *side* equals the long diameter multiplied by .382.

The *side* equals the short diameter multiplied by .415.

The *long diameter* equals diameter of enclosing circle or one side multiplied by 2.62.

The *short diameter* equals the long diameter multiplied by .93, or one side multiplied by 2.45.

The *area* equals one side multiplied by itself and by 4.8284.

The area of enclosing circle is 1.126 times area of octagon.

The area of inside circle is .972 times area of octagon.

The diameter of a circle having equal area is .953 times the long diameter of the octagon.

	0°		1°		2°		3°	
TAN.	CO-TAN.	TAN.	CO-TAN.	TAN.	CO-TAN.	TAN.	CO-TAN.	'
.00000	Infinite.	.01746	57.2900	.03492	28.6363	.05241	19.0811	60
.00029	3437.750	.01775	56.3506	.03521	28.3994	.05270	18.9755	59
.00058	1718.870	.01804	55.4415	.03550	28.1664	.05299	18.8711	58
.00087	1145.920	.01833	54.5613	.03579	27.9372	.05328	18.7678	57
.00116	859.436	.01862	53.7086	.03609	27.7117	.05357	18.6656	56
.00145	687.549	.01891	52.8821	.03638	27.4899	.05387	18.5645	55
.00175	572.957	.01920	52.0807	.03667	27.2715	.05416	18.4645	54
.00204	491.106	.01949	51.3032	.03696	27.0566	.05445	18.3655	53
.00233	429.718	.01978	50.5485	.03725	26.8450	.05474	18.2677	52
.00262	381.971	.02007	49.8157	.03754	26.6367	.05503	18.1708	51
.00291	343.774	.02036	49.1039	.03783	26.4316	.05533	18.0750	50
.00320	312.521	.02066	48.4121	.03812	26.2296	.05562	17.9802	49
.00349	286.478	.02095	47.7395	.03842	26.0307	.05591	17.8863	48
.00378	264.441	.02124	47.0853	.03871	25.8348	.05620	17.7934	47
.00407	245.552	.02153	46.4489	.03900	25.6418	.05649	17.7015	46
.00436	229.182	.02182	45.8294	.03929	25.4517	.05678	17.6106	45
.00465	214.858	.02211	45.2261	.03958	25.2644	.05708	17.5205	44
.00495	202.219	.02240	44.6386	.03987	25.0798	.05737	17.4314	43
.00524	190.984	.02269	44.0661	.04016	24.8978	.05766	17.3432	42
.00553	180.932	.02298	43.5081	.04046	24.7185	.05795	17.2558	41
.00582	171.885	.02328	42.9641	.04075	24.5418	.05824	17.1693	40
.00611	163.700	.02357	42.4335	.04104	24.3675	.05854	17.0837	39
.00640	156.259	.02386	41.9158	.04133	24.1957	.05883	16.9990	38
.00669	149.465	.02415	41.4106	.04162	24.0263	.05912	16.9150	37
.00698	143.237	.02444	40.9174	.04191	23.8593	.05941	16.8319	36
.00727	137.507	.02473	40.4358	.04220	23.6945	.05970	16.7496	35
.00756	132.219	.02502	39.9655	.04250	23.5321	.05999	16.6681	34
.00785	127.321	.02531	39.5059	.04279	23.3718	.06029	16.5874	33
.00814	122.774	.02560	39.0568	.04308	23.2137	.06058	16.5075	32
.00844	118.540	.02589	38.6177	.04337	23.0577	.06087	16.4283	31
.00873	114.589	.02619	38.1885	.04366	22.9038	.06116	16.3499	30
.00902	110.892	.02648	37.7686	.04395	22.7519	.06145	16.2722	29
.00931	107.426	.02677	37.3579	.04424	22.6020	.06175	16.1952	28
.00960	104.171	.02706	36.9560	.04454	22.4541	.06204	16.1190	27
.00989	101.107	.02735	36.5627	.04483	22.3081	.06233	16.0435	26
.01018	98.2179	.02764	36.1776	.04512	22.1640	.06262	15.9687	25
.01047	95.4895	.02793	35.8006	.04541	22.0217	.06291	15.8945	24
.01076	92.9085	.02822	35.4313	.04570	21.8813	.06321	15.8211	23
.01105	90.4633	.02851	35.0695	.04599	21.7426	.06350	15.7483	22
.01135	88.1436	.02881	34.7151	.04628	21.6056	.06379	15.6762	21
.01164	85.9398	.02910	34.3678	.04658	21.4704	.06408	15.6048	20
.01193	83.8435	.02939	34.0273	.04687	21.3369	.06437	15.5340	19
.01222	81.8470	.02968	33.6935	.04716	21.2049	.06467	15.4638	18
.01251	79.9434	.02997	33.3662	.04745	21.0747	.06496	15.3943	17
.01280	78.1263	.03026	33.0452	.04774	20.9460	.06525	15.3254	16
.01309	76.3900	.03055	32.7303	.04803	20.8188	.06554	15.2571	15
.01338	74.7292	.03084	32.4213	.04832	20.6932	.06584	15.1893	14
.01367	73.1390	.03114	32.1181	.04862	20.5691	.06613	15.1222	13
.01396	71.6151	.03143	31.8205	.04891	20.4465	.06642	15.0557	12
.01425	70.1533	.03172	31.5284	.04920	20.3253	.06671	14.9898	11
.01455	68.7501	.03201	31.2416	.04949	20.2056	.06700	14.9244	10
.01484	67.4019	.03230	30.9599	.04978	20.0872	.06730	14.8596	9
.01513	66.1055	.03259	30.6833	.05007	19.9702	.06759	14.7954	8
.01542	64.8580	.03288	30.4116	.05037	19.8546	.06788	14.7317	7
.01571	63.6567	.03317	30.1446	.05066	19.7403	.06817	14.6685	6
.01600	62.4992	.03346	29.8823	.05095	19.6273	.06847	14.6059	5
.01629	61.3829	.03376	29.6245	.05124	19.5156	.06876	14.5438	4
.01658	60.3058	.03405	29.3711	.05153	19.4051	.06905	14.4823	3
.01687	59.2659	.03434	29.1220	.05182	19.2959	.06934	14.4212	2
.01716	58.2612	.03463	28.8771	.05212	19.1879	.06963	14.3607	1
.01746	57.2900	.03492	28.6363	.05241	19.0811	.06993	14.3007	0
CO-TAN.	TAN.	CO-TAN.	TAN.	CO-TAN.	TAN.	CO-TAN.	TAN.	'
	89°		88°		87°		86°	

′	4° Tan.	Co-tan.	5° Tan.	Co-tan.	6° Tan.	Co-tan.	7° Tan.	Co-tan.
0	.06993	14.3007	.08749	11.4301	.10510	9.51436	.12278	8.14435
1	.07022	14.2411	.08778	11.3919	.10540	9.48781	.12308	8.12481
2	.07051	14.1821	.08807	11.3540	.10569	9.46141	.12338	8.10536
3	.07080	14.1235	.08837	11.3163	.10599	9.43515	.12367	8.08600
4	.07110	14.0655	.08866	11.2789	.10628	9.40904	.12397	8.06674
5	.07139	14.0079	.08895	11.2417	.10657	9.38307	.12426	8.04756
6	.07168	13.9507	.08925	11.2048	.10687	9.35724	.12456	8.02848
7	.07197	13.8940	.08954	11.1681	.10716	9.33154	.12485	8.00948
8	.07227	13.8378	.08983	11.1316	.10746	9.30599	.12515	7.99058
9	.07256	13.7821	.09013	11.0954	.10775	9.28058	.12544	7.97176
10	.07285	13.7267	.09042	11.0594	.10805	9.25530	.12574	7.95302
11	.07314	13.6719	.09071	11.0237	.10834	9.23016	.12603	7.93438
12	.07344	13.6174	.09101	10.9882	.10863	9.20516	.12633	7.91582
13	.07373	13.5634	.09130	10.9529	.10893	9.18028	.12662	7.89734
14	.07402	13.5098	.09159	10.9178	.10922	9.15554	.12692	7.87895
15	.07431	13.4566	.09189	10.8829	.10952	9.13093	.12722	7.86064
16	.07461	13.4039	.09218	10.8483	.10981	9.10646	.12751	7.84242
17	.07490	13.3515	.09247	10.8139	.11011	9.08211	.12781	7.82428
18	.07519	13.2996	.09277	10.7797	.11040	9.05789	.12810	7.80622
19	.07548	13.2480	.09306	10.7457	.11070	9.03379	.12840	7.78825
20	.07578	13.1969	.09335	10.7119	.11099	9.00983	.12869	7.77035
21	.07607	13.1461	.09365	10.6783	.11128	8.98598	.12899	7.75254
22	.07636	13.0958	.09394	10.6450	.11158	8.96227	.12929	7.73480
23	.07665	13.0458	.09423	10.6118	.11187	8.93867	.12958	7.71715
24	.07695	12.9962	.09453	10.5789	.11217	8.91520	.12988	7.69957
25	.07724	12.9469	.09482	10.5462	.11246	8.89185	.13017	7.68208
26	.07753	12.8981	.09511	10.5136	.11276	8.86862	.13047	7.66466
27	.07782	12.8496	.09541	10.4813	.11305	8.84551	.13076	7.64732
28	.07812	12.8014	.09570	10.4491	.11335	8.82252	.13106	7.63005
29	.07841	12.7536	.09600	10.4172	.11364	8.79964	.13136	7.61287
30	.07870	12.7062	.09629	10.3854	.11394	8.77689	.13165	7.59575
31	.07899	12.6591	.09658	10.3538	.11423	8.75425	.13195	7.57872
32	.07929	12.6124	.09688	10.3224	.11452	8.73172	.13224	7.56176
33	.07958	12.5660	.09717	10.2913	.11482	8.70931	.13254	7.54487
34	.07987	12.5199	.09746	10.2602	.11511	8.68701	.13284	7.52806
35	.08017	12.4742	.09776	10.2294	.11541	8.66482	.13313	7.51132
36	.08046	12.4288	.09805	10.1988	.11570	8.64275	.13343	7.49465
37	.08075	12.3838	.09834	10.1683	.11600	8.62078	.13372	7.47806
38	.08104	12.3390	.09864	10.1381	.11629	8.59803	.13402	7.46154
39	.08134	12.2946	.09893	10.1080	.11659	8.57718	.13432	7.44509
40	.08163	12.2505	.09923	10.0780	.11688	8.55555	.13461	7.42871
41	.08192	12.2067	.09952	10.0483	.11718	8.53402	.13491	7.41240
42	.08221	12.1632	.09981	10.0187	.11747	8.51259	.13521	7.39610
43	.08251	12.1201	.10011	9.98931	.11777	8.49128	.13550	7.37999
44	.08280	12.0772	.10040	9.96007	.11806	8.47007	.13580	7.36389
45	.08309	12.0346	.10069	9.93101	.11836	8.44896	.13609	7.34786
46	.08339	11.9923	.10099	9.90211	.11865	8.42795	.13639	7.33190
47	.08368	11.9504	.10128	9.87338	.11895	8.40705	.13669	7.31600
48	.08397	11.9087	.10158	9.84482	.11924	8.38625	.13698	7.30018
49	.08427	11.8673	.10187	9.81641	.11954	8.36555	.13728	7.28442
50	.08456	11.8262	.10216	9.78817	.11983	8.34496	.13758	7.26873
51	.08485	11.7853	.10246	9.76009	.12013	8.32446	.13787	7.25310
52	.08514	11.7448	.10275	9.73217	.12042	8.30406	.13817	7.23754
53	.08544	11.7045	.10305	9.70441	.12072	8.28376	.13846	7.22204
54	.08573	11.6645	.10334	9.67680	.12101	8.26355	.13876	7.20661
55	.08602	11.6248	.10363	9.64935	.12131	8.24345	.13906	7.19125
56	.08632	11.5853	.10393	9.62205	.12160	8.22344	.13935	7.17594
57	.08661	11.5461	.10422	9.59490	.12190	8.20352	.13965	7.16071
58	.08690	11.5072	.10452	9.56791	.12219	8.18370	.13995	7.14553
59	.08720	11.4685	.10481	9.54106	.12249	8.16398	.14024	7.13042
60	.08749	11.4301	.10510	9.51436	.12278	8.14435	.14054	7.11537
′	Co-tan.	Tan.	Co-tan.	Tan.	Co-tan.	Tan.	Co-tan.	Tan.
	85°		84°		83°		82°	

8° TAN.	8° CO-TAN.	9° TAN.	9° CO-TAN.	10° TAN.	10° CO-TAN.	11° TAN.	11° CO-TAN.	'
.14054	7.11537	.15838	6.31375	.17633	5.67128	.19438	5.14455	60
.14084	7.10038	.15868	6.30189	.17663	5.66165	.19468	5.13658	59
.14113	7.08546	.15898	6.29007	.17693	5.65205	.19498	5.12862	58
.14143	7.07059	.15928	6.27829	.17723	5.64248	.19529	5.12069	57
.14173	7.05579	.15958	6.26655	.17753	5.63295	.19559	5.11279	56
.14202	7.04105	.15988	6.25486	.17783	5.62344	.19589	5.10490	55
.14232	7.02637	.16017	6.24321	.17813	5.61397	.19619	5.09704	54
.14262	7.01174	.16047	6.23160	.17843	5.60452	.19649	5.08921	53
.14291	6.99718	.16077	6.22003	.17873	5.59511	.19680	5.08139	52
.14321	6.98268	.16107	6.20851	.17903	5.58573	.19710	5.07360	51
.14351	6.96823	.16137	6.19703	.17933	5.57638	.19740	5.06584	50
.14381	6.95385	.16167	6.18559	.17963	5.56706	.19770	5.05809	49
.14410	6.93952	.16196	6.17419	.17993	5.55777	.19801	5.05037	48
.14440	6.92525	.16226	6.16283	.18023	5.54851	.19831	5.04267	47
.14470	6.91104	.16256	6.15151	.18053	5.53927	.19861	5.03499	46
.14499	6.89688	.16286	6.14023	.18083	5.53007	.19891	5.02734	45
.14529	6.88278	.16316	6.12899	.18113	5.52090	.19921	5.01971	44
.14559	6.86874	.16346	6.11779	.18143	5.51176	.19952	5.01210	43
.14588	6.85475	.16376	6.10664	.18173	5.50264	.19982	5.00451	42
.14618	6.84082	.16405	6.09552	.18203	5.49356	.20012	4.99695	41
.14648	6.82694	.16435	6.08444	.18233	5.48451	.20042	4.98940	40
.14678	6.81312	.16465	6.07340	.18263	5.47548	.20073	4.98188	39
.14707	6.79936	.16495	6.06240	.18293	5.46648	.20103	4.97438	38
.14737	6.78564	.16525	6.05143	.18323	5.45751	.20133	4.96690	37
.14767	6.77199	.16555	6.04051	.18353	5.44857	.20164	4.95945	36
.14796	6.75838	.16585	6.02962	.18383	5.43966	.20194	4.95201	35
.14826	6.74483	.16615	6.01878	.18414	5.43077	.20224	4.94460	34
.14856	6.73133	.16645	6.00797	.18444	5.42192	.20254	4.93721	33
.14886	6.71789	.16674	5.99720	.18474	5.41309	.20285	4.92984	32
.14915	6.70450	.16704	5.98646	.18504	5.40429	.20315	4.92249	31
.14945	6.69116	.16734	5.97576	.18534	5.39552	.20345	4.91516	30
.14975	6.67787	.16764	5.96510	.18564	5.38677	.20376	4.90785	29
.15005	6.66463	.16794	5.95448	.18594	5.37805	.20406	4.90056	28
.15034	6.65144	.16824	5.94390	.18624	5.36936	.20436	4.89330	27
.15064	6.63831	.16854	5.93335	.18654	5.36070	.20466	4.88605	26
.15094	6.62523	.16884	5.92283	.18684	5.35206	.20497	4.87882	25
.15124	6.61219	.16914	5.91235	.18714	5.34345	.20527	4.87162	24
.15153	6.59921	.16944	5.90191	.18745	5.33487	.20557	4.86444	23
.15183	6.58627	.16974	5.89151	.18775	5.32631	.20588	4.85727	22
.15213	6.57339	.17004	5.88114	.18805	5.31778	.20618	4.85013	21
.15243	6.56055	.17033	5.87080	.18835	5.30928	.20648	4.84300	20
.15272	6.54777	.17063	5.86051	.18865	5.30080	.20679	4.83590	19
.15302	6.53503	.17093	5.85024	.18895	5.29235	.20709	4.82882	18
.15332	6.52234	.17123	5.84001	.18925	5.28393	.20739	4.82175	17
.15362	6.50970	.17153	5.82982	.18955	5.27553	.20770	4.81471	16
.15391	6.49710	.17183	5.81966	.18986	5.26715	.20800	4.80769	15
.15421	6.48456	.17213	5.80953	.19016	5.25880	.20830	4.80068	14
.15451	6.47206	.17243	5.79944	.19046	5.25048	.20861	4.79370	13
.15481	6.45961	.17273	5.78938	.19076	5.24218	.20891	4.78673	12
.15511	6.44720	.17303	5.77936	.19106	5.23391	.20921	4.77978	11
.15540	6.43484	.17333	5.76937	.19136	5.22566	.20952	4.77286	10
.15570	6.42253	.17363	5.75941	.19166	5.21744	.20982	4.76595	9
.15600	6.41026	.17393	5.74949	.19197	5.20925	.21013	4.75906	8
.15630	6.39804	.17423	5.73960	.19227	5.20107	.21043	4.75219	7
.15660	6.38587	.17453	5.72974	.19257	5.19293	.21073	4.74534	6
.15689	6.37374	.17483	5.71992	.19287	5.18480	.21104	4.73851	5
.15719	6.36165	.17513	5.71013	.19317	5.17671	.21134	4.73170	4
.15749	6.34961	.17543	5.70037	.19347	5.16863	.21164	4.72490	3
.15779	6.33761	.17573	5.69064	.19378	5.16058	.21195	4.71813	2
.15809	6.32566	.17603	5.68094	.19408	5.15256	.21225	4.71137	1
.15838	6.31375	.17633	5.67128	.19438	5.14455	.21256	4.70463	0

CO-TAN.	TAN.	CO-TAN.	TAN.	CO-TAN.	TAN.	CO-TAN.	TAN.	'
81°		80°		79°		78°		

′	12° Tan.	Co-tan.	13° Tan.	Co-tan.	14° Tan.	Co-tan.	15° Tan.	Co-tan.	′
0	.21256	4.70463	.23087	4.33148	.24933	4.01078	.26795	3.73205	60
1	.21286	4.69791	.23117	4.32573	.24964	4.00582	.26826	3.72771	59
2	.21316	4.69121	.23148	4.32001	.24995	4.00086	.26857	3.72338	58
3	.21347	4.68452	.23179	4.31430	.25026	3.99592	.26888	3.71907	57
4	.21377	4.67786	.23209	4.30860	.25056	3.99099	.26920	3.71476	56
5	.21408	4.67121	.23240	4.30291	.25087	3.98607	.26951	3.71046	55
6	.21438	4.66458	.23271	4.29724	.25118	3.98117	.26982	3.70616	54
7	.21469	4.65797	.23301	4.29159	.25149	3.97627	.27013	3.70188	53
8	.21499	4.65138	.23332	4.28595	.25180	3.97139	.27044	3.69761	52
9	.21529	4.64480	.23363	4.28032	.25211	3.96651	.27076	3.69335	51
10	.21560	4.63825	.23393	4.27471	.25242	3.96165	.27107	3.68909	50
11	.21590	4.63171	.23424	4.26911	.25273	3.95680	.27138	3.68485	49
12	.21621	4.62518	.23455	4.26352	.25304	3.95196	.27169	3.68061	48
13	.21651	4.61868	.23485	4.25795	.25335	3.94713	.27201	3.67638	47
14	.21682	4.61219	.23516	4.25239	.25366	3.94232	.27232	3.67217	46
15	.21712	4.60572	.23547	4.24685	.25397	3.93751	.27263	3.66796	45
16	.21743	4.59927	.23578	4.24132	.25428	3.93271	.27294	3.66376	44
17	.21773	4.59283	.23608	4.23580	.25459	3.92793	.27326	3.65957	43
18	.21804	4.58641	.23639	4.23030	.25490	3.92316	.27357	3.65538	42
19	.21834	4.58001	.23670	4.22481	.25521	3.91839	.27388	3.65121	41
20	.21864	4.57363	.23700	4.21933	.25552	3.91364	.27419	3.64705	40
21	.21895	4.56726	.23731	4.21387	.25583	3.90890	.27451	3.64289	39
22	.21925	4.56091	.23762	4.20842	.25614	3.90417	.27482	3.63874	38
23	.21956	4.55458	.23793	4.20298	.25645	3.89945	.27513	3.63461	37
24	.21986	4.54826	.23823	4.19756	.25676	3.89474	.27545	3.63048	36
25	.22017	4.54196	.23854	4.19215	.25707	3.89004	.27576	3.62636	35
26	.22047	4.53568	.23885	4.18675	.25738	3.88536	.27607	3.62224	34
27	.22078	4.52941	.23916	4.18137	.25769	3.88068	.27638	3.61814	33
28	.22108	4.52316	.23946	4.17600	.25800	3.87601	.27670	3.61405	32
29	.22139	4.51693	.23977	4.17064	.25831	3.87136	.27701	3.60996	31
30	.22169	4.51071	.24008	4.16530	.25862	3.86671	.27732	3.60588	30
31	.22200	4.50451	.24039	4.15997	.25893	3.86208	.27764	3.60181	29
32	.22231	4.49832	.24069	4.15465	.25924	3.85745	.27795	3.59775	28
33	.22261	4.49215	.24100	4.14934	.25955	3.85284	.27826	3.59370	27
34	.22292	4.48600	.24131	4.14405	.25986	3.84824	.27858	3.58966	26
35	.22322	4.47986	.24162	4.13877	.26017	3.84364	.27889	3.58562	25
36	.22353	4.47374	.24193	4.13350	.26048	3.83906	.27920	3.58160	24
37	.22383	4.46764	.24223	4.12825	.26079	3.83449	.27952	3.57758	23
38	.22414	4.46155	.24254	4.12301	.26110	3.82992	.27983	3.57357	22
39	.22444	4.45548	.24285	4.11778	.26141	3.82537	.28015	3.56957	21
40	.22475	4.44942	.24316	4.11256	.26172	3.82083	.28046	3.56557	20
41	.22505	4.44338	.24347	4.10736	.26203	3.81630	.28077	3.56159	19
42	.22536	4.43735	.24377	4.10216	.26235	3.81177	.28109	3.55761	18
43	.22567	4.43134	.24408	4.09699	.26266	3.80726	.28140	3.55364	17
44	.22597	4.42534	.24439	4.09182	.26297	3.80276	.28172	3.54968	16
45	.22628	4.41936	.24470	4.08666	.26328	3.79827	.28203	3.54573	15
46	.22658	4.41340	.24501	4.08152	.26359	3.79378	.28234	3.54179	14
47	.22689	4.40745	.24532	4.07639	.26390	3.78931	.28266	3.53785	13
48	.22719	4.40152	.24562	4.07127	.26421	3.78485	.28297	3.53393	12
49	.22750	4.39560	.24593	4.06616	.26452	3.78040	.28329	3.53001	11
50	.22781	4.38969	.24624	4.06107	.26483	3.77595	.28360	3.52609	10
51	.22811	4.38381	.24655	4.05599	.26515	3.77152	.28391	3.52219	9
52	.22842	4.37793	.24686	4.05092	.26546	3.76709	.28423	3.51829	8
53	.22872	4.37207	.24717	4.04586	.26577	3.76268	.28454	3.51441	7
54	.22903	4.36623	.24747	4.04081	.26608	3.75828	.28486	3.51053	6
55	.22934	4.36040	.24778	4.03578	.26639	3.75388	.28517	3.50666	5
56	.22964	4.35459	.24809	4.03075	.26670	3.74950	.28549	3.50279	4
57	.22995	4.34879	.24840	4.02574	.26701	3.74512	.28580	3.49894	3
58	.23026	4.34300	.24871	4.02074	.26733	3.74075	.28612	3.49509	2
59	.23056	4.33723	.24902	4.01576	.26764	3.73640	.28643	3.49125	1
60	.23087	4.33148	.24933	4.01078	.26795	3.73205	.28675	3.48741	0
′	Co-tan.	Tan.	Co-tan.	Tan.	Co-tan.	Tan.	Co-tan.	Tan.	′

77° | 76° | 75° | 74°

16° TAN.	16° CO-TAN.	17° TAN.	17° CO-TAN.	18° TAN.	18° CO-TAN.	19° TAN.	19° CO-TAN.	′
.28675	3.48741	.30573	3.27085	.32492	3.07768	.34433	2.90421	60
.28706	3.48359	.30605	3.26745	.32524	3.07464	.34465	2.90147	59
.28738	3.47977	.30637	3.26406	.32556	3.07160	.34498	2.89873	58
.28769	3.47596	.30669	3.26067	.32588	3.06857	.34530	2.89600	57
.28800	3.47216	.30700	3.25729	.32621	3.06554	.34563	2.89327	56
.28832	3.46837	.30732	3.25392	.32653	3.06252	.34596	2.89055	55
.28864	3.46458	.30764	3.25055	.32685	3.05950	.34628	2.88783	54
.28895	3.46080	.30796	3.24719	.32717	3.05649	.34661	2.88511	53
.28927	3.45703	.30828	3.24383	.32749	3.05349	.34693	2.88240	52
.28958	3.45327	.30860	3.24049	.32782	3.05049	.34726	2.87970	51
.28990	3.44951	.30891	3.23714	.32814	3.04749	.34758	2.87700	50
.29021	3.44576	.30923	3.23381	.32846	3.04450	.34791	2.87430	49
.29053	3.44202	.30955	3.23048	.32878	3.04152	.34824	2.87161	48
.29084	3.43829	.30987	3.22715	.32911	3.03854	.34856	2.86892	47
.29116	3.43456	.31019	3.22384	.32943	3.03556	.34889	2.86624	46
.29147	3.43084	.31051	3.22053	.32975	3.03260	.34922	2.86356	45
.29179	3.42713	.31083	3.21722	.33007	3.02963	.34954	2.86089	44
.29210	3.42343	.31115	3.21392	.33040	3.02667	.34987	2.85822	43
.29242	3.41973	.31147	3.21063	.33072	3.02372	.35019	2.85555	42
.29274	3.41604	.31178	3.20734	.33104	3.02077	.35052	2.85289	41
.29305	3.41236	.31210	3.20406	.33136	3.01783	.35085	2.85023	40
.29337	3.40869	.31242	3.20079	.33160	3.01489	.35117	2.84758	39
.29368	3.40502	.31274	3.19752	.33201	3.01196	.35150	2.84494	38
.29400	3.40136	.31306	3.19426	.33233	3.00903	.35183	2.84229	37
.29432	3.39771	.31338	3.19100	.33266	3.00611	.35216	2.83965	36
.29463	3.39406	.31370	3.18775	.33298	3.00319	.35248	2.83702	35
.29495	3.39042	.31402	3.18451	.33330	3.00028	.35281	2.83439	34
.29526	3.38679	.31434	3.18127	.33363	2.99738	.35314	2.83176	33
.29558	3.38317	.31466	3.17804	.33395	2.99447	.35346	2.82914	32
.29590	3.37955	.31498	3.17481	.33427	2.99158	.35379	2.82653	31
.29621	3.37594	.31530	3.17159	.33460	2.98868	.35412	2.82391	30
.29653	3.37234	.31562	3.16838	.33492	2.98580	.35445	2.82130	29
.29685	3.36875	.31594	3.16517	.33524	2.98292	.35477	2.81870	28
.29716	3.36516	.31626	3.16197	.33557	2.98004	.35510	2.81610	27
.29748	3.36158	.31658	3.15877	.33589	2.97717	.35543	2.81350	26
.29780	3.35800	.31690	3.15558	.33621	2.97430	.35576	2.81091	25
.29811	3.35443	.31722	3.15240	.33654	2.97144	.35608	2.80833	24
.29843	3.35087	.31754	3.14922	.33686	2.96858	.35641	2.80574	23
.29875	3.34732	.31786	3.14605	.33718	2.96573	.35674	2.80316	22
.29906	3.34377	.31818	3.14288	.33751	2.96288	.35707	2.80059	21
.29938	3.34023	.31850	3.13972	.33783	2.96004	.35740	2.79802	20
.29970	3.33670	.31882	3.13656	.33816	2.95721	.35772	2.79545	19
.30001	3.33317	.31914	3.13341	.33848	2.95437	.35805	2.79289	18
.30033	3.32965	.31946	3.13027	.33881	2.95155	.35838	2.79033	17
.30065	3.32614	.31978	3.12713	.33913	2.94872	.35871	2.78778	16
.30097	3.32264	.32010	3.12400	.33945	2.94590	.35904	2.78523	15
.30128	3.31914	.32042	3.12087	.33978	2.94309	.35937	2.78269	14
.30160	3.31565	.32074	3.11775	.34010	2.94028	.35969	2.78014	13
.30192	3.31216	.32106	3.11464	.34043	2.93748	.36002	2.77761	12
.30224	3.30868	.32139	3.11153	.34075	2.93468	.36035	2.77507	11
.30255	3.30521	.32171	3.10842	.34108	2.93189	.36068	2.77254	10
.30287	3.30174	.32203	3.10532	.34140	2.92910	.36101	2.77002	9
.30319	3.29829	.32235	3.10223	.34173	2.92632	.36134	2.76750	8
.30351	3.29483	.32267	3.09914	.34205	2.92354	.36167	2.76498	7
.30382	3.29139	.32299	3.09606	.34238	2.92076	.36199	2.76247	6
.30414	3.28795	.32331	3.09298	.34270	2.91799	.36232	2.75996	5
.30446	3.28452	.32363	3.08991	.34303	2.91523	.36265	2.75746	4
.30478	3.28109	.32396	3.08685	.34335	2.91246	.36298	2.75496	3
.30509	3.27767	.32428	3.08379	.34368	2.90971	.36331	2.75246	2
.30541	3.27426	.32460	3.08073	.34400	2.90696	.36364	2.74997	1
.30573	3.27085	.32492	3.07768	.34433	2.90421	.36397	2.74748	0

Co-tan.	Tan.	Co-tan.	Tan.	Co-tan.	Tan.	Co-tan.	Tan.	′
	73°		72°		71°		70°	

′	20° TAN.	20° CO-TAN.	21° TAN.	21° CO-TAN.	22° TAN.	22° CO-TAN.	23° TAN.	23° CO-TAN.	′
0	.36397	2.74748	.38386	2.60509	.40403	2.47509	.42447	2.35585	60
1	.36430	2.74499	.38420	2.60283	.40436	2.47302	.42482	2.35395	59
2	.36463	2.74251	.38453	2.60057	.40470	2.47095	.42516	2.35205	58
3	.36496	2.74004	.38487	2.59831	.40504	2.46888	.42551	2.35015	57
4	.36529	2.73756	.38520	2.59606	.40538	2.46682	.42585	2.34825	56
5	.36562	2.73509	.38553	2.59381	.40572	2.46476	.42619	2.34636	55
6	.36595	2.73263	.38587	2.59156	.40606	2.46270	.42654	2.34447	54
7	.36628	2.73017	.38620	2.58932	.40640	2.46065	.42688	2.34258	53
8	.36661	2.72771	.38654	2.58708	.40674	2.45860	.42722	2.34069	52
9	.36694	2.72526	.38687	2.58484	.40707	2.45655	.42757	2.33881	51
10	.36727	2.72281	.38721	2.58261	.40741	2.45451	.42791	2.33693	50
11	.36760	2.72036	.38754	2.58038	.40775	2.45246	.42826	2.33505	49
12	.36793	2.71792	.38787	2.57815	.40809	2.45043	.42860	2.33317	48
13	.36826	2.71548	.38821	2.57593	.40843	2.44839	.42894	2.33130	47
14	.36859	2.71305	.38854	2.57371	.40877	2.44636	.42929	2.32943	46
15	.36892	2.71062	.38888	2.57150	.40911	2.44433	.42963	2.32756	45
16	.36925	2.70819	.38921	2.56928	.40945	2.44230	.42998	2.32570	44
17	.36958	2.70577	.38955	2.56707	.40979	2.44027	.43032	2.32383	43
18	.36991	2.70335	.38988	2.56487	.41013	2.43825	.43067	2.32197	42
19	.37024	2.70094	.39022	2.56266	.41047	2.43623	.43101	2.32012	41
20	.37057	2.69853	.39055	2.56046	.41081	2.43422	.43136	2.31826	40
21	.37090	2.69612	.39089	2.55827	.41115	2.43220	.43170	2.31641	39
22	.37124	2.69371	.39122	2.55608	.41149	2.43019	.43205	2.31456	38
23	.37157	2.69131	.39156	2.55389	.41183	2.42819	.43239	2.31271	37
24	.37190	2.68892	.39190	2.55170	.41217	2.42618	.43274	2.31086	36
25	.37223	2.68653	.39223	2.54952	.41251	2.42418	.43308	2.30902	35
26	.37256	2.68414	.39257	2.54734	.41285	2.42218	.43343	2.30718	34
27	.37289	2.68175	.39290	2.54516	.41319	2.42019	.43378	2.30534	33
28	.37322	2.67937	.39324	2.54299	.41353	2.41819	.43412	2.30351	32
29	.37355	2.67700	.39357	2.54082	.41387	2.41620	.43447	2.30167	31
30	.37388	2.67462	.39391	2.53865	.41421	2.41421	.43481	2.29984	30
31	.37422	2.67225	.39425	2.53648	.41455	2.41223	.43516	2.29801	29
32	.37455	2.66989	.39458	2.53432	.41490	2.41025	.43550	2.29619	28
33	.37488	2.66752	.39492	2.53217	.41524	2.40827	.43585	2.29437	27
34	.37521	2.66516	.39526	2.53001	.41558	2.40629	.43620	2.29254	26
35	.37554	2.66281	.39559	2.52786	.41592	2.40432	.43654	2.29073	25
36	.37588	2.66046	.39593	2.52571	.41626	2.40235	.43689	2.28891	24
37	.37621	2.65811	.39626	2.52357	.41660	2.40038	.43724	2.28710	23
38	.37654	2.65576	.39660	2.52142	.41694	2.39841	.43758	2.28528	22
39	.37687	2.65342	.39694	2.51929	.41728	2.39645	.43793	2.28348	21
40	.37720	2.65109	.39727	2.51715	.41763	2.39449	.43828	2.28167	20
41	.37754	2.64875	.39761	2.51502	.41797	2.39253	.43862	2.27987	19
42	.37787	2.64642	.39795	2.51289	.41831	2.39058	.43897	2.27806	18
43	.37820	2.64410	.39829	2.51076	.41865	2.38862	.43932	2.27626	17
44	.37853	2.64177	.39862	2.50864	.41899	2.38668	.43966	2.27447	16
45	.37887	2.63945	.39896	2.50652	.41933	2.38473	.44001	2.27267	15
46	.37920	2.63714	.39930	2.50440	.41968	2.38279	.44036	2.27088	14
47	.37953	2.63483	.39963	2.50229	.42002	2.38084	.44071	2.26909	13
48	.37986	2.63252	.39997	2.50018	.42036	2.37891	.44105	2.26730	12
49	.38020	2.63021	.40031	2.49807	.42070	2.37697	.44140	2.26552	11
50	.38053	2.62791	.40065	2.49597	.42105	2.37504	.44175	2.26374	10
51	.38086	2.62561	.40098	2.49386	.42139	2.37311	.44210	2.26196	9
52	.38120	2.62332	.40132	2.49177	.42173	2.37118	.44244	2.26018	8
53	.38153	2.62103	.40166	2.48967	.42207	2.36925	.44279	2.25840	7
54	.38186	2.61874	.40200	2.48758	.42242	2.36733	.44314	2.25663	6
55	.38220	2.61646	.40234	2.48549	.42276	2.36541	.44349	2.25486	5
56	.38253	2.61418	.40267	2.48340	.42310	2.36349	.44384	2.25309	4
57	.38286	2.61190	.40301	2.48132	.42345	2.36158	.44418	2.25132	3
58	.38320	2.60963	.40335	2.47924	.42379	2.35967	.44453	2.24956	2
59	.38353	2.60736	.40369	2.47716	.42413	2.35776	.44488	2.24780	1
60	.38386	2.60509	.40403	2.47509	.42447	2.35585	.44523	2.24604	0
′	CO-TAN.	TAN.	CO-TAN.	TAN.	CO-TAN.	TAN.	CO-TAN.	TAN.	
	69°		68°		67°		66°		

24° Tan.	Co-tan.	25° Tan.	Co-tan.	26° Tan.	Co-tan.	27° Tan.	Co-tan.	
.44523	2.24604	.46631	2.14451	.48773	2.05030	.50953	1.96261	60
.44558	2.24428	.46666	2.14288	.48809	2.04879	.50989	1.96120	59
.44593	2.24252	.46702	2.14125	.48845	2.04728	.51026	1.95979	58
.44627	2.24077	.46737	2.13963	.48881	2.04577	.51063	1.95838	57
.44662	2.23902	.46772	2.13801	.48917	2.04426	.51099	1.95698	56
.44697	2.23727	.46808	2.13639	.48953	2.04276	.51136	1.95557	55
.44732	2.23553	.46843	2.13477	.48989	2.04125	.51173	1.95417	54
.44767	2.23378	.46879	2.13316	.49026	2.03975	.51209	1.95277	53
.44802	2.23204	.46914	2.13154	.49062	2.03825	.51246	1.95137	52
.44837	2.23030	.46950	2.12993	.49098	2.03675	.51283	1.94997	51
.44872	2.22857	.46985	2.12832	.49134	2.03526	.51319	1.94858	50
.44907	2.22683	.47021	2.12671	.49170	2.03376	.51356	1.94718	49
.44942	2.22510	.47056	2.12511	.49206	2.03227	.51393	1.94579	48
.44977	2.22337	.47092	2.12350	.49242	2.03078	.51430	1.94440	47
.45012	2.22164	.47128	2.12190	.49278	2.02929	.51467	1.94301	46
.45047	2.21992	.47163	2.12030	.49315	2.02780	.51503	1.94162	45
.45082	2.21819	.47199	2.11871	.49351	2.02631	.51540	1.94023	44
.45117	2.21647	.47234	2.11711	.49387	2.02483	.51577	1.93885	43
.45152	2.21475	.47270	2.11552	.49423	2.02335	.51614	1.93746	42
.45187	2.21304	.47305	2.11392	.49459	2.02187	.51651	1.93608	41
.45222	2.21132	.47341	2.11233	.49495	2.02039	.51688	1.93470	40
.45257	2.20961	.47377	2.11075	.49532	2.01891	.51724	1.93332	39
.45292	2.20790	.47412	2.10916	.49568	2.01743	.51761	1.93195	38
.45327	2.20619	.47448	2.10758	.49604	2.01596	.51798	1.93057	37
.45362	2.20449	.47483	2.10600	.49640	2.01449	.51835	1.92920	36
.45397	2.20278	.47519	2.10442	.49677	2.01302	.51872	1.92782	35
.45432	2.20108	.47555	2.10284	.49713	2.01155	.51909	1.92645	34
.45467	2.19938	.47590	2.10126	.49749	2.01008	.51946	1.92508	33
.45502	2.19769	.47626	2.09969	.49786	2.00862	.51983	1.92371	32
.45537	2.19599	.47662	2.09811	.49822	2.00715	.52020	1.92235	31
.45573	2.19430	.47698	2.09654	.49858	2.00569	.52057	1.92098	30
.45608	2.19261	.47733	2.09498	.49894	2.00423	.52094	1.91962	29
.45643	2.19092	.47769	2.09341	.49931	2.00277	.52131	1.91826	28
.45678	2.18923	.47805	2.09184	.49967	2.00131	.52168	1.91690	27
.45713	2.18755	.47840	2.09028	.50004	1.99986	.52205	1.91554	26
.45748	2.18587	.47876	2.08872	.50040	1.99841	.52242	1.91418	25
.45784	2.18419	.47912	2.08716	.50076	1.99695	.52279	1.91282	24
.45819	2.18251	.47948	2.08560	.50113	1.99550	.52316	1.91147	23
.45854	2.18084	.47984	2.08405	.50149	1.99406	.52353	1.91012	22
.45889	2.17916	.48019	2.08250	.50185	1.99261	.52390	1.90876	21
.45924	2.17749	.48055	2.08094	.50222	1.99116	.52427	1.90741	20
.45960	2.17582	.48091	2.07939	.50258	1.98972	.52464	1.90607	19
.45995	2.17416	.48127	2.07785	.50295	1.98828	.52501	1.90472	18
.46030	2.17249	.48163	2.07630	.50331	1.98684	.52538	1.90337	17
.46065	2.17083	.48198	2.07476	.50368	1.98540	.52575	1.90203	16
.46101	2.16917	.48234	2.07321	.50404	1.98396	.52613	1.90069	15
.46136	2.16751	.48270	2.07167	.50441	1.98253	.52650	1.89935	14
.46171	2.16585	.48306	2.07014	.50477	1.98110	.52687	1.89801	13
.46206	2.16420	.48342	2.06860	.50514	1.97966	.52724	1.89667	12
.46242	2.16255	.48378	2.06706	.50550	1.97823	.52761	1.89533	11
.46277	2.16090	.48414	2.06553	.50587	1.97680	.52798	1.89400	10
.46312	2.15925	.48450	2.06400	.50623	1.97538	.52836	1.89266	9
.46348	2.15760	.48486	2.06247	.50660	1.97395	.52873	1.89133	8
.46383	2.15596	.48521	2.06094	.50696	1.97253	.52910	1.89000	7
.46418	2.15432	.48557	2.05942	.50733	1.97111	.52947	1.88867	6
.46454	2.15268	.48593	2.05790	.50769	1.96969	.52984	1.88734	5
.46489	2.15104	.48629	2.05637	.50806	1.96827	.53022	1.88602	4
.46525	2.14940	.48665	2.05485	.50843	1.96685	.53059	1.88469	3
.46560	2.14777	.48701	2.05333	.50879	1.96544	.53096	1.88337	2
.46595	2.14614	.48737	2.05182	.50916	1.96402	.53134	1.88205	1
.46631	2.14451	.48773	2.05030	.50953	1.96261	.53171	1.88073	0
Co-tan.	Tan. 65°	Co-tan.	Tan. 64°	Co-tan.	Tan. 63°	Co-tan.	Tan. 62°	′

	28°		29°		30°		31°		
	TAN.	CO-TAN.	TAN.	CO-TAN.	TAN.	CO-TAN.	TAN.	CO-TAN.	
0	.53171	1.88073	.55431	1.80405	.57735	1.73205	.60086	1.66428	60
1	.53208	1.87941	.55469	1.80281	.57774	1.73089	.60126	1.66318	59
2	.53246	1.87809	.55507	1.80158	.57813	1.72973	.60165	1.66209	58
3	.53283	1.87677	.55545	1.80034	.57851	1.72857	.60205	1.66099	57
4	.53320	1.87546	.55583	1.79911	.57890	1.72741	.60245	1.65990	56
5	.53358	1.87415	.55621	1.79788	.57929	1.72625	.60284	1.65881	55
6	.53395	1.87283	.55659	1.79665	.57968	1.72509	.60324	1.65772	54
7	.53432	1.87152	.55697	1.79542	.58007	1.72393	.60364	1.65663	53
8	.53470	1.87021	.55736	1.79419	.58046	1.72278	.60403	1.65534	52
9	.53507	1.86891	.55774	1.79296	.58085	1.72163	.60443	1.65445	51
10	.53545	1.86760	.55812	1.79174	.58124	1.72047	.60483	1.65337	50
11	.53582	1.86630	.55850	1.79051	.58162	1.71932	.60522	1.65228	49
12	.53620	1.86499	.55888	1.78929	.58201	1.71817	.60562	1.65120	48
13	.53657	1.86369	.55926	1.78807	.58240	1.71702	.60602	1.65011	47
14	.53694	1.86239	.55964	1.78685	.58279	1.71588	.60642	1.64903	46
15	.53732	1.86109	.56003	1.78563	.58318	1.71473	.60681	1.64795	45
16	.53769	1.85979	.56041	1.78441	.58357	1.71358	.60721	1.64687	44
17	.53807	1.85850	.56079	1.78319	.58396	1.71244	.60761	1.64579	43
18	.53844	1.85720	.56117	1.78198	.58435	1.71129	.60801	1.64471	42
19	.53882	1.85591	.56156	1.78077	.58474	1.71015	.60841	1.64363	41
20	.53920	1.85462	.56194	1.77955	.58513	1.70901	.60881	1.64256	40
21	.53957	1.85333	.56232	1.77834	.58552	1.70787	.60921	1.64148	39
22	.53995	1.85204	.56270	1.77713	.58591	1.70673	.60960	1.64041	38
23	.54032	1.85075	.56309	1.77592	.58631	1.70560	.61000	1.63934	37
24	.54070	1.84946	.56347	1.77471	.58670	1.70446	.61040	1.63826	36
25	.54107	1.84818	.56385	1.77351	.58709	1.70332	.61080	1.63719	35
26	.54145	1.84689	.56424	1.77230	.58748	1.70219	.61120	1.63612	34
27	.54183	1.84561	.56462	1.77110	.58787	1.70106	.61160	1.63505	33
28	.54220	1.84433	.56500	1.76990	.58826	1.69992	.61200	1.63398	32
29	.54258	1.84305	.56539	1.76869	.58865	1.69879	.61240	1.63292	31
30	.54296	1.84177	.56577	1.76749	.58904	1.69766	.61280	1.63185	30
31	.54333	1.84049	.56616	1.76630	.58944	1.69653	.61320	1.63079	29
32	.54371	1.83922	.56654	1.76510	.58983	1.69541	.61360	1.62972	28
33	.54409	1.83794	.56693	1.76390	.59022	1.69428	.61400	1.62866	27
34	.54446	1.83667	.56731	1.76271	.59061	1.69316	.61440	1.62760	26
35	.54484	1.83540	.56769	1.76151	.59101	1.69203	.61480	1.62654	25
36	.54522	1.83413	.56808	1.76032	.59140	1.69091	.61520	1.62548	24
37	.54560	1.83286	.56846	1.75913	.59179	1.68979	.61561	1.62442	23
38	.54597	1.83159	.56885	1.75794	.59218	1.68866	.61601	1.62336	22
39	.54635	1.83033	.56923	1.75675	.59258	1.68754	.61641	1.62230	21
40	.54673	1.82906	.56962	1.75556	.59297	1.68643	.61681	1.62125	20
41	.54711	1.82780	.57000	1.75437	.59336	1.68531	.61721	1.62019	19
42	.54748	1.82654	.57039	1.75319	.59376	1.68419	.61761	1.61914	18
43	.54786	1.82528	.57078	1.75200	.59415	1.68308	.61801	1.61808	17
44	.54824	1.82402	.57116	1.75082	.59454	1.68196	.61842	1.61703	16
45	.54862	1.82276	.57155	1.74964	.59494	1.68085	.61882	1.61598	15
46	.54900	1.82150	.57193	1.74846	.59533	1.67974	.61922	1.61493	14
47	.54938	1.82025	.57232	1.74728	.59573	1.67863	.61962	1.61388	13
48	.54975	1.81899	.57271	1.74610	.59612	1.67752	.62003	1.61283	12
49	.55013	1.81774	.57309	1.74492	.59651	1.67641	.62043	1.61179	11
50	.55051	1.81649	.57348	1.74375	.59691	1.67530	.62083	1.61074	10
51	.55089	1.81524	.57386	1.74257	.59730	1.67419	.62124	1.60970	9
52	.55127	1.81399	.57425	1.74140	.59770	1.67309	.62164	1.60865	8
53	.55165	1.81274	.57464	1.74022	.59809	1.67198	.62204	1.60761	7
54	.55203	1.81150	.57503	1.73905	.59849	1.67088	.62245	1.60657	6
55	.55241	1.81025	.57541	1.73788	.59888	1.66978	.62285	1.60553	5
56	.55279	1.80901	.57580	1.73671	.59928	1.66867	.62325	1.60449	4
57	.55317	1.80777	.57619	1.73555	.59967	1.66757	.62366	1.60345	3
58	.55355	1.80653	.57657	1.73438	.60007	1.66647	.62406	1.60241	2
59	.55393	1.80529	.57696	1.73321	.60046	1.66538	.62446	1.60137	1
60	.55431	1.80405	.57735	1.73205	.60086	1.66428	.62487	1.60033	0
	CO-TAN.	TAN.	CO-TAN.	TAN.	CO-TAN.	TAN.	CO-TAN.	TAN.	
		61°		60°		59°		58°	

	32°		33°		34°		35°	
	TAN.	CO-TAN.	TAN.	CO-TAN.	TAN.	CO-TAN.	TAN.	CO-TAN.	′
.62487	1.60033	.64941	1.53986	.67451	1.48256	.70021	1.42815	60
.62527	1.59930	.64982	1.53888	.67493	1.48163	.70064	1.42726	59
.62568	1.59826	.65023	1.53791	.67536	1.48070	.70107	1.42638	58
.62608	1.59723	.65065	1.53693	.67578	1.47977	.70151	1.42550	57
.62649	1.59620	.65106	1.53595	.67620	1.47885	.70194	1.42462	56
.62689	1.59517	.65148	1.53497	.67663	1.47792	.70238	1.42374	55
.62730	1.59414	.65189	1.53400	.67705	1.47699	.70281	1.42286	54
.62770	1.59311	.65231	1.53302	.67748	1.47607	.70325	1.42198	53
.62811	1.59208	.65272	1.53205	.67790	1.47514	.70368	1.42110	52
.62852	1.59105	.65314	1.53107	.67832	1.47422	.70412	1.42022	51
.62892	1.59002	.65355	1.53010	.67875	1.47330	.70455	1.41934	50
.62933	1.58900	.65397	1.52913	.67917	1.47238	.70499	1.41847	49
.62973	1.58797	.65438	1.52816	.67960	1.47146	.70542	1.41759	48
.63014	1.58695	.65480	1.52719	.68002	1.47053	.70586	1.41672	47
.63055	1.58593	.65521	1.52622	.68045	1.46962	.70629	1.41584	46
.63095	1.58490	.65563	1.52525	.68088	1.46870	.70673	1.41497	45
.63136	1.58388	.65604	1.52429	.68130	1.46778	.70717	1.41409	44
.63177	1.58286	.65646	1.52332	.68173	1.46686	.70760	1.41322	43
.63217	1.58184	.65688	1.52235	.68215	1.46595	.70804	1.41235	42
.63258	1.58083	.65729	1.52139	.68258	1.46503	.70848	1.41148	41
.63299	1.57981	.65771	1.52043	.68301	1.46411	.70891	1.41061	40
.63340	1.57879	.65813	1.51946	.68343	1.46320	.70935	1.40974	39
.63380	1.57778	.65854	1.51850	.68386	1.46229	.70979	1.40887	38
.63421	1.57676	.65896	1.51754	.68429	1.46137	.71023	1.40800	37
.63462	1.57575	.65938	1.51658	.68471	1.46046	.71066	1.40714	36
.63503	1.57474	.65980	1.51562	.68514	1.45955	.71110	1.40627	35
.63544	1.57372	.66021	1.51466	.68557	1.45864	.71154	1.40540	34
.63584	1.57271	.66063	1.51370	.68600	1.45773	.71198	1.40454	33
.63625	1.57170	.66105	1.51275	.68642	1.45682	.71242	1.40367	32
.63666	1.57069	.66147	1.51179	.68685	1.45592	.71285	1.40281	31
.63707	1.56969	.66189	1.51084	.68728	1.45501	.71329	1.40195	30
.63748	1.56868	.66230	1.50988	.68771	1.45410	.71373	1.40109	29
.63789	1.56767	.66272	1.50893	.68814	1.45320	.71417	1.40022	28
.63830	1.56667	.66314	1.50797	.68857	1.45229	.71461	1.39936	27
.63871	1.56566	.66356	1.50702	.68900	1.45139	.71505	1.39850	26
.63912	1.56466	.66398	1.50607	.68942	1.45049	.71549	1.39764	25
.63953	1.56366	.66440	1.50512	.68985	1.44958	.71593	1.39679	24
.63994	1.56265	.66482	1.50417	.69028	1.44868	.71637	1.39593	23
.64035	1.56165	.66524	1.50322	.69071	1.44778	.71681	1.39507	22
.64076	1.56065	.66566	1.50228	.69114	1.44688	.71725	1.39421	21
.64117	1.55966	.66608	1.50133	.69157	1.44598	.71769	1.39336	20
.64158	1.55866	.66650	1.50038	.69200	1.44508	.71813	1.39250	19
.64199	1.55766	.66692	1.49944	.69243	1.44418	.71857	1.39165	18
.64240	1.55666	.66734	1.49849	.69286	1.44329	.71901	1.39079	17
.64281	1.55567	.66776	1.49755	.69329	1.44239	.71946	1.38994	16
.64322	1.55467	.66818	1.49661	.69372	1.44149	.71990	1.38909	15
.64363	1.55368	.66860	1.49566	.69416	1.44060	.72034	1.38824	14
.64404	1.55269	.66902	1.49472	.69459	1.43970	.72078	1.38738	13
.64446	1.55170	.66944	1.49378	.69502	1.43881	.72122	1.38653	12
.64487	1.55071	.66986	1.49284	.69545	1.43792	.72166	1.38568	11
.64528	1.54972	.67028	1.49190	.69588	1.43703	.72211	1.38484	10
.64569	1.54873	.67071	1.49097	.69631	1.43614	.72255	1.38399	9
.64610	1.54774	.67113	1.49003	.69675	1.43525	.72299	1.38314	8
.64652	1.54675	.67155	1.48909	.69718	1.43436	.72344	1.38229	7
.64693	1.54576	.67197	1.48816	.69761	1.43347	.72388	1.38145	6
.64734	1.54478	.67239	1.48722	.69804	1.43258	.72432	1.38060	5
.64775	1.54379	.67282	1.48629	.69847	1.43169	.72477	1.37976	4
.64817	1.54281	.67324	1.48536	.69891	1.43080	.72521	1.37891	3
.64858	1.54183	.67366	1.48442	.69934	1.42992	.72565	1.37807	2
.64899	1.54085	.67409	1.48349	.69977	1.42903	.72610	1.37722	1
.64941	1.53986	.67451	1.48256	.70021	1.42815	.72654	1.37638	0
CO-TAN.	TAN.	CO-TAN.	TAN.	CO-TAN.	TAN.	CO-TAN.	TAN.	′
	57°		56°		55°		54°	

′	36° TAN.	36° CO-TAN.	37° TAN.	37° CO-TAN.	38° TAN.	38° CO-TAN.	39° TAN.	39° CO-TAN.
0	.72654	1.37638	.75355	1.32704	.78129	1.27994	.80978	1.23490
1	.72699	1.37554	.75401	1.32624	.78175	1.27917	.81027	1.23416
2	.72743	1.37470	.75447	1.32544	.78222	1.27841	.81075	1.23343
3	.72788	1.37386	.75492	1.32464	.78269	1.27764	.81123	1.23270
4	.72832	1.37302	.75538	1.32384	.78316	1.27688	.81171	1.23196
5	.72877	1.37218	.75584	1.32304	.78363	1.27611	.81220	1.23123
6	.72921	1.37134	.75629	1.32224	.78410	1.27535	.81268	1.23050
7	.72966	1.37050	.75675	1.32144	.78457	1.27458	.81316	1.22977
8	.73010	1.36967	.75721	1.32064	.78504	1.27382	.81364	1.22904
9	.73055	1.36883	.75767	1.31984	.78551	1.27306	.81413	1.22831
10	.73100	1.36800	.75812	1.31904	.78598	1.27230	.81461	1.22758
11	.73144	1.36716	.75858	1.31825	.78645	1.27153	.81510	1.22685
12	.73189	1.36633	.75904	1.31745	.78692	1.27077	.81558	1.22612
13	.73234	1.36549	.75950	1.31666	.78739	1.27001	.81606	1.22539
14	.73278	1.36466	.75996	1.31586	.78786	1.26925	.81655	1.22467
15	.73323	1.36383	.76042	1.31507	.78834	1.26849	.81703	1.22394
16	.73368	1.36300	.76088	1.31427	.78881	1.26774	.81752	1.22321
17	.73413	1.36217	.76134	1.31348	.78928	1.26698	.81800	1.22249
18	.73457	1.36133	.76180	1.31269	.78975	1.26622	.81849	1.22176
19	.73502	1.36051	.76226	1.31190	.79022	1.26546	.81898	1.22104
20	.73547	1.35968	.76272	1.31110	.79070	1.26471	.81946	1.22031
21	.73592	1.35885	.76318	1.31031	.79117	1.26395	.81995	1.21959
22	.73637	1.35802	.76364	1.30952	.79164	1.26319	.82044	1.21886
23	.73681	1.35719	.76410	1.30873	.79212	1.26244	.82092	1.21814
24	.73726	1.35637	.76456	1.30795	.79259	1.26169	.82141	1.21742
25	.73771	1.35554	.76502	1.30716	.79306	1.26093	.82190	1.21670
26	.73816	1.35472	.76548	1.30637	.79354	1.26018	.82238	1.21598
27	.73861	1.35389	.76594	1.30558	.79401	1.25943	.82287	1.21526
28	.73906	1.35307	.76640	1.30480	.79449	1.25867	.82336	1.21454
29	.73951	1.35224	.76686	1.30401	.79496	1.25792	.82385	1.21382
30	.73996	1.35142	.76733	1.30323	.79544	1.25717	.82434	1.21310
31	.74041	1.35060	.76779	1.30244	.79591	1.25642	.82483	1.21238
32	.74086	1.34978	.76825	1.30166	.79639	1.25567	.82531	1.21166
33	.74131	1.34896	.76871	1.30087	.79686	1.25492	.82580	1.21094
34	.74176	1.34814	.76918	1.30009	.79734	1.25417	.82629	1.21023
35	.74221	1.34732	.76964	1.29931	.79781	1.25343	.82678	1.20951
36	.74267	1.34650	.77010	1.29853	.79829	1.25268	.82727	1.20879
37	.74312	1.34568	.77057	1.29775	.79877	1.25193	.82776	1.20808
38	.74357	1.34487	.77103	1.29696	.79924	1.25118	.82825	1.20736
39	.74402	1.34405	.77149	1.29618	.79972	1.25044	.82874	1.20665
40	.74447	1.34323	.77196	1.29541	.80020	1.24969	.82923	1.20593
41	.74492	1.34242	.77242	1.29463	.80067	1.24895	.82972	1.20522
42	.74538	1.34160	.77289	1.29385	.80115	1.24820	.83022	1.20451
43	.74583	1.34079	.77335	1.29307	.80163	1.24746	.83071	1.20379
44	.74628	1.33998	.77382	1.29229	.80211	1.24672	.83120	1.20308
45	.74674	1.33916	.77428	1.29152	.80258	1.24597	.83169	1.20237
46	.74719	1.33835	.77475	1.29074	.80306	1.24523	.83218	1.20166
47	.74764	1.33754	.77521	1.28997	.80354	1.24449	.83268	1.20095
48	.74810	1.33673	.77568	1.28919	.80402	1.24375	.83317	1.20024
49	.74855	1.33592	.77615	1.28842	.80450	1.24301	.83366	1.19953
50	.74900	1.33511	.77661	1.28764	.80498	1.24227	.83415	1.19882
51	.74946	1.33430	.77708	1.28687	.80546	1.24153	.83465	1.19811
52	.74991	1.33349	.77754	1.28610	.80594	1.24079	.83514	1.19740
53	.75037	1.33268	.77801	1.28533	.80642	1.24005	.83564	1.19669
54	.75082	1.33187	.77848	1.28456	.80690	1.23931	.83613	1.19599
55	.75128	1.33107	.77895	1.28379	.80738	1.23858	.83662	1.19528
56	.75173	1.33026	.77941	1.28302	.80786	1.23784	.83712	1.19457
57	.75219	1.32946	.77988	1.28225	.80834	1.23710	.83761	1.19387
58	.75264	1.32865	.78035	1.28148	.80882	1.23637	.83811	1.19316
59	.75310	1.32785	.78082	1.28071	.80930	1.23563	.83860	1.19246
60	.75355	1.32704	.78129	1.27994	.80978	1.23490	.83910	1.19175
′	CO-TAN.	TAN.	CO-TAN.	TAN.	CO-TAN.	TAN.	CO-TAN.	TAN.
	53°		52°		51°		50°	

40° Tan.	Co-tan.	41° Tan.	Co-tan.	42° Tan.	Co-tan.	43° Tan.	Co-tan.	'
.83910	1.19175	.86929	1.15037	.90040	1.11061	.93252	1.07237	60
.83960	1.19105	.86980	1.14969	.90093	1.10996	.93306	1.07174	59
.84009	1.19035	.87031	1.14902	.90146	1.10931	.93360	1.07112	58
.84059	1.18964	.87082	1.14834	.90199	1.10867	.93415	1.07049	57
.84108	1.18894	.87133	1.14767	.90251	1.10802	.93469	1.06987	56
.84158	1.18824	.87184	1.14699	.90304	1.10737	.93524	1.06925	55
.84208	1.18754	.87236	1.14632	.90357	1.10672	.93578	1.06862	54
.84258	1.18684	.87287	1.14565	.90410	1.10607	.93633	1.06800	53
.84307	1.18614	.87338	1.14498	.90463	1.10543	.93688	1.06738	52
.84357	1.18544	.87389	1.14430	.90516	1.10478	.93742	1.06676	51
.84407	1.18474	.87441	1.14363	.90569	1.10414	.93797	1.06613	50
.84457	1.18404	.87492	1.14296	.90621	1.10349	.93852	1.06551	49
.84507	1.18334	.87543	1.14229	.90674	1.10285	.93906	1.06489	48
.84556	1.18264	.87595	1.14162	.90727	1.10220	.93961	1.06427	47
.84606	1.18194	.87646	1.14095	.90781	1.10156	.94016	1.06365	46
.84656	1.18125	.87698	1.14028	.90834	1.10091	.94071	1.06303	45
.84706	1.18055	.87749	1.13961	.90887	1.10027	.94125	1.06241	44
.84756	1.17986	.87801	1.13894	.90940	1.09963	.94180	1.06179	43
.84806	1.17916	.87852	1.13828	.90993	1.09899	.94235	1.06117	42
.84856	1.17846	.87904	1.13761	.91046	1.09834	.94290	1.06056	41
.84906	1.17777	.87955	1.13694	.91099	1.09770	.94345	1.05994	40
·84956	1·17708	.88007	1.13627	.91153	1.09706	.94400	1.05932	39
.85006	1.17638	.88059	1.13561	.91206	1.09642	.94455	1.05870	38
.85057	1.17569	.88110	1.13494	.91259	1.09578	.94510	1.05809	37
.85107	1.17500	.88162	1.13428	.91313	1.09514	.94565	1.05747	36
.85157	1.17430	.88214	1.13361	.91366	1.09450	.94620	1.05685	35
.85207	1.17361	.88265	1.13295	.91419	1.09386	.94676	1.05624	34
.85257	1.17292	.88317	1.13228	.91473	1.09322	.94731	1.05562	33
.85307	1.17223	.88369	1.13162	.91526	1.09258	.94786	1.05501	32
.85358	1.17154	.88421	1.13096	.91580	1.09195	.94841	1.05439	31
.85408	1.17085	.88473	1.13029	.91633	1.09131	.94896	1.05378	30
.85458	1.17016	.88524	1.12963	.91687	1.09067	.94952	1.05317	29
.85509	1.16947	.88576	1.12897	.91740	1.09003	.95007	1.05255	28
.85559	1.16878	.88628	1.12831	.91794	1.08940	.95062	1.05194	27
.85609	1.16809	.88680	1.12765	.91847	1.08876	.95118	1.05133	26
.85660	1.16741	.88732	1.12699	.91901	1.08813	.95173	1.05072	25
.85710	1.16672	.88784	1.12633	.91955	1.08749	.95229	1.05010	24
.85761	1.16603	.88836	1.12567	.92008	1.08686	.95284	1.04949	23
.85811	1.16535	.88888	1.12501	.92062	1.08622	.95340	1.04888	22
.85862	1.16466	.88940	1.12435	.92116	1.08559	.95395	1.04827	21
.85912	1.16398	.88992	1.12369	.92170	1.08496	.95451	1.04766	20
.85963	1.16329	.89045	1.12303	.92224	1.08432	.95506	1.04705	19
.86014	1.16261	.89097	1.12238	.92277	1.08369	.95562	1.04644	18
.86064	1.16192	.89149	1.12172	.92331	1.08306	.95618	1.04583	17
.86115	1.16124	.89201	1.12106	.92385	1.08243	.95673	1.04522	16
.86166	1.16056	.89253	1.12041	.92439	1.08179	.95729	1.04461	15
.86216	1.15987	.89306	1.11975	.92493	1.08116	.95785	1.04401	14
.86267	1.15919	.89358	1.11909	.92547	1.08053	.95841	1.04340	13
.86318	1.15851	.89410	1.11844	.92601	1.07990	.95897	1.04279	12
.86368	1.15783	.89463	1.11778	.92655	1.07927	.95952	1.04218	11
.86419	1.15715	.89515	1.11713	.92709	1.07864	.96008	1.04158	10
.86470	1.15647	.89567	1.11648	.92763	1.07801	.96064	1.04097	9
.86521	1.15579	.89620	1.11582	.92817	1.07738	.96120	1.04036	8
.86572	1.15511	.89672	1.11517	.92872	1.07676	.96176	1.03976	7
.86623	1.15443	.89725	1.11452	.92926	1.07613	.96232	1.03915	6
.86674	1.15375	.89777	1.11387	.92980	1.07550	.96288	1.03855	5
.86725	1.15308	.89830	1.11321	.93034	1.07487	.96344	1.03794	4
.86776	1.15240	.89883	1.11256	.93088	1.07425	.96400	1.03734	3
.86827	1.15172	.89935	1.11191	.93143	1.07362	.96457	1.03674	2
.86878	1.15104	.89988	1.11126	.93197	1.07299	.96513	1.03613	1
.86929	1.15037	.90040	1.11061	.93252	1.07237	.96569	1.03553	0
Co-tan.	Tan.	Co-tan.	Tan.	Co-tan.	Tan.	Co-tan.	Tan.	
49°		48°		47°		46°		

44°

'	TAN.	CO-TAN.	'	'	TAN.	CO-TAN.	'	'	TAN.	CO-TAN.	'
0	.96569	1.03553	60	21	.97756	1.02295	39	41	.98901	1.01112	19
1	.96625	1.03493	59	22	.97813	1.02236	38	42	.98958	1.01053	18
2	.96681	1.03433	58	23	.97870	1.02176	37	43	.99016	1.00994	17
3	.96738	1.03372	57	24	.97927	1.02117	36	44	.99073	1.00935	16
4	.96794	1.03312	56	25	.97984	1.02057	35	45	.99131	1.00876	15
5	.96850	1.03252	55	26	.98041	1.01998	34	46	.99189	1.00818	14
6	.96907	1.03192	54	27	.98098	1.01939	33	47	.99247	1.00759	13
7	.96963	1.03132	53	28	.98155	1.01879	32	48	.99304	1.00701	12
8	.97020	1.03072	52	29	.98213	1.01820	31	49	.99362	1.00642	11
9	.97076	1.03012	51	30	.98270	1.01761	30	50	.99420	1.00583	10
10	.97133	1.02952	50	31	.98327	1.01702	29	51	.99478	1.00525	9
11	.97189	1.02892	49	32	.98384	1.01642	28	52	.99536	1.00467	8
12	.97246	1.02832	48	33	.98441	1.01583	27	53	.99594	1.00408	7
13	.97302	1.02772	47	34	.98499	1.01524	26	54	.99652	1.00350	6
14	.97359	1.02713	46	35	.98556	1.01465	25	55	.99710	1.00291	5
15	.97416	1.02653	45	36	.98613	1.01406	24	56	.99768	1.00233	4
16	.97472	1.02593	44	37	.98671	1.01347	23	57	.99826	1.00175	3
17	.97529	1.02533	43	38	.98728	1.01288	22	58	.99884	1.00116	2
18	.97586	1.02474	42	39	.98786	1.01229	21	59	.99942	1.00058	1
19	.97643	1.02414	41	40	.98843	1.01170	20	60	1	1	0
20	.97700	1.02355	40								
'	CO-TAN.	TAN.		'	CO-TAN.	TAN.		'	CO-TAN.	TAN.	'

45°

NATURAL SINES AND COSINES

0°

'	SINE	COSINE	'	'	SINE	COSINE	'	'	SINE	COSINE	'
0	.00000	1	60	21	.00611	.99998	39	41	.01193	.99993	19
1	.00029	1	59	22	.00640	.99998	38	42	.01222	.99993	18
2	.00058	1	58	23	.00669	.99998	37	43	.01251	.99992	17
3	.00087	1	57	24	.00698	.99998	36	44	.01280	.99992	16
4	.00116	1	56	25	.00727	.99997	35	45	.01309	.99991	15
5	.00145	1	55	26	.00756	.99997	34	46	.01338	.99991	14
6	.00175	1	54	27	.00785	.99997	33	47	.01367	.99991	13
7	.00204	1	53	28	.00814	.99997	32	48	.01396	.99990	12
8	.00233	1	52	29	.00844	.99996	31	49	.01425	.99990	11
9	.00262	1	51	30	.00873	.99996	30	50	.01454	.99989	10
10	.00291	1	50	31	.00902	.99996	29	51	.01483	.99989	9
11	.00320	.99999	49	32	.00931	.99996	28	52	.01513	.99989	8
12	.00349	.99999	48	33	.00960	.99995	27	53	.01542	.99988	7
13	.00378	.99999	47	34	.00989	.99995	26	54	.01571	.99988	6
14	.00407	.99999	46	35	.01018	.99995	25	55	.01600	.99987	5
15	.00436	.99999	45	36	.01047	.99995	24	56	.01629	.99987	4
16	.00465	.99999	44	37	.01076	.99994	23	57	.01658	.99986	3
17	.00495	.99999	43	38	.01105	.99994	22	58	.01687	.99986	2
18	.00524	.99999	42	39	.01134	.99994	21	59	.01716	.99985	1
19	.00553	.99998	41	40	.01164	.99993	20	60	.01745	.99985	0
10	.00582	.99998	40								
'	COSINE	SINE		'	COSINE	SINE		'	COSINE	SINE	'

89°

Sine (1°)	Cosine (1°)	Sine (2°)	Cosine (2°)	Sine (3°)	Cosine (3°)	Sine (4°)	Cosine (4°)	'
.01745	.99985	.03490	.99939	.05234	.99863	.06976	.99756	60
.01774	.99984	.03519	.99938	.05263	.99861	.07005	.99754	59
.01803	.99984	.03548	.99937	.05292	.99860	.07034	.99752	58
.01832	.99983	.03577	.99936	.05321	.99858	.07063	.99750	57
.01862	.99983	.03606	.99935	.05350	.99857	.07092	.99748	56
.01891	.99982	.03635	.99934	.05379	.99855	.07121	.99746	55
.01920	.99982	.03664	.99933	.05408	.99854	.07150	.99744	54
.01949	.99981	.03693	.99932	.05437	.99852	.07179	.99742	53
.01978	.99980	.03723	.99931	.05466	.99851	.07208	.99740	52
.02007	.99980	.03752	.99930	.05495	.99849	.07237	.99738	51
.02036	.99979	.03781	.99929	.05524	.99847	.07266	.99736	50
.02065	.99979	.03810	.99927	.05553	.99846	.07295	.99734	49
.02094	.99978	.03839	.99926	.05582	.99844	.07324	.99731	48
.02123	.99977	.03868	.99925	.05611	.99842	.07353	.99729	47
.02152	.99977	.03897	.99924	.05640	.99841	.07382	.99727	46
.02181	.99976	.03926	.99923	.05669	.99839	.07411	.99725	45
.02211	.99976	.03955	.99922	.05698	.99838	.07440	.99723	44
.02240	.99975	.03984	.99921	.05727	.99836	.07469	.99721	43
.02269	.99974	.04013	.99919	.05756	.99834	.07498	.99719	42
.02298	.99974	.04042	.99918	.05785	.99833	.07527	.99716	41
.02327	.99973	.04071	.99917	.05814	.99831	.07556	.99714	40
.02356	.99972	.04100	.99916	.05844	.99829	.07585	.99712	39
.02385	.99972	.04129	.99915	.05873	.99827	.07614	.99710	38
.02414	.99971	.04159	.99913	.05902	.99826	.07643	.99708	37
.02443	.99970	.04188	.99912	.05931	.99824	.07672	.99705	36
.02472	.99969	.04217	.99911	.05960	.99822	.07701	.99703	35
.02501	.99969	.04246	.99910	.05989	.99821	.07730	.99701	34
.02530	.99968	.04275	.99909	.06018	.99819	.07759	.99699	33
.02560	.99967	.04304	.99907	.06047	.99817	.07788	.99696	32
.02589	.99966	.04333	.99906	.06076	.99815	.07817	.99694	31
.02618	.99966	.04362	.99905	.06105	.99813	.07846	.99692	30
.02647	.99965	.04391	.99904	.06134	.99812	.07875	.99689	29
.02676	.99964	.04420	.99902	.06163	.99810	.07904	.99687	28
.02705	.99963	.04449	.99901	.06192	.99808	.07933	.99685	27
.02734	.99963	.04478	.99900	.06221	.99806	.07962	.99683	26
.02763	.99962	.04507	.99898	.06250	.99804	.07991	.99680	25
.02792	.99961	.04536	.99897	.06279	.99803	.08020	.99678	24
.02821	.99960	.04565	.99896	.06308	.99801	.08049	.99676	23
.02850	.99959	.04594	.99894	.06337	.99799	.08078	.99673	22
.02879	.99959	.04623	.99893	.06366	.99797	.08107	.99671	21
.02908	.99958	.04653	.99892	.06395	.99795	.08136	.99668	20
.02938	.99957	.04682	.99890	.06424	.99793	.08165	.99666	19
.02967	.99956	.04711	.99889	.06453	.99792	.08194	.99664	18
.02996	.99955	.04740	.99888	.06482	.99790	.08223	.99661	17
.03025	.99954	.04769	.99886	.06511	.99788	.08252	.99659	16
.03054	.99953	.04798	.99885	.06540	.99786	.08281	.99657	15
.03083	.99952	.04827	.99883	.06569	.99784	.08310	.99654	14
.03112	.99952	.04856	.99882	.06598	.99782	.08339	.99652	13
.03141	.99951	.04885	.99881	.06627	.99780	.08368	.99649	12
.03170	.99950	.04914	.99879	.06656	.99778	.08397	.99647	11
.03199	.99949	.04943	.99878	.06685	.99776	.08426	.99644	10
.03228	.99948	.04972	.99876	.06714	.99774	.08455	.99642	9
.03257	.99947	.05001	.99875	.06743	.99772	.08484	.99639	8
.03286	.99946	.05030	.99873	.06773	.99770	.08513	.99637	7
.03316	.99945	.05059	.99872	.06802	.99768	.08542	.99635	6
.03345	.99944	.05088	.99870	.06831	.99766	.08571	.99632	5
.03374	.99943	.05117	.99869	.06860	.99764	.08600	.99630	4
.03403	.99942	.05146	.99867	.06889	.99762	.08629	.99627	3
.03432	.99941	.05175	.99866	.06918	.99760	.08658	.99625	2
.03461	.99940	.05205	.99864	.06947	.99758	.08687	.99622	1
.03490	.99939	.05234	.99863	.06976	.99756	.08716	.99619	0

Cosine	Sine	Cosine	Sine	Cosine	Sine	Cosine	Sine	'
88°		87°		86°		85°		

′	5° Sine	5° Cosine	6° Sine	6° Cosine	7° Sine	7° Cosine	8° Sine	8° Cosine	′
0	.08716	.99619	.10453	.99452	.12187	.99255	.13917	.99027	60
1	.08745	.99617	.10482	.99449	.12216	.99251	.13946	.99023	59
2	.08774	.99614	.10511	.99446	.12245	.99248	.13975	.99019	58
3	.08803	.99612	.10540	.99443	.12274	.99244	.14004	.99015	57
4	.08831	.99609	.10569	.99440	.12302	.99240	.14033	.99011	56
5	.08860	.99607	.10597	.99437	.12331	.99237	.14061	.99006	55
6	.08889	.99604	.10626	.99434	.12360	.99233	.14090	.99002	54
7	.08918	.99602	.10655	.99431	.12389	.99230	.14119	.98998	53
8	.08947	.99599	.10684	.99428	.12418	.99226	.14148	.98994	52
9	.08976	.99596	.10713	.99424	.12447	.99222	.14177	.98990	51
10	.09005	.99594	.10742	.99421	.12476	.99219	.14205	.98986	50
11	.09034	.99591	.10771	.99418	.12504	.99215	.14234	.98982	49
12	.09063	.99588	.10800	.99415	.12533	.99211	.14263	.98978	48
13	.09092	.99586	.10829	.99412	.12562	.99208	.14292	.98973	47
14	.09121	.99583	.10858	.99409	.12591	.99204	.14320	.98969	46
15	.09150	.99580	.10887	.99406	.12620	.99200	.14349	.98965	45
16	.09179	.99578	.10916	.99402	.12649	.99197	.14378	.98961	44
17	.09208	.99575	.10945	.99399	.12678	.99193	.14407	.98957	43
18	.09237	.99572	.10973	.99396	.12706	.99189	.14436	.98953	42
19	.09266	.99570	.11002	.99393	.12735	.99186	.14464	.98948	41
20	.09295	.99567	.11031	.99390	.12764	.99182	.14493	.98944	40
21	.09324	.99564	.11060	.99386	.12793	.99178	.14522	.98940	39
22	.09353	.99562	.11089	.99383	.12822	.99175	.14551	.98936	38
23	.09382	.99559	.11118	.99380	.12851	.99171	.14580	.98931	37
24	.09411	.99556	.11147	.99377	.12880	.99167	.14608	.98927	36
25	.09440	.99553	.11176	.99374	.12908	.99163	.14637	.98923	35
26	.09469	.99551	.11205	.99370	.12937	.99160	.14666	.98919	34
27	.09498	.99548	.11234	.99367	.12966	.99156	.14695	.98914	33
28	.09527	.99545	.11263	.99364	.12995	.99152	.14723	.98910	32
29	.09556	.99542	.11291	.99360	.13024	.99148	.14752	.98906	31
30	.09585	.99540	.11320	.99357	.13053	.99144	.14781	.98902	30
31	.09614	.99537	.11349	.99354	.13081	.99141	.14810	.98897	29
32	.09642	.99534	.11378	.99351	.13110	.99137	.14838	.98893	28
33	.09671	.99531	.11407	.99347	.13139	.99133	.14867	.98889	27
34	.09700	.99528	.11436	.99344	.13168	.99129	.14896	.98884	26
35	.09729	.99526	.11465	.99341	.13197	.99125	.14925	.98880	25
36	.09758	.99523	.11494	.99337	.13226	.99122	.14954	.98876	24
37	.09787	.99520	.11523	.99334	.13254	.99118	.14982	.98871	23
38	.09816	.99517	.11552	.99331	.13283	.99114	.15011	.98867	22
39	.09845	.99514	.11580	.99327	.13312	.99110	.15040	.98863	21
40	.09874	.99511	.11609	.99324	.13341	.99106	.15069	.98858	20
41	.09903	.99508	.11638	.99320	.13370	.99102	.15097	.98854	19
42	.09932	.99506	.11667	.99317	.13399	.99098	.15126	.98849	18
43	.09961	.99503	.11696	.99314	.13427	.99094	.15155	.98845	17
44	.09990	.99500	.11725	.99310	.13456	.99091	.15184	.98841	16
45	.10019	.99497	.11754	.99307	.13485	.99087	.15212	.98836	15
46	.10048	.99494	.11783	.99303	.13514	.99083	.15241	.98832	14
47	.10077	.99491	.11812	.99300	.13543	.99079	.15270	.98827	13
48	.10106	.99488	.11840	.99297	.13572	.99075	.15299	.98823	12
49	.10135	.99485	.11869	.99293	.13600	.99071	.15327	.98818	11
50	.10164	.99482	.11898	.99290	.13629	.99067	.15356	.98814	10
51	.10192	.99479	.11927	.99286	.13658	.99063	.15385	.98809	9
52	.10221	.99476	.11956	.99283	.13687	.99059	.15414	.98805	8
53	.10250	.99473	.11985	.99279	.13716	.99055	.15442	.98800	7
54	.10279	.99470	.12014	.99276	.13744	.99051	.15471	.98796	6
55	.10308	.99467	.12043	.99272	.13773	.99047	.15500	.98791	5
56	.10337	.99464	.12071	.99269	.13802	.99043	.15529	.98787	4
57	.10366	.99461	.12100	.99265	.13831	.99039	.15557	.98782	3
58	.10395	.99458	.12129	.99262	.13860	.99035	.15586	.98778	2
59	.10424	.99455	.12158	.99258	.13889	.99031	.15615	.98773	1
60	.10453	.99452	.12187	.99255	.13917	.99027	.15643	.98769	0

′	Cosine	Sine	Cosine	Sine	Cosine	Sine	Cosine	Sine	′
	84°		83°		82°		81°		

9° SINE	COSINE	10° SINE	COSINE	11° SINE	COSINE	12° SINE	COSINE
.15643	.98769	.17365	.98481	.19081	.98163	.20791	.97815
.15672	.98764	.17393	.98476	.19109	.98157	.20820	.97809
.15701	.98760	.17422	.98471	.19138	.98152	.20848	.97803
.15730	.98755	.17451	.98466	.19167	.98146	.20877	.97797
.15758	.98751	.17479	.98461	.19195	.98140	.20905	.97791
.15787	.98746	.17508	.98455	.19224	.98135	.20933	.97784
.15816	.98741	.17537	.98450	.19252	.98129	.20962	.97778
.15845	.98737	.17565	.98445	.19281	.98124	.20990	.97772
.15873	.98732	.17594	.98440	.19309	.98118	.21019	.97766
.15902	.98728	.17623	.98435	.19338	.98112	.21047	.97760
.15931	.98723	.17651	.98430	.19366	.98107	.21076	.97754
.15959	.98718	.17680	.98425	.19395	.98101	.21104	.97748
.15988	.98714	.17708	.98420	.19423	.98096	.21132	.97742
.16017	.98709	.17737	.98414	.19452	.98090	.21161	.97735
.16046	.98704	.17766	.98409	.19481	.98084	.21189	.97729
.16074	.98700	.17794	.98404	.19509	.98079	.21218	.97723
.16103	.98695	.17823	.98399	.19538	.98073	.21246	.97717
.16132	.98690	.17852	.98394	.19566	.98067	.21275	.97711
.16160	.98689	.17880	.98389	.19595	.98061	.21303	.97705
.16189	.98681	.17909	.98383	.19623	.98056	.21331	.97698
.16218	.98676	.17937	.98378	.19652	.98050	.21360	.97692
.16246	.98671	.17966	.98373	.19680	.98044	.21388	.97686
.16275	.98667	.17995	.98368	.19709	.98039	.21417	.97680
.16304	.98662	.18023	.98362	.19737	.98033	.21445	.97673
.16333	.98657	.18052	.98357	.19766	.98027	.21474	.97667
.16361	.98652	.18081	.98352	.19794	.98021	.21502	.97661
.16390	.98648	.18109	.98347	.19823	.98016	.21530	.97655
.16419	.98643	.18138	.98341	.19851	.98010	.21559	.97648
.16447	.98638	.18166	.98336	.19880	.98004	.21587	.97642
.16476	.98633	.18195	.98331	.19908	.97988	.21616	.97636
.16505	.98629	.18224	.98325	.19937	.97992	.21644	.97630
.16533	.98624	.18252	.98320	.19965	.97987	.21672	.97623
.16562	.98619	.18281	.98315	.19994	.97981	.21701	.97617
.16591	.98614	.18309	.98310	.20022	.97975	.21729	.97611
.16620	.98609	.18338	.98304	.20031	.97969	.21758	.97604
.16648	.98604	.18367	.98299	.20079	.97963	.21786	.97598
.16677	.98600	.18395	.98294	.20108	.97958	.21814	.97592
.16706	.98595	.18424	.98288	.20136	.97952	.21843	.97585
.16734	.98590	.18452	.98283	.20165	.97946	.21871	.97579
.16763	.98585	.18481	.98277	.20193	.97940	.21899	.97573
.16792	.98580	.18509	.98272	.20222	.97934	.21928	.97566
.16820	.98575	.18538	.98267	.20250	.97928	.21956	.97560
.16849	.98570	.18567	.98261	.20279	.97922	.21985	.97553
.16878	.98565	.18595	.98256	.20307	.97916	.22013	.97547
.16906	.98561	.18624	.98250	.20336	.97910	.22041	.97541
.16935	.98556	.18652	.98245	.20364	.97905	.22070	.97534
.16964	.98551	.18681	.98240	.20393	.97899	.22098	.97528
.16992	.98546	.18710	.98234	.20421	.97893	.22126	.97521
.17021	.98541	.18738	.98229	.20450	.97887	.22155	.97515
.17050	.98536	.18767	.98223	.20478	.97881	.22183	.97508
.17078	.98531	.18795	.98218	.20507	.97875	.22212	.97502
.17107	.98526	.18824	.98212	.20535	.97869	.22240	.97496
.17136	.98521	.18852	.98207	.20563	.97863	.22268	.97489
.17164	.98516	.18881	.98201	.20592	.97857	.22297	.97483
.17193	.98511	.18910	.98196	.20620	.97851	.22325	.97476
.17222	.98506	.18938	.98190	.20649	.97845	.22353	.97470
.17250	.98501	.18967	.98185	.20677	.97839	.22382	.97463
.17279	.98496	.18995	.98179	.20706	.97833	.22410	.97457
.17308	.98491	.19024	.98174	.20734	.97827	.22438	.97450
.17336	.98486	.19052	.98168	.20763	.97821	.22467	.97444
.17365	.98481	.19081	.98163	.20791	.97815	.22495	.97437
COSINE	SINE	COSINE	SINE	COSINE	SINE	COSINE	SINE
80°		**79°**		**78°**		**77°**	

'	13° Sine	Cosine	14° Sine	Cosine	15° Sine	Cosine	16° Sine	Cosine	'
0	.22495	.97437	.24192	.97030	.25882	.96593	.27564	.96126	60
1	.22523	.97430	.24220	.97023	.25910	.96585	.27592	.96118	59
2	.22552	.97424	.24249	.97015	.25938	.96578	.27620	.96110	58
3	.22580	.97417	.24277	.97008	.25966	.96570	.27648	.96102	57
4	.22608	.97411	.24305	.97001	.25994	.96562	.27676	.96094	56
5	.22637	.97404	.24333	.96994	.26022	.96555	.27704	.96086	55
6	.22665	.97398	.24362	.96987	.26050	.96547	.27731	.96078	54
7	.22693	.97391	.24390	.96980	.26079	.96540	.27759	.96070	53
8	.22722	.97384	.24418	.96973	.26107	.96532	.27787	.96062	52
9	.22750	.97378	.24446	.96966	.26135	.96524	.27815	.96054	51
10	.22778	.97371	.24474	.96959	.26163	.96517	.27843	.96046	50
11	.22807	.97365	.24503	.96952	.26191	.96509	.27871	.96037	49
12	.22835	.97358	.24531	.96945	.26219	.96502	.27899	.96029	48
13	.22863	.97351	.24559	.96937	.26247	.96494	.27927	.96021	47
14	.22892	.97345	.24587	.96930	.26275	.96486	.27955	.96013	46
15	.22920	.97338	.24615	.96923	.26303	.96479	.27983	.96005	45
16	.22948	.97331	.24644	.96916	.26331	.96471	.28011	.95997	44
17	.22977	.97325	.24672	.96909	.26359	.96463	.28039	.95989	43
18	.23005	.97318	.24700	.96902	.26387	.96456	.28067	.95981	42
19	.23033	.97311	.24728	.96894	.26415	.96448	.28095	.95972	41
20	.23062	.97304	.24756	.96887	.26443	.96440	.28123	.95964	40
21	.23090	.97298	.24784	.96880	.26471	.96433	.28150	.95956	39
22	.23118	.97291	.24813	.96873	.26500	.96425	.28178	.95948	38
23	.23146	.97284	.24841	.96866	.26528	.96417	.28206	.95940	37
24	.23175	.97278	.24869	.96858	.26556	.96410	.28234	.95931	36
25	.23203	.97271	.24897	.96851	.26584	.96402	.28262	.95923	35
26	.23231	.97264	.24925	.96844	.26612	.96394	.28290	.95915	34
27	.23260	.97257	.24954	.96837	.26640	.96386	.28318	.95907	33
28	.23288	.97251	.24982	.96829	.26668	.96379	.28346	.95898	32
29	.23316	.97244	.25010	.96822	.26696	.96371	.28374	.95890	31
30	.23345	.97237	.25038	.96815	.26724	.96363	.28402	.95882	30
31	.23373	.97230	.25066	.96807	.26752	.96355	.28429	.95874	29
32	.23401	.97223	.25094	.96800	.26780	.96347	.28457	.95865	28
33	.23429	.97217	.25122	.96793	.26808	.96340	.28485	.95857	27
34	.23458	.97210	.25151	.96786	.26836	.96332	.28513	.95849	26
35	.23486	.97203	.25179	.96778	.26864	.96324	.28541	.95841	25
36	.23514	.97196	.25207	.96771	.26892	.96316	.28569	.95832	24
37	.23542	.97189	.25235	.96764	.26920	.96308	.28597	.95824	23
38	.23571	.97182	.25263	.96756	.26948	.96301	.28625	.95816	22
39	.23599	.97176	.25291	.96749	.26976	.96293	.28652	.95807	21
40	.23627	.97169	.25320	.96742	.27004	.96285	.28680	.95799	20
41	.23656	.97162	.25348	.96734	.27032	.96277	.28708	.95790	19
42	.23684	.97155	.25376	.96727	.27060	.96269	.28736	.95782	18
43	.23712	.97148	.25404	.96719	.27088	.96261	.28764	.95774	17
44	.23740	.97141	.25432	.96712	.27116	.96253	.28792	.95766	16
45	.23769	.97134	.25460	.96705	.27144	.96246	.28820	.95757	15
46	.23797	.97127	.25488	.96697	.27172	.96238	.28847	.95749	14
47	.23825	.97120	.25516	.96690	.27200	.96230	.28875	.95740	13
48	.23853	.97113	.25545	.96682	.27228	.96222	.28903	.95732	12
49	.23882	.97106	.25573	.96675	.27256	.96214	.28931	.95724	11
50	.23910	.97100	.25601	.96667	.27284	.96206	.28959	.95715	10
51	.23938	.97093	.25629	.96660	.27312	.96198	.28987	.95707	9
52	.23966	.97086	.25657	.96653	.27340	.96190	.29015	.95698	8
53	.23995	.97079	.25685	.96645	.27368	.96182	.29042	.95690	7
54	.24023	.97072	.25713	.96638	.27396	.96174	.29070	.95681	6
55	.24051	.97065	.25741	.96630	.27424	.96166	.29098	.95673	5
56	.24079	.97058	.25769	.96623	.27452	.96158	.29126	.95664	4
57	.24108	.97051	.25798	.96615	.27480	.96150	.29154	.95656	3
58	.24136	.97044	.25826	.96608	.27508	.96142	.29182	.95647	2
59	.24164	.97037	.25854	.96600	.27536	.96134	.29209	.95639	1
60	.24192	.97030	.25882	.96593	.27564	.96126	.29237	.95630	0
'	Cosine	Sine	Cosine	Sine	Cosine	Sine	Cosine	Sine	'
	76°		75°		74°		73°		

	17°		18°		19°		20°	
	Sine	Cosine	Sine	Cosine	Sine	Cosine	Sine	Cosine	′
	.29237	.95630	.30902	.95106	.32557	.94552	.34202	.93969	60
	.29265	.95622	.30929	.95097	.32584	.94542	.34229	.93959	59
	.29293	.95613	.30957	.95088	.32612	.94533	.34257	.93949	58
	.29321	.95605	.30985	.95079	.32639	.94523	.34284	.93939	57
	.29348	.95596	.31012	.95070	.32667	.94514	.34311	.93929	56
	.29376	.95588	.31040	.95061	.32694	.94504	.34339	.93919	55
	.29404	.95579	.31068	.95052	.32722	.94495	.34366	.93909	54
	.29432	.95571	.31095	.95043	.32749	.94485	.34393	.93899	53
	.29460	.95562	.31123	.95033	.32777	.94476	.34421	.93889	52
	.29487	.95554	.31151	.95024	.32804	.94466	.34448	.93879	51
	.29515	.95545	.31178	.95015	.32832	.94457	.34475	.93869	50
	.29543	.95536	.31206	.95006	.32859	.94447	.34503	.93859	49
	.29571	.95528	.31233	.94997	.32887	.94438	.34530	.93849	48
	.29599	.95519	.31261	.94988	.32914	.94428	.34557	.93839	47
	.29626	.95511	.31289	.94979	.32942	.94418	.34584	.93829	46
	.29654	.95502	.31316	.94970	.32969	.94409	.34612	.93819	45
	.29682	.95493	.31344	.94961	.32997	.94399	.34639	.93809	44
	.29710	.95485	.31372	.94952	.33024	.94390	.34666	.93799	43
	.29737	.95476	.31399	.94943	.33051	.94380	.34694	.93789	42
	.29765	.95467	.31427	.94933	.33079	.94370	.34721	.93779	41
	.29793	.95459	.31454	.94924	.33106	.94361	.34748	.93769	40
	.29821	.95450	.31482	.94915	.33134	.94351	.34775	.93759	39
	.29849	.95441	.31510	.94906	.33161	.94342	.34803	.93748	38
	.29876	.95433	.31537	.94897	.33189	.94332	.34830	.93738	37
	.29904	.95424	.31565	.94888	.33216	.94322	.34857	.93728	36
	.29932	.95415	.31593	.94878	.33244	.94313	.34884	.93718	35
	.29960	.95407	.31620	.94869	.33271	.94303	.34912	.93708	34
	.29987	.95398	.31648	.94860	.33298	.94293	.34939	.93698	33
	.30015	.95389	.31675	.94851	.33326	.94284	.34966	.93688	32
	.30043	.95380	.31703	.94842	.33353	.94274	.34993	.93677	31
	.30071	.95372	.31730	.94832	.33381	.94264	.35021	.93667	30
	.30098	.95363	.31758	.94823	.33408	.94254	.35048	.93657	29
	.30126	.95354	.31786	.94814	.33436	.94245	.35075	.93647	28
	.30154	.95345	.31813	.94805	.33463	.94235	.35102	.93637	27
	.30182	.95337	.31841	.94795	.33490	.94225	.35130	.93626	26
	.30209	.95328	.31868	.94786	.33518	.94215	.35157	.93616	25
	.30237	.95319	.31896	.94777	.33545	.94206	.35184	.93606	24
	.30265	.95310	.31923	.94768	.33573	.94196	.35211	.93596	23
	.30292	.95301	.31951	.94758	.33600	.94186	.35239	.93585	22
	.30320	.95293	.31979	.94749	.33627	.94176	.35266	.93575	21
	.30348	.95284	.32006	.94740	.33655	.94167	.35293	.93565	20
	.30376	.95275	.32034	.94730	.33682	.94157	.35320	.93555	19
	.30403	.95266	.32061	.94721	.33710	.94147	.35347	.93544	18
	.30431	.95257	.32089	.94712	.33737	.94137	.35375	.93534	17
	.30459	.95248	.32116	.94702	.33764	.94127	.35402	.93524	16
	.30486	.95240	.32144	.94693	.33792	.94118	.35429	.93514	15
	.30514	.95231	.32171	.94684	.33819	.94108	.35456	.93503	14
	.30542	.95222	.32199	.94674	.33846	.94098	.35484	.93493	13
	.30570	.95213	.32227	.94665	.33874	.94088	.35511	.93483	12
	.30597	.95204	.32254	.94656	.33901	.94078	.35538	.93472	11
	.30625	.95195	.32282	.94646	.33929	.94068	.35565	.93462	10
	.30653	.95186	.32309	.94637	.33956	.94058	.35592	.93452	9
	.30680	.95177	.32337	.94627	.33983	.94049	.35619	.93441	8
	.30708	.95168	.32364	.94618	.34011	.94039	.35647	.93431	7
	.30736	.95159	.32392	.94609	.34038	.94029	.35674	.93420	6
	.30763	.95150	.32419	.94599	.34065	.94019	.35701	.93410	5
	.30791	.95142	.32447	.94590	.34093	.94009	.35728	.93400	4
	.30819	.95133	.32474	.94580	.34120	.93999	.35755	.93389	3
	.30846	.95124	.32502	.94571	.34147	.93989	.35782	.93379	2
	.30874	.95115	.32529	.94561	.34175	.93979	.35810	.93368	1
	.30902	.95106	.32557	.94552	.34202	.93969	.35837	.93358	0
	Cosine	Sine	Cosine	Sine	Cosine	Sine	Cosine	Sine	′
		72°		71°		70°		69°	

′	21° SINE	21° COSINE	22° SINE	22° COSINE	23° SINE	23° COSINE	24° SINE	24° COSINE
0	.35837	.93358	.37461	.92718	.39073	.92050	.40674	.91355
1	.35864	.93348	.37488	.92707	.39100	.92039	.40700	.91343
2	.35891	.93337	.37515	.92697	.39127	.92028	.40727	.91331
3	.35918	.93327	.37542	.92686	.39153	.92016	.40753	.91319
4	.35945	.93316	.37569	.92675	.39180	.92005	.40780	.91307
5	.35973	.93306	.37595	.92664	.39207	.91994	.40806	.91295
6	.36000	.93295	.37622	.92653	.39234	.91982	.40833	.91283
7	.36027	.93285	.37649	.92642	.39260	.91971	.40860	.91272
8	.36054	.93274	.37676	.92631	.39287	.91959	.40886	.91260
9	.36081	.93264	.37703	.92620	.39314	.91948	.40913	.91248
10	.36108	.93253	.37730	.92609	.39341	.91936	.40939	.91236
11	.36135	.93243	.37757	.92598	.39367	.91925	.40966	.91224
12	.36162	.93232	.37784	.92587	.39394	.91914	.40992	.91212
13	.36190	.93222	.37811	.92576	.39421	.91902	.41019	.91200
14	.36217	.93211	.37838	.92565	.39448	.91891	.41045	.91188
15	.36244	.93201	.37865	.92554	.39474	.91879	.41072	.91176
16	.36271	.93190	.37892	.92543	.39501	.91868	.41098	.91164
17	.36298	.93180	.37919	.92532	.39528	.91856	.41125	.91152
18	.36325	.93169	.37946	.92521	.39555	.91845	.41151	.91140
19	.36352	.93159	.37973	.92510	.39581	.91833	.41178	.91128
20	.36379	.93148	.37999	.92499	.39608	.91822	.41204	.91116
21	.36406	.93137	.38026	.92488	.39635	.91810	.41231	.91104
22	.36434	.93127	.38053	.92477	.39661	.91799	.41257	.91092
23	.36461	.93116	.38080	.92466	.39688	.91787	.41284	.91080
24	.36488	.93106	.38107	.92455	.39715	.91775	.41310	.91068
25	.36515	.93095	.38134	.92444	.39741	.91764	.41337	.91056
26	.36542	.93084	.38161	.92432	.39768	.91752	.41363	.91044
27	.36569	.93074	.38188	.92421	.39795	.91741	.41390	.91032
28	.36596	.93063	.38215	.92410	.39822	.91729	.41416	.91020
29	.36623	.93052	.38241	.92399	.39848	.91718	.41443	.91008
30	.36650	.93042	.38268	.92388	.39875	.91706	.41469	.90996
31	.36677	.93031	.38295	.92377	.39902	.91694	.41496	.90984
32	.36704	.93020	.38322	.92366	.39928	.91683	.41522	.90972
33	.36731	.93010	.38349	.92355	.39955	.91671	.41549	.90960
34	.36758	.92999	.38376	.92343	.39982	.91660	.41575	.90948
35	.36785	.92988	.38403	.92332	.40008	.91648	.41602	.90936
36	.36812	.92978	.38430	.92321	.40035	.91636	.41628	.90924
37	.36839	.92967	.38456	.92310	.40062	.91625	.41655	.90911
38	.36867	.92956	.38483	.92299	.40088	.91613	.41681	.90899
39	.36894	.92945	.38510	.92287	.40115	.91601	.41707	.90887
40	.36921	.92935	.38537	.92276	.40141	.91590	.41734	.90875
41	.36948	.92924	.38564	.92265	.40168	.91578	.41760	.90863
42	.36975	.92913	.38591	.92254	.40195	.91566	.41787	.90851
43	.37002	.92902	.38617	.92243	.40221	.91555	.41813	.90839
44	.37029	.92892	.38644	.92231	.40248	.91543	.41840	.90826
45	.37056	.92881	.38671	.92220	.40275	.91531	.41866	.90814
46	.37083	.92870	.38698	.92209	.40301	.91519	.41892	.90802
47	.37110	.92859	.38725	.92198	.40328	.91508	.41919	.90790
48	.37137	.92849	.38752	.92186	.40355	.91496	.41945	.90778
49	.37164	.92838	.38778	.92175	.40381	.91484	.41972	.90766
50	.37191	.92827	.38805	.92164	.40408	.91472	.41998	.90753
51	.37218	.92816	.38832	.92152	.40434	.91461	.42024	.90741
52	.37245	.92805	.38859	.92141	.40461	.91449	.42051	.90729
53	.37272	.92794	.38886	.92130	.40488	.91437	.42077	.90717
54	.37299	.92784	.38912	.92119	.40514	.91425	.42104	.90704
55	.37326	.92773	.38939	.92107	.40541	.91414	.42130	.90692
56	.37353	.92762	.38966	.92096	.40567	.91402	.42156	.90680
57	.37380	.92751	.38993	.92085	.40594	.91390	.42183	.90668
58	.37407	.92740	.39020	.92073	.40621	.91378	.42209	.90655
59	.37434	.92729	.39046	.92062	.40647	.91366	.42235	.90643
60	.37461	.92718	.39073	.92050	.40674	.91355	.42262	.90631
	COSINE	SINE	COSINE	SINE	COSINE	SINE	COSINE	SINE
	68°		67°		66°		65°	

NATURAL SINES AND COSINES

25°		26°		27°		28°	
SINE	COSINE	SINE	COSINE	SINE	COSINE	SINE	COSINE
.42262	.90631	.43837	.89879	.45399	.89101	.46947	.88295
.42288	.90618	.43863	.89867	.45425	.89087	.46973	.88281
.42315	.90606	.43889	.89854	.45451	.89074	.46999	.88267
.42341	.90594	.43916	.89841	.45477	.89061	.47024	.88254
.42367	.90582	.43942	.89828	.45503	.89048	.47050	.88240
.42394	.90569	.43968	.89816	.45529	.89035	.47076	.88226
.42420	.90557	.43994	.89803	.45554	.89021	.47101	.88213
.42446	.90545	.44020	.89790	.45580	.89008	.47127	.88199
.42473	.90532	.44046	.89777	.45606	.88995	.47153	.88185
.42499	.90520	.44072	.89764	.45632	.88981	.47178	.88172
.42525	.90507	.44098	.89752	.45658	.88968	.47204	.88158
.42552	.90495	.44124	.89739	.45684	.88955	.47229	.88144
.42578	.90483	.44151	.89726	.45710	.88942	.47255	.88130
.42604	.90470	.44177	.89713	.45736	.88928	.47281	.88117
.42631	.90458	.44203	.89700	.45762	.88915	.47306	.88103
.42657	.90446	.44229	.89687	.45787	.88902	.47332	.88089
.42683	.90433	.44255	.89674	.45813	.88888	.47358	.88075
.42709	.90421	.44281	.89662	.45839	.88875	.47383	.88062
.42736	.90408	.44307	.89649	.45865	.88862	.47409	.88048
.42762	.90396	.44333	.89636	.45891	.88848	.47434	.88034
.42788	.90383	.44359	.89623	.45917	.88835	.47460	.88020
.42815	.90371	.44385	.89610	.45942	.88822	.47486	.88006
.42841	.90358	.44411	.89597	.45968	.88808	.47511	.87993
.42867	.90346	.44437	.89584	.45994	.88795	.47537	.87979
.42894	.90334	.44464	.89571	.46020	.88782	.47562	.87965
.42920	.90321	.44490	.89558	.46046	.88768	.47588	.87951
.42946	.90309	.44516	.89545	.46072	.88755	.47614	.87937
.42972	.90296	.44542	.89532	.46097	.88741	.47639	.87923
.42999	.90284	.44568	.89519	.46123	.88728	.47665	.87909
.43025	.90271	.44594	.89506	.46149	.88715	.47690	.87896
.43051	.90259	.44620	89493	.46175	.88701	.47716	.87882
.43077	.90246	.44646	.89480	.46201	.88688	.47741	.87868
.43104	.90233	.44672	.89467	.46226	.88674	.47767	.87854
.43130	.90221	.44698	.89454	.46252	.88661	.47793	.87840
.43156	.90208	.44724	.89441	.46278	.88647	.47818	.87826
.43182	.90196	.44750	.89428	.46304	.88634	.47844	.87812
.43209	.90183	.44776	.89415	.46330	.88620	.47869	.87798
.43235	.90171	.44802	.89402	.46355	.88607	.47895	.87784
.43261	.90158	.44828	.89389	.46381	.88593	.47920	.87770
.43287	.90146	.44854	.89376	.46407	.88580	.47946	.87756
.43313	.90133	.44880	.89363	.46433	.88566	.47971	.87743
.43340	.90120	.44906	.89350	.46458	.88553	.47997	.87729
.43366	.90108	.44932	.89337	.46484	.88539	.48022	.87715
.43392	.90095	.44958	.89324	.46510	.88526	.48048	.87701
.43418	.90082	.44984	.89311	.46536	.88512	.48073	.87687
.43445	.90070	.45010	.89298	.46561	.88499	.48099	.87673
.43471	.90057	.45036	.89285	.46587	.88485	.48124	.87659
.43497	.90045	.45062	.89272	.46613	.88472	.48150	.87645
.43523	.90032	.45088	.89259	.46639	.88458	.48175	.87631
.43549	.90019	.45114	.89245	.46664	.88445	.48201	.87617
.43575	.90007	.45140	.89232	.46690	.88431	.48226	.87603
.43602	.89994	.45166	.89219	.46716	.88417	.48252	.87589
.43628	.89981	45192	.89206	.46742	.88404	.48277	.87575
.43654	.89968	.45218	.89193	.46767	.88390	.48303	.87561
.43680	.89956	.45243	.89180	.46793	.88377	.48328	.87546
.43706	.89943	.45269	.89167	.46819	.88363	.48354	.87532
.43733	.89930	.45295	.89153	.46844	.88349	.48379	.87518
.43759	.89918	.45321	.89140	.46870	.88336	.48405	.87504
.43785	.89905	.45347	.89127	.46896	.88322	.48430	.87490
.43811	.89892	.45373	.89114	.46921	.88308	.48456	.87476
.43837	.89879	.45399	.89101	.46947	.88295	.48481	.87462

COSINE	SINE	COSINE	SINE	COSINE	SINE	COSINE	SINE
64°		63°		62°		61°	

′	29° Sine	Cosine	30° Sine	Cosine	31° Sine	Cosine	32° Sine	Cosine	′
0	.48481	.87462	.50000	.86603	.51504	.85717	.52992	.84805	60
1	.48506	.87448	.50025	.86588	.51529	.85702	.53017	.84789	59
2	.48532	.87434	.50050	.86573	.51554	.85687	.53041	.84774	58
3	.48557	.87420	.50076	.86559	.51579	.85672	.53066	.84759	57
4	.48583	.87406	.50101	.86544	.51604	.85657	.53091	.84743	56
5	.48608	.87391	.50126	.86530	.51628	.85642	.53115	.84728	55
6	.48634	.87377	.50151	.86515	.51653	.85627	.53140	.84712	54
7	.48659	.87363	.50176	.86501	.51678	.85612	.53164	.84697	53
8	.48684	.87349	.50201	.86486	.51703	.85597	.53189	.84681	52
9	.48710	.87335	.50227	.86471	.51728	.85582	.53214	.84666	51
10	.48735	.87321	.50252	.86457	.51753	.85567	.53238	.84650	50
11	.48761	.87306	.50277	.86442	.51778	.85551	.53263	.84635	49
12	.48786	.87292	.50302	.86427	.51803	.85536	.53288	.84619	48
13	.48811	.87278	.50327	.86413	.51828	.85521	.53312	.84604	47
14	.48837	.87264	.50352	.86398	.51852	.85506	.53337	.84588	46
15	.48862	.87250	.50377	.86384	.51877	.85491	.53361	.84573	45
16	.48888	.87235	.50403	.86369	.51902	.85476	.53386	.84557	44
17	.48913	.87221	.50428	.86354	.51927	.85461	.53411	.84542	43
18	.48938	.87207	.50453	.86340	.51952	.85446	.53435	.84526	42
19	.48964	.87193	.50478	.86325	.51977	.85431	.53460	.84511	41
20	.48989	.87178	.50503	.86310	.52002	.85416	.53484	.84495	40
21	.49014	.87164	.50528	.86295	.52026	.85401	.53509	.84480	39
22	.49040	.87150	.50553	.86281	.52051	.85385	.53534	.84464	38
23	.49065	.87136	.50578	.86266	.52076	.85370	.53558	.84448	37
24	.49090	.87121	.50603	.86251	.52101	.85355	.53583	.84433	36
25	.49116	.87107	.50628	.86237	.52126	.85340	.53607	.84417	35
26	.49141	.87093	.50654	.86222	.52151	.85325	.53632	.84402	34
27	.49166	.87079	.50679	.86207	.52175	.85310	.53656	.84386	33
28	.49192	.87064	.50704	.86192	.52200	.85294	.53681	.84370	32
29	.49217	.87050	.50729	.86178	.52225	.85279	.53705	.84355	31
30	.49242	.87036	.50754	.86163	.52250	.85264	.53730	.84339	30
31	.49268	.87021	.50779	.86148	.52275	.85249	.53754	.84324	29
32	.49293	.87007	.50804	.86133	.52299	.85234	.53779	.84308	28
33	.49318	.86993	.50829	.86119	.52324	.85218	.53804	.84292	27
34	.49344	.86978	.50854	.86104	.52349	.85203	.53828	.84277	26
35	.49369	.86964	.50879	.86089	.52374	.85188	.53853	.84261	25
36	.49394	.86949	.50904	.86074	.52399	.85173	.53877	.84245	24
37	.49419	.86935	.50929	.86059	.52423	.85157	.53902	.84230	23
38	.49445	.86921	.50954	.86045	.52448	.85142	.53926	.84214	22
39	.49470	.86906	.50979	.86030	.52473	.85127	.53951	.84198	21
40	.49495	.86892	.51004	.86015	.52498	.85111	.53975	.84182	20
41	.49521	.86878	.51029	.86000	.52522	.85096	.54000	.84167	19
42	.49546	.86863	.51054	.85985	.52547	.85081	.54024	.84151	18
43	.49571	.86849	.51079	.85970	.52572	.85066	.54049	.84135	17
44	.49596	.86834	.51104	.85956	.52597	.85051	.54073	.84120	16
45	.49622	.86820	.51129	.85941	.52621	.85035	.54097	.84104	15
46	.49647	.86805	.51154	.85926	.52646	.85020	.54122	.84088	14
47	.49672	.86791	.51179	.85911	.52671	.85005	.54146	.84072	13
48	.49697	.86777	.51204	.85896	.52696	.84989	.54171	.84057	12
49	.49723	.86762	.51229	.85881	.52720	.84974	.54195	.84041	11
50	.49748	.86748	.51254	.85866	.52745	.84959	.54220	.84025	10
51	.49773	.86733	.51279	.85851	.52770	.84943	.54244	.84009	9
52	.49798	.86719	.51304	.85836	.52794	.84928	.54269	.83994	8
53	.49824	.86704	.51329	.85821	.52819	.84913	.54293	.83978	7
54	.49849	.86690	.51354	.85806	.52844	.84897	.54317	.83962	6
55	.49874	.86675	.51379	.85792	.52869	.84882	.54342	.83946	5
56	.49899	.86661	.51404	.85777	.52893	.84866	.54366	.83930	4
57	.49924	.86646	.51429	.85762	.52918	.84851	.54391	.83915	3
58	.49950	.86632	.51454	.85747	.52943	.84836	.54415	.83899	2
59	.49975	.86617	.51479	.85732	.52967	.84820	.54440	.83883	1
60	.50000	.86603	.51504	.85717	.52992	.84805	.54464	.83867	0

	Cosine	Sine	Cosine	Sine	Cosine	Sine	Cosine	Sine	′
		60°		59°		58°		57°	

33°		34°		35°		36°		'
Sine	Cosine	Sine	Cosine	Sine	Cosine	Sine	Cosnie	
.54464	.83867	.55919	.82904	.57358	.81915	.58779	.80002	60
.54488	.83851	.55943	.82887	.57381	.81899	.58802	.80885	59
.54513	.83835	.55968	.82871	.57405	.81882	.58826	.80867	58
.54537	.83819	.55992	.82855	.57429	.81865	.58849	.80850	57
.54561	.83804	.56016	.82839	.57453	.81848	.58873	.80833	56
.54586	.83788	.56040	.82822	.57477	.81832	.58896	.80816	55
.54610	.83772	.56064	.82806	.57501	.81815	.58920	.80799	54
.54635	.83756	.56088	.82790	.57524	.81798	.58943	.80782	53
.54659	.83740	.56112	.82773	.57548	.81782	.58967	.80765	52
.54683	.83724	.56136	.82757	.57572	.81765	.58990	.80748	51
.54708	.83708	.56160	.82741	.57596	.81748	.59014	.80730	50
.54732	83692	.56184	.82724	.57619	.81731	.59037	.80713	49
.54756	.83676	.56208	.82708	.57643	.81714	.59061	.80696	48
.54781	.83660	.56232	.82692	.57667	.81698	.59084	.80679	47
.54805	.83645	.56256	.82675	.57691	.81681	.59108	.80662	46
.54829	83629	.56280	.82659	.57715	.81664	.59131	.80644	45
.54854	.83613	.56305	.82643	.57738	.81647	.59154	.80627	44
.54878	.83597	.56329	.82626	.57762	.81631	.59178	.80610	43
.54902	.83581	.56353	.82610	.57786	.81614	.59201	.80593	42
.54927	83565	.56377	.82593	.57810	.81597	.59225	.80576	41
.54951	.83549	.56401	.82577	.57833	.81580	.59248	.80558	40
.54975	.83533	.56425	.82561	.57857	.81563	.59272	.80541	39
.54999	.83517	.56449	.82544	.57881	.81546	.59295	.80524	38
.55024	.83501	.56473	.82528	.57904	.81530	.59318	.80507	37
.55048	.83485	.56497	.82511	.57928	.81513	.59342	.80489	36
.55072	.83469	.56521	.82495	.57952	.81496	.59365	.80472	35
.55097	.83453	.56545	.82478	.57976	.81479	.59389	.80455	34
.55121	.83437	.56569	.82462	.57999	.81462	.59412	.80438	33
.55145	.83421	.56593	.82446	.58023	.81445	.59436	.80420	32
.55169	.83405	.56617	.82429	.58047	.81428	.59459	.80403	31
.55194	.83389	.56641	.82413	.58070	.81412	.59482	.80386	30
.55218	.83373	.56665	.82396	.58094	.81395	.59506	.80368	29
.55242	.83356	.56689	.82380	.58118	.81378	.59529	.80351	28
.55266	.83340	.56713	.82363	.58141	.81361	.59552	.80334	27
.55291	.83324	.56736	.82347	.58165	.81344	.59576	.80316	26
.55315	.83308	.56760	.82330	.58189	.81327	.59599	.80299	25
.55339	.83292	.56784	.82314	.58212	.81310	.59622	.80282	24
.55363	.83276	.56808	.82297	.58236	.81293	.59646	.80264	23
.55388	.83260	.56832	.82281	.58260	.81276	.59669	.80247	22
.55412	.83244	.56856	.82264	.58283	.81259	.59693	.80230	21
.55436	.83228	.56880	.82248	.58307	.81242	.59716	.80212	20
.55460	.83212	.56904	.82231	.58330	.81225	.59739	.80195	19
.55484	.83195	.56928	.82214	.58354	.81208	.59763	.80178	18
.55509	.83179	.56952	.82198	.58378	.81191	.59786	.80160	17
.55533	.83163	.56976	.82181	.58401	.81174	.59809	.80143	16
.55557	.83147	.57000	.82165	.58425	.81157	.59832	.80125	15
.55581	.83131	.57024	.82148	.58449	.81140	.59856	.80108	14
.55605	.83115	.57047	.82132	.58472	.81123	.59879	.80091	13
.55630	.83098	.57071	.82115	.58496	.81106	.59902	.80073	12
.55654	.83082	.57095	.82098	.58519	.81089	.59926	.80056	11
.55678	.83066	.57119	.82082	.58543	.81072	.59949	.80038	10
.55702	.83050	.57143	.82065	.58567	.81055	.59972	.80021	9
.55726	.83034	.57167	.82048	.58590	.81038	.59995	.80003	8
.55750	.83017	.57191	.82032	.58614	.81021	.60019	.79986	7
.55775	.83001	.57215	.82015	.58637	.81004	.60042	.79968	6
.55799	.82985	.57238	.81999	.58661	.80987	.60065	.79951	5
.55823	.82969	.57262	.81982	.58684	.80970	.60089	.79934	4
.55847	.82953	.57286	.81965	.58708	.80953	.60112	.79916	3
.55871	.82936	.57310	.81949	.58731	.80936	.60135	.79899	2
.55895	.82920	.57334	.81932	.58755	.80919	.60158	.79881	1
.55919	.82904	.57358	.81915	.58779	.80902	.60182	.79864	0
Cosine	Sine	Cosine	Sine	Cosine	Sine	Cosine	Sine	

| 56° | | 55° | | 54° | | 53° | | |

′	37° Sine	Cosine	38° Sine	Cosine	39° Sine	Cosine	40° Sine	Cosine	′
0	.60182	.79864	.61566	.78801	.62932	.77715	.64279	.76604	60
1	.60205	.79846	.61589	.78783	.62955	.77696	.64301	.76586	59
2	.60228	.79829	.61612	.78765	.62977	.77678	.64323	.76567	58
3	.60251	.79811	.61635	.78747	.63000	.77660	.64346	.76548	57
4	.60274	.79793	.61658	.78729	.63022	.77641	.64368	.76530	56
5	.60298	.79776	.61681	.78711	.63045	.77623	.64390	.76511	55
6	.60321	.79758	.61704	.78694	.63068	.77605	.64412	.76492	54
7	.60344	.79741	.61726	.78676	.63090	.77586	.64435	.76473	53
8	.60367	.79723	.61749	.78658	.63113	.77568	.64457	.76455	52
9	.60390	.79706	.61772	.78640	.63135	.77550	.64479	.76436	51
10	.60414	.79688	.61795	.78622	.63158	.77531	.64501	.76417	50
11	.60437	.79671	.61818	.78604	.63180	.77513	.64524	.76398	49
12	.60460	.79653	.61841	.78586	.63203	.77494	.64546	.76380	48
13	.60483	.79635	.61864	.78568	.63225	.77476	.64568	.76361	47
14	.60506	.79618	.61887	.78550	.63248	.77458	.64590	.76342	46
15	.60529	.79600	.61909	.78532	.63271	.77439	.64612	.76323	45
16	.60553	.79583	.61932	.78514	.63293	.77421	.64635	.76304	44
17	.60576	.79565	.61955	.78496	.63316	.77402	.64657	.76286	43
18	.60599	.79547	.61978	.78478	.63338	.77384	.64679	.76267	42
19	.60622	.79530	.62001	.78460	.63361	.77366	.64701	.76248	41
20	.60645	.79512	.62024	.78442	.63383	.77347	.64723	.76229	40
21	.60668	.79494	.62046	.78424	.63406	.77329	.64746	.76210	39
22	.60691	.79477	.62069	.78405	.63428	.77310	.64768	.76192	38
23	.60714	.79459	.62092	.78387	.63451	.77292	.64790	.76173	37
24	.60738	.79441	.62115	.78369	.63473	.77273	.64812	.76154	36
25	.60761	.79424	.62138	.78351	.63496	.77255	.64834	.76135	35
26	.60784	.79406	.62160	.78333	.63518	.77236	.64856	.76116	34
27	.60807	.79388	.62183	.78315	.63540	.77218	.64878	.76097	33
28	.60830	.79371	.62206	.78297	.63563	.77199	.64901	.76078	32
29	.60853	.79353	.62229	.78279	.63585	.77181	.64923	.76059	31
30	.60876	.79335	.62251	.78261	.63608	.77162	.64945	.76041	30
31	.60899	.79318	.62274	.78243	.63630	.77144	.64967	.76022	29
32	.60922	.79300	.62297	.78225	.63653	.77125	.64989	.76003	28
33	.60945	.79282	.62320	.78206	.63675	.77107	.65011	.75984	27
34	.60968	.79264	.62342	.78188	.63698	.77088	.65033	.75965	26
35	.60991	.79247	.62365	.78170	.63720	.77070	.65055	.75946	25
36	.61015	.79229	.62388	.78152	.63742	.77051	.65077	.75927	24
37	.61038	.79211	.62411	.78134	.63765	.77033	.65100	.75908	23
38	.61061	.79193	.62433	.78116	.63787	.77014	.65122	.75889	22
39	.61084	.79176	.62456	.78098	.63810	.76996	.65144	.75870	21
40	.61107	.79158	.62479	.78079	.63832	.76977	.65166	.75851	20
41	.61130	.79140	.62502	.78061	.63854	.76959	.65188	.75832	19
42	.61153	.79122	.62524	.78043	.63877	.76940	.65210	.75813	18
43	.61176	.79105	.62547	.78025	.63899	.76921	.65232	.75794	17
44	.61199	.79087	.62570	.78007	.63922	.76903	.65254	.75775	16
45	.61222	.79069	.62592	.77988	.63944	.76884	.65276	.75756	15
46	.61245	.79051	.62615	.77970	.63966	.76866	.65298	.75738	14
47	.61268	.79033	.62638	.77952	.63989	.76847	.65320	.75719	13
48	.61291	.79016	.62660	.77934	.64011	.76828	.65342	.75700	12
49	.61314	.78998	.62683	.77916	.64033	.76810	.65364	.75680	11
50	.61337	.78980	.62706	.77897	.64056	.76791	.65386	.75661	10
51	.61360	.78962	.62728	.77879	.64078	.76772	.65408	.75642	9
52	.61383	.78944	.62751	.77861	.64100	.76754	.65430	.75623	8
53	.61406	.78926	.62774	.77843	.64123	.76735	.65452	.75604	7
54	.61429	.78908	.62796	.77824	.64145	.76717	.65474	.75585	6
55	.61451	.78891	.62819	.77806	.64167	.76698	.65496	.75566	5
56	.61474	.78873	.62842	.77788	.64190	.76679	.65518	.75547	4
57	.61497	.78855	.62864	.77769	.64212	.76661	.65540	.75528	3
58	.61520	.78837	.62887	.77751	.64234	.76642	.65562	.75509	2
59	.61543	.78819	.62909	.77733	.64256	.76623	.65584	.75490	1
60	.61566	.78801	.62932	.77715	.64279	.76604	.65606	.75471	0
′	Cosine	Sine	Cosine	Sine	Cosine	Sine	Cosine	Sine	′
	52°		51°		50°		49°		

| | 41° | | 42° | | 43° | | 44° | |
	SINE	COSINE	SINE	COSINE	SINE	COSINE	SINE	COSINE	′
	.65606	.75471	.66913	.74314	.68200	.73135	.69466	.71934	60
	.65628	.75452	.66935	.74295	.68221	.73116	.69487	.71914	59
	.65650	.75433	.66956	.74276	.68242	.73096	.69508	.71894	58
	.65672	.75414	.66978	.74256	.68264	.73076	.69529	.71873	57
	.65694	.75395	.66999	.74237	.68285	.73056	.69549	.71853	56
	.65716	.75375	.67021	.74217	.68306	.73036	.69570	.71833	55
	.65738	.75356	.67043	.74198	.68327	.73016	.69591	.71813	54
	.65759	.75337	.67064	.74178	.68349	.72996	.69612	.71792	53
	.65781	.75318	.67086	.74159	.68370	.72976	.69633	.71772	52
	.65803	.75299	.67107	.74139	.68391	.72957	.69654	.71752	51
	.65825	.75280	.67129	.74120	.68412	.72937	.69675	.71732	50
	.65847	.75261	.67151	.74100	.68434	.72917	.69696	.71711	49
	.65869	.75241	.67172	.74080	.68455	.72897	.69717	.71691	48
	.65891	.75222	.67194	.74061	.68476	.72877	.69737	.71671	47
	.65913	.75203	.67215	.74041	.68497	.72857	.69758	.71650	46
	.65935	.75184	.67237	.74022	.68518	.72837	.69779	.71630	45
	.65956	.75165	.67258	.74002	.68539	.72817	.69800	.71610	44
	.65978	.75146	.67280	.73983	.68561	.72797	.69821	.71590	43
	.66000	.75126	.67301	.73963	.68582	.72777	.69842	.71569	42
	.66022	.75107	.67323	.73944	.68603	.72757	.69862	.71549	41
	.66044	.75088	.67344	.73924	.68624	.72737	.69883	.71529	40
	.66066	.75069	.67366	.73904	.68645	.72717	.69904	.71508	39
	.66088	.75050	.67387	.73885	.68666	.72697	.69925	.71488	38
	.66109	.75030	.67409	.73865	.68688	.72677	.69946	.71468	37
	.66131	.75011	.67430	.73846	.68709	.72657	.69966	.71447	36
	.66153	.74992	.67452	.73826	.68730	.72637	.69987	.71427	35
	.66175	.74973	.67473	.73806	.68751	.72617	.70008	.71407	34
	.66197	.74953	.67495	.73787	.68772	.72597	.70029	.71386	33
	.66218	.74934	.67516	.73767	.68793	.72577	.70049	.71366	32
	.66240	.74915	.67538	.73747	.68814	.72557	.70070	.71345	31
	.66262	.74896	.67559	.73728	.68835	.72537	.70091	.71325	30
	.66284	.74876	.67580	.73708	.68857	.72517	.70112	.71305	29
	.66306	.74857	.67602	.73688	.68878	.72497	.70132	.71284	28
	.66327	.74838	.67623	.73669	.68899	.72477	.70153	.71264	27
	.66349	.74818	.67645	.73649	.68920	.72457	.70174	.71243	26
	.66371	.74799	.67666	.73629	.68941	.72437	.70195	.71223	25
	.66393	.74780	.67688	.73610	.68962	.72417	.70215	.71203	24
	.66414	.74760	.67709	.73590	.68983	.72397	.70236	.71182	23
	.66436	.74741	.67730	.73570	.69004	.72377	.70257	.71162	22
	.66458	.74722	.67752	.73551	.69025	.72357	.70277	.71141	21
	.66480	.74703	.67773	.73531	.69046	.72337	.70298	.71121	20
	.66501	.74683	.67795	.73511	.69067	.72317	.70319	.71100	19
	.66523	.74664	.67816	.73491	.69088	.72297	.70339	.71080	18
	.66545	.74644	.67837	.73472	.69109	.72277	.70360	.71059	17
	.66566	.74625	.67859	.73452	.69130	.72257	.70381	.71039	16
	.66588	.74606	.67880	.73432	.69151	.72236	.70401	.71019	15
	.66610	.74586	.67901	.73413	.69172	.72216	.70422	.70998	14
	.66632	.74567	.67923	.73393	.69193	.72196	.70443	.70978	13
	.66653	.74548	.67944	.73373	.69214	.72176	.70463	.70957	12
	.66675	.74528	.67965	.73353	.69235	.72156	.70484	.70937	11
	.66697	.74509	.67987	.73333	.69256	.72136	.70505	.70916	10
	.66718	.74489	.68008	.73314	.69277	.72116	.70525	.70896	9
	.66740	.74470	.68029	.73294	.69298	.72095	.70546	.70875	8
	.66762	.74451	.68051	.73274	.69319	.72075	.70567	.70855	7
	.66783	.74431	.68072	.73254	.69340	.72055	.70587	.70834	6
	.66805	.74412	.68093	.73234	.69361	.72035	.70608	.70813	5
	.66827	.74392	.68115	.73215	.69382	.72015	.70628	.70793	4
	.66848	.74373	.68136	.73195	.69403	.71995	.70649	.70772	3
	.66870	.74353	.68157	.73175	.69424	.71974	.70670	.70752	2
	.66891	.74334	.68179	.73155	.69445	.71954	.70690	.70731	1
	.66913	.74314	.68200	.73135	.69466	.71934	.70711	.70711	0
	COSINE	SINE	COSINE	SINE	COSINE	SINE	COSINE	SINE	′
		48°		47°		46°		45°	

′	0° Sec.	0° Co-sec.	1° Sec.	1° Co-sec.	2° Sec.	2° Co-sec.	3° Sec.	3° Co-sec.
0	1	Infinite.	1.0001	57.299	1.0006	28.654	1.0014	19.107
1	1	3437.70	1.0001	56.359	1.0006	28.417	1.0014	19.002
2	1	1718.90	1.0002	55.450	1.0006	28.184	1.0014	18.897
3	1	1145.90	1.0002	54.570	1.0006	27.955	1.0014	18.794
4	1	859.44	1.0002	53.718	1.0006	27.730	1.0014	18.692
5	1	687.55	1.0002	52.891	1.0007	27.508	1.0014	18.591
6	1	572.96	1.0002	52.090	1.0007	27.290	1.0015	18.491
7	1	491.11	1.0002	51.313	1.0007	27.075	1.0015	18.393
8	1	429.72	1.0002	50.558	1.0007	26.864	1.0015	18.295
9	1	381.97	1.0002	49.826	1.0007	26.655	1.0015	18.198
10	1	343.77	1.0002	49.114	1.0007	26.450	1.0015	18.103
11	1	312.52	1.0002	48.422	1.0007	26.249	1.0015	18.008
12	1	286.48	1.0002	47.750	1.0007	26.050	1.0016	17.914
13	1	264.44	1.0002	47.096	1.0007	25.854	1.0016	17.821
14	1	245.55	1.0002	46.460	1.0008	25.661	1.0016	17.730
.15	1	229.18	1.0002	45.840	1.0008	25.471	1.0016	17.639
16	1	214.86	1.0002	45.237	1.0008	25.284	1.0016	17.549
17	1	202.22	1.0002	44.650	1.0008	25.100	1.0016	17.460
18	1	190.99	1.0002	44.077	1.0008	24.918	1.0017	17.372
19	1	180.73	1.0003	43.520	1.0008	24.739	1.0017	17.285
20	1	171.89	1.0003	42.976	1.0008	24.562	1.0017	17.198
21	1	163.70	1.0003	42.445	1.0008	24.358	1.0017	17.113
22	1	156.26	1.0003	41.928	1.0008	24.216	1.0017	17.028
23	1	149.47	1.0003	41.423	1.0009	24.047	1.0017	16.944
24	1	143.24	1.0003	40.930	1.0009	23.880	1.0018	16.861
25	1	137.51	1.0003	40.448	1.0009	23.716	1.0018	16.779
26	1	132.22	1.0003	39.978	1.0009	23.553	1.0018	16.698
27	1	127.32	1.0003	39.518	1.0009	23.393	1.0018	16.617
28	1	122.78	1.0003	39.069	1.0009	23.235	1.0018	16.538
29	1	118.54	1.0003	38.631	1.0009	23.079	1.0018	16.459
30	1	114.59	1.0003	38.201	1.0009	22.925	1.0019	16.380
31	1	110.90	1.0003	37.782	1.0010	22.774	1.0019	16.303
32	1	107.43	1.0003	37.371	1.0010	22.624	1.0019	16.226
33	1	104.17	1.0004	36.969	1.0010	22.476	1.0019	16.150
34	1	101.11	1.0004	36.576	1.0010	22.330	1.0019	16.075
35	1	90.223	1.0004	36.191	1.0010	22.186	1.0019	16.000
36	1	95.495	1.0004	35.814	1.0010	22.044	1.0020	15.926
37	1	92.914	1.0004	35.445	1.0010	21.904	1.0020	15.853
38	1.0001	92.469	1.0004	35.084	1.0010	21.765	1.0020	15.780
39	1.0001	88.149	1.0004	34.729	1.0011	21.629	1.0020	15.708
40	1.0001	85.946	1.0004	34.382	1.0011	21.494	1.0020	15.637
41	1.0001	83.849	1.0004	34.042	1.0011	21.360	1.0021	15.566
42	1.0001	81.853	1.0004	33.708	1.0011	21.228	1.0021	15.496
43	1.0001	79.950	1.0004	33.381	1.0011	21.098	1.0021	15.427
44	1.0001	78.133	1.0004	33.060	1.0011	20.970	1.0021	15.358
45	1.0001	76.396	1.0005	32.745	1.0011	20.843	1.0021	15.290
46	1.0001	74.736	1.0005	32.437	1.0012	20.717	1.0022	15.222
47	1.0001	73.146	1.0005	32.134	1.0012	20.593	1.0022	15.155
48	1.0001	71.622	1.0005	31.836	1.0012	20.471	1.0022	15.089
49	1.0001	71.160	1.0005	31.544	1.0012	20.350	1.0022	15.023
50	1.0001	68.757	1.0005	31.257	1.0012	20.230	1.0022	14.958
51	1.0001	67.409	1.0005	30.976	1.0012	20.112	1.0023	14.893
52	1.0001	66.113	1.0005	30.699	1.0012	19.995	1.0023	14.829
53	1.0001	64.866	1.0005	30.428	1.0013	19.880	1.0023	14.765
54	1.0001	63.664	1.0005	30.161	1.0013	19.766	1.0023	14.702
55	1.0001	62.507	1.0005	29.899	1.0013	19.653	1.0023	14.640
56	1.0001	61.391	1.0006	29.641	1.0013	19.541	1.0024	14.578
57	1.0001	61.314	1.0006	29.388	1.0013	19.431	1.0024	14.517
58	1.0001	59.274	1.0006	29.139	1.0013	19.322	1.0024	14.456
59	1.0001	58.270	1.0006	28.894	1.0013	19.214	1.0024	14.395
60	1.0001	57.299	1.0006	28.654	1.0014	19.107	1.0024	14.335
′	Co-sec.	Sec.	Co-sec.	Sec.	Co-sec.	Sec.	Co-sec.	Sec.
	89°		88°		87°		86°	

4° Sec.	4° Co-sec.	5° Sec.	5° Co-sec.	6° Sec.	6° Co-sec	7° Sec.	7° Co-sec.	'
1.0024	14.335	1.0038	11.474	1.0055	9.5668	1.0075	8.2055	60
1.0025	14.276	1.0038	11.436	1.0055	9.5404	1.0075	8.1861	59
1.0025	14.217	1.0039	11.398	1.0056	9.5141	1.0076	8.1668	58
1.0025	14.159	1.0039	11.360	1.0056	9.4880	1.0076	8.1476	57
1.0025	14.101	1.0039	11.323	1.0056	9.4620	1.0076	8.1285	56
1.0025	14.043	1.0039	11.286	1.0057	9.4362	1.0077	8.1094	55
1.0026	13.986	1.0040	11.249	1.0057	9.4105	1.0077	8.0905	54
1.0026	13.930	1.0040	11.213	1.0057	9.3850	1.0078	8.0717	53
1.0026	13.874	1.0040	11.176	1.0057	9.3596	1.0078	8.0529	52
1.0026	13.818	1.0040	11.140	1.0058	9.3343	1.0078	8.0342	51
1.0026	13.763	1.0041	11.104	1.0058	9.3092	1.0079	8.0156	50
1.0027	13.708	1.0041	11.069	1.0058	9.2842	1.0079	7.9971	49
1.0027	13.654	1.0041	11.033	1.0059	9.2593	1.0079	7.9787	48
1.0027	13.600	1.0041	10.988	1.0059	9.2346	1.0080	7.9604	47
1.0027	13.547	1.0042	10.963	1.0059	9.2100	1.0080	7.9421	46
1.0027	13.494	1.0042	10.929	1.0060	9.1855	1.0080	7.9240	45
1.0028	13.441	1.0042	10.894	1.0060	9.1612	1.0081	7.9059	44
1.0028	13.389	1.0043	10.860	1.0060	9.1370	1.0081	7.8879	43
1.0028	13.337	1.0043	10.826	1.0061	9.1129	1.0082	7.8700	42
1.0028	13.286	1.0043	10.792	1.0061	9.0890	1.0082	7.8522	41
1.0029	13.235	1.0043	10.758	1.0061	9.0651	1.0082	7.8344	40
1.0029	13.184	1.0044	10.725	1.0062	9.0414	1.0083	7.8168	39
1.0029	13.134	1.0044	10.692	1.0062	9.0179	1.0083	7.7992	38
1.0029	13.084	1.0044	10.659	1.0062	8.9944	1.0084	7.7817	37
1.0029	13.034	1.0044	10.626	1.0063	8.9711	1.0084	7.7642	36
1.0030	12.985	1.0045	10.593	1.0063	8.9479	1.0084	7.7469	35
1.0030	12.937	1.0045	10.561	1.0063	8.9248	1.0085	7.7296	34
1.0030	12.888	1.0045	10.529	1.0064	8.9018	1.0085	7.7124	33
1.0030	12.840	1.0046	10.497	1.0064	8.8790	1.0085	7.6953	32
1.0031	12.793	1.0046	10.465	1.0064	8.8563	1.0086	7.6783	31
1.0031	12.745	1.0046	10.433	1.0065	8.8337	1.0086	7.6613	30
1.0031	12.698	1.0046	10.402	1.0065	8.8112	1.0087	7.6444	29
1.0031	12.652	1.0047	10.371	1.0065	8.7888	1.0087	7.6276	28
1.0032	12.606	1.0047	10.340	1.0066	8.7665	1.0088	7.6108	27
1.0032	12.560	1.0047	10.309	1.0066	8.7444	1.0088	7.5942	26
1.0032	12.514	1.0048	10.278	1.0066	8.7223	1.0088	7.5776	25
1.0032	12.469	1.0048	10.248	1.0067	8.7004	1.0089	7.5611	24
1.0032	12.424	1.0048	10.217	1.0067	8.6786	1.0089	7.5446	23
1.0033	12.379	1.0048	10.187	1.0067	8.6569	1.0089	7.5282	22
1.0033	12.335	1.0049	10.157	1.0068	8.6353	1.0090	7.5119	21
1.0033	12.291	1.0049	10.127	1.0068	8.6138	1.0090	7.4957	20
1.0033	12.248	1.0049	10.098	1.0068	8.5924	1.0090	7.4795	19
1.0034	12.204	1.0050	10.068	1.0069	8.5711	1.0091	7.4634	18
1.0034	12.161	1.0050	10.039	1.0069	8.5499	1.0091	7.4474	17
1.0034	12.118	1.0050	10.010	1.0069	8.5289	1.0092	7.4315	16
1.0034	12.076	1.0050	9.9812	1.0070	8.5079	1.0092	7.4156	15
1.0035	12.034	1.0051	9.9525	1.0070	8.4871	1.0092	7.3998	14
1.0035	11.992	1.0051	9.9239	1.0070	8.4663	1.0093	7.3840	13
1.0035	11.950	1.0051	9.8955	1.0071	8.4457	1.0093	7.3683	12
1.0035	11.909	1.0052	9.8672	1.0071	8.4251	1.0094	7.3527	11
1.0036	11.868	1.0052	9.8391	1.0071	8.4046	1.0094	7.3372	10
1.0036	11.828	1.0052	9.8112	1.0072	8.3843	1.0094	7.3217	9
1.0036	11.787	1.0053	9.7834	1.0072	8.3640	1.0095	7.3063	8
1.0036	11.747	1.0053	9.7558	1.0073	8.3439	1.0095	7.2909	7
1.0037	11.707	1.0053	9.7283	1.0073	8.3238	1.0096	7.2757	6
1.0037	11.668	1.0053	9.7010	1.0073	8.3039	1.0096	7.2604	5
1.0037	11.628	1.0054	9.6739	1.0074	8.2840	1.0097	7.2453	4
1.0037	11.589	1.0054	9.6469	1.0074	8.2642	1.0097	7.2302	3
1.0038	11.550	1.0054	9.6200	1.0074	8.2446	1.0097	7.2152	2
1.0038	11.512	1.0055	9.5933	1.0075	8.2250	1.0098	7.2002	1
1.0038	11.474	1.0055	9.5668	1.0075	8.2055	1.0098	7.1853	0
Co-sec.	Sec.	Co-sec.	Sec.	Co-sec.	Sec.	Co-sec.	Sec.	'
85°		84°		83°		82°		

′	13° SINE	COSINE	14° SINE	COSINE	15° SINE	COSINE	16° SINE	COSINE	′
0	.22495	.97437	.24192	.97030	.25882	.96593	.27564	.96126	60
1	.22523	.97430	.24220	.97023	.25910	.96585	.27592	.96118	59
2	.22552	.97424	.24249	.97015	.25938	.96578	.27620	.96110	58
3	.22580	.97417	.24277	.97008	.25966	.96570	.27648	.96102	57
4	.22608	.97411	.24305	.97001	.25994	.96562	.27676	.96094	56
5	.22637	.97404	.24333	.96994	.26022	.96555	.27704	.96086	55
6	.22665	.97398	.24362	.96987	.26050	.96547	.27731	.96078	54
7	.22693	.97391	.24390	.96980	.26079	.96540	.27759	.96070	53
8	.22722	.97384	.24418	.96973	.26107	.96532	.27787	.96062	52
9	.22750	.97378	.24446	.96966	.26135	.96524	.27815	.96054	51
10	.22778	.97371	.24474	.96959	.26163	.96517	.27843	.96046	50
11	.22807	.97365	.24503	.96952	.26191	.96509	.27871	.96037	49
12	.22835	.97358	.24531	.96945	.26219	.96502	.27899	.96029	48
13	.22863	.97351	.24559	.96937	.26247	.96494	.27927	.96021	47
14	.22892	.97345	.24587	.96930	.26275	.96486	.27955	.96013	46
15	.22920	.97338	.24615	.96923	.26303	.96479	.27983	.96005	45
16	.22948	.97331	.24644	.96916	.26331	.96471	.28011	.95997	44
17	.22977	.97325	.24672	.96909	.26359	.96463	.28039	.95989	43
18	.23005	.97318	.24700	.96902	.26387	.96456	.28067	.95981	42
19	.23033	.97311	.24728	.96894	.26415	.96448	.28095	.95972	41
20	.23062	.97304	.24756	.96887	.26443	.96440	.28123	.95964	40
21	.23090	.97298	.24784	.96880	.26471	.96433	.28150	.95956	39
22	.23118	.97291	.24813	.96873	.26500	.96425	.28178	.95948	38
23	.23146	.97284	.24841	.96866	.26528	.96417	.28206	.95940	37
24	.23175	.97278	.24869	.96858	.26556	.96410	.28234	.95931	36
25	.23203	.97271	.24897	.96851	.26584	.96402	.28262	.95923	35
26	.23231	.97264	.24925	.96844	.26612	.96394	.28290	.95915	34
27	.23260	.97257	.24954	.96837	.26640	.96386	.28318	.95907	33
28	.23288	.97251	.24982	.96829	.26668	.96379	.28346	.95898	32
29	.23316	.97244	.25010	.96822	.26696	.96371	.28374	.95890	31
30	.23345	.97237	.25038	.96815	.26724	.96363	.28402	.95882	30
31	.23373	.97230	.25066	.96807	.26752	.96355	.28429	.95874	29
32	.23401	.97223	.25094	.96800	.26780	.96347	.28457	.95865	28
33	.23429	.97217	.25122	.96793	.26808	.96340	.28485	.95857	27
34	.23458	.97210	.25151	.96786	.26836	.96332	.28513	.95849	26
35	.23486	.97203	.25179	.96778	.26864	.96324	.28541	.95841	25
36	.23514	.97196	.25207	.96771	.26892	.96316	.28569	.95832	24
37	.23542	.97189	.25235	.96764	.26920	.96308	.28597	.95824	23
38	.23571	.97182	.25263	.96756	.26948	.96301	.28625	.95816	22
39	.23599	.97176	.25291	.96749	.26976	.96293	.28652	.95807	21
40	.23627	.97169	.25320	.96742	.27004	.96285	.28680	.95799	20
41	.23656	.97162	.25348	.96734	.27032	.96277	.28708	.95791	19
42	.23684	.97155	.25376	.96727	.27060	.96269	.28736	.95782	18
43	.23712	.97148	.25404	.96719	.27088	.96261	.28764	.95774	17
44	.23740	.97141	.25432	.96712	.27116	.96253	.28792	.95766	16
45	.23769	.97134	.25460	.96705	.27144	.96246	.28820	.95757	15
46	.23797	.97127	.25488	.96697	.27172	.96238	.28847	.95749	14
47	.23825	.97120	.25516	.96690	.27200	.96230	.28875	.95740	13
48	.23853	.97113	.25545	.96682	.27228	.96222	.28903	.95732	12
49	.23882	.97106	.25573	.96675	.27256	.96214	.28931	.95724	11
50	.23910	.97100	.25601	.96667	.27284	.96206	.28959	.95715	10
51	.23938	.97093	.25629	.96660	.27312	.96198	.28987	.95707	9
52	.23966	.97086	.25657	.96653	.27340	.96190	.29015	.95698	8
53	.23995	.97079	.25685	.96645	.27368	.96182	.29042	.95690	7
54	.24023	.97072	.25713	.96638	.27396	.96174	.29070	.95681	6
55	.24051	.97065	.25741	.96630	.27424	.96166	.29098	.95673	5
56	.24079	.97058	.25769	.96623	.27452	.96158	.29126	.95664	4
57	.24108	.97051	.25798	.96615	.27480	.96150	.29154	.95656	3
58	.24136	.97044	.25826	.96608	.27508	.96142	.29182	.95647	2
59	.24164	.97037	.25854	.96600	.27536	.96134	.29209	.95639	1
60	.24192	.97030	.25882	.96593	.27564	.96126	.29237	.95630	0
′	COSINE	SINE	COSINE	SINE	COSINE	SINE	COSINE	SINE	′
	76°		75°		74°		73°		

17° Sine	17° Cosine	18° Sine	18° Cosine	19° Sine	19° Cosine	20° Sine	20° Cosine	'
.29237	.95630	.30902	.95106	.32557	.94552	.34202	.93969	60
.29265	.95622	.30929	.95097	.32584	.94542	.34229	.93959	59
.29293	.95613	.30957	.95088	.32612	.94533	.34257	.93949	58
.29321	.95605	.30985	.95079	.32639	.94523	.34284	.93939	57
.29348	.95596	.31012	.95070	.32667	.94514	.34311	.93929	56
.29376	.95588	.31040	.95061	.32694	.94504	.34339	.93919	55
.29404	.95579	.31068	.95052	.32722	.94495	.34366	.93909	54
.29432	.95571	.31095	.95043	.32749	.94485	.34393	.93899	53
.29460	.95562	.31123	.95033	.32777	.94476	.34421	.93889	52
.29487	.95554	.31151	.95024	.32804	.94466	.34448	.93879	51
.29515	.95545	.31178	.95015	.32832	.94457	.34475	.93869	50
.29543	.95536	.31206	.95006	.32859	.94447	.34503	.93859	49
.29571	.95528	.31233	.94997	.32887	.94438	.34530	.93849	48
.29599	.95519	.31261	.94988	.32914	.94428	.34557	.93839	47
.29626	.95511	.31289	.94979	.32942	.94418	.34584	.93829	46
.29654	.95502	.31316	.94970	.32969	.94409	.34612	.93819	45
.29682	.95493	.31344	.94961	.32997	.94399	.34639	.93809	44
.29710	.95485	.31372	.94952	.33024	.94390	.34666	.93799	43
.29737	.95476	.31399	.94943	.33051	.94380	.34694	.93789	42
.29765	.95467	.31427	.94933	.33079	.94370	.34721	.93779	41
.29793	.95459	.31454	.94924	.33106	.94361	.34748	.93769	40
.29821	.95450	.31482	.94915	.33134	.94351	.34775	.93759	39
.29849	.95441	.31510	.94906	.33161	.94342	.34803	.93748	38
.29876	.95433	.31537	.94897	.33189	.94332	.34830	.93738	37
.29904	.95424	.31565	.94888	.33216	.94322	.34857	.93728	36
.29932	.95415	.31593	.94878	.33244	.94313	.34884	.93718	35
.29960	.95407	.31620	.94869	.33271	.94303	.34912	.93708	34
.29987	.95398	.31648	.94860	.33298	.94293	.34939	.93698	33
.30015	.95389	.31675	.94851	.33326	.94284	.34966	.93688	32
.30043	.95380	.31703	.94842	.33353	.94274	.34993	.93677	31
.30071	.95372	.31730	.94832	.33381	.94264	.35021	.93667	30
.30098	.95363	.31758	.94823	.33408	.94254	.35048	.93657	29
.30126	.95354	.31786	.94814	.33436	.94245	.35075	.93647	28
.30154	.95345	.31813	.94805	.33463	.94235	.35102	.93637	27
.30182	.95337	.31841	.94795	.33490	.94225	.35130	.93626	26
.30209	.95328	.31868	.94786	.33518	.94215	.35157	.93616	25
.30237	.95319	.31896	.94777	.33545	.94206	.35184	.93606	24
.30265	.95310	.31923	.94768	.33573	.94196	.35211	.93596	23
.30292	.95301	.31951	.94758	.33600	.94186	.35239	.93585	22
.30320	.95293	.31979	.94749	.33627	.94176	.35266	.93575	21
.30348	.95284	.32006	.94740	.33655	.94167	.35293	.93565	20
.30376	.95275	.32034	.94730	.33682	.94157	.35320	.93555	19
.30403	.95266	.32061	.94721	.33710	.94147	.35347	.93544	18
.30431	.95257	.32089	.94712	.33737	.94137	.35375	.93534	17
.30459	.95248	.32116	.94702	.33764	.94127	.35402	.93524	16
.30486	.95240	.32144	.94693	.33792	.94118	.35429	.93514	15
.30514	.95231	.32171	.94684	.33819	.94108	.35456	.93503	14
.30542	.95222	.32199	.94674	.33846	.94098	.35484	.93493	13
.30570	.95213	.32227	.94665	.33874	.94088	.35511	.93483	12
.30597	.95204	.32254	.94656	.33901	.94078	.35538	.93472	11
.30625	.95195	.32282	.94646	.33929	.94068	.35565	.93462	10
.30653	.95186	.32309	.94637	.33956	.94058	.35592	.93452	9
.30680	.95177	.32337	.94627	.33983	.94049	.35619	.93441	8
.30708	.95168	.32364	.94618	.34011	.94039	.35647	.93431	7
.30736	.95159	.32392	.94609	.34038	.94029	.35674	.93420	6
.30763	.95150	.32419	.94599	.34065	.94019	.35701	.93410	5
.30791	.95142	.32447	.94590	.34093	.94009	.35728	.93400	4
.30819	.95133	.32474	.94580	.34120	.93999	.35755	.93389	3
.30846	.95124	.32502	.94571	.34147	.93989	.35782	.93379	2
.30874	.95115	.32529	.94561	.34175	.93979	.35810	.93368	1
.30902	.95106	.32557	.94552	.34202	.93969	.35837	.93358	0
Cosine	Sine	Cosine	Sine	Cosine	Sine	Cosine	Sine	'
72°		71°		70°		69°		

'	21° Sine	Cosine	22° Sine	Cosine	23° Sine	Cosine	24° Sine	Cosine
0	.35837	.93358	.37461	.92718	.39073	.92050	.40674	.91355
1	.35864	.93348	.37488	.92707	.39100	.92039	.40700	.91343
2	.35891	.93337	.37515	.92697	.39127	.92028	.40727	.91331
3	.35918	.93327	.37542	.92686	.39153	.92016	.40753	.91319
4	.35945	.93316	.37569	.92675	.39180	.92005	.40780	.91307
5	.35973	.93306	.37595	.92664	.39207	.91994	.40806	.91295
6	.36000	.93295	.37622	.92653	.39234	.91982	.40833	.91283
7	.36027	.93285	.37649	.92642	.39260	.91971	.40860	.91272
8	.36054	.93274	.37676	.92631	.39287	.91959	.40886	.91260
9	.36081	.93264	.37703	.92620	.39314	.91948	.40913	.91248
10	.36108	.93253	.37730	.92609	.39341	.91936	.40939	.91236
11	.36135	.93243	.37757	.92598	.39367	.91925	.40966	.91224
12	.36162	.93232	.37784	.92587	.39394	.91914	.40992	.91212
13	.36190	.93222	.37811	.92576	.39421	.91902	.41019	.91200
14	.36217	.93211	.37838	.92565	.39448	.91891	.41045	.91188
15	.36244	.93201	.37865	.92554	.39474	.91879	.41072	.91176
16	.36271	.93190	.37892	.92543	.39501	.91868	.41098	.91164
17	.36298	.93180	.37919	.92532	.39528	.91856	.41125	.91152
18	.36325	.93169	.37946	.92521	.39555	.91845	.41151	.91140
19	.36352	.93159	.37973	.92510	.39581	.91833	.41178	.91128
20	.36379	.93148	.37999	.92499	.39608	.91822	.41204	.91116
21	.36406	.93137	.38026	.92488	.39635	.91810	.41231	.91104
22	.36434	.93127	.38053	.92477	.39661	.91799	.41257	.91092
23	.36461	.93116	.38080	.92466	.39688	.91787	.41284	.91080
24	.36488	.93106	.38107	.92455	.39715	.91775	.41310	.91068
25	.36515	.93095	.38134	.92444	.39741	.91764	.41337	.91056
26	.36542	.93084	.38161	.92432	.39768	.91752	.41363	.91044
27	.36569	.93074	.38188	.92421	.39795	.91741	.41390	.91032
28	.36596	.93063	.38215	.92410	.39822	.91729	.41416	.91020
29	.36623	.93052	.38241	.92399	.39848	.91718	.41443	.91008
30	.36650	.93042	.38268	.92388	.39875	.91706	.41469	.90990
31	.36677	.93031	.38295	.92377	.39902	.91694	.41496	.90984
32	.36704	.93020	.38322	.92366	.39928	.91683	.41522	.90972
33	.36731	.93010	.38349	.92355	.39955	.91671	.41549	.90960
34	.36758	.92999	.38376	.92343	.39982	.91660	.41575	.90948
35	.36785	.92988	.38403	.92332	.40008	.91648	.41602	.90936
36	.36812	.92978	.38430	.92321	.40035	.91636	.41628	.90924
37	.36839	.92967	.38456	.92310	.40062	.91625	.41655	.90911
38	.36867	.92956	.38483	.92299	.40088	.91613	.41681	.90899
39	.36894	.92945	.38510	.92287	.40115	.91601	.41707	.90887
40	.36921	.92935	.38537	.92276	.40141	.91590	.41734	.90875
41	.36948	.92924	.38564	.92265	.40168	.91578	.41760	.90863
42	.36975	.92913	.38591	.92254	.40195	.91566	.41787	.90851
43	.37002	.92902	.38617	.92243	.40221	.91555	.41813	.90830
44	.37029	.92892	.38644	.92231	.40248	.91543	.41840	.90820
45	.37056	.92881	.38671	.92220	.40275	.91531	.41866	.90814
46	.37083	.92870	.38698	.92209	.40301	.91519	.41892	.90802
47	.37110	.92859	.38725	.92198	.40328	.91508	.41919	.90700
48	.37137	.92849	.38752	.92186	.40355	.91496	.41945	.90778
49	.37164	.92838	.38778	.92175	.40381	.91484	.41972	.90760
50	.37191	.92827	.38805	.92164	.40408	.91472	.41998	.90753
51	.37218	.92816	.38832	.92152	.40434	.91461	.42024	.90741
52	.37245	.92805	.38859	.92141	.40461	.91449	.42051	.90720
53	.37272	.92794	.38886	.92130	.40488	.91437	.42077	.90717
54	.37299	.92784	.38912	.92119	.40514	.91425	.42104	.90704
55	.37326	.92773	.38939	.92107	.40541	.91414	.42130	.90602
56	.37353	.92762	.38966	.92096	.40567	.91402	.42156	.90680
57	.37380	.92751	.38993	.92085	.40594	.91390	.42183	.90668
58	.37407	.92740	.39020	.92073	.40621	.91378	.42209	.90655
59	.37434	.92729	.39046	.92062	.40647	.91366	.42235	.90643
60	.37461	.92718	.39073	.92050	.40674	.91355	.42262	.90631
'	Cosine	Sine	Cosine	Sine	Cosine	Sine	Cosine	Sine
	68°		67°		66°		65°	

NATURAL SINES AND COSINES

25° Sine	Cosine	26° Sine	Cosine	27° Sine	Cosine	28° Sine	Cosine
.42262	.90631	.43837	.89879	.45399	.89101	.46947	.88295
.42288	.90618	.43863	.89867	.45425	.89087	.46973	.88281
.42315	.90606	.43889	.89854	.45451	.89074	.46999	.88267
.42341	.90594	.43916	.89841	.45477	.89061	.47024	.88254
.42367	.90582	.43942	.89828	.45503	.89048	.47050	.88240
.42394	.90569	.43968	.89816	.45529	.89035	.47076	.88226
.42420	.90557	.43994	.89803	.45554	.89021	.47101	.88213
.42446	.90545	.44020	.89790	.45580	.89008	.47127	.88199
.42473	.90532	.44046	.89777	.45606	.88995	.47153	.88185
.42499	.90520	.44072	.89764	.45632	.88981	.47178	.88172
.42525	.90507	.44098	.89752	.45658	.88968	.47204	.88158
.42552	.90495	.44124	.89739	.45684	.88955	.47229	.88144
.42578	.90483	.44151	.89726	.45710	.88942	.47255	.88130
.42604	.90470	.44177	.89713	.45736	.88928	.47281	.88117
.42631	.90458	.44203	.89700	.45762	.88915	.47306	.88103
.42657	.90446	.44229	.89687	.45787	.88902	.47332	.88089
.42683	.90433	.44255	.89674	.45813	.88888	.47358	.88075
.42709	.90421	.44281	.89662	.45839	.88875	.47383	.88062
.42736	.90408	.44307	.89649	.45865	.88862	.47409	.88048
.42762	.90396	.44333	.89636	.45891	.88848	.47434	.88034
.42788	.90383	.44359	.89623	.45917	.88835	.47460	.88020
.42815	.90371	.44385	.89610	.45942	.88822	.47486	.88006
.42841	.90358	.44411	.89597	.45968	.88808	.47511	.87993
.42867	.90346	.44437	.89584	.45994	.88795	.47537	.87979
.42894	.90334	.44464	.89571	.46020	.88782	.47562	.87965
.42920	.90321	.44490	.89558	.46046	.88768	.47588	.87951
.42946	.90309	.44516	.89545	.46072	.88755	.47614	.87937
.42972	.90296	.44542	.89532	.46097	.88741	.47639	.87923
.42999	.90284	.44568	.89519	.46123	.88728	.47665	.87909
.43025	.90271	.44594	.89506	.46149	.88715	.47690	.87896
.43051	.90259	.44620	.89493	.46175	.88701	.47716	.87882
.43077	.90246	.44646	.89480	.46201	.88688	.47741	.87868
.43104	.90233	.44672	.89467	.46226	.88674	.47767	.87854
.43130	.90221	.44698	.89454	.46252	.88661	.47793	.87840
.43156	.90208	.44724	.89441	.46278	.88647	.47818	.87826
.43182	.90196	.44750	.89428	.46304	.88634	.47844	.87812
.43209	.90183	.44776	.89415	.46330	.88620	.47869	.87798
.43235	.90171	.44802	.89402	.46355	.88607	.47895	.87784
.43261	.90158	.44828	.89389	.46381	.88593	.47920	.87770
.43287	.90146	.44854	.89376	.46407	.88580	.47946	.87756
.43313	.90133	.44880	.89363	.46433	.88566	.47971	.87743
.43340	.90120	.44906	.89350	.46458	.88553	.47997	.87729
.43366	.90108	.44932	.89337	.46484	.88539	.48022	.87715
.43392	.90095	.44958	.89324	.46510	.88526	.48048	.87701
.43418	.90082	.44984	.89311	.46536	.88512	.48073	.87687
.43445	.90070	.45010	.89298	.46561	.88499	.48099	.87673
.43471	.90057	.45036	.89285	.46587	.88485	.48124	.87659
.43497	.90045	.45062	.89272	.46613	.88472	.48150	.87645
.43523	.90032	.45088	.89259	.46639	.88458	.48175	.87631
.43549	.90019	.45114	.89245	.46664	.88445	.48201	.87617
.43575	.90007	.45140	.89232	.46690	.88431	.48226	.87603
.43602	.89994	.45166	.89219	.46716	.88417	.48252	.87589
.43628	.89981	.45192	.89206	.46742	.88404	.48277	.87575
.43654	.89968	.45218	.89193	.46767	.88390	.48303	.87561
.43680	.89956	.45243	.89180	.46793	.88377	.48328	.87546
.43706	.89943	.45269	.89167	.46819	.88363	.48354	.87532
.43733	.89930	.45295	.89153	.46844	.88349	.48379	.87518
.43759	.89918	.45321	.89140	.46870	.88336	.48405	.87504
.43785	.89905	.45347	.89127	.46896	.88322	.48430	.87490
.43811	.89892	.45373	.89114	.46921	.88308	.48456	.87476
.43837	.89879	.45399	.89101	.46947	.88295	.48481	.87462

Cosine	Sine	Cosine	Sine	Cosine	Sine	Cosine	Sine
64°		63°		62°		61°	

'	29° SINE	29° COSINE	30° SINE	30° COSINE	31° SINE	31° COSINE	32° SINE	32° COSINE
0	.48481	.87462	.50000	.86603	.51504	.85717	.52992	.84805
1	.48506	.87448	.50025	.86588	.51529	.85702	.53017	.84789
2	.48532	.87434	.50050	.86573	.51554	.85687	.53041	.84774
3	.48557	.87420	.50076	.86559	.51579	.85672	.53066	.84759
4	.48583	.87406	.50101	.86544	.51604	.85657	.53091	.84743
5	.48608	.87391	.50126	.86530	.51628	.85642	.53115	.84728
6	.48634	.87377	.50151	.86515	.51653	.85627	.53140	.84712
7	.48659	.87363	.50176	.86501	.51678	.85612	.53164	.84697
8	.48684	.87349	.50201	.86486	.51703	.85597	.53189	.84681
9	.48710	.87335	.50227	.86471	.51728	.85582	.53214	.84666
10	.48735	.87321	.50252	.86457	.51753	.85567	.53238	.84650
11	.48761	.87306	.50277	.86442	.51778	.85551	.53263	.84635
12	.48786	.87292	.50302	.86427	.51803	.85536	.53288	.84619
13	.48811	.87278	.50327	.86413	.51828	.85521	.53312	.84604
14	.48837	.87264	.50352	.86398	.51852	.85506	.53337	.84588
15	.48862	.87250	.50377	.86384	.51877	.85491	.53361	.84573
16	.48888	.87235	.50403	.86369	.51902	.85476	.53386	.84557
17	.48913	.87221	.50428	.86354	.51927	.85461	.53411	.84542
18	.48938	.87207	.50453	.86340	.51952	.85446	.53435	.84526
19	.48964	.87193	.50478	.86325	.51977	.85431	.53460	.84511
20	.48989	.87178	.50503	.86310	.52002	.85416	.53484	.84495
21	.49014	.87164	.50528	.86295	.52026	.85401	.53509	.84480
22	.49040	.87150	.50553	.86281	.52051	.85385	.53534	.84464
23	.49065	.87136	.50578	.86266	.52076	.85370	.53558	.84448
24	.49090	.87121	.50603	.86251	.52101	.85355	.53583	.84433
25	.49116	.87107	.50628	.86237	.52126	.85340	.53607	.84417
26	.49141	.87093	.50654	.86222	.52151	.85325	.53632	.84402
27	.49166	.87079	.50679	.86207	.52175	.85310	.53656	.84386
28	.49192	.87064	.50704	.86192	.52200	.85294	.53681	.84370
29	.49217	.87050	.50729	.86178	.52225	.85279	.53705	.84355
30	.49242	.87036	.50754	.86163	.52250	.85264	.53730	.84339
31	.49268	.87021	.50779	.86148	.52275	.85249	.53754	.84324
32	.49293	.87007	.50804	.86133	.52299	.85234	.53779	.84308
33	.49318	.86993	.50829	.86119	.52324	.85218	.53804	.84292
34	.49344	.86978	.50854	.86104	.52349	.85203	.53828	.84277
35	.49369	.86964	.50879	.86089	.52374	.85188	.53853	.84261
36	.49394	.86949	.50904	.86074	.52399	.85173	.53877	.84245
37	.49419	.86935	.50929	.86059	.52423	.85157	.53902	.84230
38	.49445	.86921	.50954	.86045	.52448	.85142	.53926	.84214
39	.49470	.86906	.50979	.86030	.52473	.85127	.53951	.84198
40	.49495	.86892	.51004	.86015	.52498	.85112	.53975	.84182
41	.49521	.86878	.51029	.86000	.52522	.85096	.54000	.84167
42	.49546	.86863	.51054	.85985	.52547	.85081	.54024	.84151
43	.49571	.86849	.51079	.85970	.52572	.85066	.54049	.84135
44	.49596	.86834	.51104	.85956	.52597	.85051	.54073	.84120
45	.49622	.86820	.51129	.85941	.52621	.85035	.54097	.84104
46	.49647	.86805	.51154	.85926	.52646	.85020	.54122	.84088
47	.49672	.86791	.51179	.85911	.52671	.85005	.54146	.84072
48	.49697	.86777	.51204	.85896	.52696	.84989	.54171	.84057
49	.49723	.86762	.51229	.85881	.52720	.84974	.54195	.84041
50	.49748	.86748	.51254	.85866	.52745	.84959	.54220	.84025
51	.49773	.86733	.51279	.85851	.52770	.84943	.54244	.84009
52	.49798	.86719	.51304	.85836	.52794	.84928	.54269	.83994
53	.49824	.86704	.51329	.85821	.52819	.84913	.54293	.83978
54	.49849	.86690	.51354	.85806	.52844	.84897	.54317	.83962
55	.49874	.86675	.51379	.85792	.52869	.84882	.54342	.83946
56	.49899	.86661	.51404	.85777	.52893	.84866	.54366	.83930
57	.49924	.86646	.51429	.85762	.52918	.84851	.54391	.83915
58	.49950	.86632	.51454	.85747	.52943	.84836	.54415	.83899
59	.49975	.86617	.51479	.85732	.52967	.84820	.54440	.83883
60	.50000	.86603	.51504	.85717	.52992	.84805	.54464	.83867
	COSINE	SINE	COSINE	SINE	COSINE	SINE	COSINE	SINE
	60°		59°		58°		57°	

33°		34°		35°		36°		
Sine	Cosine	Sine	Cosine	Sine	Cosine	Sine	Cosnie	′
.54464	.83867	.55919	.82904	.57358	.81915	.58779	.80902	60
.54488	.83851	.55943	.82887	.57381	.81899	.58802	.80885	59
.54513	.83835	.55968	.82871	.57405	.81882	.58826	.80867	58
.54537	.83819	.55992	.82855	.57429	.81865	.58849	.80850	57
.54561	.83804	.56016	.82839	.57453	.81848	.58873	.80833	56
.54586	.83788	.56040	.82822	.57477	.81832	.58896	.80816	55
.54610	.83772	.56064	.82806	.57501	.81815	.58920	.80799	54
.54635	.83756	.56088	.82790	.57524	.81798	.58943	.80782	53
.54659	.83740	.56112	.82773	.57548	.81782	.58967	.80765	52
.54683	.83724	.56136	.82757	.57572	.81765	.58990	.80748	51
.54708	.83708	.56160	.82741	.57596	.81748	.59014	.80730	50
.54732	.83692	.56184	.82724	.57619	.81731	.59037	.80713	49
.54756	.83676	.56208	.82708	.57643	.81714	.59061	.80696	48
.54781	.83660	.56232	.82692	.57667	.81698	.59084	.80679	47
.54805	.83645	.56256	.82675	.57691	.81681	.59108	.80662	46
.54829	.83629	.56280	.82659	.57715	.81664	.59131	.80644	45
.54854	.83613	.56305	.82643	.57738	.81647	.59154	.80627	44
.54878	.83597	.56329	.82626	.57762	.81631	.59178	.80610	43
.54902	.83581	.56353	.82610	.57786	.81614	.59201	.80593	42
.54927	.83565	.56377	.82593	.57810	.81597	.59225	.80576	41
.54951	.83549	.56401	.82577	.57833	.81580	.59248	.80558	40
.54975	.83533	.56425	.82561	.57857	.81563	.59272	.80541	39
.54999	.83517	.56449	.82544	.57881	.81546	.59295	.80524	38
.55024	.83501	.56473	.82528	.57904	.81530	.59318	.80507	37
.55048	.83485	.56497	.82511	.57928	.81513	.59342	.80489	36
.55072	.83469	.56521	.82495	.57952	.81496	.59365	.80472	35
.55097	.83453	.56545	.82478	.57976	.81479	.59389	.80455	34
.55121	.83437	.56569	.82462	.57999	.81462	.59412	.80438	33
.55145	.83421	.56593	.82446	.58023	.81445	.59436	.80420	32
.55169	.83405	.56617	.82429	.58047	.81428	.59459	.80403	31
.55194	.83389	.56641	.82413	.58070	.81412	.59482	.80386	30
.55218	.83373	.56665	.82396	.58094	.81395	.59506	.80368	29
.55242	.83356	.56689	.82380	.58118	.81378	.59529	.80351	28
.55266	.83340	.56713	.82363	.58141	.81361	.59552	.80334	27
.55291	.83324	.56736	.82347	.58165	.81344	.59576	.80316	26
.55315	.83308	.56760	.82330	.58189	.81327	.59599	.80299	25
.55339	.83292	.56784	.82314	.58212	.81310	.59622	.80282	24
.55363	.83276	.56808	.82297	.58236	.81293	.59646	.80264	23
.55388	.83260	.56832	.82281	.58260	.81276	.59669	.80247	22
.55412	.83244	.56856	.82264	.58283	.81259	.59693	.80230	21
.55436	.83228	.56880	.82248	.58307	.81242	.59716	.80212	20
.55460	.83212	.56904	.82231	.58330	.81225	.59739	.80195	19
.55484	.83195	.56928	.82214	.58354	.81208	.59763	.80178	18
.55509	.83179	.56952	.82198	.58378	.81191	.59786	.80160	17
.55533	.83163	.56976	.82181	.58401	.81174	.59809	.80143	16
.55557	.83147	.57000	.82165	.58425	.81157	.59832	.80125	15
.55581	.83131	.57024	.82148	.58449	.81140	.59856	.80108	14
.55605	.83115	.57047	.82132	.58472	.81123	.59879	.80091	13
.55630	.83098	.57071	.82115	.58496	.81106	.59902	.80073	12
.55654	.83082	.57095	.82098	.58519	.81089	.59926	.80056	11
.55678	.83066	.57119	.82082	.58543	.81072	.59949	.80038	10
.55702	.83050	.57143	.82065	.58567	.81055	.59972	.80021	9
.55726	.83034	.57167	.82048	.58590	.81038	.59995	.80003	8
.55750	.83017	.57191	.82032	.58614	.81021	.60019	.79986	7
.55775	.83001	.57215	.82015	.58637	.81004	.60042	.79968	6
.55799	.82985	.57238	.81999	.58661	.80987	.60065	.79951	5
.55823	.82969	.57262	.81982	.58684	.80970	.60089	.79934	4
.55847	.82953	.57286	.81965	.58708	.80953	.60112	.79916	3
.55871	.82936	.57310	.81949	.58731	.80936	.60135	.79899	2
.55895	.82920	.57334	.81932	.58755	.80919	.60158	.79881	1
.55919	.82904	.57358	.81915	.58779	.80902	.60182	.79864	0
Cosine	Sine	Cosine	Sine	Cosine	Sine	Cosine	Sine	′
56°		55°		54°		53°		

′	37° SINE	COSINE	38° SINE	COSINE	39° SINE	COSINE	40° SINE	COSINE	′
0	.60182	.79864	.61566	.78801	.62932	.77715	.64279	.76604	60
1	.60205	.79846	.61589	.78783	.62955	.77696	.64301	.76586	59
2	.60228	.79829	.61612	.78765	.62977	.77678	.64323	.76567	58
3	.60251	.79811	.61635	.78747	.63000	.77660	.64346	.76548	57
4	.60274	.79793	.61658	.78729	.63022	.77641	.64368	.76530	56
5	.60298	.79776	.61681	.78711	.63045	.77623	.64390	.76511	55
6	.60321	.79758	.61704	.78694	.63068	.77605	.64412	.76492	54
7	.60344	.79741	.61726	.78676	.63090	.77586	.64435	.76473	53
8	.60367	.79723	.61749	.78658	.63113	.77568	.64457	.76455	52
9	.60390	.79706	.61772	.78640	.63135	.77550	.64479	.76436	51
10	.60414	.79688	.61795	.78622	.63158	.77531	.64501	.76417	50
11	.60437	.79671	.61818	.78604	.63180	.77513	.64524	.76398	49
12	.60460	.79653	.61841	.78586	.63203	.77494	.64546	.76380	48
13	.60483	.79635	.61864	.78568	.63225	.77476	.64568	.76361	47
14	.60506	.79618	.61887	.78550	.63248	.77458	.64590	.76342	46
15	.60529	.79600	.61909	.78532	.63271	.77439	.64612	.76323	45
16	.60553	.79583	.61932	.78514	.63293	.77421	.64635	.76304	44
17	.60576	.79565	.61955	.78496	.63316	.77402	.64657	.76286	43
18	.60599	.79547	.61978	.78478	.63338	.77384	.64679	.76267	42
19	.60622	.79530	.62001	.78460	.63361	.77366	.64701	.76248	41
20	.60645	.79512	.62024	.78442	.63383	.77347	.64723	.76229	40
21	.60668	.79494	.62046	.78424	.63406	.77329	.64746	.76210	39
22	.60691	.79477	.62069	.78405	.63428	.77310	.64768	.76192	38
23	.60714	.79459	.62092	.78387	.63451	.77292	.64790	.76173	37
24	.60738	.79441	.62115	.78369	.63473	.77273	.64812	.76154	36
25	.60761	.79424	.62138	.78351	.63496	.77255	.64834	.76135	35
26	.60784	.79406	.62160	.78333	.63518	.77236	.64856	.76116	34
27	.60807	.79388	.62183	.78315	.63540	.77218	.64878	.76097	33
28	.60830	.79371	.62206	.78297	.63563	.77199	.64901	.76078	32
29	.60853	.79353	.62229	.78279	.63585	.77181	.64923	.76059	31
30	.60876	.79335	.62251	.78261	.63608	.77162	.64945	.76041	30
31	.60899	.79318	.62274	.78243	.63630	.77144	.64967	.76022	29
32	.60922	.79300	.62297	.78225	.63653	.77125	.64989	.76003	28
33	.60945	.79282	.62320	.78206	.63675	.77107	.65011	.75984	27
34	.60968	.79264	.62342	.78188	.63698	.77088	.65033	.75965	26
35	.60991	.79247	.62365	.78170	.63720	.77070	.65055	.75946	25
36	.61015	.79229	.62388	.78152	.63742	.77051	.65077	.75927	24
37	.61038	.79211	.62411	.78134	.63765	.77033	.65100	.75908	23
38	.61061	.79193	.62433	.78116	.63787	.77014	.65122	.75889	22
39	.61084	.79176	.62456	.78098	.63810	.76996	.65144	.75870	21
40	.61107	.79158	.62479	.78079	.63832	.76977	.65166	.75851	20
41	.61130	.79140	.62502	.78061	.63854	.76959	.65188	.75832	19
42	.61153	.79122	.62524	.78043	.63877	.76940	.65210	.75813	18
43	.61176	.79105	.62547	.78025	.63899	.76921	.65232	.75794	17
44	.61199	.79087	.62570	.78007	.63922	.76903	.65254	.75775	16
45	.61222	.79069	.62592	.77988	.63944	.76884	.65276	.75756	15
46	.61245	.79051	.62615	.77970	.63966	.76866	.65298	.75738	14
47	.61268	.79033	.62638	.77952	.63989	.76847	.65320	.75719	13
48	.61291	.79016	.62660	.77934	.64011	.76828	.65342	.75700	12
49	.61314	.78998	.62683	.77916	.64033	.76810	.65364	.75680	11
50	.61337	.78980	.62706	.77897	.64056	.76791	.65386	.75661	10
51	.61360	.78962	.62728	.77879	.64078	.76772	.65408	.75642	9
52	.61383	.78944	.62751	.77861	.64100	.76754	.65430	.75623	8
53	.61406	.78926	.62774	.77843	.64123	.76735	.65452	.75604	7
54	.61429	.78908	.62796	.77824	.64145	.76717	.65474	.75585	6
55	.61451	.78891	.62819	.77806	.64167	.76698	.65496	.75566	5
56	.61474	.78873	.62842	.77788	.64190	.76679	.65518	.75547	4
57	.61497	.78855	.62864	.77769	.64212	.76661	.65540	.75528	3
58	.61520	.78837	.62887	.77751	.64234	.76642	.65562	.75509	2
59	.61543	.78819	.62909	.77733	.64256	.76623	.65584	.75490	1
60	.61566	.78801	.62932	.77715	.64279	.76604	.65606	.75471	0
′	COSINE	SINE	COSINE	SINE	COSINE	SINE	COSINE	SINE	′
	52°		51°		50°		49°		

41°		42°		43°		44°		
SINE	COSINE	SINE	COSINE	SINE	COSINE	SINE	COSINE	'
.65606	.75471	.66913	.74314	.68200	.73135	.69466	.71934	60
.65628	.75452	.66935	.74295	.68221	.73116	.69487	.71914	59
.65650	.75433	.66956	.74276	.68242	.73096	.69508	.71894	58
.65672	.75414	.66978	.74256	.68264	.73076	.69529	.71873	57
.65694	.75395	.66999	.74237	.68285	.73056	.69549	.71853	56
.65716	.75375	.67021	.74217	.68306	.73036	.69570	.71833	55
.65738	.75356	.67043	.74198	.68327	.73016	.69591	.71813	54
.65759	.75337	.67064	.74178	.68349	.72996	.69612	.71792	53
.65781	.75318	.67086	.74159	.68370	.72976	.69633	.71772	52
.65803	.75299	.67107	.74139	.68391	.72957	.69654	.71752	51
.65825	.75280	.67129	.74120	.68412	.72937	.69675	.71732	50
.65847	.75261	.67151	.74100	.68434	.72917	.69696	.71711	49
.65869	.75241	.67172	.74080	.68455	.72897	.69717	.71691	48
.65891	.75222	.67194	.74061	.68476	.72877	.69737	.71671	47
.65913	.75203	.67215	.74041	.68497	.72857	.69758	.71650	46
.65935	.75184	.67237	.74022	.68518	.72837	.69779	.71630	45
.65956	.75165	.67258	.74002	.68539	.72817	.69800	.71610	44
.65978	.75146	.67280	.73983	.68561	.72797	.69821	.71590	43
.66000	.75126	.67301	.73963	.68582	.72777	.69842	.71569	42
.66022	.75107	.67323	.73944	.68603	.72757	.69862	.71549	41
.66044	.75088	.67344	.73924	.68624	.72737	.69883	.71529	40
.66066	.75069	.67366	.73904	.68645	.72717	.69904	.71508	39
.66088	.75050	.67387	.73885	.68666	.72697	.69925	.71488	38
.66109	.75030	.67409	.73865	.68688	.72677	.69946	.71468	37
.66131	.75011	.67430	.73846	.68709	.72657	.69966	.71447	36
.66153	.74992	.67452	.73826	.68730	.72637	.69987	.71427	35
.66175	.74973	.67473	.73806	.68751	.72617	.70008	.71407	34
.66197	.74953	.67495	.73787	.68772	.72597	.70029	.71386	33
.66218	.74934	.67516	.73767	.68793	.72577	.70049	.71366	32
.66240	.74915	.67538	.73747	.68814	.72557	.70070	.71345	31
.66262	.74896	.67559	.73728	.68835	.72537	.70091	.71325	30
.66284	.74876	.67580	.73708	.68857	.72517	.70112	.71305	29
.66306	.74857	.67602	.73688	.68878	.72497	.70132	.71284	28
.66327	.74838	.67623	.73669	.68899	.72477	.70153	.71264	27
.66349	.74818	.67645	.73649	.68920	.72457	.70174	.71243	26
.66371	.74799	.67666	.73629	.68941	.72437	.70195	.71223	25
.66393	.74780	.67688	.73610	.68962	.72417	.70215	.71203	24
.66414	.74760	.67709	.73590	.68983	.72397	.70236	.71182	23
.66436	.74741	.67730	.73570	.69004	.72377	.70257	.71162	22
.66458	.74722	.67752	.73551	.69025	.72357	.70277	.71141	21
.66480	.74703	.67773	.73531	.69046	.72337	.70298	.71121	20
.66501	.74683	.67795	.73511	.69067	.72317	.70319	.71100	19
.66523	.74664	.67816	.73491	.69088	.72297	.70339	.71080	18
.66545	.74644	.67837	.73472	.69109	.72277	.70360	.71059	17
.66566	.74625	.67859	.73452	.69130	.72257	.70381	.71039	16
.66588	.74606	.67880	.73432	.69151	.72236	.70401	.71019	15
.66610	.74586	.67901	.73413	.69172	.72216	.70422	.70998	14
.66632	.74567	.67923	.73393	.69193	.72196	.70443	.70978	13
.66653	.74548	.67944	.73373	.69214	.72176	.70463	.70957	12
.66675	.74528	.67965	.73353	.69235	.72156	.70484	.70937	11
.66697	.74509	.67987	.73333	.69256	.72136	.70505	.70916	10
.66718	.74489	.68008	.73314	.69277	.72116	.70525	.70896	9
.66740	.74470	.68029	.73294	.69298	.72095	.70546	.70875	8
.66762	.74451	.68051	.73274	.69319	.72075	.70567	.70855	7
.66783	.74431	.68072	.73254	.69340	.72055	.70587	.70834	6
.66805	.74412	.68093	.73234	.69361	.72035	.70608	.70813	5
.66827	.74392	.68115	.73215	.69382	.72015	.70628	.70793	4
.66848	.74373	.68136	.73195	.69403	.71995	.70649	.70772	3
.66870	.74353	.68157	.73175	.69424	.71974	.70670	.70752	2
.66891	.74334	.68179	.73155	.69445	.71954	.70690	.70731	1
.66913	.74314	.68200	.73135	.69466	.71934	.70711	.70711	0
COSINE	SINE	COSINE	SINE	COSINE	SINE	COSINE	SINE	'
	48°		47°		46°		45°	

	0°		1°		2°		3°		
′	Sec.	Co-sec.	Sec.	Co-sec.	Sec.	Co-sec.	Sec.	Co-sec.	′
0	1	Infinite.	1.0001	57.299	1.0006	28.654	1.0014	19.107	60
1	1	3437.70	1.0001	56.359	1.0006	28.417	1.0014	19.002	59
2	1	1718.90	1.0002	55.450	1.0006	28.184	1.0014	18.897	58
3	1	1145.90	1.0002	54.570	1.0006	27.955	1.0014	18.794	57
4	1	859.44	1.0002	53.718	1.0006	27.730	1.0014	18.692	56
5	1	687.55	1.0002	52.891	1.0007	27.508	1.0014	18.591	55
6	1	572.96	1.0002	52.090	1.0007	27.290	1.0015	18.491	54
7	1	491.11	1.0002	51.313	1.0007	27.075	1.0015	18.393	53
8	1	429.72	1.0002	50.558	1.0007	26.864	1.0015	18.295	52
9	1	381.97	1.0002	49.826	1.0007	26.655	1.0015	18.198	51
10	1	343.77	1.0002	49.114	1.0007	26.450	1.0015	18.103	50
11	1	312.52	1.0002	48.422	1.0007	26.249	1.0015	18.008	49
12	1	286.48	1.0002	47.750	1.0007	26.050	1.0016	17.914	48
13	1	264.44	1.0002	47.096	1.0007	25.854	1.0016	17.821	47
14	1	245.55	1.0002	46.460	1.0008	25.661	1.0016	17.730	46
15	1	229.18	1.0002	45.840	1.0008	25.471	1.0016	17.639	45
16	1	214.86	1.0002	45.237	1.0008	25.284	1.0016	17.549	44
17	1	202.22	1.0002	44.650	1.0008	25.100	1.0016	17.460	43
18	1	190.99	1.0002	44.077	1.0008	24.918	1.0017	17.372	42
19	1	180.73	1.0003	43.520	1.0008	24.739	1.0017	17.285	41
20	1	171.89	1.0003	42.976	1.0008	24.562	1.0017	17.198	40
21	1	163.70	1.0003	42.445	1.0008	24.358	1.0017	17.113	39
22	1	156.26	1.0003	41.928	1.0008	24.216	1.0017	17.028	38
23	1	149.47	1.0003	41.423	1.0009	24.047	1.0017	16.944	37
24	1	143.24	1.0003	40.930	1.0009	23.880	1.0018	16.861	36
25	1	137.51	1.0003	40.448	1.0009	23.716	1.0018	16.779	35
26	1	132.22	1.0003	39.978	1.0009	23.553	1.0018	16.698	34
27	1	127.32	1.0003	39.518	1.0009	23.393	1.0018	16.617	33
28	1	122.78	1.0003	39.069	1.0009	23.235	1.0018	16.538	32
29	1	118.54	1.0003	38.631	1.0009	23.079	1.0018	16.459	31
30	1	114.59	1.0003	38.201	1.0009	22.925	1.0019	16.380	30
31	1	110.90	1.0003	37.782	1.0010	22.774	1.0019	16.303	29
32	1	107.43	1.0003	37.371	1.0010	22.624	1.0019	16.226	28
33	1	104.17	1.0004	36.969	1.0010	22.476	1.0019	16.150	27
34	1	101.11	1.0004	36.576	1.0010	22.330	1.0019	16.075	26
35	1	98.223	1.0004	36.191	1.0010	22.186	1.0019	16.000	25
36	1	95.495	1.0004	35.814	1.0010	22.044	1.0020	15.926	24
37	1	92.914	1.0004	35.445	1.0010	21.904	1.0020	15.853	23
38	1.0001	92.469	1.0004	35.084	1.0010	21.765	1.0020	15.780	22
39	1.0001	88.149	1.0004	34.729	1.0011	21.629	1.0020	15.708	21
40	1.0001	85.946	1.0004	34.382	1.0011	21.494	1.0020	15.637	20
41	1.0001	83.849	1.0004	34.042	1.0011	21.360	1.0021	15.566	19
42	1.0001	81.853	1.0004	33.708	1.0011	21.228	1.0021	15.496	18
43	1.0001	79.950	1.0004	33.381	1.0011	21.098	1.0021	15.427	17
44	1.0001	78.133	1.0004	33.060	1.0011	20.970	1.0021	15.358	16
45	1.0001	76.396	1.0005	32.745	1.0011	20.843	1.0021	15.290	15
46	1.0001	74.736	1.0005	32.437	1.0012	20.717	1.0022	15.222	14
47	1.0001	73.146	1.0005	32.134	1.0012	20.593	1.0022	15.155	13
48	1.0001	71.622	1.0005	31.836	1.0012	20.471	1.0022	15.089	12
49	1.0001	71.160	1.0005	31.544	1.0012	20.350	1.0022	15.023	11
50	1.0001	68.757	1.0005	31.257	1.0012	20.230	1.0022	14.958	10
51	1.0001	67.409	1.0005	30.976	1.0012	20.112	1.0023	14.893	9
52	1.0001	66.113	1.0005	30.699	1.0012	19.995	1.0023	14.829	8
53	1.0001	64.866	1.0005	30.428	1.0013	19.880	1.0023	14.765	7
54	1.0001	63.664	1.0005	30.161	1.0013	19.766	1.0023	14.702	6
55	1.0001	62.507	1.0005	29.899	1.0013	19.653	1.0023	14.640	5
56	1.0001	61.391	1.0006	29.641	1.0013	19.541	1.0024	14.578	4
57	1.0001	61.314	1.0006	29.388	1.0013	19.431	1.0024	14.517	3
58	1.0001	59.274	1.0006	29.139	1.0013	19.322	1.0024	14.456	2
59	1.0001	58.270	1.0006	28.894	1.0013	19.214	1.0024	14.395	1
60	1.0001	57.299	1.0006	28.654	1.0014	19.107	1.0024	14.335	0
′	Co-sec.	Sec.	Co-sec.	Sec.	Co-sec.	Sec.	Co-sec.	Sec.	
		89°		88°		87°		86°	

| | 4° | | 5° | | 6° | | 7° | |
SEC.	CO-SEC.	SEC.	CO-SEC.	SEC.	CO-SEC.	SEC.	CO-SEC.	'
1.0024	14.335	1.0038	11.474	1.0055	9.5668	1.0075	8.2055	60
1.0025	14.276	1.0038	11.436	1.0055	9.5404	1.0075	8.1861	59
1.0025	14.217	1.0039	11.398	1.0056	9.5141	1.0076	8.1668	58
1.0025	14.159	1.0039	11.360	1.0056	9.4880	1.0076	8.1476	57
1.0025	14.101	1.0039	11.323	1.0056	9.4620	1.0076	8.1285	56
1.0025	14.043	1.0039	11.286	1.0057	9.4362	1.0077	8.1094	55
1.0026	13.986	1.0040	11.249	1.0057	9.4105	1.0077	8.0905	54
1.0026	13.930	1.0040	11.213	1.0057	9.3850	1.0078	8.0717	53
1.0026	13.874	1.0040	11.176	1.0057	9.3596	1.0078	8.0529	52
1.0026	13.818	1.0040	11.140	1.0058	9.3343	1.0078	8.0342	51
1.0026	13.763	1.0041	11.104	1.0058	9.3092	1.0079	8.0156	50
1.0027	13.708	1.0041	11.069	1.0058	9.2842	1.0079	7.9971	49
1.0027	13.654	1.0041	11.033	1.0059	9.2593	1.0079	7.9787	48
1.0027	13.600	1.0041	10.988	1.0059	9.2346	1.0080	7.9604	47
1.0027	13.547	1.0042	10.963	1.0059	9.2100	1.0080	7.9421	46
1.0027	13.494	1.0042	10.929	1.0060	9.1855	1.0080	7.9240	45
1.0028	13.441	1.0042	10.894	1.0060	9.1612	1.0081	7.9059	44
1.0028	13.389	1.0043	10.860	1.0060	9.1370	1.0081	7.8879	43
1.0028	13.337	1.0043	10.826	1.0061	9.1129	1.0082	7.8700	42
1.0028	13.286	1.0043	10.792	1.0061	9.0890	1.0082	7.8522	41
1.0029	13.235	1.0043	10.758	1.0061	9.0651	1.0082	7.8344	40
1.0029	13.184	1.0044	10.725	1.0062	9.0414	1.0083	7.8168	39
1.0029	13.134	1.0044	10.692	1.0062	9.0179	1.0083	7.7992	38
1.0029	13.084	1.0044	10.659	1.0062	8.9944	1.0084	7.7817	37
1.0029	13.034	1.0044	10.626	1.0063	8.9711	1.0084	7.7642	36
1.0030	12.985	1.0045	10.593	1.0063	8.9479	1.0084	7.7469	35
1.0030	12.937	1.0045	10.561	1.0063	8.9248	1.0085	7.7296	34
1.0030	12.888	1.0045	10.529	1.0064	8.9018	1.0085	7.7124	33
1.0030	12.840	1.0046	10.497	1.0064	8.8790	1.0085	7.6953	32
1.0031	12.793	1.0046	10.465	1.0064	8.8563	1.0086	7.6783	31
1.0031	12.745	1.0046	10.433	1.0065	8.8337	1.0086	7.6613	30
1.0031	12.698	1.0046	10.402	1.0065	8.8112	1.0087	7.6444	29
1.0031	12.652	1.0047	10.371	1.0065	8.7888	1.0087	7.6276	28
1.0032	12.606	1.0047	10.340	1.0066	8.7665	1.0087	7.6108	27
1.0032	12.560	1.0047	10.309	1.0066	8.7444	1.0088	7.5942	26
1.0032	12.514	1.0048	10.278	1.0066	8.7223	1.0088	7.5776	25
1.0032	12.469	1.0048	10.248	1.0067	8.7004	1.0089	7.5611	24
1.0032	12.424	1.0048	10.217	1.0067	8.6786	1.0089	7.5446	23
1.0033	12.379	1.0048	10.187	1.0067	8.6569	1.0089	7.5282	22
1.0033	12.335	1.0049	10.157	1.0068	8.6353	1.0090	7.5119	21
1.0033	12.291	1.0049	10.127	1.0068	8.6138	1.0090	7.4957	20
1.0033	12.248	1.0049	10.098	1.0068	8.5924	1.0090	7.4795	19
1.0034	12.204	1.0050	10.068	1.0069	8.5711	1.0091	7.4634	18
1.0034	12.161	1.0050	10.039	1.0069	8.5499	1.0091	7.4474	17
1.0034	12.118	1.0050	10.010	1.0069	8.5289	1.0092	7.4315	16
1.0034	12.076	1.0050	9.9812	1.0070	8.5079	1.0092	7.4156	15
1.0035	12.034	1.0051	9.9525	1.0070	8.4871	1.0092	7.3998	14
1.0035	11.992	1.0051	9.9239	1.0070	8.4663	1.0093	7.3840	13
1.0035	11.950	1.0051	9.8955	1.0071	8.4457	1.0093	7.3683	12
1.0035	11.909	1.0052	9.8672	1.0071	8.4251	1.0094	7.3527	11
1.0036	11.868	1.0052	9.8391	1.0071	8.4046	1.0094	7.3372	10
1.0036	11.828	1.0052	9.8112	1.0072	8.3843	1.0094	7.3217	9
1.0036	11.787	1.0053	9.7834	1.0072	8.3640	1.0095	7.3063	8
1.0036	11.747	1.0053	9.7558	1.0073	8.3439	1.0095	7.2909	7
1.0037	11.707	1.0053	9.7283	1.0073	8.3238	1.0096	7.2757	6
1.0037	11.668	1.0053	9.7010	1.0073	8.3039	1.0096	7.2604	5
1.0037	11.628	1.0054	9.6739	1.0074	8.2840	1.0097	7.2453	4
1.0037	11.589	1.0054	9.6469	1.0074	8.2642	1.0097	7.2302	3
1.0038	11.550	1.0054	9.6200	1.0074	8.2446	1.0097	7.2152	2
1.0038	11.512	1.0055	9.5933	1.0075	8.2250	1.0098	7.2002	1
1.0038	11.474	1.0055	9.5668	1.0075	8.2055	1.0098	7.1853	0
CO-SEC.	SEC.	CO-SEC.	SEC.	CO-SEC.	SEC.	CO-SEC.	SEC.	'
	85°		84°		83°		82°	

′	8° SEC.	CO-SEC.	9° SEC.	CO-SEC.	10° SEC.	CO-SEC.	11° SEC.	CO-SEC.	′
0	1.0098	7.1853	1.0125	6.3924	1.0154	5.7588	1.0187	5.2408	60
1	1.0099	7.1704	1.0125	6.3807	1.0155	5.7493	1.0188	5.2330	59
2	1.0099	7.1557	1.0125	6.3690	1.0155	5.7398	1.0188	5.2252	58
3	1.0099	7.1409	1.0126	6.3574	1.0156	5.7304	1.0189	5.2174	57
4	1.0100	7.1263	1.0126	6.3458	1.0156	5.7210	1.0189	5.2097	56
5	1.0100	7.1117	1.0127	6.3343	1.0157	5.7117	1.0190	5.2019	55
6	1.0101	7.0972	1.0127	6.3228	1.0157	5.7023	1.0191	5.1942	54
7	1.0101	7.0827	1.0128	6.3113	1.0158	5.6930	1.0191	5.1865	53
8	1.0102	7.0683	1.0128	6.2999	1.0158	5.6838	1.0192	5.1788	52
9	1.0102	7.0539	1.0129	6.2885	1.0159	5.6745	1.0192	5.1712	51
10	1.0102	7.0396	1.0129	6.2772	1.0159	5.6653	1.0193	5.1636	50
11	1.0103	7.0254	1.0130	6.2659	1.0160	5.6561	1.0193	5.1560	49
12	1.0103	7.0112	1.0130	6.2546	1.0160	5.6470	1.0194	5.1484	48
13	1.0104	6.9971	1.0131	6.2434	1.0161	5.6379	1.0195	5.1409	47
14	1.0104	6.9830	1.0131	6.2322	1.0162	5.6288	1.0195	5.1333	46
15	1.0104	6.9690	1.0132	6.2211	1.0162	5.6197	1.0196	5.1258	45
16	1.0105	6.9550	1.0132	6.2100	1.0163	5.6107	1.0196	5.1183	44
17	1.0105	6.9411	1.0133	6.1990	1.0163	5.6017	1.0197	5.1109	43
18	1.0106	6.9273	1.0133	6.1880	1.0164	5.5928	1.0198	5.1034	42
19	1.0106	6.9135	1.0134	6.1770	1.0164	5.5838	1.0198	5.0960	41
20	1.0107	6.8998	1.0134	6.1661	1.0165	5.5749	1.0199	5.0886	40
21	1.0107	6.8861	1.0135	6.1552	1.0165	5.5660	1.0199	5.0812	39
22	1.0107	6.8725	1.0135	6.1443	1.0166	5.5572	1.0200	5.0739	38
23	1.0108	6.8589	1.0136	6.1335	1.0166	5.5484	1.0201	5.0666	37
24	1.0108	6.8454	1.0136	6.1227	1.0167	5.5396	1.0201	5.0593	36
25	1.0109	6.8320	1.0136	6.1120	1.0167	5.5308	1.0202	5.0520	35
26	1.0109	6.8185	1.0137	6.1013	1.0168	5.5221	1.0202	5.0447	34
27	1.0110	6.8052	1.0137	6.0906	1.0169	5.5134	1.0203	5.0375	33
28	1.0110	6.7919	1.0138	6.0800	1.0169	5.5047	1.0204	5.0302	32
29	1.0111	6.7787	1.0138	6.0694	1.0170	5.4960	1.0204	5.0230	31
30	1.0111	6.7655	1.0139	6.0588	1.0170	5.4874	1.0205	5.0158	30
31	1.0111	6.7523	1.0139	6.0483	1.0171	5.4788	1.0205	5.0087	29
32	1.0112	6.7392	1.0140	6.0379	1.0171	5.4702	1.0206	5.0015	28
33	1.0112	6.7262	1.0140	6.0274	1.0172	5.4617	1.0207	4.9944	27
34	1.0113	6.7132	1.0141	6.0170	1.0172	5.4532	1.0207	4.9873	26
35	1.0113	6.7003	1.0141	6.0066	1.0173	5.4447	1.0208	4.9802	25
36	1.0114	6.6874	1.0142	5.9963	1.0174	5.4362	1.0208	4.9732	24
37	1.0114	6.6745	1.0142	5.9860	1.0174	5.4278	1.0209	4.9661	23
38	1.0115	6.6617	1.0143	5.9758	1.0175	5.4194	1.0210	4.9591	22
39	1.0115	6.6490	1.0143	5.9655	1.0175	5.4110	1.0210	4.9521	21
40	1.0115	6.6363	1.0144	5.9554	1.0176	5.4026	1.0211	4.9452	20
41	1.0116	6.6237	1.0144	5.9452	1.0176	5.3943	1.0211	4.9382	19
42	1.0116	6.6111	1.0145	5.9351	1.0177	5.3860	1.0212	4.9313	18
43	1.0117	6.5985	1.0145	5.9250	1.0177	5.3777	1.0213	4.9243	17
44	1.0117	6.5860	1.0146	5.9150	1.0178	5.3695	1.0213	4.9175	16
45	1.0118	6.5736	1.0146	5.9049	1.0179	5.3612	1.0214	4.9106	15
46	1.0118	6.5612	1.0147	5.8950	1.0179	5.3530	1.0215	4.9037	14
47	1.0119	6.5488	1.0147	5.8850	1.0180	5.3449	1.0215	4.8969	13
48	1.0119	6.5365	1.0148	5.8751	1.0180	5.3367	1.0216	4.8901	12
49	1.0119	6.5243	1.0148	5.8652	1.0181	5.3286	1.0216	4.8833	11
50	1.0120	6.5121	1.0149	5.8554	1.0181	5.3205	1.0217	4.8765	10
51	1.0120	6.4999	1.0150	5.8456	1.0182	5.3124	1.0218	4.8697	9
52	1.0121	6.4878	1.0150	5.8358	1.0182	5.3044	1.0218	4.8630	8
53	1.0121	6.4757	1.0151	5.8261	1.0183	5.2963	1.0219	4.8563	7
54	1.0122	6.4637	1.0151	5.8163	1.0184	5.2883	1.0220	4.8496	6
55	1.0122	6.4517	1.0152	5.8067	1.0184	5.2803	1.0220	4.8429	5
56	1.0123	6.4398	1.0152	5.7970	1.0185	5.2724	1.0221	4.8362	4
57	1.0123	6.4279	1.0153	5.7874	1.0185	5.2645	1.0221	4.8296	3
58	1.0124	6.4160	1.0153	5.7778	1.0186	5.2566	1.0222	4.8229	2
59	1.0124	6.4042	1.0154	5.7683	1.0186	5.2487	1.0223	4.8163	1
60	1.0125	6.3924	1.0154	5.7588	1.0187	5.2408	1.0223	4.8097	0

	CO-SEC.	SEC.	CO-SEC.	SEC.	CO-SEC.	SEC.	CO-SEC.	SEC.	′
	81°		80°		79°		78°		

12° Sec.	Co-sec.	13° Sec.	Co-sec.	14° Sec.	Co-sec.	15° Sec.	Co-sec.	'
1.0223	4.8097	1.0263	4.4454	1.0306	4.1336	1.0353	3.8637	60
1.0224	4.8032	1.0264	4.4398	1.0307	4.1287	1.0353	3.8595	59
1.0225	4.7966	1.0264	4.4342	1.0308	4.1239	1.0354	3.8553	58
1.0225	4.7901	1.0265	4.4287	1.0308	4.1191	1.0355	3.8512	57
1.0226	4.7835	1.0266	4.4231	1.0309	4.1144	1.0356	3.8470	56
1.0226	4.7770	1.0266	4.4176	1.0310	4.1096	1.0357	3.8428	55
1.0227	4.7706	1.0267	4.4121	1.0311	4.1048	1.0358	3.8387	54
1.0228	4.7641	1.0268	4.4065	1.0311	4.1001	1.0358	3.8346	53
1.0228	4.7576	1.0268	4.4011	1.0312	4.0953	1.0359	3.8304	52
1.0229	4.7512	1.0269	4.3956	1.0313	4.0906	1.0360	3.8263	51
1.0230	4.7448	1.0270	4.3910	1.0314	4.0859	1.0361	3.8222	50
1.0230	4.7384	1.0271	4.3847	1.0314	4.0812	1.0362	3.8181	49
1.0231	4.7320	1.0271	4.3792	1.0315	4.0765	1.0362	3.8140	48
1.0232	4.7257	1.0272	4.3738	1.0316	4.0718	1.0363	3.8100	47
1.0232	4.7193	1.0273	4.3684	1.0317	4.0672	1.0364	3.8059	46
1.0233	4.7130	1.0273	4.3630	1.0317	4.0625	1.0365	3.8018	45
1.0234	4.7067	1.0274	4.3576	1.0318	4.0579	1.0366	3.7978	44
1.0234	4.7004	1.0275	4.3522	1.0319	4.0532	1.0367	3.7937	43
1.0235	4.6942	1.0276	4.3469	1.0320	4.0486	1.0367	3.7897	42
1.0235	4.6879	1.0276	4.3415	1.0320	4.0440	1.0368	3.7857	41
1.0236	4.6817	1.0277	4.3362	1.0321	4.0394	1.0369	3.7816	40
1.0237	4.6754	1.0278	4.3309	1.0322	4.0348	1.0370	3.7776	39
1.0237	4.6692	1.0278	4.3256	1.0323	4.0302	1.0371	3.7736	38
1.0238	4.6631	1.0279	4.3203	1.0323	4.0256	1.0371	3.7697	37
1.0239	4.6569	1.0280	4.3150	1.0324	4.0211	1.0372	3.7657	36
1.0239	4.6507	1.0280	4.3098	1.0325	4.0165	1.0373	3.7617	35
1.0240	4.6446	1.0281	4.3045	1.0326	4.0120	1.0374	3.7577	34
1.0241	4.6385	1.0282	4.2993	1.0327	4.0074	1.0375	3.7538	33
1.0241	4.6324	1.0283	4.2941	1.0327	4.0029	1.0376	3.7498	32
1.0242	4.6263	1.0283	4.2888	1.0328	3.9984	1.0376	3.7459	31
1.0243	4.6202	1.0284	4.2836	1.0329	3.9939	1.0377	3.7420	30
1.0243	4.6142	1.0285	4.2785	1.0330	3.9894	1.0378	3.7380	29
1.0244	4.6081	1.0285	4.2733	1.0330	3.9850	1.0379	3.7341	28
1.0245	4.6021	1.0286	4.2681	1.0331	3.9805	1.0380	3.7302	27
1.0245	4.5961	1.0287	4.2630	1.0332	3.9760	1.0381	3.7263	26
1.0246	4.5901	1.0288	4.2579	1.0333	3.9716	1.0382	3.7224	25
1.0247	4.5841	1.0288	4.2527	1.0334	3.9672	1.0382	3.7186	24
1.0247	4.5782	1.0289	4.2476	1.0334	3.9627	1.0383	3.7147	23
1.0248	4.5722	1.0290	4.2425	1.0335	3.9583	1.0384	3.7108	22
1.0249	4.5663	1.0291	4.2375	1.0336	3.9539	1.0385	3.7070	21
1.0249	4.5604	1.0291	4.2324	1.0337	3.9495	1.0386	3.7031	20
1.0250	4.5545	1.0292	4.2273	1.0338	3.9451	1.0387	3.6993	19
1.0251	4.5486	1.0293	4.2223	1.0338	3.9408	1.0387	3.6955	18
1.0251	4.5428	1.0293	4.2173	1.0339	3.9364	1.0388	3.6917	17
1.0252	4.5369	1.0294	4.2122	1.0340	3.9320	1.0389	3.6878	16
1.0253	4.5311	1.0295	4.2072	1.0341	3.9277	1.0390	3.6840	15
1.0253	4.5253	1.0296	4.2022	1.0341	3.9234	1.0391	3.6802	14
1.0254	4.5195	1.0296	4.1972	1.0342	3.9199	1.0392	3.6765	13
1.0255	4.5137	1.0297	4.1923	1.0343	3.9147	1.0393	3.6727	12
1.0255	4.5079	1.0298	4.1873	1.0344	3.9104	1.0393	3.6689	11
1.0256	4.5021	1.0299	4.1824	1.0345	3.9061	1.0394	3.6651	10
1.0257	4.4964	1.0299	4.1774	1.0345	3.9018	1.0395	3.6614	9
1.0257	4.4907	1.0300	4.1725	1.0346	3.8976	1.0396	3.6576	8
1.0258	4.4850	1.0301	4.1676	1.0347	3.8933	1.0397	3.6539	7
1.0259	4.4793	1.0302	4.1627	1.0348	3.8990	1.0398	3.6502	6
1.0260	4.4736	1.0302	4.1578	1.0349	3.8848	1.0399	3.6464	5
1.0260	4.4679	1.0303	4.1529	1.0349	3.8805	1.0399	3.6427	4
1.0261	4.4623	1.0304	4.1481	1.0350	3.8763	1.0400	3.6390	3
1.0262	4.4566	1.0305	4.1432	1.0351	3.8721	1.0401	3.6353	2
1.0262	4.4510	1.0305	4.1384	1.0352	3.8679	1.0402	3.6316	1
1.0263	4.4454	1.0306	4.1336	1.0353	3.8637	1.0403	3.6279	0
Co-sec. 77°	Sec.	Co-sec. 76°	Sec.	Co-sec. 75°	Sec.	Co-sec. 74°	Sec.	

′	16° Sec.	Co-sec.	17° Sec.	Co-sec.	18° Sec.	Co-sec.	19° Sec.	Co-sec.	
0	1.0403	3.6279	1.0457	3.4203	1.0515	3.2361	1.0576	3.0715	
1	1.0404	3.6243	1.0458	3.4170	1.0516	3.2332	1.0577	3.0690	
2	1.0405	3.6206	1.0459	3.4138	1.0517	3.2303	1.0578	3.0664	
3	1.0406	3.6169	1.0460	3.4106	1.0518	3.2274	1.0579	3.0638	
4	1.0406	3.6133	1.0461	3.4073	1.0519	3.2245	1.0580	3.0612	
5	1.0407	5.6096	1.0461	3.4041	1.0520	3.2216	1.0581	3.0586	
6	1.0408	3.6060	1.0462	3.4009	1.0521	3.2188	1.0582	3.0561	
7	1.0409	3.6024	1.0463	3.3977	1.0522	3.2159	1.0584	3.0535	
8	1.0410	3.5987	1.0464	3.3945	1.0523	3.2131	1.0585	3.0509	
9	1.0411	3.5951	1.0465	3.3913	1.0524	3.2102	1.0586	3.0484	
10	1.0412	3.5915	1.0466	3.3881	1.0525	3.2074	1.0587	3.0458	
11	1.0413	3.5879	1.0467	3.3849	1.0526	3.2045	1.0588	3.0433	
12	1.0413	3.5843	1.0468	3.3817	1.0527	3.2017	1.0589	3.0407	
13	1.0414	3.5807	1.0469	3.3785	1.0528	3.1989	1.0590	3.0382	
14	1.0415	3.5772	1.0470	3.3754	1.0529	3.1960	1.0591	3.0357	
15	1.0416	3.5736	1.0471	3.3722	1.0530	3.1932	1.0592	3.0331	
16	1.0417	3.5700	1.0472	3.3690	1.0531	3.1904	1.0593	3.0306	
17	1.0418	3.5665	1.0473	3.3659	1.0532	3.187′	1.0594	3.0281	
18	1.0419	3.5629	1.0474	3.3627	1.0533	3.1848	1.0595	3.0256	
19	1.0420	3.5594	1.0475	3.3596	1.0534	3.1820	1.0596	3.0231	
20	1.0420	3.5559	1.0476	3.3565	1.0535	3.1792	1.0598	3.0206	
21	1.0421	3.5523	1.0477	3.3534	1.0536	3.1764	1.0599	3.0181	
22	1.0422	3.5488	1.0478	3.3502	1.0537	3.1736	1.0600	3.0156	
23	1.0423	3.5453	1.0478	3.3471	1.0538	3.1708	1.0601	3.0131	
24	1.0424	3.5418	1.0479	3.3440	1.0539	3.1681	1.0602	3.0106	
25	1.0425	3.5383	1.0480	3.3409	1.0540	3.1653	1.0603	3.0081	
26	1.0426	3.5348	1.0481	3.3378	1.0541	3.1625	1.0604	3.0056	
27	1.0427	3.5313	1.0482	3.3347	1.0542	3.1598	1.0605	3.0031	
28	1.0428	3.5279	1.0483	3.3316	1.0543	3.1570	1.0606	3.0007	
29	1.0428	3.5244	1.0484	3.3286	1.0544	3.1543	1.0607	2.9982	
30	1.0429	3.5209	1.0485	3.3255	1.0545	3.1515	1.0608	2.9957	
31	1.0430	3.5175	1.0486	3.3224	1.0546	3.1488	1.0609	2.9933	
32	1.0431	3.5140	1.0487	3.3194	1.0547	3.1461	1.0611	2.9908	
33	1.0432	3.5106	1.0488	3.3163	1.0548	3.1433	1.0612	2.9884	
34	1.0433	3.5072	1.0489	3.3133	1.0549	3.1406	1.0613	2.9859	
35	1.0434	3.5037	1.0490	3.3102	1.0550	3.1379	1.0614	2.9835	
36	1.0435	3.5003	1.0491	3.3072	1.0551	3.1352	1.0615	2.9810	
37	1.0436	3.4969	1.0492	3.3042	1.0552	3.1325	1.0616	2.9786	
38	1.0437	3.4935	1.0493	3.3011	1.0553	3.1298	1.0617	2.9762	
39	1.0438	3.4901	1.0494	3.2981	1.0554	3.1271	1.0618	2.9738	
40	1.0438	3.4867	1.0495	3.2951	1.0555	3.1244	1.0619	2.9713	
41	1.0439	3.4833	1.0496	3.2921	1.0556	3.1217	1.0620	2.9689	
42	1.0440	3.4799	1.0497	3.2891	1.0557	3.1190	1.0622	2.9665	
43	1.0441	3.4766	1.0498	3.2861	1.0558	3.1163	1.0623	2.9641	
44	1.0442	3.4732	1.0499	3.2831	1.0559	3.1137	1.0624	2.9617	
45	1.0443	3.4698	1.0500	3.2801	1.0560	3.1110	1.0625	2.9593	
46	1.0444	3.4665	1.0501	3.2772	1.0561	3.1083	1.0626	2.9569	
47	1.0445	3.4632	1.0502	3.2742	1.0562	3.1057	1.0627	2.9545	
48	1.0446	3.4598	1.0503	3.2712	1.0563	3.1030	1.0628	2.9521	
49	1.0447	3.4565	1.0504	3.2683	1.0565	3.1004	1.0629	2.9497	
50	1.0448	3.4532	1.0505	3.2653	1.0566	3.0977	1.0630	2.9474	
51	1.0448	3.4498	1.0506	3.2624	1.0567	3.0951	1.0632	2.9450	
52	1.0449	3.4465	1.0507	3.2594	1.0568	3.0925	1.0633	2.9426	
53	1.0450	3.4432	1.0508	3.2565	1.0569	3.0898	1.0634	2.9402	
54	1.0451	3.4399	1.0509	3.2535	1.0570	3.0872	1.0635	2.9379	
55	1.0452	3.4366	1.0510	3.2506	1.0571	3.0846	1.0636	2.9355	
56	1.0453	3.4334	1.0511	3.2477	1.0572	3.0820	1.0637	2.9332	
57	1.0454	3.4301	1.0512	3.2448	1.0573	3.0793	1.0638	2.9308	
58	1.0455	3.4268	1.0513	3.2419	1.0574	3.0767	1.0639	2.9285	
59	1.0456	3.4236	1.0514	3.2390	1.0575	3.0741	1.0641	2.9261	
60	1.0457	3.4203	1.0515	3.2361	1.0576	3.0715	1.0642	2.9238	

	Co-sec.	Sec.	Co-sec.	Sec.	Co-sec.	Sec.	Co-sec.	Sec.	
	73°		72°		71°		70°		

20°		21°		22°		23°		
Sec.	Co-sec.	Sec.	Co-sec.	Sec.	Co-sec.	Sec.	Co-sec.	'
1.0642	2.9238	1.0711	2.7904	1.0785	2.6695	1.0864	2.5593	60
1.0643	2.9215	1.0713	2.7883	1.0787	2.6675	1.0865	2.5575	59
1.0644	2.9191	1.0714	2.7862	1.0788	2.6656	1.0866	2.5558	58
1.0645	2.9168	1.0715	2.7841	1.0789	2.6637	1.0868	2.5540	57
1.0646	2.9145	1.0716	2.7820	1.0790	2.6618	1.0869	2.5523	56
1.0647	2.9122	1.0717	2.7799	1.0792	2.6599	1.0870	2.5506	55
1.0648	2.9098	1.0719	2.7778	1.0793	2.6580	1.0872	2.5488	54
1.0650	2.9075	1.0720	2.7757	1.0794	2.6561	1.0873	2.5471	53
1.0651	2.9052	1.0721	2.7736	1.0795	2.6542	1.0874	2.5453	52
1.0652	2.9029	1.0722	2.7715	1.0797	2.6523	1.0876	2.5436	51
1.0653	2.9006	1.0723	2.7694	1.0798	2.6504	1.0877	2.5419	50
1.0654	2.8983	1.0725	2.7674	1.0799	2.6485	1.0878	2.5402	49
1.0655	2.8960	1.0726	2.7653	1.0801	2.6466	1.0880	2.5384	48
1.0656	2.8937	1.0727	2.7632	1.0802	2.6447	1.0881	2.5367	47
1.0658	2.8915	1.0728	2.7611	1.0803	2.6428	1.0882	2.5350	46
1.0659	2.8892	1.0729	2.7591	1.0804	2.6410	1.0884	2.5333	45
1.0660	2.8869	1.0731	2.7570	1.0806	2.6391	1.0885	2.5316	44
1.0661	2.8846	1.0732	2.7550	1.0807	2.6372	1.0886	2.5299	43
1.0662	2.8824	1.0733	2.7529	1.0808	2.6353	1.0888	2.5281	42
1.0663	2.8801	1.0734	2.7509	1.0810	2.6335	1.0889	2.5264	41
1.0664	2.8778	1.0736	2.7488	1.0811	2.6316	1.0891	2.5247	40
1.0666	2.8756	1.0737	2.7468	1.0812	2.6297	1.0892	2.5230	39
1.0667	2.8733	1.0738	2.7447	1.0813	2.6279	1.0893	2.5213	38
1.0668	2.8711	1.0739	2.7427	1.0815	2.6260	1.0895	2.5196	37
1.0669	2.8688	1.0740	2.7406	1.0816	2.6242	1.0896	2.5179	36
1.0670	2.8666	1.0742	2.7386	1.0817	2.6223	1.0897	2.5163	35
1.0671	2.8644	1.0743	2.7366	1.0819	2.6205	1.0899	2.5146	34
1.0673	2.8621	1.0744	2.7346	1.0820	2.6186	1.0900	2.5129	33
1.0674	2.8599	1.0745	2.7325	1.0821	2.6168	1.0902	2.5112	32
1.0675	2.8577	1.0747	2.7305	1.0823	2.6150	1.0903	2.5095	31
1.0676	2.8554	1.0748	2.7285	1.0824	2.6131	1.0904	2.5078	30
1.0677	2.8532	1.0749	2.7265	1.0825	2.6113	1.0906	2.5062	29
1.0678	2.8510	1.0750	2.7245	1.0826	2.6095	1.0907	2.5045	28
1.0679	2.8488	1.0751	2.7225	1.0828	2.6076	1.0908	2.5028	27
1.0681	2.8466	1.0753	2.7205	1.0829	2.6058	1.0910	2.5011	26
1.0682	2.8444	1.0754	2.7185	1.0830	2.6040	1.0911	2.4995	25
1.0683	2.8422	1.0755	2.7165	1.0832	2.6022	1.0913	2.4978	24
1.0684	2.8400	1.0756	2.7145	1.0833	2.6003	1.0914	2.4961	23
1.0685	2.8378	1.0758	2.7125	1.0834	2.5985	1.0915	2.4945	22
1.0686	2.8356	1.0759	2.7105	1.0836	2.5967	1.0917	2.4928	21
1.0688	2.8334	1.0760	2.7085	1.0837	2.5949	1.0918	2.4912	20
1.0689	2.8312	1.0761	2.7065	1.0838	2.5931	1.0920	2.4895	19
1.0690	2.8290	1.0763	2.7045	1.0840	2.5913	1.0921	2.4879	18
1.0691	2.8269	1.0764	2.7026	1.0841	2.5895	1.0922	2.4862	17
1.0692	2.8247	1.0765	2.7006	1.0842	2.5877	1.0924	2.4846	16
1.0694	2.8225	1.0766	2.6986	1.0844	2.5859	1.0925	2.4829	15
1.0695	2.8204	1.0768	2.6967	1.0845	2.5841	1.0927	2.4813	14
1.0696	2.8182	1.0769	2.6947	1.0846	2.5823	1.0928	2.4797	13
1.0697	2.8160	1.0770	2.6927	1.0847	2.5805	1.0929	2.4780	12
1.0698	2.8139	1.0771	2.6908	1.0849	2.5787	1.0931	2.4764	11
1.0699	2.8117	1.0773	2.6888	1.0850	2.5770	1.0932	2.4748	10
1.0701	2.8096	1.0774	2.6869	1.0851	2.5752	1.0934	2.4731	9
1.0702	2.8074	1.0775	2.6849	1.0853	2.5734	1.0935	2.4715	8
1.0703	2.8053	1.0776	2.6830	1.0854	2.5716	1.0936	2.4699	7
1.0704	2.8032	1.0778	2.6810	1.0855	2.5699	1.0938	2.4683	6
1.0705	2.8010	1.0779	2.6791	1.0857	2.5681	1.0939	2.4666	5
1.0707	2.7989	1.0780	2.6772	1.0858	2.5663	1.0941	2.4650	4
1.0708	2.7968	1.0781	2.6752	1.0859	2.5646	1.0942	2.4634	3
1.0709	2.7947	1.0783	2.6733	1.0861	2.5628	1.0943	2.4618	2
1.0710	2.7925	1.0784	2.6714	1.0862	2.5610	1.0945	2.4602	1
1.0711	2.7904	1.0785	2.6695	1.0864	2.5593	1.0946	2.4586	0
Co-sec.	Sec.	Co-sec.	Sec.	Co-sec.	Sec.	Co-sec.	Sec.	
	69°		68°		67°		66°	

′	24° Sec.	24° Co-sec.	25° Sec.	25° Co-sec.	26° Sec.	26° Co-sec.	27° Sec.	27° Co-sec.
0	1.0946	2.4586	1.1034	2.3662	1.1126	2.2812	1.1223	2.2027
1	1.0948	2.4570	1.1035	2.3647	1.1127	2.2798	1.1225	2.2014
2	1.0949	2.4554	1.1037	2.3632	1.1129	2.2784	1.1226	2.2002
3	1.0951	2.4538	1.1038	2.3618	1.1131	2.2771	1.1228	2.1989
4	1.0952	2.4522	1.1040	2.3603	1.1132	2.2757	1.1230	2.1977
5	1.0953	2.4506	1.1041	2.3588	1.1134	2.2744	1.1231	2.1964
6	1.0955	2.4490	1.1043	2.3574	1.1135	2.2730	1.1233	2.1952
7	1.0956	2.4474	1.1044	2.3559	1.1137	2.2717	1.1235	2.1939
8	1.0958	2.4458	1.1046	2.3544	1.1139	2.2703	1.1237	2.1927
9	1.0959	2.4442	1.1047	2.3530	1.1140	2.2690	1.1238	2.1914
10	1.0961	2.4426	1.1049	2.3515	1.1142	2.2676	1.1240	2.1902
11	1.0962	2.4411	1.1050	2.3501	1.1143	2.2663	1.1242	2.1889
12	1.0963	2.4395	1.1052	2.3486	1.1145	2.2650	1.1243	2.1877
13	1.0965	2.4379	1.1053	2.3472	1.1147	2.2636	1.1245	2.1865
14	1.0966	2.4363	1.1055	2.3457	1.1148	2.2623	1.1247	2.1852
15	1.0968	2.4347	1.1056	2.3443	1.1150	2.2610	1.1248	2.1840
16	1.0969	2.4332	1.1058	2.3428	1.1151	2.2596	1.1250	2.1828
17	1.0971	2.4316	1.1059	2.3414	1.1153	2.2583	1.1252	2.1815
18	1.0972	2.4300	1.1061	2.3399	1.1155	2.2570	1.1253	2.1803
19	1.0973	2.4285	1.1062	2.3385	1.1156	2.2556	1.1255	2.1791
20	1.0975	2.4269	1.1064	2.3371	1.1158	2.2543	1.1257	2.1778
21	1.0976	2.4254	1.1065	2.3356	1.1159	2.2530	1.1258	2.1766
22	1.0978	2.4238	1.1067	2.3342	1.1161	2.2517	1.1260	2.1754
23	1.0979	2.4222	1.1068	2.3328	1.1163	2.2503	1.1262	2.1742
24	1.0981	2.4207	1.1070	2.3313	1.1164	2.2490	1.1264	2.1730
25	1.0982	2.4191	1.1072	2.3299	1.1166	2.2477	1.1265	2.1717
26	1.0984	2.4176	1.1073	2.3285	1.1167	2.2464	1.1267	2.1705
27	1.0985	2.4160	1.1075	2.3271	1.1169	2.2451	1.1269	2.1693
28	1.0986	2.4145	1.1076	2.3256	1.1171	2.2438	1.1270	2.1681
29	1.0988	2.4130	1.1078	2.3242	1.1172	2.2425	1.1272	2.1669
30	1.0989	2.4114	1.1079	2.3228	1.1174	2.2411	1.1274	2.1657
31	1.0991	2.4099	1.1081	2.3214	1.1176	2.2398	1.1275	2.1645
32	1.0992	2.4083	1.1082	2.3200	1.1177	2.2385	1.1277	2.1633
33	1.0994	2.4068	1.1084	2.3186	1.1179	2.2372	1.1279	2.1620
34	1.0995	2.4053	1.1085	2.3172	1.1180	2.2359	1.1281	2.1608
35	1.0997	2.4037	1.1087	2.3158	1.1182	2.2346	1.1282	2.1596
36	1.0998	2.4022	1.1088	2.3143	1.1184	2.2333	1.1284	2.1584
37	1.1000	2.4007	1.1090	2.3129	1.1185	2.2320	1.1286	2.1572
38	1.1001	2.3992	1.1092	2.3115	1.1187	2.2307	1.1287	2.1560
39	1.1003	2.3976	1.1093	2.3101	1.1189	2.2294	1.1289	2.1548
40	1.1004	2.3961	1.1095	2.3087	1.1190	2.2282	1.1291	2.1535
41	1.1005	2.3946	1.1096	2.3073	1.1192	2.2269	1.1293	2.1525
42	1.1007	2.3931	1.1098	2.3059	1.1193	2.2256	1.1294	2.1513
43	1.1008	2.3916	1.1099	2.3046	1.1195	2.2243	1.1296	2.1501
44	1.1010	2.3901	1.1101	2.3032	1.1197	2.2230	1.1298	2.1489
45	1.1011	2.3886	1.1102	2.3018	1.1198	2.2217	1.1299	2.1477
46	1.1013	2.3871	1.1104	2.3004	1.1200	2.2204	1.1301	2.1465
47	1.1014	2.3856	1.1106	2.2990	1.1202	2.2192	1.1303	2.1453
48	1.1016	2.3841	1.1107	2.2976	1.1203	2.2179	1.1305	2.1441
49	1.1017	2.3826	1.1109	2.2962	1.1205	2.2166	1.1306	2.1430
50	1.1019	2.3811	1.1110	2.2949	1.1207	2.2153	1.1308	2.1418
51	1.1020	2.3796	1.1112	2.2935	1.1208	2.2141	1.1310	2.1406
52	1.1022	2.3781	1.1113	2.2921	1.1210	2.2128	1.1312	2.1394
53	1.1023	2.3766	1.1115	2.2907	1.1212	2.2115	1.1313	2.1382
54	1.1025	2.3751	1.1116	2.2894	1.1213	2.2103	1.1315	2.1371
55	1.1026	2.3736	1.1118	2.2880	1.1215	2.2090	1.1317	2.1359
56	1.1028	2.3721	1.1120	2.2866	1.1217	2.2077	1.1319	2.1347
57	1.1029	2.3706	1.1121	2.2853	1.1218	2.2065	1.1320	2.1335
58	1.1031	2.3691	1.1123	2.2839	1.1220	2.2052	1.1322	2.1324
59	1.1032	2.3677	1.1124	2.2825	1.1222	2.2039	1.1324	2.1312
60	1.1034	2.3662	1.1126	2.2812	1.1223	2.2027	1.1326	2.1300

′	Co-sec.	Sec.	Co-sec.	Sec.	Co-sec.	Sec.	Co-sec.	Sec.
		65°		64°		63°		62°

28° Sec.	28° Co-sec.	29° Sec.	29° Co-sec.	30° Sec.	30° Co-sec.	31° Sec.	31° Co-sec.	′
1.1326	2.1300	1.1433	2.0627	1.1547	2.0000	1.1666	1.9416	60
1.1327	2.1289	1.1435	2.0616	1.1549	1.9990	1.1668	1.9407	59
1.1329	2.1277	1.1437	2.0605	1.1551	1.9980	1.1670	1.9397	58
1.1331	2.1266	1.1439	2.0594	1.1553	1.9970	1.1672	1.9388	57
1.1333	2.1254	1.1441	2.0583	1.1555	1.9960	1.1674	1.9378	56
1.1334	2.1242	1.1443	2.0573	1.1557	1.9950	1.1676	1.9369	55
1.1336	2.1231	1.1445	2.0562	1.1559	1.9940	1.1678	1.9360	54
1.1338	2.1219	1.1446	2.0551	1.1561	1.9930	1.1681	1.9350	53
1.1340	2.1208	1.1448	2.0540	1.1562	1.9920	1.1683	1.9341	52
1.1341	2.1196	1.1450	2.0530	1.1564	1.9910	1.1685	1.9332	51
1.1343	2.1185	1.1452	2.0519	1.1566	1.9900	1.1687	1.9322	50
1.1345	2.1173	1.1454	2.0508	1.1568	1.9890	1.1689	1.9313	49
1.1347	2.1162	1.1456	2.0498	1.1570	1.9880	1.1691	1.9304	48
1.1349	2.1150	1.1458	2.0487	1.1572	1.9870	1.1693	1.9295	47
1.1350	2.1139	1.1459	2.0476	1.1574	1.9860	1.1695	1.9285	46
1.1352	2.1127	1.1461	2.0466	1.1576	1.9850	1.1697	1.9276	45
1.1354	2.1116	1.1463	2.0455	1.1578	1.9840	1.1699	1.9267	44
1.1356	2.1104	1.1465	2.0444	1.1580	1.9830	1.1701	1.9258	43
1.1357	2.1093	1.1467	2.0434	1.1582	1.9820	1.1703	1.9248	42
1.1359	2.1082	1.1469	2.0423	1.1584	1.9811	1.1705	1.9239	41
1.1361	2.1070	1.1471	2.0413	1.1586	1.9801	1.1707	1.9230	40
1.1363	2.1059	1.1473	2.0402	1.1588	1.9791	1.1709	1.9221	39
1.1365	2.1048	1.1474	2.0392	1.1590	1.9781	1.1712	1.9212	38
1.1366	2.1036	1.1476	2.0381	1.1592	1.9771	1.1714	1.9203	37
1.1368	2.1025	1.1478	2.0370	1.1594	1.9761	1.1716	1.9193	36
1.1370	2.1014	1.1480	2.0360	1.1596	1.9752	1.1718	1.9184	35
1.1372	2.1002	1.1482	2.0349	1.1598	1.9742	1.1720	1.9175	34
1.1373	2.0991	1.1484	2.0339	1.1600	1.9732	1.1722	1.9166	33
1.1375	2.0980	1.1486	2.0329	1.1602	1.9722	1.1724	1.9157	32
1.1377	2.0969	1.1488	2.0318	1.1604	1.9713	1.1726	1.9148	31
1.1379	2.0957	1.1489	2.0308	1.1606	1.9703	1.1728	1.9139	30
1.1381	2.0946	1.1491	2.0297	1.1608	1.9693	1.1730	1.9130	29
1.1382	2.0935	1.1493	2.0287	1.1610	1.9683	1.1732	1.9121	28
1.1384	2.0924	1.1495	2.0276	1.1612	1.9674	1.1734	1.9112	27
1.1386	2.0912	1.1497	2.0266	1.1614	1.9664	1.1737	1.9102	26
1.1388	2.0901	1.1499	2.0256	1.1616	1.9654	1.1739	1.9093	25
1.1390	2.0890	1.1501	2.0245	1.1618	1.9645	1.1741	1.9084	24
1.1391	2.0879	1.1503	2.0235	1.1620	1.9635	1.1743	1.9075	23
1.1393	2.0868	1.1505	2.0224	1.1622	1.9625	1.1745	1.9066	22
1.1395	2.0857	1.1507	2.0214	1.1624	1.9616	1.1747	1.9057	21
1.1397	2.0846	1.1508	2.0204	1.1626	1.9606	1.1749	1.9048	20
1.1399	2.0835	1.1510	2.0194	1.1628	1.9596	1.1751	1.9039	19
1.1401	2.0824	1.1512	2.0183	1.1630	1.9587	1.1753	1.9030	18
1.1402	2.0812	1.1514	2.0173	1.1632	1.9577	1.1756	1.9021	17
1.1404	2.0801	1.1516	2.0163	1.1634	1.9568	1.1758	1.9013	16
1.1406	2.0790	1.1518	2.0152	1.1636	1.9558	1.1760	1.9004	15
1.1408	2.0779	1.1520	2.0142	1.1638	1.9549	1.1762	1.8995	14
1.1410	2.0768	1.1522	2.0132	1.1640	1.9539	1.1764	1.8986	13
1.1411	2.0757	1.1524	2.0122	1.1642	1.9530	1.1766	1.8977	12
1.1413	2.0746	1.1526	2.0111	1.1644	1.9520	1.1768	1.8968	11
1.1415	2.0735	1.1528	2.0101	1.1646	1.9510	1.1770	1.8959	10
1.1417	2.0725	1.1530	2.0091	1.1648	1.9501	1.1772	1.8950	9
1.1419	2.0714	1.1531	2.0081	1.1650	1.9491	1.1775	1.8941	8
1.1421	2.0703	1.1533	2.0071	1.1652	1.9482	1.1777	1.8932	7
1.1422	2.0692	1.1535	2.0061	1.1654	1.9473	1.1779	1.8924	6
1.1424	2.0681	1.1537	2.0050	1.1656	1.9463	1.1781	1.8915	5
1.1426	2.0670	1.1539	2.0040	1.1658	1.9454	1.1783	1.8906	4
1.1428	2.0659	1.1541	2.0030	1.1660	1.9444	1.1785	1.8897	3
1.1430	2.0648	1.1543	2.0020	1.1662	1.9435	1.1787	1.8888	2
1.1432	2.0637	1.1545	2.0010	1.1664	1.9425	1.1790	1.8879	1
1.1433	2.0627	1.1547	2.0000	1.1666	1.9416	1.1792	1.8871	0
Co-sec.	Sec.	Co-sec.	Sec.	Co-sec.	Sec.	Co-sec.	Sec.	
61°		60°		59°		58°		

′	32° Sec.	Co-sec.	33° Sec.	Co-sec.	34° Sec.	Co-sec.	35° Sec.	Co-sec.
0	1.1792	1.8871	1.1924	1.8361	1.2062	1.7883	1.2208	1.7434
1	1.1794	1.8862	1.1926	1.8352	1.2064	1.7875	1.2210	1.7427
2	1.1796	1.8853	1.1928	1.8344	1.2067	1.7867	1.2213	1.7420
3	1.1798	1.8844	1.1930	1.8336	1.2069	1.7860	1.2215	1.7413
4	1.1800	1.8836	1.1933	1.8328	1.2072	1.7852	1.2218	1.7405
5	1.1802	1.8827	1.1935	1.8320	1.2074	1.7844	1.2220	1.7398
6	1.1805	1.8818	1.1937	1.8311	1.2076	1.7837	1.2223	1.7391
7	1.1807	1.8809	1.1939	1.8303	1.2079	1.7829	1.2225	1.7384
8	1.1809	1.8801	1.1942	1.8295	1.2081	1.7821	1.2228	1.7377
9	1.1811	1.8792	1.1944	1.8287	1.2083	1.7814	1.2230	1.7369
10	1.1813	1.8783	1.1946	1.8279	1.2086	1.7806	1.2233	1.7362
11	1.1815	1.8775	1.1948	1.8271	1.2088	1.7798	1.2235	1.7355
12	1.1818	1.8766	1.1951	1.8263	1.2091	1.7791	1.2238	1.7348
13	1.1820	1.8757	1.1953	1.8255	1.2093	1.7783	1.2240	1.7341
14	1.1822	1.8749	1.1955	1.8246	1.2095	1.7776	1.2243	1.7334
15	1.1824	1.8740	1.1958	1.8238	1.2098	1.7768	1.2245	1.7327
16	1.1826	1.8731	1.1960	1.8230	1.2100	1.7760	1.2248	1.7319
17	1.1828	1.8723	1.1962	1.8222	1.2103	1.7753	1.2250	1.7312
18	1.1831	1.8714	1.1964	1.8214	1.2105	1.7745	1.2253	1.7305
19	1.1833	1.8706	1.1967	1.8206	1.2107	1.7738	1.2255	1.7298
20	1.1835	1.8697	1.1969	1.8198	1.2110	1.7730	1.2258	1.7291
21	1.1837	1.8688	1.1971	1.8190	1.2112	1.7723	1.2260	1.7284
22	1.1839	1.8680	1.1974	1.8182	1.2115	1.7715	1.2263	1.7277
23	1.1841	1.8671	1.1976	1.8174	1.2117	1.7708	1.2265	1.7270
24	1.1844	1.8663	1.1978	1.8166	1.2119	1.7700	1.2268	1.7263
25	1.1846	1.8654	1.1980	1.8158	1.2122	1.7693	1.2270	1.7256
26	1.1848	1.8646	1.1983	1.8150	1.2124	1.7685	1.2273	1.7249
27	1.1850	1.8637	1.1985	1.8142	1.2127	1.7678	1.2276	1.7242
28	1.1852	1.8629	1.1987	1.8134	1.2129	1.7670	1.2278	1.7234
29	1.1855	1.8620	1.1990	1.8126	1.2132	1.7663	1.2281	1.7227
30	1.1857	1.8611	1.1992	1.8118	1.2134	1.7655	1.2283	1.7220
31	1.1859	1.8603	1.1994	1.8110	1.2136	1.7648	1.2286	1.7213
32	1.1861	1.8595	1.1997	1.8102	1.2139	1.7640	1.2288	1.7206
33	1.1863	1.8586	1.1999	1.8094	1.2141	1.7633	1.2291	1.7199
34	1.1866	1.8578	1.2001	1.8086	1.2144	1.7625	1.2293	1.7192
35	1.1868	1.8569	1.2004	1.8078	1.2146	1.7618	1.2296	1.7185
36	1.1870	1.8561	1.2006	1.8070	1.2149	1.7610	1.2298	1.7178
37	1.1872	1.8552	1.2008	1.8062	1.2151	1.7603	1.2301	1.7171
38	1.1874	1.8544	1.2010	1.8054	1.2153	1.7596	1.2304	1.7164
39	1.1877	1.8535	1.2013	1.8047	1.2156	1.7588	1.2306	1.7157
40	1.1879	1.8527	1.2015	1.8039	1.2158	1.7581	1.2309	1.7151
41	1.1881	1.8519	1.2017	1.8031	1.2161	1.7573	1.2311	1.7144
42	1.1883	1.8510	1.2020	1.8023	1.2163	1.7566	1.2314	1.7137
43	1.1886	1.8502	1.2022	1.8015	1.2166	1.7559	1.2316	1.7130
44	1.1888	1.8493	1.2024	1.8007	1.2168	1.7551	1.2319	1.7123
45	1.1890	1.8485	1.2027	1.7999	1.2171	1.7544	1.2322	1.7116
46	1.1892	1.8477	1.2029	1.7992	1.2173	1.7537	1.2324	1.7109
47	1.1894	1.8468	1.2031	1.7984	1.2175	1.7529	1.2327	1.7102
48	1.1897	1.8460	1.2034	1.7976	1.2178	1.7522	1.2329	1.7095
49	1.1899	1.8452	1.2036	1.7968	1.2180	1.7514	1.2332	1.7088
50	1.1901	1.8443	1.2039	1.7960	1.2183	1.7507	1.2335	1.7081
51	1.1903	1.8435	1.2041	1.7953	1.2185	1.7500	1.2337	1.7075
52	1.1906	1.8427	1.2043	1.7945	1.2188	1.7493	1.2340	1.7068
53	1.1908	1.8418	1.2046	1.7937	1.2190	1.7485	1.2342	1.7061
54	1.1910	1.8410	1.2048	1.7929	1.2193	1.7478	1.2345	1.7054
55	1.1912	1.8402	1.2050	1.7921	1.2195	1.7471	1.2348	1.7047
56	1.1915	1.8394	1.2053	1.7914	1.2198	1.7463	1.2350	1.7040
57	1.1917	1.8385	1.2055	1.7906	1.2200	1.7456	1.2353	1.7033
58	1.1919	1.8377	1.2057	1.7898	1.2203	1.7449	1.2355	1.7027
59	1.1921	1.8369	1.2060	1.7891	1.2205	1.7442	1.2358	1.7020
60	1.1922	1.8361	1.2062	1.7883	1.2208	1.7434	1.2361	1.7013
′	Co-sec.	Sec.	Co-sec.	Sec.	Co-sec.	Sec.	Co-sec.	Sec.
	57°		56°		55°		54°	

36° Sec.	Co-sec.	37° Sec.	Co-sec.	38° Sec.	Co-sec.	39° Sec.	Co-sec.	′
1.2361	1.7013	1.2521	1.6616	1.2690	1.6243	1.2867	1.5890	60
1.2363	1.7006	1.2524	1.6610	1.2693	1.6237	1.2871	1.5884	59
1.2366	1.6999	1.2527	1.6603	1.2696	1.6231	1.2874	1.5879	58
1.2368	1.6993	1.2530	1.6597	1.2699	1.6224	1.2877	1.5873	57
1.2371	1.6986	1.2532	1.6591	1.2702	1.6218	1.2880	1.5867	56
1.2374	1.6979	1.2535	1.6584	1.2705	1.6212	1.2883	1.5862	55
1.2376	1.6972	1.2538	1.6578	1.2707	1.6206	1.2886	1.5856	54
1.2379	1.6965	1.2541	1.6572	1.2710	1.6200	1.2889	1.5850	53
1.2382	1.6959	1.2543	1.6565	1.2713	1.6194	1.2892	1.5845	52
1.2384	1.6952	1.2546	1.6559	1.2716	1.6188	1.2895	1.5839	51
1.2387	1.6945	1.2549	1.6552	1.2719	1.6182	1.2898	1.5833	50
1.2389	1.6938	1.2552	1.6546	1.2722	1.6176	1.2901	1.5828	49
1.2392	1.6932	1.2554	1.6540	1.2725	1.6170	1.2904	1.5822	48
1.2395	1.6925	1.2557	1.6533	1.2728	1.6164	1.2907	1.5816	47
1.2397	1.6918	1.2560	1.6527	1.2731	1.6159	1.2910	1.5811	46
1.2400	1.6912	1.2563	1.6521	1.2734	1.6153	1.2913	1.5805	45
1.2403	1.6905	1.2565	1.6514	1.2737	1.6147	1.2916	1.5799	44
1.2405	1.6898	1.2568	1.6508	1.2739	1.6141	1.2919	1.5794	43
1.2408	1.6891	1.2571	1.6502	1.2742	1.6135	1.2922	1.5788	42
1.2411	1.6885	1.2574	1.6496	1.2745	1.6129	1.2926	1.5783	41
1.2413	1.6878	1.2577	1.6489	1.2748	1.6123	1.2929	1.5777	40
1.2416	1.6871	1.2579	1.6483	1.2751	1.6117	1.2932	1.5771	39
1.2419	1.6865	1.2582	1.6477	1.2754	1.6111	1.2935	1.5766	38
1.2421	1.6858	1.2585	1.6470	1.2757	1.6105	1.2938	1.5760	37
1.2424	1.6851	1.2588	1.6464	1.2760	1.6099	1.2941	1.5755	36
1.2427	1.6845	1.2591	1.6458	1.2763	1.6093	1.2944	1.5749	35
1.2429	1.6838	1.2593	1.6452	1.2766	1.6087	1.2947	1.5743	34
1.2432	1.6831	1.2596	1.6445	1.2769	1.6081	1.2950	1.5738	33
1.2435	1.6825	1.2599	1.6439	1.2772	1.6077	1.2953	1.5732	32
1.2437	1.6818	1.2602	1.6433	1.2775	1.6070	1.2956	1.5727	31
1.2440	1.6812	1.2605	1.6427	1.2778	1.6064	1.2960	1.5721	30
1.2443	1.6805	1.2607	1.6420	1.2781	1.6058	1.2963	1.5716	29
1.2445	1.6798	1.2610	1.6414	1.2784	1.6052	1.2966	1.5710	28
1.2448	1.6792	1.2613	1.6408	1.2787	1.6046	1.2969	1.5705	27
1.2451	1.6785	1.2616	1.6402	1.2790	1.6040	1.2972	1.5699	26
1.2453	1.6779	1.2619	1.6396	1.2793	1.6034	1.2975	1.5694	25
1.2456	1.6772	1.2622	1.6389	1.2795	1.6029	1.2978	1.5688	24
1.2459	1.6766	1.2624	1.6383	1.2798	1.6023	1.2981	1.5683	23
1.2461	1.6759	1.2627	1.6377	1.2801	1.6017	1.2985	1.5677	22
1.2464	1.6752	1.2630	1.6371	1.2804	1.6011	1.2988	1.5672	21
1.2467	1.6746	1.2633	1.6365	1.2807	1.6005	1.2991	1.5666	20
1.2470	1.6739	1.2636	1.6359	1.2810	1.6000	1.2994	1.5661	19
1.2472	1.6733	1.2639	1.6352	1.2813	1.5994	1.2997	1.5655	18
1.2475	1.6726	1.2641	1.6346	1.2816	1.5988	1.3000	1.5650	17
1.2478	1.6720	1.2644	1.6340	1.2819	1.5982	1.3003	1.5644	16
1.2480	1.6713	1.2647	1.6334	1.2822	1.5976	1.3006	1.5639	15
1.2483	1.6707	1.2650	1.6328	1.2825	1.5971	1.3010	1.5633	14
1.2486	1.6700	1.2653	1.6322	1.2828	1.5965	1.3013	1.5628	13
1.2488	1.6694	1.2656	1.6316	1.2831	1.5959	1.3016	1.5622	12
1.2490	1.6687	1.2659	1.6309	1.2834	1.5953	1.3019	1.5617	11
1.2494	1.6681	1.2661	1.6303	1.2837	1.5947	1.3022	1.5611	10
1.2497	1.6674	1.2664	1.6297	1.2840	1.5942	1.3025	1.5606	9
1.2499	1.6668	1.2667	1.6291	1.2843	1.5936	1.3029	1.5600	8
1.2502	1.6661	1.2670	1.6285	1.2846	1.5930	1.3032	1.5595	7
1.2505	1.6655	1.2673	1.6279	1.2849	1.5924	1.3035	1.5590	6
1.2508	1.6648	1.2676	1.6273	1.2852	1.5919	1.3038	1.5584	5
1.2510	1.6642	1.2679	1.6267	1.2855	1.5913	1.3041	1.5570	4
1.2513	1.6636	1.2681	1.6261	1.2858	1.5907	1.3044	1.5573	3
1.2516	1.6629	1.2684	1.6255	1.2861	1.5901	1.3048	1.5568	2
1.2519	1.6623	1.2687	1.6249	1.2864	1.5896	1.3051	1.5563	1
1.2521	1.6616	1.2690	1.6243	1.2867	1.5890	1.3054	1.5557	0
Co-sec.	Sec.	Co-sec.	Sec.	Co-sec.	Sec.	Co-sec.	Sec.	′
53°		52°		51°		50°		

′	Sec.	Co-sec.	Sec.	Co-sec.	Sec.	Co-sec.	Sec.	Co-sec.
		0°		**1°**		**2°**		**3°**
0	1	Infinite.	1.0001	57.299	1.0006	28.654	1.0014	19.107
1	1	3437.70	1.0001	56.359	1.0006	28.417	1.0014	19.002
2	1	1718.90	1.0002	55.450	1.0006	28.184	1.0014	18.897
3	1	1145.90	1.0002	54.570	1.0006	27.955	1.0014	18.794
4	1	859.44	1.0002	53.718	1.0006	27.730	1.0014	18.692
5	1	687.55	1.0002	52.891	1.0007	27.508	1.0014	18.591
6	1	572.96	1.0002	52.090	1.0007	27.290	1.0015	18.491
7	1	491.11	1.0002	51.313	1.0007	27.075	1.0015	18.393
8	1	429.72	1.0002	50.558	1.0007	26.864	1.0015	18.295
9	1	381.97	1.0002	49.826	1.0007	26.655	1.0015	18.198
10	1	343.77	1.0002	49.114	1.0007	26.450	1.0015	18.103
11	1	312.52	1.0002	48.422	1.0007	26.249	1.0015	18.008
12	1	286.48	1.0002	47.750	1.0007	26.050	1.0016	17.914
13	1	264.44	1.0002	47.096	1.0007	25.854	1.0016	17.821
14	1	245.55	1.0002	46.460	1.0008	25.661	1.0016	17.730
.15	1	229.18	1.0002	45.840	1.0008	25.471	1.0016	17.639
16	1	214.86	1.0002	45.237	1.0008	25.284	1.0016	17.549
17	1	202.22	1.0002	44.650	1.0008	25.100	1.0016	17.460
18	1	190.99	1.0002	44.077	1.0008	24.918	1.0017	17.372
19	1	180.73	1.0003	43.520	1.0008	24.739	1.0017	17.285
20	1	171.89	1.0003	42.976	1.0008	24.562	1.0017	17.198
21	1	163.70	1.0003	42.445	1.0008	24.358	1.0017	17.113
22	1	156.26	1.0003	41.928	1.0008	24.216	1.0017	17.028
23	1	149.47	1.0003	41.423	1.0009	24.047	1.0017	16.944
24	1	143.24	1.0003	40.930	1.0009	23.880	1.0018	16.861
25	1	137.51	1.0003	40.448	1.0009	23.716	1.0018	16.779
26	1	132.22	1.0003	39.978	1.0009	23.553	1.0018	16.698
27	1	127.32	1.0003	39.518	1.0009	23.393	1.0018	16.617
28	1	122.78	1.0003	39.069	1.0009	23.235	1.0018	16.538
29	1	118.54	1.0003	38.631	1.0009	23.079	1.0018	16.459
30	1	114.59	1.0003	38.201	1.0009	22.925	1.0019	16.380
31	1	110.90	1.0003	37.782	1.0010	22.774	1.0019	16.303
32	1	107.43	1.0003	37.371	1.0010	22.624	1.0019	16.220
33	1	104.17	1.0004	36.969	1.0010	22.476	1.0019	16.150
34	1	101.11	1.0004	36.576	1.0010	22.330	1.0019	16.075
35	1	98.223	1.0004	36.191	1.0010	22.186	1.0019	16.000
36	1	95.495	1.0004	35.814	1.0010	22.044	1.0020	15.920
37	1	92.914	1.0004	35.445	1.0010	21.904	1.0020	15.853
38	1.0001	90.469	1.0004	35.084	1.0010	21.765	1.0020	15.780
39	1.0001	88.149	1.0004	34.729	1.0011	21.629	1.0020	15.708
40	1.0001	85.946	1.0004	34.382	1.0011	21.494	1.0020	15.637
41	1.0001	83.849	1.0004	34.042	1.0011	21.360	1.0021	15.566
42	1.0001	81.853	1.0004	33.708	1.0011	21.228	1.0021	15.496
43	1.0001	79.950	1.0004	33.381	1.0011	21.098	1.0021	15.427
44	1.0001	78.133	1.0004	33.060	1.0011	20.970	1.0021	15.358
45	1.0001	76.396	1.0005	32.745	1.0011	20.843	1.0021	15.290
46	1.0001	74.736	1.0005	32.437	1.0012	20.717	1.0022	15.222
47	1.0001	73.146	1.0005	32.134	1.0012	20.593	1.0022	15.155
48	1.0001	71.622	1.0005	31.836	1.0012	20.471	1.0022	15.089
49	1.0001	70.160	1.0005	31.544	1.0012	20.350	1.0022	15.023
50	1.0001	68.757	1.0005	31.257	1.0012	20.230	1.0022	14.958
51	1.0001	67.409	1.0005	30.976	1.0012	20.112	1.0023	14.893
52	1.0001	66.113	1.0005	30.699	1.0012	19.995	1.0023	14.829
53	1.0001	64.866	1.0005	30.428	1.0013	19.880	1.0023	14.765
54	1.0001	63.664	1.0005	30.161	1.0013	19.766	1.0023	14.702
55	1.0001	62.507	1.0005	29.899	1.0013	19.653	1.0023	14.640
56	1.0001	61.391	1.0006	29.641	1.0013	19.541	1.0024	14.578
57	1.0001	60.314	1.0006	29.388	1.0013	19.431	1.0024	14.517
58	1.0001	59.274	1.0006	29.139	1.0013	19.322	1.0024	14.450
59	1.0001	58.270	1.0006	28.894	1.0013	19.214	1.0024	14.395
60	1.0001	57.299	1.0006	28.654	1.0014	19.107	1.0024	14.335
′	Co-sec.	Sec.	Co-sec.	Sec.	Co-sec.	Sec.	Co-sec.	Sec.
		89°		**88°**		**87°**		**86°**

4°		5°		6°		7°		'
Sec.	Co-sec.	Sec.	Co-sec.	Sec.	Co-sec	Sec.	Co-sec.	
1.0024	14.335	1.0038	11.474	1.0055	9.5668	1.0075	8.2055	60
1.0025	14.276	1.0038	11.436	1.0055	9.5404	1.0075	8.1861	59
1.0025	14.217	1.0039	11.398	1.0056	9.5141	1.0076	8.1668	58
1.0025	14.159	1.0039	11.360	1.0056	9.4880	1.0076	8.1476	57
1.0025	14.101	1.0039	11.323	1.0056	9.4620	1.0076	8.1285	56
1.0025	14.043	1.0039	11.286	1.0057	9.4362	1.0077	8.1094	55
1.0026	13.986	1.0040	11.249	1.0057	9.4105	1.0077	8.0905	54
1.0026	13.930	1.0040	11.213	1.0057	9.3850	1.0078	8.0717	53
1.0026	13.874	1.0040	11.176	1.0057	9.3596	1.0078	8.0529	52
1.0026	13.818	1.0040	11.140	1.0058	9.3343	1.0078	8.0342	51
1.0026	13.763	1.0041	11.104	1.0058	9.3092	1.0079	8.0156	50
1.0027	13.708	1.0041	11.069	1.0058	9.2842	1.0079	7.9971	49
1.0027	13.654	1.0041	11.033	1.0059	9.2593	1.0079	7.9787	48
1.0027	13.600	1.0041	10.988	1.0059	9.2346	1.0080	7.9604	47
1.0027	13.547	1.0042	10.963	1.0059	9.2100	1.0080	7.9421	46
1.0027	13.494	1.0042	10.929	1.0060	9.1855	1.0080	7.9240	45
1.0028	13.441	1.0042	10.894	1.0060	9.1612	1.0081	7.9059	44
1.0028	13.389	1.0043	10.860	1.0060	9.1370	1.0081	7.8879	43
1.0028	13.337	1.0043	10.826	1.0061	9.1129	1.0082	7.8700	42
1.0028	13.286	1.0043	10.792	1.0061	9.0890	1.0082	7.8522	41
1.0029	13.235	1.0043	10.758	1.0061	9.0651	1.0082	7.8344	40
1.0029	13.184	1.0044	10.725	1.0062	9.0414	1.0083	7.8168	39
1.0029	13.134	1.0044	10.692	1.0062	9.0179	1.0083	7.7992	38
1.0029	13.084	1.0044	10.659	1.0062	8.9944	1.0084	7.7817	37
1.0029	13.034	1.0044	10.626	1.0063	8.9711	1.0084	7.7642	36
1.0030	12.985	1.0045	10.593	1.0063	8.9479	1.0084	7.7469	35
1.0030	12.937	1.0045	10.561	1.0063	8.9248	1.0085	7.7296	34
1.0030	12.888	1.0045	10.529	1.0064	8.9018	1.0085	7.7124	33
1.0030	12.840	1.0046	10.497	1.0064	8.8790	1.0085	7.6953	32
1.0031	12.793	1.0046	10.465	1.0064	8.8563	1.0086	7.6783	31
1.0031	12.745	1.0046	10.433	1.0065	8.8337	1.0086	7.6613	30
1.0031	12.698	1.0046	10.402	1.0065	8.8112	1.0087	7.6444	29
1.0031	12.652	1.0047	10.371	1.0065	8.7888	1.0087	7.6276	28
1.0032	12.606	1.0047	10.340	1.0066	8.7665	1.0087	7.6108	27
1.0032	12.560	1.0047	10.309	1.0066	8.7444	1.0088	7.5942	26
1.0032	12.514	1.0048	10.278	1.0066	8.7223	1.0088	7.5776	25
1.0032	12.469	1.0048	10.248	1.0067	8.7004	1.0089	7.5611	24
1.0032	12.424	1.0048	10.217	1.0067	8.6786	1.0089	7.5446	23
1.0033	12.379	1.0048	10.187	1.0067	8.6569	1.0089	7.5282	22
1.0033	12.335	1.0049	10.157	1.0068	8.6353	1.0090	7.5119	21
1.0033	12.291	1.0049	10.127	1.0068	8.6138	1.0090	7.4957	20
1.0033	12.248	1.0049	10.098	1.0068	8.5924	1.0090	7.4795	19
1.0034	12.204	1.0050	10.068	1.0069	8.5711	1.0091	7.4634	18
1.0034	12.161	1.0050	10.039	1.0069	8.5499	1.0091	7.4474	17
1.0034	12.118	1.0050	10.010	1.0069	8.5289	1.0092	7.4315	16
1.0034	12.076	1.0050	9.9812	1.0070	8.5079	1.0092	7.4156	15
1.0035	12.034	1.0051	9.9525	1.0070	8.4871	1.0092	7.3998	14
1.0035	11.992	1.0051	9.9239	1.0070	8.4663	1.0093	7.3840	13
1.0035	11.950	1.0051	9.8955	1.0071	8.4457	1.0093	7.3683	12
1.0035	11.909	1.0052	9.8672	1.0071	8.4251	1.0094	7.3527	11
1.0036	11.868	1.0052	9.8391	1.0071	8.4046	1.0094	7.3372	10
1.0036	11.828	1.0052	9.8112	1.0072	8.3843	1.0094	7.3217	9
1.0036	11.787	1.0053	9.7834	1.0072	8.3640	1.0095	7.3063	8
1.0036	11.747	1.0053	9.7558	1.0073	8.3439	1.0095	7.2909	7
1.0037	11.707	1.0053	9.7283	1.0073	8.3238	1.0096	7.2757	6
1.0037	11.668	1.0053	9.7010	1.0073	8.3039	1.0096	7.2604	5
1.0037	11.628	1.0054	9.6739	1.0074	8.2840	1.0097	7.2453	4
1.0037	11.589	1.0054	9.6469	1.0074	8.2642	1.0097	7.2302	3
1.0038	11.550	1.0054	9.6200	1.0074	8.2446	1.0097	7.2152	2
1.0038	11.512	1.0055	9.5933	1.0075	8.2250	1.0098	7 2002	1
1.0038	11.474	1.0055	9.5668	1.0075	8.2055	1.0098	7.1853	0
Co-sec.	Sec.	Co-sec.	Sec.	Co-sec.	Sec.	Co-sec.	Sec.	
	85°		84°		83°		82°	

′	8° Sec.	Co-sec.	9° Sec.	Co-sec.	10° Sec.	Co-sec.	11° Sec.	Co-sec.
0	1.0098	7.1853	1.0125	6.3924	1.0154	5.7588	1.0187	5.2408
1	1.0099	7.1704	1.0125	6.3807	1.0155	5.7493	1.0188	5.2330
2	1.0099	7.1557	1.0125	6.3690	1.0155	5.7398	1.0188	5.2252
3	1.0099	7.1409	1.0126	6.3574	1.0156	5.7304	1.0189	5.2174
4	1.0100	7.1263	1.0126	6.3458	1.0156	5.7210	1.0189	5.2097
5	1.0100	7.1117	1.0127	6.3343	1.0157	5.7117	1.0190	5.2019
6	1.0101	7.0972	1.0127	6.3228	1.0157	5.7023	1.0191	5.1942
7	1.0101	7.0827	1.0128	6.3113	1.0158	5.6930	1.0191	5.1865
8	1.0102	7.0683	1.0128	6.2999	1.0158	5.6838	1.0192	5.1788
9	1.0102	7.0539	1.0129	6.2885	1.0159	5.6745	1.0192	5.1712
10	1.0102	7.0396	1.0129	6.2772	1.0159	5.6653	1.0193	5.1636
11	1.0103	7.0254	1.0130	6.2659	1.0160	5.6561	1.0193	5.1560
12	1.0103	7.0112	1.0130	6.2546	1.0160	5.6470	1.0194	5.1484
13	1.0104	6.9971	1.0131	6.2434	1.0161	5.6379	1.0195	5.1409
14	1.0104	6.9830	1.0131	6.2322	1.0162	5.6288	1.0195	5.1333
15	1.0104	6.9690	1.0132	6.2211	1.0162	5.6197	1.0196	5.1258
16	1.0105	6.9550	1.0132	6.2100	1.0163	5.6107	1.0196	5.1183
17	1.0105	6.9411	1.0133	6.1990	1.0163	5.6017	1.0197	5.1109
18	1.0106	6.9273	1.0133	6.1880	1.0164	5.5928	1.0198	5.1034
19	1.0106	6.9135	1.0134	6.1770	1.0164	5.5838	1.0198	5.0960
20	1.0107	6.8998	1.0134	6.1661	1.0165	5.5749	1.0199	5.0886
21	1.0107	6.8861	1.0135	6.1552	1.0165	5.5660	1.0199	5.0812
22	1.0107	6.8725	1.0135	6.1443	1.0166	5.5572	1.0200	5.0739
23	1.0108	6.8589	1.0136	6.1335	1.0166	5.5484	1.0201	5.0666
24	1.0108	6.8454	1.0136	6.1227	1.0167	5.5396	1.0201	5.0593
25	1.0109	6.8320	1.0136	6.1120	1.0167	5.5308	1.0202	5.0520
26	1.0109	6.8185	1.0137	6.1013	1.0168	5.5221	1.0202	5.0447
27	1.0110	6.8052	1.0137	6.0906	1.0169	5.5134	1.0203	5.0375
28	1.0110	6.7919	1.0138	6.0800	1.0169	5.5047	1.0204	5.0302
29	1.0111	6.7787	1.0138	6.0694	1.0170	5.4960	1.0204	5.0230
30	1.0111	6.7655	1.0139	6.0588	1.0170	5.4874	1.0205	5.0158
31	1.0111	6.7523	1.0139	6.0483	1.0171	5.4788	1.0205	5.0087
32	1.0112	6.7392	1.0140	6.0379	1.0171	5.4702	1.0206	5.0015
33	1.0112	6.7262	1.0140	6.0274	1.0172	5.4617	1.0207	4.9944
34	1.0113	6.7132	1.0141	6.0170	1.0172	5.4532	1.0207	4.9873
35	1.0113	6.7003	1.0141	6.0066	1.0173	5.4447	1.0208	4.9802
36	1.0114	6.6874	1.0142	5.9963	1.0174	5.4362	1.0208	4.9732
37	1.0114	6.6745	1.0142	5.9860	1.0174	5.4278	1.0209	4.9661
38	1.0115	6.6617	1.0143	5.9758	1.0175	5.4194	1.0210	4.9591
39	1.0115	6.6490	1.0143	5.9655	1.0175	5.4110	1.0210	4.9521
40	1.0115	6.6363	1.0144	5.9554	1.0176	5.4026	1.0211	4.9452
41	1.0116	6.6237	1.0144	5.9452	1.0176	5.3943	1.0211	4.9382
42	1.0116	6.6111	1.0145	5.9351	1.0177	5.3860	1.0212	4.9313
43	1.0117	6.5985	1.0145	5.9250	1.0177	5.3777	1.0213	4.9243
44	1.0117	6.5860	1.0146	5.9150	1.0178	5.3695	1.0213	4.9175
45	1.0118	6.5736	1.0146	5.9049	1.0179	5.3612	1.0214	4.9106
46	1.0118	6.5612	1.0147	5.8950	1.0179	5.3530	1.0215	4.9037
47	1.0119	6.5488	1.0147	5.8850	1.0180	5.3449	1.0215	4.8969
48	1.0119	6.5365	1.0148	5.8751	1.0180	5.3367	1.0216	4.8901
49	1.0119	6.5243	1.0148	5.8652	1.0181	5.3286	1.0216	4.8833
50	1.0120	6.5121	1.0149	5.8554	1.0181	5.3205	1.0217	4.8765
51	1.0120	6.4999	1.0150	5.8456	1.0182	5.3124	1.0218	4.8697
52	1.0121	6.4878	1.0150	5.8358	1.0182	5.3044	1.0218	4.8630
53	1.0121	6.4757	1.0151	5.8261	1.0183	5.2963	1.0219	4.8563
54	1.0122	6.4637	1.0151	5.8163	1.0184	5.2883	1.0220	4.8496
55	1.0122	6.4517	1.0152	5.8067	1.0184	5.2803	1.0220	4.8429
56	1.0123	6.4398	1.0152	5.7970	1.0185	5.2724	1.0221	4.8362
57	1.0123	6.4279	1.0153	5.7874	1.0185	5.2645	1.0221	4.8296
58	1.0124	6.4160	1.0153	5.7778	1.0186	5.2566	1.0222	4.8229
59	1.0124	6.4042	1.0154	5.7683	1.0186	5.2487	1.0223	4.8163
60	1.0125	6.3924	1.0154	5.7588	1.0187	5.2408	1.0223	4.8097

	Co-sec.	Sec.	Co-sec.	Sec.	Co-sec.	Sec.	Co-sec.	Sec.	′
		81°		80°		79°		78°	

12°		13°		14°		15°		′
SEC.	CO-SEC.	SEC.	CO-SEC.	SEC.	CO-SEC.	SEC.	CO-SEC.	
1.0223	4.8097	1.0263	4.4454	1.0306	4.1336	1.0353	3.8637	60
1.0224	4.8032	1.0264	4.4398	1.0307	4.1287	1.0353	3.8595	59
1.0225	4.7966	1.0264	4.4342	1.0308	4.1239	1.0354	3.8553	58
1.0225	4.7901	1.0265	4.4287	1.0308	4.1191	1.0355	3.8512	57
1.0226	4.7835	1.0266	4.4231	1.0309	4.1144	1.0356	3.8470	56
1.0226	4.7770	1.0266	4.4176	1.0310	4.1096	1.0357	3.8428	55
1.0227	4.7706	1.0267	4.4121	1.0311	4.1048	1.0358	3.8387	54
1.0228	4.7641	1.0268	4.4065	1.0311	4.1001	1.0358	3.8346	53
1.0228	4.7576	1.0268	4.4011	1.0312	4.0953	1.0359	3.8304	52
1.0229	4.7512	1.0269	4.3956	1.0313	4.0906	1.0360	3.8263	51
1.0230	4.7448	1.0270	4.3910	1.0314	4.0859	1.0361	3.8222	50
1.0230	4.7384	1.0271	4.3847	1.0314	4.0812	1.0362	3.8181	49
1.0231	4.7320	1.0271	4.3792	1.0315	4.0765	1.0362	3.8140	48
1.0232	4.7257	1.0272	4.3738	1.0316	4.0718	1.0363	3.8100	47
1.0232	4.7193	1.0273	4.3684	1.0317	4.0672	1.0364	3.8059	46
1.0233	4.7130	1.0273	4.3630	1.0317	4.0625	1.0365	3.8018	45
1.0234	4.7067	1.0274	4.3576	1.0318	4.0579	1.0366	3.7978	44
1.0234	4.7004	1.0275	4.3522	1.0319	4.0532	1.0367	3.7937	43
1.0235	4.6942	1.0276	4.3469	1.0320	4.0486	1.0367	3.7897	42
1.0235	4.6879	1.0276	4.3415	1.0320	4.0440	1.0368	3.7857	41
1.0236	4.6817	1.0277	4.3362	1.0321	4.0394	1.0369	3.7816	40
1.0237	4.6754	1.0278	4.3309	1.0322	4.0348	1.0370	3.7776	39
1.0237	4.6692	1.0278	4.3256	1.0323	4.0302	1.0371	3.7736	38
1.0238	4.6631	1.0279	4.3203	1.0323	4.0256	1.0371	3.7697	37
1.0239	4.6569	1.0280	4.3150	1.0324	4.0211	1.0372	3.7657	36
1.0239	4.6507	1.0280	4.3098	1.0325	4.0165	1.0373	3.7617	35
1.0240	4.6446	1.0281	4.3045	1.0326	4.0120	1.0374	3.7577	34
1.0241	4.6385	1.0282	4.2993	1.0327	4.0074	1.0375	3.7538	33
1.0241	4.6324	1.0283	4.2941	1.0327	4.0029	1.0376	3.7498	32
1.0242	4.6263	1.0283	4.2888	1.0328	3.9984	1.0376	3.7459	31
1.0243	4.6202	1.0284	4.2836	1.0329	3.9939	1.0377	3.7420	30
1.0243	4.6142	1.0285	4.2785	1.0330	3.9894	1.0378	3.7380	29
1.0244	4.6081	1.0285	4.2733	1.0330	3.9850	1.0379	3.7341	28
1.0245	4.6021	1.0286	4.2681	1.0331	3.9805	1.0380	3.7302	27
1.0245	4.5961	1.0287	4.2630	1.0332	3.9760	1.0381	3.7263	26
1.0246	4.5901	1.0288	4.2579	1.0333	3.9716	1.0382	3.7224	25
1.0247	4.5841	1.0288	4.2527	1.0334	3.9672	1.0382	3.7186	24
1.0247	4.5782	1.0289	4.2476	1.0334	3.9627	1.0383	3.7147	23
1.0248	4.5722	1.0290	4.2425	1.0335	3.9583	1.0384	3.7108	22
1.0249	4.5663	1.0291	4.2375	1.0336	3.9539	1.0385	3.7070	21
1.0249	4.5604	1.0291	4.2324	1.0337	3.9495	1.0386	3.7031	20
1.0250	4.5545	1.0292	4.2273	1.0338	3.9451	1.0387	3.6993	19
1.0251	4.5486	1.0293	4.2223	1.0338	3.9408	1.0387	3.6955	18
1.0251	4.5428	1.0293	4.2173	1.0339	3.9364	1.0388	3.6917	17
1.0252	4.5369	1.0294	4.2122	1.0340	3.9320	1.0389	3.6878	16
1.0253	4.5311	1.0295	4.2072	1.0341	3.9277	1.0390	3.6840	15
1.0253	4.5253	1.0296	4.2022	1.0341	3.9234	1.0391	3.6802	14
1.0254	4.5195	1.0296	4.1972	1.0342	3.9199	1.0392	3.6765	13
1.0255	4.5137	1.0297	4.1923	1.0343	3.9147	1.0393	3.6727	12
1.0255	4.5079	1.0298	4.1873	1.0344	3.9104	1.0393	3.6689	11
1.0256	4.5021	1.0299	4.1824	1.0345	3.9061	1.0394	3.6651	10
1.0257	4.4964	1.0299	4.1774	1.0345	3.9018	1.0395	3.6614	9
1.0257	4.4907	1.0300	4.1725	1.0346	3.8976	1.0396	3.6576	8
1.0258	4.4850	1.0301	4.1676	1.0347	3.8933	1.0397	3.6539	7
1.0259	4.4793	1.0302	4.1627	1.0348	3.8890	1.0398	3.6502	6
1.0260	4.4736	1.0302	4.1578	1.0349	3.8848	1.0399	3.6464	5
1.0260	4.4679	1.0303	4.1529	1.0349	3.8805	1.0399	3.6427	4
1.0261	4.4623	1.0304	4.1481	1.0350	3.8763	1.0400	3.6390	3
1.0262	4.4566	1.0305	4.1432	1.0351	3.8721	1.0401	3.6353	2
1.0262	4.4510	1.0305	4.1384	1.0352	3.8679	1.0402	3.6316	1
1.0263	4.4454	1.0306	4.1336	1.0353	3.8637	1.0403	3.6279	0
CO-SEC.	SEC.	CO-SEC.	SEC.	CO-SEC.	SEC.	CO-SEC.	SEC.	
77°		76°		75°		74°		

′	16° SEC.	Co-SEC.	17° SEC.	Co-SEC.	18° SEC.	Co-SEC.	19° SEC.	Co-SEC.	
0	1.0403	3.6279	1.0457	3.4203	1.0515	3.2361	1.0576	3.0715	
1	1.0404	3.6243	1.0458	3.4170	1.0516	3.2332	1.0577	3.0690	
2	1.0405	3.6206	1.0459	3.4138	1.0517	3.2303	1.0578	3.0664	
3	1.0406	3.6169	1.0460	3.4106	1.0518	3.2274	1.0579	3.0638	
4	1.0406	3.6133	1.0461	3.4073	1.0519	3.2245	1.0580	3.0612	
5	1.0407	5.6096	1.0461	3.4041	1.0520	3.2216	1.0581	3.0586	
6	1.0408	3.6060	1.0462	3.4009	1.0521	3.2188	1.0582	3.0561	
7	1.0409	3.6024	1.0463	3.3977	1.0522	3.2159	1.0584	3.0535	
8	1.0410	3.5987	1.0464	3.3945	1.0523	3.2131	1.0585	3.0509	
9	1.0411	3.5951	1.0465	3.3913	1.0524	3.2102	1.0586	3.0484	
10	1.0412	3.5915	1.0466	3.3881	1.0525	3.2074	1.0587	3.0458	
11	1.0413	3.5879	1.0467	3.3849	1.0526	3.2045	1.0588	3.0433	
12	1.0413	3.5843	1.0468	3.3817	1.0527	3.2017	1.0589	3.0407	
13	1.0414	3.5807	1.0469	3.3785	1.0528	3.1989	1.0590	3.0382	
14	1.0415	3.5772	1.0470	3.3754	1.0529	3.1960	1.0591	3.0357	
15	1.0416	3.5736	1.0471	3.3722	1.0530	3.1932	1.0592	3.0331	
16	1.0417	3.5700	1.0472	3.3690	1.0531	3.1904	1.0593	3.0306	
17	1.0418	3.5665	1.0473	3.3659	1.0532	3.187*	1.0594	3.0281	
18	1.0419	3.5629	1.0474	3.3627	1.0533	3.1848	1.0595	3.0256	
19	1.0420	3.5594	1.0475	3.3596	1.0534	3.1820	1.0596	3.0231	
20	1.0420	3.5559	1.0476	3.3565	1.0535	3.1792	1.0598	3.0206	
21	1.0421	3.5523	1.0477	3.3534	1.0536	3.1764	1.0599	3.0181	
22	1.0422	3.5488	1.0478	3.3502	1.0537	3.1736	1.0600	3.0156	
23	1.0423	3.5453	1.0478	3.3471	1.0538	3.1708	1.0601	3.0131	
24	1.0424	3.5418	1.0479	3.3440	1.0539	3.1681	1.0602	3.0106	
25	1.0425	3.5383	1.0480	3.3409	1.0540	3.1653	1.0603	3.0081	
26	1.0426	3.5348	1.0481	3.3378	1.0541	3.1625	1.0604	3.0056	
27	1.0427	3.5313	1.0482	3.3347	1.0542	3.1598	1.0605	3.0031	
28	1.0428	3.5279	1.0483	3.3316	1.0543	3.1570	1.0606	3.0007	
29	1.0428	3.5244	1.0484	3.3286	1.0544	3.1543	1.0607	2.9982	
30	1.0429	3.5209	1.0485	3.3255	1.0545	3.1515	1.0608	2.9957	
31	1.0430	3.5175	1.0486	3.3224	1.0546	3.1488	1.0609	2.9933	
32	1.0431	3.5140	1.0487	3.3194	1.0547	3.1461	1.0611	2.9908	
33	1.0432	3.5106	1.0488	3.3163	1.0548	3.1433	1.0612	2.9884	
34	1.0433	3.5072	1.0489	3.3133	1.0549	3.1406	1.0613	2.9859	
35	1.0434	3.5037	1.0490	3.3102	1.0550	3.1379	1.0614	2.9835	
36	1.0435	3.5003	1.0491	3.3072	1.0551	3.1352	1.0615	2.9810	
37	1.0436	3.4969	1.0492	3.3042	1.0552	3.1325	1.0616	2.9786	
38	1.0437	3.4935	1.0493	3.3011	1.0553	3.1298	1.0617	2.9762	
39	1.0438	3.4901	1.0494	3.2981	1.0554	3.1271	1.0618	2.9738	
40	1.0438	3.4867	1.0495	3.2951	1.0555	3.1244	1.0619	2.9713	
41	1.0439	3.4833	1.0496	3.2921	1.0556	3.1217	1.0620	2.9689	
42	1.0440	3.4799	1.0497	3.2891	1.0557	3.1190	1.0622	2.9665	
43	1.0441	3.4766	1.0498	3.2861	1.0558	3.1163	1.0623	2.9641	
44	1.0442	3.4732	1.0499	3.2831	1.0559	3.1137	1.0624	2.9617	
45	1.0443	3.4698	1.0500	3.2801	1.0560	3.1110	1.0625	2.9593	
46	1.0444	3.4665	1.0501	3.2772	1.0561	3.1083	1.0626	2.9569	
47	1.0445	3.4632	1.0502	3.2742	1.0562	3.1057	1.0627	2.9545	
48	1.0446	3.4598	1.0503	3.2712	1.0563	3.1030	1.0628	2.9521	
49	1.0447	3.4565	1.0504	3.2683	1.0565	3.1004	1.0629	2.9497	
50	1.0448	3.4532	1.0505	3.2653	1.0566	3.0977	1.0630	2.9474	
51	1.0448	3.4498	1.0506	3.2624	1.0567	3.0951	1.0632	2.9450	
52	1.0449	3.4465	1.0507	3.2594	1.0568	3.0925	1.0633	2.9426	
53	1.0450	3.4432	1.0508	3.2565	1.0569	3.0898	1.0634	2.9402	
54	1.0451	3.4399	1.0509	3.2535	1.0570	3.0872	1.0635	2.9379	
55	1.0452	3.4366	1.0510	3.2506	1.0571	3.0846	1.0636	2.9355	
56	1.0453	3.4334	1.0511	3.2477	1.0572	3.0820	1.0637	2.9332	
57	1.0454	3.4301	1.0512	3.2448	1.0573	3.0793	1.0638	2.9308	
58	1.0455	3.4268	1.0513	3.2419	1.0574	3.0767	1.0639	2.9285	
59	1.0456	3.4236	1.0514	3.2390	1.0575	3.0741	1.0641	2.9261	
60	1.0457	3.4203	1.0515	3.2361	1.0576	3.0715	1.0642	2.9238	
	Co-SEC.	SEC.	Co-SEC.	SEC.	Co-SEC.	SEC.	Co-SEC.	SEC.	
		73°		72°		71°		70°	

Sec.	Co-sec.	Sec.	Co-sec.	Sec.	Co-sec.	Sec.	Co-sec.	'
20°		**21°**		**22°**		**23°**		
1.0642	2.9238	1.0711	2.7004	1.0785	2.6695	1.0864	2.5593	60
1.0643	2.9215	1.0713	2.7883	1.0787	2.6675	1.0865	2.5575	59
1.0644	2.9191	1.0714	2.7862	1.0788	2.6656	1.0866	2.5558	58
1.0645	2.9168	1.0715	2.7841	1.0789	2.6637	1.0868	2.5540	57
1.0646	2.9145	1.0716	2.7820	1.0790	2.6618	1.0869	2.5523	56
1.0647	2.9122	1.0717	2.7799	1.0792	2.6599	1.0870	2.5506	55
1.0648	2.9098	1.0719	2.7778	1.0793	2.6580	1.0872	2.5488	54
1.0650	2.9075	1.0720	2.7757	1.0794	2.6561	1.0873	2.5471	53
1.0651	2.9052	1.0721	2.7736	1.0795	2.6542	1.0874	2.5453	52
1.0652	2.9029	1.0722	2.7715	1.0797	2.6523	1.0876	2.5436	51
1.0653	2.9006	1.0723	2.7694	1.0798	2.6504	1.0877	2.5419	50
1.0654	2.8983	1.0725	2.7674	1.0799	2.6485	1.0878	2.5402	49
1.0655	2.8960	1.0726	2.7653	1.0801	2.6466	1.0880	2.5384	48
1.0656	2.8937	1.0727	2.7632	1.0802	2.6447	1.0881	2.5367	47
1.0658	2.8915	1.0728	2.7611	1.0803	2.6428	1.0882	2.5350	46
1.0659	2.8892	1.0729	2.7591	1.0804	2.6410	1.0884	2.5333	45
1.0660	2.8869	1.0731	2.7570	1.0806	2.6391	1.0885	2.5316	44
1.0661	2.8846	1.0732	2.7550	1.0807	2.6372	1.0886	2.5299	43
1.0662	2.8824	1.0733	2.7529	1.0808	2.6353	1.0888	2.5281	42
1.0663	2.8801	1.0734	2.7509	1.0810	2.6335	1.0889	2.5264	41
1.0664	2.8778	1.0736	2.7488	1.0811	2.6316	1.0891	2.5247	40
1.0666	2.8756	1.0737	2.7468	1.0812	2.6297	1.0892	2.5230	39
1.0667	2.8733	1.0738	2.7447	1.0813	2.6279	1.0893	2.5213	38
1.0668	2.8711	1.0739	2.7427	1.0815	2.6260	1.0895	2.5196	37
1.0669	2.8688	1.0740	2.7406	1.0816	2.6242	1.0896	2.5179	36
1.0670	2.8666	1.0742	2.7386	1.0817	2.6223	1.0897	2.5163	35
1.0671	2.8644	1.0743	2.7366	1.0819	2.6205	1.0899	2.5146	34
1.0673	2.8621	1.0744	2.7346	1.0820	2.6186	1.0900	2.5129	33
1.0674	2.8599	1.0745	2.7325	1.0821	2.6168	1.0902	2.5112	32
1.0675	2.8577	1.0747	2.7305	1.0823	2.6150	1.0903	2.5095	31
1.0676	2.8554	1.0748	2.7285	1.0824	2.6131	1.0904	2.5078	30
1.0677	2.8532	1.0749	2.7265	1.0825	2.6113	1.0906	2.5062	29
1.0678	2.8510	1.0750	2.7245	1.0826	2.6095	1.0907	2.5045	28
1.0679	2.8488	1.0751	2.7225	1.0828	2.6076	1.0908	2.5028	27
1.0681	2.8466	1.0753	2.7205	1.0829	2.6058	1.0910	2.5011	26
1.0682	2.8444	1.0754	2.7185	1.0830	2.6040	1.0911	2.4995	25
1.0683	2.8422	1.0755	2.7165	1.0832	2.6022	1.0913	2.4978	24
1.0684	2.8400	1.0756	2.7145	1.0833	2.6003	1.0914	2.4961	23
1.0685	2.8378	1.0758	2.7125	1.0834	2.5985	1.0915	2.4945	22
1.0686	2.8356	1.0759	2.7105	1.0836	2.5967	1.0917	2.4928	21
1.0688	2.8334	1.0760	2.7085	1.0837	2.5949	1.0918	2.4912	20
1.0689	2.8312	1.0761	2.7065	1.0838	2.5931	1.0920	2.4895	19
1.0690	2.8290	1.0763	2.7045	1.0840	2.5913	1.0921	2.4879	18
1.0691	2.8269	1.0764	2.7026	1.0841	2.5895	1.0922	2.4862	17
1.0692	2.8247	1.0765	2.7006	1.0842	2.5877	1.0924	2.4846	16
1.0694	2.8225	1.0766	2.6986	1.0844	2.5859	1.0927	2.4829	15
1.0695	2.8204	1.0768	2.6967	1.0845	2.5841	1.0927	2.4813	14
1.0696	2.8182	1.0769	2.6947	1.0846	2.5823	1.0928	2.4797	13
1.0697	2.8160	1.0770	2.6927	1.0847	2.5805	1.0929	2.4780	12
1.0698	2.8139	1.0771	2.6908	1.0849	2.5787	1.0931	2.4764	11
1.0699	2.8117	1.0773	2.6888	1.0850	2.5770	1.0932	2.4748	10
1.0701	2.8096	1.0774	2.6869	1.0851	2.5752	1.0934	2.4731	9
1.0702	2.8074	1.0775	2.6849	1.0853	2.5734	1.0935	2.4715	8
1.0703	2.8053	1.0776	2.6830	1.0854	2.5716	1.0936	2.4699	7
1.0704	2.8032	1.0778	2.6810	1.0855	2.5699	1.0938	2.4683	6
1.0705	2.8010	1.0779	2.6791	1.0857	2.5681	1.0939	2.4666	5
1.0707	2.7989	1.0780	2.6772	1.0858	2.5663	1.0941	2.4650	4
1.0708	2.7968	1.0781	2.6752	1.0859	2.5646	1.0942	2.4634	3
1.0709	2.7947	1.0783	2.6733	1.0861	2.5628	1.0943	2.4618	2
1.0710	2.7925	1.0784	2.6714	1.0862	2.5610	1.0945	2.4602	1
1.0711	2.7904	1.0785	2.6695	1.0864	2.5593	1.0946	2.4586	0
Co-sec.	Sec.	Co-sec.	Sec.	Co-sec.	Sec.	Co-sec.	Sec.	'
69°		**68°**		**67°**		**66°**		

′	21° SINE	COSINE	22° SINE	COSINE	23° SINE	COSINE	24° SINE	COSINE	′
0	.35837	.93358	.37461	.92718	.39073	.92050	.40674	.91355	60
1	.35864	.93348	.37488	.92707	.39100	.92039	.40700	.91343	59
2	.35891	.93337	.37515	.92697	.39127	.92028	.40727	.91331	58
3	.35918	.93327	.37542	.92686	.39153	.92016	.40753	.91319	57
4	.35945	.93316	.37569	.92675	.39180	.92005	.40780	.91307	56
5	.35973	.93306	.37595	.92664	.39207	.91994	.40806	.91295	55
6	.36000	.93295	.37622	.92653	.39234	.91982	.40833	.91283	54
7	.36027	.93285	.37649	.92642	.39260	.91971	.40860	.91272	53
8	.36054	.93274	.37676	.92631	.39287	.91959	.40886	.91260	52
9	.36081	.93264	.37703	.92620	.39314	.91948	.40913	.91248	51
10	.36108	.93253	.37730	.92609	.39341	.91936	.40939	.91236	50
11	.36135	.93243	.37757	.92598	.39367	.91925	.40966	.91224	49
12	.36162	.93232	.37784	.92587	.39394	.91914	.40992	.91212	48
13	.36190	.93222	.37811	.92576	.39421	.91902	.41019	.91200	47
14	.36217	.93211	.37838	.92565	.39448	.91891	.41045	.91188	46
15	.36244	.93201	.37865	.92554	.39474	.91879	.41072	.91176	45
16	.36271	.93190	.37892	.92543	.39501	.91868	.41098	.91164	44
17	.36298	.93180	.37919	.92532	.39528	.91856	.41125	.91152	43
18	.36325	.93169	.37946	.92521	.39555	.91845	.41151	.91140	42
19	.36352	.93159	.37973	.92510	.39581	.91833	.41178	.91128	41
20	.36379	.93148	.37999	.92499	.39608	.91822	.41204	.91116	40
21	.36406	.93137	.38026	.92488	.39635	.91810	.41231	.91104	39
22	.36434	.93127	.38053	.92477	.39661	.91799	.41257	.91092	38
23	.36461	.93116	.38080	.92466	.39688	.91787	.41284	.91080	37
24	.36488	.93106	.38107	.92455	.39715	.91775	.41310	.91068	36
25	.36515	.93095	.38134	.92444	.39741	.91764	.41337	.91056	35
26	.36542	.93084	.38161	.92432	.39768	.91752	.41363	.91044	34
27	.36569	.93074	.38188	.92421	.39795	.91741	.41390	.91032	33
28	.36596	.93063	.38215	.92410	.39822	.91729	.41416	.91020	32
29	.36623	.93052	.38241	.92399	.39848	.91718	.41443	.91008	31
30	.36650	.93042	.38268	.92388	.39875	.91706	.41469	.90996	30
31	.36677	.93031	.38295	.92377	.39902	.91694	.41496	.90984	29
32	.36704	.93020	.38322	.92366	.39928	.91683	.41522	.90972	28
33	.36731	.93010	.38349	.92355	.39955	.91671	.41549	.90960	27
34	.36758	.92999	.38376	.92343	.39982	.91660	.41575	.90948	26
35	.36785	.92988	.38403	.92332	.40008	.91648	.41602	.90936	25
36	.36812	.92978	.38430	.92321	.40035	.91636	.41628	.90924	24
37	.36839	.92967	.38456	.92310	.40062	.91625	.41655	.90911	23
38	.36867	.92956	.38483	.92299	.40088	.91613	.41681	.90899	22
39	.36894	.92945	.38510	.92287	.40115	.91601	.41707	.90887	21
40	.36921	.92935	.38537	.92276	.40141	.91590	.41734	.90875	20
41	.36948	.92924	.38564	.92265	.40168	.91578	.41760	.90863	19
42	.36975	.92913	.38591	.92254	.40195	.91566	.41787	.90851	18
43	.37002	.92902	.38617	.92243	.40221	.91555	.41813	.90839	17
44	.37029	.92892	.38644	.92231	.40248	.91543	.41840	.90826	16
45	.37056	.92881	.38671	.92220	.40275	.91531	.41866	.90814	15
46	.37083	.92870	.38698	.92209	.40301	.91519	.41892	.90802	14
47	.37110	.92859	.38725	.92198	.40328	.91508	.41919	.90790	13
48	.37137	.92849	.38752	.92186	.40355	.91496	.41945	.90778	12
49	.37164	.92838	.38778	.92175	.40381	.91484	.41972	.90766	11
50	.37191	.92827	.38805	.92164	.40408	.91472	.41998	.90753	10
51	.37218	.92816	.38832	.92152	.40434	.91461	.42024	.90741	9
52	.37245	.92805	.38859	.92141	.40461	.91449	.42051	.90729	8
53	.37272	.92794	.38886	.92130	.40488	.91437	.42077	.90717	7
54	.37299	.92784	.38912	.92119	.40514	.91425	.42104	.90704	6
55	.37326	.92773	.38939	.92107	.40541	.91414	.42130	.90692	5
56	.37353	.92762	.38966	.92096	.40567	.91402	.42156	.90680	4
57	.37380	.92751	.38993	.92085	.40594	.91390	.42183	.90668	3
58	.37407	.92740	.39020	.92073	.40621	.91378	.42209	.90655	2
59	.37434	.92729	.39046	.92062	.40647	.91366	.42235	.90643	1
60	.37461	.92718	.39073	.92050	.40674	.91355	.42262	.90631	0
	COSINE	SINE	COSINE	SINE	COSINE	SINE	COSINE	SINE	′
	68°		67°		66°		65°		

'	25° Sine	Cosine	26° Sine	Cosine	27° Sine	Cosine	28° Sine	Cosine	'
0	.42262	.90631	.43837	.89879	.45399	.89101	.46947	.88295	60
1	.42288	.90618	.43863	.89867	.45425	.89087	.46973	.88281	59
2	.42315	.90606	.43889	.89854	.45451	.89074	.46999	.88267	58
3	.42341	.90594	.43916	.89841	.45477	.89061	.47024	.88254	57
4	.42367	.90582	.43942	.89828	.45503	.89048	.47050	.88240	56
5	.42394	.90569	.43968	.89816	.45529	.89035	.47076	.88226	55
6	.42420	.90557	.43994	.89803	.45554	.89021	.47101	.88213	54
7	.42446	.90545	.44020	.89790	.45580	.89008	.47127	.88199	53
8	.42473	.90532	.44046	.89777	.45606	.88995	.47153	.88185	52
9	.42499	.90520	.44072	.89764	.45632	.88981	.47178	.88172	51
10	.42525	.90507	.44098	.89752	.45658	.88968	.47204	.88158	50
11	.42552	.90495	.44124	.89739	.45684	.88955	.47229	.88144	49
12	.42578	.90483	.44151	.89726	.45710	.88942	.47255	.88130	48
13	.42604	.90470	.44177	.89713	.45736	.88928	.47281	.88117	47
14	.42631	.90458	.44203	.89700	.45762	.88915	.47306	.88103	46
15	.42657	.90446	.44229	.89687	.45787	.88902	.47332	.88089	45
16	.42683	.90433	.44255	.89674	.45813	.88888	.47358	.88075	44
17	.42709	.90421	.44281	.89662	.45839	.88875	.47383	.88062	43
18	.42736	.90408	.44307	.89649	.45865	.88862	.47409	.88048	42
19	.42762	.90396	.44333	.89636	.45891	.88848	.47434	.88034	41
20	.42788	.90383	.44359	.89623	.45917	.88835	.47460	.88020	40
21	.42815	.90371	.44385	.89610	.45942	.88822	.47486	.88006	39
22	.42841	.90358	.44411	.89597	.45968	.88808	.47511	.87993	38
23	.42867	.90346	.44437	.89584	.45994	.88795	.47537	.87979	37
24	.42894	.90334	.44464	.89571	.46020	.88782	.47562	.87965	36
25	.42920	.90321	.44490	.89558	.46046	.88768	.47588	.87951	35
26	.42946	.90309	.44516	.89545	.46072	.88755	.47614	.87937	34
27	.42972	.90296	.44542	.89532	.46097	.88741	.47639	.87923	33
28	.42999	.90284	.44568	.89519	.46123	.88728	.47665	.87909	32
29	.43025	.90271	.44594	.89506	.46149	.88715	.47690	.87896	31
30	.43051	.90259	.44620	89493	.46175	.88701	.47716	.87882	30
31	.43077	.90246	.44646	.89480	.46201	.88688	.47741	.87868	29
32	.43104	.90233	.44672	.89467	.46226	.88674	.47767	.87854	28
33	.43130	.90221	.44698	.89454	.46252	.88661	.47793	.87840	27
34	.43156	.90208	.44724	.89441	.46278	.88647	.47818	.87826	26
35	.43182	.90196	.44750	.89428	.46304	.88634	.47844	.87812	25
36	.43209	.90183	.44776	.89415	.46330	.88620	.47869	.87798	24
37	.43235	.90171	.44802	.89402	.46355	.88607	.47895	.87784	23
38	.43261	.90158	.44828	.89389	.46381	.88593	.47920	.87770	22
39	.43287	.90146	.44854	.89376	.46407	.88580	.47946	.87756	21
40	.43313	.90133	.44880	.89363	.46433	.88566	.47971	.87743	20
41	.43340	.90120	.44906	.89350	.46458	.88553	.47997	.87729	19
42	.43366	.90108	.44932	.89337	.46484	.88539	.48022	.87715	18
43	.43392	.90095	.44958	.89324	.46510	.88526	.48048	.87701	17
44	.43418	.90082	.44984	.89311	.46536	.88512	.48073	.87687	16
45	.43445	.90070	.45010	.89298	.46561	.88499	.48099	.87673	15
46	.43471	.90057	.45036	.89285	.46587	.88485	.48124	.87659	14
47	.43497	.90045	.45062	.89272	.46613	.88472	.48150	.87645	13
48	.43523	.90032	.45088	.89259	.46639	.88458	.48175	.87631	12
49	.43549	.90019	.45114	.89245	.46664	.88445	.48201	.87617	11
50	.43575	.90007	.45140	.89232	.46690	.88431	.48226	.87603	10
51	.43602	.89994	.45166	.89219	.46716	.88417	.48252	.87589	9
52	.43628	.89981	45192	.89206	.46742	.88404	.48277	.87575	8
53	.43654	.89968	.45218	.89193	.46767	.88390	.48303	.87561	7
54	.43680	.89956	.45243	.89180	.46793	.88377	.48328	.87546	6
55	.43706	.89943	.45269	.89167	.46819	.88363	.48354	.87532	5
56	.43733	.89930	.45295	.89153	.46844	.88349	.48379	.87518	4
57	.43759	.89918	.45321	.89140	.46870	.88336	.48405	.87504	3
58	.43785	.89905	.45347	.89127	.46896	.88322	.48430	.87490	2
59	.43811	.89892	.45373	.89114	.46921	.88308	.48456	.87476	1
60	.43837	.89879	.45399	.89101	.46947	.88295	.48481	.87462	0
'	Cosine	Sine	Cosine	Sine	Cosine	Sine	Cosine	Sine	

| 64° | 63° | 62° | 61° |

'	24° SEC.	24° CO-SEC.	25° SEC.	25° CO-SEC.	26° SEC.	26° CO-SEC.	27° SEC.	27° CO-SEC.	'
0	1.0946	2.4586	1.1034	2.3662	1.1126	2.2812	1.1223	2.2027	60
1	1.0948	2.4570	1.1035	2.3647	1.1127	2.2798	1.1225	2.2014	59
2	1.0949	2.4554	1.1037	2.3632	1.1129	2.2784	1.1226	2.2002	58
3	1.0951	2.4538	1.1038	2.3618	1.1131	2.2771	1.1228	2.1989	57
4	1.0952	2.4522	1.1040	2.3603	1.1132	2.2757	1.1230	2.1977	56
5	1.0953	2.4506	1.1041	2.3588	1.1134	2.2744	1.1231	2.1964	55
6	1.0955	2.4490	1.1043	2.3574	1.1135	2.2730	1.1233	2.1951	54
7	1.0956	2.4474	1.1044	2.3559	1.1137	2.2717	1.1235	2.1939	53
8	1.0958	2.4458	1.1046	2.3544	1.1139	2.2703	1.1237	2.1927	52
9	1.0959	2.4442	1.1047	2.3530	1.1140	2.2690	1.1238	2.1914	51
10	1.0961	2.4426	1.1049	2.3515	1.1142	2.2676	1.1240	2.1902	50
11	1.0962	2.4411	1.1050	2.3501	1.1143	2.2663	1.1242	2.1889	49
12	1.0963	2.4395	1.1052	2.3486	1.1145	2.2650	1.1243	2.1877	48
13	1.0965	2.4379	1.1053	2.3472	1.1147	2.2636	1.1245	2.1865	47
14	1.0966	2.4363	1.1055	2.3457	1.1148	2.2623	1.1247	2.1852	46
15	1.0968	2.4347	1.1056	2.3443	1.1150	2.2610	1.1248	2.1840	45
16	1.0969	2.4332	1.1058	2.3428	1.1151	2.2596	1.1250	2.1828	44
17	1.0971	2.4316	1.1059	2.3414	1.1153	2.2583	1.1252	2.1815	43
18	1.0972	2.4300	1.1061	2.3399	1.1155	2.2570	1.1253	2.1803	42
19	1.0973	2.4285	1.1062	2.3385	1.1156	2.2556	1.1255	2.1791	41
20	1.0975	2.4269	1.1064	2.3371	1.1158	2.2543	1.1257	2.1778	40
21	1.0976	2.4254	1.1065	2.3356	1.1159	2.2530	1.1258	2.1766	39
22	1.0978	2.4238	1.1067	2.3342	1.1161	2.2517	1.1260	2.1754	38
23	1.0979	2.4222	1.1068	2.3328	1.1163	2.2503	1.1262	2.1742	37
24	1.0981	2.4207	1.1070	2.3313	1.1164	2.2490	1.1264	2.1730	36
25	1.0982	2.4191	1.1072	2.3299	1.1166	2.2477	1.1265	2.1717	35
26	1.0984	2.4176	1.1073	2.3285	1.1167	2.2464	1.1267	2.1705	34
27	1.0985	2.4160	1.1075	2.3271	1.1169	2.2451	1.1269	2.1693	33
28	1.0986	2.4145	1.1076	2.3256	1.1171	2.2438	1.1270	2.1681	32
29	1.0988	2.4130	1.1078	2.3242	1.1172	2.2425	1.1272	2.1669	31
30	1.0989	2.4114	1.1079	2.3228	1.1174	2.2411	1.1274	2.1657	30
31	1.0991	2.4099	1.1081	2.3214	1.1176	2.2398	1.1275	2.1645	29
32	1.0992	2.4083	1.1082	2.3200	1.1177	2.2385	1.1277	2.1633	28
33	1.0994	2.4068	1.1084	2.3186	1.1179	2.2372	1.1279	2.1620	27
34	1.0995	2.4053	1.1085	2.3172	1.1180	2.2359	1.1281	2.1608	26
35	1.0997	2.4037	1.1087	2.3158	1.1182	2.2346	1.1282	2.1596	25
36	1.0998	2.4022	1.1088	2.3143	1.1184	2.2333	1.1284	2.1584	24
37	1.1000	2.4007	1.1090	2.3129	1.1185	2.2320	1.1286	2.1572	23
38	1.1001	2.3992	1.1092	2.3115	1.1187	2.2307	1.1287	2.1560	22
39	1.1003	2.3976	1.1093	2.3101	1.1189	2.2294	1.1289	2.1548	21
40	1.1004	2.3961	1.1095	2.3087	1.1190	2.2282	1.1291	2.1536	20
41	1.1005	2.3946	1.1096	2.3073	1.1192	2.2269	1.1293	2.1525	19
42	1.1007	2.3931	1.1098	2.3059	1.1193	2.2256	1.1294	2.1513	18
43	1.1008	2.3916	1.1099	2.3046	1.1195	2.2243	1.1296	2.1501	17
44	1.1010	2.3901	1.1101	2.3032	1.1197	2.2230	1.1298	2.1489	16
45	1.1011	2.3886	1.1102	2.3018	1.1198	2.2217	1.1299	2.1477	15
46	1.1013	2.3871	1.1104	2.3004	1.1200	2.2204	1.1301	2.1465	14
47	1.1014	2.3856	1.1106	2.2990	1.1202	2.2192	1.1303	2.1453	13
48	1.1016	2.3841	1.1107	2.2976	1.1203	2.2179	1.1305	2.1441	12
49	1.1017	2.3826	1.1109	2.2962	1.1205	2.2166	1.1306	2.1430	11
50	1.1019	2.3811	1.1110	2.2949	1.1207	2.2153	1.1308	2.1418	10
51	1.1020	2.3796	1.1112	2.2935	1.1208	2.2141	1.1310	2.1406	9
52	1.1022	2.3781	1.1113	2.2921	1.1210	2.2128	1.1312	2.1394	8
53	1.1023	2.3766	1.1115	2.2907	1.1212	2.2115	1.1313	2.1382	7
54	1.1025	2.3751	1.1116	2.2894	1.1213	2.2103	1.1315	2.1371	6
55	1.1026	2.3736	1.1118	2.2880	1.1215	2.2090	1.1317	2.1359	5
56	1.1028	2.3721	1.1120	2.2866	1.1217	2.2077	1.1319	2.1347	4
57	1.1029	2.3706	1.1121	2.2853	1.1218	2.2065	1.1320	2.1335	3
58	1.1031	2.3691	1.1123	2.2839	1.1220	2.2052	1.1322	2.1324	2
59	1.1032	2.3677	1.1124	2.2825	1.1222	2.2039	1.1324	2.1312	1
60	1.1034	2.3662	1.1126	2.2812	1.1223	2.2027	1.1326	2.1300	0
'	CO-SEC.	SEC.	CO-SEC.	SEC.	CO-SEC.	SEC.	CO-SEC.	SEC.	
	65°		64°		63°		62°		

28° Sec.	Co-sec.	29° Sec.	Co-sec.	30° Sec.	Co-sec.	31° Sec.	Co-sec.	'
1.1326	2.1300	1.1433	2.0627	1.1547	2.0000	1.1666	1.9416	60
1.1327	2.1289	1.1435	2.0616	1.1549	1.9990	1.1668	1.9407	59
1.1329	2.1277	1.1437	2.0605	1.1551	1.9980	1.1670	1.9397	58
1.1331	2.1266	1.1439	2.0594	1.1553	1.9970	1.1672	1.9388	57
1.1333	2.1254	1.1441	2.0583	1.1555	1.9960	1.1674	1.9378	56
1.1334	2.1242	1.1443	2.0573	1.1557	1.9950	1.1676	1.9369	55
1.1336	2.1231	1.1445	2.0562	1.1559	1.9940	1.1678	1.9360	54
1.1338	2.1219	1.1446	2.0551	1.1561	1.9930	1.1681	1.9350	53
1.1340	2.1208	1.1448	2.0540	1.1562	1.9920	1.1683	1.9341	52
1.1341	2.1196	1.1450	2.0530	1.1564	1.9910	1.1685	1.9332	51
1.1343	2.1185	1.1452	2.0519	1.1566	1.9900	1.1687	1.9322	50
1.1345	2.1173	1.1454	2.0508	1.1568	1.9890	1.1689	1.9313	49
1.1347	2.1162	1.1456	2.0498	1.1570	1.9880	1.1691	1.9304	48
1.1349	2.1150	1.1458	2.0487	1.1572	1.9870	1.1693	1.9295	47
1.1350	2.1139	1.1459	2.0476	1.1574	1.9860	1.1695	1.9285	46
1.1352	2.1127	1.1461	2.0466	1.1576	1.9850	1.1697	1.9276	45
1.1354	2.1116	1.1463	2.0455	1.1578	1.9840	1.1699	1.9267	44
1.1356	2.1104	1.1465	2.0444	1.1580	1.9830	1.1701	1.9258	43
1.1357	2.1093	1.1467	2.0434	1.1582	1.9820	1.1703	1.9248	42
1.1359	2.1082	1.1469	2.0423	1.1584	1.9811	1.1705	1.9239	41
1.1361	2.1070	1.1471	2.0413	1.1586	1.9801	1.1707	1.9230	40
1.1363	2.1059	1.1473	2.0402	1.1588	1.9791	1.1709	1.9221	39
1.1365	2.1048	1.1474	2.0392	1.1590	1.9781	1.1712	1.9212	38
1.1366	2.1036	1.1476	2.0381	1.1592	1.9771	1.1714	1.9203	37
1.1368	2.1025	1.1478	2.0370	1.1594	1.9761	1.1716	1.9193	36
1.1370	2.1014	1.1480	2.0360	1.1596	1.9752	1.1718	1.9184	35
1.1372	2.1002	1.1482	2.0349	1.1598	1.9742	1.1720	1.9175	34
1.1373	2.0991	1.1484	2.0339	1.1600	1.9732	1.1722	1.9166	33
1.1375	2.0980	1.1486	2.0329	1.1602	1.9722	1.1724	1.9157	32
1.1377	2.0969	1.1488	2.0318	1.1604	1.9713	1.1726	1.9148	31
1.1379	2.0957	1.1489	2.0308	1.1606	1.9703	1.1728	1.9139	30
1.1381	2.0946	1.1491	2.0297	1.1608	1.9693	1.1730	1.9130	29
1.1382	2.0935	1.1493	2.0287	1.1610	1.9683	1.1732	1.9121	28
1.1384	2.0924	1.1495	2.0276	1.1612	1.9674	1.1734	1.9112	27
1.1386	2.0912	1.1497	2.0266	1.1614	1.9664	1.1737	1.9102	26
1.1388	2.0901	1.1499	2.0256	1.1616	1.9654	1.1739	1.9093	25
1.1390	2.0890	1.1501	2.0245	1.1618	1.9645	1.1741	1.9084	24
1.1391	2.0879	1.1503	2.0235	1.1620	1.9635	1.1743	1.9075	23
1.1393	2.0868	1.1505	2.0224	1.1622	1.9625	1.1745	1.9066	22
1.1395	2.0857	1.1507	2.0214	1.1624	1.9616	1.1747	1.9057	21
1.1397	2.0846	1.1508	2.0204	1.1626	1.9606	1.1749	1.9048	20
1.1399	2.0835	1.1510	2.0194	1.1628	1.9596	1.1751	1.9039	19
1.1401	2.0824	1.1512	2.0183	1.1630	1.9587	1.1753	1.9030	18
1.1402	2.0812	1.1514	2.0173	1.1632	1.9577	1.1756	1.9021	17
1.1404	2.0801	1.1516	2.0163	1.1634	1.9568	1.1758	1.9013	16
1.1406	2.0790	1.1518	2.0152	1.1636	1.9558	1.1760	1.9004	15
1.1408	2.0779	1.1520	2.0142	1.1638	1.9549	1.1762	1.8995	14
1.1410	2.0768	1.1522	2.0132	1.1640	1.9539	1.1764	1.8986	13
1.1411	2.0757	1.1524	2.0122	1.1642	1.9530	1.1766	1.8977	12
1.1413	2.0746	1.1526	2.0111	1.1644	1.9520	1.1768	1.8968	11
1.1415	2.0735	1.1528	2.0101	1.1646	1.9510	1.1770	1.8959	10
1.1417	2.0725	1.1530	2.0091	1.1648	1.9501	1.1772	1.8950	9
1.1419	2.0714	1.1531	2.0081	1.1650	1.9491	1.1775	1.8941	8
1.1421	2.0703	1.1533	2.0071	1.1652	1.9482	1.1777	1.8932	7
1.1422	2.0692	1.1535	2.0061	1.1654	1.9473	1.1779	1.8924	6
1.1424	2.0681	1.1537	2.0050	1.1656	1.9463	1.1781	1.8915	5
1.1426	2.0670	1.1539	2.0040	1.1658	1.9454	1.1783	1.8906	4
1.1428	2.0659	1.1541	2.0030	1.1660	1.9444	1.1785	1.8897	3
1.1430	2.0648	1.1543	2.0020	1.1662	1.9435	1.1787	1.8888	2
1.1432	2.0637	1.1545	2.0010	1.1664	1.9425	1.1790	1.8879	1
1.1433	2.0627	1.1547	2.0000	1.1666	1.9416	1.1792	1.8871	'
Co-sec.	Sec.	Co-sec.	Sec.	Co-sec.	Sec.	Co-sec.	Sec.	
61°		60°		59°		58°		

′	37° Sine	Cosine	38° Sine	Cosine	39° Sine	Cosine	40° Sine	Cosine	′
0	.60182	.79864	.61566	.78801	.62932	.77715	.64279	.76604	60
1	.60205	.79846	.61589	.78783	.62955	.77696	.64301	.76586	59
2	.60228	.79829	.61612	.78765	.62977	.77678	.64323	.76567	58
3	.60251	.79811	.61635	.78747	.63000	.77660	.64346	.76548	57
4	.60274	.79793	.61658	.78729	.63022	.77641	.64368	.76530	56
5	.60298	.79776	.61681	.78711	.63045	.77623	.64390	.76511	55
6	.60321	.79758	.61704	.78694	.63068	.77605	.64412	.76492	54
7	.60344	.79741	.61726	.78676	.63090	.77586	.64435	.76473	53
8	.60367	.79723	.61749	.78658	.63113	.77568	.64457	.76455	52
9	.60390	.79706	.61772	.78640	.63135	.77550	.64479	.76436	51
10	.60414	.79688	.61795	.78622	.63158	.77531	.64501	.76417	50
11	.60437	.79671	.61818	.78604	.63180	.77513	.64524	.76398	49
12	.60460	.79653	.61841	.78586	.63203	.77494	.64546	.76380	48
13	.60483	.79635	.61864	.78568	.63225	.77476	.64568	.76361	47
14	.60506	.79618	.61887	.78550	.63248	.77458	.64590	.76342	46
15	.60529	.79600	.61909	.78532	.63271	.77439	.64612	.76323	45
16	.60553	.79583	.61932	.78514	.63293	.77421	.64635	.76304	44
17	.60576	.79565	.61955	.78496	.63316	.77402	.64657	.76286	43
18	.60599	.79547	.61978	.78478	.63338	.77384	.64679	.76267	42
19	.60622	.79530	.62001	.78460	.63361	.77366	.64701	.76248	41
20	.60645	.79512	.62024	.78442	.63383	.77347	.64723	.76229	40
21	.60668	.79494	.62046	.78424	.63406	.77329	.64746	.76210	39
22	.60691	.79477	.62069	.78405	.63428	.77310	.64768	.76192	38
23	.60714	.79459	.62092	.78387	.63451	.77292	.64790	.76173	37
24	.60738	.79441	.62115	.78369	.63473	.77273	.64812	.76154	36
25	.60761	.79424	.62138	.78351	.63496	.77255	.64834	.76135	35
26	.60784	.79406	.62160	.78333	.63518	.77236	.64856	.76116	34
27	.60807	.79388	.62183	.78315	.63540	.77218	.64878	.76097	33
28	.60830	.79371	.62206	.78297	.63563	.77199	.64901	.76078	32
29	.60853	.79353	.62229	.78279	.63585	.77181	.64923	.76059	31
30	.60876	.79335	.62251	.78261	.63608	.77162	.64945	.76041	30
31	.60899	.79318	.62274	.78243	.63630	.77144	.64967	.76022	29
32	.60922	.79300	.62297	.78225	.63653	.77125	.64989	.76003	28
33	.60945	.79282	.62320	.78206	.63675	.77107	.65011	.75984	27
34	.60968	.79264	.62342	.78188	.63698	.77088	.65033	.75965	26
35	.60991	.79247	.62365	.78170	.63720	.77070	.65055	.75946	25
36	.61015	.79229	.62388	.78152	.63742	.77051	.65077	.75927	24
37	.61038	.79211	.62411	.78134	.63765	.77033	.65100	.75908	23
38	.61061	.79193	.62433	.78116	.63787	.77014	.65122	.75889	22
39	.61084	.79176	.62456	.78098	.63810	.76996	.65144	.75870	21
40	.61107	.79158	.62479	.78079	.63832	.76977	.65166	.75851	20
41	.61130	.79140	.62502	.78061	.63854	.76959	.65188	.75832	19
42	.61153	.79122	.62524	.78043	.63877	.76940	.65210	.75813	18
43	.61176	.79105	.62547	.78025	.63899	.76921	.65232	.75794	17
44	.61199	.79087	.62570	.78007	.63922	.76903	.65254	.75775	16
45	.61222	.79069	.62592	.77988	.63944	.76884	.65276	.75756	15
46	.61245	.79051	.62615	.77970	.63966	.76866	.65298	.75738	14
47	.61268	.79033	.62638	.77952	.63989	.76847	.65320	.75719	13
48	.61291	.79016	.62660	.77934	.64011	.76828	.65342	.75700	12
49	.61314	.78998	.62683	.77916	.64033	.76810	.65364	.75680	11
50	.61337	.78980	.62706	.77897	.64056	.76791	.65386	.75661	10
51	.61360	.78962	.62728	.77879	.64078	.76772	.65408	.75642	9
52	.61383	.78944	.62751	.77861	.64100	.76754	.65430	.75623	8
53	.61406	.78926	.62774	.77843	.64123	.76735	.65452	.75604	7
54	.61429	.78908	.62796	.77824	.64145	.76717	.65474	.75585	6
55	.61451	.78891	.62819	.77806	.64167	.76698	.65496	.75566	5
56	.61474	.78873	.62842	.77788	.64190	.76679	.65518	.75547	4
57	.61497	.78855	.62864	.77769	.64212	.76661	.65540	.75528	3
58	.61520	.78837	.62887	.77751	.64234	.76642	.65562	.75509	2
59	.61543	.78819	.62909	.77733	.64256	.76623	.65584	.75490	1
60	.61566	.78801	.62932	.77715	.64279	.76604	.65606	.75471	0
′	Cosine	Sine	Cosine	Sine	Cosine	Sine	Cosine	Sine	′
	52°		51°		50°		49°		

41° Sine	41° Cosine	42° Sine	42° Cosine	43° Sine	43° Cosine	44° Sine	44° Cosine	′
.65606	.75471	.66913	.74314	.68200	.73135	.69466	.71934	60
.65628	.75452	.66935	.74295	.68221	.73116	.69487	.71914	59
.65650	.75433	.66956	.74276	.68242	.73096	.69508	.71894	58
.65672	.75414	.66978	.74256	.68264	.73076	.69529	.71873	57
.65694	.75395	.66999	.74237	.68285	.73056	.69549	.71853	56
.65716	.75375	.67021	.74217	.68306	.73036	.69570	.71833	55
.65738	.75356	.67043	.74198	.68327	.73016	.69591	.71813	54
.65759	.75337	.67064	.74178	.68349	.72996	.69612	.71792	53
.65781	.75318	.67086	.74159	.68370	.72976	.69633	.71772	52
.65803	.75299	.67107	.74139	.68391	.72957	.69654	.71752	51
.65825	.75280	.67129	.74120	.68412	.72937	.69675	.71732	50
.65847	.75261	.67151	.74100	.68434	.72917	.69696	.71711	49
.65869	.75241	.67172	.74080	.68455	.72897	.69717	.71691	48
.65891	.75222	.67194	.74061	.68476	.72877	.69737	.71671	47
.65913	.75203	.67215	.74041	.68497	.72857	.69758	.71650	46
.65935	.75184	.67237	.74022	.68518	.72837	.69779	.71630	45
.65956	.75165	.67258	.74002	.68539	.72817	.69800	.71610	44
.65978	.75146	.67280	.73983	.68561	.72797	.69821	.71590	43
.66000	.75126	.67301	.73963	.68582	.72777	.69842	.71569	42
.66022	.75107	.67323	.73944	.68603	.72757	.69862	.71549	41
.66044	.75088	.67344	.73924	.68624	.72737	.69883	.71529	40
.66066	.75069	.67366	.73904	.68645	.72717	.69904	.71508	39
.66088	.75050	.67387	.73885	.68666	.72697	.69925	.71488	38
.66109	.75030	.67409	.73865	.68688	.72677	.69946	.71468	37
.66131	.75011	.67430	.73846	.68709	.72657	.69966	.71447	36
.66153	.74992	.67452	.73826	.68730	.72637	.69987	.71427	35
.66175	.74973	.67473	.73806	.68751	.72617	.70008	.71407	34
.66197	.74953	.67495	.73787	.68772	.72597	.70029	.71386	33
.66218	.74934	.67516	.73767	.68793	.72577	.70049	.71366	32
.66240	.74915	.67538	.73747	.68814	.72557	.70070	.71345	31
.66262	.74896	.67559	.73728	.68835	.72537	.70091	.71325	30
.66284	.74876	.67580	.73708	.68857	.72517	.70112	.71305	29
.66306	.74857	.67602	.73688	.68878	.72497	.70132	.71284	28
.66327	.74838	.67623	.73669	.68899	.72477	.70153	.71264	27
.66349	.74818	.67645	.73649	.68920	.72457	.70174	.71243	26
.66371	.74799	.67666	.73629	.68941	.72437	.70195	.71223	25
.66393	.74780	.67688	.73610	.68962	.72417	.70215	.71203	24
.66414	.74760	.67709	.73590	.68983	.72397	.70236	.71182	23
.66436	.74741	.67730	.73570	.69004	.72377	.70257	.71162	22
.66458	.74722	.67752	.73551	.69025	.72357	.70277	.71141	21
.66480	.74703	.67773	.73531	.69046	.72337	.70298	.71121	20
.66501	.74683	.67795	.73511	.69067	.72317	.70319	.71100	19
.66523	.74664	.67816	.73491	.69088	.72297	.70339	.71080	18
.66545	.74644	.67837	.73472	.69109	.72277	.70360	.71059	17
.66566	.74625	.67859	.73452	.69130	.72257	.70381	.71039	16
.66588	.74606	.67880	.73432	.69151	.72236	.70401	.71019	15
.66610	.74586	.67901	.73413	.69172	.72216	.70422	.70998	14
.66632	.74567	.67923	.73393	.69193	.72196	.70443	.70978	13
.66653	.74548	.67944	.73373	.69214	.72176	.70463	.70957	12
.66675	.74528	.67965	.73353	.69235	.72156	.70484	.70937	11
.66697	.74509	.67987	.73333	.69256	.72136	.70505	.70916	10
.66718	.74489	.68008	.73314	.69277	.72116	.70525	.70896	9
.66740	.74470	.68029	.73294	.69298	.72095	.70546	.70875	8
.66762	.74451	.68051	.73274	.69319	.72075	.70567	.70855	7
.66783	.74431	.68072	.73254	.69340	.72055	.70587	.70834	6
.66805	.74412	.68093	.73234	.69361	.72035	.70608	.70813	5
.66827	.74392	.68115	.73215	.69382	.72015	.70628	.70793	4
.66848	.74373	.68136	.73195	.69403	.71995	.70649	.70772	3
.66870	.74353	.68157	.73175	.69424	.71974	.70670	.70752	2
.66891	.74334	.68179	.73155	.69445	.71954	.70690	.70731	1
.66913	.74314	.68200	.73135	.69466	.71934	.70711	.70711	0
Cosine	Sine	Cosine	Sine	Cosine	Sine	Cosine	Sine	′
48°		47°		46°		45°		

′	0° SEC.	0° CO-SEC.	1° SEC.	1° CO-SEC.	2° SEC.	2° CO-SEC.	3° SEC.	3° CO-SEC.	′
0	1	Infinite.	1.0001	57.299	1.0006	28.654	1.0014	19.107	60
1	1	3437.70	1.0001	56.359	1.0006	28.417	1.0014	19.002	59
2	1	1718.90	1.0002	55.450	1.0006	28.184	1.0014	18.897	58
3	1	1145.90	1.0002	54.570	1.0006	27.955	1.0014	18.794	57
4	1	859.44	1.0002	53.718	1.0006	27.730	1.0014	18.692	56
5	1	687.55	1.0002	52.891	1.0007	27.508	1.0014	18.591	55
6	1	572.96	1.0002	52.090	1.0007	27.290	1.0015	18.491	54
7	1	491.11	1.0002	51.313	1.0007	27.075	1.0015	18.393	53
8	1	429.72	1.0002	50.558	1.0007	26.864	1.0015	18.295	52
9	1	381.97	1.0002	49.826	1.0007	26.655	1.0015	18.198	51
10	1	343.77	1.0002	49.114	1.0007	26.450	1.0015	18.103	50
11	1	312.52	1.0002	48.422	1.0007	26.249	1.0015	18.008	49
12	1	286.48	1.0002	47.750	1.0007	26.050	1.0016	17.914	48
13	1	264.44	1.0002	47.096	1.0007	25.854	1.0016	17.821	47
14	1	245.55	1.0002	46.460	1.0008	25.661	1.0016	17.730	46
.15	1	229.18	1.0002	45.840	1.0008	25.471	1.0016	17.639	45
16	1	214.86	1.0002	45.237	1.0008	25.284	1.0016	17.549	44
17	1	202.22	1.0002	44.650	1.0008	25.100	1.0016	17.460	43
18	1	190.99	1.0002	44.077	1.0008	24.918	1.0017	17.372	42
19	1	180.73	1.0003	43.520	1.0008	24.739	1.0017	17.285	41
20	1	171.89	1.0003	42.976	1.0008	24.562	1.0017	17.198	40
21	1	163.70	1.0003	42.445	1.0008	24.358	1.0017	17.113	39
22	1	156.26	1.0003	41.928	1.0008	24.216	1.0017	17.028	38
23	1	149.47	1.0003	41.423	1.0009	24.047	1.0017	16.944	37
24	1	143.24	1.0003	40.930	1.0009	23.880	1.0018	16.861	36
25	1	137.51	1.0003	40.448	1.0009	23.716	1.0018	16.779	35
26	1	132.22	1.0003	39.978	1.0009	23.553	1.0018	16.698	34
27	1	127.32	1.0003	39.518	1.0009	23.393	1.0018	16.617	33
28	1	122.78	1.0003	39.069	1.0009	23.235	1.0018	16.538	32
29	1	118.54	1.0003	38.631	1.0009	23.079	1.0018	16.459	31
30	1	114.59	1.0003	38.201	1.0009	22.925	1.0019	16.380	30
31	1	110.90	1.0003	37.782	1.0010	22.774	1.0019	16.303	29
32	1	107.43	1.0003	37.371	1.0010	22.624	1.0019	16.226	28
33	1	104.17	1.0004	36.969	1.0010	22.476	1.0019	16.150	27
34	1	101.11	1.0004	36.576	1.0010	22.330	1.0019	16.075	26
35	1	90.223	1.0004	36.191	1.0010	22.186	1.0019	16.000	25
36	1	95.495	1.0004	35.814	1.0010	22.044	1.0020	15.926	24
37	1	92.914	1.0004	35.445	1.0010	21.904	1.0020	15.853	23
38	1.0001	90.469	1.0004	35.084	1.0010	21.765	1.0020	15.780	22
39	1.0001	88.149	1.0004	34.729	1.0011	21.629	1.0020	15.708	21
40	1.0001	85.946	1.0004	34.382	1.0011	21.494	1.0020	15.637	20
41	1.0001	83.849	1.0004	34.042	1.0011	21.360	1.0021	15.566	19
42	1.0001	81.853	1.0004	33.708	1.0011	21.228	1.0021	15.496	18
43	1.0001	79.950	1.0004	33.381	1.0011	21.098	1.0021	15.427	17
44	1.0001	78.133	1.0004	33.060	1.0011	20.970	1.0021	15.358	16
45	1.0001	76.396	1.0005	32.745	1.0011	20.843	1.0021	15.290	15
46	1.0001	74.736	1.0005	32.437	1.0012	20.717	1.0022	15.222	14
47	1.0001	73.146	1.0005	32.134	1.0012	20.593	1.0022	15.155	13
48	1.0001	71.622	1.0005	31.836	1.0012	20.471	1.0022	15.089	12
49	1.0001	70.160	1.0005	31.544	1.0012	20.350	1.0022	15.023	11
50	1.0001	68.757	1.0005	31.257	1.0012	20.230	1.0022	14.958	10
51	1.0001	67.409	1.0005	30.976	1.0012	20.112	1.0023	14.893	9
52	1.0001	66.113	1.0005	30.699	1.0012	19.995	1.0023	14.829	8
53	1.0001	64.866	1.0005	30.428	1.0013	19.880	1.0023	14.765	7
54	1.0001	63.664	1.0005	30.161	1.0013	19.766	1.0023	14.702	6
55	1.0001	62.507	1.0005	29.899	1.0013	19.653	1.0023	14.640	5
56	1.0001	61.391	1.0006	29.641	1.0013	19.541	1.0024	14.578	4
57	1.0001	60.314	1.0006	29.388	1.0013	19.431	1.0024	14.517	3
58	1.0001	59.274	1.0006	29.139	1.0013	19.322	1.0024	14.456	2
59	1.0001	58.270	1.0006	28.894	1.0013	19.214	1.0024	14.395	1
60	1.0001	57.299	1.0006	28.654	1.0014	19.107	1.0024	14.335	0
′	CO-SEC.	SEC.	CO-SEC.	SEC.	CO-SEC.	SEC.	CO-SEC.	SEC.	
		89°		88°		87°		86°	

4°		5°		6°		7°		
SEC.	CO-SEC.	SEC.	CO-SEC.	SEC.	CO-SEC.	SEC.	CO-SEC.	′
1.0024	14.335	1.0038	11.474	1.0055	9.5668	1.0075	8.2055	60
1.0025	14.276	1.0038	11.436	1.0055	9.5404	1.0075	8.1861	59
1.0025	14.217	1.0039	11.398	1.0056	9.5141	1.0076	8.1668	58
1.0025	14.159	1.0039	11.360	1.0056	9.4880	1.0076	8.1476	57
1.0025	14.101	1.0039	11.323	1.0056	9.4620	1.0076	8.1285	56
1.0025	14.043	1.0039	11.286	1.0057	9.4362	1.0077	8.1094	55
1.0026	13.986	1.0040	11.249	1.0057	9.4105	1.0077	8.0905	54
1.0026	13.930	1.0040	11.213	1.0057	9.3850	1.0078	8.0717	53
1.0026	13.874	1.0040	11.176	1.0057	9.3596	1.0078	8.0529	52
1.0026	13.818	1.0040	11.140	1.0058	9.3343	1.0078	8.0342	51
1.0026	13.763	1.0041	11.104	1.0058	9.3092	1.0079	8.0156	50
1.0027	13.708	1.0041	11.069	1.0058	9.2842	1.0079	7.9971	49
1.0027	13.654	1.0041	11.033	1.0059	9.2593	1.0079	7.9787	48
1.0027	13.600	1.0041	10.988	1.0059	9.2346	1.0080	7.9604	47
1.0027	13.547	1.0042	10.963	1.0059	9.2100	1.0080	7.9421	46
1.0027	13.494	1.0042	10.929	1.0060	9.1855	1.0080	7.9240	45
1.0028	13.441	1.0042	10.894	1.0060	9.1612	1.0081	7.9059	44
1.0028	13.389	1.0043	10.860	1.0060	9.1370	1.0081	7.8879	43
1.0028	13.337	1.0043	10.826	1.0061	9.1129	1.0082	7.8700	42
1.0028	13.286	1.0043	10.792	1.0061	9.0890	1.0082	7.8522	41
1.0029	13.235	1.0043	10.758	1.0061	9.0651	1.0082	7.8344	40
1.0029	13.184	1.0044	10.725	1.0062	9.0414	1.0083	7.8168	39
1.0029	13.134	1.0044	10.692	1.0062	9.0179	1.0083	7.7992	38
1.0029	13.084	1.0044	10.659	1.0062	8.9944	1.0084	7.7817	37
1.0029	13.034	1.0044	10.626	1.0063	8.9711	1.0084	7.7642	36
1.0030	12.985	1.0045	10.593	1.0063	8.9479	1.0084	7.7469	35
1.0030	12.937	1.0045	10.561	1.0063	8.9248	1.0085	7.7296	34
1.0030	12.888	1.0045	10.529	1.0064	8.9018	1.0085	7.7124	33
1.0030	12.840	1.0046	10.497	1.0064	8.8790	1.0085	7.6953	32
1.0031	12.793	1.0046	10.465	1.0064	8.8563	1.0086	7.6783	31
1.0031	12.745	1.0046	10.433	1.0065	8.8337	1.0086	7.6613	30
1.0031	12.698	1.0046	10.402	1.0065	8.8112	1.0087	7.6444	29
1.0031	12.652	1.0047	10.371	1.0065	8.7888	1.0087	7.6276	28
1.0032	12.606	1.0047	10.340	1.0066	8.7665	1.0087	7.6108	27
1.0032	12.560	1.0047	10.309	1.0066	8.7444	1.0088	7.5942	26
1.0032	12.514	1.0048	10.278	1.0066	8.7223	1.0088	7.5776	25
1.0032	12.469	1.0048	10.248	1.0067	8.7004	1.0089	7.5611	24
1.0032	12.424	1.0048	10.217	1.0067	8.6786	1.0089	7.5446	23
1.0033	12.379	1.0048	10.187	1.0067	8.6569	1.0089	7.5282	22
1.0033	12.335	1.0049	10.157	1.0068	8.6353	1.0090	7.5119	21
1.0033	12.291	1.0049	10.127	1.0068	8.6138	1.0090	7.4957	20
1.0033	12.248	1.0049	10.098	1.0068	8.5924	1.0091	7.4795	19
1.0034	12.204	1.0050	10.068	1.0069	8.5711	1.0091	7.4634	18
1.0034	12.161	1.0050	10.039	1.0069	8.5499	1.0091	7.4474	17
1.0034	12.118	1.0050	10.010	1.0069	8.5289	1.0092	7.4315	16
1.0034	12.076	1.0050	9.9812	1.0070	8.5079	1.0092	7.4156	15
1.0035	12.034	1.0051	9.9525	1.0070	8.4871	1.0092	7.3998	14
1.0035	11.992	1.0051	9.9239	1.0070	8.4663	1.0093	7.3840	13
1.0035	11.950	1.0051	9.8955	1.0071	8.4457	1.0093	7.3683	12
1.0035	11.909	1.0052	9.8672	1.0071	8.4251	1.0094	7.3527	11
1.0036	11.868	1.0052	9.8391	1.0071	8.4046	1.0094	7.3372	10
1.0036	11.828	1.0052	9.8112	1.0072	8.3843	1.0094	7.3217	9
1.0036	11.787	1.0053	9.7834	1.0072	8.3640	1.0095	7.3063	8
1.0036	11.747	1.0053	9.7558	1.0073	8.3439	1.0095	7.2909	7
1.0037	11.707	1.0053	9.7283	1.0073	8.3238	1.0096	7.2757	6
1.0037	11.668	1.0053	9.7010	1.0073	8.3039	1.0096	7.2604	5
1.0037	11.628	1.0054	9.6739	1.0074	8.2840	1.0097	7.2453	4
1.0037	11.589	1.0054	9.6469	1.0074	8.2642	1.0097	7.2302	3
1.0038	11.550	1.0054	9.6200	1.0074	8.2446	1.0097	7.2152	2
1.0038	11.512	1.0055	9.5933	1.0075	8.2250	1.0098	7.2002	1
1.0038	11.474	1.0055	9.5668	1.0075	8.2055	1.0098	7.1853	0
CO-SEC.	SEC.	CO-SEC.	SEC.	CO-SEC.	SEC.	CO-SEC.	SEC.	′
	85°		84°		83°		82°	

′	13° SINE	COSINE	14° SINE	COSINE	15° SINE	COSINE	16° SINE	COSINE
0	.22495	.97437	.24192	.97030	.25882	.96593	.27564	.96126
1	.22523	.97430	.24220	.97023	.25910	.96585	.27592	.96118
2	.22552	.97424	.24249	.97015	.25938	.96578	.27620	.96110
3	.22580	.97417	.24277	.97008	.25966	.96570	.27648	.96102
4	.22608	.97411	.24305	.97001	.25994	.96562	.27676	.96094
5	.22637	.97404	.24333	.96994	.26022	.96555	.27704	.96086
6	.22665	.97398	.24362	.96987	.26050	.96547	.27731	.96078
7	.22693	.97391	.24390	.96980	.26079	.96540	.27759	.96070
8	.22722	.97384	.24418	.96973	.26107	.96532	.27787	.96062
9	.22750	.97378	.24446	.96966	.26135	.96524	.27815	.96054
10	.22778	.97371	.24474	.96959	.26163	.96517	.27843	.96046
11	.22807	.97365	.24503	.96952	.26191	.96509	.27871	.96037
12	.22835	.97358	.24531	.96945	.26219	.96502	.27899	.96029
13	.22863	.97351	.24559	.96937	.26247	.96494	.27927	.96021
14	.22892	.97345	.24587	.96930	.26275	.96486	.27955	.96013
15	.22920	.97338	.24615	.96923	.26303	.96479	.27983	.96005
16	.22948	.97331	.24644	.96916	.26331	.96471	.28011	.95997
17	.22977	.97325	.24672	.96909	.26359	.96463	.28039	.95989
18	.23005	.97318	.24700	.96902	.26387	.96456	.28067	.95981
19	.23033	.97311	.24728	.96894	.26415	.96448	.28095	.95972
20	.23062	.97304	.24756	.96887	.26443	.96440	.28123	.95964
21	.23090	.97298	.24784	.96880	.26471	.96433	.28150	.95956
22	.23118	.97291	.24813	.96873	.26500	.96425	.28178	.95948
23	.23146	.97284	.24841	.96866	.26528	.96417	.28206	.95940
24	.23175	.97278	.24869	.96858	.26556	.96410	.28234	.95931
25	.23203	.97271	.24897	.96851	.26584	.96402	.28262	.95923
26	.23231	.97264	.24925	.96844	.26612	.96394	.28290	.95915
27	.23260	.97257	.24954	.96837	.26640	.96386	.28318	.95907
28	.23288	.97251	.24982	.96829	.26668	.96379	.28346	.95898
29	.23316	.97244	.25010	.96822	.26696	.96371	.28374	.95890
30	.23345	.97237	.25038	.96815	.26724	.96363	.28402	.95882
31	.23373	.97230	.25066	.96807	.26752	.96355	.28429	.95874
32	.23401	.97223	.25094	.96800	.26780	.96347	.28457	.95865
33	.23429	.97217	.25122	.96793	.26808	.96340	.28485	.95857
34	.23458	.97210	.25151	.96786	.26836	.96332	.28513	.95849
35	.23486	.97203	.25179	.96778	.26864	.96324	.28541	.95841
36	.23514	.97196	.25207	.96771	.26892	.96316	.28569	.95832
37	.23542	.97189	.25235	.96764	.26920	.96308	.28597	.95824
38	.23571	.97182	.25263	.96756	.26948	.96301	.28625	.95816
39	.23599	.97176	.25291	.96749	.26976	.96293	.28652	.95807
40	.23627	.97169	.25320	.96742	.27004	.96285	.28680	.95799
41	.23656	.97162	.25348	.96734	.27032	.96277	.28708	.95791
42	.23684	.97155	.25376	.96727	.27060	.96269	.28736	.95782
43	.23712	.97148	.25404	.96719	.27088	.96261	.28764	.95774
44	.23740	.97141	.25432	.96712	.27116	.96253	.28792	.95766
45	.23769	.97134	.25460	.96705	.27144	.96246	.28820	.95757
46	.23797	.97127	.25488	.96697	.27172	.96238	.28847	.95749
47	.23825	.97120	.25516	.96690	.27200	.96230	.28875	.95740
48	.23853	.97113	.25545	.96682	.27228	.96222	.28903	.95732
49	.23882	.97106	.25573	.96675	.27256	.96214	.28931	.95724
50	.23910	.97100	.25601	.96667	.27284	.96206	.28959	.95715
51	.23938	.97093	.25629	.96660	.27312	.96198	.28987	.95707
52	.23966	.97086	.25657	.96653	.27340	.96190	.29015	.95698
53	.23995	.97079	.25685	.96645	.27368	.96182	.29042	.95690
54	.24023	.97072	.25713	.96638	.27396	.96174	.29070	.95681
55	.24051	.97065	.25741	.96630	.27424	.96166	.29098	.95673
56	.24079	.97058	.25769	.96623	.27452	.96158	.29126	.95664
57	.24108	.97051	.25798	.96615	.27480	.96150	.29154	.95656
58	.24136	.97044	.25826	.96608	.27508	.96142	.29182	.95647
59	.24164	.97037	.25854	.96600	.27536	.96134	.29209	.95639
60	.24192	.97030	.25882	.96593	.27564	.96126	.29237	.95630

′	COSINE	SINE	COSINE	SINE	COSINE	SINE	COSINE	SINE
	76°		75°		74°		73°	

	17°		18°		19°		20°	
	Sine	Cosine	Sine	Cosine	Sine	Cosine	Sine	Cosine	′
	.29237	.95630	.30902	.95106	.32557	.94552	.34202	.93969	60
	.29265	.95622	.30929	.95097	.32584	.94542	.34229	.93959	59
	.29293	.95613	.30957	.95088	.32612	.94533	.34257	.93949	58
	.29321	.95605	.30985	.95079	.32639	.94523	.34284	.93939	57
	.29348	.95596	.31012	.95070	.32667	.94514	.34311	.93929	56
	.29376	.95588	.31040	.95061	.32694	.94504	.34339	.93919	55
	.29404	.95579	.31068	.95052	.32722	.94495	.34366	.93909	54
	.29432	.95571	.31095	.95043	.32749	.94485	.34393	.93899	53
	.29460	.95562	.31123	.95033	.32777	.94476	.34421	.93889	52
	.29487	.95554	.31151	.95024	.32804	.94466	.34448	.93879	51
	.29515	.95545	.31178	.95015	.32832	.94457	.34475	.93869	50
	.29543	.95536	.31206	.95006	.32859	.94447	.34503	.93859	49
	.29571	.95528	.31233	.94997	.32887	.94438	.34530	.93849	48
	.29599	.95519	.31261	.94988	.32914	.94428	.34557	.93839	47
	.29626	.95511	.31289	.94979	.32942	.94418	.34584	.93829	46
	.29654	.95502	.31316	.94970	.32969	.94409	.34612	.93819	45
	.29682	.95493	.31344	.94961	.32997	.94399	.34639	.93809	44
	.29710	.95485	.31372	.94952	.33024	.94390	.34666	.93799	43
	.29737	.95476	.31399	.94943	.33051	.94380	.34694	.93789	42
	.29765	.95467	.31427	.94933	.33079	.94370	.34721	.93779	41
	.29793	.95459	.31454	.94924	.33106	.94361	.34748	.93769	40
	.29821	.95450	.31482	.94915	.33134	.94351	.34775	.93759	39
	.29849	.95441	.31510	.94906	.33161	.94342	.34803	.93748	38
	.29876	.95433	.31537	.94897	.33189	.94332	.34830	.93738	37
	.29904	.95424	.31565	.94888	.33216	.94322	.34857	.93728	36
	.29932	.95415	.31593	.94878	.33244	.94313	.34884	.93718	35
	.29960	.95407	.31620	.94869	.33271	.94303	.34912	.93708	34
	.29987	.95398	.31648	.94860	.33298	.94293	.34939	.93698	33
	.30015	.95389	.31675	.94851	.33326	.94284	.34966	.93688	32
	.30043	.95380	.31703	.94842	.33353	.94274	.34993	.93677	31
	.30071	.95372	.31730	.94832	.33381	.94264	.35021	.93667	30
	.30098	.95363	.31758	.94823	.33408	.94254	.35048	.93657	29
	.30126	.95354	.31786	.94814	.33436	.94245	.35075	.93647	28
	.30154	.95345	.31813	.94805	.33463	.94235	.35102	.93637	27
	.30182	.95337	.31841	.94795	.33490	.94225	.35130	.93626	26
	.30209	.95328	.31868	.94786	.33518	.94215	.35157	.93616	25
	.30237	.95319	.31896	.94777	.33545	.94206	.35184	.93606	24
	.30265	.95310	.31923	.94768	.33573	.94196	.35211	.93596	23
	.30292	.95301	.31951	.94758	.33600	.94186	.35239	.93585	22
	.30320	.95293	.31979	.94749	.33627	.94176	.35266	.93575	21
	.30348	.95284	.32006	.94740	.33655	.94167	.35293	.93565	20
	.30376	.95275	.32034	.94730	.33682	.94157	.35320	.93555	19
	.30403	.95266	.32061	.94721	.33710	.94147	.35347	.93544	18
	.30431	.95257	.32089	.94712	.33737	.94137	.35375	.93534	17
	.30459	.95248	.32116	.94702	.33764	.94127	.35402	.93524	16
	.30486	.95240	.32144	.94693	.33792	.94118	.35429	.93514	15
	.30514	.95231	.32171	.94684	.33819	.94108	.35456	.93503	14
	.30542	.95222	.32199	.94674	.33846	.94098	.35484	.93493	13
	.30570	.95213	.32227	.94665	.33874	.94088	.35511	.93483	12
	.30597	.95204	.32254	.94656	.33901	.94078	.35538	.93472	11
	.30625	.95195	.32282	.94646	.33929	.94068	.35565	.93462	10
	.30653	.95186	.32309	.94637	.33956	.94058	.35592	.93452	9
	.30680	.95177	.32337	.94627	.33983	.94049	.35619	.93441	8
	.30708	.95168	.32364	.94618	.34011	.94039	.35647	.93431	7
	.30736	.95159	.32392	.94609	.34038	.94029	.35674	.93420	6
	.30763	.95150	.32419	.94599	.34065	.94019	.35701	.93410	5
	.30791	.95142	.32447	.94590	.34093	.94009	.35728	.93400	4
	.30819	.95133	.32474	.94580	.34120	.93999	.35755	.93389	3
	.30846	.95124	.32502	.94571	.34147	.93989	.35782	.93379	2
	.30874	.95115	.32529	.94561	.34175	.93979	.35810	.93368	1
	.30902	.95106	.32557	.94552	.34202	.93969	.35837	.93358	0
	Cosine	Sine	Cosine	Sine	Cosine	Sine	Cosine	Sine	′
		72°		71°		70°		69°	

′	21° SINE	COSINE	22° SINE	COSINE	23° SINE	COSINE	24° SINE	COSINE	′
0	.35837	.93358	.37461	.92718	.39073	.92050	.40674	.91355	60
1	.35864	.93348	.37488	.92707	.39100	.92039	.40700	.91343	59
2	.35891	.93337	.37515	.92697	.39127	.92028	.40727	.91331	58
3	.35918	.93327	.37542	.92686	.39153	.92016	.40753	.91319	57
4	.35945	.93316	.37569	.92675	.39180	.92005	.40780	.91307	56
5	.35973	.93306	.37595	.92664	.39207	.91994	.40806	.91295	55
6	.36000	.93295	.37622	.92653	.39234	.91982	.40833	.91283	54
7	.36027	.93285	.37649	.92642	.39260	.91971	.40860	.91272	53
8	.36054	.93274	.37676	.92631	.39287	.91959	.40886	.91260	52
9	.36081	.93264	.37703	.92620	.39314	.91948	.40913	.91248	51
10	.36108	.93253	.37730	.92609	.39341	.91936	.40939	.91236	50
11	.36135	.93243	.37757	.92598	.39367	.91925	.40966	.91224	49
12	.36162	.93232	.37784	.92587	.39394	.91914	.40992	.91212	48
13	.36190	.93222	.37811	.92576	.39421	.91902	.41019	.91200	47
14	.36217	.93211	.37838	.92565	.39448	.91891	.41045	.91188	46
15	.36244	.93201	.37865	.92554	.39474	.91879	.41072	.91176	45
16	.36271	.93190	.37892	.92543	.39501	.91868	.41098	.91164	44
17	.36298	.93180	.37919	.92532	.39528	.91856	.41125	.91152	43
18	.36325	.93169	.37946	.92521	.39555	.91845	.41151	.91140	42
19	.36352	.93159	.37973	.92510	.39581	.91833	.41178	.91128	41
20	.36379	.93148	.37999	.92499	.39608	.91822	.41204	.91116	40
21	.36406	.93137	.38026	.92488	.39635	.91810	.41231	.91104	39
22	.36434	.93127	.38053	.92477	.39661	.91799	.41257	.91092	38
23	.36461	.93116	.38080	.92466	.39688	.91787	.41284	.91080	37
24	.36488	.93106	.38107	.92455	.39715	.91775	.41310	.91068	36
25	.36515	.93095	.38134	.92444	.39741	.91764	.41337	.91056	35
26	.36542	.93084	.38161	.92432	.39768	.91752	.41363	.91044	34
27	.36569	.93074	.38188	.92421	.39795	.91741	.41390	.91032	33
28	.36596	.93063	.38215	.92410	.39822	.91729	.41416	.91020	32
29	.36623	.93052	.38241	.92399	.39848	.91718	.41443	.91008	31
30	.36650	.93042	.38268	.92388	.39875	.91706	.41469	.90996	30
31	.36677	.93031	.38295	.92377	.39902	.91694	.41496	.90984	29
32	.36704	.93020	.38322	.92366	.39928	.91683	.41522	.90972	28
33	.36731	.93010	.38349	.92355	.39955	.91671	.41549	.90960	27
34	.36758	.92999	.38376	.92343	.39982	.91660	.41575	.90948	26
35	.36785	.92988	.38403	.92332	.40008	.91648	.41602	.90936	25
36	.36812	.92978	.38430	.92321	.40035	.91636	.41628	.90924	24
37	.36839	.92967	.38456	.92310	.40062	.91625	.41655	.90911	23
38	.36867	.92956	.38483	.92299	.40088	.91613	.41681	.90899	22
39	.36894	.92945	.38510	.92287	.40115	.91601	.41707	.90887	21
40	.36921	.92935	.38537	.92276	.40141	.91590	.41734	.90875	20
41	.36948	.92924	.38564	.92265	.40168	.91578	.41760	.90863	19
42	.36975	.92913	.38591	.92254	.40195	.91566	.41787	.90851	18
43	.37002	.92902	.38617	.92243	.40221	.91555	.41813	.90839	17
44	.37029	.92892	.38644	.92231	.40248	.91543	.41840	.90826	16
45	.37056	.92881	.38671	.92220	.40275	.91531	.41866	.90814	15
46	.37083	.92870	.38698	.92209	.40301	.91519	.41892	.90802	14
47	.37110	.92859	.38725	.92198	.40328	.91508	.41919	.90790	13
48	.37137	.92849	.38752	.92186	.40355	.91496	.41945	.90778	12
49	.37164	.92838	.38778	.92175	.40381	.91484	.41972	.90766	11
50	.37191	.92827	.38805	.92164	.40408	.91472	.41998	.90753	10
51	.37218	.92816	.38832	.92152	.40434	.91461	.42024	.90741	9
52	.37245	.92805	.38859	.92141	.40461	.91449	.42051	.90729	8
53	.37272	.92794	.38886	.92130	.40488	.91437	.42077	.90717	7
54	.37299	.92784	.38912	.92119	.40514	.91425	.42104	.90704	6
55	.37326	.92773	.38939	.92107	.40541	.91414	.42130	.90692	5
56	.37353	.92762	.38966	.92096	.40567	.91402	.42156	.90680	4
57	.37380	.92751	.38993	.92085	.40594	.91390	.42183	.90668	3
58	.37407	.92740	.39020	.92073	.40621	.91378	.42209	.90655	2
59	.37434	.92729	.39046	.92062	.40647	.91366	.42235	.90643	1
60	.37461	.92718	.39073	.92050	.40674	.91355	.42262	.90631	0
′	COSINE	SINE	COSINE	SINE	COSINE	SINE	COSINE	SINE	
	68°		67°		66°		65°		

	25°		26°		27°		28°	
	SINE	COSINE	SINE	COSINE	SINE	COSINE	SINE	COSINE	'
.42262	.90631	.43837	.89879	.45399	.89101	.46947	.88295	60
.42288	.90618	.43863	.89867	.45425	.89087	.46973	.88281	59
.42315	.90606	.43889	.89854	.45451	.89074	.46999	.88267	58
.42341	.90594	.43916	.89841	.45477	.89061	.47024	.88254	57
.42367	.90582	.43942	.89828	.45503	.89048	.47050	.88240	56
.42394	.90569	.43968	.89816	.45529	.89035	.47076	.88226	55
.42420	.90557	.43994	.89803	.45554	.89021	.47101	.88213	54
.42446	.90545	.44020	.89790	.45580	.89008	.47127	.88199	53
.42473	.90532	.44046	.89777	.45606	.88995	.47153	.88185	52
.42499	.90520	.44072	.89764	.45632	.88981	.47178	.88172	51
.42525	.90507	.44098	.89752	.45658	.88968	.47204	.88158	50
.42552	.90495	.44124	.89739	.45684	.88955	.47229	.88144	49
.42578	.90483	.44151	.89726	.45710	.88942	.47255	.88130	48
.42604	.90470	.44177	.89713	.45736	.88928	.47281	.88117	47
.42631	.90458	.44203	.89700	.45762	.88915	.47306	.88103	46
.42657	.90446	.44229	.89687	.45787	.88902	.47332	.88089	45
.42683	.90433	.44255	.89674	.45813	.88888	.47358	.88075	44
.42709	.90421	.44281	.89662	.45839	.88875	.47383	.88062	43
.42736	.90408	.44307	.89649	.45865	.88862	.47409	.88048	42
.42762	.90396	.44333	.89636	.45891	.88848	.47434	.88034	41
.42788	.90383	.44359	.89623	.45917	.88835	.47460	.88020	40
.42815	.90371	.44385	.89610	.45942	.88822	.47486	.88006	39
.42841	.90358	.44411	.89597	.45968	.88808	.47511	.87993	38
.42867	.90346	.44437	.89584	.45994	.88795	.47537	.87979	37
.42894	.90334	.44464	.89571	.46020	.88782	.47562	.87965	36
.42920	.90321	.44490	.89558	.46046	.88768	.47588	.87951	35
.42946	.90309	.44516	.89545	.46072	.88755	.47614	.87937	34
.42972	.90296	.44542	.89532	.46097	.88741	.47639	.87923	33
.42999	.90284	.44568	.89519	.46123	.88728	.47665	.87909	32
.43025	.90271	.44594	.89506	.46149	.88715	.47690	.87896	31
.43051	.90259	.44620	.89493	.46175	.88701	.47716	.87882	30
.43077	.90246	.44646	.89480	.46201	.88688	.47741	.87868	29
.43104	.90233	.44672	.89467	.46226	.88674	.47767	.87854	28
.43130	.90221	.44698	.89454	.46252	.88661	.47793	.87840	27
.43156	.90208	.44724	.89441	.46278	.88647	.47818	.87826	26
.43182	.90196	.44750	.89428	.46304	.88634	.47844	.87812	25
.43209	.90183	.44776	.89415	.46330	.88620	.47869	.87798	24
.43235	.90171	.44802	.89402	.46355	.88607	.47895	.87784	23
.43261	.90158	.44828	.89389	.46381	.88593	.47920	.87770	22
.43287	.90146	.44854	.89376	.46407	.88580	.47946	.87756	21
.43313	.90133	.44880	.89363	.46433	.88566	.47971	.87743	20
.43340	.90120	.44906	.89350	.46458	.88553	.47997	.87729	19
.43366	.90108	.44932	.89337	.46484	.88539	.48022	.87715	18
.43392	.90095	.44958	.89324	.46510	.88526	.48048	.87701	17
.43418	.90082	.44984	.89311	.46536	.88512	.48073	.87687	16
.43445	.90070	.45010	.89298	.46561	.88499	.48099	.87673	15
.43471	.90057	.45036	.89285	.46587	.88485	.48124	.87659	14
.43497	.90045	.45062	.89272	.46613	.88472	.48150	.87645	13
.43523	.90032	.45088	.89259	.46639	.88458	.48175	.87631	12
.43549	.90019	.45114	.89245	.46664	.88445	.48201	.87617	11
.43575	.90007	.45140	.89232	.46690	.88431	.48226	.87603	10
.43602	.89994	.45166	.89219	.46716	.88417	.48252	.87589	9
.43628	.89981	.45192	.89206	.46742	.88404	.48277	.87575	8
.43654	.89968	.45218	.89193	.46767	.88390	.48303	.87561	7
.43680	.89956	.45243	.89180	.46793	.88377	.48328	.87546	6
.43706	.89943	.45269	.89167	.46819	.88363	.48354	.87532	5
.43733	.89930	.45295	.89153	.46844	.88349	.48379	.87518	4
.43759	.89918	.45321	.89140	.46870	.88336	.48405	.87504	3
.43785	.89905	.45347	.89127	.46896	.88322	.48430	.87490	2
.43811	.89892	.45373	.89114	.46921	.88308	.48456	.87476	1
.43837	.89879	.45399	.89101	.46947	.88295	.48481	.87462	0
COSINE	SINE	COSINE	SINE	COSINE	SINE	COSINE	SINE	'
	64°		63°		62°		61°	

′	29° Sine	Cosine	30° Sine	Cosine	31° Sine	Cosine	32° Sine	Cosine
0	.48481	.87462	.50000	.86603	.51504	.85717	.52992	.84805
1	.48506	.87448	.50025	.86588	.51529	.85702	.53017	.84789
2	.48532	.87434	.50050	.86573	.51554	.85687	.53041	.84774
3	.48557	.87420	.50076	.86559	.51579	.85672	.53066	.84759
4	.48583	.87406	.50101	.86544	.51604	.85657	.53091	.84743
5	.48608	.87391	.50126	.86530	.51628	.85642	.53115	.84728
6	.48634	.87377	.50151	.86515	.51653	.85627	.53140	.84712
7	.48659	.87363	.50176	.86501	.51678	.85612	.53164	.84697
8	.48684	.87349	.50201	.86486	.51703	.85597	.53189	.84681
9	.48710	.87335	.50227	.86471	.51728	.85582	.53214	.84666
10	.48735	.87321	.50252	.86457	.51753	.85567	.53238	.84650
11	.48761	.87306	.50277	.86442	.51778	.85551	.53263	.84635
12	.48786	.87292	.50302	.86427	.51803	.85536	.53288	.84619
13	.48811	.87278	.50327	.86413	.51828	.85521	.53312	.84604
14	.48837	.87264	.50352	.86398	.51852	.85506	.53337	.84588
15	.48862	.87250	.50377	.86384	.51877	.85491	.53361	.84573
16	.48888	.87235	.50403	.86369	.51902	.85476	.53386	.84557
17	.48913	.87221	.50428	.86354	.51927	.85461	.53411	.84542
18	.48938	.87207	.50453	.86340	.51952	.85446	.53435	.84526
19	.48964	.87193	.50478	.86325	.51977	.85431	.53460	.84511
20	.48989	.87178	.50503	.86310	.52002	.85416	.53484	.84495
21	.49014	.87164	.50528	.86295	.52026	.85401	.53509	.84480
22	.49040	.87150	.50553	.86281	.52051	.85385	.53534	.84464
23	.49065	.87136	.50578	.86266	.52076	.85370	.53558	.84448
24	.49090	.87121	.50603	.86251	.52101	.85355	.53583	.84433
25	.49116	.87107	.50628	.86237	.52126	.85340	.53607	.84417
26	.49141	.87093	.50654	.86222	.52151	.85325	.53632	.84402
27	.49166	.87079	.50679	.86207	.52175	.85310	.53656	.84386
28	.49192	.87064	.50704	.86192	.52200	.85294	.53681	.84370
29	.49217	.87050	.50729	.86178	.52225	.85279	.53705	.84355
30	.49242	.87036	.50754	.86163	.52250	.85264	.53730	.84339
31	.49268	.87021	.50779	.86148	.52275	.85249	.53754	.84324
32	.49293	.87007	.50804	.86133	.52299	.85234	.53779	.84308
33	.49318	.86993	.50829	.86119	.52324	.85218	.53804	.84292
34	.49344	.86978	.50854	.86104	.52349	.85203	.53828	.84277
35	.49369	.86964	.50879	.86089	.52374	.85188	.53853	.84261
36	.49394	.86949	.50904	.86074	.52399	.85173	.53877	.84245
37	.49419	.86935	.50929	.86059	.52423	.85157	.53902	.84230
38	.49445	.86921	.50954	.86045	.52448	.85142	.53926	.84214
39	.49470	.86906	.50979	.86030	.52473	.85127	.53951	.84198
40	.49495	.86892	.51004	.86015	.52498	.85112	.53975	.84182
41	.49521	.86878	.51029	.86000	.52522	.85096	.54000	.84167
42	.49546	.86863	.51054	.85985	.52547	.85081	.54024	.84151
43	.49571	.86849	.51079	.85970	.52572	.85066	.54049	.84135
44	.49596	.86834	.51104	.85956	.52597	.85051	.54073	.84120
45	.49622	.86820	.51129	.85941	.52621	.85035	.54097	.84104
46	.49647	.86805	.51154	.85926	.52646	.85020	.54122	.84088
47	.49672	.86791	.51179	.85911	.52671	.85005	.54146	.84072
48	.49697	.86777	.51204	.85896	.52696	.84989	.54171	.84057
49	.49723	.86762	.51229	.85881	.52720	.84974	.54195	.84041
50	.49748	.86748	.51254	.85866	.52745	.84959	.54220	.84025
51	.49773	.86733	.51279	.85851	.52770	.84943	.54244	.84009
52	.49798	.86719	.51304	.85836	.52794	.84928	.54269	.83994
53	.49824	.86704	.51329	.85821	.52819	.84913	.54293	.83978
54	.49849	.86690	.51354	.85806	.52844	.84897	.54317	.83962
55	.49874	.86675	.51379	.85792	.52869	.84882	.54342	.83946
56	.49899	.86661	.51404	.85777	.52893	.84866	.54366	.83930
57	.49924	.86646	.51429	.85762	.52918	.84851	.54391	.83915
58	.49950	.86632	.51454	.85747	.52943	.84836	.54415	.83899
59	.49975	.86617	.51479	.85732	.52967	.84820	.54440	.83883
60	.50000	.86603	.51504	.85717	.52992	.84805	.54464	.83867

′	Cosine	Sine	Cosine	Sine	Cosine	Sine	Cosine	Sine
	60°		59°		58°		57°	

33° Sine	Cosine	34° Sine	Cosine	35° Sine	Cosine	36° Sine	Cosnie	'
.54464	.83867	.55919	.82904	.57358	.81915	.58779	.80902	60
.54488	.83851	.55943	.82887	.57381	.81899	.58802	.80885	59
.54513	.83835	.55968	.82871	.57405	.81882	.58826	.80867	58
.54537	.83819	.55992	.82855	.57429	.81865	.58849	.80850	57
.54561	.83804	.56016	.82839	.57453	.81848	.58873	.80833	56
.54586	.83788	.56040	.82822	.57477	.81832	.58896	.80816	55
.54610	.83772	.56064	.82806	.57501	.81815	.58920	.80799	54
.54635	.83756	.56088	.82790	.57524	.81798	.58943	.80782	53
.54659	.83740	.56112	.82773	.57548	.81782	.58967	.80765	52
.54683	.83724	.56136	.82757	.57572	.81765	.58990	.80748	51
.54708	.83708	.56160	.82741	.57596	.81748	.59014	.80730	50
.54732	83692	.56184	.82724	.57619	.81731	.59037	.80713	49
.54756	.83676	.56208	.82708	.57643	.81714	.59061	.80696	48
.54781	.83660	.56232	.82692	.57667	.81698	.59084	.80679	47
.54805	.83645	.56256	.82675	.57691	.81681	.59108	.80662	46
.54829	.83629	.56280	.82659	.57715	.81664	.59131	.80644	45
.54854	.83613	.56305	.82643	.57738	.81647	.59154	.80627	44
.54878	.83597	.56329	.82626	.57762	.81631	.59178	.80610	43
.54902	.83581	.56353	.82610	.57786	.81614	.59201	.80593	42
.54927	.83565	.56377	.82593	.57810	.81597	.59225	.80576	41
.54951	.83549	.56401	.82577	.57833	.81580	.59248	.80558	40
.54975	.83533	.56425	.82561	.57857	.81563	.59272	.80541	39
.54999	.83517	.56449	.82544	.57881	.81546	.59295	.80524	38
.55024	.83501	.56473	.82528	.57904	.81530	.59318	.80507	37
.55048	.83485	.56497	.82511	.57928	.81513	.59342	.80489	36
.55072	.83469	.56521	.82495	.57952	.81496	.59365	.80472	35
.55097	.83453	.56545	.82478	.57976	.81479	.59389	.80455	34
.55121	.83437	.56569	.82462	.57999	.81462	.59412	.80438	33
.55145	.83421	.56593	.82446	.58023	.81445	.59436	.80420	32
.55169	.83405	.56617	.82429	.58047	.81428	.59459	.80403	31
.55194	.83389	.56641	.82413	.58070	.81412	.59482	.80386	30
.55218	.83373	.56665	.82396	.58094	.81395	.59506	.80368	29
.55242	.83356	.56689	.82380	.58118	.81378	.59529	.80351	28
.55266	.83340	.56713	.82363	.58141	.81361	.59552	.80334	27
.55291	.83324	.56736	.82347	.58165	.81344	.59576	.80316	26
.55315	.83308	.56760	.82330	.58189	.81327	.59599	.80299	25
.55339	.83292	.56784	.82314	.58212	.81310	.59622	.80282	24
.55363	.83276	.56808	.82297	.58236	.81293	.59646	.80264	23
.55388	.83260	.56832	.82281	.58260	.81276	.59669	.80247	22
.55412	.83244	.56856	.82264	.58283	.81259	.59693	.80230	21
.55436	.83228	.56880	.82248	.58307	.81242	.59716	.80212	20
.55460	.83212	.56904	.82231	.58330	.81225	.59739	.80195	19
.55484	.83195	.56928	.82214	.58354	.81208	.59763	.80178	18
.55509	.83179	.56952	.82198	.58378	.81191	.59786	.80160	17
.55533	.83163	.56976	.82181	.58401	.81174	.59809	.80143	16
.55557	.83147	.57000	.82165	.58425	.81157	.59832	.80125	15
.55581	.83131	.57024	.82148	.58449	.81140	.59856	.80108	14
.55605	.83115	.57047	.82132	.58472	.81123	.59879	.80091	13
.55630	.83098	.57071	.82115	.58496	.81106	.59902	.80073	12
.55654	.83082	.57095	.82098	.58519	.81089	.59926	.80056	11
.55678	.83066	.57119	.82082	.58543	.81072	.59949	.80038	10
.55702	.83050	.57143	.82065	.58567	.81055	.59972	.80021	9
.55726	.83034	.57167	.82048	.58590	.81038	.59995	.80003	8
.55750	.83017	.57191	.82032	.58614	.81021	.60019	.79986	7
.55775	.83001	.57215	.82015	.58637	.81004	.60042	.79968	6
.55799	.82985	.57238	.81999	.58661	.80987	.60065	.79951	5
.55823	.82969	.57262	.81982	.58684	.80970	.60089	.79934	4
.55847	.82953	.57286	.81965	.58708	.80953	.60112	.79916	3
.55871	.82936	.57310	.81949	.58731	.80936	.60135	.79899	2
.55895	.82920	.57334	.81932	.58755	.80919	.60158	.79881	1
.55919	.82904	.57358	.81915	.58779	.80902	.60182	.79864	0
Cosine	Sine	Cosine	Sine	Cosine	Sine	Cosine	Sine	'
	56°		55°		54°		53°	

′	37° Sine	Cosine	38° Sine	Cosine	39° Sine	Cosine	40° Sine	Cosine	′
0	.60182	.79864	.61566	.78801	.62932	.77715	.64279	.76604	60
1	.60205	.79846	.61589	.78783	.62955	.77696	.64301	.76586	59
2	.60228	.79829	.61612	.78765	.62977	.77678	.64323	.76567	58
3	.60251	.79811	.61635	.78747	.63000	.77660	.64346	.76548	57
4	.60274	.79793	.61658	.78729	.63022	.77641	.64368	.76530	56
5	.60298	.79776	.61681	.78711	.63045	.77623	.64390	.76511	55
6	.60321	.79758	.61704	.78694	.63068	.77605	.64412	.76492	54
7	.60344	.79741	.61726	.78676	.63090	.77586	.64435	.76473	53
8	.60367	.79723	.61749	.78658	.63113	.77568	.64457	.76455	52
9	.60390	.79706	.61772	.78640	.63135	.77550	.64479	.76436	51
10	.60414	.79688	.61795	.78622	.63158	.77531	.64501	.76417	50
11	.60437	.79671	.61818	.78604	.63180	.77513	.64524	.76398	49
12	.60460	.79653	.61841	.78586	.63203	.77494	.64546	.76380	48
13	.60483	.79635	.61864	.78568	.63225	.77476	.64568	.76361	47
14	.60506	.79618	.61887	.78550	.63248	.77458	.64590	.76342	46
15	.60529	.79600	.61909	.78532	.63271	.77439	.64612	.76323	45
16	.60553	.79583	.61932	.78514	.63293	.77421	.64635	.76304	44
17	.60576	.79565	.61955	.78496	.63316	.77402	.64657	.76286	43
18	.60599	.79547	.61978	.78478	.63338	.77384	.64679	.76267	42
19	.60622	.79530	.62001	.78460	.63361	.77366	.64701	.76248	41
20	.60645	.79512	.62024	.78442	.63383	.77347	.64723	.76229	40
21	.60668	.79494	.62046	.78424	.63406	.77329	.64746	.76210	39
22	.60691	.79477	.62069	.78405	.63428	.77310	.64768	.76192	38
23	.60714	.79459	.62092	.78387	.63451	.77292	.64790	.76173	37
24	.60738	.79441	.62115	.78369	.63473	.77273	.64812	.76154	36
25	.60761	.79424	.62138	.78351	.63496	.77255	.64834	.76135	35
26	.60784	.79406	.62160	.78333	.63518	.77236	.64856	.76116	34
27	.60807	.79388	.62183	.78315	.63540	.77218	.64878	.76097	33
28	.60830	.79371	.62206	.78297	.63563	.77199	.64901	.76078	32
29	.60853	.79353	.62229	.78279	.63585	.77181	.64923	.76059	31
30	.60876	.79335	.62251	.78261	.63608	.77162	.64945	.76041	30
31	.60899	.79318	.62274	.78243	.63630	.77144	.64967	.76022	29
32	.60922	.79300	.62297	.78225	.63653	.77125	.64989	.76003	28
33	.60945	.79282	.62320	.78206	.63675	.77107	.65011	.75984	27
34	.60968	.79264	.62342	.78188	.63698	.77088	.65033	.75965	26
35	.60991	.79247	.62365	.78170	.63720	.77070	.65055	.75946	25
36	.61015	.79229	.62388	.78152	.63742	.77051	.65077	.75927	24
37	.61038	.79211	.62411	.78134	.63765	.77033	.65100	.75908	23
38	.61061	.79193	.62433	.78116	.63787	.77014	.65122	.75889	22
39	.61084	.79176	.62456	.78098	.63810	.76996	.65144	.75870	21
40	.61107	.79158	.62479	.78079	.63832	.76977	.65166	.75851	20
41	.61130	.79140	.62502	.78061	.63854	.76959	.65188	.75832	19
42	.61153	.79122	.62524	.78043	.63877	.76940	.65210	.75813	18
43	.61176	.79105	.62547	.78025	.63899	.76921	.65232	.75794	17
44	.61199	.79087	.62570	.78007	.63922	.76903	.65254	.75775	16
45	.61222	.79069	.62592	.77988	.63944	.76884	.65276	.75756	15
46	.61245	.79051	.62615	.77970	.63966	.76866	.65298	.75738	14
47	.61268	.79033	.62638	.77952	.63989	.76847	.65320	.75719	13
48	.61291	.79016	.62660	.77934	.64011	.76828	.65342	.75700	12
49	.61314	.78998	.62683	.77916	.64033	.76810	.65364	.75680	11
50	.61337	.78980	.62706	.77897	.64056	.76791	.65386	.75661	10
51	.61360	.78962	.62728	.77879	.64078	.76772	.65408	.75642	9
52	.61383	.78944	.62751	.77861	.64100	.76754	.65430	.75623	8
53	.61406	.78926	.62774	.77843	.64123	.76735	.65452	.75604	7
54	.61429	.78908	.62796	.77824	.64145	.76717	.65474	.75585	6
55	.61451	.78891	.62819	.77806	.64167	.76698	.65496	.75566	5
56	.61474	.78873	.62842	.77788	.64190	.76679	.65518	.75547	4
57	.61497	.78855	.62864	.77769	.64212	.76661	.65540	.75528	3
58	.61520	.78837	.62887	.77751	.64234	.76642	.65562	.75509	2
59	.61543	.78819	.62909	.77733	.64256	.76623	.65584	.75490	1
60	.61566	.78801	.62932	.77715	.64279	.76604	.65606	.75471	0
′	Cosine	Sine	Cosine	Sine	Cosine	Sine	Cosine	Sine	′

52°	51°	50°	49°

41° SINE	41° COSINE	42° SINE	42° COSINE	43° SINE	43° COSINE	44° SINE	44° COSINE	'
.65606	.75471	.66913	.74314	.68200	.73135	.69466	.71934	60
.65628	.75452	.66935	.74295	.68221	.73116	.69487	.71914	59
.65650	.75433	.66956	.74276	.68242	.73096	.69508	.71894	58
.65672	.75414	.66978	.74256	.68264	.73076	.69529	.71873	57
.65694	.75395	.66999	.74237	.68285	.73056	.69549	.71853	56
.65716	.75375	.67021	.74217	.68306	.73036	.69570	.71833	55
.65738	.75356	.67043	.74198	.68327	.73016	.69591	.71813	54
.65759	.75337	.67064	.74178	.68349	.72996	.69612	.71792	53
.65781	.75318	.67086	.74159	.68370	.72976	.69633	.71772	52
.65803	.75299	.67107	.74139	.68391	.72957	.69654	.71752	51
.65825	.75280	.67129	.74120	.68412	.72937	.69675	.71732	50
.65847	.75261	.67151	.74100	.68434	.72917	.69696	.71711	49
.65869	.75241	.67172	.74080	.68455	.72897	.69717	.71691	48
.65891	.75222	.67194	.74061	.68476	.72877	.69737	.71671	47
.65913	.75203	.67215	.74041	.68497	.72857	.69758	.71650	46
.65935	.75184	.67237	.74022	.68518	.72837	.69779	.71630	45
.65956	.75165	.67258	.74002	.68539	.72817	.69800	.71610	44
.65978	.75146	.67280	.73983	.68561	.72797	.69821	.71590	43
.66000	.75126	.67301	.73963	.68582	.72777	.69842	.71569	42
.66022	.75107	.67323	.73044	.68603	.72757	.69862	.71549	41
.66044	.75088	.67344	.73924	.68624	.72737	.69883	.71529	40
.66066	.75069	.67366	.73904	.68645	.72717	.69904	.71508	39
.66088	.75050	.67387	.73885	.68666	.72697	.69925	.71488	38
.66109	.75030	.67409	.73865	.68688	.72677	.69946	.71468	37
.66131	.75011	.67430	.73846	.68709	.72657	.69966	.71447	36
.66153	.74992	.67452	.73826	.68730	.72637	.69987	.71427	35
.66175	.74973	.67473	.73806	.68751	.72617	.70008	.71407	34
.66197	.74953	.67495	.73787	.68772	.72597	.70029	.71386	33
.66218	.74934	.67516	.73767	.68793	.72577	.70049	.71366	32
.66240	.74915	.67538	.73747	.68814	.72557	.70070	.71345	31
.66262	.74896	.67559	.73728	.68835	.72537	.70091	.71325	30
.66284	.74876	.67580	.73708	.68857	.72517	.70112	.71305	29
.66306	.74857	.67602	.73688	.68878	.72497	.70132	.71284	28
.66327	.74838	.67623	.73669	.68899	.72477	.70153	.71264	27
.66349	.74818	.67645	.73649	.68920	.72457	.70174	.71243	26
.66371	.74799	.67666	.73629	.68941	.72437	.70195	.71223	25
.66393	.74780	.67688	.73610	.68962	.72417	.70215	.71203	24
.66414	.74760	.67709	.73590	.68983	.72397	.70236	.71182	23
.66436	.74741	.67730	.73570	.69004	.72377	.70257	.71162	22
.66458	.74722	.67752	.73551	.69025	.72357	.70277	.71141	21
.66480	.74703	.67773	.73531	.69046	.72337	.70298	.71121	20
.66501	.74683	.67795	.73511	.69067	.72317	.70319	.71100	19
.66523	.74664	.67816	.73491	.69088	.72297	.70339	.71080	18
.66545	.74644	.67837	.73472	.69109	.72277	.70360	.71059	17
.66566	.74625	.67859	.73452	.69130	.72257	.70381	.71039	16
.66588	.74606	.67880	.73432	.69151	.72236	.70401	.71019	15
.66610	.74586	.67901	.73413	.69172	.72216	.70422	.70998	14
.66632	.74567	.67923	.73393	.69193	.72196	.70443	.70978	13
.66653	.74548	.67944	.73373	.69214	.72176	.70463	.70957	12
.66675	.74528	.67965	.73353	.69235	.72156	.70484	.70937	11
.66697	.74509	.67987	.73333	.69256	.72136	.70505	.70916	10
.66718	.74489	.68008	.73314	.69277	.72116	.70525	.70896	9
.66740	.74470	.68029	.73294	.69298	.72095	.70546	.70875	8
.66762	.74451	.68051	.73274	.69319	.72075	.70567	.70855	7
.66783	.74431	.68072	.73254	.69340	.72055	.70587	.70834	6
.66805	.74412	.68093	.73234	.69361	.72035	.70608	.70813	5
.66827	.74392	.68115	.73215	.69382	.72015	.70628	.70793	4
.66848	.74373	.68136	.73195	.69403	.71995	.70649	.70772	3
.66870	.74353	.68157	.73175	.69424	.71974	.70670	.70752	2
.66891	.74334	.68179	.73155	.69445	.71054	.70690	.70731	1
.66913	.74314	.68200	.73135	.69466	.71934	.70711	.70711	0
COSINE	SINE	COSINE	SINE	COSINE	SINE	COSINE	SINE	'
48°		47°		46°		45°		

′	SEC. (0°)	Co-sec. (0°)	SEC. (1°)	Co-sec. (1°)	SEC. (2°)	Co-sec. (2°)	SEC. (3°)	Co-sec. (3°)	′
0	1	Infinite.	1.0.01	57.299	1.0006	28.654	1.0014	19.107	60
1	1	3437.70	1.0001	56.359	1.0006	28.417	1.0014	19.002	59
2	1	1718.90	1.0002	55.450	1.0006	28.184	1.0014	18.897	58
3	1	1145.90	1.0002	54.570	1.0006	27.955	1.0014	18.794	57
4	1	859.44	1.0002	53.718	1.0006	27.730	1.0014	18.692	56
5	1	687.55	1.0002	52.891	1.0007	27.508	1.0014	18.591	55
6	1	572.96	1.0002	52.090	1.0007	27.290	1.0015	18.491	54
7	1	491.11	1.0002	51.313	1.0007	27.075	1.0015	18.393	53
8	1	429.72	1.0002	50.558	1.0007	26.864	1.0015	18.295	52
9	1	381.97	1.0002	49.826	1.0007	26.655	1.0015	18.198	51
10	1	343.77	1.0002	49.114	1.0007	26.450	1.0015	18.103	50
11	1	312.52	1.0002	48.422	1.0007	26.249	1.0015	18.008	49
12	1	286.48	1.0002	47.750	1.0007	26.050	1.0016	17.914	48
13	1	264.44	1.0002	47.096	1.0007	25.854	1.0016	17.821	47
14	1	245.55	1.0002	46.460	1.0008	25.661	1.0016	17.730	46
.15	1	229.18	1.0002	45.840	1.0008	25.471	1.0016	17.639	45
16	1	214.86	1.0002	45.237	1.0008	25.284	1.0016	17.549	44
17	1	202.22	1.0002	44.650	1.0008	25.100	1.0016	17.460	43
18	1	190.99	1.0002	44.077	1.0008	24.918	1.0017	17.372	42
19	1	180.73	1.0003	43.520	1.0008	24.739	1.0017	17.285	41
20	1	171.89	1.0003	42.976	1.0008	24.562	1.0017	17.198	40
21	1	163.70	1.0003	42.445	1.0008	24.358	1.0017	17.113	39
22	1	156.26	1.0003	41.928	1.0008	24.216	1.0017	17.028	38
23	1	149.47	1.0003	41.423	1.0009	24.047	1.0017	16.944	37
24	1	143.24	1.0003	40.930	1.0009	23.880	1.0018	16.861	36
25	1	137.51	1.0003	40.448	1.0009	23.716	1.0018	16.779	35
26	1	132.22	1.0003	39.978	1.0009	23.553	1.0018	16.698	34
27	1	127.32	1.0003	39.518	1.0009	23.393	1.0018	16.617	33
28	1	122.78	1.0003	39.069	1.0009	23.235	1.0018	16.538	32
29	1	118.54	1.0003	38.631	1.0009	23.079	1.0018	16.459	31
30	1	114.59	1.0003	38.201	1.0009	22.925	1.0019	16.380	30
31	1	110.90	1.0003	37.782	1.0010	22.774	1.0019	16.303	29
32	1	107.43	1.0003	37.371	1.0010	22.624	1.0019	16.226	28
33	1	104.17	1.0004	36.969	1.0010	22.476	1.0019	16.150	27
34	1	101.11	1.0004	36.576	1.0010	22.330	1.0019	16.075	26
35	1	90.223	1.0004	36.191	1.0010	22.186	1.0019	16.000	25
36	1	95.495	1.0004	35.814	1.0010	22.044	1.0020	15.926	24
37	1	92.914	1.0004	35.445	1.0010	21.904	1.0020	15.853	23
38	1.0001	92.469	1.0004	35.084	1.0010	21.765	1.0020	15.780	22
39	1.0001	88.149	1.0004	34.729	1.0011	21.629	1.0020	15.708	21
40	1.0001	85.946	1.0004	34.382	1.0011	21.494	1.0020	15.637	20
41	1.0001	83.849	1.0004	34.042	1.0011	21.360	1.0021	15.566	19
42	1.0001	81.853	1.0004	33.708	1.0011	21.228	1.0021	15.496	18
43	1.0001	79.950	1.0004	33.381	1.0011	21.098	1.0021	15.427	17
44	1.0001	78.133	1.0004	33.060	1.0011	20.970	1.0021	15.358	16
45	1.0001	76.396	1.0005	32.745	1.0011	20.843	1.0021	15.290	15
46	1.0001	74.736	1.0005	32.437	1.0012	20.717	1.0022	15.222	14
47	1.0001	73.146	1.0005	32.134	1.0012	20.593	1.0022	15.155	13
48	1.0001	71.622	1.0005	31.836	1.0012	20.471	1.0022	15.089	12
49	1.0001	71.160	1.0005	31.544	1.0012	20.350	1.0022	15.023	11
50	1.0001	68.757	1.0005	31.257	1.0012	20.230	1.0022	14.958	10
51	1.0001	67.409	1.0005	30.976	1.0012	20.112	1.0023	14.893	9
52	1.0001	66.113	1.0005	30.699	1.0012	19.995	1.0023	14.829	8
53	1.0001	64.866	1.0005	30.428	1.0013	19.880	1.0023	14.765	7
54	1.0001	63.664	1.0005	30.161	1.0013	19.766	1.0023	14.702	6
55	1.0001	62.507	1.0005	29.899	1.0013	19.653	1.0023	14.640	5
56	1.0001	61.391	1.0006	29.641	1.0013	19.541	1.0024	14.578	4
57	1.0001	61.314	1.0006	29.388	1.0013	19.431	1.0024	14.517	3
58	1.0001	59.274	1.0006	29.139	1.0013	19.322	1.0024	14.456	2
59	1.0001	58.270	1.0006	28.894	1.0013	19.214	1.0024	14.395	1
60	1.0001	57.299	1.0006	28.654	1.0014	19.107	1.0024	14.335	
′	Co-sec.	Sec. (89°)	Co-sec.	Sec. (88°)	Co-sec.	Sec. (87°)	Co-sec.	Sec. (86°)	

	4°		5°		6°		7°		'
	Sec.	Co-sec.	Sec.	Co-sec.	Sec.	Co-sec	Sec.	Co-sec.	
	1.0024	14.335	1.0038	11.474	1.0055	9.5668	1.0075	8.2055	60
	1.0025	14.276	1.0038	11.436	1.0055	9.5404	1.0075	8.1861	59
	1.0025	14.217	1.0039	11.398	1.0056	9.5141	1.0076	8.1668	58
	1.0025	14.159	1.0039	11.360	1.0056	9.4880	1.0076	8.1476	57
	1.0025	14.101	1.0039	11.323	1.0056	9.4620	1.0076	8.1285	56
	1.0025	14.043	1.0039	11.286	1.0057	9.4362	1.0077	8.1094	55
	1.0026	13.986	1.0040	11.249	1.0057	9.4105	1.0077	8.0905	54
	1.0026	13.930	1.0040	11.213	1.0057	9.3850	1.0078	8.0717	53
	1.0026	13.874	1.0040	11.176	1.0057	9.3596	1.0078	8.0529	52
	1.0026	13.818	1.0040	11.140	1.0058	9.3343	1.0078	8.0342	51
	1.0026	13.763	1.0041	11.104	1.0058	9.3092	1.0079	8.0156	50
	1.0027	13.708	1.0041	11.069	1.0058	9.2842	1.0079	7.9971	49
	1.0027	13.654	1.0041	11.033	1.0059	9.2593	1.0079	7.9787	48
	1.0027	13.600	1.0041	10.988	1.0059	9.2346	1.0080	7.9604	47
	1.0027	13.547	1.0042	10.963	1.0059	9.2100	1.0080	7.9421	46
	1.0027	13.494	1.0042	10.929	1.0060	9.1855	1.0080	7.9240	45
	1.0028	13.441	1.0042	10.894	1.0060	9.1612	1.0081	7.9059	44
	1.0028	13.389	1.0043	10.860	1.0060	9.1370	1.0081	7.8879	43
	1.0028	13.337	1.0043	10.826	1.0061	9.1129	1.0082	7.8700	42
	1.0028	13.286	1.0043	10.792	1.0061	9.0890	1.0082	7.8522	41
	1.0029	13.235	1.0043	10.758	1.0061	9.0651	1.0082	7.8344	40
	1.0029	13.184	1.0044	10.725	1.0062	9.0414	1.0083	7.8168	39
	1.0029	13.134	1.0044	10.692	1.0062	9.0179	1.0083	7.7992	38
	1.0029	13.084	1.0044	10.659	1.0062	8.9944	1.0084	7.7817	37
	1.0029	13.034	1.0044	10.626	1.0063	8.9711	1.0084	7.7642	36
	1.0030	12.985	1.0045	10.593	1.0063	8.9479	1.0084	7.7469	35
	1.0030	12.937	1.0045	10.561	1.0063	8.9248	1.0085	7.7296	34
	1.0030	12.888	1.0045	10.529	1.0064	8.9018	1.0085	7.7124	33
	1.0030	12.840	1.0046	10.497	1.0064	8.8790	1.0085	7.6953	32
	1.0031	12.793	1.0046	10.465	1.0064	8.8563	1.0086	7.6783	31
	1.0031	12.745	1.0046	10.433	1.0065	8.8337	1.0086	7.6613	30
	1.0031	12.698	1.0046	10.402	1.0065	8.8112	1.0087	7.6444	29
	1.0031	12.652	1.0047	10.371	1.0065	8.7888	1.0087	7.6276	28
	1.0032	12.606	1.0047	10.340	1.0066	8.7665	1.0087	7.6108	27
	1.0032	12.560	1.0047	10.309	1.0066	8.7444	1.0088	7.5942	26
	1.0032	12.514	1.0048	10.278	1.0066	8.7223	1.0088	7.5776	25
	1.0032	12.469	1.0048	10.248	1.0067	8.7004	1.0089	7.5611	24
	1.0032	12.424	1.0048	10.217	1.0067	8.6786	1.0089	7.5446	23
	1.0033	12.379	1.0048	10.187	1.0067	8.6569	1.0089	7.5282	22
	1.0033	12.335	1.0049	10.157	1.0068	8.6353	1.0090	7.5119	21
	1.0033	12.291	1.0049	10.127	1.0068	8.6138	1.0090	7.4957	20
	1.0033	12.248	1.0049	10.098	1.0068	8.5924	1.0090	7.4795	19
	1.0034	12.204	1.0050	10.068	1.0069	8.5711	1.0091	7.4634	18
	1.0034	12.161	1.0050	10.039	1.0069	8.5499	1.0091	7.4474	17
	1.0034	12.118	1.0050	10.010	1.0069	8.5289	1.0092	7.4315	16
	1.0034	12.076	1.0051	9.9812	1.0070	8.5079	1.0092	7.4156	15
	1.0035	12.034	1.0051	9.9525	1.0070	8.4871	1.0092	7.3998	14
	1.0035	11.992	1.0051	9.9239	1.0070	8.4663	1.0093	7.3840	13
	1.0035	11.950	1.0051	9.8955	1.0071	8.4457	1.0093	7.3683	12
	1.0035	11.909	1.0052	9.8672	1.0071	8.4251	1.0094	7.3527	11
	1.0036	11.868	1.0052	9.8391	1.0071	8.4046	1.0094	7.3372	10
	1.0036	11.828	1.0052	9.8112	1.0072	8.3843	1.0094	7.3217	9
	1.0036	11.787	1.0053	9.7834	1.0072	8.3640	1.0095	7.3063	8
	1.0036	11.747	1.0053	9.7558	1.0073	8.3439	1.0095	7.2909	7
	1.0037	11.707	1.0053	9.7283	1.0073	8.3238	1.0096	7.2757	6
	1.0037	11.668	1.0053	9.7010	1.0073	8.3039	1.0096	7.2604	5
	1.0037	11.628	1.0054	9.6739	1.0074	8.2840	1.0097	7.2453	4
	1.0037	11.589	1.0054	9.6469	1.0074	8.2642	1.0097	7.2302	3
	1.0038	11.550	1.0054	9.6200	1.0074	8.2446	1.0097	7.2152	2
	1.0038	11.512	1.0055	9.5933	1.0075	8.2250	1.0098	7.2002	1
	1.0038	11.474	1.0055	9.5668	1.0075	8.2055	1.0098	7.1853	0
'	Co-sec.	Sec.	Co-sec.	Sec.	Co-sec.	Sec.	Co-sec.	Sec.	
		85°		84°		83°		82°	

′	8° Sec.	8° Co-sec.	9° Sec.	9° Co-sec.	10° Sec.	10° Co-sec.	11° Sec.	11° Co-sec.	′
0	1.0098	7.1853	1.0125	6.3924	1.0154	5.7588	1.0187	5.2408	60
1	1.0099	7.1704	1.0125	6.3807	1.0155	5.7493	1.0188	5.2330	59
2	1.0099	7.1557	1.0125	6.3690	1.0155	5.7398	1.0188	5.2252	58
3	1.0099	7.1409	1.0126	6.3574	1.0156	5.7304	1.0189	5.2174	57
4	1.0100	7.1263	1.0126	6.3458	1.0156	5.7210	1.0189	5.2097	56
5	1.0100	7.1117	1.0127	6.3343	1.0157	5.7117	1.0190	5.2019	55
6	1.0101	7.0972	1.0127	6.3228	1.0157	5.7023	1.0191	5.1942	54
7	1.0101	7.0827	1.0128	6.3113	1.0158	5.6930	1.0191	5.1865	53
8	1.0102	7.0683	1.0128	6.2999	1.0158	5.6838	1.0192	5.1788	52
9	1.0102	7.0539	1.0129	6.2885	1.0159	5.6745	1.0192	5.1711	51
10	1.0102	7.0396	1.0129	6.2772	1.0159	5.6653	1.0193	5.1636	50
11	1.0103	7.0254	1.0130	6.2659	1.0160	5.6561	1.0193	5.1560	49
12	1.0103	7.0112	1.0130	6.2546	1.0160	5.6470	1.0194	5.1484	48
13	1.0104	6.9971	1.0131	6.2434	1.0161	5.6379	1.0195	5.1409	47
14	1.0104	6.9830	1.0131	6.2322	1.0162	5.6288	1.0195	5.1333	46
15	1.0104	6.9690	1.0132	6.2211	1.0162	5.6197	1.0196	5.1258	45
16	1.0105	6.9550	1.0132	6.2100	1.0163	5.6107	1.0196	5.1183	44
17	1.0105	6.9411	1.0133	6.1990	1.0163	5.6017	1.0197	5.1109	43
18	1.0106	6.9273	1.0133	6.1880	1.0164	5.5928	1.0198	5.1034	42
19	1.0106	6.9135	1.0134	6.1770	1.0164	5.5838	1.0198	5.0960	41
20	1.0107	6.8998	1.0134	6.1661	1.0165	5.5749	1.0199	5.0886	40
21	1.0107	6.8861	1.0135	6.1552	1.0165	5.5660	1.0199	5.0812	39
22	1.0107	6.8725	1.0135	6.1443	1.0166	5.5572	1.0200	5.0739	38
23	1.0108	6.8589	1.0136	6.1335	1.0166	5.5484	1.0201	5.0666	37
24	1.0108	6.8454	1.0136	6.1227	1.0167	5.5396	1.0201	5.0593	36
25	1.0109	6.8320	1.0136	6.1120	1.0167	5.5308	1.0202	5.0520	35
26	1.0109	6.8185	1.0137	6.1013	1.0168	5.5221	1.0202	5.0447	34
27	1.0110	6.8052	1.0137	6.0906	1.0169	5.5134	1.0203	5.0375	33
28	1.0110	6.7919	1.0138	6.0800	1.0169	5.5047	1.0204	5.0302	32
29	1.0111	6.7787	1.0138	6.0694	1.0170	5.4960	1.0204	5.0230	31
30	1.0111	6.7655	1.0139	6.0588	1.0170	5.4874	1.0205	5.0158	30
31	1.0111	6.7523	1.0139	6.0483	1.0171	5.4788	1.0205	5.0087	29
32	1.0112	6.7392	1.0140	6.0379	1.0171	5.4702	1.0206	5.0015	28
33	1.0112	6.7262	1.0140	6.0274	1.0172	5.4617	1.0207	4.9944	27
34	1.0113	6.7132	1.0141	6.0170	1.0172	5.4532	1.0207	4.9873	26
35	1.0113	6.7003	1.0141	6.0066	1.0173	5.4447	1.0208	4.9802	25
36	1.0114	6.6874	1.0142	5.9963	1.0174	5.4362	1.0208	4.9732	24
37	1.0114	6.6745	1.0142	5.9860	1.0174	5.4278	1.0209	4.9661	23
38	1.0115	6.6617	1.0143	5.9758	1.0175	5.4194	1.0210	4.9591	22
39	1.0115	6.6490	1.0143	5.9655	1.0175	5.4110	1.0210	4.9521	21
40	1.0115	6.6363	1.0144	5.9554	1.0176	5.4026	1.0211	4.9452	20
41	1.0116	6.6237	1.0144	5.9452	1.0176	5.3943	1.0211	4.9382	19
42	1.0116	6.6111	1.0145	5.9351	1.0177	5.3860	1.0212	4.9313	18
43	1.0117	6.5985	1.0145	5.9250	1.0177	5.3777	1.0213	4.9243	17
44	1.0117	6.5860	1.0146	5.9150	1.0178	5.3695	1.0213	4.9175	16
45	1.0118	6.5736	1.0146	5.9049	1.0179	5.3612	1.0214	4.9106	15
46	1.0118	6.5612	1.0147	5.8950	1.0179	5.3530	1.0215	4.9037	14
47	1.0119	6.5488	1.0147	5.8850	1.0180	5.3449	1.0215	4.8969	13
48	1.0119	6.5365	1.0148	5.8751	1.0180	5.3367	1.0216	4.8901	12
49	1.0119	6.5243	1.0148	5.8652	1.0181	5.3286	1.0216	4.8833	11
50	1.0120	6.5121	1.0149	5.8554	1.0181	5.3205	1.0217	4.8765	10
51	1.0120	6.4999	1.0150	5.8456	1.0182	5.3124	1.0218	4.8697	9
52	1.0121	6.4878	1.0150	5.8358	1.0182	5.3044	1.0218	4.8630	8
53	1.0121	6.4757	1.0151	5.8261	1.0183	5.2963	1.0219	4.8563	7
54	1.0122	6.4637	1.0151	5.8163	1.0184	5.2883	1.0220	4.8496	6
55	1.0122	6.4517	1.0152	5.8067	1.0184	5.2803	1.0220	4.8429	5
56	1.0123	6.4398	1.0152	5.7970	1.0185	5.2724	1.0221	4.8362	4
57	1.0123	6.4279	1.0153	5.7874	1.0185	5.2645	1.0221	4.8296	3
58	1.0124	6.4160	1.0153	5.7778	1.0186	5.2566	1.0222	4.8229	2
59	1.0124	6.4042	1.0154	5.7683	1.0186	5.2487	1.0223	4.8163	1
60	1.0125	6.3924	1.0154	5.7588	1.0187	5.2408	1.0223	4.8097	0
′	Co-sec.	Sec.	Co-sec.	Sec.	Co-sec.	Sec.	Co-sec.	Sec.	′
	81°		80°		79°		78°		

12° Sec.	Co-sec.	13° Sec.	Co-sec.	14° Sec.	Co-sec.	15° Sec.	Co-sec.	'
1.0223	4.8097	1.0263	4.4454	1.0306	4.1336	1.0353	3.8637	60
1.0224	4.8032	1.0264	4.4398	1.0307	4.1287	1.0353	3.8595	59
1.0225	4.7966	1.0264	4.4342	1.0308	4.1239	1.0354	3.8553	58
1.0225	4.7901	1.0265	4.4287	1.0308	4.1191	1.0355	3.8512	57
1.0226	4.7835	1.0266	4.4231	1.0309	4.1144	1.0356	3.8470	56
1.0226	4.7770	1.0266	4.4176	1.0310	4.1096	1.0357	3.8428	55
1.0227	4.7706	1.0267	4.4121	1.0311	4.1048	1.0358	3.8387	54
1.0228	4.7641	1.0268	4.4065	1.0311	4.1001	1.0358	3.8346	53
1.0228	4.7576	1.0268	4.4011	1.0312	4.0953	1.0359	3.8304	52
1.0229	4.7512	1.0269	4.3956	1.0313	4.0906	1.0360	3.8263	51
1.0230	4.7448	1.0270	4.3910	1.0314	4.0859	1.0361	3.8222	50
1.0230	4.7384	1.0271	4.3847	1.0314	4.0812	1.0362	3.8181	49
1.0231	4.7320	1.0271	4.3792	1.0315	4.0765	1.0362	3.8140	48
1.0232	4.7257	1.0272	4.3738	1.0316	4.0718	1.0363	3.8100	47
1.0232	4.7193	1.0273	4.3684	1.0317	4.0672	1.0364	3.8059	46
1.0233	4.7130	1.0273	4.3630	1.0317	4.0625	1.0365	3.8018	45
1.0234	4.7067	1.0274	4.3576	1.0318	4.0579	1.0366	3.7978	44
1.0234	4.7004	1.0275	4.3522	1.0319	4.0532	1.0367	3.7937	43
1.0235	4.6942	1.0276	4.3469	1.0320	4.0486	1.0367	3.7897	42
1.0235	4.6879	1.0276	4.3415	1.0320	4.0440	1.0368	3.7857	41
1.0236	4.6817	1.0277	4.3362	1.0321	4.0394	1.0369	3.7816	40
1.0237	4.6754	1.0278	4.3309	1.0322	4.0348	1.0370	3.7776	39
1.0237	4.6692	1.0278	4.3256	1.0323	4.0302	1.0371	3.7736	38
1.0238	4.6631	1.0279	4.3203	1.0323	4.0256	1.0371	3.7697	37
1.0239	4.6569	1.0280	4.3150	1.0324	4.0211	1.0372	3.7657	36
1.0239	4.6507	1.0280	4.3098	1.0325	4.0165	1.0373	3.7617	35
1.0240	4.6446	1.0281	4.3045	1.0326	4.0120	1.0374	3.7577	34
1.0241	4.6385	1.0282	4.2993	1.0327	4.0074	1.0375	3.7538	33
1.0241	4.6324	1.0283	4.2941	1.0327	4.0029	1.0376	3.7498	32
1.0242	4.6263	1.0283	4.2888	1.0328	3.9984	1.0377	3.7459	31
1.0243	4.6202	1.0284	4.2836	1.0329	3.9939	1.0377	3.7420	30
1.0243	4.6142	1.0285	4.2785	1.0330	3.9894	1.0378	3.7380	29
1.0244	4.6081	1.0285	4.2733	1.0330	3.9850	1.0379	3.7341	28
1.0245	4.6021	1.0286	4.2681	1.0331	3.9805	1.0380	3.7302	27
1.0245	4.5961	1.0287	4.2630	1.0332	3.9760	1.0381	3.7263	26
1.0246	4.5901	1.0288	4.2579	1.0333	3.9716	1.0382	3.7224	25
1.0247	4.5841	1.0288	4.2527	1.0334	3.9672	1.0382	3.7186	24
1.0247	4.5782	1.0289	4.2476	1.0334	3.9627	1.0383	3.7147	23
1.0248	4.5722	1.0290	4.2425	1.0335	3.9583	1.0384	3.7108	22
1.0249	4.5663	1.0291	4.2375	1.0336	3.9539	1.0385	3.7070	21
1.0249	4.5604	1.0291	4.2324	1.0337	3.9495	1.0386	3.7031	20
1.0250	4.5545	1.0292	4.2273	1.0338	3.9451	1.0387	3.6993	19
1.0251	4.5486	1.0293	4.2223	1.0338	3.9408	1.0387	3.6955	18
1.0251	4.5428	1.0293	4.2173	1.0339	3.9364	1.0388	3.6917	17
1.0252	4.5369	1.0294	4.2122	1.0340	3.9320	1.0389	3.6878	16
1.0253	4.5311	1.0295	4.2072	1.0341	3.9277	1.0390	3.6840	15
1.0253	4.5253	1.0296	4.2022	1.0341	3.9234	1.0391	3.6802	14
1.0254	4.5195	1.0296	4.1972	1.0342	3.9199	1.0392	3.6765	13
1.0255	4.5137	1.0297	4.1923	1.0343	3.9147	1.0393	3.6727	12
1.0255	4.5079	1.0298	4.1873	1.0344	3.9104	1.0393	3.6689	11
1.0256	4.5021	1.0299	4.1824	1.0345	3.9061	1.0394	3.6651	10
1.0257	4.4964	1.0299	4.1774	1.0345	3.9018	1.0395	3.6614	9
1.0257	4.4907	1.0300	4.1725	1.0346	3.8976	1.0396	3.6576	8
1.0258	4.4850	1.0301	4.1676	1.0347	3.8933	1.0397	3.6539	7
1.0259	4.4793	1.0302	4.1627	1.0348	3.8990	1.0398	3.6502	6
1.0260	4.4736	1.0302	4.1578	1.0349	3.8848	1.0399	3.6464	5
1.0260	4.4679	1.0303	4.1529	1.0349	3.8805	1.0399	3.6427	4
1.0261	4.4623	1.0304	4.1481	1.0350	3.8763	1.0400	3.6390	3
1.0262	4.4566	1.0305	4.1432	1.0351	3.8721	1.0401	3.6353	2
1.0262	4.4510	1.0305	4.1384	1.0352	3.8679	1.0402	3.6316	1
1.0263	4.4454	1.0306	4.1336	1.0353	3.8637	1.0403	3.6279	0

Co-sec.	Sec.	Co-sec.	Sec.	Co-sec.	Sec.	Co-sec.	Sec.
77°		76°		75°		74°	

′	16° SEC.	16° CO-SEC.	17° SEC.	17° CO-SEC.	18° SEC.	18° CO-SEC.	19° SEC.	19° CO-SEC.
0	1.0403	3.6279	1.0457	3.4203	1.0515	3.2361	1.0576	3.0715
1	1.0404	3.6243	1.0458	3.4170	1.0516	3.2332	1.0577	3.0690
2	1.0405	3.6206	1.0459	3.4138	1.0517	3.2303	1.0578	3.0664
3	1.0406	3.6169	1.0460	3.4106	1.0518	3.2274	1.0579	3.0638
4	1.0406	3.6133	1.0461	3.4073	1.0519	3.2245	1.0580	3.0612
5	1.0407	5.6096	1.0461	3.4041	1.0520	3.2216	1.0581	3.0586
6	1.0408	3.6060	1.0462	3.4009	1.0521	3.2188	1.0582	3.0561
7	1.0409	3.6024	1.0463	3.3977	1.0522	3.2159	1.0584	3.0535
8	1.0410	3.5987	1.0464	3.3945	1.0523	3.2131	1.0585	3.0509
9	1.0411	3.5951	1.0465	3.3913	1.0524	3.2102	1.0586	3.0484
10	1.0412	3.5915	1.0466	3.3881	1.0525	3.2074	1.0587	3.0458
11	1.0413	3.5879	1.0467	3.3849	1.0526	3.2045	1.0588	3.0433
12	1.0413	3.5843	1.0468	3.3817	1.0527	3.2017	1.0589	3.0407
13	1.0414	3.5807	1.0469	3.3785	1.0528	3.1989	1.0590	3.0382
14	1.0415	3.5772	1.0470	3.3754	1.0529	3.1960	1.0591	3.0357
15	1.0416	3.5736	1.0471	3.3722	1.0530	3.1932	1.0592	3.0331
16	1.0417	3.5700	1.0472	3.3690	1.0531	3.1904	1.0593	3.0306
17	1.0418	3.5665	1.0473	3.3659	1.0532	3.187′	1.0594	3.0281
18	1.0419	3.5629	1.0474	3.3627	1.0533	3.1848	1.0595	3.0256
19	1.0420	3.5594	1.0475	3.3596	1.0534	3.1820	1.0596	3.0231
20	1.0420	3.5559	1.0476	3.3565	1.0535	3.1792	1.0598	3.0206
21	1.0421	3.5523	1.0477	3.3534	1.0536	3.1764	1.0599	3.0181
22	1.0422	3.5488	1.0478	3.3502	1.0537	3.1736	1.0600	3.0156
23	1.0423	3.5453	1.0478	3.3471	1.0538	3.1708	1.0601	3.0131
24	1.0424	3.5418	1.0479	3.3440	1.0539	3.1681	1.0602	3.0106
25	1.0425	3.5383	1.0480	3.3409	1.0540	3.1653	1.0603	3.0081
26	1.0426	3.5348	1.0481	3.3378	1.0541	3.1625	1.0604	3.0056
27	1.0427	3.5313	1.0482	3.3347	1.0542	3.1598	1.0605	3.0031
28	1.0428	3.5279	1.0483	3.3316	1.0543	3.1570	1.0606	3.0007
29	1.0428	3.5244	1.0484	3.3286	1.0544	3.1543	1.0607	2.9982
30	1.0429	3.5209	1.0485	3.3255	1.0545	3.1515	1.0608	2.9957
31	1.0430	3.5175	1.0486	3.3224	1.0546	3.1488	1.0609	2.9933
32	1.0431	3.5140	1.0487	3.3194	1.0547	3.1461	1.0611	2.9908
33	1.0432	3.5106	1.0488	3.3163	1.0548	3.1433	1.0612	2.9884
34	1.0433	3.5072	1.0489	3.3133	1.0549	3.1406	1.0613	2.9859
35	1.0434	3.5037	1.0490	3.3102	1.0550	3.1379	1.0614	2.9835
36	1.0435	3.5003	1.0491	3.3072	1.0551	3.1352	1.0615	2.9810
37	1.0436	3.4969	1.0492	3.3042	1.0552	3.1325	1.0616	2.9786
38	1.0437	3.4935	1.0493	3.3011	1.0553	3.1298	1.0617	2.9762
39	1.0438	3.4901	1.0494	3.2981	1.0554	3.1271	1.0618	2.9738
40	1.0438	3.4867	1.0495	3.2951	1.0555	3.1244	1.0619	2.9713
41	1.0439	3.4833	1.0496	3.2921	1.0556	3.1217	1.0620	2.9689
42	1.0440	3.4799	1.0497	3.2891	1.0557	3.1190	1.0622	2.9665
43	1.0441	3.4766	1.0498	3.2861	1.0558	3.1163	1.0623	2.9641
44	1.0442	3.4732	1.0499	3.2831	1.0559	3.1137	1.0624	2.9617
45	1.0443	3.4698	1.0500	3.2801	1.0560	3.1110	1.0625	2.9593
46	1.0444	3.4665	1.0501	3.2772	1.0561	3.1083	1.0626	2.9569
47	1.0445	3.4632	1.0502	3.2742	1.0562	3.1057	1.0627	2.9545
48	1.0446	3.4598	1.0503	3.2712	1.0563	3.1030	1.0628	2.9521
49	1.0447	3.4565	1.0504	3.2683	1.0565	3.1004	1.0629	2.9497
50	1.0448	3.4532	1.0505	3.2653	1.0566	3.0977	1.0630	2.9474
51	1.0448	3.4498	1.0506	3.2624	1.0567	3.0951	1.0632	2.9450
52	1.0449	3.4465	1.0507	3.2594	1.0568	3.0925	1.0633	2.9426
53	1.0450	3.4432	1.0508	3.2565	1.0569	3.0898	1.0634	2.9402
54	1.0451	3.4399	1.0509	3.2535	1.0570	3.0872	1.0636	2.9379
55	1.0452	3.4366	1.0510	3.2506	1.0571	3.0846	1.0636	2.9355
56	1.0453	3.4334	1.0511	3.2477	1.0572	3.0820	1.0637	2.9332
57	1.0454	3.4301	1.0512	3.2448	1.0573	3.0793	1.0638	2.9308
58	1.0455	3.4268	1.0513	3.2419	1.0574	3.0767	1.0639	2.9285
59	1.0456	3.4236	1.0514	3.2390	1.0575	3.0741	1.0641	2.9261
60	1.0457	3.4203	1.0515	3.2361	1.0576	3.0715	1.0642	2.9238

| ′ | Co-SEC. | SEC. 73° | Co-SEC. | SEC. 72° | Co-SEC. | SEC. 71° | Co-SEC. | SEC. 70° |

20° Sec.	20° Co-sec.	21° Sec.	21° Co-sec.	22° Sec.	22° Co-sec.	23° Sec.	23° Co-sec.	'
1.0642	2.9238	1.0711	2.7904	1.0785	2.6695	1.0864	2.5593	60
1.0643	2.9215	1.0713	2.7883	1.0787	2.6675	1.0865	2.5575	59
1.0644	2.9191	1.0714	2.7862	1.0788	2.6656	1.0866	2.5558	58
1.0645	2.9168	1.0715	2.7841	1.0789	2.6637	1.0868	2.5540	57
1.0646	2.9145	1.0716	2.7820	1.0790	2.6618	1.0869	2.5523	56
1.0647	2.9122	1.0717	2.7799	1.0792	2.6599	1.0870	2.5506	55
1.0648	2.9098	1.0719	2.7778	1.0793	2.6580	1.0872	2.5488	54
1.0650	2.9075	1.0720	2.7757	1.0794	2.6561	1.0873	2.5471	53
1.0651	2.9052	1.0721	2.7736	1.0795	2.6542	1.0874	2.5453	52
1.0652	2.9029	1.0722	2.7715	1.0797	2.6523	1.0876	2.5436	51
1.0653	2.9006	1.0723	2.7694	1.0798	2.6504	1.0877	2.5419	50
1.0654	2.8983	1.0725	2.7674	1.0799	2.6485	1.0878	2.5402	49
1.0655	2.8960	1.0726	2.7653	1.0801	2.6466	1.0880	2.5384	48
1.0656	2.8937	1.0727	2.7632	1.0802	2.6447	1.0881	2.5367	47
1.0658	2.8915	1.0728	2.7611	1.0803	2.6428	1.0882	2.5350	46
1.0659	2.8892	1.0729	2.7591	1.0804	2.6410	1.0884	2.5333	45
1.0660	2.8869	1.0731	2.7570	1.0806	2.6391	1.0885	2.5316	44
1.0661	2.8846	1.0732	2.7550	1.0807	2.6372	1.0886	2.5299	43
1.0662	2.8824	1.0733	2.7529	1.0808	2.6353	1.0888	2.5281	42
1.0663	2.8801	1.0734	2.7509	1.0810	2.6335	1.0889	2.5264	41
1.0664	2.8778	1.0736	2.7488	1.0811	2.6316	1.0891	2.5247	40
1.0666	2.8756	1.0737	2.7468	1.0812	2.6297	1.0892	2.5230	39
1.0667	2.8733	1.0738	2.7447	1.0813	2.6279	1.0893	2.5213	38
1.0668	2.8711	1.0739	2.7427	1.0815	2.6260	1.0895	2.5196	37
1.0669	2.8688	1.0740	2.7406	1.0816	2.6242	1.0896	2.5179	36
1.0670	2.8666	1.0742	2.7386	1.0817	2.6223	1.0897	2.5163	35
1.0671	2.8644	1.0743	2.7366	1.0819	2.6205	1.0899	2.5146	34
1.0673	2.8621	1.0744	2.7346	1.0820	2.6186	1.0900	2.5129	33
1.0674	2.8599	1.0745	2.7325	1.0821	2.6168	1.0902	2.5112	32
1.0675	2.8577	1.0747	2.7305	1.0823	2.6150	1.0903	2.5095	31
1.0676	2.8554	1.0748	2.7285	1.0824	2.6131	1.0904	2.5078	30
1.0677	2.8532	1.0749	2.7265	1.0825	2.6113	1.0906	2.5062	29
1.0678	2.8510	1.0750	2.7245	1.0826	2.6095	1.0907	2.5045	28
1.0679	2.8488	1.0751	2.7225	1.0828	2.6076	1.0908	2.5028	27
1.0681	2.8466	1.0753	2.7205	1.0829	2.6058	1.0910	2.5011	26
1.0682	2.8444	1.0754	2.7185	1.0830	2.6040	1.0911	2.4995	25
1.0683	2.8422	1.0755	2.7165	1.0832	2.6022	1.0913	2.4978	24
1.0684	2.8400	1.0756	2.7145	1.0833	2.6003	1.0914	2.4961	23
1.0685	2.8378	1.0758	2.7125	1.0834	2.5985	1.0915	2.4945	22
1.0686	2.8356	1.0759	2.7105	1.0836	2.5967	1.0917	2.4928	21
1.0688	2.8334	1.0760	2.7085	1.0837	2.5949	1.0918	2.4912	20
1.0689	2.8312	1.0761	2.7065	1.0838	2.5931	1.0920	2.4895	19
1.0690	2.8290	1.0763	2.7045	1.0840	2.5913	1.0921	2.4879	18
1.0691	2.8269	1.0764	2.7026	1.0841	2.5895	1.0922	2.4862	17
1.0692	2.8247	1.0765	2.7006	1.0842	2.5877	1.0924	2.4846	16
1.0694	2.8225	1.0766	2.6986	1.0844	2.5859	1.0925	2.4829	15
1.0695	2.8204	1.0768	2.6967	1.0845	2.5841	1.0927	2.4813	14
1.0696	2.8182	1.0769	2.6947	1.0846	2.5823	1.0928	2.4797	13
1.0697	2.8160	1.0770	2.6927	1.0847	2.5805	1.0929	2.4780	12
1.0698	2.8139	1.0771	2.6908	1.0849	2.5787	1.0931	2.4764	11
1.0699	2.8117	1.0773	2.6888	1.0850	2.5770	1.0932	2.4748	10
1.0701	2.8096	1.0774	2.6869	1.0851	2.5752	1.0934	2.4731	9
1.0702	2.8074	1.0775	2.6849	1.0853	2.5734	1.0935	2.4715	8
1.0703	2.8053	1.0776	2.6830	1.0855	2.5716	1.0936	2.4699	7
1.0704	2.8032	1.0778	2.6810	1.0855	2.5699	1.0938	2.4683	6
1.0705	2.8010	1.0779	2.6791	1.0857	2.5681	1.0939	2.4666	5
1.0707	2.7989	1.0780	2.6772	1.0858	2.5663	1.0941	2.4650	4
1.0708	2.7968	1.0781	2.6752	1.0859	2.5646	1.0942	2.4634	3
1.0709	2.7947	1.0783	2.6733	1.0861	2.5628	1.0943	2.4618	2
1.0710	2.7925	1.0784	2.6714	1.0862	2.5610	1.0945	2.4602	1
1.0711	2.7904	1.0785	2.6695	1.0864	2.5593	1.0946	2.4586	0
Co-sec.	Sec.	Co-sec.	Sec.	Co-sec.	Sec.	Co-sec.	Sec.	
69°		68°		67°		66°		

′	24° Sec.	Co-sec.	25° Sec.	Co-sec.	26° Sec.	Co-sec.	27° Sec.	Co-sec.
0	1.0946	2.4586	1.1034	2.3662	1.1126	2.2812	1.1223	2.2027
1	1.0948	2.4570	1.1035	2.3647	1.1127	2.2798	1.1225	2.2014
2	1.0949	2.4554	1.1037	2.3632	1.1129	2.2784	1.1226	2.2002
3	1.0951	2.4538	1.1038	2.3618	1.1131	2.2771	1.1228	2.1989
4	1.0952	2.4522	1.1040	2.3603	1.1132	2.2757	1.1230	2.1977
5	1.0953	2.4506	1.1041	2.3588	1.1134	2.2744	1.1231	2.1964
6	1.0955	2.4490	1.1043	2.3574	1.1135	2.2730	1.1233	2.1952
7	1.0956	2.4474	1.1044	2.3559	1.1137	2.2717	1.1235	2.1939
8	1.0958	2.4458	1.1046	2.3544	1.1139	2.2703	1.1237	2.1927
9	1.0959	2.4442	1.1047	2.3530	1.1140	2.2690	1.1238	2.1914
10	1.0961	2.4426	1.1049	2.3515	1.1142	2.2676	1.1240	2.1902
11	1.0962	2.4411	1.1050	2.3501	1.1143	2.2663	1.1242	2.1889
12	1.0963	2.4395	1.1052	2.3486	1.1145	2.2650	1.1243	2.1877
13	1.0965	2.4379	1.1053	2.3472	1.1147	2.2636	1.1245	2.1865
14	1.0966	2.4363	1.1055	2.3457	1.1148	2.2623	1.1247	2.1852
15	1.0968	2.4347	1.1056	2.3443	1.1150	2.2610	1.1248	2.1840
16	1.0969	2.4332	1.1058	2.3428	1.1151	2.2596	1.1250	2.1828
17	1.0971	2.4316	1.1059	2.3414	1.1153	2.2583	1.1252	2.1815
18	1.0972	2.4300	1.1061	2.3399	1.1155	2.2570	1.1253	2.1803
19	1.0973	2.4285	1.1062	2.3385	1.1156	2.2556	1.1255	2.1791
20	1.0975	2.4269	1.1064	2.3371	1.1158	2.2543	1.1257	2.1778
21	1.0976	2.4254	1.1065	2.3356	1.1159	2.2530	1.1258	2.1766
22	1.0978	2.4238	1.1067	2.3342	1.1161	2.2517	1.1260	2.1754
23	1.0979	2.4222	1.1068	2.3328	1.1163	2.2503	1.1262	2.1742
24	1.0981	2.4207	1.1070	2.3313	1.1164	2.2490	1.1264	2.1730
25	1.0982	2.4191	1.1072	2.3299	1.1166	2.2477	1.1265	2.1717
26	1.0984	2.4176	1.1073	2.3285	1.1167	2.2464	1.1267	2.1705
27	1.0985	2.4160	1.1075	2.3271	1.1169	2.2451	1.1269	2.1693
28	1.0986	2.4145	1.1076	2.3256	1.1171	2.2438	1.1270	2.1681
29	1.0988	2.4130	1.1078	2.3242	1.1172	2.2425	1.1272	2.1669
30	1.0989	2.4114	1.1079	2.3228	1.1174	2.2411	1.1274	2.1657
31	1.0991	2.4099	1.1081	2.3214	1.1176	2.2398	1.1275	2.1645
32	1.0992	2.4083	1.1082	2.3200	1.1177	2.2385	1.1277	2.1633
33	1.0994	2.4068	1.1084	2.3186	1.1179	2.2372	1.1279	2.1620
34	1.0995	2.4053	1.1085	2.3172	1.1180	2.2359	1.1281	2.1608
35	1.0997	2.4037	1.1087	2.3158	1.1182	2.2346	1.1282	2.1596
36	1.0998	2.4022	1.1088	2.3143	1.1184	2.2333	1.1284	2.1584
37	1.1000	2.4007	1.1090	2.3129	1.1185	2.2320	1.1286	2.1572
38	1.1001	2.3992	1.1092	2.3115	1.1187	2.2307	1.1287	2.1560
39	1.1003	2.3976	1.1093	2.3101	1.1189	2.2294	1.1289	2.1548
40	1.1004	2.3961	1.1095	2.3087	1.1190	2.2282	1.1291	2.1536
41	1.1005	2.3946	1.1096	2.3073	1.1192	2.2269	1.1293	2.1525
42	1.1007	2.3931	1.1098	2.3059	1.1193	2.2256	1.1294	2.1513
43	1.1008	2.3916	1.1099	2.3046	1.1195	2.2243	1.1296	2.1501
44	1.1010	2.3901	1.1101	2.3032	1.1197	2.2230	1.1298	2.1489
45	1.1011	2.3886	1.1102	2.3018	1.1198	2.2217	1.1299	2.1477
46	1.1013	2.3871	1.1104	2.3004	1.1200	2.2204	1.1301	2.1465
47	1.1014	2.3856	1.1106	2.2990	1.1202	2.2192	1.1303	2.1453
48	1.1016	2.3841	1.1107	2.2976	1.1203	2.2179	1.1305	2.1441
49	1.1017	2.3826	1.1109	2.2962	1.1205	2.2166	1.1306	2.1430
50	1.1019	2.3811	1.1110	2.2949	1.1207	2.2153	1.1308	2.1418
51	1.1020	2.3796	1.1112	2.2935	1.1208	2.2141	1.1310	2.1406
52	1.1022	2.3781	1.1113	2.2921	1.1210	2.2128	1.1312	2.1394
53	1.1023	2.3766	1.1115	2.2907	1.1212	2.2115	1.1313	2.1382
54	1.1025	2.3751	1.1116	2.2894	1.1213	2.2103	1.1315	2.1371
55	1.1026	2.3736	1.1118	2.2880	1.1215	2.2090	1.1317	2.1359
56	1.1028	2.3721	1.1120	2.2866	1.1217	2.2077	1.1319	2.1347
57	1.1029	2.3706	1.1121	2.2853	1.1218	2.2065	1.1320	2.1335
58	1.1031	2.3691	1.1123	2.2839	1.1220	2.2052	1.1322	2.1324
59	1.1032	2.3677	1.1124	2.2825	1.1222	2.2039	1.1324	2.1312
60	1.1034	2.3662	1.1126	2.2812	1.1223	2.2027	1.1326	2.1300

'	28° Sec.	Co-sec.	29° Sec.	Co-sec.	30° Sec.	Co-sec.	31° Sec.	Co-sec.	'
0	1.1326	2.1300	1.1433	2.0627	1.1547	2.0000	1.1666	1.9416	60
1	1.1327	2.1289	1.1435	2.0616	1.1549	1.9990	1.1668	1.9407	59
2	1.1329	2.1277	1.1437	2.0605	1.1551	1.9980	1.1670	1.9397	58
3	1.1331	2.1266	1.1439	2.0594	1.1553	1.9970	1.1672	1.9388	57
4	1.1333	2.1254	1.1441	2.0583	1.1555	1.9960	1.1674	1.9378	56
5	1.1334	2.1242	1.1443	2.0573	1.1557	1.9950	1.1676	1.9369	55
6	1.1336	2.1231	1.1445	2.0562	1.1559	1.9940	1.1678	1.9360	54
7	1.1338	2.1219	1.1446	2.0551	1.1561	1.9930	1.1681	1.9350	53
8	1.1340	2.1208	1.1448	2.0540	1.1562	1.9920	1.1683	1.9341	52
9	1.1341	2.1196	1.1450	2.0530	1.1564	1.9910	1.1685	1.9332	51
0	1.1343	2.1185	1.1452	2.0519	1.1566	1.9900	1.1687	1.9322	50
1	1.1345	2.1173	1.1454	2.0508	1.1568	1.9890	1.1689	1.9313	49
2	1.1347	2.1162	1.1456	2.0498	1.1570	1.9880	1.1691	1.9304	48
3	1.1349	2.1150	1.1458	2.0487	1.1572	1.9870	1.1693	1.9295	47
4	1.1350	2.1139	1.1459	2.0476	1.1574	1.9860	1.1695	1.9285	46
5	1.1352	2.1127	1.1461	2.0466	1.1576	1.9850	1.1697	1.9276	45
6	1.1354	2.1116	1.1463	2.0455	1.1578	1.9840	1.1699	1.9267	44
7	1.1356	2.1104	1.1465	2.0444	1.1580	1.9830	1.1701	1.9258	43
8	1.1357	2.1093	1.1467	2.0434	1.1582	1.9820	1.1703	1.9248	42
9	1.1359	2.1082	1.1469	2.0423	1.1584	1.9811	1.1705	1.9239	41
0	1.1361	2.1070	1.1471	2.0413	1.1586	1.9801	1.1707	1.9230	40
1	1.1363	2.1059	1.1473	2.0402	1.1588	1.9791	1.1709	1.9221	39
2	1.1365	2.1048	1.1474	2.0392	1.1590	1.9781	1.1712	1.9212	38
3	1.1366	2.1036	1.1476	2.0381	1.1592	1.9771	1.1714	1.9203	37
4	1.1368	2.1025	1.1478	2.0370	1.1594	1.9761	1.1716	1.9193	36
5	1.1370	2.1014	1.1480	2.0360	1.1596	1.9752	1.1718	1.9184	35
6	1.1372	2.1002	1.1482	2.0349	1.1598	1.9742	1.1720	1.9175	34
7	1.1373	2.0991	1.1484	2.0339	1.1600	1.9732	1.1722	1.9166	33
8	1.1375	2.0980	1.1486	2.0329	1.1602	1.9722	1.1724	1.9157	32
9	1.1377	2.0969	1.1488	2.0318	1.1604	1.9713	1.1726	1.9148	31
0	1.1379	2.0957	1.1489	2.0308	1.1606	1.9703	1.1728	1.9139	30
1	1.1381	2.0946	1.1491	2.0297	1.1608	1.9693	1.1730	1.9130	29
2	1.1382	2.0935	1.1493	2.0287	1.1610	1.9683	1.1732	1.9121	28
3	1.1384	2.0924	1.1495	2.0276	1.1612	1.9674	1.1734	1.9112	27
4	1.1386	2.0912	1.1497	2.0266	1.1614	1.9664	1.1737	1.9102	26
5	1.1388	2.0901	1.1499	2.0256	1.1616	1.9654	1.1739	1.9093	25
6	1.1390	2.0890	1.1501	2.0245	1.1618	1.9645	1.1741	1.9084	24
7	1.1391	2.0879	1.1503	2.0235	1.1620	1.9635	1.1743	1.9075	23
8	1.1393	2.0868	1.1505	2.0224	1.1622	1.9625	1.1745	1.9066	22
9	1.1395	2.0857	1.1507	2.0214	1.1624	1.9616	1.1747	1.9057	21
0	1.1397	2.0846	1.1508	2.0204	1.1626	1.9606	1.1749	1.9048	20
1	1.1399	2.0835	1.1510	2.0194	1.1628	1.9596	1.1751	1.9039	19
2	1.1401	2.0824	1.1512	2.0183	1.1630	1.9587	1.1753	1.9030	18
3	1.1402	2.0812	1.1514	2.0173	1.1632	1.9577	1.1756	1.9021	17
4	1.1404	2.0801	1.1516	2.0163	1.1634	1.9568	1.1758	1.9013	16
5	1.1406	2.0790	1.1518	2.0152	1.1636	1.9558	1.1760	1.9004	15
6	1.1408	2.0779	1.1520	2.0142	1.1638	1.9549	1.1762	1.8995	14
7	1.1410	2.0768	1.1522	2.0132	1.1640	1.9539	1.1764	1.8986	13
8	1.1411	2.0757	1.1524	2.0122	1.1642	1.9530	1.1766	1.8977	12
9	1.1413	2.0746	1.1526	2.0111	1.1644	1.9520	1.1768	1.8968	11
0	1.1415	2.0735	1.1528	2.0101	1.1646	1.9510	1.1770	1.8959	10
1	1.1417	2.0725	1.1530	2.0091	1.1648	1.9501	1.1772	1.8950	9
2	1.1419	2.0714	1.1531	2.0081	1.1650	1.9491	1.1775	1.8941	8
3	1.1421	2.0703	1.1533	2.0071	1.1652	1.9482	1.1777	1.8932	7
4	1.1422	2.0692	1.1535	2.0061	1.1654	1.9473	1.1779	1.8924	6
5	1.1424	2.0681	1.1537	2.0050	1.1656	1.9463	1.1781	1.8915	5
6	1.1426	2.0670	1.1539	2.0040	1.1658	1.9454	1.1783	1.8906	4
7	1.1428	2.0659	1.1541	2.0030	1.1660	1.9444	1.1785	1.8897	3
8	1.1430	2.0648	1.1543	2.0020	1.1662	1.9435	1.1787	1.8888	2
9	1.1432	2.0637	1.1545	2.0010	1.1664	1.9425	1.1790	1.8879	1
0	1.1433	2.0627	1.1547	2.0000	1.1666	1.9416	1.1792	1.8871	0
'	Co-sec.	Sec.	Co-sec.	Sec.	Co-sec.	Sec.	Co-sec.	Sec.	'
	61°		60°		59°		58°		

′	32° Sec.	Co-sec.	33° Sec.	Co-sec.	34° Sec.	Co-sec.	35° Sec.	Co-sec.	
0	1.1792	1.8871	1.1924	1.8361	1.2062	1.7883	1.2208	1.7434	60
1	1.1794	1.8862	1.1926	1.8352	1.2064	1.7875	1.2210	1.7427	59
2	1.1796	1.8853	1.1928	1.8344	1.2067	1.7867	1.2213	1.7420	58
3	1.1798	1.8844	1.1930	1.8336	1.2069	1.7860	1.2215	1.7413	57
4	1.1800	1.8836	1.1933	1.8328	1.2072	1.7852	1.2218	1.7405	56
5	1.1802	1.8827	1.1935	1.8320	1.2074	1.7844	1.2220	1.7398	55
6	1.1805	1.8818	1.1937	1.8311	1.2076	1.7837	1.2223	1.7391	54
7	1.1807	1.8809	1.1939	1.8303	1.2079	1.7829	1.2225	1.7384	53
8	1.1809	1.8801	1.1942	1.8295	1.2081	1.7821	1.2228	1.7377	52
9	1.1811	1.8792	1.1944	1.8287	1.2083	1.7814	1.2230	1.7369	51
10	1.1813	1.8783	1.1946	1.8279	1.2086	1.7806	1.2233	1.7362	50
11	1.1815	1.8785	1.1948	1.8271	1.2088	1.7798	1.2235	1.7355	49
12	1.1818	1.8766	1.1951	1.8263	1.2091	1.7791	1.2238	1.7348	48
13	1.1820	1.8757	1.1953	1.8255	1.2093	1.7783	1.2240	1.7341	47
14	1.1822	1.8749	1.1955	1.8246	1.2095	1.7776	1.2243	1.7334	46
15	1.1824	1.8740	1.1958	1.8238	1.2098	1.7768	1.2245	1.7327	45
16	1.1826	1.8731	1.1960	1.8230	1.2100	1.7760	1.2248	1.7319	44
17	1.1828	1.8723	1.1962	1.8222	1.2103	1.7753	1.2250	1.7312	43
18	1.1831	1.8714	1.1964	1.8214	1.2105	1.7745	1.2253	1.7305	42
19	1.1833	1.8706	1.1967	1.8206	1.2107	1.7738	1.2255	1.7298	41
20	1.1835	1.8697	1.1969	1.8198	1.2110	1.7730	1.2258	1.7291	40
21	1.1837	1.8688	1.1971	1.8190	1.2112	1.7723	1.2260	1.7284	39
22	1.1839	1.8680	1.1974	1.8182	1.2115	1.7715	1.2263	1.7277	38
23	1.1841	1.8671	1.1976	1.8174	1.2117	1.7708	1.2265	1.7270	37
24	1.1844	1.8663	1.1978	1.8166	1.2119	1.7700	1.2268	1.7263	36
25	1.1846	1.8654	1.1980	1.8158	1.2122	1.7693	1.2270	1.7256	35
26	1.1848	1.8646	1.1983	1.8150	1.2124	1.7685	1.2273	1.7249	34
27	1.1850	1.8637	1.1985	1.8142	1.2127	1.7678	1.2276	1.7242	33
28	1.1852	1.8629	1.1987	1.8134	1.2129	1.7670	1.2278	1.7234	32
29	1.1855	1.8620	1.1990	1.8126	1.2132	1.7663	1.2281	1.7227	31
30	1.1857	1.8611	1.1992	1.8118	1.2134	1.7655	1.2283	1.7220	30
31	1.1859	1.8603	1.1994	1.8110	1.2136	1.7648	1.2286	1.7213	29
32	1.1861	1.8595	1.1997	1.8102	1.2139	1.7640	1.2288	1.7206	28
33	1.1863	1.8586	1.1999	1.8094	1.2141	1.7633	1.2291	1.7199	27
34	1.1866	1.8578	1.2001	1.8086	1.2144	1.7625	1.2293	1.7192	26
35	1.1868	1.8569	1.2004	1.8078	1.2146	1.7618	1.2296	1.7185	25
36	1.1870	1.8561	1.2006	1.8070	1.2149	1.7610	1.2298	1.7178	24
37	1.1872	1.8552	1.2008	1.8062	1.2151	1.7603	1.2301	1.7171	23
38	1.1874	1.8544	1.2010	1.8054	1.2153	1.7596	1.2304	1.7164	22
39	1.1877	1.8535	1.2013	1.8047	1.2156	1.7588	1.2306	1.7157	21
40	1.1879	1.8527	1.2015	1.8039	1.2158	1.7581	1.2309	1.7151	20
41	1.1881	1.8519	1.2017	1.8031	1.2161	1.7573	1.2311	1.7144	19
42	1.1883	1.8510	1.2020	1.8023	1.2163	1.7566	1.2314	1.7137	18
43	1.1886	1.8502	1.2022	1.8015	1.2166	1.7559	1.2316	1.7130	17
44	1.1888	1.8493	1.2024	1.8007	1.2168	1.7551	1.2319	1.7123	16
45	1.1890	1.8485	1.2027	1.7999	1.2171	1.7544	1.2322	1.7116	15
46	1.1892	1.8477	1.2029	1.7992	1.2173	1.7537	1.2324	1.7109	14
47	1.1894	1.8468	1.2031	1.7984	1.2175	1.7529	1.2327	1.7102	13
48	1.1897	1.8460	1.2034	1.7976	1.2178	1.7522	1.2329	1.7095	12
49	1.1899	1.8452	1.2036	1.7968	1.2180	1.7514	1.2332	1.7088	11
50	1.1901	1.8443	1.2039	1.7960	1.2183	1.7507	1.2335	1.7081	10
51	1.1903	1.8435	1.2041	1.7953	1.2185	1.7500	1.2337	1.7075	9
52	1.1906	1.8427	1.2043	1.7945	1.2188	1.7493	1.2340	1.7068	8
53	1.1908	1.8418	1.2046	1.7937	1.2190	1.7485	1.2342	1.7061	7
54	1.1910	1.8410	1.2048	1.7929	1.2193	1.7478	1.2345	1.7054	6
55	1.1912	1.8402	1.2050	1.7921	1.2195	1.7471	1.2348	1.7047	5
56	1.1915	1.8394	1.2053	1.7914	1.2198	1.7463	1.2350	1.7040	4
57	1.1917	1.8385	1.2055	1.7906	1.2200	1.7456	1.2353	1.7033	3
58	1.1919	1.8377	1.2057	1.7898	1.2203	1.7449	1.2355	1.7027	2
59	1.1921	1.8369	1.2060	1.7891	1.2205	1.7442	1.2358	1.7020	1
60	1.1922	1.8361	1.2062	1.7883	1.2208	1.7434	1.2361	1.7013	0

′	Co-sec.	Sec.	Co-sec.	Sec.	Co-sec.	Sec.	Co-sec.	Sec.
	57°		56°		55°		54°	

36°		37°		38°		39°		
Sec.	Co-sec.	Sec.	Co-sec.	Sec.	Co-sec.	Sec.	Co-sec.	'
1.2361	1.7013	1.2521	1.6616	1.2690	1.6243	1.2867	1.5890	60
1.2363	1.7006	1.2524	1.6610	1.2693	1.6237	1.2871	1.5884	59
1.2366	1.6999	1.2527	1.6603	1.2696	1.6231	1.2874	1.5879	58
1.2368	1.6993	1.2530	1.6597	1.2699	1.6224	1.2877	1.5873	57
1.2371	1.6986	1.2532	1.6591	1.2702	1.6218	1.2880	1.5867	56
1.2374	1.6979	1.2535	1.6584	1.2705	1.6212	1.2883	1.5862	55
1.2376	1.6972	1.2538	1.6578	1.2707	1.6206	1.2886	1.5856	54
1.2379	1.6965	1.2541	1.6572	1.2710	1.6200	1.2889	1.5850	53
1.2382	1.6959	1.2543	1.6565	1.2713	1.6194	1.2892	1.5845	52
1.2384	1.6952	1.2546	1.6559	1.2716	1.6188	1.2895	1.5839	51
1.2387	1.6945	1.2549	1.6552	1.2719	1.6182	1.2898	1.5833	50
1.2389	1.6938	1.2552	1.6546	1.2722	1.6176	1.2901	1.5828	49
1.2392	1.6932	1.2554	1.6540	1.2725	1.6170	1.2904	1.5822	48
1.2395	1.6925	1.2557	1.6533	1.2728	1.6164	1.2907	1.5816	47
1.2397	1.6918	1.2560	1.6527	1.2731	1.6159	1.2910	1.5811	46
1.2400	1.6912	1.2563	1.6521	1.2734	1.6153	1.2913	1.5805	45
1.2403	1.6905	1.2565	1.6514	1.2737	1.6147	1.2916	1.5799	44
1.2405	1.6898	1.2568	1.6508	1.2739	1.6141	1.2919	1.5794	43
1.2408	1.6891	1.2571	1.6502	1.2742	1.6135	1.2922	1.5788	42
1.2411	1.6885	1.2574	1.6496	1.2745	1.6129	1.2926	1.5783	41
1.2413	1.6878	1.2577	1.6489	1.2748	1.6123	1.2929	1.5777	40
1.2416	1.6871	1.2579	1.6483	1.2751	1.6117	1.2932	1.5771	39
1.2419	1.6865	1.2582	1.6477	1.2754	1.6111	1.2935	1.5766	38
1.2421	1.6858	1.2585	1.6470	1.2757	1.6105	1.2938	1.5760	37
1.2424	1.6851	1.2588	1.6464	1.2760	1.6099	1.2941	1.5755	36
1.2427	1.6845	1.2591	1.6458	1.2763	1.6093	1.2944	1.5749	35
1.2429	1.6838	1.2593	1.6452	1.2766	1.6087	1.2947	1.5743	34
1.2432	1.6831	1.2596	1.6445	1.2769	1.6081	1.2950	1.5738	33
1.2435	1.6825	1.2599	1.6439	1.2772	1.6077	1.2953	1.5732	32
1.2437	1.6818	1.2602	1.6433	1.2775	1.6070	1.2956	1.5727	31
1.2440	1.6812	1.2605	1.6427	1.2778	1.6064	1.2960	1.5721	30
1.2443	1.6805	1.2607	1.6420	1.2781	1.6058	1.2963	1.5716	29
1.2445	1.6798	1.2610	1.6414	1.2784	1.6052	1.2966	1.5710	28
1.2448	1.6792	1.2613	1.6408	1.2787	1.6046	1.2969	1.5705	27
1.2451	1.6785	1.2616	1.6402	1.2790	1.6040	1.2972	1.5699	26
1.2453	1.6779	1.2619	1.6396	1.2793	1.6034	1.2975	1.5694	25
1.2456	1.6772	1.2622	1.6389	1.2795	1.6029	1.2978	1.5688	24
1.2459	1.6766	1.2624	1.6383	1.2798	1.6023	1.2981	1.5683	23
1.2461	1.6759	1.2627	1.6377	1.2801	1.6017	1.2985	1.5677	22
1.2464	1.6752	1.2630	1.6371	1.2804	1.6011	1.2988	1.5672	21
1.2467	1.6746	1.2633	1.6365	1.2807	1.6005	1.2991	1.5666	20
1.2470	1.6739	1.2636	1.6359	1.2810	1.6000	1.2994	1.5661	19
1.2472	1.6733	1.2639	1.6352	1.2813	1.5994	1.2997	1.5655	18
1.2475	1.6726	1.2641	1.6346	1.2816	1.5988	1.3000	1.5650	17
1.2478	1.6720	1.2644	1.6340	1.2819	1.5982	1.3003	1.5644	16
1.2480	1.6713	1.2647	1.6334	1.2822	1.5976	1.3006	1.5639	15
1.2483	1.6707	1.2650	1.6328	1.2825	1.5971	1.3010	1.5633	14
1.2486	1.6700	1.2653	1.6322	1.2828	1.5965	1.3013	1.5628	13
1.2488	1.6694	1.2656	1.6316	1.2831	1.5959	1.3016	1.5622	12
1.2490	1.6687	1.2659	1.6309	1.2834	1.5953	1.3019	1.5617	11
1.2494	1.6681	1.2661	1.6303	1.2837	1.5947	1.3022	1.5611	10
1.2497	1.6674	1.2664	1.6297	1.2840	1.5942	1.3025	1.5606	9
1.2499	1.6668	1.2667	1.6291	1.2843	1.5936	1.3029	1.5600	8
1.2502	1.6661	1.2670	1.6285	1.2846	1.5930	1.3032	1.5595	7
1.2505	1.6655	1.2673	1.6279	1.2849	1.5924	1.3035	1.5590	6
1.2508	1.6648	1.2676	1.6273	1.2852	1.5919	1.3038	1.5584	5
1.2510	1.6642	1.2679	1.6267	1.2855	1.5913	1.3041	1.5579	4
1.2513	1.6636	1.2681	1.6261	1.2858	1.5907	1.3044	1.5573	3
1.2516	1.6629	1.2684	1.6255	1.2861	1.5901	1.3048	1.5568	2
1.2519	1.6623	1.2687	1.6249	1.2864	1.5896	1.3051	1.5563	1
1.2521	1.6616	1.2690	1.6243	1.2867	1.5890	1.3054	1.5557	0
Co-sec.	Sec.	Co-sec.	Sec.	Co-sec.	Sec.	Co-sec.	Sec.	
53°		52°		51°		50°		

′	40° Sec.	40° Co-sec.	41° Sec.	41° Co-sec.	42° Sec.	42° Co-sec.	43° Sec.	43° Co-sec.	′
0	1.3054	1.5557	1.3250	1.5242	1.3456	1.4945	1.3673	1.4663	60
1	1.3057	1.5552	1.3253	1.5237	1.3460	1.4940	1.3677	1.4658	59
2	1.3060	1.5546	1.3257	1.5232	1.3463	1.4935	1.3681	1.4654	58
3	1.3064	1.5541	1.3260	1.5227	1.3467	1.4930	1.3684	1.4649	57
4	1.3067	1.5536	1.3263	1.5222	1.3470	1.4925	1.3688	1.4644	56
5	1.3070	1.5530	1.3267	1.5217	1.3474	1.4921	1.3692	1.4640	55
6	1.3073	1.5525	1.3270	1.5212	1.3477	1.4916	1.3695	1.4635	54
7	1.3076	1.5520	1.3274	1.5207	1.3481	1.4911	1.3699	1.4631	53
8	1.3080	1.5514	1.3277	1.5202	1.3485	1.4906	1.3703	1.4626	52
9	1.3083	1.5509	1.3280	1.5197	1.3488	1.4901	1.3707	1.4622	51
10	1.3086	1.5503	1.3284	1.5192	1.3492	1.4897	1.3710	1.4617	50
11	1.3089	1.5498	1.3287	1.5187	1.3495	1.4892	1.3714	1.4613	49
12	1.3092	1.5493	1.3290	1.5182	1.3499	1.4887	1.3718	1.4608	48
13	1.3096	1.5487	1.3294	1.5177	1.3502	1.4882	1.3722	1.4604	47
14	1.3099	1.5482	1.3297	1.5171	1.3506	1.4877	1.3725	1.4599	46
15	1.3102	1.5477	1.3301	1.5166	1.3509	1.4873	1.3729	1.4595	45
16	1.3105	1.5471	1.3304	1.5161	1.3513	1.4868	1.3733	1.4590	44
17	1.3109	1.5466	1.3307	1.5156	1.3517	1.4863	1.3737	1.4586	43
18	1.3112	1.5461	1.3311	1.5151	1.3520	1.4858	1.3740	1.4581	42
19	1.3115	1.5456	1.3314	1.5146	1.3524	1.4854	1.3744	1.4577	41
20	1.3118	1.5450	1.3318	1.5141	1.3527	1.4849	1.3748	1.4572	40
21	1.3121	1.5445	1.3321	1.5136	1.3531	1.4844	1.3752	1.4568	39
22	1.3125	1.5440	1.3324	1.5131	1.3534	1.4839	1.3756	1.4563	38
23	1.3128	1.5434	1.3328	1.5126	1.3538	1.4835	1.3759	1.4559	37
24	1.3131	1.5429	1.3331	1.5121	1.3542	1.4830	1.3763	1.4554	36
25	1.3134	1.5424	1.3335	1.5116	1.3545	1.4825	1.3767	1.4550	35
26	1.3138	1.5419	1.3338	1.5111	1.3549	1.4821	1.3771	1.4545	34
27	1.3141	1.5413	1.3342	1.5106	1.3552	1.4816	1.3774	1.4541	33
28	1.3144	1.5408	1.3345	1.5101	1.3556	1.4811	1.3778	1.4536	32
29	1.3148	1.5403	1.3348	1.5096	1.3560	1.4806	1.3782	1.4532	31
30	1.3151	1.5398	1.3352	1.5092	1.3563	1.4802	1.3786	1.4527	30
31	1.3154	1.5392	1.3355	1.5087	1.3567	1.4797	1.3790	1.4523	29
32	1.3157	1.5387	1.3359	1.5082	1.3571	1.4792	1.3794	1.4518	28
33	1.3161	1.5382	1.3362	1.5077	1.3574	1.4788	1.3797	1.4514	27
34	1.3164	1.5377	1.3366	1.5072	1.3578	1.4783	1.3801	1.4510	26
35	1.3167	1.5371	1.3369	1.5067	1.3581	1.4778	1.3805	1.4505	25
36	1.3170	1.5366	1.3372	1.5062	1.3585	1.4774	1.3809	1.4501	24
37	1.3174	1.5361	1.3376	1.5057	1.3589	1.4769	1.3813	1.4496	23
38	1.3177	1.5356	1.3379	1.5052	1.3592	1.4764	1.3816	1.4492	22
39	1.3180	1.5351	1.3383	1.5047	1.3596	1.4760	1.3820	1.4487	21
40	1.3184	1.5345	1.3386	1.5042	1.3600	1.4755	1.3824	1.4483	20
41	1.3187	1.5340	1.3390	1.5037	1.3603	1.4750	1.3828	1.4479	19
42	1.3190	1.5335	1.3393	1.5032	1.3607	1.4746	1.3832	1.4474	18
43	1.3193	1.5330	1.3397	1.5027	1.3611	1.4741	1.3836	1.4470	17
44	1.3197	1.5325	1.3400	1.5022	1.3614	1.4736	1.3839	1.4465	16
45	1.3200	1.5319	1.3404	1.5018	1.3618	1.4732	1.3843	1.4461	15
46	1.3203	1.5314	1.3407	1.5013	1.3622	1.4727	1.3847	1.4457	14
47	1.3207	1.5309	1.3411	1.5008	1.3625	1.4723	1.3851	1.4452	13
48	1.3210	1.5304	1.3414	1.5003	1.3629	1.4718	1.3855	1.4448	12
49	1.3213	1.5299	1.3418	1.4998	1.3633	1.4713	1.3859	1.4443	11
50	1.3217	1.5294	1.3421	1.4993	1.3636	1.4709	1.3863	1.4439	10
51	1.3220	1.5289	1.3425	1.4988	1.3640	1.4704	1.3867	1.4435	9
52	1.3223	1.5283	1.3428	1.4983	1.3644	1.4699	1.3870	1.4430	8
53	1.3227	1.5278	1.3432	1.4979	1.3647	1.4695	1.3874	1.4426	7
54	1.3230	1.5273	1.3435	1.4974	1.3651	1.4690	1.3878	1.4422	6
55	1.3233	1.5268	1.3439	1.4969	1.3655	1.4686	1.3882	1.4417	5
56	1.3237	1.5263	1.3442	1.4964	1.3658	1.4681	1.3886	1.4413	4
57	1.3240	1.5258	1.3446	1.4959	1.3662	1.4676	1.3890	1.4408	3
58	1.3243	1.5253	1.3449	1.4954	1.3666	1.4672	1.3894	1.4404	2
59	1.3247	1.5248	1.3453	1.4949	1.3669	1.4667	1.3898	1.4400	1
60	1.3250	1.5242	1.3456	1.4945	1.3673	1.4663	1.3902	1.4395	0
′	Co-sec.	Sec.	Co-sec.	Sec.	Co-sec.	Sec.	Co-sec.	Sec.	′
	49°		48°		47°		46°		

'	44° Sec.	44° Co-sec.	'	'	44° Sec.	44° Co-sec.	'	'	44° Sec	44° Co-sec.	'
0	1.3902	1.4395	60	21	1.3984	1.4305	39	41	1.4065	1.4221	19
1	1.3905	1.4391	59	22	1.3988	1.4301	38	42	1.4069	1.4217	18
2	1.3909	1.4387	58	23	1.3992	1.4297	37	43	1.4073	1.4212	17
3	1.3913	1.4382	57	24	1.3996	1.4292	36	44	1.4077	1.4208	16
4	1.3917	1.4378	56	25	1.4000	1.4288	35	45	1.4081	1.4204	15
5	1.3921	1.4374	55	26	1.4004	1.4284	34	46	1.4085	1.4200	14
6	1.3925	1.4370	54	27	1.4008	1.4280	33	47	1.4089	1.4196	13
7	1.3929	1.4365	53	28	1.4012	1.4276	32	48	1.4093	1.4192	12
8	1.3933	1.4361	52	29	1.4016	1.4271	31	49	1.4097	1.4188	11
9	1.3937	1.4357	51	30	1.4020	1.4267	30	50	1.4101	1.4183	10
10	1.3941	1.4352	50	31	1.4024	1.4263	29	51	1.4105	1.4179	9
11	1.3945	1.4348	49	32	1.4028	1.4259	28	52	1.4109	1.4175	8
12	1.3949	1.4344	48	33	1.4032	1.4254	27	53	1.4113	1.4171	7
13	1.3953	1.4339	47	34	1.4036	1.4250	26	54	1.4117	1.4167	6
14	1.3957	1.4335	46	35	1.4040	1.4246	25	55	1.4122	1.4163	5
15	1.3960	1.4331	45	36	1.4044	1.4242	24	56	1.4126	1.4159	4
16	1.3964	1.4327	44	37	1.4048	1.4238	23	57	1.4130	1.4154	3
17	1.3968	1.4322	43	38	1.4052	1.4233	22	58	1.4134	1.4150	2
18	1.3972	1.4318	42	39	1.4056	1.4229	21	59	1.4138	1.4146	1
19	1.3976	1.4314	41	40	1.4060	1.4225	20	60	1.4142	1.4142	0
20	1.3980	1.4310	40								
'	Co-sec.	Sec.	'	'	Co-sec.	Sec.	'	'	Co-sec.	Sec	'

45°

DICTIONARY OF MACHINE SHOP TERMS

This has been compiled to assist in a definite understanding of the names of tools, appliances and shop terms which are used in various parts of the country, and will, we trust, prove of value in this way. Cross references have been used in many cases, and we believe that no trouble will be experienced in finding the definition desired even where it may not be under the letter expected. Cutters of all kinds are under "cutters," twist drills under "drills," and by bearing this in mind no delay will be experienced. Practical suggestions as to additions to this section will be appreciated.

DICTIONARY OF SHOP TERMS

A

Ampere — The unit of electric current. The amount of current which one volt can force through a resistance of one ohm.

Ampere Hour. — One ampere flowing for one hour.

Ampere Turns. — Used in magnet work to represent the number of turns times the number of amperes.

Angle Irons — Pieces, usually castings, for holding work at an angle with the face-plate of a lathe, the platen of a planer or other similar work. Usually at right angles but can be anything desired.

Angle Plate — A cast-iron plate with two surfaces at right angles to each other; one side is bolted to a machine table, the other carries the work.

Annealing — Softening steel, rolled brass or copper by heating to a low heat and allowing to cool gradually.

Annealing Boxes — Boxes, usually of cast iron, in which steel is packed with lime or sand to retard the cooling as much as possible.

Anode — The positive terminal of any source of electricity as a battery, or where the current goes into a plating bath.

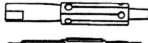

Anvils — Blocks of iron or steel on which metals are hammered or forged. Usually have a steel face. A square hole is usually provided for holding hardies, fuller blocks, etc.

Apron — A protecting or covering piece which encloses or covers any mechanism, as the apron of a lathe.

Arbor — Shaft or bar to hold work while it is being turned or otherwise worked on. Usually made with a 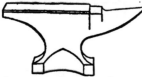 slight taper (about .010 inch per foot) to drive into work and hold by friction. Also applied to shaft for holding circular saws, milling cutters, etc. Often called mandrel.

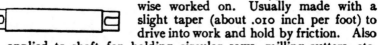

Arbor, Expansion — Arbor which can be varied in diameter to hold different sized work. These vary greatly in design, as shown. The first and last are spring sleeves of different types, the second has blades which can be adjusted to size.

Arc — The passage of current across the space between two separated points.

Armature. — Usually the revolving part of a dynamo or motor or the movable part of any magnetic device.

B

Babbitt Metal — A good mixture for bearings where the load is not too heavy. Consists of varying proportions of tin, antimony, and copper, and sometimes lead. Tin is the base.

Back-lash — Usually applied to lost motion in gears, sometimes to screw in a nut.

Backing-off — Removing metal behind the cutting edge to relieve friction in taps, reamers, drills, etc. Also called "relieving."

Back Rest — A rest attached to the lathe ways for supporting long, slender shafts or other work being turned.

Balance, Running — High-speed pulleys require balancing by running at speed and seeing that they run without tremor or vibration. This is called running balance.

Balance, Standing — When a pulley has been balanced on the balancing ways it is called a standing balance. See Balance-running.

Balancing Ways — Level strips on which the shaft carrying the pulley or other revolving body is placed. If the pulley is unbalanced the heavy side will roll to bottom.

Ball Reamer — See Reamer, Ball.

Bastard — Not regular. The term is usually applied to a file, meaning a cut between the rough and second cut, or to a thread, meaning one that is not of the standard proportions.

Battery. — A combination of chemicals which will give off an electric current.

Bearings, Ball — Made to reduce friction by interposing balls between the shaft and the bearing. They are made in various ways but all aim to have a rolling instead of a sliding action.

Bearings, Roller — Similar to ball bearings except rollers are used instead of balls. In some cases the rollers are practically hollow round springs from square stock. These are known as flexible roller bearings (Hyatt).

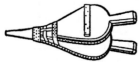

Bellows — Devices of wood and leather for producing a current of air for fanning fires or blowing dust.

Bearing, Base Plate — For supporting pillow blocks or journal boxes.

Belt Carriers — Pulleys for supporting a long belt between driving and driven pulleys. May or may not have flanges.

Belt Dressing — Preparation for preserving or cleaning a belt or making it cling to pulleys.

Belt Fastener — Hooks or other device for joining the ends of belt.

Belt Lacing — Methods of fastening ends of belt with a more or less flexible joint by means of leather or wire lacing.

Belt, Muley — A belt running around a corner guided by idler pulleys on a muley shaft.

elt Polisher or Strap — A belt covered with glue and emery or other abrasive is driven over pulleys and work held against it.

elt Shifter — Device for shifting belt or belts on countershaft or elsewhere, from loose to tight pulleys and vice-versa. These are made in many varieties. Not used where clutches are employed.

elt Tightener — Loose pulleys arranged for taking up slack of belts; often called idlers.

ench — Usual hight is 34 to 35 inches from floor to top of bench, width about 29 inches. Should be 3 inches from wall to allow circulation of air, in order to give sprinklers a chance at a fire underneath.

ench, Leveling — Bench with a level surface so that work can be laid on it to test. Made of iron.

ending Machine — For bending rods, beams, rails, plate, etc. Run by hydraulic or other power.

Bevel — A tool for measuring or laying off bevels as shown. Also a surface not at right angles to the main surface; may be any angle. When at 45 degrees sometimes called a miter.

Blocks, Differential — Hoisting apparatus consisting of differential gears for lifting heavy loads.

Blocks, Tackle — Sheaves or pulleys mounted in a shell or case, used with hoisting ropes or chains to raise heavy weights.

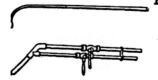

Blow Pipe — A pipe for blowing a jet of air into a flame for heating work locally, such as soldering. The upper picture is a plain one for use with an alcohol lamp, the other has a gas and an air tube. Each is regulated by the small valve so as to make the hottest flame.

BOLTS

Agricultural Bolt. Agricultural bolts, as indicated by the name, are used in farm machines and appliances. The body of the bolt has a series of helical lands and grooves which are formed by a rolling process.

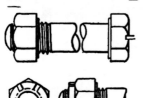

A. L. A. M. Bolt — This bolt is adopted by the Association of Licensed Automobile Manufacturers. It has a slotted head and castellated nut.

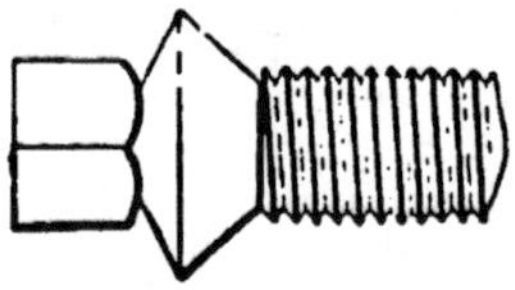

Boiler Patch Bolt — A bolt used in fastening patches on boilers. The patch is countersunk for the cone head, and boiler shell tapped for bolt thread. The square head is knocked off after bolt is screwed in place.

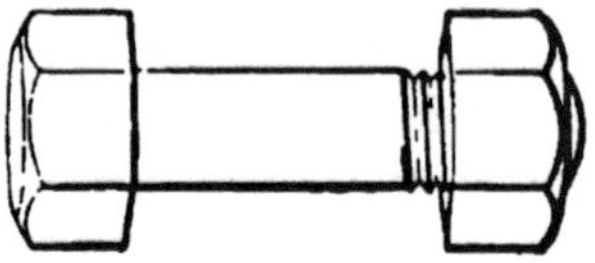

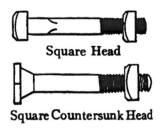

Coupling Bolt — Bolts for shaft couplings are finished all over and must be a close fit in the hole reamed in the two flanges of the coupling, so that the sections shall be rigidly secured together.

Expansion Bolt — In attaching parts to brick, stone or concrete walls and floors, expansion bolts are frequently employed. The

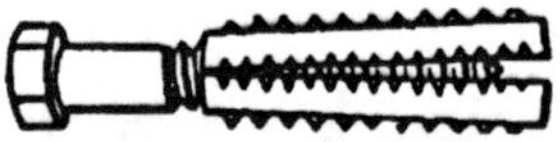

"Star" bolt in the illustration has an internally threaded, split sleeve which is slipped into a hole made in the wall and then expanded by running in the screw. The projections on the surface of the shell, and the fact that the hole receiving it is made larger at the rear, assure the device holding fast when the expander is in place.

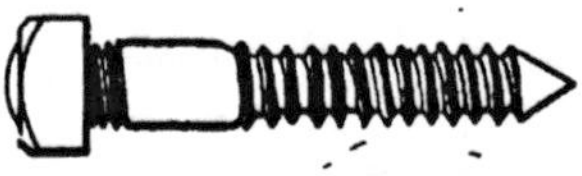

Hanger Bolt — This bolt is used for attaching hangers to woodwork and consists of a lag screw at one end with a machine bolt thread and nut at the other.

Machine Bolts

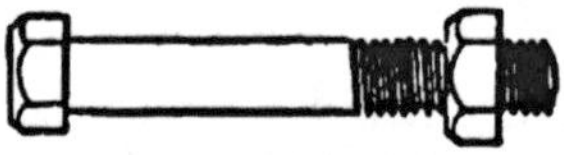

Hexagon Head

Square Head

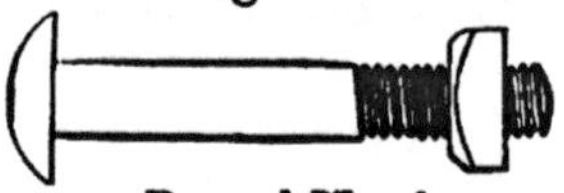

Round Head

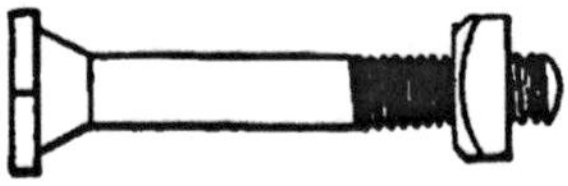

Square Countersunk Head

Miscellaneous Bolts

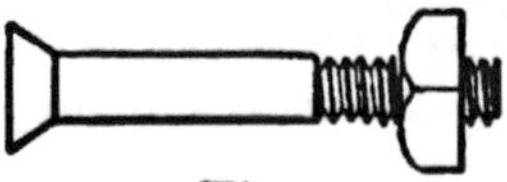

Tire

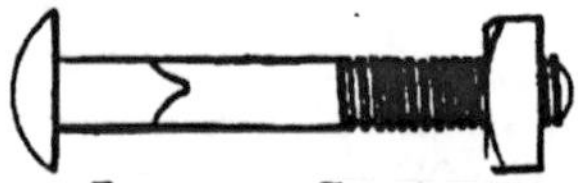

Loom or Carriage

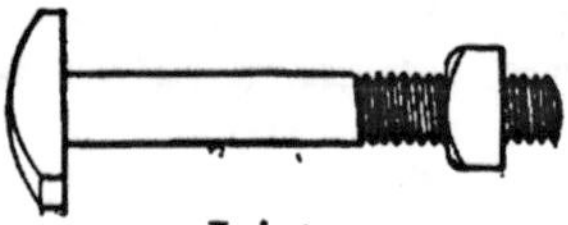

Joint

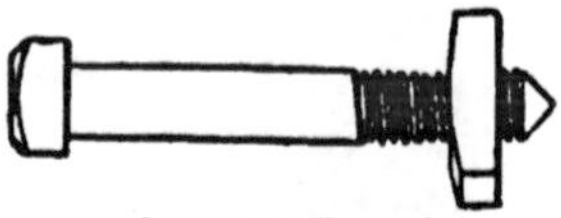

Oval T-Head

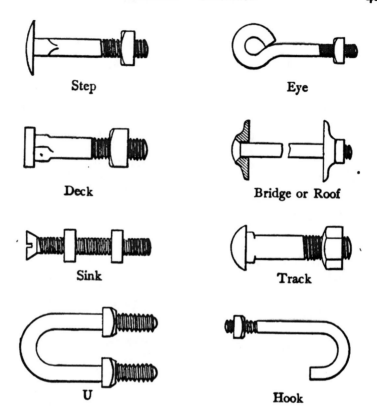

Step

Eye

Deck

Bridge or Roof

Sink

Track

U

Hook

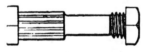

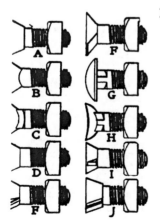

"North" Bolt — The "North" bolt is used in agricultural machinery and appliances and has a series of longitudinal lands rolled on the body to the same diameter as the bolt.

Plow Bolt — Several types of plow and cultivator bolts are shown in the accompanying engravings, the forms illustrated being typical of a variety of bolts manufactured for agricultural apparatus.

A — Large Round Head
B — Square Head
C — Round Head, Square Shank
D — Round Head
E — Key Head
F — Tee Head
G — Button Head
H — Concave Head
I — Reverse Key Head
J — Large Key Head.

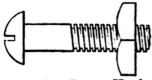

Round or Button Head

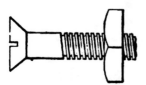

Flat or Countersunk
Head

Stove Bolt — Stove bolts are made in sizes ranging from ⅛ to ⅜ inch. There is no standard form of thread for these bolts to which all makers adhere, and even the same makers in some cases have a different shape of thread for different sizes of bolts. The heads commonly formed are the round, or button head, and the flat or countersunk head.

Hexagon Head

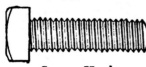

Square Head

Tap Bolt — Tap bolts are usually threaded the full length of the body, which is not machined prior to running on the die. Only the point and the under side of the head are finished. They are not hardened and are used as a rule for the rougher classes of machine work. The heads are the same width as machine bolt heads.

T-Head Planer Bolt

T-Head Planer Bolt — A bolt with a T-head having oblique ends which may be dropped into the T-slot of a planer and locked by giving it a quarter turn, until the sloping ends strike against the sides of the slot. Commonly employed for holding work on the planer table.

Bolt Cutter — Machine for threading bolts, cutting threads on them.

Bolt Header — Machine for upsetting the bolt body to form the head.

Bolster — A block sometimes called the die block, in which a punch press die is held. It is attached to the bed by bolts at either end.

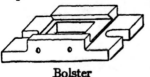

Bolster

Boring and Turning Mill — Machine having a rotating horizontal table for the work with one or more stationary vertical tools for boring, turning or facing; a turret is often provided for holding a number of tools in one of the heads. Often called "vertical mill." Horizontal boring machines are not called "mills."

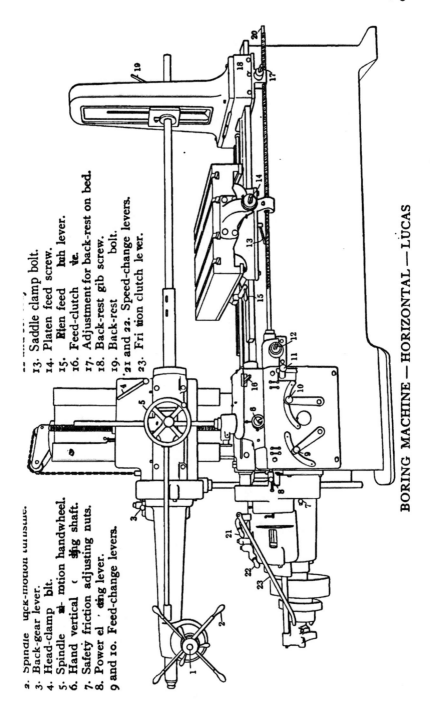

2. Spindle quick-motion turnstile.
3. Back-gear lever.
4. Head-clamp bolt.
5. Spindle at- motion handwheel.
6. Hand vertical (ding shaft.
7. Safety friction adjusting nuts.
8. Power el · ding lever.
9 and 10. Feed-change levers.

13. Saddle clamp bolt.
14. Platen feed screw.
15. Elen feed hoh lever.
16. Feed-clutch le.
17. Adjustment for back-rest on bed.
18. Back-rest gib screw.
19. Back-rest bolt.
21 and 22. Speed-change levers.
23. Fri tion clutch le ver.

BORING MACHINE — HORIZONTAL — LUCAS

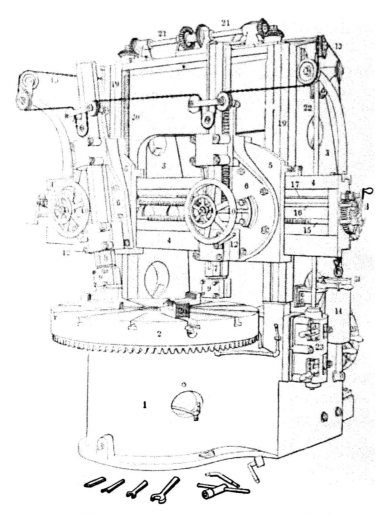

BORING MILL — VERTICAŁ — NILES

1. Base.
2. Table.
3. Housing.
4. Cross-rail.
5. Saddle.
6. Swivel.
7. Right spindle.
8. Left spindle.
9. Tool heads.
10. Vertical feed wheels.
11. Power feed lock.
12. Spindle bearings.
13. Counterweight chain.

14. Counterweight.
15. Cross-feed screw.
16. Cross-feed screw.
17. Vertical feed rod.
18. Power feed gears.
19. Housing slides.
20. Vertical cross-rail screw.
21. Cross-rail hoist.
22. Vertical power rod.
23. Gear box.
24. Power control handle.
25. Driving pulleys.
26. Chuck jaws.

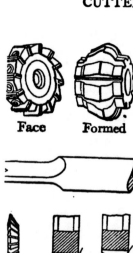

Face   Formed

Face and Formed Cutters — The face cutter to the left, of Brown & Sharpe inserted tooth type is made in large sizes and cuts on the periphery and ends of teeth.

The formed cutter to the right may be sharpened by grinding on the face without changing the shape. For milling wide forms several cutters are often placed side by side in a gang.

Fish Tail Cutter — A simple cutter for milling a seat or groove in a shaft or other piece. Usually operated at rapid speed and light cut and feed.

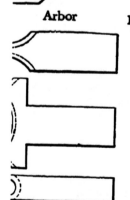

A   *B*   *C*

Fluting Cutters — Cutter *A* is an angular mill for cutting the teeth in spiral mills, cutter *B* is for tap fluting and *C* for milling reamer flutes. In each case the cutter is shown with one face set radial to the center of the work.

Arbor

Fly Cutters — Fly cutters are simple formed cutters which may be held in an arbor like that shown at the top of the group. The arbor is placed in the miller spindle and the tool or other work to be formed is given a slow feed past the revolving cutter. After roughing out, the cutter can be held stationary and used like a planer tool for finishing the work which is fed past it and so given a scraping cut.

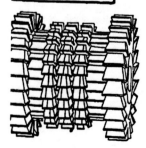

Gang Cutters — Cutters are used in a gang on an arbor for milling a broad surface of any desired form. The cutters shown have interlocking and overlapping teeth so that proper spacing may be maintained. In extensive manufacturing operation the gangs of cutters are usually kept set up on their arbor and never removed except for grinding.

Button — A steel bushing, hardened and ground, used for locating a jig plate or some similar piece in which holes have to be bored in exact position. The button is attached to the work by a small screw and is then adjusted by a micrometer or otherwise until it is central at the exact point where it is desired to bore the hole. The work is then placed on the face plate of the lathe, and with a test indicator resting on the outside of the button, the piece is readily set central. It is then clamped fast to the face plate, the button is removed and the hole bored. Frequently, several buttons are used on the same piece of work, their relative positions being adjusted to conform to the center distances required between holes. The work is then indicated true by each button in succession, and one hole after another bored.

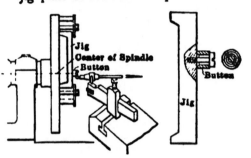

C

CALIPERS

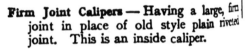

Firm Joint Calipers — Having a large, firm joint in place of old style plain riveted joint. This is an inside caliper.

Gear Tooth Caliper — A caliper with two beams at right angles. The vertical beam gives tooth depth to pitch line and the other the thickness at pitch line. Both have verniers. Used in measuring teeth for accuracy.

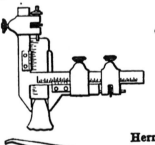

Hermaphrodite Caliper — A combination of one leg of a divider and one leg of a caliper. Used in testing centered work and in laying off distances from the edge of a piece.

Keyhole Caliper — Has one straight leg and the other curved.

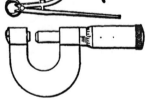

Micrometer Caliper — A measuring instrument consisting of a screw and having its barrel divided into small parts so as to measure slight degrees of rotation. Usually measure to thousandths, sometimes to ten-thousandths.

Odd Leg Caliper — Calipers having both legs pointing in the same direction. Used in measuring shoulder distances on flat work, boring half round, boxes etc.

Outside, Spring Caliper — Tool for measuring the outside diameter of work. Controlled by spring and threaded nut. Nuts are sometimes split or otherwise designed to allow rapid movement when desired, final adjustment being made by screw.

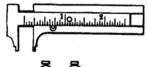

Slide Caliper — A beam caliper made with a graduated slide. Generally made small for carrying in the pocket.

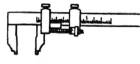

Square-micrometer Caliper — A beam caliper having jaws square with the blade, and having a micrometer adjustment to read to thousandths of an inch.

Thread Caliper — Similar to outside calipers except it has broad points to go over the tops of several threads.

Transfer Caliper. — Caliper which can be set to a given size, the auxiliary arm set, and the calipers opened at will, as they can be reset to the auxiliary arm at any time. Used to caliper recesses and places where they must be moved to get them out.

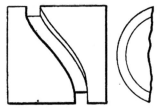

Cam, Drum or Barrel — The drum cam has a path for the roll cut around the periphery, and imparts a to-and-for motion to a slide or lever in a plane parallel to the axis of the cam. Sometimes these cams are built up of a plain drum with cam plates attached.

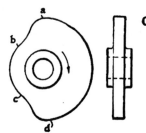

Cam, Edge — Edge or peripheral cams (also called disk cams) operate a mechanism in one direction only, gravity, or a spring, being relied upon to hold the cam roll in contact with the edge of the cam. On the cam shown, *a* to *b* is the drop; *b* to *c* the dwell; *c* to *d*, rise; *d* to *a*, dwell.

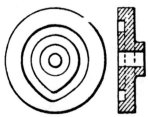

Cam, Face — Face cams have a groove or roll path cut in the face and operate a lever or other mechanism positively in both directions, as the roll is always guided by the sides of the slot.

"C" Clamp — See Clamp "C."

"C" Washer. — See Washer, Open.

Carbonizing — The heat treatment of steel so that the outer surface will be hard. The surface absorbs carbon from the material used.

Card Patterns — Patterns made on a gate so as to be all molded at once and to provide gates for the metal to flow.

Case-hardening — A surface hardening by which the outer skin of a piece of iron or steel absorbs carbon or carbonizes so as to harden when cooled in water. The piece is usually packed in an iron box with bone, leather or charcoal, or all three, and heated slowly several hours, then quenched.

Cat Head — A collar or sleeve which fits loosely over a shaft and is clamped to it by set screws. Used for steady rest to run on where it is not desired to run it on the work.

Same name is also given to the head carrying cutters on boring bars.

Cat Head Chuck — A chuck in which the end of a shaft or other piece is driven by a number of set screws tapped through the wall of the chuck.

Cathode — The negative terminal of an electric bath or battery.

Center, Dead — The back center or the stationary center on which the work revolves. On many grinding machines both centers are dead.

Center, Live — The center in the revolving spindle of a lathe or similar machine. It is highly important that this should run true or it will cause the work to move in an eccentric path.

Center Punch — Punch for marking points on metal. Made of steel with a sharp point and hardened. Often called a prick punch.

Center Punch, Automatic — Has a spring actuated hammer in the handle, which is released when the handle is pressed way down.

The point can be placed where desired and the blow given by a pressure of the hand. In some cases the blow can be varied.

Center Punch, Bell or Self-centering — A center punch sliding in a bell or cone mouthed casing so when placed square over the end of any bar it will locate the center with sufficient accuracy for most purposes.

Center Punch, Locating — Having an extra leg which has a spring point and is adjustable. The spring point is placed in the first punch mark and so locates the next punch mark at the right distance from the first.

Centering Machines — For drilling and reaming center of work for the lathe or grinder.

Chamber — A long recess. See Recess.

Chasers — Tools used for cutting threads by chasing. Usually have several teeth of right pitch, but name is sometimes applied to a single point tool used in brass work on a Fox lathe. Chasers are made circular or flat and in the old days many were used by hand.

Chasing Threads — Cutting threads by moving a tool along the work at the right speed to give the proper pitch. Distinguishes between threads cut with a die and those cut with a threading tool.

Chattering — A slight jumping of the tool away from the work or vice-versa, and which leaves little ridges in same direction as the teeth. Occurs at times in any class of work and with any kind of tool. Due to springing of some parts of the machine.

Cherry — See Cutters, Milling.

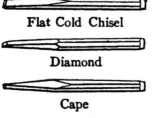

Chisel, Blacksmith's Hot — A chisel for cutting hot metal. Has a handle so that it can be used without getting the hand too near the heated metal.

Chipping — The cutting of metal with cold chisel and hammer. Also used when a piece "chips" or breaks out of a piece or punch.

Chisel, Cape — Chisel with a narrow blade for cutting keyways and similar work.

Flat Cold Chisel

Diamond

Cape

Round

Chisel, Cold — The usual machinists' chisel for cutting or "chipping" metal with a plain cutting edge as in illustration.

Chisel, Diamond or Lozenge — Similar to a cape chisel but with square end and cutting edge at one corner. Used for cutting a sharp-bottomed groove.

Chisel, Round — A round end chisel with the cutting edge ground back at an angle. Used for cutting oil grooves and similar work.

Chuck, Draw — Operated by moving longitudinally in a taper bearing. Used on precision work.

Chuck, Drill — A chuck made especially for holding drills in drilling machines. Sizes run from the smallest up to one inch.

Chuck, Eccentric — For turning eccentrics or other work in which hole is not concentric with outside. Usually made adjustable to suit varying degrees of eccentricity.

Chuck, Expanding — For turning hollow work which must be held on inside. Jaws go inside of work.

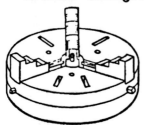

Chucks, Lathe — Devices for holding work. Usually screw on spindle and have two, three or four jaws. These may be independent or move together by screws only (in case of two jawed) or screws and gears in case of more than two. There is also a spiral or scroll chuck without gears or screws of the ordinary kind.

Chuck, Magnetic — Has no jaws but holds iron and steel by magnetism.

Chuck, Master — The main body of a screw chuck which screws on the nose of the lathe spindle and which carries the sub- or screwchuck for holding the work. Mostly used in brass work.

Chuck, Nipple — For holding short piece of pipe to be threaded.

Chuck, Oval — Chuck designed to move the work to and from the tool so as to produce an oval instead of a round. Sometimes called an elliptic chuck.

Plain Base

Swivel Base

Chuck, Planer — For holding work on bed or platen of planer, shaper or milling machine. Sometimes called a vise. They are made with both plain and swivel bases as shown, and usually have locking strips which hold the piece carrying the set screws.

Chuck, Screw — Chucks made with internal or external thread to hold work which has been already threaded. These very often screw into a master chuck. Mostly used in brass work.

Chuck, Spring — See Screw Machine Tools.

Chucking Machines — Usually have a turret for tools, a revolving chuck or table for work, and generally used for boring and reaming. May be either vertical or horizontal.

Circuit — The path in which an electric current flows.

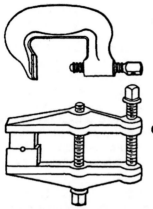

Clamp, "C" — Clamp shaped like a letter "C" for holding work in various ways. Are sometimes cast but more often drop forged for heavier work.

Clamps, Machinist — Clamps for holding work together, holding jigs or templets on work, etc.

Clash Gears — Gears which are thrown into mesh by moving the centers together and sometimes by sliding the gears on parallel shafts till the teeth get a full bearing. The latter are sometimes called sliding gears.

Clutch — Any device which permits one shaft to engage and drive another, may be either friction or positive, usually the former. Made of all sorts of combinations of cams, levers and toggles.

Clutch, Friction — A device whereby motion of loose pulley is transmitted to shaft to be driven. Various methods are employed but all depend on forcing some kind of friction surfaces together so that one drives the other.

Ratchet Tooth Clutch

Clutches, Positive — Devices for connecting machines to a constantly running shaft or one part to another, at will. There are many kinds, both positive and friction. The illustrations show two of the most common of the positive clutches.

Square Jaw Clutch

Collar — A ring used for holding shafting, loose pulleys, in proper position or for fastening to boring tools to prevent them going in too deep.

Collar, Safety — Having a clamping device instead of set screw or having set screw below surface or so covered as not to catch anything brought in contact with it.

Commutator — The part of a dynamo or motor which takes off or leads current into the machine.

Compound Rest — An auxiliary tool slide on lathe carriage arranged to swivel so as to turn at any desired angle with the lathe centers or with cross slide. Usually graduated into degrees.

Cotter, Spring — Also called split cotter, split pin, etc., is used in a hole drilled crosswise of a stud, shaft or some similar member, and its ends spread apart to retain it in place and keep some member carried by the shaft from slipping off.

Counterbore — Has a pilot to fit a hole already drilled, or drilled and reamed, and its body with cutting edges on the end is used to enlarge the hole to receive a screw head or body or for some similar purpose.

Countershaft — The shaft for driving a machine which is itself driven by the main or line shaft.

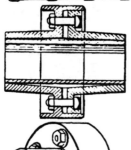

Coupling, Clamp — Couplings made in two or more parts, clamping around the shafts by transverse bolts. Hold either by friction or have dowels in shaft. Sometimes called compression although this is confusing.

Coupling, Compression — Grips shafting by drawing together tapered parts. This forces them against shaft and holds it firmly. Bolts parallel with shaft draw parts together.

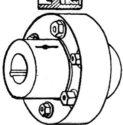

Coupling, Flanged — A flange is keyed to each shaft and these flanges are bolted together. Also called "plate" coupling.

Coupling, Friction — Couplings which depend on frictional contact.

Coupling, Jaw or Clutch — Positively engaged by jaws or projections on the face of opposing parts.

Coupling, Shaft — Devices for fastening ends of shafting together so that both may be driven as one shaft. These are made in a great variety of ways, from plain set screw coupling to elaborate compression devices.

Coupling, Wedge — Coupling that clamps the shaft with a wedging action. Practically like a compression coupling. Generally enclosed in a sleeve. Also called vise coupling.

Cope — The upper part of a flask.

Coping Machine — For cutting away the flanges and corners of beams and bending the ends.

Counter, Revolution — Device for counting the revolutions of a shaft. Generally made with a worm and a gear having 100 teeth so that one turn of dial equals 100 revolutions.

ɔuntershaft — Shaft carrying tight and loose pulleys (or friction clutch pulleys) for starting and stopping machines or reversing their motion.

rane, Gantry — Traveling crane mounted on posts or legs for yard use.

rane, Jib — Crane with a swinging boom or arm.

rane, Locomotive — Crane mounted on a car with an engine so as to be self-propelling on a track.

rane, Pillar — Having the boom or moving arm fastened to pillar or post.

rane, Portable — Hoisting frame on wheels which can be run around to the work and used to handle work in and out of lathes and other machines.

rane, Post — See Crane, Pillar.

rane, Swing — Same as Jib Crane.

rane, Traveling — Crane with a bridge or cross beam having wheels at each end so it can be run on overhead tracks to any point in the shop.

rimping — Fluting, corrugating or compressing metal ring to reduce its diameter.

ross-rail — The part of a planer, boring mill or similar machine on which the tool heads or slides move and are supported.

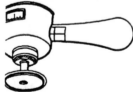

Cut Meter — Instrument for measuring the surface speed of work either in lathe or planer. A wheel is pressed against the moving surface and the speed is shown in feet per minute.

utters, Flue Sheet — Special cutters for making holes as for flues, in flue sheets or in other sheet metal or structural work.

CUTTERS, MILLING

Angular Cutters — Such cutters are used for milling straight and spiral mills, ratchet teeth, etc. Cutters for spiral mill grooving are commonly made with an angle of 12 degrees on one side and 40, 48 or 53 degree angle on the other.

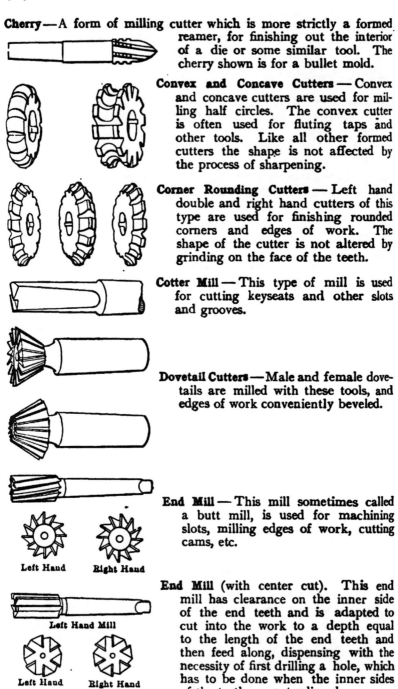

Cherry—A form of milling cutter which is more strictly a formed reamer, for finishing out the interior of a die or some similar tool. The cherry shown is for a bullet mold.

Convex and Concave Cutters — Convex and concave cutters are used for milling half circles. The convex cutter is often used for fluting taps and other tools. Like all other formed cutters the shape is not affected by the process of sharpening.

Corner Rounding Cutters — Left hand double and right hand cutters of this type are used for finishing rounded corners and edges of work. The shape of the cutter is not altered by grinding on the face of the teeth.

Cotter Mill — This type of mill is used for cutting keyseats and other slots and grooves.

Dovetail Cutters —Male and female dovetails are milled with these tools, and edges of work conveniently beveled.

Left Hand   Right Hand

End Mill — This mill sometimes called a butt mill, is used for machining slots, milling edges of work, cutting cams, etc.

Left Hand Mill

End Mill (with center cut). This end mill has clearance on the inner side of the end teeth and is adapted to cut into the work to a depth equal to the length of the end teeth and then feed along, dispensing with the necessity of first drilling a hole, which has to be done when the inner sides of the teeth are not relieved.

Left Hand   Right Hand

The mills are often used for heavy cuts particularly in cast iron.

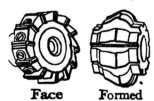

Face   Formed

Face and Formed Cutters — The face cutter to the left, of Brown & Sharpe inserted tooth type is made in large sizes and cuts on the periphery and ends of teeth.

The formed cutter to the right may be sharpened by grinding on the face without changing the shape. For milling wide forms several cutters are often placed side by side in a gang.

Fish Tail Cutter — A simple cutter for milling a seat or groove in a shaft or other piece. Usually operated at rapid speed and light cut and feed.

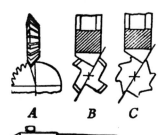

A   B   C

Fluting Cutters — Cutter *A* is an angular mill for cutting the teeth in spiral mills, cutter *B* is for tap fluting and *C* for milling reamer flutes. In each case the cutter is shown with one face set radial to the center of the work.

Arbor

Fly Cutters — Fly cutters are simple formed cutters which may be held in an arbor like that shown at the top of the group. The arbor is placed in the miller spindle and the tool or other work to be formed is given a slow feed past the revolving cutter. After roughing out, the cutter can be held stationary and used like a planer tool for finishing the work which is fed past it and so given a scraping cut.

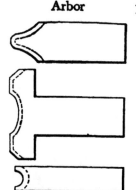

Gang Cutters — Cutters are used in a gang on an arbor for milling a broad surface of any desired form. The cutters shown have interlocking and overlapping teeth so that proper spacing may be maintained. In extensive manufacturing operation the gangs of cutters are usually kept set up on their arbor and never removed except for grinding.

Gear Cutter (Involute).   In the Brown & Sharpe system of involute gear cutters, eight cutters are regularly made for each pitch, as follows:

No. 1 will cut wheels from 135 teeth to a rack.

No. 2 will cut wheels from 55 teeth to 134 teeth.

No. 3 will cut wheels from 35 teeth to 54 teeth.

No. 4 will cut wheels from 26 teeth to 34 teeth.

No. 5 will cut wheels from 21 teeth to 25 teeth.

No. 6 will cut wheels from 17 teeth to 20 teeth.

No. 7 will cut wheels from 14 teeth to 16 teeth.

No. 8 will cut wheels from 12 teeth to 13 teeth.

Such cutters are always accurately formed and can be sharpened without affecting the shape of the teeth.

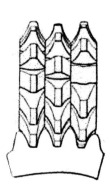

Gear Cutters, Duplex — The Gould & Eberhart duplex cutters are used in gangs of two or more; the number of cutters in the gang depending on the number of teeth in the gear to be cut.   The following table gives the number of cutters which may be used in cutting different numbers of teeth.

Under	30 teeth	1 cutter
Over	30 teeth	2 cutters
Over	50 teeth	3 cutters
Over	70 teeth	4 cutters
Over	95 teeth	5 cutters
Over	120 teeth	6 cutters
Over	150 teeth	7 cutters
Over	180 teeth	8 cutters
Over	230 teeth	10 cutters
Over	260 teeth	12 cutters.

Gear Stocking Cutter — The object of stocking cutters is to rough out the teeth in gears, leaving a smaller amount of metal to be removed by the finishing cutter.   They increase the accuracy with which gears may be cut, and save the finishing cutter as well.

In all cases where accuracy and smooth running are necessary the gears should first be roughed out.   One stocking cutter answers for all gears of the same pitch.

Hob — A form of milling cutter with teeth spirally arranged like a thread on a screw and with flutes to give cutting edges as indicated. Used for cutting the teeth of worm gears to suit the worm which is to operate the gear. Hobs are formed and backed off so that the faces of the teeth may be ground without changing the shape.

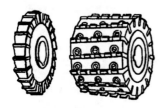

Inserted Tooth Cutter — Brown & Sharpe inserted tooth cutters have taper bushings and screws for holding the blades in position in the bodies. Inserted tooth construction is generally recommended for cutters 6 inches or larger in diameter. There are many types of inserted tooth cutters and in most cases the blades are readily removed and replaced when broken or worn out.

Inserted Tooth Cutter (Pratt & Whitney). In this type of cutter the teeth or blades are secured in position by taper pins driven into holes between every other pair of blades; the cutter head being slotted as shown to allow the metal at each side of the taper pin to be pressed firmly against the inserted blades.

Interlocking Side Cutters — These cutters have overlapping teeth and may be adjusted apart to maintain a definite width for milling slots, etc., by using packing between the inner faces.

Plain Cutters — These cutters are for milling flat surfaces. When over $\frac{3}{4}$ inch wide the teeth are usually cut spirally at an angle of from 10 to 15 degrees, to give an easy shearing cut. When of considerable length relative to diameter they are called slabbing mills.

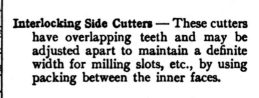

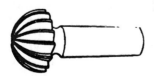

Rose Cutter — The hemispherical cutter known as a rose mill is one of a large variety of forms employed for working out dies and other parts in the profiler. Cutters of this form are also used for making spherical seats for ball joints, etc.

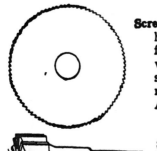

Screw Slotting Cutter — Screw slotting cutters have fine pitch teeth especially adapted for the slotting of screw heads and similar work. The cutters are not ground on the sides. They are made of various thicknesses corresponding to the numbers of the American Wire Gage.

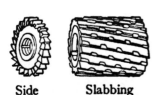

Shank Cutter — Shank milling cutters are made in all sorts of forms with shanks which can be conveniently held true in miller or profiler while in operation.

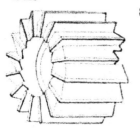

Shell End Cutter — Shell end mills are designed to do heavier work than that for which the regular type of end mills are suited. They are made to be used on an arbor and are secured by a screw in the end of the arbor. The end of the cutter is counterbored to receive the head of the screw and the back end is slotted for driving as indicated.

Side or Straddle, and Slabbing Cutters — Side cutters like that to the left cut on the periphery and sides, are suitable for milling slots and when used in pairs are called straddle mills. May be packed out to mill any desired width of slot or opposite faces of a piece of any thickness.

Side   Slabbing

Slabbing cutters are frequently made with nicked teeth to break up the chip and so give an easier cut than would be possible with a plain tooth.

Slitting Saw, Metal — Metal slitting saws are thin milling cutters. The sides are finished true by grinding, and a little thicker at the outside edge than near the center, for proper clearance. Coarse teeth are best adapted for brass work and deep slots and fine teeth for cutting thin metal.

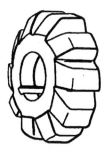

Sprocket Wheel Cutter — Cutters for milling the teeth on sprocket wheels for chains are formed to the necessary outline and admit of grinding on the face like regular gear cutters, without changing the form of the tooth.

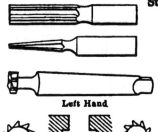

Straight Shank Cutter — Straight shank cutters of small size are extensively used in profilers and vertical millers for die sinking, profiling, routing, etc. They are held in spring chucks or collets.

Left Hand

T-Slot Cutter — Slots for bolts in miller and other tables are milled with T-slot cutters. They are made to standard dimensions to suit bolts of various sizes. The narrow part of the slot is first milled in the casting, then the bottom portion is widened out with the T-slot cutter.

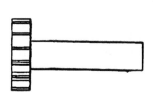

Left Hand
Cutter

Right Hand
Cutter

Woodruff Key Cutter — The Whitney Mfg. Co's keys are semi-circular in form and for cutting the seats in the shaft to receive them a cutter of the type shown is used. These cutters are made of right diameter and thickness to suit all the different sizes of keys in the Woodruff system.

Cutting-off Machine — For cutting desired lengths from commercial bars of iron, steel or other material, usually has stationary tools and revolves the work. The latter is gripped by the rotating chuck; and as the tools are fed toward the center, the spindle in some types of machines is driven at an accelerated speed so that as the diameter of the cut is reduced, the speed of rotation is increased to maintain a practically uniform surface speed of work. In cold-saw cutting-off machines, the work is held in a vise and a rotating circular cutter is fed against it. Such machines are used not only for severing round stock but also for cutting off square and rectangular bars, rails, I-beams, channels and other structural steel shapes.

Cutting-off Tool Post — The tool block used on the cross slide of a turret lathe or other machine for carrying the tool for cutting off the completed piece of work from the bar of material held in the chuck. The tool post may be made to receive a straight tool or a circular cutter.

D

Daniels Planer — See Planer.

Dead Center — The center in the tail spindle of a lathe or grinder which does not revolve.

Derrick — Structure consisting of a fixed upright and an arm hinged at bottom, which is raised and lowered and usually swings around to handle heavy loads.

Dial Feed — A revolving disk which carries blanks between the punch and die.

Diamond Hand Tool — Used for dressing grinding wheels after they

have been roughed out with the cheaper forms of cutters. Fixed diamonds are usually considered better than those held by hand.

Die Chasers — Threaded sections inserted in a die head for cutting bolts and screws.

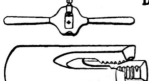

Die, Screw Plate or Stock — A frame or handle for holding a threading die. Sometimes die and handles are of one piece.

Die, Spring or Prong — Die with cutting portions in the end of prongs; can be adjusted somewhat by springing prongs together with a collar on outside.

Dies, Bolt — Dies for cutting bolts. Some are solid, others adjustable. Some for hand die stocks or plates but mostly for machine bolt cutters.

DIES, PUNCH PRESS

Bending Dies, Compound — In compound bending dies of the type
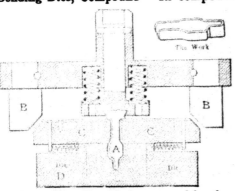
shown the work is carried down into the die by punch A, and held there while the beveled fingers B act upon the slides C and cause them to move forward in the top of die D and bend the material to the outline of the punch. Upon the up-stroke of the punch the slides C are pressed outwardly by their springs and the bent piece of work is removed by the punch from the die. It will be seen that the holder for punch A, upon which depends the interior form of the piece being bent, is not positively secured in its holder, but is instead adapted to slide up and down in its seat.

although prevented from turning by a small pin at the upper end of the shank which is engaged by a slot in the punch carrier. The springs shown above the punch proper tend to hold the punch in its lower position and at the same time after the punch has passed down into the die allow the punch carrier to descend still further to press fingers *B* into operation against slides *C* which bend the work to the outline of the punch.

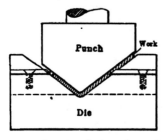

Bending Dies, Plain — Simple bending dies are made with the upper face of the die and the bottom of the punch shaped to conform to the bend it is desired to give the blank. A common type is shown in the engraving.

In simple bending dies the upper face of the die is cut out to the desired form and the piece of work formed to required shape by being pressed directly down into the die by the punch.

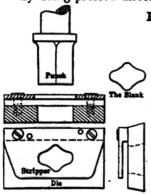

Blanking Dies — Blanking dies are about the most commonly used of all the varieties of press tools. A simple form of die is seen in the illustration. The strip of sheet metal is fed under the stripper and is prevented by that member from lifting with the punch upon the up-stroke, following the punching out of the blank. Where several punches are combined in one hole for blanking as many pieces simultaneously, they are known as multiple blanking tools.

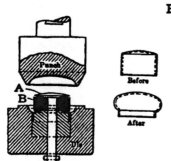

Bulging Dies — The "before and after" sketches show the character of the work handled in bulging dies. The shell after being drawn up straight is placed over mushroom plunger *A* in the bulging die, and when the punch descends the rubber disk *B* is forced out, expanding the shell into the curved chamber formed by the punch and die. Upon the punch ascending, the rubber returns to its original form and the expanded work is then removed.

Burnishing Dies — Burnishing dies are made a little smaller at the bottom than at the top and when the work is forced down throug the die, the edge of the piece is given a very high finish makir polishing and hand finishing operations unnecessary. The burnishing process forms a very accurate sizing method also.

Coining Dies — Coining dies are operated in very powerful presses of the embossing type similar to those used for forming designs on silverware, medals, jewelry, etc. The position of the work, the retaining coller and the dies are indicated at *A*, *B*, and *C*.

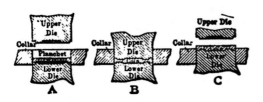

In the latter section the coin is shown delivered at the top of the die.

Combination Dies — Combination dies are used in single acting presses for such work as cutting a blank and at the same stroke turning down the edge and drawing the piece into the required shape. In most cases the work is pushed out of the dies by the action of a spring. Such a set of dies is shown in the engraving, for making a box cover and body. The work is blanked by cutting-punch *A*, and formed to the right shape by *B* and *C*, the former holding the piece by spring pressure to the block *C* while punch *A*, continuing to descend, draws the box to the required shape. Ring *D* acts as an ejector or shedder and is pressed down, compressing the rubber at *E* during the drawing operation, and upon the up-stroke of the punch, ascends and ejects the work from the dies.

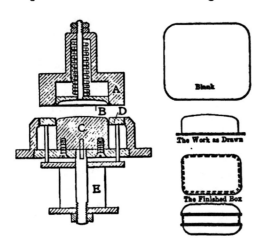

Dies of this type, with a spring actuated punch or die inside the regular blanking tool, are often used for simultaneously blanking and piercing, blanking and bending, etc.

It will be noticed that the lower view in the group showing the work at the right-hand side of the sectional illustration of the die, represents the box cover and body as they appear when assembled after the superfluous metal in the flange or " fin " has been removed in a trimming die. This fin as left on the piece when coming from the combination die is shown in the view immediately under the blank. A trimming die for finishing such work evenly is shown on page 439.

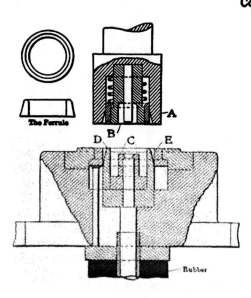

Compound Dies — Compound dies have a die in the upper punch and a punch in the lower die. The ferrule-making tools shown have a blanking and outer drawing punch *A*, with a central die *B*, to receive lower punch *C* which cuts out the center of the ferrule blank and allows the metal to be drawn down inside as well as outside of the bevel edged member *D*. As the work is drawn down ring *E* descends compressing rubber cushion *F* below and upon the return movement the ferrule is ejected from the die.

Cupping Dies — Used for drawing up a cup from a disk or planchet. Same as drawing dies.

Curling Dies — Curling dies are used for producing a curled edge

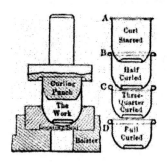

around the top of a piece drawn up from sheet metal. When the top is to be stiffened by a wire ring around which the metal is curled, a wiring die is used, the construction of which is practically the same as the curling die. The illustration shows a curling die and the appearance of a shell at various stages during the operation of curling over the edge. The diagrams *A*, *B*, *C*, *D*, show the progress in the curling process as the punch descends, pressing down on the edge of the straight drawn shell.

Dinking Dies — Dinking dies, or hollow cutters, although not usually classed with regular dies are used

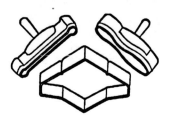

so commonly as to entitle them to be listed in that class. They are adapted to punching out all sorts of shapes from leather, cloth, or paper. The edges of the dies (a few specimens of which are shown in the engraving) are usually beveled about 20 degrees outside. Where made for press use t' handle is omitted. As a surface í the cutting edge of the die to strike on, a block is built up of seasoned rock maple, set endwise of the grain.

Double Action Dies — This type of die is used in a press which has a double acting ram; that is, there are two slides, one inside the other, which have different strokes.

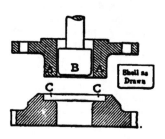

To the outer slide is fastened the combined cutting punch and blank holder *A*, which is operated slightly in advance of the drawing punch *B* actuated by the inner slide. The blank upon being cut from the stock by *A*, drops into the top of the die at *C* and is kept under pressure by the flat end of cutting punch *A* to prevent its wrinkling, while punch *B* continues downward drawing the metal from between the pressure surfaces and into the shape required.

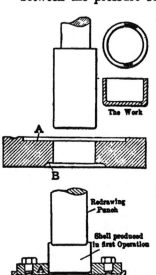

Drawing Dies, Plain — Dies of the type shown can be used for shallow drawing only, as there is no pressure on the blank to prevent its wrinkling when forced down into the die by the punch. The blank fits the recess *A* in the upper face of the die and the die itself which is slightly tapering is made the diameter of the punch plus twice the thickness of the wall required for the shell. The bottom edge *B* of the die strips the shell from the punch when the latter ascends.

Re-drawing dies are used for drawing out a shell or cup already formed from the sheet metal. In the illustration, a shell ready for re-drawing is shown in position in the dies, which need little explanation. The work is located in the upper plate *A*, and after being forced through the die *B*, is stripped from the punch by edge *C*. Ordinarily, a shell which is to be given considerable elongation, is passed through a number of re-drawing dies.

Re-drawing dies are sometimes referred to as reducing dies, although the latter, as explained under the proper heading on page 436, are used for drawing down the end of a shell only, as in the case of a cartridge shell, which is made with a neck somewhat smaller in diameter than the body.

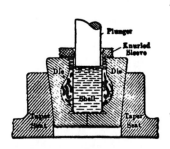

Fluid Dies — Water or fluid dies are used for forming artistic hollow ware of silver, and other soft metals, in exact reproduction of chased work. The die as shown is a hinged mold cast from carved models and finished with all details clean and sharp. The shell to be worked is filled with liquid and enclosed in the die and a plunger in the press ram then descends and causes the fluid to force the metal out into the design in the die.

Follow Dies — Follow tools consist of two or more punches and dies in one punch holder and die body, these being arranged in

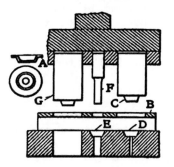

tandem fashion so that after the first operation the stock is fed to the next point and a second operation performed; and so on. In the die shown, which is for making piece *A*, the strip of metal is first entered beneath stripper *B* far enough to allow the first shell to be drawn at the first stroke by punch *C* and die *D*. The metal is then moved along one space and the shell drawn at the first stroke is centered and located within the locating portion of piercing die *E*. At the next stroke the hole in the center is pierced with punch *F*, and a second shell drawn in the stock by punch and die *C* and *D*. The stock is then fed forward another space and the blanking punch *G* cuts out the piece from the metal. At the same stroke a third piece is being formed on the end of the stock and a second hole pierced. Thus three operations are carried on simultaneously.

Gang Dies — Gang tools have two or more punches and dies in one

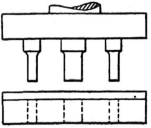

holder for making as many openings in a blank at one stroke of the press. Sometimes dies which perform a number of operations on a piece which is fed along successively under one punch after another are called "gang" dies; strictly speaking, however, such tools are "follow" dies. Where a large number of punches are combined they are called a multiple punch, or if they are of quite small diameter for piercing are sometimes known as perforating punches.

Heading Dies — Heading dies strike up the heads on cartridges and other shells, and are generally operated in a horizontal heading machine.

Index Dies — For certain classes of work such as notching the edges of large disks or armature punchings, an index die is sometimes used consisting of a rotary index plate adapted to carry the work step by step past the punches which cut out one notch or a series of notches at each stroke of the press.

Perforating Dies — Perforating tools consist of a number of piercing tools in one set of dies and may be called also multiple piercing tools. In the example shown, which punches a large number of holes in a disk, the work is held by the spring-controlled pressure-pad *A* against the face of the die *B* while the punches at *C* are forced down through the sheet metal. In this case the punches are easily replaced when broken, by unscrewing the holder from the shank and slipping the small punch out from the back.

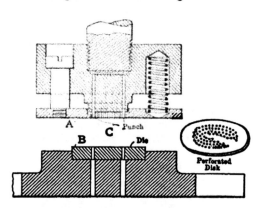

Piercing Dies — Piercing tools are used for punching small holes through sheet metal. Where arranged for punching a large number of holes simultaneously they are often called perforating dies.

Piercing Dies, Compound — Compound piercing tools have, in addition to the regular punches carried by the holder in the ram, a set of horizontal punches for making holes through the sides of the work. These side punches are operated by slides moved inward by wedge-shaped fingers, the arrangement being the same as in the case of the compound bending dies, an illustration of which is given under that head.

Reducing Dies — Reducing dies are re-drawing dies for reducing a portion of the shell only, whereas the regular re-drawing die reduces the whole length. Reducing dies for cartridge shells form the familiar "bottle neck" shell now so commonly manufactured, with a larg body for the powder and a smaller neck into which the bullet is secured. In dial feed presses ordinarly employed for cartridge making operations, two or more reducing dies are often used for shaping the neck of the shell to the required dimensions, each die operating in turn upon the shell as it is carried around step by step under the press tools by the intermittent rotary movement of the feeding dial.

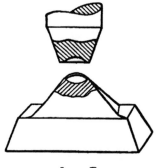

Riveting Dies — Riveting dies for the punch press are provided with cavities in the working faces to suit the shape of the head it is desired to produce on the ends of the rivets.

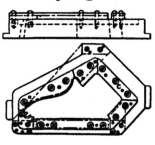

Sectional Dies — Frequently dies of complicated outline are built up in sections to enable them to be more easily constructed and kept in order. This form is resorted to often in the case of large dies where a break at one point would mean considerable expense for a new die. Also the difficulties of hardening are reduced with the sectional construction. As shown, the various parts are secured to a common base or holder.

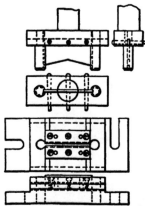

Shearing Dies — Shearing dies are used for cutting-off operations, and are frequently combined with other press tools so that after certain operations on a piece it can be severed from the end of the stock. The shearing tools in the engraving are arranged for simply cutting up stock into pieces of the required length and the punch itself is of the inserted type secured by pins in its holder.

Split Dies — Split dies form one type of sectional die — the simplest; they are made in halves to facilitate working out to shape, hardening, and economical maintainance.

Sub-press Dies — A sub-press and its tools are represented on the following page. Such tools are used for small parts which have to be made accurately and are very common in watch and typewriter shops and similar places. The tools are held positively in line in this press and as a result their efficiency is maintained indefinitely. The press is slipped bodily into the regular power press with the base clamped to the press bed and the neck of the sub-press plunger connected with the ram of the press.

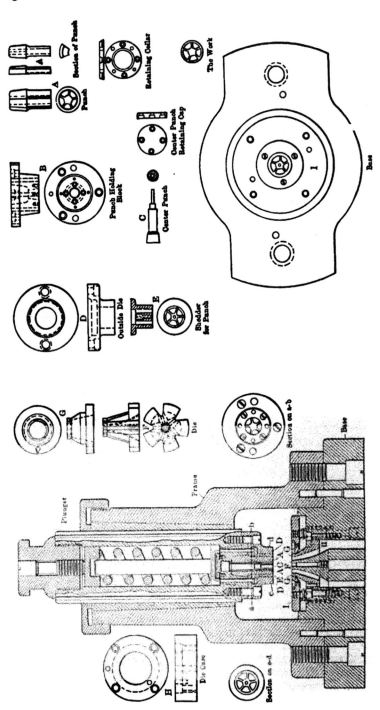

Sub-Press with Tools for a Watch Wheel

Swaging Dies — Swaging operations are resorted to where it is desired to shape up or round over the edges of work already blanked out. Thus in watch wheel work the arms and inside edges of the rims are sometimes swaged to a nicely rounded form subsequently to the blanking out of the wheel in the sub-press. Swaging dies for such work are of course made with shallow impressions which correspond to a split mold between the two halves of which the blank is properly shaped.

Bullet swaging dies receive the slug as it comes from the bullet mold and shape the end to the required cone point.

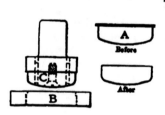

Trimming Dies — Trimming dies remove the superfluous metal left around the edges or ends of various classes of drawn and formed work. In the case shown, the box body *A* has been drawn up and a fin left all the way round; this is dropped into the trimming die *B* and the punch *C* in carrying it through the die trims the edge off evenly, as indicated. Work of the nature shown in this illustration is blanked, drawn up and formed ready for trimming, by means of combination tools, a typical example of which will be found under the combination die heading on page 432. The box body illustrated here as it appears before and after trimming is shown in connection with the combination dies as it appears in the blank, after it is formed, and after assembling with its cover.

Triple-action Dies — These dies are used in triple-acting presses,

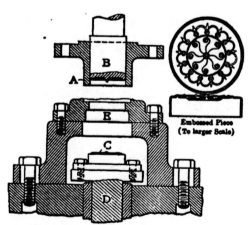

where in addition to the double-action slides which take the place of the regular single-acting ram, there is also a third slide or plunger which operates under the table or die bed. Thus a piece like that shown which has to be blanked, drawn and embossed, is operated upon from above by the cutting and drawing punches *A* and *B*, and upon the latter carrying the drawn work down to the face of the embossing die *C*, that die is forced upward by the plunger *D* beneath and gives the piece the desired impression. On the up-stroke of the punch the work is stripped from it by edge *E* and falls out of the press.

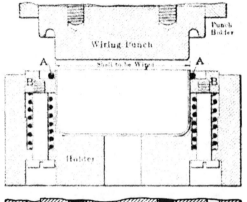

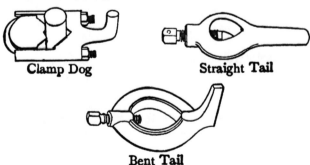

Wiring Dies— Wiring dies are much the same in construction as plain curling dies. In the engraving, the wire ring is shown at *A* around the top of the shell to be wired and in a channel at the top of the spring-supported ring *B*. As indicated in the lower illustration, the punch as it descends, depresses the ring *B* and curls the edge of the shell around the wire ring *A*.

Disks, Reference — Accurate disks of standard dimensions for setting calipers and measuring with. Usually of hardened steel.

Divider, Spring — The spring tends to force the points apart and adjustments are made by the nurled nut on the screw.

Doctor — Local term for adjuster or adapter so that chucks from one lathe can be used on another. Sometimes used same as "dutchman."

Dog — Name given to any projecting piece which strikes and moves some other part, as the reversing dogs or stops on a planer or milling machine. Sometimes applied to the pawl of a ratchet.

Dog, Clamp — Grips work by clamping with the two parts of the dog. There are many types both home-made and for sale.

Dog, Lathe — Devices for clamping on work so that it can be revolved by face-plate. Straight tail dogs are driven by a stud on face-plate. Curved tail (usual way) dogs have the end bent to go into a slot in face-plate.

Clamp Dog   Straight Tail

Bent Tail

Drag — The bottom part of a flask, sometimes called the nowell.

Draw Bench — Place where wire is drawn from rods, being drawn through plates or bull blocks with successively smaller openings.

Drift — A tool for cutting out the sides of an opening while driven through with a hammer.

DRILLS

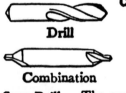

Drill

Combination

Center Drill — The short drills used for centering shafts before facing and turning are called center drills. The drill and reamer or countersink for the 60 degree center hole when combined as shown allow the centering to be done more readily than when separate tools are used.

Core Drill — The core drill is a hollow tool which cuts out a core instead of removing the metal in the form of chips. Such drills are generally used to procure a core from the center of castings or forgings for the determination of the tensile strength or other physical properties of the metal.

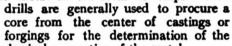

Gun Barrel Drill — Gun barrel drills are run at high speed and under very light feed, oil being forced through a hole in the drill to clear the chips and cool the cutting point and work. The drill itself is short and fastened to a shank of suitable length.

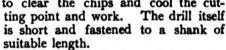

Hog Nose Drill — More like a boring tool. Mostly used for boring out cored holes. Must be very stiff to be effective but when made right and used to advantage, does lots of hard work.

Hollow Drill — The hollow drill is for deep-hole drilling. It has an opening through the body and is attached to a shank of the necessary length for the depth of hole to be drilled.

Oil-drill (Morse) — These drills convey lubricant to the point, through holes formed in the solid metal. Where the drills are larger than $2\frac{1}{2}$ inches an inserted copper tube is employed to carry the oil to the drill point and wash out the chips and keep the drill cool. The oil enters through the hollow shank or through a connection at the side as shown.

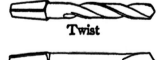

Twist

Ratchet Drill — The square taper shanks of these drills are made to fit a ratchet for drilling holes by hand.

Flat

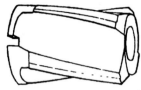

Shell Drill — Shell drills are fitted to a taper shank and used for chucking out cored holes and enlarging holes drilled with a two-flute twist drill. The angle of the spiral lips is about 15 degrees.

Straight Flute Drill — The straight flute, or "Farmer" drill as it is frequently called after its inventor, does not clear itself as well

as the twist drill does, but is stiffer, and does not "run" or follow blow-holes or soft spots as readily as the twist drill. It is also better for drilling brass and other soft metals.

Three and Four-groove Drills — Where large holes are to be made in solid stock, it is advisable to use a three or four groove drill after running the required two-flute drill through the piece. These drills will enlarge the hole to the size required and are also useful in boring out cored holes in castings.

Three Groove

Four Groove

Twist Drill — Usually made with two flutes or grooves, running around the body. This furnishes cutting edges and the chips follow the flutes out of the hole being drilled.

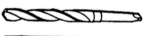

Wood Drill (Bit) — Bits for wood drilling are made in various forms. The pod drill is cut out hollow at the working end; the double flute spiral drill has a regular bit point; the single flute drill is full diameter for a short distance only and is cleared the rest of the length as indicated.

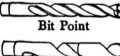

Pod

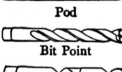

Bit Point

Single Flute

Drill, Chain — Device to be used in connection with a brace or breast drill in many places where it is not convenient to bring a ratchet drill into use.

Drill Speeder — Device which goes on drill spindle and gears up the speed of drills so that small drills can be used economically on large drill presses.

Drill Vise — See Vise, Drill.

Drill, Radial — Parts of

1. Vertical driving-shaft gear.
2. Center driving-shaft gear.
3. Elevating tumble-plate segment.
4. Elevating-screw gear.
5. Column cap.

6. Vertical driving shaft.
7. Column sleeve.
8. Elevating-lever shaft.
9. Elevating screw.
10. Arm girdle.
11. Arm-binder handle.

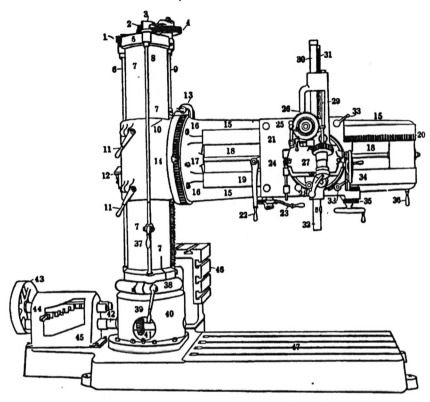

RADIAL DRILL — FULL UNIVERSAL — BICKFORD

12. Arm-miter gear guard.
13. Arm-worm box.
14. Arm pointer.
15. Full universal arm.
16. Arm-clamping nuts.
17. Arm-dowel pin.
18. Arm shaft.
19. Arm ways.
20. Arm rack.
21. Saddle.
22. Reversing lever.
23. Back-gear lever.
24. Head-swiveling worm.
25. Feed-trip lever.
26. Index gear.
27. Universal head.
28. Quick-return lever.
29. Feed-rack worm shaft.

30. Spindle sleeve.
31. Feed rack.
32. Spindle.
33. Saddle-binding lever.
34. Feed hand wheel.
35. Head-moving gear.
36. Arm-swinging handle.
37. Elevating lever.
38. Clamping ring.
39. Clamping-ring handle.
40. Column.
41. Column driving-miters.
42. Driving-shaft coupling.
43. Driving pulley.
44. Speed-change lever.
45. Speed-box case.
46. Box table.
47. Base.

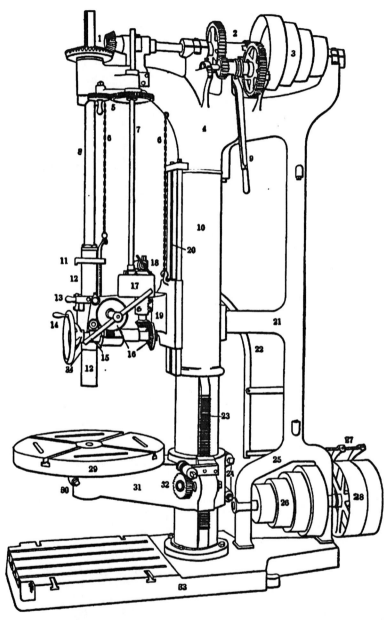

DRILL   PRESS — CINCINNATI   MACHINE
TOOL   COMPANY

Drill Press — Parts of

1. Main driving gears, bevel.
2. Back gears.
3. Upper cone pulley.
4. Yoke to frame.
5. Feed gears.
6. Counterweight chains.
7. Feed shaft.
8. Spindle.
9. Back-gear lever.
10. Column.
11. Automatic stop.
12. Spindle sleeve.
13. Feed-trip lever.
14. Hand-feed wheel.
15. Quick-return lever.
16. Feed gearing.
17. Feed box.
18. Feed-change handle.
19. Sliding head.
20. Face of column.
21. Back brace.
22. Belt shifter.
23. Rack for elevating table.
24. Table-arm clamping screws.
25. Pulley stand.
26. Lower cone pulley.
27. Belt-shifting fingers.
28. Tight and loose pulleys
29. Table.
30. Table-clamp screw.
31. Table arm.
32. Table-adjusting gear.
33. Base.
34. Ball-thrust bearing.

Drive or Force Fit — See Fit.

Dry Sand Molds — Molds made of green sand and baked dry in ovens or otherwise dried out before pouring.

Dutchman — Local term for a wedge or liner to make a piece fit. Used to make a poor job useable. A round key or pin fitting endwise in a hole drilled half in a shaft and half in the piece to be attached thereto.

E

Ejector — An ejector on punch press work is a ring, collar or disk actuated by spring pressure or by pressure of a rubber disk, to remove blanks from the interior of compound and other dies. It is often called a shedder.

Elliptic Chuck — See Chuck, Oval.

Emery Jointer — Grinder for making a close joint between the share and mold board of steel plows.

Emery Wheel Dressers — See Grinding Wheels and Diamonds.

End Measuring Rod — Arranged for internal measurements similar to the internal cylindrical gages.

Expanding Arbor or Mandrel — See Arbor.

Extractor, Oil — Machine for extracting oil from iron and metal chips. Revolves rapidly and throws out the oil by centrifugal force.

F

Face Cam — See Cam, Face.

Face Plate — The plate or disk which screws on the nose of a lathe spindle and drives or carries work to be turned or bored. Sometimes applied to table of vertical boring mill.

Face Reamer — See Reamer, Face.

Feather — Might be called a sliding key — sometimes called a spline. Used to prevent a pulley, gear or other part, from turning on the shaft but allows it to move lengthwise as in the feed shaft used on most lathes and other tools. Feather is nearly always fastened to the sliding piece.

Field — Usually the stationary part of a dynamo or motor.

Files — Tools of hardened steel having sharp cutting points or teeth across their surface. These are forced up by a chisel and hammer.

Filing Machine — Runs a file by power, usually vertically. Useful in many kinds of small work.

Fin — The thin edge or mark left by the parting of a mold or die. In drop forge work this is called the "flash."

Fit, Drive or Force or Press — Fitting a shaft to a hole by making the hole so the shaft can be driven or forced in with a sledge or some power press, often requiring many tons pressure.

Fit, Running or Sliding — Enough allowance between shaft and hole to allow it to run or slide without sticking or heating.

Fit, Shrink — Fitting a shaft to a hole by making the hole slightly smaller than the shaft, then heating the piece with the hole till it expands enough to allow shaft to enter. When cool the shaft is very tightly seized if the allowance is right.

Fit, Wringing — A smaller allowance than for running but so that the shaft can be twisted into the hole by hand. Usually applied to some such work as a boring bar in a horizontal boring machine. Sometimes used in connection with twisting two flat surfaces together to exclude the air.

Flask — The frame which holds the sand mold for the casting. Includes both the cope and drag.

Flat Reamer — See Reamer, Flat.

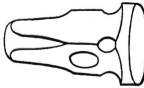 **Flatter** — Round face. A blacksmith's tool which is held on the work and struck by a sledge. Used to take out hammer marks and smooth up a forging.

Flute — Shop name for a groove. Applied to taps, reamers, drills and other tools.

Fly-wheel — Heavy wheel for steadying motion of machinery. On an engine it carries the crank past the center and produces a uniform rotation.

Follower Rest — A back rest for supporting long lathe work; attached to the carriage and following immediately behind the turning tool.

Foot Stock — The tail stock or tail block of a lathe, grinder, etc.

Force — A master punch which is used under a powerful press to form an impression in a die. Forces are commonly employed in the making of coining and other embossing dies. A similar tool used by jewelers is called a "hub." It is sometimes referred to incorrectly as a "hob."

Forge — Open fireplace for heating metals for welding, forging, etc. Has forced draft by fan or bellows.

Forging Press — Heavy machine for shaping metal by forcing into dies by a steady pressure instead of a sudden blow as in drop forging. Similar to a bulldozer.

Fork Center — A center for driving woodwork in the lathe. Also used in hand or fox lathes for driving special work.

Fox Lathe — Lathe for brass workers having a "chasing bar" for cutting threads and often has a turret on the tail stock.

Franklin Metal — An alloy having zinc as a base, used for casting in metal molds.

Fuller — Blacksmith's tool something like a hammer, having a round nose for spreading or fulling the iron under hammer blowers.

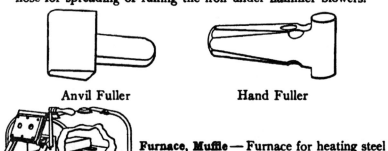

Anvil Fuller   Hand Fuller

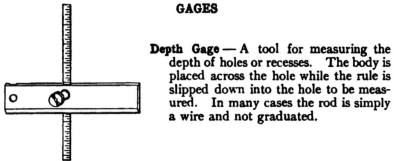

Furnace, Muffle — Furnace for heating steel to harden, in which the flame does not come in contact with the metals.

Furniture — In machine shops applies to tool racks, lathe pans, tote boxes, etc.

Fuse — A piece of metal which melts when too much current passes and acts as a safety valve.

GAGES

Depth Gage — A tool for measuring the depth of holes or recesses. The body is placed across the hole while the rule is slipped down into the hole to be measured. In many cases the rod is simply a wire and not graduated.

Drill Gage — Flat steel plate drilled with different size drills and each hole marked with correct size or number.

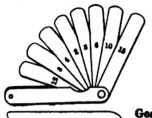

Feeler or Thickness Gage — Has blades of different thicknesses, in thousandths of an inch, so that slight variations can be felt or measured.

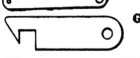

Gear Tooth Depth Gage — A gage for measuring the depth of gear teeth. Requires a different gage for each pitch.

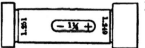

Limit Gage — A plug or other gage having one end larger and the other smaller than the nominal size. If the small end of the plug goes in but the large end does not, the size is between the two and within the limits of the gage. Similarly, in the case of a female limit gage, if the large end of the gage goes over the piece of work and the small end does not go over it, the work is within the established limits. Ordinarily, one end of a limit gage is marked "Go," and the other end "Not Go," or else they are stamped + and −.

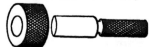

Plug and Ring Gage — Gages for use in measuring inside and outside work or for use in setting calipers.

Radius or Curve Gage — Made like a feeler or thread gage but has each blade with a given outside radius on one end and inside radius on the other for gaging small fillets or round edges.

Scratch Gage — For scratching a line at a given distance from one side of a piece. Adjustable for different lengths.

Snap Gage — A solid caliper used for either inside or outside measurement. This shows a combined gage for outside and inside work. Sizes can be the same or give the allowance for any kind of fit desired.

Splining or Key-seat Gage — Gage for laying out key-seats on shafts.

Surface Gage — A tool for gaging the hight between a flat surface such as a planer table or a surface plate and some point on the work. This can then be transferred to any other point.

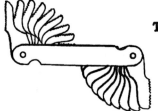

Thread Gage — Tool with a number of blades, each having the same number of notches per inch as the thread it represents. Made for different kinds of threads and in various forms.

Wire Gage — Gage for measuring sizes of wire. The wire fits between the sides of the opening, not in the holes. Sometimes made in the form of a circular disk.

Worm Thread Tool Gage — For grinding thread tool for worm threads — 29 degree angle.

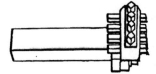

Gang Tool — A holder with a number of tools, generally used in the planer but sometimes in the lathe. Each tool cuts a little deeper than the one ahead of it.

GEARS

Angular Gears — Sometimes applied to bevel gears and also to spur gears with helical or skew teeth. See those terms for definition.

Annular Gear — Toothed ring for use in universal chucks and similar places. Teeth can be on any of the four faces although when inside it is usually called an internal gear.

Bevel Gears — Gears cut on conical surfaces to transmit power with shafts at an angle to each other. When made for shafts at right angles and with both gears of the same size are often called "miter" gears. Teeth may be either straight, skew or herring bone.

Crown Gear — A gear with teeth on the side of rim. Used before facilities for cutting bevel gears existed. Seldom found now.

Elliptical or Eccentric Gears — Gears in which the shaft is not in the center. May be of almost any shape, oval, heart-shape, etc. Printing presses usually have good examples of this.

Helical Gears — Gears having teeth at an angle across the face to give a more constant pull. Also give side thrust. More often called "skew" teeth.

Herring-bone Gears — Gears having teeth cut at a double angle. Made by putting two helical or "skew" tooth gears together. Does away with side or end thrust.

Intermittent Gears — Gears where the teeth are not continuous but have plain surfaces between. On the driven gear these plain surfaces are concave to fit the plain surface of the driver and the driven wheel is stationary while the plain surfaces are in contact.

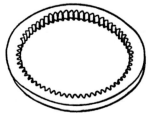

Internal Gears — Gears having teeth on the inside of a ring or shell.

Module or Metric Gears — French system of making gears with metric measurement. Pitch diameter in millimeters divided by the number of teeth in the gear.

Pin Gear — Gear with teeth formed by pins such as the old lantern pinion. Also formed by short projecting pins or knobs and only used now in some feeding devices.

Quill Gears — Gears or pinions cut on a quill or sleeve.

Skew Gears — See Helical.

Spiral Gears — Spur gears with spiral teeth which run together at an angle and do the work of bevel gears.

Spur Gears — Wheels or cylinders whose shafts are parallel, having teeth across face. Teeth can be straight, helical or skew or herring bone.

Staggered Tooth Gears — Made up of two or more straight tooth spur gears, teeth set so that teeth and spaces break joints instead of presenting a continuous pull.

Worm Gears — Spur gears with teeth cut on angle to be driven by a worm. Teeth are usually cut out with a hob to fit the worm.

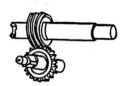  

A   B

Sprocket Gears—Toothed wheels for chain driving. A is the regular and B is a hook tooth for running one way only.

Gear Teeth — The projections which, meshing together, transmit a positive motion. The involute curve tooth is now almost universal. The older form has a 14½ degree pressure angle but some are using a shorter tooth, known as a "stub" tooth, with 20 degrees pressure angle. An involute tooth rack has straight sides to the ·teeth.

Gears, Pitch of — *Chordal,* distance from center of one tooth to center of next in a direct line.

Circular, distance from center of one tooth to center of next along the pitch line.

Diametral, number of teeth per inch of diameter.

Geneva Motion — A device which gives a positive but intermittent motion to the driven wheel but prevents its moving in either direction without the driver. The driver may have one tooth as shown or a number if desired. Also made so as to prevent a complete revolution of the driven wheel.

German Silver — An alloy of copper 60 parts, zinc 20 parts, nickel 20 parts.

Gib — A piece located alongside a sliding member to take up wear.

Gland — A cylindrical piece enveloping a stem and used in a stuffing-box to make a tight joint.

Green Sand Molds — Molds made of sand that is moistened for molding and not dried out or baked before pouring.

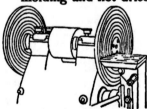

Grinder, Disk — A grinding machine having steel disks which are covered with emery cloth. Some disks have spiral grooves to give cushions under the emery cloth.

Grinder Wheel Dresser — A tool consisting of pointed or corrugated disks of hard metal which really break or pry off small particles of the grinding wheel when held against its rapidly revolving surface.

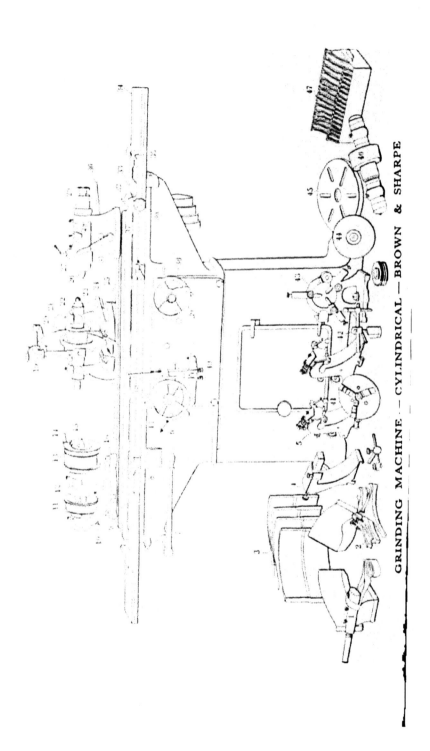

GRINDING MACHINE.—CYLINDRICAL—BROWN & SHARPE

Grinding Machine — Parts of

1. Internal grinding fixture.
2. Water guard supports.
3. Water guards.
4. Plain back rests.
5. Universal back rest.
6. Automatic cross-feed pawl.
7. Starting and stopping lever.
8. Table-reversing dogs.
9. Headstock index finger.
10. Live spindle-locking pin.
11. Live spindle-driving pulley.
12. Headstock.
13. Dead center pulley.
14. Work driving arm and pin.
15. Headstock center.
16. Headstock base.
17. Cross-feed hand wheel.
18. Reversing lever.
19. Water piping.
20. Wheel-driving pulley.
21. Wheel guards.
22. Spindle box.
23. Wheel stand.
24. Wheel-stand platen.
25. Wheel-stand slide.
26. Footstock center.
27. Diamond tool-holder.
28. Footstock.
29. Tension adjusting knob.
30. Quick-adjusting lever.
31. Clamping lever
32. Clamping bolt.
33. Table scale.
34. Bed water guard.
35. Sliding table.
36. Swivel table knob.
37. Swivel table.
38. Hand wheel.
39. Table travel control.
40. Automatic cross-feed.
41. Universal chuck.
42. Tooth rest.
43. Center rest.
44. Face-grinding chuck.
45. Face plate.
46. Internal grinding counter.
47. Work-driving dogs.

Grinding Wheels — Common types of grinding wheels made of emery, corundum, carborundum and alundum, are the disk, ring, saucer, cup and cylinder. Disk and ring wheels are used on the periphery; saucer wheels on the thin edge; cup and cylinder wheels on the end. The latter are commonly used for surface grinding.

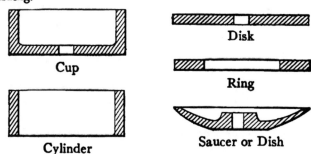

Cup

Disk

Ring

Cylinder

Saucer or Dish

Gripe — Local name for machine clamp.

Ground joint — A joint finished by grinding the two parts together with emery and oil or by other abrasives.

Ground — A connection between the electric circuit and the earth.

Gudgeon — Local name for a trunnion or bearing which projects from a piece as a cannon.

Guide Liner — A tool for use in locomotive work for lining up guides and cross heads.

H

Half Nut — A nut which is split lengthwise. Sometimes half is used and rides on screw, in others both halves clamp around screw as in the half nut of a lathe carriage.

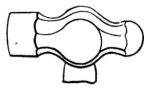

Hammer — The common types of machinists' hammers are the ball peen, straight peen and cross peen, as shown. The so-called engineer's and the riveting hammers have cross peens.

Ball Peen

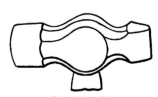

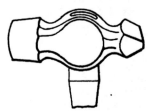

Straight Peen   Cross Peen

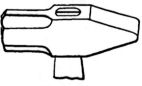

Riveting   Engineers

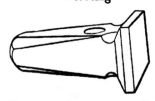

Hammer, Blacksmith's Flatter — A flat-faced hammer used to smooth the surfaces of forgings. Is held on the work and struck by a helper with a sledge.

Hammer, Bumping or Horning — For closing seams on large cans, buckets, etc.

Hammer, Drop — Hammer head or "monkey" or "drop" is raised by hand or power and falls by gravity. Sometimes raised by a board attached to top of hammer head and running between pulleys. Others use a belt.

Hammer, Helve — Power hammer in which there is an arm pivoted in the center and power applied at the back end while the hammer is at the other and strikes the work on an anvil.

Hammer, Lever Trip — Trips the hammer by a cam or lever and allows it to fall.

Hammer, Spring — Comparatively small hammer giving a great variety in the force of blow. This is controlled by pressure of foot on lever.

Hand Wheels, Clutched — Hand wheels connected to shaft by a clutch which can be thrown out by a knob or otherwise so that accidental movement of wheel will not disturb setting. Used on milling machines and similar places.

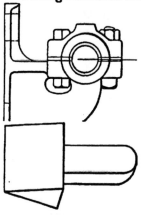

Hanger, Drop — Shaft hanger to be fastened to ceiling with bearing held in lower end.

Hanger, Post — Shafting hanger for fastening to posts or other vertical structures.

Hardie — Blacksmith's cutting chisel which fits a hole in the anvil and forms the lower tool in cutting off work.

Harveyizing — The surface hardening of steel armor plates by using a bed of charcoal over the work and then gas turned on so it will soak in from the top. Not adapted to small work.

Hindley Worm — See Worm.

Hoist, Chain — Hoist with chain passing through pulley block used for hoisting.

Holder, Drill — Device for holding drill stationary while work is revolved by lathe chuck, or face place. Not a drill chuck.

Hooks, Twin or Sister — Double crane hook which resembles an anchor and allows load to be carried on either side.

Hub — A master punch used in making jewelry dies for fancy embossing, and various forms to which articles of gold and silver are to be struck.

Hunting Tooth — An extra tooth in a wheel to give it one more tooth than its mate in order to prevent the same teeth from meshing together all the time.

I

Idler or Idler Pulley — See Pulley, Idler.

Incandescent — A substance heated to white heat as in the bulb of a lamp.

Indexing, Compound — Indexing by combination of two settings of index, either by adding or subtracting.

Indexing, Differential — Indexing with the index plate geared to the spindle, thus giving a differential motion that allows the indexing to be done with one circle of holes and with the index crank turned in the same direction, as in plain indexing.

Indexing, Direct — Indexing work by direct use of dividing head of milling machine.

Indicator, Lathe Test — Instrument with multiplying levers which shows slight variations in the truth of revolving work. Used for setting work in lathe or on face-plate.

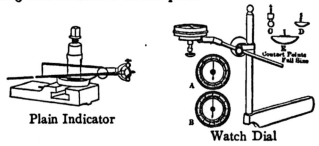

Plain Indicator   Watch Dial

Induction Motor — A motor which runs by the magnetic pull through the air without contact. Usually a constant-speed motor.

J

Jack, Hydraulic — Device for raising weight or exerting pressure by pumping oil or other liquid under a piston or ram.

Jack, Leveling — Small jacks (usually screw jacks) for leveling and holding work on planer beds and similar places. Practically adjustable blocking.

Jack, Screw — Device for elevating weights by means of a screw.

Jack Shaft — See Shaft, Jack.

Jam Plates — Old name for screw plates and in many cases a true one as the thread was jammed instead of cut.

Jig, Drill — A device for holding work while drilling, having bushings through which the drill is guided so that the holes are correctly located in the piece. Milling and planing jigs (commonly called fixtures) hold work while it is machined in the milling machine and planer. Parts produced in jigs and fixtures are interchangeable.

Joint, Universal — Shaft connection which allows freedom in any direction and still conveys a positive motion. Most of them can transmit power through any angle up to 45 degrees.

Journal Box — The part of a bearing in which the shaft revolves.

K

Kerf — The slot or passageway cut by a saw.

Key — The piece used to fasten any hollow object to a shaft or rod. Usually applied to fastening pulleys and fly-wheels to shafts; or locomotive driving wheels to their axles. Keys may be square, rectangular, round or other shape and fasten in any way. Are usually rectangular and run lengthwise of shaft.

Key, Barth — This key was invented several years ago by Carl G. Barth. It is simply a rectangular key with one-half of both sides beveled off at 45 degrees. The key need not fit tightly, as the pressure tends to drive it better into its seat. As a feather key this key has been used in a great many cases to replace rectangular feather keys which have given trouble. It has also been used to replace keys which were sheared off under heavy load.

Key, Center — A flat piece of steel, with tapered sides, for removing taper shank drills from drill spindle or similar work.

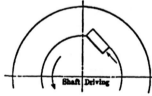

Key, Lewis — A key invented by Wilfred Lewis about 20 years ago. Its position is such that it is subjected to compression only.

Key, Round End — Is fitted into a shaft by end milling a seat into which the key is secured. Where a key of some length is fixed in the shaft and a member arranged to slide thereon it is called a feather or feather key.

Key Seater — Machine for cutting keyways in shafts or hubs of pulleys or gears.

Key, Taper — The taper key is made with and without head. The taper is commonly $\frac{1}{8}$ or $\frac{3}{16}$ inch per foot.

Key, Woodruff — A semi-circular key used in various kinds of shafts, studs, etc. It is fitted in place by merely sinking a seat with a shank mill such as the Whitney cutter.

Keyway — A groove, usually square or rectangular, in which the key is driven or in which a "feather" slides. The groove in both the shaft and piece which is to be fastened to it, or guided on it, is called a keyway.

Knurling — See Nurling.

L

Land — Space between flutes or grooves in drills, taps, reamers or other tools.

Lap — Applied to seams which lap each other. To the distance a valve must move before opening its port when valve is central on seat. To a tool usually consisting of lead, iron or copper charged with abrasive for fine grinding. See Lap, Lead.

Lap Cutter — For preparing the ends of band-saws with bands for brazing. Uses milling cutters.

Lap Grinder — This prepares the laps of band-saws by grinding.

Lap, Lead — Usually a bar of lead or covered with lead, a trifle smaller than the hole to be ground. Emery or some fine abrasive is used which gives a fine surface. Laps are sometimes held in the hand or are run in a machine and the work held stationary. Also consists at times of a lead-covered disk, revolving horizontally, which is used for grinding flat surfaces. Very similar in action to a potter's wheel.

Lathe, Double Spindle — Has two working spindles, so located that one gives a much larger swing than the other, and both can be used to advantage. Especially good for repair shops.

Lathe, Engine — The ordinary form of lathe with lead screw, power feed, etc.

Lathe, Engine — Parts of

1. Rear bearing.	20. Cross-feed screw.
2. Back-gear case.	21. Cross-power feed.
3. Cone pulley.	22. Half-nut handle.
4. Face-gear guard.	23. Regular power feed.
5. Front bearing.	24. Feed reverse.
6. Face plate.	25. Gear stud.
7. Live center.	26. Hand feed.
8. Dead center.	27. Front apron.
9. Tail spindle.	28. Rear apron.
10. Tail-spindle lock.	29. Lead screw.
11. Tailstock slide.	30. Feed rod.
12. Locking bolts.	31. Feed gears.
13. Tailstock base.	32. Feed box.
14. Tailstock pinion.	33. Change gear handle.
15. Tailstock hand wheel.	34. Compound gears.
16. Steady rest.	35. Change-gear handle.
17. Tool post.	36. Change-gear handle.
18. Compound rest.	37. Change-gear handle.
19. Cross-slide.	38. Bed.

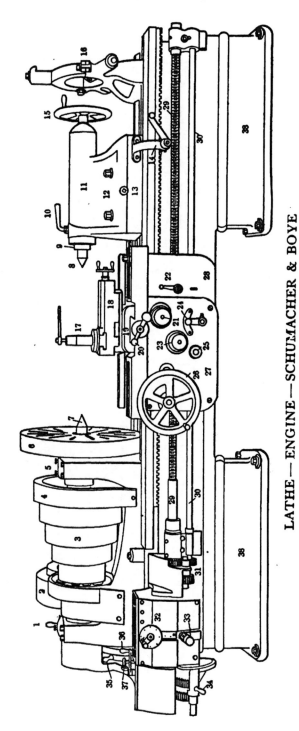

LATHE — ENGINE — SCHUMACHER & BOYE

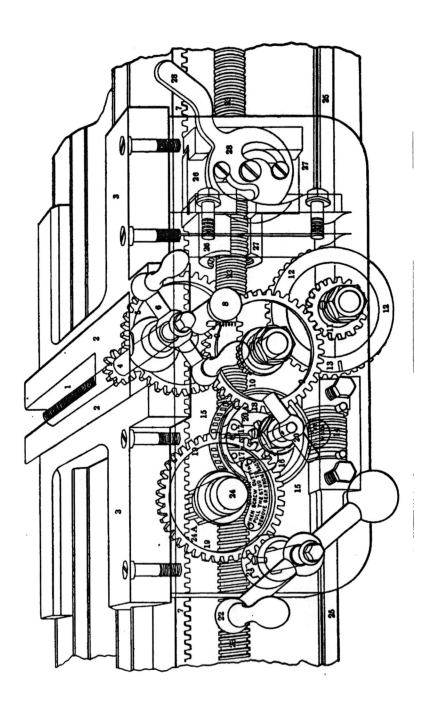

Lathe Apron, Reed — Parts of

1. Cross-feed screw.
2. Cross-slides.
3. Wing of saddle.
4. Cross-feed pinion.
5. Cross-feed gear.
6. Cross-feed handle.
7. Rack.
8. Power cross-feed and control.
9. Gear in train.
10. Pinion for cross-feed.
11. Main driving pinion.
12. Bevel gear.
13. Bevel pinion.
14. Feed-worm.
15. Feed-worm wheel.
16. Clutch ring.
17. Clutch levers.
18. Pinion.
19. Gear in train.
20. Feed-clutch handle.
20A. Clutch spreader.
21. Hand pinion.
22. Carriage handle.
23. Lead screw.
24. Rack pinion knob.
24A. Rack pinion.
25. Feed rod.
26. Upper-half nut.
27. Lower-half nut.
28. Half-nut cam.

NOTE:—Cross-feed is from bevel pinion 13, through gears 12, 11, 9, 10, and 4. Regular feed is through worm 14, worm wheel 15, clutch 16, pinion 18, gears 19 and 24A. Hand movement is through handle 22, pinion 21, 19 and 24A.

Lathe, Extension — So made that bed can be lengthened or shortened. When bed is made longer, there is a gap near head, increasing the swing for face-plate work.

Lathe, Fox — Brass workers' lathe having a "fox" or chasing bar for cutting threads. The bar has a "leader" which acts as a nut on a short lead screw or "hob" of the desired pitch (or half the pitch if the hob is geared down 2 to 1) and carries a tool along at the right feed for the thread. Sometimes has a turret on the back head.

Lathe, Gap — Has V-shaped gap in front of head stock to increase swing for face-plate work.

Lathe, Gun — For boring and turning cannons and rapid-fire guns.

Lathe, Ingot — For boring, turning and cutting off steel ingots.

Lathe, Precision — Bench lathe made especially for small and very accurate die, jig or model work.

Lathe, Projectile.— Simply a heavy lathe for turning up projectiles. Sometimes has attachment for pointing them.

Lathe, Pulley — Especially designed for turning pulleys, can turn them crowning or straight.

Lathe, Roll Turning — For turning rolling mill, steel mill and calendar rolls.

Lathe, Screw Cutting — Having lead screw and change gears for cutting threads.

Lathe, Shafting — For turning long shafts or similar work.

Lathe, Speed — A simple lathe with no mechanically actuated carriage or attachments.

Lathe, Spinning — For forming sheet metal into various hollow shapes, all circular. Done by forcing against a form of some kind (with a single round ended tool) while it is revolving.

Lathe, Stone Turning — Specially designed for turning stone columns or similar shapes.

Lathe, Turret — Having a multiple tool holder which revolves. This is the turret. Usually takes place of tail or foot stock but not always. Usually has automatic devices for turning turret and sometimes for feeding tools against work.

Lathe, Vertical — Name given one type of Bullard boring mill on account of a side head which acts very much like a lathe carriage and does a large variety of work that would ordinarily be done on the face-plate of a lathe.

Lathe, Watchmaker's — A very small precision lathe.

Lead — The advance made by one turn of a screw. Often confused with pitch of thread but not the same unless in the case of a single thread. With a double thread the lead is twice as much as the pitch.

Level — Instrument with a glass tube or vial containing a liquid which does not quite fill it. The tube is usually ground on an arc so that bubble can easily get to the center. Alcohol is generally used as it does not freeze at ordinary temperatures.

Level, Engineers' — Level mounted on tripod and having telescope for leveling distant objects.

Level, Pocket — Small level to be carried in pocket.

Level, Quartering — A tool for testing driving wheels to see if crank-pins are set 90 degrees apart. The level has a forked end and with the angles shown. Placing this on the crank-pin and lining the edge with the center of axle should bring the bubble of level in the center. If the same result is obtained on the other wheel the crank-pins are 90 degrees apart.

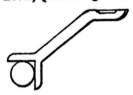

1st Class

Power   Fulcrum   Weight

2nd Class

Fulcrum   Weight   Power

3rd Class

Fulcrum   Power   Weight

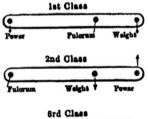

Levers — Arms pivoted or bearing on points called fulcrums. Divided into three classes as shown. First has fulcrum or bearing point between power and weight, second has weight between power and fulcrum and third has power between weight and fulcrum.

Line Shaft — See Shaft, Line.

Liner — A piece for separating pieces a desired distance; also called shim.

Live Center — See Center, Live.

Loam Mold — Made with a mixture of coarse sand and loam into a sort of plaster which is spread over brick or other framework to make the mold. Used on large castings to produce a smoother finish than is to be had with green sand.

M

Machinists' Clamp — See Clamp.

Magnet Electro — Usually a bar of iron having coils of insulated wire around it which carry current. Permanent magnets are of hardened steel with no wire or current around them.

Mandrel — See Arbor.

Marking Machine — For stamping trade-marks, patent dates, etc., on cutlery, gun barrels, etc. Stamps are usually on rolls and rolled into work.

Master Die — A die made standard and used only for reference purposes or for threading taps.

Master Plate — See Plate, Master.

Master Tap — A tap cut to standard dimensions and used only for reference purposes or for tapping master dies.

Match Board — The board used to hold patterns, half on each side, while being molded on some types of molding machines.

Measuring Machine — Practically a large bench caliper of any desired form to measure work such as taps, reamers, gages, etc.

Measuring Rods — See End Measuring Rods.

Milling Cutters — See Cutters, Milling.

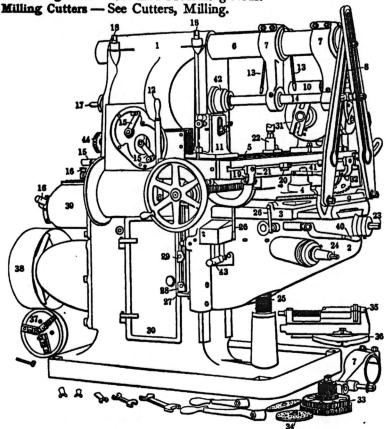

MILLING MACHINE — UNIVERSAL — MILWAUKEE

Milling Machine — Universal — Parts of

1. Column.
2. Knee.
3. Saddle.
4. Swivel carriage.
5. Work table.
6. Over arm.
7. Arm brackets (arbor supports.)
8. Arm braces (harness).
9. Knee clamp (for arm braces).
10. Spiral dividing head.
11. Tailstock.
12. Starting lever.
13. Oil tubes.
14. Cutter arbor.
15. Speed-changing lever.
16. Feed-changing lever.
17. Draw-in rod for arbor.
18. Arm-clamp screws.
19. Table stops.
20. Table-feed trip block.
21. Fixed table-feed trips.
22. Steady rest.
23. Cross-feed screw.
24. Elevating shaft.
25. Elevating screw (telescopic).
26. Saddle-clamp levers.
27. Knee-clamp levers.
28. Fixed vertical feed trip.
29. Vertical feed-trip blocks.
30. Door.
31. Dog driver.
32. Change-gear bracket.
33. Change gears.
34. Index plates.
35. Vise.
36. Swivel base.
37. Universal chuck.
38. Driving pulley.
39. Feed box.
40. Cross and vertical feed handle.
41. Table-feed handle.
42. Clutch-drive collar.
43. Interlocking lever to prevent the engagement of more than one feed at a time.

Milling Machine — Operating tool is a revolving cutter. Has table for carrying work and moving it so as to feed against cutter.

Milling Machine, Universal — Has work table and feeds so arranged that all classes of plane, circular, helical, index, or other milling may be done. Equipped with index centers, chuck, etc.

Milling Machine, Vertical — Has a vertical spindle for carrying cutter.

Milling Machine — Vertical — Parts of

1. Spindle drawbar cap.
2. Back-gear pull pin.
3. Spindle-driving pulley.
4. Spindle head.
5. Back-gear sliding pinion and stem gear.
6. Spindle upper box.
6. Spindle lower box.
7. Spindle head bearing.
8. Head-feed gear.
9. Idler pulleys.
10. Standard.
11. Spindle.
12. Rotary attachment.
13. Rotary attachment feed-trip dog and lever.
14. Rotary attachment feed clutch.
15. Rotary attachment base.
16. Table and table oil pans.
17. Rotary attachment binder.
18. Feed-trip dogs, right and left.
19. Feed-trip plate.
20. Cross-feed screw.
21. Feed-clutch lever.
22. Carriage.
23. Feed clutch.
24. Table-feed screw.
25. Rotary attachment feed gears.
26. Rotary attachment hand wheel.
27. Rotary attachment feed rod.
28. Feed-driving cone.
29. Feed bracket.
30. Universal joint.
31. Telescopic feed shaft.
32. Driving cone.
33. Driving pulley.
34. Knee-elevating shaft.
35. Knee-elevating telescopic screw.
36. Face of standard.
37. Base.

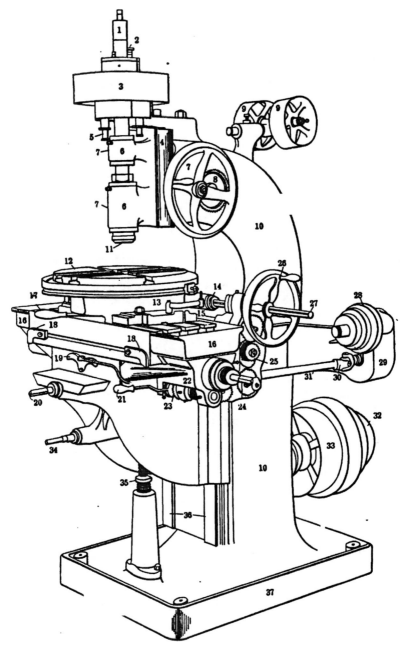

MILLING MACHINE — VERTICAL — BECKER

Miter — A bevel of 45 degrees.

Mold — The mold consists of the cope and the drag or nowel, with the sand inside molded to pattern and ready to pour.

Mold Board — The board used to put over a flask to keep sand from falling when being handled and sometimes used to clamp on when fastening molds together.

Mufflers — Ovens or furnaces, usually of clay, where direct heat is not required.

Muley Belt — See Belt, Muley.

Muley Shaft — See Shaft, Muley.

N

Necking Tool — Tool for turning a groove or neck in a piece of work.

Nose — In shop work applied to the business end of tools or things. The threaded end of a lathe or milling-machine spindle or the end of "hog nose drill" or similar tool.

Nowel — Same as Drag.

Nurling — The rolling of depressions of various kinds into metal by the use of revolving hardened steel wheels pressed against the work. The design on the nurl will be reproduced on the work. Generally used to give a roughened surface for turning a nut or screw by hand.

Nut, Cold Punched — A nut punched from flat bar stock. The hole is usually reamed to size before tapping.

Nut, Hot Pressed — A nut formed hot in a forging machine.

Nut, Castellated or Castle — A nut with slot across the face to admit a cotter pin for locking in place.

Nut Machine — For cutting, drilling and tapping nuts from a bar or rod.

Nut Tapper — For tapping hole in nuts.

Nut, Wing — A nut operated by hand and very commonly used where a light and quick clamping action is required.

Nuts — See Bolts.

O

Ogee — Name given to a finish or beading consisting of a reverse curve. Applied to work of any class, wood or iron.

Ohm — The unit of electrical resistance. One volt will force one ampere through a resistance of one ohm.

Oval — Continuously curved but not round, as a circle which has been more or less flattened.

P

Pawl — A hinged piece which engages teeth in a gear, rack or ratchet for moving it or for arresting its motion. Sometimes used to designate a piece such as a reversing dog on a planer or milling machine.

Peening — The stretching of metal by hammering or rolling the surface. Used to stretch babbitt to fit tightly in a bearing, to straighten bars by stretching the short or concave side, etc.

Pickling Castings — Dipping into acid solution to soften scale and remove sand. Solution of three or four parts of water to one of sulphuric acid is used for iron. For brass use five parts water to one of nitric acid.

Pin, Collar — A collar pin is driven tight into a machine frame or member and adapted to carry a roll, gear, or other part at the 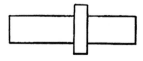 outer end. It differs from the collar stud in not having a thread at the inner end. When drilled through the end for a cotter pin it is known as a fulcrum pin, as it is then especially suited for carrying rocker arms, etc.

Pin, Dowel, Screw — Dowel pins are customarily made straight, or plain taper and fitted into reamed holes. When applied in such a position in a mechanism that it is impossible to remove them by driving out, they are sometimes threaded and screwed into place.

Pin, Taper — Taper pins for dowels and other purposes are regularly manufactured with a taper of $\frac{1}{4}$ inch to the foot and from $\frac{3}{4}$ to  6 inches long, the diameters at the large end of the sizes in the series ranging from about $\frac{6}{32}$ to $\frac{7}{8}$ inch. The reamers for these pins are so proportioned that each size "overlaps" the next smaller size by about $\frac{1}{2}$ inch.

Pickling Forgings — Putting in bath of 1 part sulphuric acid to 25 parts boiling water to remove scale. Can be done in 10 minutes. Rinse in boiling water and they will dry before rusting.

Pitch — The distance from the center of one screw thread, or gear tooth or serration of any kind to the center of the next. In screws with a single thread the pitch is the same as the lead but not otherwise.

 **Pillow Blocks** — Low shaft bearings, resting on foundations, or floors or other supports.

Pitman — A connecting rod; term used more commonly in connection with agricultural implements.

Planchet — Blank piece of metal punched out of a sheet before being finished by further work. Such as the blanks from which coins are made.

Planer — For producing plane surfaces on metals. Work is held on table or platen which runs back and forth under the tool which is stationary.

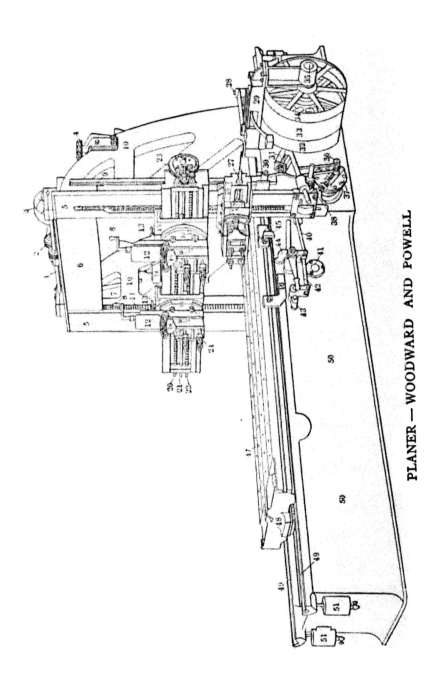

PLANER — WOODWARD AND POWELL

Planer — Parts of

1. Shaft for raising cross-rail.
2. Gears for raising cross-rail.
3. Pulley drive for raising cross-rail.
4. Chain for counterweighting the cross-rail.
5. Face of uprights.
6. Tie piece between uprights.
7. Handle controlling.
8. Crank handle for raising tool block.
9. Rack for moving feed screw in cross-rail.
10. Upright or housing.
11. Screw for elevating cross-rail.
12. Tool slide.
13. Screw to clamp saddle to cross-rail.
14. Counterweight for left side of cross-rail.
15. Saddle.
16. Swivel.
17. Clamping bolt.
18. Clapper box.
19. Clapper block.
20. Feed screw for left-hand head.
21. Vertical feed rod.
22. Feed screw for right-hand head.
23. Feed mechanism on end of cross-rail.
24. Tool-holding straps.
25. Clapper block pin.
26. Side head.
27. Side head for feed screw.
28. Belt shifter.
29. Drive-pulley support.
30. Connection to feed rack.
31. Regulator for vertical feed.
32. Forward driving pulley.
33. Loose pulley.
34. Return driving pulley.
35. Driving shaft.
36. Regulator cross-feed.
37. Cross-feed drive.
38. Vertical feed pinion.
39. Vertical feed rod.
40. Rod to belt shifter for reversing.
41. Bull or driving-wheel shaft.
42. Connections to safety lock.
43. Lock to prevent table being moved.
44. Reversing latch or trip.
45. Forward stop dog.
46. Backward stop dog.
47. Platen or table.
48. Rack under platen.
49. Ways or V's.
50. Bed.
51. Oil reservoirs.

Planer Centers — A pair of index centers to hold work for planing. Similar to plain milling machine centers.

Planer, Daniels — Wood planer with a table for carrying work under a two-armed knife swung horizontally from a vertical spindle. Very similar to a vertical spindle-milling machine of the planer type. Excellent for taking warp or wind out of lumber.

Planer, Open Side — A planer with only one upright or housing, supporting an overhanging arm which takes the place of the usual cross rail. Useful in planing work too wide to go in the ordinary planer.

Planer, Radius — For planing parts of circles such as links for locomotive or stationary engine valve motion.

Planer, Rotary — Really a large milling machine in which the work is carried past a rotary cutter by the platen.

Planer Tools — See Tools, Planer.

Planer, Traveling Head — Planer in which work is stationary and tool moves over it.

Planishing — The finishing of sheet metal by hammering with smooth faced hammers or their equivalents.

Plate, Master — A steel plate serving as a model by which holes in jigs, fixtures and other tools are accurately located for boring. In the illustration the piece to be bored is shown dowelled to the master plate which is mounted on the face plate of the lathe. In the master plate there are as many holes as are to be bored in the work, and the center distances are correct. The plate is located on a center plug fitting the lathe spindle, and after a given hole is bored in the work, the master plate and work are shifted and relocated with the center plug in the next hole in the plate and the corresponding hole in the work is then bored out. This is one of the most accurate methods employed by the toolmaker on precision work.

Platen — A work holding table on miller, planer or drill.

Plumb Bob, Mercury — Plumb bob filled with mercury to secure weight in a small space.

Potter's Wheel — Probably the oldest machine known. Consists of a vertical shaft with a disk mounted horizontally at the top. The potter puts a lump of clay in the center, revolves the wheel with his foot or by power, and shapes the revolving clay as desired. See Lap, Lead for modern application of this in machine work.

Press, Blanking — For heavy punching or swaging.

Press, Broaching — A press for forcing broaches through holes in metal work. Usually cleans out or forms holes that are not round.

Press, Cabbaging — For compressing loose sheet metal scrap into convenient form for handling and remelting.

Press, Coining — For making metal planchets from which coins are stamped.

Press, Double Action — Has a telescoping ram or one ram inside the other, each driven by an independent cam so that one motion follows the other and performs two operations for each revolution of the press.

Press Fit — See Fit.

Press, Forcing — For forcing one piece into another, such as a rod brass into a rod.

Press, Forging — For forging metal by subjecting it to heavy pressure between formers or dies.

Press, Horning — For closing side seams on pieced tinware.

Press, Inclinable — One that can be used in vertical or inclined position.

Press, Screw — Pressure is applied by screw. Heavier work is done in this way than by foot or hand press.

Press, Straight-sided — Made with perfectly straight sides so as to give great strength and rigidity for heavy work.

Press, Pendulum — Foot press having a pendulum like lever for applying power to the ram.

Profiling Machine — Has rotary cutter that can be made to follow outline or pattern in shaping small parts of machines. Practically a vertical milling machine.

Protractor, Bevel — Graduated semicircular protractor having a pivoted arm for measuring off angles.

Pull-pin — A means of locking or unlocking two parts of machinery. Sometimes slides gears in or out of mesh and at others operates a sliding key which engages any desired gear of a number on stud.

Pulley, Gallow or Guide — Loose pulley mounted in movable frames to guide and tighten belts.

Pulley, Idler — A pulley running loose on a shaft and driving no machinery, merely guiding the belt. Practically same as a "loose pulley."

Pulley, Loose — Pulley running loosely on shaft doing no work. Carries belt when not driving tight (or fast, or working) pulley. Used on countershafts, planers, grinders, etc., where machine is idle part of time. Belt is then on the loose pulley but when shifted to tight pulley the machine starts up. See Belt Shifter, Friction Clutch.

Punching — A piece cut out of sheet stock by punch and die; the same as blank.

Punch, Belt — Hollow, round or elliptical punch for cutting holes for belt lacing.

Q

Quadrant — A piece forming a quarter circle. A segment of a circle. The swinging plate carrying the change gears in the feed train at the end of the lathe.

Quick Return — A mechanism employed in various machine tools to give a table, ram or other member a rapid movement during the return or non-cutting stroke.

Quill — A hollow shaft which revolves on a solid shaft, carrying pulleys or clutches. When clutches are closed the quill and shaft revolve together.

R

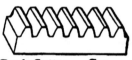 **Rack** — A strip cut with teeth so that a gear can mesh with it to convert rotary into reciprocating motion or vice versa.

Rack Cutter — Cuts regularly spaced teeth in a straight line. Cutting tool is either a milling cutter or a single point tool.

Ratchet — A gear with triangular shaped teeth adapted to be engaged by a pawl which either imparts intermittent motion to the ratchet or else locks it against backward movement when operated otherwise.

 **Ratchet Drill** — Device for turning a drill when the handle cannot make a complete revolution. A pawl on the handle drops into a ratchet wheel on the barrel so that it can be turned one or more teeth.

Recess — A groove below the normal surface of work. On flat work a groove to allow tool to run into as a planer, or a slide to run over as a cross-head on a guide. In boring a groove inside a hole. If long it is often called a chamber. ·

Relief or Relieving — The removing of, or the amount removed to reduce friction back of cutting edge of a drill, reamer, tap, etc. Also applied to other than cutting tools. See "backing off."

REAMERS

Reamer — A tool to enlarge a hole already existing, whether a cast or cored hole or one made by a drill or boring bar. Reamers are of many kinds and shape as indicated below. Usually a reamer gives the finishing touch to a hole.

Ball Reamer — Usually a fluted or rose reamer for making the female portion of a ball joint. It is considered advisable to space the teeth irregularly as this tends to prevent chattering.

Bridge Reamer — A reamer used by boiler-makers, bridge builders, ship-builders, etc., has a straight body from A to B and tapered end from B to C. This reamer has a  taper shank and can be used in an air drill.

Flat

Fluted

Center Reamer — Center reamers, or countersinks for centering the ends of shafts, etc., are usually made 60 degrees included angle.

Fluted

Rose

Chucking Reamer — Chucking reamers are used in turret machines. The plain, fluted type has teeth relieved the whole length; while the rose reamer cuts only on the end as there is no peripheral clearance. Where possible reamers used in the turret should be mounted in floating holders which allow the reamer to play sidewise sufficiently to line up with the hole in the work which may be so drilled or bored as not to run perfectly true prior to the reaming operation.

Chucking Reamer (Three-groove) — Spiral fluted chucking reamers with three and four grooves are employed for enlarging cored holes, etc. They are also made with oil passages through them and in this form are adapted to operating in steel.

Flat Reamer — A reamer made of a flat piece of steel. Not much used except on brass work and then usually packed with wooden strips to fit the hole tightly. Flat reamers are not much used except for taper work.

Half-round Reamer — Used considerably in some classes of work, particularly in small sizes and taper work when taper is slight. Not much used in large sizes. Somewhat resembles the "hog-nose drill" in general appearance except that this is always quite short on cutting edge.

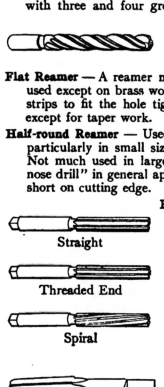

Straight

Threaded End

Spiral

Hand Reamer — Reamers enlarge and finish a hole produced by drilling, boring, etc. The cut should be light for hand reamers and the reamer held straight to avoid ruining the hole. The threaded end reamer has a fine thread to assist in drawing the reamer into the work. The spiral reamer is cut left-hand to prevent its drawing into the hole too rapidly. Reamers are slightly tapered at the point to enable them to enter.

Pipe Reamer (Briggs) — Pipe reamers to the Briggs standard taper of ¾-inch per foot are used for reaming out work prior to tapping with the pipe tap.

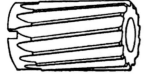

Shell Reamer — Shell reamers have taper holes to fit the end of an arbor on which they are held in the chucking machine. They are made with both straight and spiral flutes.

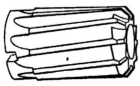

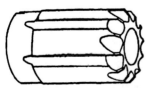

Shell Reamer (Rose) — Rose reamers cut on the end only as there is no peripheral clearance. They are very accurate tools for finishing holes. The shell reamers are made with taper holes to fit an arbor for holding them in the turret machine.

Taper Reamer — For finishing taper holes two or more reamers are sometimes used. The roughing reamer is often provided with nicked or stepped teeth to break up the chip. Taper reamers are also made with spiral teeth. Where the taper is slight the spiral should be left-hand to prevent the reamer from drawing in too fast; where the taper is abrupt the teeth, if cut with right-hand spiral, will help hold the reamer to the cut and make the operation more satisfactory.

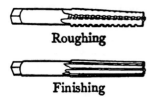

Roughing

Finishing

Taper Pin Reamer — Standard taper pin reamers are made $\frac{1}{4}$-inch taper per foot and each size in the series will overlap the next size smaller by about one-half inch.

Taper Reamer (Locomotive) — Reamers for locomotive taper pins have a taper of $\frac{1}{8}$ inch per foot.

Rests, Slide — Detachable rests capable of being clamped to brass lathe bed at any desired point and usually arranged to give motion to tool in two ways; across the bed to reduce diameter or cut-off, and with the bed for turning. Invented by Henry Maudsley.

Rheostat — An adjustable resistance box so that part of the current can be cut out of the motor.

Riddle — Name given to a sieve used in foundries for siftings and for the molds.

Riffle — Name given a small file used by die sinkers and on similar work.

Rivet — A pin for holding two or more plates or pieces together. A head is formed on one end when made; the other end is upset after the rivet is put in place and draws the riveted members close together.

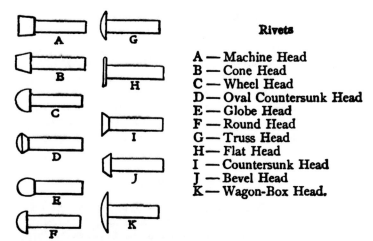

Rivets

A — Machine Head
B — Cone Head
C — Wheel Head
D — Oval Countersunk Head
E — Globe Head
F — Round Head
G — Truss Head
H — Flat Head
I — Countersunk Head
J — Bevel Head
K — Wagon-Box Head.

Rivet Machine — For making rivets from metal rods.

Roller Bearing — See Bearing.

Rule, Hook — Rule with a hook on the end for measuring through pulley holes and in similar places.

Rule, Key-seat — For laying out key-seats on shaft or in hubs.

Rule, Shrink — Graduated to allow for shrinkage in casting. Used by pattern-makers and varies with metal to be cast.

Run — Applied to drilling or reaming when the tool shows a tendency to leave the straight or direct path. Caused by one lip or cutting edge being less sharp than the other, being ground so one lip leads the other, or from uneven hardness of material being drilled.

Running or Sliding Fit — See Fits.

Rust Joint — A joint made by application of cast-iron turnings mixed with sal-ammoniac and sulphur to cause the turnings to rust and become a solid body.

S

Salamander — The mass of iron left in the hearth when a furnace is blown out for repairs.

Sand Blast — Sand is blown by compressed air through a hose as desired. Used to clean castings, stonework, etc.

Sanding Machine — A machine in which woodwork is finished by means of rolls or wheels covered with sandpaper.

Sanding Belt — Endless belt of some strong fabric, charged with glue and sand. For sandpapering wood held in hand or by clamps.

Saw, Band — Continuous metal band, toothed on one edge and guided between rolls. Mostly used on woodwork, but occasionally on metal work, especially in European shops.

Saw Bench, Universal — Bench on which lumber is brought to the saw for ripping, cross-cutting, dadoing, mitering, etc.

Saw, Cold — For sawing metal. Circular saws are generally used though not always, band saws being occasionally employed.

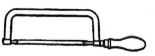

Saw, Hack — Close-toothed saw for cutting metal. Usually held in a hand frame but power hack saws are now becoming very common in shops.

Scarf — The bevel edge formed on a piece of metal which is to be lap-welded.

SCREW MACHINE TOOLS

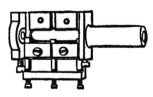

Box Tool, Bushing — The cutters in this tool are placed with edges radial to the stock and may be adjusted to turn the required diameter by the screws in the rear. The stock is supported in a bushing and must therefore be very true and accurate as to size.

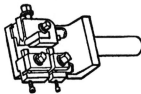

Box Tool, Finishing — The material turned in this box tool is supported by adjustable back rest jaws and the cutters are also adjustable in and out as well as lengthwise of the tool body.

Box Tool, Roughing — This tool has one or more cutters inverted over the work and with cutting edges tangent to the material.

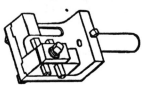

The back rest is bored out the size the screw or other piece is to be turned and the cutter turns the end of the piece to size before it enters the back rest. Sometimes a pointing tool is inserted in the shank for finishing the end of the work.

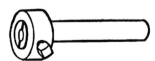

Drill Holder — The end of the drill holder is split and provided with a clamp collar by which the holder is closed on the drill.

Feed Tube — The screw-machine feed tube or feed finger is closed prior to hardening and maintains at all times a grip on the stock.

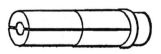

The rear end is threaded and screwed into the tube by which it is operated. It is drawn back over the stock and when the chuck opens is moved forward feeding the stock the right distance for the next piece.

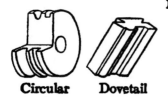

Circular   **Dovetail**

Forming Tools — Circular forming cutters are generally cut below center to give proper clearance and the tool post is bored a corresponding amount above center to bring the tool on the center line. Dovetail cutters are made at an angle of about 10 degrees for clearance.

Hollow Mill — Usually made with 3 prongs or cutting edges and with a slight taper inside toward the rear. A clamp collar is used on mill like a spring die collar and a reasonable amount of adjustment may be obtained by this collar. Hollow mills are frequently used in place of box tools for turning work in the screw machine.

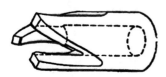

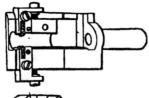

Nurling Tool — The two nurls in this box are adjustable to suit different diameters of work.

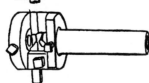

Pointing Tool — The bushing in this tool receives and supports the end of the round stock and the cutters carried in the frame form and point the end.

Revolving Die Holder — The common type of revolving die holder which is generally used with spring dies, has a pair of driving pins behind the head and in the flange of the sleeve which fits into the turret hole. At the rear end of the sleeve is a cam surface which engages a pin in the shank of the head when the die is reversed. The die is run on to the work with the driving pins engaged.

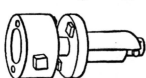

When the work is reversed, the cam at the rear engages the pin in the shank and holds the die from turning so that it must draw off the work.

Spotting Tool — This tool spots a center in the end of the bar of stock to allow the drill to start properly, and also faces the end of the piece true. Sometimes called "centering and facing" tool. It is desirable to have the included angle of the cutting point less than that of the drill which follows it in order that the latter may start true by cutting at the corners first.

Spring Collet — Spring collets or chucks are made to receive round,

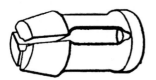

square, hexagonal or other stock worked in the screw machine. The collet is hardened and is closed in operation by being pressed into the conical cap into which it fits. When released it springs open sufficiently to free the stock and allow it to be fed through the collet.

Spring Die and Extension — Spring dies

or prong dies are provided with a collar at the end for adjusting and are easily sharpened by grinding in the flutes.

Screw Plates — Holders for dies for cutting threads on bolts or screws. Dies are usually separate but sometimes cut in the piece which forms the holder.

SCREWS

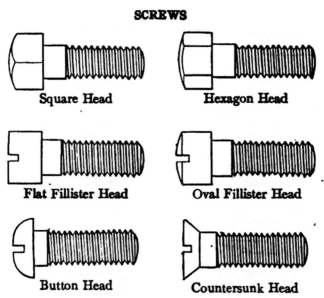

Square Head   Hexagon Head

Flat Fillister Head   Oval Fillister Head

Button Head   Countersunk Head

Cap Screws — Cap screws are machined straight from point to head, have finished heads and up to 4 inches in length are usually threaded three-fourths of the length. When longer than 4 inches they are threaded one half the length, which is measured under the head, except in the case of countersunk head screws which are measured over all. Cap-screw sizes vary by 16ths and 8ths and are regularly made up to 1 or $1\frac{1}{4}$ inch diameter, while machine screws with which they are frequently confused are made to the machine-screw gage sizes.

Flat fillister heads on cap screws are often called "round" heads; oval fillister heads are frequently designated as "fillister" heads, and countersunk heads as "flat" heads. When a countersunk or flat head has an oval top it is called a "French" head. Fillister heads are also made with rounded corners as well as with the oval head shown above. Fillister head screws are known in England as cheese-head screws. The included angle of the countersunk or flat head is 70 degrees.

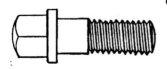

Collar Screw — Collar or collar head screws are used for much the same purposes as regular cap screws, and, in fact, are sometimes designated as "collar" cap or "collar head" cap screws.

Lag Screw — Lag screws, or coach screws, as they are often called, have a thread like a wood screw and a square or a hexagonal head. They are used for attaching countershaft hangers to over-head joists for fastening machines to wood floors and for many other purposes where a heavy wood screw is required.

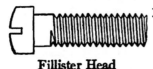

Machine Screws — Machine screws are made to the sizes of the machine-screw gage instead of running like cap screws in even fractions of an inch.

Fillister Head

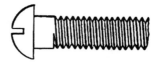

Counter Sunk or Flat Head

Button or Round

Set Screws — Set screws are threaded the full length of body and may or may not be necked under the head. They are usually case-hardened. Ordinarily the width of head across flats and the length of head are equal to the diameter of the screw. The headless set screw is known in England as a "grub" screw.

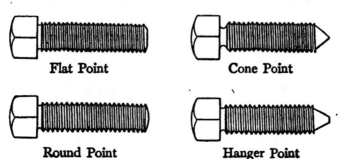

Flat Point   Cone Point

Round Point   Hanger Point

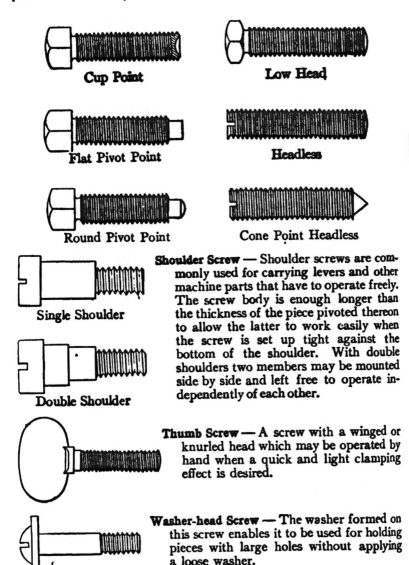

Cup Point

Low Head

Flat Pivot Point

Headless

Round Pivot Point

Cone Point Headless

Single Shoulder

Double Shoulder

Shoulder Screw — Shoulder screws are commonly used for carrying levers and other machine parts that have to operate freely. The screw body is enough longer than the thickness of the piece pivoted thereon to allow the latter to work easily when the screw is set up tight against the bottom of the shoulder. With double shoulders two members may be mounted side by side and left free to operate independently of each other.

Thumb Screw — A screw with a winged or knurled head which may be operated by hand when a quick and light clamping effect is desired.

Washer-head Screw — The washer formed on this screw enables it to be used for holding pieces with large holes without applying a loose washer.

Wood Screws — Wood screws are made in an endless variety of forms, a number of which are shown on the following page. They range in size from No. 0 to No. 30 by the American Screw Company's gage and are regularly made in lengths from ¼ inch to 6 inches. Generally the thread is cut about seven tenths of the total length of the screw. The flat-head wood screw has an included angle of head of 82 degrees.

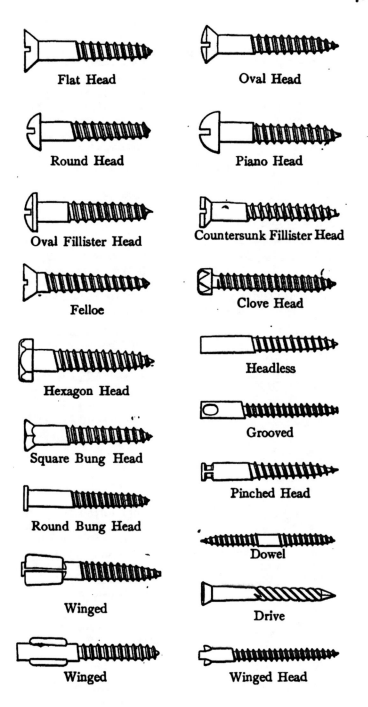

Flat Head

Oval Head

Round Head

Piano Head

Oval Fillister Head

Countersunk Fillister Head

Felloe

Clove Head

Hexagon Head

Headless

Square Bung Head

Grooved

Round Bung Head

Pinched Head

Winged

Dowel

Drive

Winged

Winged Head

Screw Thread, Acme 29 degree Standard —

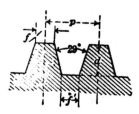

$$p = \text{pitch} = \frac{1}{\text{no. threads per inch}}$$

d = depth = $\frac{1}{2}$ p + .010.

f = flat on top of thread = p x .3707

The Acme screw thread is practically the same depth as the square thread and much stronger. It is used extensively for lead screws, feed screws, etc.

Screw Thread, British Association Standard—

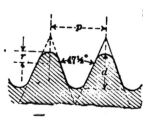

p = pitch

d = depth = p x .6

$$r = \text{radius} = \frac{2 \text{ x p}}{11}$$

This thread has been adopted in England for small screws used by opticians and in telegraph work, upon recommendations made by the Committee of the British Association. The diameter and pitches in this system are in millimeters.

Screw Thread, Buttress —

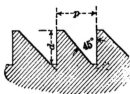

$$p = \text{pitch} = \frac{1}{\text{no. threads per inch}}$$

d = depth = $\frac{3}{4}$ p

The buttress thread takes a bearing on one side only and is very strong in that direction. The ratchet thread is of practically the same form but sharper.

Screw Thread, International (Metric) Standard—

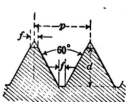

p = pitch

d = depth = p x .6495

$$f = \text{flat} = \frac{p}{8}$$

The International thread is of the same form as the Sellers or U. S. Standard. This system was recommended by a Congress held at Zurich in 1898, and is much the same as the metric system of threads generally used in France. The sizes and pitches in the system are in millimeters.

Screw Thread, Square —

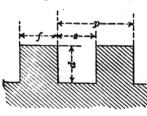

$$p = \text{pitch} = \frac{1}{\text{no. threads per inch}}$$

d = depth = $\frac{1}{2}$ p

f = width of flat = $\frac{1}{2}$ p

s = width of space = $\frac{1}{2}$ p.

While theoretically depth, width of space and thread are each one half the pitch, in practice the groove is cut slightly wider and deeper.

Screw Thread, United States Standard —

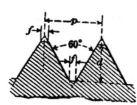

$$p = pitch = \frac{1}{no.\ threads\ per\ inch}$$

$$d = depth = p \times .6495$$

$$f = flat = \frac{p}{8}$$

This thread was devised by Wm. Sellers, and recommended by the Franklin Institute in 1869. It is called the U. S. Standard, the Franklin Institute, and the Sellers thread. The advantages of this thread are, that it is not easily injured, tap and dies will retain their size longer, and bolts and screws with this thread are stronger and better appearing. The system has been adopted by the United States Government, Master Mechanics and Master Car Builders' Associations, Machine Bolt Makers, and by many manufacturing establishments.

Screw Thread, V, 60 degree Sharp —

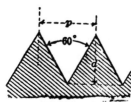

$$p = pitch = \frac{1}{no.\ threads\ per\ inch}$$

$$d = depth = p \times .8660$$

While the sharp V form gives a deeper thread than the U. S. Standard, the objections urged against the thread are, that the sharp top is injured by the slightest accident, and, in the case of taps and dies, the fine edge is quickly lost, causing constant variation in fitting.

Screw Thread, Whitworth Standard —

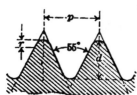

$$p = pitch = \frac{1}{no.\ threads\ per\ inch}$$

$$d = depth = p \times .64033$$

$$r = radius = p \times .1373$$

The Whitworth thread is the standard in use in England. It was devised by Sir Joseph Whitworth in 1841, the system then proposed by him being slightly modified in 1857 and 1861.

Worm Thread, Brown & Sharpe 29 degree —

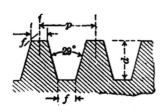

$$p = pitch = \frac{1}{no.\ threads\ per\ inch}$$

$$d = depth = p \times .6866$$

$f = $ flat on top of thread $= p \times .335$
This thread is commonly used in America for worms. It is considerably deeper than the Acme screw thread of the same angle, namely 29 degrees.

Sector — A device used on an index plate of a dividing head for indicating the number of holes to be included at each advance of the index crank in dividing circles. The sector can be opened or closed to form as large or small an arc as necessary to cover the desired number of holes for each movement of the crank.

Set — The bend to one side of the teeth of a saw.

Set Screw — See Screws.

Shaft-bearing Stand — Shaft bearing which is fastened to floor.

Shaft Coupling — See Coupling.

Shaft, Flexible — Shaft made of a helical spring or of jointed parts, usually confined in a leather or fabric casing, to transmit power in varying directions.

Shaft, Jack — A secondary or auxiliary shaft, driven by the engine and in turn driving the dynamos or other machinery. Jack shafts are often introduced between a regular machine counter-shaft and the line shaft.

Shaft, Line — The shafting driving the machinery of a shop or section of a shop by means of pulleys and belts.

Shaft, Muley — A vertical shaft carrying two idler pulleys for carrying a belt around a corner. To be avoided where possible.

Shaper — Work is held on table or knee and tool moves across it, held by a tool post on the moving ram. Table adjustable for depth of cut, etc.

Shaper — Parts of

1. Tool post.
2. Clapper block.
3. Clapper box.
4. Clamping bolts.
5. Down-feed screw.
6. Tool slide.
7. Tool head.
8. Binder for head.
9. Stop for down feed.
10. Down-feed adjustment.
11. Ram adjuster.
12. Ram.
13. Position lever.
14. Clamp for down feed.
15. Ram slide.
16. Face of column.
17. Ram guide.
18. Frame or body.
19. Feed box.
20. Feed regulator.
21. Cone-driving pulley.
22. Lever bearing
23. Power elevation of table.
24. Vise.
25. Swiveling base.
26. Table.
27. Saddle.
28. Cross-feed screw.
29. Cross-feed dog.
30. Cross-feed handle.
31. Elevating screw.

Shaper, Crank — Ram is driven by a crank motion.

Shaper, Draw Cut — Cutting stroke takes place when tool is moving toward frame of machine. This tends to draw the parts together.

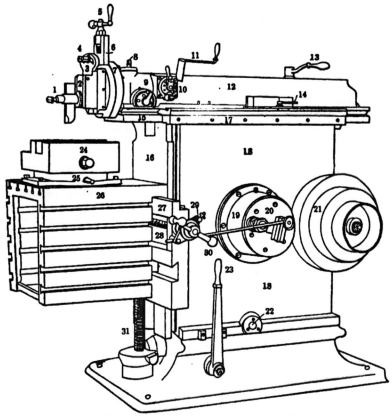

SHAPER — POTTER AND JOHNSTON

Shaper, Friction — Ram is driven by rack and pinion through friction clutches. Ram is reversed by simultaneous release and engagement of these clutches. These are driven by open and crossed belts in opposite directions.

Shaper, Gear — Planes gear teeth by using a hardened cutter, shaped like a pinion gear, and moving across the face of the gear with a planing or shaping cut.

Shaper, Geared — Ram is driven by rack and pinion with a slow cutting stroke and a quick return by shifting open and crossed belts the same as on a planer.

Shapers, Traverse or Traveling Head — Ram feeds across work, which is stationary.

Shear — Tool for cutting metals between two blades. The name given to the way or V of a lathe or planer. A hoisting apparatus used on wharves or docks, consisting of two heavy struts like a long inverted V.

Shears — The ways on which the lathe carriage and tail stock move are called "shears" by some, "ways" by others. They may be either V, flat or any other shape.

Shears, Alligator — Name given to machine where moving knife or cutter works on a pivot.

Shears, Squaring — Has cutter bar guided at both ends.

Shears, Slitting — Arranged for slitting sheet metal. Rotary cutters are usually employed.

Shearing Machine — For cutting off rods, bars or plates.

Shedder — A plate or ring operated by springs or by a rubber pad to eject a blank from a compound die. It acts as an internal stripper, and is sometimes known as an ejector.

Sherardizing — The name given to a new process of dry galvanizing of any iron product.

Shifter Forks — Arms to guide belt from tight to loose pulley or vice versa, by pressing the sides.

Shim — A liner or piece to place between surfaces to secure proper adjustment.

Shrink Fit — See Fits.

Slip Washer — See Open Washer.

Slotted Washer — See Open Washer.

Slotter — A machine for planing vertical surfaces or cutting slots. Tool travels vertically.

Socket, Grip — A device for driving drills and other tools with either a straight or taper shank.

Sow — In foundry work, the gate or central channel which feeds iron into the pigs when making pig iron.

Sow or Sow-block — Local name for a chuck for holding work, such as dies. A ball chuck.

Spinning — The forming of sheet metal by rolling it against forms such as lamp bodies. Lathes are made especially for this work.

Spline — Used in some sections in place of "key" and in others the same as "feather." See Key and Feather.

Split Nut — Nut split lengthwise so as to open for quick adjustment.

Spot or Spotting — Spotting is making a spot or flat surface for a set-screw point or to lay out from.

 **Spring, Compression** — A helical spring which tends to shorten in action.

Spring, Helical — A spring coiled lengthwise of its axis like a screw thread. Often incorrectly called a spiral spring.

Spring, Leaf — A built up spring made of flat stock like a carriage spring or locomotive driving spring.

Spring, Spiral — A spring wound with one coil over the other as in a clock spring. Usually of flat stock, but not always.

Spring, Tension — A helical spring which tends to lengthen in action.

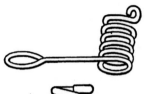

Spring, Torsion — A helical spring which operates with a coiling or uncoiling action as a door spring.

Spring, Valve — A helical spring used on valve stems and similar places; each coil being smaller than the one below, in order that the spring may close up into a very small space and then have a considerable range of action.

Spring Cotter — See Cotter.

Sprue Cutter — A cutting punch for trimming sprues from soft metal castings.

Square, Caliper — A square with a caliper adjustment for laying out work.

Square, Combination — A tool combining square, level and protractor in one tool.

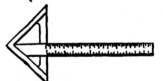

Square, Center — For finding the center of a round bar by placing across the end and scribing lines in two different positions. Also used as a T-square. Not so much used as formerly.

T-Square — A straight edge with a head at one end commonly used on the drawing board for drawing straight lines. It forms a guide also, along which triangles are slid. Generally made of wood, although sometimes of metal and often provided with a swiveling head which serves as a protractor when graduated in degrees.

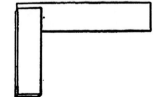

Square, Try — Small square for testing work as to its being at right angles.

Stand, Vise — Stand, usually of metal, for holding a vise firmly in any desired part of the shop, making it a portable vise.

Steady Rest — A rest attached to the lathe ways for supporting long, slender work.

Steel, High Speed — A name given to steels which do not lose their hardness by being heated under high speed cuts. Alloy steels which depend on tungsten, chromium, manganese, molybdenum, etc., for their hardness.

Stocks, Ratchet — Die stocks with ratchet handles.

Straight Edge — A piece of metal having one edge ground and scraped flat and true. Small ones are sometimes made of steel but large, straight edges are usually of cast iron, proportioned to resist bending, and are used for testing the truth of flat surfaces such as plane ways.

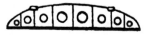

Strap — See Belt Polisher.

Strapping — A method of buffing by the use of a flexible strap or belt, usually made of cloth and covered with abrasive held in place by glue. Runs over two pulleys or one pulley and a rod or plate at high speed.

String Jig — Fixture for holding a row of pieces to be milled or planed.

Stripper — A thin plate placed over the die, in a punch press, with a gap beneath to admit the sheet stock and an opening to allow the punch to pass freely; upon the up-stroke of the punch it prevents the strip of metal from lifting with the punch.

Stripping-plate — A plate containing holes of the same outline as the pattern and used to prevent sand following the patterns when drawn out on some molding machines.

Stud, Collar — The collar stud forms a satisfactory device for carrying gears, cam rolls, rocker levers, etc. It is often provided with a hole at the end for a cutter pin or is slotted for a split washer, to retain the gear, or other part in place.

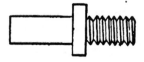

Stud, Shoulder — A stud of this form is used for mounting levers and other parts which could be operated on a plain, unthreaded stud, which stud, however, cannot be conveniently set or removed when necessary. It is also a form of post or guide sometimes employed in machine construction for carrying one or more sliding parts.

Stud, Threaded — Studs are threaded on both ends to lengths required and screwed tight into place. A nut is run on the outer end. They are commonly used for holding cylinder heads in place and for other purposes where it is desirable that the screw shall remain stationary to prevent injury of threads tapped in the main piece.

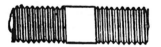

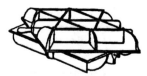

Surface Plates — Cast-iron plates have surfaces scraped flat for use in testing work. Should be made in sets of three and so scraped that each one has a perfect bearing with the other two.

Swaging — Changing the sectional shape of a piece of metal by hammering, rolling or otherwise forcing the particles to change shape without cutting.

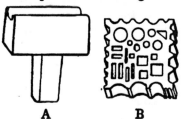

Swaging Blocks — Blocks of cast or wrought iron to assist blacksmith in swaging and bending iron to various shapes. *A* is for use in the hardy hole in the anvil, *B* can be used anywhere but is usually on or beside the anvil.

A   B

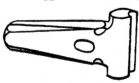

Swaging Hammer — A connection with the swaging block to swage metal to the desired size and shape.

Swaging Machine — For reducing tapering or pointing wire or tubing either between rolling dies or by hammering with rapid blows between dies of suitable shape.

Sweating — Another name for soldering.

Swing of a Lathe — In the United States the swing of a lathe means the largest diameter of the work that can be swung over the ways or shears. In England it means the distance from lathe center to the ways or one half the U. S. measurements.

T

Take-up — Name given to device for taking up slack in belt or rope drives.

Tap — Hardened and tempered steel tool for cutting internal threads. Has a thread cut on it and flutes to give cutting edges.

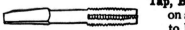

Tap, Bit-brace — Tap of any kind, usually on all bolt taps, with shank made square to be driven by bit-brace.

Tap, Echols Thread — This form of tap has every other thread cut away on each land, but these are staggered so that a space on one land has a tooth behind it on the next land. This is done for chip clearance.

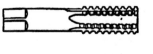

Tap, Hand, First or Taper — Bolt tap usually for hand use. The first or taper tap has the front end tapered to enter easily.

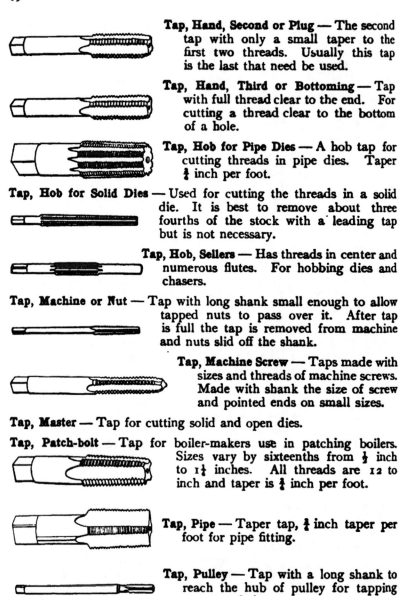

Tap, Hand, Second or Plug — The second tap with only a small taper to the first two threads. Usually this tap is the last that need be used.

Tap, Hand, Third or Bottoming — Tap with full thread clear to the end. For cutting a thread clear to the bottom of a hole.

Tap, Hob for Pipe Dies — A hob tap for cutting threads in pipe dies. Taper ¾ inch per foot.

Tap, Hob for Solid Dies — Used for cutting the threads in a solid die. It is best to remove about three fourths of the stock with a leading tap but is not necessary.

Tap, Hob, Sellers — Has threads in center and numerous flutes. For hobbing dies and chasers.

Tap, Machine or Nut — Tap with long shank small enough to allow tapped nuts to pass over it. After tap is full the tap is removed from machine and nuts slid off the shank.

Tap, Machine Screw — Taps made with sizes and threads of machine screws. Made with shank the size of screw and pointed ends on small sizes.

Tap, Master — Tap for cutting solid and open dies.

Tap, Patch-bolt — Tap for boiler-makers use in patching boilers. Sizes vary by sixteenths from ½ inch to 1¼ inches. All threads are 12 to inch and taper is ¾ inch per foot.

Tap, Pipe — Taper tap, ¾ inch taper per foot for pipe fitting.

Tap, Pulley — Tap with a long shank to reach the hub of pulley for tapping set-screw holes.

Tap Remover — Device for removing broken taps. Usually have prongs which go down in the flutes and around the central portion.

Tap, Staybolt — Tap for threading boiler sheets for staybolts. *A* reams the hole, *B* is a taper thread, *C* is straight thread of right size. *D* square for driving tap. All standard staybolt taps have 12 threads per inch.

D   C   B   A

Tap, Step — Tap made with "steps" or varying diameters. Front end cuts part of thread, next step takes out more and so on to the end. Only used for heavy threads, usually square or Acme.

Tap, Stove-bolt — Made same way as machine-screw taps but in only six standard sizes.

Tap, Tapper — Similar to a machine tap except that it has no square on the end.

Taper Reamer — See Reamer, Taper.

Tapped Face-plate — Having a number of tapped holes instead of slots. Studs screw in at any desired point.

Tapping Machine — For cutting threads with taps (tapping) in nuts or other holes.

Threads, Screw — See Screw Threads.

Threading Tool, Rivet-Dock — The tool is a rotary cutter with cutting teeth of different depths. The first tooth starts the cut, then instead of feeding the carriage into the work, the cutter is turned and the next tooth takes the next cut.

Toggle — Arrangement of levers to multiply pressure obtained by making movement given to work very much less than movement of applied power.

Tongs — Tools for holding hot or cold metals.

Tool, Boring — For operating on internal surface of holes.

Tool, Cutting-off — For cutting work apart on lathe or cutting-off machine.

Tool, Diamond — Black diamond set in metal for tracing emery or other abrasive wheels. Also used to some extent for truing up hardened steel or iron.

Tool Holder, Lathe or Planer — A body or shank, adopted to hold  small pieces of tool steel for cutting tools. These can be removed for sharpening or renewal without moving the holder and saves resetting the tool to the work.

Tools, Inserted Cutter — Holders in which are held small steel cutting tools. These are usually removed for grinding or replacing when broken or worn out. Usually made of self-hardening steel.

Tool, Nurling — For roughing or checking the outside of turned work so it can be readily grasped by hand. The tool is a wheel

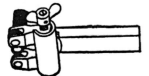

with the desired markings cut in the edge and hardened. It is forced against the work and actually forces the metal up into the depressions in the wheel. Most nurls are held in the end of a hand tool but for heavy nurling they are made to go in the tool post as shown.

TOOLS, LATHE — WM. SELLERS & CO.

Lathe

Square Thread	60° V Thread	Bent Side	Side	Bent Roughing	Roughing	Kind of Tool
Right	Right	Right	Right	Top Rake 8°	Top Rake 8°	
Top d / End e / Side f / Side a	Top d / End e / Side b	Top a / End e / Side a	Top d / End e / Side a	Top d / Side b / Side a	Side a	Face
0° / 0° / 0° / 10°	0° / 15° / 1° / 15°	15° / 6° / 6°	15° / 6° / 6°	15° / 6°	6°	Clearance
Square Thread	60° V Thread	Bent Side	Side	Nicking	Finishing	Kind of Tool
Left	Left	Left	Left			
Top d / End e / Side b / Side a	Top d / End e / Side b / Side a	Top d / End e / Side a	Top d / End e / Side b / Side a	Top d / End e / Side b / Side a	Top d / End e / Side b / Side a	Face
0° / 0° / 10° / 0°	0° / 15° / 12° / 1°	15° / 6° / 6°	15° / 6° / 6° / 5°	1° / 0° / 6° / 5°	15° / 6° / 6° / 4°	Clearance
Square Thread Bent	60° V Thread Bent	Inside Bent	Bent Brass	Bent Nicking	Bent Finishing	Kind of Tool
Right	Right	Left	Right	Left	Left	
Top d / End e / Side b / Side a	Top d / End e / Side b	Top d / End e / Side a	Top d / End e / Side a	Top d / End e / Side b / Side a	Top d / End e / Side b / Side a	Face
0° / 0° / 0° / 10°	0° / 0° / 1° / 15°	15° / 6° / 6°	0° / 10° / 0° / 10°	1° / 0° / 6° / 6°	15° / 6° / 6° / 4°	Clearance
Square Thread Bent	60° V Thread Bent	Inside Bent	Brass	Bent Nicking	Bent Finishing	Kind of Tool
Left	Left	Right	Right	Right	Right	
Top d / End e / Side b	Top d / End e / Side b / Side a	Top d / End e / Side b / Side a	Top d / End e / Side b / Side a	Top d / End e / Side b / Side a	Top d / End e / Side b / Side a	Face
0° / 0° / 10° / 0°	0° / 15° / 15°	12° / 6° / 6° / 6°	0° / 10° / 0° / 10°	1° / 0° / 6° / 6°	15° / 6° / 6° / 4°	Clearance

TOOLS, LATHE

Left-hand Side Tool

Right-hand Side Tool

Left-hand Bent Side Tool

Right-hand Bent Side Tool

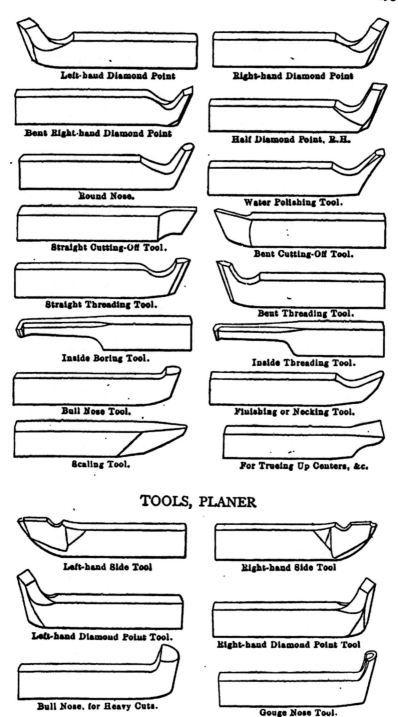

Left-hand Diamond Point

Right-hand Diamond Point

Bent Right-hand Diamond Point

Half Diamond Point, R.H.

Round Nose.

Water Polishing Tool.

Straight Cutting-Off Tool.

Bent Cutting-Off Tool.

Straight Threading Tool.

Bent Threading Tool.

Inside Boring Tool.

Inside Threading Tool.

Bull Nose Tool.

Finishing or Necking Tool.

Scaling Tool.

For Trueing Up Centers, &c.

TOOLS, PLANER

Left-hand Side Tool

Right-hand Side Tool

Left-hand Diamond Point Tool.

Right-hand Diamond Point Tool

Bull Nose, for Heavy Cuts.

Gouge Nose Tool.

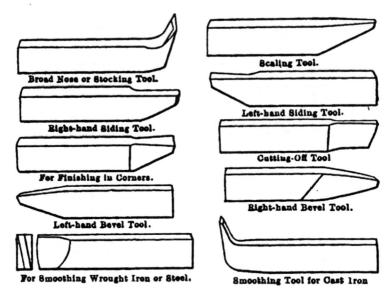

Broad Nose or Stocking Tool.

Scaling Tool.

Right-hand Siding Tool.

Left-hand Siding Tool.

For Finishing in Corners.

Cutting-Off Tool

Left-hand Bevel Tool.

Right-hand Bevel Tool.

For Smoothing Wrought Iron or Steel.

Smoothing Tool for Cast Iron

TOOLS, PLANER AND SLOTTER

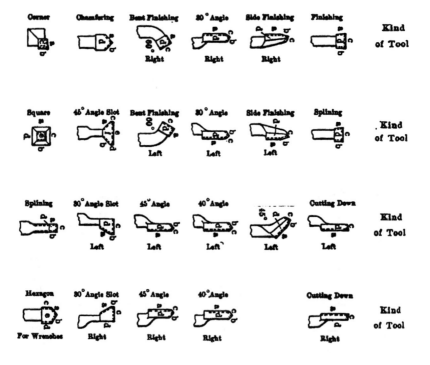

| Corner | Chamfering | Bent Finishing | 30° Angle | Side Finishing | Finishing | |
| Right | Right | Right | | | | Kind of Tool |

| Square | 45° Angle Slot | Bent Finishing | 30° Angle | Side Finishing | Splining | |
| | | Left | Left | Left | | Kind of Tool |

| Splining | 30° Angle Slot | 45° Angle | 40° Angle | | Cutting Down | |
| | Left | Left | Left | Left | Left | Kind of Tool |

| Hexagon | 30° Angle Slot | 45° Angle | 40° Angle | | Cutting Down | |
| For Wrenches | Right | Right | Right | | Right | Kind of Tool |

Tote Boxes — See Tote Pans.

Tote Pans — Pans or trays of steel for carrying small parts from one part of shop to another.

Train — A series of gears, as the feed train of a lathe connecting spindle with lead screw.

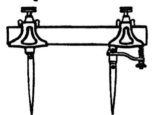

Trammels — For drawing large circles. Fit on a beam and their capacity depends on the length of the beam.

Trepanning Tool — Tool for cutting an annular groove outside or around a bored hole.

Tripper — Device that trips any piece of mechanism at the desired time. An example is found in conveyers where the tripper dumps the material at the desired point.

Tumbler Gear — An intermediate gear which meshes in between other gears to reverse the direction of the driven gear of the train.

Stub End

Hook and Eye

Turn Buckle — Turn buckles are for connecting and tightening truss rods, tie rods, etc., used in construction work.

Tuyere — The pipe or opening into forge through which air is forced.

V

Veeder Metal — An alloy with tin as a base, used for casting in metal molds.

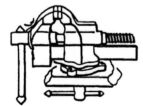

Vise, Chipping or Filing — Heavy bench vise used for holding work to chip. Vises for filing only are similar but lighter.

Vise, Drill — Vise for use on drill press to hold work being drilled.

Vise, Hand—A small vise to be held in the hand. For small work that requires turning frequently to get at different sides.

Vise, Jig — A drill vise with arms which carry bushings so that pieces can be drilled in duplicate without special jigs for them.

 **Vise, Pin** — Small hand vise for holding small wire rods.

Vise Stands — See Stands, Vise.

V's — Ways shaped like a V, either raised above the bed as on a lathe or cut below as in a planer, for guiding the travel of a carriage or table.

Volt — The unit of electrical pressure.

W

Washer, Open — Washers with one side open so as to be removed or slipped under the nut to avoid necessity of taking the nut entirely off. Also called a "C" washer.

Watt — The unit of electrical power and equals volts multiplied by amperes. 746 watts are equal to one horse-power.

Ways — The guiding or bearing surfaces on which moving parts slide, as in a lathe plane or milliug machine. The ways may be of any form, flat, V or any other shape.

Welding — The joining of metals by heating the parts to be joined to the fusing point and making a union by hammering or forcing them together. Welding in an open fire is usually confined to iron and steel but nearly all metals can be joined in this way by electric heating.

Wind — Pronounced with a long *i* as in "mind" and refers to a twist or warping away from straightness and parallelism.

Wrench, Bridge Builders' — Large heavy wrench with a hole in end for a tackle to apply power.

WRENCHES, MACHINE

15 degree angle wrenches have an opening milled at an angle of 15 degrees with the handle, which permits the turning of a hexagon nut completely around where the swing of the handle is limited to 30 degrees.

22½ degree angle wrenches have an opening which forms an angle of 22½ degrees with the handle, which permits the turning of any square head bolt or screw completely around where the swing of the handle is limited to 45 degrees.

Unfinished drop-forged wrenches are plain forgings, with openings milled to fit the nut or screw on which they are to be used.

Semi-finished wrenches are milled to fit the nut or screw on which they are to be used and case-hardened all over.

Finished wrenches are milled to fit the nut or screw on which they are to be used and are ground, polished, case-hardened all over, lacquered, with heads bright.

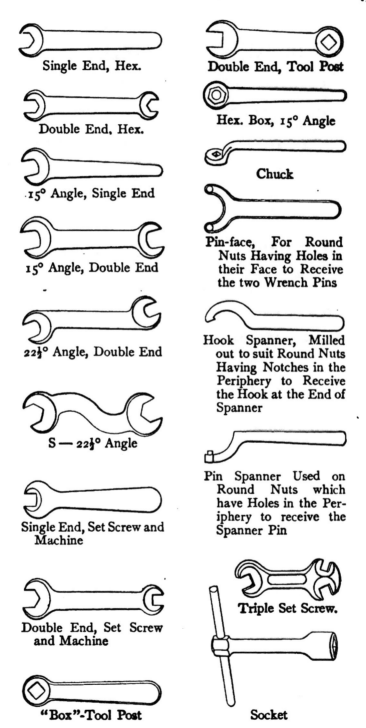

Single End, Hex.

Double End, Hex.

.15° Angle, Single End

15° Angle, Double End

22½° Angle, Double End

S — 22½° Angle

Single End, Set Screw and Machine

Double End, Set Screw and Machine

"Box"-Tool Post

Double End, Tool Post

Hex. Box, 15° Angle

Chuck

Pin-face, For Round Nuts Having Holes in their Face to Receive the two Wrench Pins

Hook Spanner, Milled out to suit Round Nuts Having Notches in the Periphery to Receive the Hook at the End of Spanner

Pin Spanner Used on Round Nuts which have Holes in the Periphery to receive the Spanner Pin

Triple Set Screw.

Socket

WRENCHES, MISCELLANEOUS

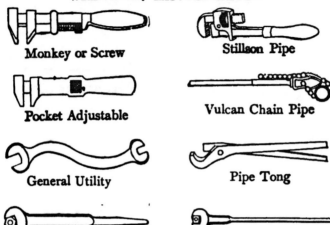

Monkey or Screw

Stillson Pipe

Pocket Adjustable

Vulcan Chain Pipe

General Utility

Pipe Tong

Construction

Track

Wrench, Tap — Wrench for holding and turning taps. Usually made adjustable for different sizes.

Wringing Fits — See Fits.

INDEX

A

PAGE

Abrasive wheels, grading 156
Abrasives, commercial 148
Accurate taper gage 245
Acme twenty-nine degree screw
  thread, and Brown & Sharpe
  worm thread compared 10, 11
Acme twenty-nine degree screw
  thread, measuring with micrometer
  and wires 35
Acme twenty-nine degree screw
  thread, table of parts 22
Addendum of gear teeth 83
A. L. A. M. standard bolts and nuts. 216
Alining shafting with a steel wire. 312, 313
Allowances for drive fits 229, 232
  for hand fits 229
  for grinding 153
  for press fits 229, 232
  for shrink fits 231
  for threading in screw
    machine 173
  for upsets............. 292
Allowances in plug gages for standard
  holes 231
Allowances in shop gages for various
  kinds of fits 231, 232
Allowances for punch and die, for
  accurate work 184–187
Allowances with the calipers for running, shrink and press fits.... 233–236
Alloys 325, 326
  case hardening of 319, 323
  for coinage 324
Aluminum bars, weight of 307
Aluminum, burnishing 327
  frosting 327
  melting............... 327
  polishing 327
  properties of 327
  shrinkage of 327
  spinning 327
  turning............... 327
Alundum 148
American or Brown & Sharpe wire-
  gage sizes 297, 299
Angle of helix for screw-threads,
  table 4, 5
Angles, calculation of 362, 367
  corresponding to given tapers
    per foot 259
  laying out by table of chords 281
  obtained by opening a two-
    foot rule 294
  of V-tools, measurement of. 11, 13
  rules for 266
Angular cutters, cutters for fluting.. 140

B

Angular cutters, number of teeth in.. 140
Annealed steel, properties of....... 322
Annealing 317
Antimony, properties of 324
Areas and circumferences of circles
  from ¼ to 1 inch........... 340, 341
Areas and circumferences of circles
  from 1 to 100 .... 342, 347
Areas and circumferences of circles
  from 1 to 520............... 348, 360
A. S. M. E. machine screw taps.... 63
A. S. M. E. machine screws, standard
  proportions of 200, 209
Automobile bolt and nut standards.. 216

B

Babbitt metal 326
Back lash of machine molded gears. 93
Ball bearings, four point 280
  two point 278, 279
Ball handles 272
  single end 273
  lever handles 273
Bar brass, copper and aluminum,
  weight of 307
Bar steel and iron, weight of 306
  flat sizes, weight of 308
Barth key dimensions 239
Baths for hardening steel 317
  for heating steel........ 315, 316
Bearing metals 325
Bearings, ball, data for four point.. 280
  data for two point,
    278, 279
Bell metal...................... 326
Belt fastenings 310
  hooks 310
    Bristol 310, 311
    lacing, Jackson.......... 310, 311
      leather........... 310, 311
    strength of 310, 311
    wire.............. 310, 311
  splice 310
  stud, Blake........... 310, 311
  working strain of 311
Belts and shafting............. 310, 314
Benches, *see* Work benches.
Bevel gear cutters 107
  table.............. 108, 109
  gears, laying out 106
  names of parts 106
Binder handles 272
Binding screws for jigs 268
Birmingham or Stubbs' iron wire
  gage sizes 297–299

PAGE

Bismuth alloy, fusible 325
        properties of............. 324
Blake belt studs 310, 311
Blanks, shell, finding diameters of.. 184
Block and roller chain sprocket
        wheels 104, 105
Block indexing in cutting gear teeth,
                                101, 102
Boiler rivet heads 224
Bolt circles, spacing 365
        heads for standard T-slots.... 218
        stock required for 293
Bolts and nuts, U. S. standard,
        finished 191
Bolts and nuts, U. S. standard,
        rough....................... 190
Bolts and nuts, A. L. A. M. standard 216
        coupling 217
Bolts, eye 219
        hook.................... 275
        lengths of 214
        lengths of threads on 214
        machine, carriage and loom,
                button head 214
Bolts, machine, with manufacturers'
        standard heads 192
Bolts, nuts, and screws 189-225
        planer head 217
        square and round countersunk
                heads 215
        stove 215
        tap....................... 215
Bone for case hardening........... 319
Box tool for screw machines, finish-
        ing 170
Box tool for screw machines, rough-
        ing 169
Box tool for screw machines, speed
        and feeds for 180
Brass bars, weight of 307
        cast, properties of 324
        plates, weight of........302, 303
        tubing, weight of 309
        wire, weight of 304, 305
Briggs standard pipe ends, drill sizes
        for..........................41, 44
Briggs standard pipe ends, gages
        for 42-46
Briggs standard pipe, joint 42
Briggs standard pipe, threads, form
        and general system 39-43
Bristol belt hooks 310, 311
British Association screw threads... 19
Bronze, properties of 324
Bronzes, Navy Department 325
Brown & Sharpe standard tapers 250, 251
Brown & Sharpe thread micrometer
        readings for U. S. threads........ 24
Brown & Sharpe thread micrometer
        readings for V threads.......... 25
Brown & Sharpe thread micrometer
        readings for Whitworth threads.. 26
Brown & Sharpe wire gage sizes. 297, 299
Brown & Sharpe worm thread and
        Acme screw thread compared.. 10, 11
Brown & Sharpe worm threads,
        measuring with micrometer and
        wires 37

PAGE

Brown & Sharpe worm threads,
        table of parts 36
Bushings, for jigs, fixed 266
                loose 266, 267
Button head bolts, machine, carriage
                and loom 214
        cap screws........... 197
Button head machine screws, Ameri-
        can Screw Co. standard...... 199
Button head machine screws, A. S.
        M. E. standard 209

C

Calculations by trigonometry 364
Calipering and fitting 225-244
Calipers, axial inclination of in
        measuring for shrink or press
        fits 234
Calipers, side play of in measuring
        for running fits.............233, 235
Cams, heart shape, milling 121
        milling by gearing up the
                dividing head 122
Cap screws, button head 197
        flat and oval counter-
                sunk head 197
        flat, round and oval
                fillister head.....195, 196
        hexagon and square
                head 194
Carbonizing 318-323
        material 319-321
        rate of...........319, 321
        with gas 321
Carborundum 148
        Company's wheel
                grade marks 156
Carriage bolts 214
Case hardening 318-323
        rate of........319, 321
        with gas 321
Castellated nuts for A. L. A. M.
        standard bolts 216
Cast gear teeth 92
        iron soldering 81
        washers 221
Castings, shrinkage of 326
        weight of 294
Centigrade thermometer scale 323
Change gears for cutting diametral
        pitch worms in the lathe........ 9
Change gears for cutting screw, how
        to find 3
Charcoal for case hardening 319
Check and jam nuts, cold punched. 210
Chemical symbols of metals....... 324
Chordal pitch 83
Chords, table of 280, 281
        table for laying out holes in
                circles 284-286
        use of for laying out angles. 281
Chromium, properties of 324
Chucking reamers, rose, cutter for
        fluting 144
Circles, areas of, from ¼ to 1 inch,
                                340, 341

PAGE

Circles, areas of, from 1 to 100 .342- 347
　　　　　　from 1 to 520 .348–360
Circles, tables of sides, angles and
　sines for laying out work in. .286–291
Circles, tables for spacing holes in 284–291
Circular forming tools 174–178
　　　　　　diameter of. . 176
Circular forming tools for conical
　points 178
Circular pitch 83
Circumferences and areas of cir-
　cles 340–360
Circumferences and diameters of
　circles from 1 to 200 361
Circumferential distances, divisions
　corresponding to 147
Clearance angle for milling cutters. . 120
　　　for punches and dies. .184–187
　　　for running fits 232
　　　of gear teeth 83
　　　of reamers, grinding. .164, 167
　　　of thread tool at side, find-
　　　　ing 4
　　　with cup and disk grind-
　　　　ing wheels 168
Coach and lag screws 222
Coach and lag screws, length of
　thread on 222
Cobalt, properties of 324
Coinage alloys 324
Cold punched nuts, check and jam. . 210
Cold punched nuts, manufacturers'
　standard 212
Cold soldering 82
Collar head screws 195
　　　　for jigs 267
Comparison of wire gages used in
　the United States 297
Complementary angles 364
Composition of bearing metals 325
　　　of steel, effect on hard-
　　　　ening 321–323
Compound gearing, train for screw
　cutting 2
Computing tapers, table for.....260–262
Concave and convex cutters, cutter
　for fluting 140
Concave and convex cutters, num-
　ber of teeth in 140
Constants for dovetail measurements 265
Conversion factors, English.....329, 330
Conversion factors, English and
　Metric 331, 332
Cooling steel 316
Copper bars, weight of 307
　　　plates, weight of 302, 303
　　　properties of 324
　　　tubing, weight of 309
　　　wire, weight of 304, 305
Corner rounding cutters, cutter for
　fluting 140
Corner rounding cutters, number of
　teeth in 140
Corundum 148
Co-secant of angle 363
Co-secants, secants and, table of .394–405
Cosine of angle 363
Cosines, sines and, table of.....382–393

PAGE

Co-tangent of angle 363
Co-tangents, tangents and, table of
　　　　　　　　　371–382
Cotter pins 219
Counterbore, depth of for fillister
　head cap screws 195, 196
Counterbores with inserted pilots... 277
Counterboring on the screw machine,
　speeds for 183
Countersunk head bolts, square and
　round 215
Coupling bolts 217
Co-versed sine of angle 363
Cubes, squares and roots of fractions
　from 1/64 to 1 inch............340, 341
Cubes, squares and roots of numbers
　from 1 to 520 348–360
Curves, finding radius without center 369
Cutters, for box tool, radial........ 170
　　　　　　sizes of steel
　　　　　　　for 170
　　　　　　tangent 169
　　　for spur gears 101
　　　milling, advantages of coarse
　　　　pitch 120
　　　milling angle of clearance.. 120
　　　cutters for fluting,
　　　　　　　　139, 140
　　　fine and coarse pitch 119
　　　keyway dimensions 145
　　　number of teeth in,
　　　　　　　　138–140
　　　power required for. 119
　　　speeds and feeds
　　　　for 118
　　　T-slot, dimensions of 146
　　　Whitney, for Woodruff keys 240
Cutting diametral pitch worms in the
　lathe 8–10
Cutting double threads 7
　　　fractional threads 1
　　　multiple threads with special
　　　　face plate 8
　　　quadruple threads 7
　　　screw threads 1–4
　　　speeds and feeds for screw
　　　　machine work 179
　　　of planers 292
　　　threads with compound gear-
　　　　ing 2
　　　threads with simple gearing. 2
　　　triple threads 7
Cycloidal gear teeth............. 83

D

Decagon, properties of 367
Decimal equivalents of fractions of
　an inch 335–337
Decimal equivalents of millimeters
　and fractions 332, 333
Decimal equivalents of squares,
　cubes, roots, etc., of fractions .340, 341
Decimals of a foot, equivalents of
　inches in 338, 339
Dedendum of gear teeth 83
Degrees obtained by opening a two-
　foot rule 294

	PAGE
Depth of keyways, total, table for finding	242, 243
Depth to drill and tap for studs	218
Diameters of circles of circumferences 1 to 200	361
Diameters of circular forming tools,	174, 176
of shell blanks, finding	184
Diametral pitch	83
worms, change gears for cutting	9
worms, cutting in the lathe	8
Diamond laps	163
powder in the machine shop	162–164
settling in oil	162
tools for charging	163
used on box wood laps	164
Diamonds, setting	154, 155
using on wheels	154
Die clearance, punch and	184
taps, taper	67
Dies and taps for screw machine work	171
Dies, screw machine, speeds for	183
spring, sizes of	172
Differential indexing on the milling machine, tables for	128–137
Dimensioning dovetail slides and gibs	262, 263
Disks used with accurate taper gage,	245–249
Dividing head, gearing for plain and differential indexing on	128–137
Dividing head, milling cams by setting	122
Dodecagon, properties of	367
Double threads, cutting	7
Dovetail forming tools	174, 175
depth of	177
slides and gibs, dimensioning	263
Dovetails, measuring with plugs in the angles	264, 265
Dowel pins, drills for	59
Drawing room standards, shop and,	266–295
Drill and wire gage sizes	296, 298, 299
Drill and wire gage sizes arranged consecutively	298, 299
Drill bushings for jigs, fixed	266
Drill bushings for jigs, fixed for stop collars	267
Drill bushings for jigs, loose	266
Drill jig screws, binding	268
collar head	267
headless	268
locking	268–269
nurled head	269
square head	268
supporting	268, 269
winged	269
jigs, straps for	269
Drills for Briggs pipe reamers	41, 44
for dowel pins	59
for drive fits	59

	PAGE
Drills for running fits	59
for taps, see Tap drills.	
for Whitworth pipe taps	41
also see Twist Drills.	
Drilling and tapping for studs, depths for	218
Drilling feeds and speeds in screw machine	182
Drive fits, allowing with the calipers for	234
Drive fits, drills and reamers for	59
limits for	229, 232

E

Emery	148
End mills, number of teeth in	138
English and Metric conversion tables	331, 332
English weights and measures	328–330
Equivalents of inches in decimals of a foot	338, 339
Equivalents of inches in millimeters.	334
Equivalents of inches in millimeters, decimal	332, 333
Erecting perpendiculars by triangles	281
Estimating lumber for patterns	294
Eye bolts	219

F

Face plate for cutting multiple threads	8
Fahrenheit thermometer scale	323
Feather key, square, dimensions of	239
Feeds and speeds for screw machines	179
milling machine speeds and	118
File makers' terms	70
teeth, cut of, actual sizes	71
Files	70–75
die sinkers or riffle	75
measurement of	70
shapes and sizes of	72
teeth per inch on	72
tests of	74
Filing, hight of work for	70
Fillets, areas, weight and volume of	295
Fillister head cap screws	195, 196
Fillister head machine screws, American Screw Co.	199
Fillister head machine screws, flat, A. S. M. E. standard	207
Fillister head machine screws, oval, A. S. M. E. Standard	206
Finding the diameters of shell blanks	184
Fine pitch screw thread diameters, measuring with wires	34
Finishing box tool for screw machine,	170
cutting speeds and feeds for	180
Fits, limits for drive	229, 232
for hand	229, 232
for press	229, 232
for running	230, 232
in shop gages for	231, 232
press and shrink, calipering for,	234

PAGE

Fits, running, side play in calipering
for 233, 235
    shrink, allowances for 231
Fittings, Briggs pipe, gages for.... 42-46
              reamer sizes for   44
              tap drill sizes
                for 41, 44
    Whitworth pipe, tap drill
        sizes for 41
Fixed bushings for jigs 266
            for jig tools with stop
                collars 267
Flat and oval countersunk head cap
    screws 197
Flat bar steel, weight of 308
    fillister head cap screws 195, 196
                machine screws,
    A. S. M. E. standard 207
Flat head machine screws, American
    Screw Co................... 198
Flat head machine screws, A. S. M. E
    standard 208
Flat on tools for U. S. form of thread,
    grinding 13, 15
Fluting cutters for hobs 142
            for milling cutters,
                139, 140
            for reamers 143
            for taps 141
Fluxes, — see Soldering 78
Foot, decimal parts of in inch
    equivalents 338, 339
Form of thread, A. S. M. E. standard
    machine screws 200, 201
Forming tools, circular and dove-
    tail 174-178
Forming tools, circular diameters of,
                174, 176
            dovetail, depth of... 177
            speeds and feeds for. 181
Four-point ball-bearing data....... 280
Fractions, decimal equivalents of,
                335, 337
    of a foot, decimal....338, 339
    squares, cubes, and roots
        of 340, 341
French (Metric) standard screw
    threads 20
Furnaces for heating steel 315
        gas 316
        liquid................... 315
Fusible plug 325

G

Gage, machine and wood screw,
    sizes of 221
Gage, sizes, drill and steel wire....
                296, 298, 299
    taper 245
        applications of 245
        applications of formu-
            ulas for 247-249
        formulas for...........245
    wire and drill, sizes arranged
        consecutively 298, 299
Gages, pipe and fittings Briggs...42-46

PAGE

Gages, plug and ring, dimensions of. 276
            laps for ..160, 161
    shop, limits in for various
        kinds of fits 231, 232
    Stubs'.................296-300
    wire, different standards used
        in the U. S.............. 297
Gas, case hardening with.......... 321
Gear blanks, laying out 93
        sizes of 99
        tables of 99, 100
    pressure angles 96
    teeth 83
        actual sizes of......... 95
    tooth parts, proportions of ...88-91
Gears and pulleys, speeds of 314
        bevel 106
            cutters for 107
            table of 108
        block indexing for 101
        cutters for 101
        for screw cutting, finding.... 3
        miter, table of 110
        sprocket 104
        spur, cutters for 101
        with cast teeth 92
Gearing...................83-117
    chordal pitch 87
    circular pitch 84, 85
    compound, for screw cut-
        ting................... 3
    constants for chordal pitch.. 87
    corresponding diametral and
        circular pitches 86
    diametral pitch 84, 85
    face of worm 115
    for cutting diametral pitch
        worms, table of 9
    metric pitch 103
    module 103
    parts of 83
    simple for screw cutting 2
    single curve tooth 96
    spiral, rules for 111, 112
        table of 113
    standard tooth 97
    stub tooth 97
    worm 115
        proportions of....... 117
German silver 326
Gold, properties of 324
Graduations on micrometer 226
            on ten thousandth mi-
                crometer 227
            on vernier 225
Grinding allowances 153
        and lapping 148-168
    flats on U. S. form of
        thread tool 13-15
    hardened work 153
    use of water in 154
    wheels and grinding 148
        combination grit .. 149
        contact of 152
        cup, clearance table
            for 168
        disk, clearance
            table for 168

PAGE

Grinding wheels, grades for different
kinds of work... 157
grading...148, 152, 156
hard 149
speeds of......150, 155
Gun metal 326

H

Hand fits, limits for 232
taps 64
cutter for fluting 141
number of flutes in 141
wheel dimensions 270
Handles, ball..................... 272
lever 273
binder................... 272
for hand wheels.......... 271
machine 274
Hardened steel, properties of....... 322
work, grinding 153
Hardening, case 318–323
effect of 321–323
Harveyizing 318
Headless jig screws 268
Heads, key, proportions of 241
of bolts, stock required for... 293
Heart-shape cams, milling......... 121
Heat treatment of steel 315
Heating steel in liquids 315
methods of 315
Helix angles of screw threads...... 4, 5
Heptagon, properties of 367
Hexagon stock, brass, weight of.... 307
steel, weight of...... 306
head cap screws 194
properties of 367, 370
High-speed steel 318
Hight of work for filing 70
Hob taps 68
cutter for fluting 142
number of flutes in 142
Hobs, cutters for fluting 142
die, number of flutes in...... 142
for worm wheels 38
pipe, cutter for fluting 142
number of flutes in.... 142
Sellers, cutter for fluting 142
number of flutes in.. 142
Holes in circles, tables for spacing,
284–291
Hollow mills for screw machine.... 170
Hook bolts 275
Hooks, belt 310
Hot pressed and cold punched nuts,
U. S. standard sizes 210
Hot pressed and forged nuts, manu-
facturers' standard sizes........ 211
Hot pressed nuts, narrow gage 213

I

Imperial wire gage sizes 297
Inclination of calipers in making
shrink or press fits............. 234
Indexing on milling machine, plain
and differential 128–137

PAGE

Inserted tooth cutters, number of
teeth in 138
Integral right-angle triangles, table of 281
International standard screw threads, 21
Involute gear tooth................ 83
Iridium, properties of............. 324
Iron bars, weight of 306
plates, weight of 301, 302
properties of 324
wire, weight of 304, 305

J

Jackson belt lacing 310, 311
Jam and check nuts, cold punched.. 210
Jarno taper 254, 255
Jig bushings, fixed 266
for stop collars 267
loose 266
parts, standard 266–269
screws, binding 268
collar head 267
headless 268
locking 268, 269
nurled head 269
square head 268
supporting 268, 269
winged 267
straps...................... 269
Jigs, laying out by trigonometry.... 365
Joint, Briggs pipe 42

K

Key head dimensions 241
Keys, amount of taper for various
lengths 244
Keys and key-seats, rules and di-
mensions for 236, 237
Keys, Barth, ... 239
feather 239
straight 237–239
taper 236, 237
Woodruff 240, 241
Keyway depth, total, finding 242, 243
Knobs, machine, sizes of.......... 271

L

Lacing, belt 310
Lag screws 222
length of thread on 222
test of................. 222
Laps, diamond 163
for holes 159, 160
for plug and ring gages ...160, 161
Lapping 158
flat surfaces 158, 159
Laying out angles by table of chords 281
holes in circles, tables
for 284–291
Lead bath for steel 316
properties of 324
Leather for case hardening 319
Lengths of bolts 314

PAGE

Lengths of threads cut on bolts 214
    of threads cut on lag screws, 222
Limits for drive fits 229, 232
    for hand fits 229, 232
    for press fits............229, 232
    for running fits 230, 232
    for shop gages, for various
      classes of fits 231, 232
    for shrink fits 230
Linear pitch of racks 83
Linear, square, and cubic measure,
    English 328
Linear, square, and cubic measure,
    Metric 330
Lining bushings for jigs........... 266
Locking jig screws 268, 269
Loom bolts ....: 214
Loose bushings for jigs............ 266
Lubricants for press tools 188
Lubricants for working copper, brass,
    steel, etc., in the punch press.... 188
Lubrication in milling steel........ 120
Lumber for patterns, estimating 294

M

Machine and wood screw gage sizes, 221
    bolts with button heads... 214
    with manufacturers'
      standard heads .. 192
    with U. S. standard
      heads 190, 191
    hand wheels 270
      handles for ... 271
    handles ....:...272–274
    knobs 271
    molded gears 93
    or nut taps, number of flutes
      in 141
    screw taps 63
Machine screw taps, cutters for flu-
    ting 141
Machine screw taps, number of flutes
    in 141
Machine screws, American Screw
    Co., fillister head 199
Machine screws, American Screw
    Co., flat and round head........ 198
Machine screws, American Screw
    Co., threads per inch 200
Machine screw, A. S. M. E. standard,
    flat fillister head 207
Machine screws, A. S. M. E. stand-
    ard, flat head 208
Machine screws, A. S. M. E. stand-
    ard, oval head 206
Machine screws, A. S. M. E. stand-
    ard, round head 209
Machine screws, A. S. M. E. stand-
    ard, special sizes 204
Machine screws, A. S. M. E. stand-
    ard, special taps for 205
Machine screws, A. S. M. E. stand-
    ard, standard proportions 200–209
Machine screws, A. S. M. E. stand-
    ard, standard taps for 203
Machine screws, A. S. M. E. stand-
    ard, thread diagram 201

PAGE

Manganese, properties of 324
Manufacturers' standard heads for
    machine bolts 192
Manufacturers' standard cold
    punched nuts 212
Manufacturers' standard hot pressed
    nuts 211
Manufacturers' standard hot pressed
    and forged nuts 211
Manufacturers' standard narrow gage
    nuts 213
Measurement of V-tools.......... 11–13
Measures and weights, English.. 328–330
              Metric ..330–332
Measuring external and internal
    dovetails 264, 265
Measuring screw threads with mi-
    crometer and wires 27–34
Measuring screw threads, Acme 29
    degree 35
Measuring screw threads, fine pitch. 34
Measuring screw threads, U. S. stand-
    ard 28, 29
Measuring screw threads, V standard,
                  30, 31
Measuring screw threads Whitworth,
                  32, 33
Measuring tapers 245
    three-fluted tools....227, 228
    worm threads.......... 37
Melting points of metals 324
Mercury, properties of 324
Metal, fusible 325
    slitting cutters, pitch of...... 138
Metals, chemical symbols of 324
    for bearings 325
    melting points of 324
    properties of 315, 327
    specific gravity of 324
    tensile strength of 324
    weight of 324
Metric and English conversion tables,
                331, 332
    pitch gears 103
    weights and measures .. 330–332
Micrometer and how it is read..... 226
    application in measuring
      three-fluted tools 228
Micrometer readings for U. S.
    threads 24
Micrometer readings for V Threads 25
Micrometer readings for Whitworth
    threads 26
Micrometer, ten thousandth........ 227
Millimeters, decimal equivalents of,
                332, 333
    equivalents of inches in. 334
Milling cams by gearing the dividing
    head 122
Milling cutters, angle of clearance for 120
Milling cutters, clearance tables for
    grinding 168
Milling cutters, coarse pitch, advan-
    tages of 120
Milling cutters, cutters for fluting,
               139, 140
Milling cutters, fine and coarse pitch,
               119, 120

PAGE

Milling cutters, for Woodruff keys . . 240
    keyways for 145
    number of teeth in,
         138-140
    power required to
      drive 119
    T-slot 146
    heart-shaped cams 121
    lubrication for 120
Milling machine, plain and differen-
    tial indexing on 128-137
Milling machine speeds and feeds . . 118
    table for cutting
      spirals 124-127
Mills, end, number of teeth in 138
    hollow, for screw machine 170
Miter gear table 110
Module, metric pitch 103
Morse standard tapers 252, 253
Multiple threads, cutting 6
    face plate for . . . 8
Muntz metal 326
Music wire sizes 301

N

Narrow gage nuts, manufacturers'
    standard 213
Narrow gage washers 220
Naval bearing metals 325
Nickel, properties of 324
Nonagon, properties of 367
Norton Company's wheel grade num-
    bers 156
Nurled head jig screws 269
Nuts, castellated, for A. L. A. M.
    standard bolts 216
Nuts, cold punched, check and jam . 210
    cold punched, manufacturers'
      standard 212
    for A. L. A. M. standard bolts . 216
    for coupling bolts 217
    for U. S. standard finished
      bolts 191
    for U. S. standard rough bolts . 190
    for planer head bolts 217
    hot pressed and forged, manu-
      facturers' standard 211
    hot pressed, manufacturers'
      standard 211
    narrow gage, manufacturers'
      standard 213
    planer 217
    see Bolts also.
    thumb, dimensions of 275
    U. S. standard hot pressed and
      cold punched 210
    wing, dimensions of 274

O

Octagon bar steel, weight of 306
    properties of 367-370
Oval, countersunk head, cap screw,
    flat end 197
Oval fillister head cap screws 196

PAGE

Oval fillister head machine screws,
    A. S. M. E. standard 206

P

Packfong metal 326
Patterns, estimating lumber for 294
    weight of castings propor-
      tionate to 294
Penetration of carbon in case harden-
    ing 319-323
Penetration tests 320, 321
Pennsylvania R. R. bearing metal . . 325
Pentagon, properties of 367
Perpendiculars, erection of, by tri-
    angles 281
Pickling bath for cast iron 70
Pilots for counterbores, table of di-
    mensions 277
Pins and reamers, taper 257
Pipe and pipe threads 39-46
    Briggs standard dimensions . . 39-43
      drill sizes for . 41, 44
    taps 66
Pipe hobs, cutters for fluting 142
    number of flutes 142
    joint, Briggs 42
    taps, cutters for fluting 142
    number of flutes 142
    threads, cutting with tools 40
    Whitworth standard dimen-
      sions, table of 41
    Whitworth standard, drill sizes
      for 41
Pitch line of gears 83
Plain milling cutters, cutter for flut-
    ing 139
Planer head bolts, nuts and washers 217
    nuts 217
Planers, actual cutting speed of 292
Plates, brass and copper, weight of,
      302, 303
    steel and iron, weights of . 301, 303
Platinum, properties of 324
Play of calipers endwise in measur-
    ing for press and shrink fits 234
Play of calipers sidewise in measur-
    ing for running fits 233, 235
Plug and ring gages, standard dimen-
    sions 276
Plug gages for Briggs standard pipe
    fittings 42-46
Plug, taper, gage for measuring 245
Plugs, use of in measuring dovetails . 264
Polygons, table of 367
Potassium cyanide for case harden-
    ing 320
Press and running fits 228, 232
    fits, inclination of calipers in
      measuring for 234
    limits for 229, 232
    tools, lubricants for 188
    punch 184
Pressure angle of gear teeth 96
Properties of metals 315-327
Proportions of machine screw heads,
    A. S. M. E. standard 206-209

PAGE

Protractor, use in testing thread tool
clearance angle 6
Pulley, finding size of 369
Pulleys and gears, speed of 314
Punch and die, allowance for accu-
rate work 184-187
Punch and die, clearance for various
metals 184-187
Punch press tools 184

Q

Quadruple thread cutting 7

R

Radius, finding without center 368
Reading the micrometer 226
Reading the ten-thousandth microm-
eter 227
Reading the vernier 225
Readings of screw thread micrometer
for U. S. threads.............. 24
Readings of screw thread micrometer
for V threads 25
Readings of screw thread micrometer
for Whitworth threads 26
Reamer and cutter grinding........ 164
Briggs pipe 44
chucking and hand, method
of grinding 163
Reamer, chucking, rose, cutter for
fluting...................... 144
Reamer, chucking, rose, number of
flutes 144
Reamer clearances, grinding, 164, 166, 167
cutters for fluting 143, 144
for taper pins 257
number of flutes 143
shell, cutters for fluting.... 143
taper, number of flutes.... 144
Reaming, feeds and speeds for screw
machine 183
Rectangular section bar stock,
weight of 308
Reed taper 254
Removable bushings for jigs........ 266
Riffles, see files, die sinkers......... 75
Right angle triangles, integral, table
of 281
Ring gages for Briggs pipe ends...42-46
Rivet heads, boiler 224
tank 224
Riveting washers 221
Roller and block chain sprocket
wheels 104-105
Rose chucking reamer, cutter for
fluting....................... 144
Rose chucking reamer, number of
flutes........................ 144
Roughing tool for screw machine... 169
Round and square countersunk head
bolts 215
Round bar stock, brass, copper and
aluminum, weight of............ 307
Round bar stock, steel and iron,
weight of 306

PAGE

Round fillister head cap screws..... 196
Round or button head machine
screws, American Screw Co...... 198
Round or button head machine
screws, A. S. M. E. standard..... 209
Rule, two foot, angle obtained by
opening 294
Rules for solving angles 366
speed of pulleys and gear 314
Running fits, drills and reamers for. 59
limits for 230, 232
side play of calipers for,
233, 235

S

Safety Emery Wheel Co.'s, wheel
grade numbers 156
Sag of wire for alining shafting..312, 313
Screw cutting, arrangement of gears
for 2
Screw cutting, examples in 1
gears for 3
multiple 6
rules for 3
Screw machine, allowance for thread-
ing........................... 173
Screw machine box tools, cutters,
etc........................... 169
Screw machine dies and taps 171
Screw machine, feeds for counter-
boring in 183
Screw machine, hollow mills....... 170
Screw machine, sizing work for
threading 172
Screw machine speeds and feeds ... 180
Screw machine, speeds and feeds for
drilling 182
Screw machine, speeds and feeds for
forming 181
Screw machine, speeds and feeds for
reaming 183
Screw machines, speed and feeds for
turning 179, 180
Screw machine, speeds for dies 183
Screw machine tools, speeds and
feeds 169-183
Screw slotting cutters, pitch of..... 138
thread angle table..........4, 5
Screw thread diameter, measuring
with wires 27-34
Screw thread micrometer caliper
readings for U. S. threads........ 24
Screw thread micrometer caliper
readings for V threads.......... 25
Screw thread micrometer caliper
readings for Whitworth threads.. 26
Screw thread tools, grinding flat for
U. S. form 13-15
Screw threads 1-38
Screw threads, Acme 29 degree, table
of 22
Screw threads, Acme, standard and
Brown & Sharpe worm threads
compared 10, 11
Screw threads, British Association,
table of 19

	PAGE
Screw threads, cutting	1
fractional	1
Screw threads, French (metric) standard, table of	20
Screw threads, International standard, table of	21
Screw threads, sharp V, table of	17
Screw threads, 60 degree, measuring with wires	30, 31
Screw threads, U. S. standard, measuring with wires	28, 29
Screw threads, U. S. standard, table of	16
Screw threads, Whitworth standard, measuring with wires	32, 33
Screw threads, Whitworth standard, table of	18
Screws, cap, button head	197
countersunk head	197
fillister head	195, 196
hexagon and square head	194
Screws, coach and lag	222
collar head	195
jig, binding	268
collar head	267
headless	268
locking	268, 269
nurled head	269
square head	268
supporting	268, 269
winged	267
Screws, machine, A. S. M. E. standard flat head	208
Screws, machine, A. S. M. E. standard flat fillister head	207
Screws, machine, A. S. M. E. standard oval fillister head	206
Screws, machine, A. S. M. E. standard round head	209
Screws, machine, A. S. M. E. standard proportions	200–209
Screws, machine, diagram of form of thread, A. S. M. E. standard	201
Screws, machine, fillister head, American Screw Co.	199
Screws, machine, flat and round head, American Screw Co.	198
Screws, machine, table of A. S. M. E., special sizes	204
Screws, machine, table of A. S. M. E., standard sizes	202
Screws, machine, threads per inch, American Screw Co., standard	200
Screws, set	193
wood	223
Secants and co-secants, table of	394–405
of angles	363
Sellers hob, cutter for fluting	142
taps	68
number of flutes	142
tapers	256, 257
Set screws	193
Setting diamonds for truing wheels	154, 155
Settling diamond powder	162
Shafting, alining by steel wire	312, 313
and belting	310–314

	PAGE
Sheet brass and copper, weight of	302, 303
steel, American or B. & S. gage	302
U. S. standard gage, weight of	301
Shell blanks, finding diameter of	184
reamer, cutter for fluting	143
number of flutes	143
Shop and drawing room standards	266–295
gages, limits for various kinds of fits	231, 232
trigonometry	362–367
Shrink fit allowances	231
fits, inclination of calipers in measuring for	234
Shrinkage of castings	326
Side clearance of thread tool, table of	4
Side or straddle mills, cutters for fluting	139
Side or straddle mills, number of teeth in	138
Side play of calipers in making running fits	233, 235
Sides, angles and sines for spacing holes, etc., in circles, table of	286–291
Silver, properties of	324
Simple gearing train for screw cutting	2
Sines and cosines, table of	382–393
Sines of angles	363
Single end ball handles, dimensions of	273
Sizing work for threading in screw machines	172
Slides and gibs, dovetail, table for dimensioning	263
Slip bushings for jigs	266
Slotted straps for jigs	269
Soldering	78–82
cast iron	81
cleaning and holding work	81
cold	82
fluxes for different metals	78
glass and porcelain	82
making the fluxes	80
strength of joint	78
Spacing bolt circles	365
holes in circles, tables for	284–291
Specific gravity of metals	324
Speculum	326
Speeds and feeds for drills in screw machine	182
Speeds and feeds for finishing box tools	180
Speeds and feeds for forming in screw machine	181
Speeds and feeds for grinding wheels	150, 155
Speeds and feeds for milling machines	118
Speeds and feeds for reaming in screw machine	183
Speeds and feeds for screw machine work	169
Speeds and feeds for turning in screw machine	179, 180

	PAGE
Speeds for threading in screw machine	183
Speeds of planers, actual cutting	292
Speeds of pulleys and gears	314
Spiral gears	111
Spirals, table for cutting on milling machine	124-127
Splice for belts	310
Spring cotterpins	219
dies, sizes of	172
Sprocket wheels	104-105
Spur gears, laying out blanks	93
Square bar stock, brass, copper and aluminum, weight of	307
Square bar stock, steel and iron, weight of	306
Square countersunk head bolts	215
head cap screws	194
jig screws	268
properties of	367, 370
thread taps	69
washers	220
Squares, cubes, roots, etc., of fractions from 1/4 to 1 inch	340, 341
Squares, cubes, roots, etc., of numbers from 1 to 520	348-360
Squares, largest that can be milled on round stock	146
Standard jig parts	266-269
Stationary bushings for jigs	266, 267
Steel and other metals	315-327
annealing	317
bars, flat sizes, weight of	308
bars, round, square, hexagon and octagon, weight of	306
cooling, apparatus for	316
heat of	315
high speed	318
lubricant for milling	120
methods of heating	315
plates, American or B. & S. gage, weight of	302
Birmingham or Stubs gage, weight of	303
U. S. standard gage, weight of	301
properties of	324
wire and twist drill gage sizes,	296, 298, 299
Brown & Sharpe gage, weight of	304
Stubs, sizes and weights	300
Stock allowed for upsets	292
flat sizes, steel, weight of	308
sheet brass and copper, weight of	302, 303
sheet steel and wire, weight of,	301-303
required for bolt heads	293
round and square, copper and aluminum, weight of	307
round and square, iron, weight of	306
square and hexagon, brass, weight of	307
round, square, hexagon and octagon, steel, weight of	306
for box tool cutters	170

	PAGE
Stove bolt taps	66
bolts	215
Straight key dimensions	237-239
Straps for jigs	269
Strength of belt lacing	310, 311
bolts, U. S. standard	189
Stub tooth gears	97
Stubs' gages	296-300
or Birmingham wire gage sizes	297-299
steel wire sizes	297-300
and weights	300
Studs, depth to drill and tap for	218
Sugar for case hardening	319
Supporting jig screws	268, 269
Symbols, chemical, of metals	324

T

T-slots, bolt heads for	218
Tangents and co-tangents, table of,	371-382
Tangents of angles	363
Tank rivet heads	224
Tap bolts	215
drills	57-61
Taper gage	245
application of	246
application of formulas for	247-249
formulas for using	247
key heads, dimensions of	241
and keyseats, dimensions of	236, 237
pins and reamers	257
reamers, number of flutes	144
Tapers and dovetails	245-265
for keys, etc., from 1/8 to 1 inch per foot	244
Tapers, measuring	243-249
per foot and corresponding angles, table of	259
standard, Brown & Sharpe,	250, 251
Jarno	254, 255
Morse	252, 253
Reed	254
Sellers	256, 257
table for computing	260-262
Tapers, table of lengths up to 24 inches	258
Taper taps	65
Tap, screw machine, length and number of lands	173
Tap threads, Acme, 29-degree, table of	23
Tapping for studs, depth of	218
Taps, Acme, 29 degrees, measuring with wires	37
Taps and dies for screw machine	171
cutters for fluting	141, 142
dimensions of	62-69
for A. S. M. E. special machine screws, table of	205
for A. S. M. E. standard machine screws, table of	203
hand	64
hob. cutters for fluting	142

PAGE

Taps, hob, number of flutes........ 142
   machine screw 62
   machine screw, A. S. M. E.... 63
   number of flutes 141, 142
   pipe 66
      cutters for fluting 142
      number of flutes 142
      Sellers hob............. 68
      square thread 69
      stove bolt 66
      taper die 67
      tapper.................... 65
Temperature of steel.............. 315
Tensile strength of metals......... 324
Ten-thousandth micrometer 227
Test of lag screws................ 222
Thermometers, Fahrenheit and Cen-
   tigrade 323
Thread angle table 4, 5
   cutting, arrangement of gears 2
        fractional 1
        multiple 6
        multiple, face-plate
         for 8
        rules for 3
   diameter, measuring with
      wires 27-34
   form of, for A. S. M. E. stand-
     ard machine screws..200, 201
Thread micrometer, readings for
   U. S. thread.................. 24
Thread micrometer, readings for V
   thread 25
Thread micrometer, readings for
   Whitworth thread 26
Thread, pipe, Briggs standard 39
   setting tools for cutting 40
   tool, use of protractor in get-
     ting clearance angle. 6
   table for angle of clear-
     ance 7
   U. S. form, grinding the
     flat for 13-15
Threading in screw machine, sizing
   work for 172
Threading in screw machine, speeds
   for 183
Threads, Acme 29 degree, measur-
   ing with wires................. 35
Threads, Acme 29 degree, table of.. 22
Threads, Acme 29 degree tap, table of 23
Threads, British Association, table
   of 19
Threads, Brown & Sharpe 29 degree
   worm, table of 36
Threads, Brown & Sharpe worm
   and Acme compared 10, 11
Threads, Brown & Sharpe worm
   measuring with wire 37
Threads, cutting.................. 1
      French (metric) standard,
       table of 20
      International standard,
       table of 21
      of worms 115
      sharp V, table of 17
      60 degree V, measuring
       with wires 30, 31

PAGE

Threads, U S. standard, measuring
   with wires ....:..28, 29
   U. S. standard, table of... 16
   Whitworth standard, meas-
     ing with wires 32, 33
   Whitworth standard, table
     of 18
Three fluted tools, measuring...227, 228
Thumb nuts, dimensions of 275
Tin, properties of................. 324
Tools, box, for screw machine 169
   for U. S. form of thread, grind-
     ing the flat 13-15
   forming, circular and dovetail,
                174, 175
   V, measurement of 11-13
Trenton Iron Co.'s wire gage sizes .. 297
Triangle, properties of.........367, 369
Triangles, integral right angle, table
   of 281
Trigonometrical table 364
Trigonometry for the shop......362-367
Triple threads, cutting............ 7
Tubes, Briggs, standard dimensions
   of............................ 40
Tubes, Whitworth, standard dimen-
   sions of 41
Tubing, brass and copper, weight of. 309
Tungsten, properties of 324
Turning screw machine work, speeds
   and feeds for 179, 180
Twenty-nine degree thread, Acme,
   table of 22
Twenty-nine degree thread, Acme,
   measuring with wires 37
Twenty-nine degree thread, Acme,
   tap, table of 23
Twenty-nine degree worm thread,
   Brown & Sharpe, table of 36
Twenty-nine degree worm thread,
   Brown & Sharpe, measuring with
   wires 37
Twist drill and steel wire gage sizes,
            296, 298, 299
   drills 47
      angle of spiral 47
      clearance of 47
  -   feed of.............. 49
      grinding or sharpening. 48
      letter and decimal sizes. 55
      sizes of 53-56
      special 50
      speed of 49
      troubles 50
Two-point ball-bearing data 278, 279
Type metal 326

U

Undecagon, properties of.......... 367
Upsets, stock allowed for.......... 292
United States form of thread, grind-
   ing flat on 13-15
United States bolts and nuts....189-191
United States bolts and nuts, finished, 191
United States bolts and nuts, rough . 190
United States bolts and nuts, strength
   of 189

PAGE

nited States standard gage for plates 297, 301

nited States standard nuts, hot pressed and cold punched 210

nited States standard screw threads, table of 16

nited States standard screw threads, measuring with wires...........28, 29

nited States standard washers.... 220

nited States threads, micrometer readings for 24

V

V-block used with micrometer to measure 3 fluted taps 228

V-threads, micrometer reading for.. 25

60 degree, measuring with wires 30, 31

table of 17

tools, measurements of, 11-13

Vanadium, properties of 324

Vernier and how to read it 225

Versed sine of angle.............. 363

W

Washburn & Moen music wire sizes, 301

Washburn & Moen wire gage sizes. 297

Washers, cast iron................ 221

for planer head bolts.. .. 217

narrow gage 220

riveting 221

square 220

U. S. standard 220

Water conversion factors 329

used in grinding 154

Webster & Horsefall music wire sizes 301

Weight of brass and copper tubing. 309

copper and aluminum bars....... 307

of castings in proportion to patterns 294

of fillets.................. 295

of iron, brass and copper wire............304, 305

of metals................. 324

of sheet steel and iron, U. S. standard gage 301

of steel and iron bars 306

bars, flat sizes 308

iron, brass and copper plates 302, 303

of steel wire, Brown & Sharpe gage 304

of steel wire, Stubs' gage... 300

Weights and measures, English. 328-330

metric ..330-332

Wheels, grades for different work... 157

combination grit 157

contact of 152

grading of, 148, 152, 156

Wheels, hard 149

speeds of 150, 155

White metal 326

Whitworth standard pipe ends, drill sizes for 41

Whitworth standard pipe threads ... 41

Whitworth standard screw threads, table of 18

Whitworth standard screw threads, measuring with wires 32, 33

Whitworth thread, micrometer readings for 26

Wing nuts, dimensions of.......... 274

Winged jig screws 267

Wire and drill gage sizes, arranged consecutively 298, 299

Wire for alining shafting 312, 313

gage sizes, steel 296, 298, 299

gages and stock weights..296-309

in use in the United States.............. 297

iron, brass and copper, weight of 304, 305

lacing 310, 311

steel, Brown & Sharpe gage, weight of 304

steel, Stubs', sizes and weights. 300

Wires, measuring Acme 29-degree thread with 35

Wires, measuring Brown & Sharpe 29-degree worm thread with 37

Wires measuring fine pitch thread diameters with................ 34

Wires, measuring 60-degree V thread with 30, 31

Wires, measuring U. S. standard thread with 28, 29

Wires, measuring Whitworth thread with 32, 33

Wood screws 223

Woodruff or Whitney key and cutter dimensions 240, 241

Work bench legs................. 78

Work benches 76

construction of...... 76

hight of 78

material for 77

Worm gearing 115

threads 115

wheel hobs 38

Worms, diametral pitch, cutting in the lathe 8

Worms, diametral pitch, table of change gears for.............. 9

Worm thread, Brown & Sharpe and Acme screw thread compared ..10, 11

Worm thread, Brown & Sharpe, table of 36

Worm thread, Brown & Sharpe, measuring with wires 37

Z

Zinc, properties of 324

LIST OF AUTHORITIES

	PAGE
Acme Machinery Co.	292, 293
Alford, L. P.	59
Almond, R. A.	13-15
American Screw Co.	198-200
American Swiss File & Tool Co.	75
Armes, F. W.	325
Atkins, H. F.	11-13
Baker Bros.	236, 237
Bardons & Oliver.	115
Becker Milling Machine Co.	108, 124-127
Brown & Sharpe Mfg. Co.	24-27, 88, 96, 103, 128-137, 250, 251
Brownstein, Benj.	295
Cantello, Walter	11, 28-33, 35, 37
Carborundum Co.	156, 157
Carstensen, Fred R	121
Cincinnati Milling Machine Co.	168
Cleveland Twist Drill Co.	47
Colburn, Geo. L.	4-6
Cook, Asa	99
Corbin Screw Corporation	200-209
Cregar, J. W.	78
Dangerfield, Jas.	34
Darbyshire, H.	149-155
Dean, C.	186, 187
Disston & Son, Henry	72
Ellis, M. E.	266-269
Fellows Gear Shaper Co.	97
Fraser, Jas.	147
Garford Co.	278-280
Goodrich, C. L.	169-183
Hartford Machine Screw Co.	193, 217
Haskell Mfg. Co., Wm. H.	214
Hedglon, M. J.	271
Hoagland, F. O.	227
Holz, Fred	118-120, 164-167
Hoopes & Townsend,	192, 214, 215, 222, 224
Hunt Co., C. W.	228-230
Johnson, E. A.	158-161
Jones & Laughlin,	239
Lachman, Robt.	146
Lake, E. F.	315-323
Milton Mfg. Co.	220, 221
Morse Twist Drill & Machine Co.	53, 252, 253
Newall Engineering Co.	231, 232
New Britain Machine Co.	76, 282, 283
Norton Co.	156, 157
Noyes, H. F.	262, 263
Nuttall & Co.	84
Pratt & Whitney Co.	45, 46, 255, 257, 270-272
Press, A. P.	188
Ranstch, E. J.	8-10
Reed Co., F. E.	254
Ryder, T.	8
Safety Emery Wheel Co.	156
Seidensticker, F. W.	287-291
Sellers & Co., Inc., Wm.	256, 257
Stabel, Jos.	277
Standard Gage Steel Co.	241
Stutz, C. C.	245
Trebert, A.	6-7
Upson Nut Co.	210-213
Valentine, A. L.	138-145
Vernon, P. V.	236
Walcott & Wood Machine Tool Co.	273
Waltham Watch Co.	31
Welsh, T. E.	264, 265
Whitney Mfg. Co.	240, 241
Woodworth, J. V.	188
Zeh, E. W.	184, 185

CPSIA information can be obtained
at www.ICGtesting.com
Printed in the USA
LVOW10s0612131117

556078LV00026B/353/P